THE PAPERS OF THOMAS A. EDISON

FINANCIAL CONTRIBUTORS

We thankfully acknowledge the vision and support of Rutgers University and the Thomas A. Edison Papers Board of Sponsors.

This edition was made possible by grant funds provided from the New Jersey Historical Commission, National Historical Publications and Records Commission, and The National Endowment for the Humanities. Major underwriting of the present volume has been provided by the Barkley Fund, through the National Trust for the Humanities. We also thank The Charles Edison Foundation and Phillips Lighting as well as many individual donors. For the assistance of all these organizations and individuals, as well as for the indispensable aid of archivists, librarians, scholars, and collectors, the editors are most grateful.

THE PAPERS OF THOMAS A. EDISON

Volume 7

Thomas A. Edison c. 1884.

Volume 7

The Papers of Thomas A. Edison

LOSSES AND LOYALTIES

April 1883–December 1884

VOLUME EDITORS
Paul B. Israel
Louis Carlat
Theresa M. Collins
David Hochfelder

ASSISTANT EDITORS
Scott Bruton
Alexandra R. Rimer

POST-GRADUATE FELLOWS
Lindsay Frederick Braun
Catherine Nisbett

GRADUATE ASSISTANTS
Kelly Enright
Dennis Halpin
Clare Hilliard

Patrick McGrath
Kristopher Shields
Daniel J. Weeks

SPONSORS
Rutgers, The State University of New Jersey
National Park Service, Thomas Edison National Historical Park
New Jersey Historical Commission
Smithsonian Institution

THE JOHNS HOPKINS UNIVERSITY PRESS
BALTIMORE

The Johns Hopkins University Press
2715 North Charles Street
Baltimore, Maryland 21218-4363
www.press.jhu.edu

The paper used in this book meets the minimum requirements of the American National Standard for Information Sciences—Permanence of Paper for Printed Library Materials, ANSI Z 39.48-1984.

Library of Congress Cataloging-in-Publication Data
(Revised for volume 3)

Edison, Thomas A. (Thomas Alva), 1847–1931
    The Papers of Thomas A. Edison

    Includes bibliographical references and index.
    Contents: v. 1. The making of an inventor, February 1847–June 1873—v. 2. From workshop to laboratory, June 1873–March 1876—v. 3. Menlo Park. The early years, April 1876–December 1877.
    1. Edison, Thomas A. (Thomas Alva), 1847–1931. 2. Edison, Thomas A. (Thomas Alva), 1847–1931—Archives. 3. Inventors—United States—Biography. I. Jenkins, Reese.
TK140.E3A2   1989   600   88-9017
ISBN 0-8018-3100-8 (v. 1. : alk. paper)
ISBN 0-8018-3101-6 (v. 2. : alk. paper)
ISBN 0-8018-3102-4 (v. 3. : alk. paper)
ISBN 0-8018-5819-4 (v. 4. : alk. paper)
ISBN 0-8018-3104-0 (v. 5. : alk. paper)
ISBN-10: 0-8018-8640-6 (v. 6. : alk. paper)
ISBN-13: 978-0-8018-8640-9 (v. 6. : alk. paper)
ISBN-10: 1-4214-0090-1 (v. 7. : alk. paper)
ISBN-13: 978-1-4214-0090-7 (v. 7. : alk. paper)

A catalog record for this book is available from the British Library.

Edison signature on case used with permission of the McGraw-Edison Company.

TO THE MEMORY OF PHILIP J. PAULY

Contents

Calendar of Documents

List of Editorial Headnotes

Preface

This volume chronicles Edison's audacious attempt to replicate the success of his New York central station in scores of U.S. towns and cities, as well as in Europe and Latin America. It also encompasses profound changes in his working and personal life, including the unexpected death of his wife. It concludes with Edison returning to the laboratory and refocusing his energies on new inventions.

Having anticipated for a year the construction of a second district in New York City and orders for central stations in other cities and towns, Edison had grown frustrated as officers of the Edison Electric Light Company focused on orders for isolated plants. He now tried to develop the market for central stations by focusing on the less expensive "village plant" system he had conceived in the preceding months. In contrast to the Pearl Street model, the village plant system required less labor and less material to build. In addition, wires for most of these plants were placed overhead on poles, thus obviating the time and expense of burying underground lines. In order to develop the business of selling and building central stations in the United States, Edison established the Thomas A. Edison Construction Department with his trusted private secretary, Samuel Insull, managing its day-to-day business affairs while Edison focused on the design and construction of the plants.

Working eighteen- to twenty-hour days, Edison was traveling more widely and frequently in the United States and Canada in 1883 and 1884. On 4 July 1883, he opened the first three-wire incandescent electric light central station in Sunbury, Pennsylvania. Thereafter, Edison traveled to several other towns to oversee the building of central stations; sixteen stations were underway before the end of 1884. He additionally visited electrical exhibitions in Chicago, Worcester (Mass.), Philadelphia, and Boston.

The unwillingness of company officials and financiers to support the central station business had prompted Edison to become, as he put it, "a business man for a year" as a builder of small central station plants. However, the generally depressed capital markets in New York and London affected his plans at home and abroad. The continuing scarcity of capital, aggravated by changes rippling through the network of investors associated with Drexel, Morgan & Company, led Edison to abandon that experiment with many of his plans unfulfilled. Similar economic forces led to his grudging acceptance of a merger with a rival in Great Britain and the virtual abandonment of that once-promising market for electric lighting. While the Italian Edison company successfully inaugurated central station service in Milan, efforts to establish central stations stalled elsewhere in Europe. This led Edison to try to develop new markets in South America, but, here too, he achieved only modest success.

Although Edison is famous for devoting more time to work than to his family, by the winter of 1884 he had grown tired of the long hours and relentless financial pressures of the central station business. Needing a break, Edison and his wife, Mary Stilwell Edison, took a long vacation in Florida during February and March. While enjoying the resorts and fishing grounds of St. Augustine and the St. Johns River in Florida, Edison seemed to be the typical tourist as he tallied up the costs of travel expenses. But he was also uniquely Edison as he took up ideas for new experiments in the midst of his leisure, if not on account of it. During the trip, Edison filled a small pocket notebook with a steady burst of new ideas, whether on problems that occupied him for years, like lamp filaments and multi-signal telegraphy, or on research that took him in new directions, like the development of fuel cells and artificial materials.

Edison's return to business at the start of April, following the vacation with Mary in Florida, marked the beginning of a turbulent period in his personal and professional life. Soon after their return, Mary's father, Nicholas Stilwell, died after what seems to have been a chronic illness. About the same time, Edison decided to close up the Construction Department due to its demands on his personal finances and because of worsening general economic conditions. Edison was having trouble getting payment from the Edison Electric Light Company for more than $11,000 that he had paid from his own pocket for canvassing and estimating towns for central stations, most of which were never built. Edison Electric was itself feeling financial strain, exacerbated by the recent financial failures of two of its principals, Henry Villard and Egisto Fabbri, and by having to take stock in the local illuminating companies to which it sold operating licenses, rather than getting all the cash it originally expected. Edison faced similar problems in collecting monies from the local companies. In an effort to improve its financial outlook, Edison Electric also sought an interest in the manufacturing shops owned by Edison and his close associates. In order to resolve these and other questions concerning the future of the Edison lighting business, a series of negotiations took place that led to a consensus by mid-June for the Edison Company for Isolated Lighting to take over the central station business and for the Edison Electric Light Company to share in the profits of the manufacturing shops. A series of agreements formalized this arrangement on 1 September.

By that time, Edison was dealing with the personal crisis of his wife's death and his desire to return full time to invention. Little is known of the circumstances, but Mary Edison died unexpectedly at their Menlo Park home on the morning of 9 August. The doctor in attendance reported only that she died of congestion of the brain, a general diagnosis based on symptoms that could result from several more specific causes. Newspaper reports later claimed that Edison sought to revive his wife using electricity, but to no avail. Daughter Marion remembered that she "found my father shaking with grief, weeping and sobbing so he could hardly tell me that mother had died in the night." Marion also recalled driving her father around the countryside at least once a week during the rest of the summer, and for some time she became her father's almost constant companion.

In the months preceding Mary's death, Edison had been spending increasing amounts of time in his laboratory. And, just before she died, he told a reporter for the *New York Daily Tribune*, "I am going into original experimenting again. I'll get out a new crop of inventions during the next year in the electrical line." His interest in new inventions was further spurred in September, during his visits to the International Electrical Exhibition in Philadelphia with Marion. While there, he reconnected with his old friend Ezra Gilliland, to whom he "mentioned that his electric light was completed and practically off his hands and he was talking of what would be a good thing to take up next." Gilliland was then in charge of American Bell Telephone's experimental shop in Boston and during their discussions, Edison decided to turn his attention to improving transmitters for the company's new long-distance lines.

As Edison worked on this and other problems for American Bell, he was also negotiating the terms of his contract with the company, which included resolving Western Union's rights to his telephone inventions. The contract issues remained unresolved at the end of the year even though Edison traveled to Boston just before Christmas in an effort to finalize his agreement with the company. This was one of several trips Edison made to Boston during the fall in connection with his telephone work. It was also in December that Edison and Gilliland began to discuss developing the system of railway telegraphy that would become a subject of sustained research for Edison in the next year.

Determined to remove himself from active participation in the business of electric lighting, Edison, aided by Samuel Insull, sought to elect a new board of directors for Edison Electric. Toward this end, they hoped to acquire enough proxies from other stockholders to enable Edison, the company's largest stockholder, to elect his slate of candidates in place of those proposed by the existing board. Before the vote took place on 28 October, a compromise was reached on the makeup of the board, and Sherburne Eaton resigned as president, although he remained as general counsel. Eugene Crowell was elected president in place of Eaton, but it was Edward Johnson, re-elected as vice president of Edison Electric and newly elected as president of the Edison Company for Isolated Lighting, who took charge of the day-to-day affairs of the company.

Having put Johnson and others that he trusted in charge of the Edison lighting business, Edison spent little of his own time on either the business or technology of electric light and power during the next year. Freed of these concerns, he returned full time to the laboratory to work on other inventions.

The progress of the Thomas A. Edison Papers depends on the support of many individuals and organizations, including the Sponsors, other financial contributors, academic scholars, Edison specialists, librarians, archivists, curators, and students. Representatives of the four Sponsors have assisted with this volume and the editors thank them for their continuing concern and attention. The strong support of public and private foundations and of their program officers has sustained the project and helped it remain editorially productive.

Preparation of this volume was made possible in part by grants from the Division of Research Programs (Scholarly Editions) of the National Endowment for the Humanities, an independent federal agency; the National Historical Publications and Records Commission; the New Jersey Historical Commission; the Charles Edison Fund; the National Trust for the Humanities; as well as through the support of Rutgers, The State University of New Jersey, and the National Park Service (Thomas Edison National Historical Park). The editors appreciate the interest and support of the many program officers and trustees, especially Elizabeth Arndt, Timothy Connolly, Sara Cureton, John P. Keegan, Marc Mappen, Peter Mickulas, and Malcolm Richardson. Any opinions, findings, conclusions, or recommendations expressed in this publication are solely those of the editors and do not necessarily reflect the views of any of the above federal foundations or agencies, the United States Government, or any other financial contributor.

The Edison Papers project is indebted to the National Park Service for its multifaceted support. The editors express particular thanks to Marie Rust, Maryanne Gerbauckas, John Maounis, and Nancy Waters of the Northeast Region, and to Greg A. Marshall, Theresa Jung, Randy Turner, Michelle Ortwein, Leonard DeGraaf, Edward Wirth, Karen Sloat-Olsen, Linda Deveau, Gerald Fabris, Joseph De Monte, Charles Magale of the Thomas Edison National Historical Park in West Orange, New Jersey.

Many associates at Rutgers University have contributed significantly to the Edison Papers. The editors are grateful to President Richard L. McCormick; Philip Furmanski, Executive Vice President for Academic Affairs; and Douglas Greenberg and Ziva Galili, Deans of the School of Arts and Sciences, along with their dedicated staffs, especially Ann Fabian, Dean for Humanities; Peter Klein, Executive Vice Dean; Todd Bristol, Vice Dean of Administration; Barbara Lemanski, Associate Dean for Policy and Personnel; Rosemary Lane, Personnel Manager; and Jason Diapaolo, Director of Business Affairs. We would especially like to thank Paul Kuznekoff, Director of Development for the Faculty of Arts and Sciences. In addition we appreciate support provided by Thomas Vosseler, David

Motovidlak, Wade Olsson, and the staff of the School of Arts and Sciences IT Services. The editors value the support of colleagues and staff in the History Department, especially Michael Adas, Rudy Bell, Alastair Bellany, Paul Clemens, James Delbourgo, Mary DeMeo, Ziva Galili, Ann Gordon, Allen Howard, Reese V. Jenkins, Jennifer Jones, Seth Koven, Norman Markowitz, Matt Matsuda, James Reed, Susan Schrepfer, Dawn Ruskai, Candace Walcott-Shepherd, and Virginia Yans. We also want to thank Michael Geselowitz and his staff at the IEEE History Center as well as members of the Rutgers University Libraries, notably Marianne Gaunt, Thomas Frusciano, Ron Becker, Tom Glynn, Jim Niessen, Rhonda Marker, and the Interlibrary Loan Office. A special thanks is due Michael Siegel, staff cartographer in the Department of Geography, who prepared the maps in this volume, and our web designer Bonnie Wasielewski. At the Rutgers University Foundation we have had the assistance of Joyce Hendricks and Willis Richie. We also want to thank Ronald Thompson, Steve Miller, and Lori DeMartino of the Division of Grant and Contract Accounting; Daphne Evans, Donna Foster, and Maryellen O'Brien of the Office of Research and Sponsored Programs; and Stephen J. DiPaolo of the Controller's Office. Additional thanks are due to Peter Shergalis, Supervisor of Material Services, Susan O'Brien, Material Services Administrator, and Jeanne Schaab, Foreperson, Custodial Services.

Many scholars have shared their insights and assisted the editors in a variety of ways. For this volume, notable help came from Stathis Arapostathis, Christopher Beauchamp, Brian Bowers, Jane Mork Gibson, Lisa Gitelman, Richard Guth, Cathy Moran Hajo, Sheldon Hochheiser, Richard John, Esther Katz, Alvin J. Salkind, Harold Wallace, and Mira Wilkins.

Institutions and their staff have provided documents, photographs, photocopies, and research assistance. The editors gratefully acknowledge Chris Baer and Max Moeller, both at the Hagley Museum and Library; William Caughlin, Joan Keegan, and George Kupczak at the AT&T Archives and History Centers in San Antonio, Texas, and Warren, New Jersey; Marc Greuther at the Henry Ford Museum; Cindy Inkrote, director of the Northumberland County (Pa.) Historical Society; Margaret Leary at the University of Michigan Law School library; Anne Locker at the Institution of Engineering and Technology, London; Lauris Olson at the Van Pelt Library of the University of Pennsylvania; Craig Orr and David

Weisz, both at the Archives Center of the National Museum of American History; Mark Reeves of the Wire Rope Works in Sunbury, Pa.; Jay Satterfield at the Rauner Special Collections Library of Dartmouth College; Mike Sweeney at the Jenkins Law Library, Philadelphia; and Laura Yeoman, archivist at the Royal Bank of Scotland Group, Edinburgh. The editors also gratefully acknowledge assistance from the Boston Public Library microform and serials reading room; the Milton S. Eisenhower Library of the Johns Hopkins University; the Metropolitan Opera; and the Wisconsin Historical Society.

Staff members, interns, students, and visiting editors not mentioned on the title page but who have contributed to this volume include Thomas E. Jeffrey, Rachel M. Weissenburger, Grace Kurkowski, Patrick McGrath, David Ranzan, Jeremy Sam, and Colin Dobie.

As always, the project has had the benefit of the superb staff of the Johns Hopkins University Press. For this volume, the editors are indebted to Robert J. Brugger, Mary Lou Kenney, and Julie McCarthy.

Chronology of
Thomas A. Edison

April 1883–December 1884

1883

4–5 April	Travels to Albany and electrical exhibition in Cornwall, Ontario.
16 April	Conducts Porfirio Díaz and his wife on a tour of Edison lighting plants and manufacturing shops in New York City.
20 April	Agrees in principle to combine electric railroad patents with those of Stephen Field in a new company.
3 May	Gives Samuel Insull power of attorney to sign contracts and conduct all business related to his Central Station Construction Department.
3–5 May	Travels to Shamokin, Pennsylvania, and signs contract to construct first village plant central station in Sunbury, Pennsylvania.
15–16 May	Requests several agents for Edison electric light to provide information regarding the use of the Maxim light in their region.
18 May	Appoints William Rich as superintendent of construction for central station plants.
1 June	Discharges all domestic servants at family home in New York City.
c. 1 June	Hires Frank Sprague.
18–19 June	Travels to Chicago for the Electric Railway Exhibition.
Spring	Designs short-core dynamo.
28 June	Comitato per Applicazioni Dell Elettricità Edison inaugurates central station service in Milan, Italy.
June	Spends substantial but indeterminate period supervising construction of first three-wire village plant in Sunbury, Pennsylvania.
c. 3–12 July	Supervises completion and 4 July start of Sunbury station.

17 July	Places Charles Campbell in charge of Engineering Department making estimates for central stations.
23 July	Donates to telegraphers' strike relief fund.
24 July	Gives conditional approval to merger of his British electric light interests with those of Joseph Swan.
1 August	New York *Evening Post* publishes interview in which Edison pledges to become "simply a business man for a year."
4–13 August	Mary Edison, daughter Marion, and sister Alice vacation in Long Beach, Long Island, with Edison and Alice's husband, William Holzer, joining them for two days.
18 August	Fires superintendent Charles Dean and bookkeeper Charles Rocap from Edison Machine Works.
18–26 September	Supervises completion and c. 20 September start of Shamokin, Pennsylvania, central station.
Summer	Adopts Frank Sprague's mathematical method of determining conductors for central stations.
29 September	Travels to Fall River and Brockton, Massachusetts.
1 October	Inaugurates Brockton central station.
c. 1 October	Edison family moves to Clarendon Hotel, New York.
2 October	Final board approval for organization of Edison and Swan United Electric Light Co., Ltd., retroactive to 30 June 1883.
c. 6 October	Returns to New York.
8 October	U.S. Commissioner of Patents rules in favor of Sawyer-Man patent on carbon loop filament.
c. 13 October	Devises new voltage regulator based on "Edison Effect" lamp.
13 November	Retains attorney John Tomlinson for lamp patent infringement lawsuits.
14 November?	Testifies as plaintiff's witness in lawsuit against overhead electric lines in New York City.
11 December	Repays more than $35,000 borrowed from Drexel, Morgan & Co.
	Sends Samuel Insull to Great Britain and continental Europe.
17–c. 18 December	Travels to Fall River, Massachusetts, to inaugurate central station there.
c. 24 December	Begins experimenting with gelatinous materials for lamp filaments.
December	Makes preliminary efforts to license and sell duplicating ink.
1884	
c. 5 January	Visits Henry Villard to console him after his financial collapse.
24 January	German patent office rules all carbon-filament incandescent lamps subject to Edison's patent.

c. 27 January	Samuel Insull returns.
29 January	Anthony Thomas proposes to Edison and other Edison Electric directors that all the Edison lighting companies and shops be reorganized into a single company.
30 January	Puts William Gilmore in charge of Engineering Department to replace Charles Campbell.
January	Directs research on electrodeposition of metallic films. Incorporates Edison Lamp Co. and Edison Machine Works.
1 February	Appoints Willis Stewart as his electric lighting agent in Chile.
	Charles Clarke officially resigns as chief engineer of Edison Electric Light Co.
c. 5 February	Drafts contract for central station engineers.
10 February	Leaves on Florida vacation with Mary Edison.
c. 23 February	Swedish patent office voids several Edison patents for lack of use in the country.
February	Withdraws "Edison Effect" voltage indicator.
	Loses retrial of civil suit brought by Lucy Seyfert.
February–March	Fills pocket notebook with plans for experiments while on Florida vacation.
24 March	Is sent copies of the "Dot and Dash" Polka, named after Edison's children, Marion and Thomas, Jr., by composer and former telegrapher John Milliken.
31 March	Returns to New York with Mary Edison.
9 April	Father-in-law Nicholas Stilwell dies at Menlo Park.
c. 22–26 April	Edison unwell and cancels planned trip to Newburgh, New York.
24 April	Announces intention to disband Construction Department.
c. 28 April	Drafts letter with Edward Johnson, Samuel Insull, and Francis Upton to Henry Villard regarding Edison Electic Light Co.'s Committee on Manufacturing and Reorganization.
April–May	Works to pass bill in New Jersey legislature authorizing electric lighting companies to put up poles or lay wires under the streets.
1 May	Moves family back to Gramercy Park house.
	Frank Sprague resigns.
1–2 May	Demonstrates electric lighting system at public exhibition in Worcester, Massachusetts.
3 May	Charles Batchelor sails back to New York.
7 May	Accepts nomination as one of the vice presidents of the newly formed American Institute of Electrical Engineers.
c. 15 May	Prepares caveat and has John Ott conduct experiments to use an electromagnet to separate nonferrous ores.

c. 21 May	Appoints Charles Batchelor general superintendent of Edison Machine Works.
22 May	Has John Ott conduct experiments on using dynamos for telegraphy in preparation for promoting their use for that purpose.
24 May	Asks Edison Electric Light Co. for authority over South American lighting business and subsequently receives it.
28 May	Negotiates retainer with Richard Dyer and Henry Seely to serve as his patent attorneys.
1 June	Article about Mary Edison published in *New York World*.
18 June	Charles Coster and Grosvenor Lowrey propose reorganization of Edison lighting companies.
June–early July	Organizes Edison Shafting Co.
22 July	Gives up office space at 65 Fifth Avenue.
	Begins week of sustained experimentation on direct conversion of coal into electricity.
	Sheriff's sale of Menlo Park property held to satisfy judgment in Seyfert lawsuit.
9 August	Mary Edison dies at Menlo Park.
23 August	Accepts appointment as a member of the "National Conference of Electricians."
1 September	Signs stockholding agreements with manufacturing shops as part of transfer of the Edison Construction Department's business to Edison Co. for Isolated Lighting.
	Signs lease for top floor of residence at 39 E. 18th Street, New York.
4–6 September	In Philadelphia with Marion to attend International Electrical Exhibition.
c. 15 September	Gives up lease on Gramercy Park home.
16–19 September	Attends International Electrical Exhibition in Philadelphia with daughter Marion.
24 September	Returns to Philadelphia for several days.
8 October	Asks Francis Upton to have lamp factory prepare special stage effect lamps for Koster & Bial's Concert Hall in New York.
16–25 October	Seeks proxies from Edison Electric stockholders.
21 October	Second sheriff's sale of Edison property at Menlo Park held to satisfy judgment in Seyfert lawsuit.
28 October	Edison Electric stockholders elect combination ticket to board of directors.
10 November	Attends Sullivan-Laflin fight at Madison Square Garden.
c. 15 November	Proposes to William Brewster to establish a syndicate for the electric light business outside of Europe and the United States.

28 November	Attends meeting of the Edison Electric Light Co. of Europe at his laboratory to discuss reorganization of French companies.
Fall	Experiments on telephone technology for and negotiates contract with American Bell Telephone Co.
c. 20–24 December	Visits Boston in connection with telephone experiments and contract negotiation with American Bell.
23 December	Attends Boston Electrical Exhibition.
November–December	Designs special theater lighting for use at Steele MacKaye's Lyceum Theater and the Bijou Theatre in Boston.

Editorial Policy and User's Guide

The editorial policy for the book edition of Thomas Edison's papers remains essentially as stated in Volume 1, with the modifications described in Volumes 2–6. The additions that follow stem from new editorial situations presented by documents in Volume 7. A comprehensive statement of the editorial policy will be published later on the Edison Papers website (http://edison.rutgers.edu).

Selection

The fifteen-volume book edition of Thomas Edison's papers will include nearly 6,500 documents selected from an estimated 5 million pages of extant Edison-related materials. For the period covered in Volume 7 (April 1883–December 1884), the editors have selected 352 documents from approximately 12,700 extant Edison-related documents. Most of the available documents from this period detail Edison's business relations or his inventive work, but some directly concern his family life. While still small, the subset of family or personal documents is notably larger than in previous volumes. This relative change is a consequence of Edison's continuing celebrity and the nature of events involving his family during this period, particularly the death of Mary Edison. The editors have selected for this volume several newspaper articles about Edison or his family, having picked them from the rapidly expanding pool of published sources available digitally through proprietary databases. Because such articles were often republished or adapted in numerous cities, and because digital collections change so quickly, the editors have not attempted to estimate the number of such printed sources available.

The editors have sought to select documents that illuminate the full range of Edison's thought, activities, and events
in his life during this period. Those documents published here
are primarily by or to Edison, his surrogate Samuel Insull, or
others working in concert with him or on his behalf. Some
third-party correspondence has been selected to highlight key
events or illustrate the context in which Edison worked. The
majority of Edison's correspondence from this time details
his business relationships, particularly his efforts to construct
central station electric systems throughout the United States.
Edison seems to have read most, if not all, of his voluminous
correspondence, and his marginal notes on incoming letters
frequently served as guides for the official responses that secretary Samuel Insull typically prepared in his name. Where
feasible, the editors have given priority to such incoming letters with Edison's comments over Insull's formal replies.
There are fewer available technical records from this time period than in periods covered by Volumes 1–6. This anomaly
is a consequence of both the nature of Edison's work at this
time and the fact that technical records created at his factories,
some of which would have documented research done directly
under his orders, did not become part of his personal papers
and were subsequently destroyed. The character of Edison's
work is also reflected in the editors' decision not to select any
artifacts for inclusion in this volume.

Note on Digital Sources

The number and scope of historical sources available electronically has increased dramatically during the preparation of
this volume. The editors have done their best to present state-
of-the-art research as of the completion of this manuscript in
mid-2010, but the tasks of culling available documents and
revising annotations with new references are potentially end-
less. In the future, significant additions and corrections can be
made in a forthcoming online version of these volumes that will
be produced by a collaborative effort of the Thomas Edison
Papers and the Johns Hopkins University Press. For this volume, the editors have made abundant use of proprietary collections of digitized newspapers, principally ProQuest databases,
JSTOR, and NewspaperArchive.com. They have also used
Ancestry.com (www.ancestry.com) for access to census records
and other genealogical and biographical information, and this
service is acknowledged within specific bibliographic records
at the end of this volume. Many of the contemporary sources

listed in the bibliography, as well as many journal articles cited in annotation, have been viewed electronically through Google Books and similar collections.

Transcription

The transcription policies used in preceding volumes have been followed in the present case, with the following additions and emendations. Edison often punctuated his sentences in very idiosyncratic ways, including the use of wholly nonstandard marks and identical marks for obviously different purposes. For the sake of intelligibility, the editors have transcribed Edison's punctuation with conventional typographic characters according to their understanding of his intent. Samuel Insull often made shorthand remarks on Edison's correspondence, but like other docket notes, these have not been transcribed or noted. Because of the abundance of documents in this volume created by secretaries, those written by an unidentified scribe are not described as "in an unknown hand"; however, where the editors can determine this identity, it is so indicated. For clarity, expressions of time are routinely transcribed in standard form.

This volume differs from its predecessors in the profusion of typed documents, many of which contain typing errors. Where these mistakes are merely artifacts of maladroit fingers or stuck keys, with no significance to the author's intended text, the editors have corrected them without comment, a departure from their past practice. The creation of carbon copies is another source of obviously mechanical problems, sometimes resulting in the duplication of blocks of text. The editors have also chosen not to reproduce or note these clerical errors.

Annotation

In the endnotes following each document, citations are generally given in the order in which the material is discussed. However, when there are several pieces of correspondence from the same person or a run of notebook entries, these are often listed together rather than in the order they are discussed to simplify the reference.

Updated Biographical Information
The proliferation of digitized primary sources has afforded relatively easy access to a wealth of new biographical information in recent years. For this reason, the editors provide up-

dated or expanded biographical entries for a number of individuals who were identified in previous volumes.

References to the Digital Edition

The editors have not provided a comprehensive calendar of Edison documents because the vastness of the archive makes preparation of such an aid impractical. Their annotations include, however, references to relevant documents in the Edison Papers digital edition; the volume may therefore serve as an entree into that publication.

The Edison Papers website (http://edison.rutgers.edu) contains approximately 161,000 images from the first three parts of the microfilm edition of documents at the Thomas Edison National Historical Park. There are also nearly 25,000 additional images not found on the microfilm that come from outside repositories. Citations to images in the digital edition are indicated by the acronym *TAED*. The citations are in an alphanumeric code unique to each document (e.g., *TAED* D8314N). In this volume, for the first time, this rule of thumb generally applies to documents found in bound volumes such as notebooks. There are cases, however, such as account books, in which unique identifiers can be assigned only with great difficulty to particular entries in a book. In those few instances, the citation gives both the general identifier for the entire book and a specific image number or numbers (e.g., AB004 [image 60]). Image numbers are also used on occasion to direct the reader to a particular point in an unusually lengthy document (e.g., W100DEC002 [image 7]). All of these images can be seen by going to the Edison Papers homepage and clicking on the link for "Single Document or Folder" under "Search Methods." This will take the user to http://edison.rutgers.edu/singldoc .htm, where the images can be seen by putting the appropriate alpha-numeric code in one of the two boxes to retrieve either a document or a folder/volume. If retrieving a folder/volume, the user should click on "List Documents" and then "Show Documents" in the introductory "target" for that folder/volume. Then click on any of the links to specific documents in the folder/volume and put the appropriate image number in the box under the "Go to Image" link. Entering an image number in that box when viewing any document will take the reader to the specific image number. Similarly, citations are sometimes made to an entire folder of documents (e.g., *TAED* D8416). Pasting the alpha-numeric folder code into the "Folder/Volume

ID" box on the search page, as above, will give the reader the option to display the contents of the entire folder.

The digital edition contains a number of other features not available in the book or microfilm editions, including lists of all of Edison's U.S. patents by execution date, issue date, and subject; links to pdf files; and a comprehensive chronology and bibliography. Other materials, such as chapter introductions from the book edition and biographical sketches, will eventually be added. Material from outside repositories, including items cited in this volume, will continue to be added. Under arrangements being made at press time, this volume and its predecessors will eventually be published in a searchable, interactive format on the web.

This volume refers to letterbooks used not only for Edison's general correspondence but also for specific purposes or by particular individuals or companies. These are grouped in a subseries of Miscellaneous Letterbooks within the Letterbook Series. As with other letterbooks, citations to the Miscellaneous Letterbooks are to the book and page, with two exceptions. Each page of the Cable Books (LM 1 and LM 2) typically contains transcriptions of several telegraphic messages. For these books, a letter designation after the page number indicates the relative position of each message (e.g., LM 2:59A).

The general availability of the Edison Papers digital edition through any standard web browser has led to the redefinition of the microfilm edition as a medium for archival preservation instead of a principal research tool. For this reason, and to simplify references in the endnotes, this volume does not give document-level citations to the Edison Papers microfilm edition. The exception to this new practice is for documents not scheduled to be added to the digital edition until well after press time, for which references to the reel and frame are given following the *TAEM* acronym, as in previous volumes (e.g., *TAEM* 72:878).

Headnotes
Volumes of the Edison Papers typically include more introductory headnotes than is common in historical documentary editions, and the present one is no exception. Each chapter begins with a brief introduction outlining Edison's personal, technical, and business activities during that period. Within chapters, occasional headnotes appear before particular documents (see List of Editorial Headnotes) for several purposes. Some present specific technical issues (e.g., "Short-Core Dynamo

Design") or describe the characteristics of a set of documents (e.g., "Edison's Florida Notebook"). A headnote may also provide a coherent narrative of related activities that otherwise appear only in scattered documents and endnote references (e.g., "Village Plant Construction"). Several headnotes in this volume present original research on major events in Edison's life that were under-represented in his personal papers (e.g., "Mary Edison's Death").

Just as chapter introductions and headnotes serve as guides for the general reader, discursive endnotes often contain annotation of interest to the general reader. Some of these discussions are broadly contextual. Other information of a biographical character suggests the tight and overlapping character of the business, technical, and social circles in which Edison's name circulated, especially within the metropolis of New York and, to a lesser extent, of London. The endnotes also include business or technical details likely to be of more concern to the specialized reader. In general, the editors provide more detailed information for technical issues that have received little scholarly attention than for topics that are already well treated in the secondary literature.

Citations
This volume follows the citation practices used in Volume 6, some of which were new. Of particular note is the format for referring to extensive correspondence over a long period. Such citations are made to the author name(s) in the digital edition as *TAED*, s.v. "name."

Appendixes

As in Volumes 1–6, we include relevant selections from the autobiographical notes that Edison prepared in 1908 and 1909 for Frank Dyer and Thomas Martin's biography of Edison (see App. 1). Following the precedent established in Volume 6, there is no longer a comprehensive list of Edison's U.S. patents, which are available on the Edison Papers website. Appendix 4 represents Edison's patenting activity in several ways.

There are two other appendixes in this volume. Appendix 2 consists of separate lists of central stations built or planned by the Edison Construction Department in 1883–1884. It is intended to indicate the financial scale and geographical scope of Edison's ambitions for this business, both of which can be hard to discern from the mass of correspondence that the Construction Department generated. Appendix 3 is a table of specifications of the short-core dynamos produced by the Edison Machine Works. It is provided as a technical reference and as a companion to a similar table in Volume 6.

Errata

Errata for previous volumes can be found on the Edison Papers website at http://edison.rutgers.edu/berrata.htm.

Editorial Symbols

~~Newark~~   Overstruck letters
  Legible manuscript cancellations; crossed-out or overwritten letters are placed before corrections

[Newark]   Text in brackets
  Material supplied by editors

[Newark?]   Text with a question mark in brackets
  Conjecture

[Newark?]ᵃ   Text with a question mark in brackets followed by a superscript letter to reference a textnote
  Conjecture of illegible text

⟨Newark⟩   Text in angle brackets
  Marginalia; in Edison's hand unless otherwise noted

[   ]   Empty brackets
  Text missing from damaged manuscript

[---]   One or more hyphens in brackets
  Conjecture of number of characters in illegible material

Superscript numbers in editors' headnotes and in the documents refer to endnotes, which are grouped at the end of each headnote and after the textnote of each document.

Superscript lowercase letters in the documents refer to textnotes, which appear collectively at the end of each document.

List of Abbreviations

ABBREVIATIONS USED TO DESCRIBE DOCUMENTS

The following abbreviations describe the basic nature of the documents included in the seventh volume of *The Papers of Thomas A. Edison*:

AD	Autograph Document
ADf	Autograph Draft
ADfS	Autograph Draft Signed
ADS	Autograph Document Signed
AL	Autograph Letter
ALS	Autograph Letter Signed
D	Document
Df	Draft
DS	Document Signed
L	Letter
LS	Letter Signed
M	Model
PD	Printed Document
PL	Printed Letter
TD	Typed Document
TL	Typed Letter
TLS	Typed Letter Signed
X	Experimental Note

In these descriptions the following meanings are assumed:

Document   Accounts, agreements and contracts, bills and receipts, legal documents, memoranda, patent applications, and published material, but excluding letters, models, and experimental notes

Draft   A preliminary or unfinished version of a document or letter

Experimental Note   Technical notes or drawings not included in letters, legal documents, and the like

Letter   Correspondence, including telegrams

Model   An artifact, whether a patent model, production model, structure, or other

The symbols may be followed in parentheses by one of these descriptive terms:

abstract   A condensation of a document

carbon copy   A mechanical copy of a document made by the author or stenographer using an intervening carbonized sheet at the time of the creation of the document

copy   A version of a document made by the author or other associated party at the time of the creation of the document

fragment   Incomplete document, the missing part of which has not been not found by the editors

historic drawing   A drawing of an artifact no longer extant or no longer in its original form

letterpress copy   A transfer copy made by pressing the original under a sheet of damp tissue paper

photographic transcript   A transcript of a document made photographically

telegram   A telegraph message

transcript   A version of a document made at a substantially later date than that of the original, by someone not directly associated with the creation of the document

STANDARD REFERENCES AND JOURNALS

Standard References

ABF	*Archives Biographiques Françaises*
American Cyclopaedia	*American Cyclopaedia: A Popular Dictionary of General Knowledge*
ANB	*American National Biography*
BDUSC	*Biographical Directory of the United States Congress*
BJHS	*British Journal for the History of Science*
Black's	*Black's Law Dictionary*, 6th ed.
Columbia	*Columbia Gazetteer of the World Online* (http://www.columbiagazetteer.org/main/Home.html)

DAB	*Dictionary of American Biography*
DBB	*Dictionary of Business Biography*
DCBO	*Dictionary of Canadian Biography Online* (http://www.biographi.ca/)
DPC	*Diccionario Politico de Chile (1810–1966)*
DSB	*Dictionary of Scientific Biography*
DSUE	*A Dictionary of Slang and Unconventional English*, 8th ed.
Ency. Brit. 1911	*Encyclopaedia Britannica*, 11th ed.
Ency. Chgo.	*Encyclopedia of Chicago* (www.encyclopedia.chicagohistory.org)
Ency. NJ	*Encyclopedia of New Jersey*
Ency. NYC	*Encyclopedia of New York City*
Ency. World Bio.	*Encyclopedia of World Biography*
Gde. Ency.	*Grande Encyclopédie Inventaire Raissonné des Sciences, des Lettres et des Arts*
GMO	*Grove Music Online* (http://www.oxfordmusiconline.com)
HDC	*Historical Dictionary of Chile*, 3rd ed.
London Ency.	*London Encyclopedia*, 1st ed. [1983]
NCAB	*National Cyclopedia of American Biography*
PMJHF	*Portraits of Modern Japanese Historical Figures*, online reference of National Diet Library, Japan (http://www.ndl.go.jp/portrait/e/index.html)
NDB	*Neue Deutsche Biographie*
NGDO	*New Grove Dictionary of Opera*
NPDS	*The New Partridge Dictionary of Slang and Unconventional English*
OED	*Oxford English Dictionary*
Oxford CBRH	*Oxford Companion to British Railway History*
Oxford DNB	*Oxford Dictionary of National Biography*
TAEB	*The Papers of Thomas A. Edison* (book edition)
TAED	*The Papers of Thomas A. Edison* (digital edition, http://edison.rutgers.edu)
TAEM	*Thomas A. Edison Papers: A Selective Microfilm Edition*
WGD	*Webster's Geographical Dictionary*
WWW	*Who's Who in America* (1899)
WWW-1	*Who Was Who in America, vol. 1*

Journals

NYT	*New York Times*
Phil. Mag.	*London, Edinburgh, and Dublin Philosophical Magazine and Journal of Science* [fifth series]
Pioneers News	*News of the Edison Pioneers,* Pioneers Printed
Sci. Am.	*Scientific American*
Sci. Am. Supp.	*Scientific American Supplement*
Teleg. J. and Elec. Rev.	*Telegraphic Journal and Electrical Review* (formerly *Telegraphic Journal*)
Trans. ASME	*Transactions of the American Society of Mechanical Engineers*

ARCHIVES AND REPOSITORIES

In general, repositories are identified according to the Library of Congress MARC code list for organizations (http://www.loc.gov/marc/organizations). Parenthetical letters added to Library of Congress abbreviations were supplied by the editors. Abbreviations contained entirely within parentheses were created by the editors and appear without parentheses in citations.

DeGH	Hagley Museum and Library, Greenville, Del.
DSI-AC	Archives Center, National Museum of American History, Smithsonian Institution, Washington, D.C.
DSI-MAH	National Museum of American History, Smithsonian Institution, Washington, D.C.
Guth	Richard N. Guth, Georgetown, Del.
Hummel	Charles Hummel, Wayne, N.J.
ICL	Cudahy Memorial Library, Loyola University, Chicago, Ill.
(LMA)	London Metropolitan Archives, London
MdBJ	Special Collections, Milton S. Eisenhower Library, Johns Hopkins University, Baltimore, Md.
MdCpNA	National Archives and Records Administration, College Park, Md.
MdHi	Maryland Historical Society, Baltimore, Md.

MH-BA	Baker Library Historical Collections, Harvard Business School
MH-HT	Harvard Theater Collection, Harvard College Library, Cambridge, Mass.
MiDbEI	Library and Archives, Henry Ford Museum & Greenfield Village, Dearborn, Mich.
(MiPhM)	Port Huron Museum of Arts and History, Port Huron, Mich.
NhD	Rauner Special Collections Library, Dartmouth College, Hanover, N.H.
Nj-Ar	Division of Archives and Records Management, New Jersey Dept. of State, Trenton
NjPwIE	Institute of Electrical and Electronics Engineers, Piscataway, N.J.
(NjWAT)	AT&T Archives and History Center, Warren, N.J.
NjWOE	Thomas Edison National Historical Park, West Orange, N.J.
NN	New York Public Library, Manuscripts and Archives Division
NNC-RB	Columbia University, Rare Book and Manuscript Library, New York
(NNMetOp)	Metropolitan Opera Archives, New York
NNNCC-AR	Division of Old Records, New York County Clerk, New York City Archives, N.Y.
NNPM	Pierpont Morgan Library, New York, N.Y.
NyBlHS	Brooklyn Historical Society, Brooklyn, N.Y.
PHi	Historical Society of Pennsylvania, Philadelphia, Pa.
PSuNCH	Northumberland County Historical Society, Sunbury, Pa.

MANUSCRIPT COLLECTIONS AND COURT CASES

Accts.	Accounts, NjWOE
Batchelor	Charles Batchelor Collection, NjWOE
Boston Bijou	Boston Bijou Theatre Co. Records, MH-HT

Brit. Pat.	British Patent
Can. Pat.	Canadian Patent
CEFC	Charles Edison Fund Collection
CPF	Clerk of the Peace File, City of London Sessions, LMA
CR	Company Records, NjWOE
DF	Document File, NjWOE
Diary	Thomas A. Edison Diary, NjWOE
Dun	R. G. Dun & Co. Collection, MH–BA
Duncan	Papers of Louis Duncan, NjPwIE
Edison and Gilliland v. Phelps	*Edison and Gilliland v. Phelps*, Patent Interference File 8028, RG-241, MdCpNA
Edison Assoc.	Edison Associates Biographical Collection, NjWOE
Edison Coll.	Edison Manuscript Collection, NNC-RB
Edison MS	Edison Manuscripts, Dibner Library, DSI-MAH
Edison v. Thomson	*Edison v. Thomson*, Patent Interference File 12332, RG 241, MdCpNA
EGF	Edison General File, NjWOE
EP&RI	Edison Papers & Related Items, MiDbEI
ESP Scraps.	Edison Speaking Phonograph Co. Scrapbooks (Incoming Correspondence), Uriah Hunt Painter Papers, HSP
FJS	Frank J. Sprague Papers, NN
Garrett	Garrett Papers (ms. 979), MdHi
HAR	Henry Augustus Rowland Collection, MdBJ
Hodgdon	Ernest Hodgdon, Derry, N.H.
Insull	Samuel Insull Papers, ICL
Kellow	Richard W. Kellow File, Legal Series, NjWOE
Kreusi	John Kruesi Collection, NjWOE
Lab.	Laboratory notebooks and scrapbooks, NjWOE
Lbk.	Letterbooks, NjWOE
Lit.	Litigation Series, NjWOE
LM	Miscellaneous Letterbook, NjWOE
	1 Cable Book (1881–1883)
	2 Cable Book (1883–1885)
	3 Insull Letterbook (1882–1883)
	5 Ore Milling Co. Letterbook (1881–1887)

	14 Construction Department (1883)
	15 Construction Department (1883)
	16 Construction Department (1883)
	17 Construction Department (1883–1884)
	20 Construction Department (1884)
	21 Construction Department (1884)
MacKaye	Percy MacKaye Papers, MacKaye Family Papers, ML-5, NhD
Meadowcroft	William H. Meadowcroft Collection, Special Collections Series, NjWOE
Miller	Harry F. Miller File, Legal Series, NjWOE
Misc. Scraps.	Miscellaneous Scrapbooks, NjWOE
NCB	Naval Consulting Board and Related Wartime Research Papers, NjWOE
Pat. App.	Patent Application Files, RG-241, MdCpNA
Pioneers Bio.	Edison Pioneers Biographical Files, NjWOE
Pioneers Printed	Edison Pioneers Printed Material [Series 5], NjWOE
PPC	Primary Printed Collection, CR, NjWOE
PP&L	Pennsylvania Power & Light Co. Papers, DeGH
PS	Patent Series, NjWOE
Scraps.	Scrapbooks, NjWOE
Sprague	Frank J. Sprague Papers, NN
TI	Telephone Interferences (Vols. 1–5), NjWOE; a printed, bound subset of the full Telephone Interferences
UHP	Uriah Hunt Painter Papers, PHi
Upton	Francis Upton Collection, NjWOE
Villard	Henry Villard Papers, MH-BA
Voucher	Laboratory Vouchers, NjWOE; voucher number and year are given in citations
Weston v. Latimer v. Edison	*Weston v. Latimer v. Edison*, Patent Interference File, RG-241, MdCpNA
WJH	William J. Hammer Collection, DSI-AC
WUTAE	Envelope of Edison Letters, Letterbox 8, Western Union Collection, DSI-MAH

LOSSES AND LOYALTIES
APRIL 1883–DECEMBER 1884

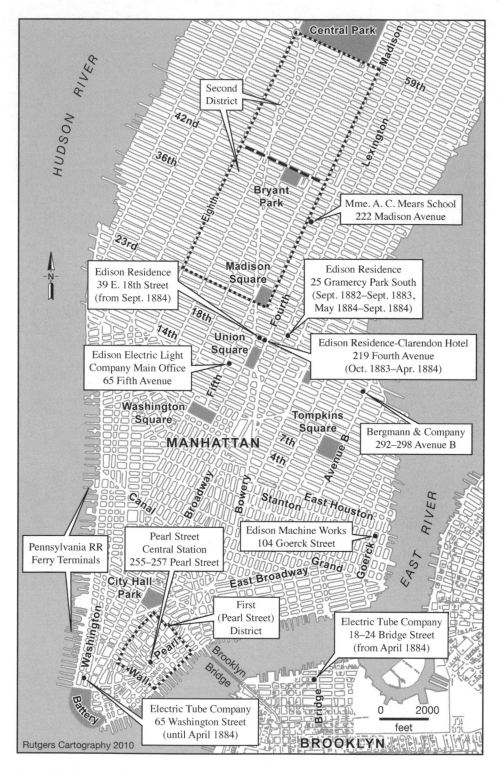

Lower Manhattan with points of significance to Edison's life and work, 1883–84.

Within the map image:

Central Park
HUDSON RIVER
Madison
59th
42nd
Second District
36th
Lexington
Bryant Park
Mme. A. C. Mears School 222 Madison Avenue
Eighth
23rd
Edison Residence 39 E. 18th Street (from Sept. 1884)
Madison Square
Edison Residence 25 Gramercy Park South (Sept. 1882–Sept. 1883, May 1884–Sept. 1884)
Fourth
18th
14th
Union Square
Edison Electric Light Company Main Office 65 Fifth Avenue
Edison Residence-Clarendon Hotel 219 Fourth Avenue (Oct. 1883–Apr. 1884)
Fifth
Washington Square
Tompkins Square
MANHATTAN
7th
4th
Avenue B
Bergmann & Company 292–298 Avenue B
Broadway
Canal
Bowery
Stanton
East Houston
EAST RIVER
Pennsylvania RR Ferry Terminals
Pearl Street Central Station 255–257 Pearl Street
Edison Machine Works 104 Goerck Street
Grand
Goerck
East Broadway
City Hall Park
First (Pearl Street) District
Washington
Pearl
Electric Tube Company 18–24 Bridge Street (from April 1884)
Wall
Brooklyn Bridge
Bridge
Battery
0   2000
feet
Electric Tube Company 65 Washington Street (until April 1884)
Rutgers Cartography 2010
BROOKLYN
-N-

April–June 1883

During the spring of 1883, Edison took steps to make his incandescent lighting system more commercially attractive. One major step involved replacing the distinctively long field magnets of his dynamo with short cores in order to make the machine both more economical in operation and cheaper to manufacture. This was a collaborative effort in which Edison followed the theoretical and practical advances made for the British Edison company by John Hopkinson. The initial impetus came from the requirements of shipboard lighting, which represented his only significant market in Britain, but the alterations also made the dynamos more attractive for small central stations in the United States. Work to improve components of the electrical system and to cheapen manufacturing processes continued at the Edison Machine Works Testing Room, at the Edison Lamp Company's factory in Harrison, New Jersey, and in Edison's laboratory at Bergmann & Company.

Edison's most significant change was to take personal charge of the effort to promote the Edison central station system. Having anticipated for a year the construction of a second district in New York City and orders for central stations in other cities and towns, Edison had grown frustrated as company officials focused on orders for isolated plants. He now tried to develop the market for central stations by focusing on the less expensive "village plant" system he had conceived in the preceding months. In contrast to the Pearl Street model, the village plant system required both less labor and less material to build. In addition, wires for most of these plants were placed overhead on poles, thus obviating the time and expense of burying underground lines.

In May, Edison established the Thomas A. Edison Construction Department to carry out the business of selling and building central stations in the United States, and he signed the first central station contracts. Although the Construction Department was a personal enterprise, it allowed Edison to work in concert with the Edison Electric Light Company to develop a market for central stations. While Edison focused on the design and construction of the plants, Samuel Insull, his private secretary, managed the day-to-day business affairs. Edison also delegated significant engineering responsibilities to William Andrews, a trusted assistant, and Frank Sprague, whom he had recently hired on the recommendation of Edward Johnson. Anticipating the need for electrical engineers, Edison also offered to provide equipment for an electrical engineering program at Columbia, Cornell, or the Stevens Institute.

As efforts to establish central stations stalled in Europe, Edison sought to expand the business in South America.[1] In April, he invited Porfirio Díaz of Mexico to tour the Edison manufacturing plants, the Pearl Street station, and some isolated plants in New York City. The following month, he began efforts to promote electric lighting in Argentina. The glum outlook in Britain led the Edison company there to propose a merger with the Swan United Electric Light Company, which Edison initially resisted. More happily, at the end of June, the Comitato per Applicazioni dell'Elettricita Edison inaugurated central station service in Milan, Italy.

Most of Edison's working days were spent in the New York offices of the Edison Electric Light Company at 65 Fifth Avenue, although business travel punctuated his calendar throughout these months. In early April, he briefly visited Albany, New York, with Insull and Frank Hastings, in an attempt to facilitate the formation of a central station company there. Edison then proceeded to Canada for the exhibition of his electric light in Cornwall, Ontario.[2] In early May, he briefly visited Shamokin to conduct negotiations for a central station plant.[3] In April, he and inventor Stephen Field had combined their patents to form the Electric Railway Company of the United States, and in June they exhibited at the National Exposition of Railway Appliances in Chicago, which Edison attended. At the beginning of June, Edison's weekly schedule was adjusted to also include the seasonal pleasures of weekends at Menlo Park with this family.

1. TAE memorandum, Apr. 1884, PN-82-04-01, Lab. (*TAED* NP016M).

2. Doc. 2418 n. 10.

3. Insull to Spencer Borden, 6 May 1883, Lbk. 16:233 (*TAED* LB016233).

–2418–

Draft Patent Application: Electric Light and Power

[New York,] April 1 1883

The object of this invention is to produce a cheap and reliable meter for measuring the Electric Current or Energy in a ~~gener~~ system of general distribution of Electricity for Light heat & power[1]

The invention consist in combining a peculiar constructed Electromotor with the line upon which the current is to be measured & providing such motor with proper counting or registering apparatus and also devices for causing the motor to perform a definite amount of work.[2]

~~If from these~~

Fig 1 shews a section of the meter—

fig 1

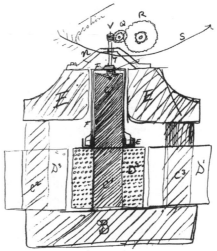

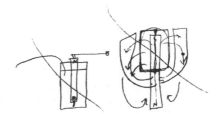

B is the iron base which forms the back of the Electromagnet C^1 being one pole, & C^2 C^3 forming with the end peices E E the other pole. C^1 C^2 C^3 [are provided with][a] bobbins of wire D[1] 2 & 3 & are all placed in one circuit[3] & this circuit forms a multiple arc circuit across the mains conductors upon the consumers premises

The direction of the Current is such E E are ~~pBoth~~ positive which C[1] opposite is negative, thus the lines of force pass from C[1] to E from the Entire Circumference of C[1]

On the top of C[1] is a depression with a bearing into which a pivotal point S runs, this depression is partially filled with mercury. The point S is made of platinum or platinum Iridium alloy so as not to amalgmate with the mercury in the depression. The mercury being used to make good & sufficient contact with the revolving pivot to carry without heating a powerful Current   F is a copper cylinder to which the pivot S is secured

fig 2[4]

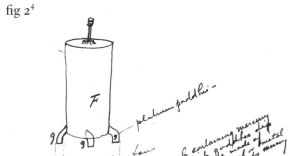

This cylinder is shewn in fig 2   at the bottom are several paddles of platinum [------][b] which rotate in the circular trough E fig 2 which is partially filled with mercury which acts both to make Electrical Contact with the cylinder which is to be rotated upon its pivot .S. and also to cause a definite amount of work to be done by the cylinder in the act of rotating.[5]

fig 4   Electrical Contact once every revolution

Another bearing .n supports the Extension of the pivot .S. & is marked T. upon the extreme end of T is a worm which acting upon a worm wheel Q gives motion to the train of wheels[c] which serves to register or indicate the number of revolutions of the cylinder .F. the[c] same as the counter used in a gas meter the revolutions being proporti[on]ate to the strength of the Current   Instead of a worm & worm wheel the rotating cylinder & shaft may at Each revolution close momentarily a local circuit & Energize an Electro magnet which serves to give motion[c] to the indicating mechanism   This is shewn in fig 4. T is the shaft W   a point   p a platinum tipped spring, R a Limiting point. at every revolution the point W comes into Contact with the spring p & closes the circuit to the magnet X.

The method of connecting the mechanism to the line is shewn in fig

fig 5

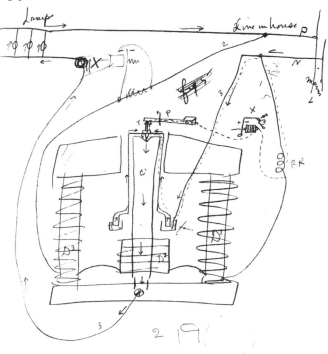

Connections of the meter   The Constant field is taken across the house mains in multiple arc. The resistance of the Constant field magnets is very high so that only sufficient Current to Energize the fields is used—[d]

L are the mains of the street P N the mains in the house. Wire 1 runs to RR which is a resistance coil, thence to the helix D^1 thence through helix D^2 thence through helix D^3 back by wire 2 to the other pole   Thus keeping a Current constantly circulating around the coils of the field magnet—

The mainline passes by wire 3 to the mercury in <u>F</u> thence up through the cylinder around its whole circumference to the pivot S down through the iron core C^1 to the base thence by wire 3 to the lamps—

When no lamps are in circuit only the field of force magnets are energized but the moment a lamp is placed across the
circuit or poles a powerful Current passes through the meter
as described & causes the Cylinder to rotate   if now a 2nd
Lamp be placed across the poles twice as much Current passes
& the speed of rotation of the cylinder is doubled   In fig 5 the
electrical contact method of working the indicating apparatus is shewn   The shaft T in rotating makes momentary contact with P once every revolution   This closes a local circuit
around the Resistances RR in which circuit is the magnet X.

That the rotating shell may so perform work so that the
Current is directly proportional to the rotation, the paddles
upon the cylinder must be arranged in a proper manner, ~~and~~
& retarding pieces may in some designs be found necessary the
retarding pieces being connected permanently with the circular trough so as to prevent the mercury rotating as a whole—
again owing to the large size of a meter & the consequent large
initial friction of the parts. It may be necessary to make the
paddles connected with the cylinder loose ~~so~~but so arranged
as to be thrown in or out of[c] the ~~mer~~ mercury more or less so
as to cause the rotations of the shell to be directly proportional[c] to the strength of the current. Paddles mounted on ring &
controlled by centrifugal apparatus[6c]

fig 6

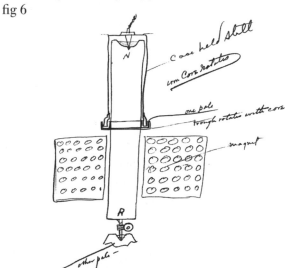

case held still   iron core rotates[7]

It is not necessary that the shell should rotate as It may be
held still & the magnet rotate as in fig 6

~~Even C′ may be a permanent~~ Even permanent magnets
may be used & Electromagnets dispensed with.

To prevent waste of Electrical Energy by keeping the field of force magnets constantly charged when there is no Lamps connected, an Electromagnet may be placed in circuit at X which having a lever & point may serve to close the same when a lamp is connected. The point & lever being interpolated in the field of force circuit. The circuit is opened by the magnet only when [-----][b] there are no lamps across the line.

Thus by the ~~used~~ of a non commutator motor or rather monodynamic motor[8] I am enabled to convey powerful Currents through the apparatus without loss by ~~rigid~~[f] solid[f] metallic or multiple contacts and also to obtain slow rotations with powerful currents, and at the same time attain this result with exceedingly small indefinite[g] friction ~~thus enabling motor other than that provided~~[h]

fig 7

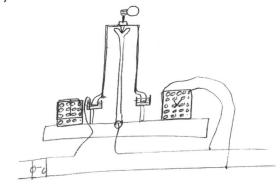

no iron   X coil wire
fig 7

It is not essential to have an Electromagnet field as the rotation of the cylinder will take place if a helix alone is used as in fig 7. but by the use of iron a much less current is required to perform the work=

Claims

A monodynamic motor substantially as described

A monodynamic motor sub as described for giving motion to recording or indicating apparatus

A monodynamoc motor sub as described having its moving or inductive parts interpolated in the circuit with the translating devices, and a constant magnetic field produced by the action of the Current from a multiple arc ckt across the line.

A monodynamic motor without a commutator interpolated within the circuit the[c] current of which circuit it is to measure in combination with a magnetic or electric[i] field of force, and recording or indicating apparatus

A monodynamic motor arranged to perform a known & definite work in excess of the f̶r̶ normal friction of the motor a̶l̶s̶o̶ & actuated mechanism so as to cause the number of revolutions of such motor to bear a known relation to the strength of the Current.

The Combination with Electric Lamps worked in multiple, with the motor in the omnibus wire of all the Lamps[j]

The use with a monodynamic motor of a contact whereby at each revolution 1 or more closings of a separate circuit is made such ckt Containing an Electromagnet [actuating?][k] recording or indicating mechanism—[j]

The combin with the monodynamic motor of worm & worm wheel.

The Electromagnet in main circuit closing when lamp put on to close the field circuit.

Mention my other patents where commutator used & dynamo, difculty wrk multiple[f] contacts without great friction & wear parts & heating & n̶e̶c̶e̶s̶s̶i̶t̶y̶ ̶v̶e̶r̶y̶ ̶l̶o̶w̶ ̶r̶e̶s̶i̶s̶t̶a̶n̶c̶e̶ ̶i̶n̶ [s̶m̶a̶?][b] difficulty attaining sufficient low resistance Etc—[9]

We tried this at Laboratory & it rotates about once a minute with 1 Lamp on— E[dison]

Be sure & get this patent up in good shape by my return Thursday[10]   E

ADfS, NjWOE, Lab., Cat. 1149 (*TAED* NM018AAL). All figures drawn on separate unsigned pages; some later renumbered in an unknown hand. [a]Paper damaged; text from patent specification. [b]Canceled. [c]Obscured overwritten text. [d]Paragraph written on same page as figure. [e]"Paddles mounted . . . apparatus" written in left margin. [f]Interlined above. [g]Written in left margin. [h]Several additional words written and canceled in left margin. [i]"magnetic or electric" interlined above. [j]Followed by dividing mark. [k]Illegible.

1. Edison's assistant John Ott made several drawings of the device described here on 17 May. It was likely "Edisons new meter" that Charles Batchelor saw work "perfectly" two days later. Unbound Notes and Drawings (1883); N-82-12-21:91–95; Lab. (*TAED* NS83ABZ; N150091); Batchelor journal, 19 May 1883, Cat. 1343:71, Batchelor (*TAED* MBJ002071).

Edison had experimented extensively over several years to design a practical electric meter. In the instrument used commercially in his light and power system at this time, current passing through an electrolytic cell deposited copper from the solution onto metal plates. The weight of the plates provided the basis for calculating the amount of current supplied to each consumer in a given period (see Doc. 2163 [headnote]). Continually trying to improve the meter, he considered at least two rotary devices in 1881, one of them a small Gramme motor, to turn the recording apparatus (Docs. 2075, 2143 esp. n. 4). In the preceding year, Edison had directed limited research on entirely different forms of meter (see Docs. 2324, 2328, and 2412).

Ott continued to work sporadically on meters, including a motor meter, during the late spring and summer. In July, he made what he called a "make and brake [break]" instrument, with unsatisfactory initial results. Ott's few surviving drawings of this device included no explanatory text, and its operation is unclear. They show no recording instrument, but the object may have been a mechanical way to register current consumption without the need to weigh metallic plates, a process requiring some skill and time but which would be crucial to the commercial success of the Edison system beyond New York City (Ott to TAE, 11 May, 12 and 22 June, 24 July, and 1 Aug. 1883; all DF [*TAED* D8369F, D8369G, D8369H, D8369I, D8369M, D8369N]; N-82-12-21:103, 105; Unbound Notes and Drawings [1883]; Lab. [*TAED* N150103, N150105; NS83ACC]). In June, experimental assistant Thomas Conant had made a series of trials "on the weighing of zinc meter plates, the object being to devise a balance whose action shall be extremely rapid and which shall at the same time weigh within a few say 5 to 10 mgs. of the correct weight" (N-82-11-14:237–73, Lab. [*TAED* N143237–N143271]).

2. This draft was the basis for a patent application that Edison executed on 6 April. The application was rejected on the grounds that this form of electric motor was anticipated by other U.S. and British patents. Edison excised eight claims and the patent issued, without significant changes to the text of the specification, in 1887. Pat. App. 370,123.

3. That is, D^1, D^2, and D^3.

4. Text is "platinum paddles—," "fan," and "trough containing mercury into which paddles dip   trough may be made of insulating material or metal not amalgamated by mercury."

5. Edison drew "fig 3" on the same page as the above figure. Similar to figure 2 in the published specification, this drawing was a cross sec-

Edison made this cross-sectional view of the meter's rotating cylinder for the patent application but did not refer to it in the draft text.

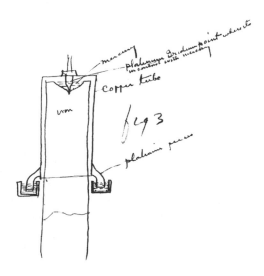

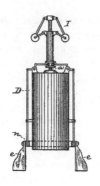

Patent drawing for a centrifugal governor to regulate the rotating cylinder's speed by varying the depth of the paddles in a trough of mercury.

tion of the cylinder and trough showing the "platinum paddles," iron cylinder inside a "copper tube," atop which was the "platinum Iridium point where its in contact with mercury." He also made an incomplete sketch below figure 3.

6. Edison started, but did not complete, a drawing of this idea on a separate page, marking it "Fig. 7." It is represented in the patent as figure 5.

7. Figure labels are "one pole," "trough rotates with core," "magnet," and "other pole."

8. Edison explained in the patent specification that the "peculiar electro dynamic motor" was "a non-commutator or uni-polar machine of such character that I apply to it the term 'mono-electro-dynamic motor,' its inductive or rotating part being a straight or one-part conductor, the current passing through it in one direction only." He experimented with this type of machine as a generator in early 1878. U.S. Pat. 370,123; see Doc. 1683.

9. Edison presumably intended his closing directives for Richard Dyer, his principal patent attorney.

The resulting patent specification referred to Edison's U.S. Patent 242,901, issued in 1881, for a meter operated by an electric motor: "but such motor is shown as a machine provided with a commutator, which machines, when used as the operative or controlling elements of electrical meters, are subject to the objection that the multiple contacts of the commutator cannot be made without considerable indefinite or variable friction, which increases the liability to error in registering, the wear upon the parts is great, and there is difficulty experienced in attaining sufficiently-low resistance." Edison had lowered resistance at the commutator while designing his large central station dynamo in 1881 and had a related patent application pending at this time. See Docs. 2131, 2134, and 2149 esp. n. 4.

10. Edison planned to leave New York on Wednesday, 4 April, on electric light business in Albany. He attended an exhibition of his system that day in Cornwall, Ontario, some two hundred fifty miles further from New York City. The editors have been unable to verify the date of his return. A. J. Lawson to Insull, 6 Apr. 1883, DF (*TAED* D8367J); Docs. 2420 and 2428; "Edison's Light in Canada," *NYT,* 5 Apr. 1883, 4.

SHORT-CORE DYNAMO DESIGN   Docs. 2419 and 2432

Docs. 2419 and 2432 are representative of the fragmentary record of major modifications to Edison dynamos. Spurred by the work of John Hopkinson in Great Britain, Edison committed substantial time and money to this effort in 1883. While he made some indisputably original alterations, notably to the armature, much of his work seems to have followed that of Hopkinson. Edison's effort neither drew upon nor contributed to the active scientific search at this time for a general theoretical understanding of dynamo performance.[1] It is uncertain whether his alterations (largely involving proportions) could have been patented in the United States, and Edison evidently did not try; nevertheless, they informed the design of every subsequent Edison dynamo.

Edison's standard dynamo design, originated in 1879 as the "long-legged Mary-Ann," used distinctively elongated magnet cores (about 42 inches) to generate a strong magnetic field around the armature. This design helped the machines attain a level of efficiency that astonished contemporary experts, and it formed the basis for every model built at the Edison Machine Works for the U.S. and foreign markets before 1883.[2]

The initial push to alter the machines came from the British marine market, where the Edison Electric Light Co., Ltd., was finding its greatest commercial success. The company had secured contracts for shipboard lighting, but marine engineers objected to the machines' large dimensions, especially the belt transmissions needed to multiply the relatively low speed of the steam engines. John Hopkinson, a talented and ambitious young consultant to the company, set out to transform the Edison dynamo into a slower machine that could be connected directly to the engine.[3] He largely succeeded in late September 1882 by connecting the field magnet pairs in parallel rather than series. This arrangement increased the current through the magnets (and consequently the strength of the field) enough to compensate for the slower rotation, but it consumed more current and lowered the machine's overall efficiency.[4]

Edison seems to have learned of Hopkinson's redesign in November 1882 when Arnold White, secretary of the British company, visited New York. He had already adapted a dynamo—the central station Jumbo—to turn at a steam engine's speed by concentrating the magnetic fields of a dozen elephantine cores onto two pole pieces and greatly increasing the number of armature windings, but this was a costly and bulky design.[5] On 1 December, he ordered tests at the Edison Machine Works of a dynamo with magnets in parallel, "in order to see at how slow a speed it was possible to run an L machine with usual load with a view to running direct from a Brotherhood Engine" (the British company's engine of choice for marine plants).[6] He also paid some attention to the magnetic circuit in December. He had the Machine Works test a "350 light K dynamo with field Surface reduced" in such a way as to further concentrate the magnetic lines of force, but this effort may have been transitory and unsystematic.[7] In any case, Edison turned his attention during the winter to different types of commutator brushes.[8] High rotational speeds exacerbated wasteful and destructive sparking at the commutator, and Edison's efforts to improve brushes at this time suggests that he was not interested in lower speeds as a possible solution. He continued to see his primary market for isolated plants on dry land and consequently, unlike the British company, did not feel constrained by space.

Hopkinson saw that changing the armature speed opened the way for other improvements. He continued to work into early 1883 with the expectation that reducing the machines' first cost would be a major advantage in the British market, even with the prospect of higher operating expenses. A gifted mathematician, he began a methodical analysis of the magnetic circuit in small laboratory models, eventually realizing that the same mass of iron employed in shorter but thicker field coils would significantly increase the machine's output (in part by increasing the cross sectional area without requiring additional wire). He constructed an armature less prone to eddy currents and internal heating, and he incorporated these changes in a provisional patent specification filed on 22 February 1883.[9] Soon afterward, Hopkinson further substituted wrought iron for cast iron in the limbs and made an incremental improvement of the sort usually seized on by Edison: winding the magnets with square wire to fill the available space more fully.[10]

Edison found himself in the unfamiliar position of playing catch-up with his own invention. Edward Johnson warned from London, just prior to 22 February, that Hopkinson had "made some very considerable improvement on the big dynamo." He reportedly thought it "very essential that we should know what it is so as to secure it in some way."[11] Though it is not clear what Edison learned through the British company or from Hopkinson's provisional specification, it probably was enough to alert him to the short field cores. Edison had no claim on Hopkinson by contract or personal loyalty, unlike employees in his laboratory or the Machine Works. Even his influence over the British company was mitigated by his grant of manufacturing rights to Mather & Platt, a Manchester engineering firm headed by Hopkinson's brother Edward.[12]

Edison took up the general problem posed by Hopkinson's work: how to increase electrical output without committing either space or electrical energy to additional magnets. For several months, beginning in late February or early March, he closely directed William Andrews in several lines of experimentation. Most details of this work (including nearly all of Edison's instructions) have been obliterated by the loss of records from the Machine Works, but it is possible to discern general outlines from Andrews's replies and reports.[13]

Edison's inclination seems to have been to increase the inductive capacity of the armature and the strength of the magnetic field. On 6 March, he claimed to have a new model that could operate 350 16-candlepower lamps, good enough for 325 lights in actual practice. Though this machine struggled to produce adequate voltage, Edison was confident—falsely, it turned out—that Hopkinson could do no better.[14] Andrews tried modifications to the armatures and field coils of Edison E and Z dynamos and small laboratory models of larger machines, like the L and K.[15] He bored out the pole pieces, at least in part to accommodate more windings on the armature, a change he believed was antithetical to short magnets.[16] He also slightly enlarged the field cores and put additional wire on them. Any of these changes could have raised the voltage, and by early April he was getting more than 100 volts (see Doc. 2419).

Edison began experimenting with short field cores about the middle of March. In mid-March, Andrews tested "the Z Dynamo with 4 short cores connected like an 'L'" model, that is, the poles of each magnet in series but the pairs in parallel with each other. Despite a "heavy loss of magnetism," it produced enough current at 107 volts for 120 A lamps and seemed to offer the chance for further improvements.[17] Samuel Insull boasted on 1 April that Edison was getting "even better results than Hopkinson & in an entirely different manner" and planned to spend $10,000 on experiments.[18] Edison promised in early April to send a "new K" 350-light dynamo for exhibit at the National Exposition of Railway Appliances in Chicago; by the time it was operating a few weeks later, it was referred to as the "H" machine. At least nine others were built for sale.[19]

Edison continued to get uneven results while working along lines similar to Hopkinson. In Doc. 2419, Andrews reported lower voltage from short cores than from other machines. A short dynamo (not necessarily the same one) gave more favorable results in Doc. 2432. Edison referred in late April to the "new H machine," one of which was running in Chicago by that time.[20] That design was probably similar to 450-light machines tested in the following weeks. Andrews tried "the 450 Light Dynamo with short cores" on 12 May with disappointing results: "The dynamo does not show as great an increase of EMF as I expected, and after about an hour's run . . . the armature was quite hot."[21] There was some breakthrough in mid-May, however, and on the 14th, a cable went to Edward Johnson announcing "Newest K running six hours five hundred thirty lamps" and instructing him to inform Edison's business associates in France and Germany. Although the trial armature had continued to run unacceptably hot, by 21 May Insull reported that the Edison Machine Works had started building "twenty 450 light machines," a number of which were installed during the summer at the Southern Exposition in Louisville, Kentucky.[22]

Although exactly what happened remains unclear, these developments coincided with a visit from Paris by Charles Batchelor, Edison's closest experimental colleague and a veteran of his earlier dynamo research. Batchelor had met Hopkinson in London on 12 March, and two days later he "Commenced to make a Z Dynamo with short cores." He reached New York by 1 May and soon was "looking very carefully" into the subject; whatever he had been able to learn of Hopkinson's work was surely useful now.[23] Johnson also provided specific help on 17 May by cabling from London the electrical resistance in Hopkinson's armature and "shorter thick" magnets.[24] (That very day, someone—possibly Batchelor—produced a comprehensive set of "Notes on Dynamos" giving the design specifications of the H and four older models.)[25] Francis Upton, a university-trained mathematician who had played a major role in developing the first Edison dynamos, was consulted during the development process, but he apparently did not leave his post at the Edison lamp factory.[26]

Details about the new H machine are scant. It used an armature that was essentially an elongated version of the 55-volt B armature configuration of the L dynamo, with two additional wires per induction loop to help carry the current. To create the required voltage, it ran at 1,200 revolutions, about one-third faster than the antecedent K, in a much stronger magnet field. Designed mainly for central station service, it was available only in the 110-volt A configuration.[27] It was conservatively rated for just 350 lamps in isolated plants to "guard against any possible breakdown," though it was warranted for the full 450 lights under the care of a central station engineer.[28] Its six twenty-five inch field cores were connected (like those of the K) in series in three pairs, each pair wired in parallel with the other two.[29] Three of the first five numbered machines were installed at the central station in Shamokin, Pennsylvania, and the H became a workhorse of the Edison central station business.[30] Edison apparently supplied a picture of the machine to Nathaniel Keith, who had it published with a brief description in Schellen 1884 (357).[31]

The new H became the general template for redesigning all the Edison isolated and village plant dynamo models during the summer and early fall. Edison had enough confidence in the modifications by early June to prepare a price list, though most new models were not ready for months. The Machine Works made at least a prototype of a four core 100-light T machine by late July,[32] about the time Edison began designing a 300-light machine, later called the Y model. The Y seems to represent a further departure from past practice in that Edison attained its capacity with a single field magnet (two cores). Breaking with Hopkinson's emphasis on reducing construction costs, he used an extraordinary amount of copper in the magnet: 207 pounds of heavy-gauge wire in each core, compared with 65 pounds in the H, L, and K machines.[33] (Charles Batchelor, who had begun his own redesign process soon after returning to Paris, made a different 300-light machine by modifying the proportions of the L dynamo's four cores, much as Edison had transformed the K into the H model.)[34] Late in the year, Edison planned to remake the 100-light T machine into a single magnet (two-core) dynamo.[35] In any case, the design process was not a straightforward one. Charles Rocap, secretary of the Machine Works, noted during the summer that "we get more out of some dynamos in proportion to the materials and cost than out of others."[36] The new generation of Edison belt-driven dynamos eventually produced substantially more current for roughly the same weight and size as their predecessors. They incorporated the shorter cores (if not necessarily the elliptical shape adopted by Hopkinson), Edison's armature modified for heavier current, and, presumably, the new brushes as well. Edison spent about $25,000 on the design and testing processes.[37] A complement of the new machines, including several H examples, was displayed at the 1884 International Exhibition in Philadelphia (by which time Edison had also developed a short-core C or "Jumbo" central station machine).[38] These particular models had a short commercial life, however, and were succeeded in the Edison catalog by a new generation of dynamos in 1885.[39]

The British technical press quickly gave the ambiguous title "Edison-Hopkinson dynamo" to the short-core design.[40] Schellen 1884, published in the United States, applied that name to Edison's new H model. In Britain, however, it was reserved specifically for machines made by Mather & Platt. Hopkinson himself differentiated between American and British dynamos, pointing out that Edison did not adopt other improvements that distinguished the British machines: namely, wrought-iron pole pieces and an armature with more iron and better internal insulation.[41] Edison also did not entirely do away with multiple magnets or use oblong cores, two changes which Hopkinson found to increase the output per mass and area and permit the low-speed operation originally sought by the British company.[42]

The short-core Edison-Hopkinson dynamo, drawn in 1885.

The new Edison
H dynamo, illustrated
in 1884.

1. Jordan 1990 identifies participants (in addition to Hopkinson) in the effort to develop a full theoretical understanding of the dynamo between 1883 and 1885. It also explains the major points at issue, particularly with respect to the magnetic circuit.

2. Edison did not seek patent protection on the distinctive form of his basic dynamo. On the development and patent status of that form see Friedel and Israel 1987, 70–72; Docs. 1702, 1727 esp. n. 2, 1735, and *TAEB* 5 chap. 2 introduction; on claims about its efficiency see Docs. 1832, 1895, 1910, and 1916 esp. n. 5; and on the design specifications of production models see *TAEB* 6 App. 3 and Andrews 1924, 166.

3. The British company had completed (or had contracts for) lighting on eleven ships in May 1883. It obtained at least £10,000 in naval contracts for slow-speed plants in August, leading company secretary Arnold White to believe "the whole of the English ironclad fleet will now fall into our hands." By late 1884, electric lighting was considered "comparatively common on board [British] ships," and the Edison company reportedly had outfitted more than fifty passenger vessels. Edison Electric Light Co., Ltd., report, 25 May 1883; Arnold White to TAE, 25 Aug. 1883; both DF (*TAED* D8338ZAF, D8338ZBK); "Notes—The S.S. 'San Martin,'" *Electrician* 13 (4 Oct. 1884): 467–68; see also *TAEB* 6 App. 2.

4. Edison was not unaware of the physical challenges posed by marine installations. Advised in April 1883 that the Edison Co. for Isolated Lighting was having trouble securing contracts for steamships because of weight constraints, he endorsed a recommendation to reduce the mass of the steam engine. He also suggested using ten candlepower lamps and substituting for two dynamos a single one with slightly enlarged field cores. John Hopkinson to Edison Electric Light Co., Ltd., 30 Sept. 1882; Joseph Hutchinson to TAE (with TAE marginalia), undated Apr. 1883; both DF (*TAED* D8239ZFI, D8325V); Arapostathis 2007, 13.

5. Edison discussed the future of the British electric lighting business during White's visit from late October or early November to mid-December. White to Samuel Insull, 13 Dec. 1882, 1 and 11 Jan. 1883, all DF (*TAED* D8239ZGC, D8338A, D8338C); see also Docs. 2374 and 2375. On the Jumbo dynamo design, see Docs. 2122 and 2238 (headnotes).

6. Charles Clarke notebook of "Electrical Experiments and Tests," [p. 4], acc. no. 1630, Box 30, Folder 2, EP&RI (*TAED* X001K3); Arapostathis 2007, 12–13.

7. Edison's attempt to reduce the area of the pole faces was reminiscent of a similar effort in connection with a disk armature in 1881, although the geometry in that case was different. Charles Clarke notebook of "Electrical Experiments and Tests," p. 21, acc. no. 1630, Box 30, Folder 2, EP&RI (*TAED* X001K3); see Doc. 2082 n. 2.

8. See Doc. 2420 esp. n. 14.

9. Brit. Pat. 973 (1883); Arapostathis 2007, 14; Thompson 1886, 166–68; Kapp 1886, 229–30; Kapp 1891, 262–63. A number of L dynamos so modified to run at 500 rpm were sold by the British company for isolated shipboard plants in 1883 ("The Edison Hopkinson Dynamo," *Electrician* 11 [20 Oct. 1883]: 546–47). One is illustrated in Doc. 2126 (headnote).

10. Arapostathis 2007, 14; Thompson 1886, 166–68.

11. Insull acknowledged an 18 March letter from Johnson (not found), which apparently provided more dynamo information. Charles Batchelor to Joshua Bailey, 20 and 22 Feb. 1883, Cat. 1239:470, 473, Batchelor (*TAED* MBLB4470, MBLB4473); TAE to Edward Johnson, 6 Mar. 1883, Lbk. 15:423 (*TAED* LB015423); Insull to Johnson, 1 Apr. 1883, LM 3:115 (*TAED* LM003115).

12. Grieg 1970, 18.

13. Edison Electric Light Co.—Testing Department, DF (*TAED* D8330); see also Doc. 2416.

14. TAE to Johnson, 6 Mar. 1883, Lbk. 15:423 (*TAED* LB015423); see also Doc. 2420. The Edison Machine Works produced about twenty of these 350-light, long-core "H" models (see Doc. 2553). At least one was sent to London by late March, and another to Boston by early May. Judging by its serial number, the latter was manufactured some months earlier, indicating that it was not originally designed or built to this capacity. The model letter designations during this transitional period confused even Edison agents. Sherburne Eaton defect reports, 28 Apr. and 12 May 1883; Spencer Borden to TAE, 15 May 1883; both DF (*TAED* D8328J, D8328K).

15. Andrews's imprecise nomenclature and sketchy descriptions of these machines, some of which were entirely experimental, make it difficult to determine exactly what he was doing at any specific time.

16. Insull reported (with Edison's approval) to Johnson on 1 April that "By simply boring out the K Field a little more Edison has got 350 lights out of this machine. He says in these experiments he is simply going back to old Menlo Park practices [where?] they always got 100 lights out of what is now the Z machine. . . . Edison tells me as you lengthen your magnets (up to a certain point) the more you can bore out your field. with short magnets you require very slight space with longer magnets you can increase the space." Insull to Johnson, 1 Apr. 1883, LM3:115 (*TAED* LM003115).

17. Andrews to TAE, 14 Mar. 1883, DF (*TAED* D8330N).

18. Although the method Insull described was technically feasible it would have increased the cost of the dynamo, which Hopkinson was trying to lower for the British market. Insull to Johnson, 1 Apr. 1883, LM 3:115 (*TAED* LM003115); cf. Doc. 2357; Hopkinson to Edison Electric Light Co., Ltd., 30 Sept. 1882, DF (*TAED* D8239ZFI).

19. The Edison Co. for Isolated Lighting had nine of the machines in stock on 16 May. Edison considered them safe for 350 lamps except in village plants where, under the care of a professional engineer, they could be run for 400 lights. TAE marginalia on George Bliss to TAE, 9 Apr. 1883; Bliss to TAE, 24 Apr. 1883; Bliss to Insull, 30 Apr. 1883; all DF (*TAED* D8372B, D8372C, D8364G); TAE to Bliss, 12 Apr. 1883; TAE to Spencer Borden, 16 May 1883; Lbk. 16:137, 354 (*TAED* LB016137, LB016354).

20. TAE to Frederick Clarke, 28 Apr. 1883, Lbk. 16:209 (*TAED* LB016209); George Bliss to TAE, 28 Apr. 1883, DF (*TAED* D8364F).

21. Andrews to TAE, 12 May 1883; Andrews test report, 12 May 1883; both DF (D8330ZAH, D8330ZAI).

22. Batchelor to Johnson, 14 May 1883, Insull to Johnson, 21 May 1883; LM 1:311A, 3:150 (*TAED* LM001311A, LM003150); Andrews to TAE, 18 May 1883, DF (*TAED* D8330ZAJ); see also Doc. 2501.

23. Batchelor sailed from Liverpool on 21 April, landing in New York on the evening of 30 April. He remained there until 19 May. Batchelor journal, 12 and 14 Mar., 21 and 30 Apr., 19 May 1883; Cat. 1343:37–38, 57, 61, 71, Batchelor (*TAED* MBJ002037, MBJ002038, MBJ002057, MBJ002061, MBJ002071).

24. Johnson also reported that the machine would run "two hundred [lamps] easy efficiency higher" (Johnson to TAE, 17 May 1883, LM 1:311C [*TAED* LM001311C]). The specifications had been provided by William Mather, of the Manchester firm, to Frank Sprague, who had tested one of the machines and submitted a formal report to Johnson the same day (Sprague 1883a). Sprague was in New York by early June and could possibly have influenced the design before the Machine Works finished its order (see Doc. 2456).

25. Notes on Dynamos, 17 May 1883, Unbound Notes and Drawings (1883), Lab. (*TAED* NS83ABZ1).

26. See Doc. 2420.

27. Notes on Dynamos, 17 May 1883, Unbound Notes and Drawings (1883), Lab. (*TAED* NS83ABZ1); Cat. 1235:46–47, Batchelor (*TAED* MBN012046); Charles Rocap to Samuel Insull, 6 July 1883; DF (*TAED* D8334W); Schellen 1884, 353; see also App. 3 and *TAEB* 6 App. 3.

28. An undated 1883 catalog of the Edison Co. for Isolated Lighting, illustrated with an engraving of the H, rated the machine at 400 lamps. TAE to Edison's Indian & Colonial Electric Co., 22 June 1883, DF (*TAED* D8316ABW); Edison Co. for Isolated Lighting booklet, undated 1883, PPC (*TAED* CA002B).

29. Unlike Hopkinson, Edison did not at this time increase the field cores' diameter. "Old Edison Type Dynamos," 59; William Hammer Notebook 8:41–42 (1882), Ser. 1, Box 13, Folder 2, WJH (*TAED* X098F02); Cat. 1235:46, Batchelor (*TAED* MBN012046).

30. Statement of Shamokin expenses, undated [November 1883?];

Edison Electric Illuminating Co. of Shamokin report, n.d. [June 1883?]; both DF (*TAED* D8360ZDG, D8457H3).

31. Nathaniel Keith to TAE, 11 May 1883, DF (*TAED* D8303ZDE).

32. See Doc. 2499. The T, developed as an experimental eighteen-inch armature model, was among those whose test results were recorded on a standardized form at the Edison Machine Works. The form also included blanks for some design specifications, which generally were filled in. Edison Electric Light Co.—Testing Dept., DF (*TAED* D8330).

33. Frank Sprague to TAE, 16 Apr. 1884; Edison Electric Light Co. test report, 21 Sept. 1883; both DF (*TAED* D8442ZDG, D8330ZAX); Cat. 1235:46–47, Batchelor (*TAED* MBN012046); Andrews 1924, 166. Gustav Soldan began testing a prototype Y machine on 19 September (see Doc. 2529).

34. By October, Batchelor was manufacturing a four-core 300-light machine that turned at a slower speed than Edison's (Batchelor to TAE, 4 Oct. 1883, DF [*TAED* D8337ZEP]). His specifications, notes, and calculations for dynamo experiments throughout the summer and fall are in Cats. 1237, 1235:4–51; Batchelor (*TAED* MBN006, MBN012004–MBN012051); see also Doc. 2501.

35. See Doc. 2553.

36. TAE to Rocap, 5 June 1883, Lbk. 17:58 (*TAED* LB017058); cf. Doc. 2499; Rocap to Insull, 6 July 1883, DF (*TAED* D8334W).

37. See Doc. 2569.

38. Franklin Institute 1884, 26, 61; on the experimental C model, see Doc. 2580.

39. See, e.g., Edison Co. for Isolated Lighting list of plants, undated September 1885, DF (*TAED* D8522J).

40. The name "Edison-Hopkinson" was in public use by August 1883. "Report on the Edison-Hopkinson Dynamo," *Electrician* 11 (11 Aug. 1883): 296–98.

41. Hopkinson was later paraphrased as telling a meeting of the Institution of Civil Engineers that:

> Mr. Edison had followed the improvements of the Edison-Hopkinson machine by shortening the magnets and increasing their sectional-area, with the result that the more recent Edison machines had fields of an intensity of nearly 4,000 per square centimeter, being nearly double of the older machines with the long magnets, but still only half the Edison-Hopkinson. By also adopting the improvements of the armature, and using either wrought-iron pole pieces or letting the wrought iron into the cast-iron, Mr. Edison would further greatly improve his machine. [Kapp 1885–86, 233]

The British-made dynamos quickly developed a reputation with the Edison Swan United Co. for unreliability, and the firm preferred those built by Edison in New York (Samuel Flood Page to TAE, 27 Mar. 1884, DF [*TAED* D8437M]).

42. "The Edison-Hopkinson Dynamo," *Electrician* 11 (20 Oct. 1883): 546–48; "The Philadelphia Exhibition.—III," ibid., 13 (11 Oct. 1884): 496; Thompson 1884a, 46–51; Thompson 1884b, 475; Arapostathis 2007, 15–18.

New York, April 3rd *1883*[a]

Dear Sir

I have just tried the experiment with the Z Dynamo that has been bored out to give ¼ inch clearance between field and armature— Load 75 "a" Lamps (8½ to HP)— With field in series, Resis 40 ohms   1200 Revs gave 108 Volts[1]

With field multiple arced & resis= 10 ohms   1200 Revs gave 117 Volts[b]

The 4 short cores Dynamo with a resis in field of 10 ohms and 120 Lamps gave with   1200 Revs. 98½ Volts— Yours truly

W. S. Andrews[2]

ALS, NjWOE, DF (*TAED* D8330Q). Letterhead of Edison Electric Light Co. [a]"*New York*," and "*188*" preprinted. [b]Followed by dividing mark.

1. See headnote above.
2. William Symes Andrews (1847–1929) was a skilled instrumentalist and mechanic who began working for Edison as a laboratory assistant at Menlo Park in 1879. He took charge of the Testing Room, a department of the Edison Electric Light Co., at the Edison Machine Works in early 1882. Doc. 2223 n. 2.

[New York,] 3rd April [188]3

My Dear Johnson[1]

The powers are against me. They seem to object to my writing to you at any length. I started in a letter to you at the office Sunday.[2] I was obliged to leave it to attend to some matters for Edison   I went back to it Monday (yesterday) & something turned up that made me drop it. I essayed to finish it this morning & the mail would close at a certain time so I closed up so much of it as was written & sent it off.

I started to [–][a] indite this over at the office & the fates would go against me ~~in~~ in the form of the wretched Engineer who put out the Light at 10 p.m. so I have repaired to "my own fireside" (that sounds strange in April) & here I propose to spin out the yarn to you t~~o~~ill I have said my say—my whole say[3]

My last letter of any length to you was dictated to Mac-Gowan[4] & I could not speak as freely as I can in this.

To go back to Hopkinsons improvements[5] Edison thinks H. wants to bulldoze the London Co[6] into buying his Dynamo improvements. I thinks my letters will show you that Edison has done as much as Hopkinson[7]   I showed your letters on this point to Upton.[8] He has promised me that he will write you on the Technical points. He says when Hopkinsons comes to run the Z machine[9] at its present speed then he will begin to find his trouble. Edison gets 120 "A" lights[10] out of a Z at 1200 rev. of the armature & no heating & not more than 20 lbs more wire on the armature. Can Hopkinson do that   Edison gets 350 lights out of the K easily at 1040 revolutions of the Armature & is building 12 machines which we will ship at $1350 each f.o.b.[11] New York. He will get more lights yet out of the K. Edison says his 40 light machine will give 100 before he gets through and it was on that, that he cabled you that he would deliver 50 machines of 100 lights f.o. Dock at London for £75 each.[12] I do not know about this price but Edison insisted to me he could do it & sent the cable as he said to show you that he was getting good results as well as Hopkinson. Edison says you must not let Hopkinson bulldoze the London Co into agreeing to pay him (H.) for his Dynamo improvements. Multiplicity of cores is not new. The big Dynamo[13] the K. & the L. have a number of Cores; and then he says Hopkinson cannot get a patent on the exact length of a Core

Furthermore Edison has been experimenting on Brushes. He has got a brush that can be put on a Commutator at any point and will not spark & he says that is going to make a big difference. It will he says enable him to get 1500 lights out of the Central Station machine without any mercury or anything else on the Commutator & without troubling his head about the non-sparking point.[14] He says that with some further experimenting he can make this brush for use on the Z, K, & L machines and he looks upon this as a tremendous point gained   Keep stirring Edison up on these matters & he will then keep his experiments going.

As to his lamp he has made a good deal of progress but he has nothing to ~~tell~~ send[b] you at present & in as much as he is now in his laboratory but little (reasons hereafter) I doubt whether you will see that lamp for some time to come. If you are coming home on May 1st (as Mrs Harrison[15] says—although I put it near July 1st) you will not see it at all before you get here. Upton has promised to touch on this subject also in his letter to you.

Now as to affairs here.

In a letter to Edison you ask him why he is so blue as you take it he must be from his letter to you & Batch[16] asking you both to return. You imagine that something has gone wrong. Edisons end of the line is all right & if those who are supposed to run the business kept their end up Edison would not write for you & Batch. The trouble is as I told you about a year ago in my letters to you when you were then in London the Edison Electric Light business is not run well[17] & so much impressed is Edison by this fact that he has practically left his Laboratory & now makes my Office his Headquarters & is attending to purely business matters in onethe way he used to when I first came over here. He said he could see that this Village Plant business[18] would go to ruin unless he came up & attended to matters. He says he has been told again & again by people that they could getting nothing satisfactory from the Executive Officers of the Light Co.[19] So Edison decided to "drop science & pursue business" The trouble first came over financial matters. The Isolated Co[20] were away behind in their collections. The Home Office[21] was gone for & they collected nearly) $100,000 between 1st Feby and 31st March. This is bang up & they are doing splendidly so far as collections are concerned. But the New England Dept have only collected $28,000 in that period. Edison came up to my office early last week & for three days Moore[22] Hutchinson[23] Hastings,[24] Edison, & myself went into all the accounts & the result is that when Borden[25] came here last Saturday he was just lectured by everyone (but the President) for money money money & we are hoping[c] he will go home & benefit by what was said to him. The Isolated Co had out lots of notes & they have all they can do to meet these notes I can tell you. They have now out about $70,000 of notes. As a consequence we have been unable to get hardly a cent from them either for Lamp Co,[26] Bergmann,[27] Machine Works[28] & Tube Co[29] & last Saturday Edison had to meet the pay Roll for every one[d] of those concerns out of his own resources although the shops aggregated bills amounting to $25,000 against the Isolated Co. The Illuminating Co[30] has no money & the Parent Co[31] had no money until they made a Call of 20% which should have been made two months ago. Eaton[32] does not face these financial problems. He ties everybodys hands; will not allow the Treasurer to do anything (although Hastings so far as he has been allowed to has worked like a horse) & simply sets down thinking things will right themselves or else calls Conferences the result of which is to delay things weeks & weeks & make us all suffer. Take for instance the machine wks we were told to build $5000 of Dynamo a week some time back and were to have six weeks notice to[c] stop. We judged tight times were coming & did not do a quarter of that amount of work, but[c] later arranged that we would only build 12 K machines which was afterwards increased to 16 & when the Isolated Co. got in a tight pinch all Eaton did was to countermand the order & leaves us with 16 Ks partly finished, our bills for material all owing for those machines. Eaton thinks it is a very simple matter to countermand goods ordered or in other words finance your business by repudiating your obligations

There are some amusing sides to this business. You should see Bergmann.[33] Almost every morning he telephones me for money and what ever we do for him he always replies with[c] his [bad?][a] Dutch accent "That will do me no good." On Saturday he told me he did not know where to get his Pay Roll so I gave him a check for $700 which considerably overdrew our Bank Account. When I gave it him he promptly said "That will do me no good." Then he tried to get Hastings. He was out & would not be back till after the Bank closed. Then Bergmann raved. I can tell you it was rich. The little Dutchman has not had any backbone in him in a month. He has gone under completely. Of course it was out of the question to let his Pay Roll go unpaid so I determined to get Edison to sign a check for $1000 on Drexel Morgan & Co[34] & trust to my being able to make it good by the Monday after. Fortunately just about two o'clock I found I could get $1000 downtown for some goods shipped that day & so I was easily able to give Bergmann the $1000 to meet his Pay Roll & when I did so I really think he had the greatest difficulty to restrain his habit of saying "That will do me no good." Things dont look much better this week. The Lamp Coys Pay Roll comes first being due tomorrow (Wednesday) & Hastings cannot give any money for it so we had to meet it. How the Tube Co, Machine Works, & Bergmann will fare I do not know   Their Pay Rolls come on Saturday. So far we have not missed a Pay Roll nor has Edison had to make any sacrifices & no doubt we shall all pull through all right. Hastings & myself are going to Albany tomorrow[c] morning & we are going to see on our way up there whether we cannot improve matters financial somewhat.

Village Plants. This business looks very promising & it is this that Edison and myself are going to give our attention to this Summer. Edison is going to take contracts for the erection of these plants. We shall close the first contract probably on Thursday & there are plenty of other places which are just crying for these Plants. Edison saw that only a few Plants would be erected in the course of a year at the rate the Light Co were setting about the business & hence his taking it up himself. He proposes that we shall give practically all our time to this business & he wants you to give some time to it when you get back.

Talking of Village Plants reminds me ~~with~~ of the arrangement Edison has made with me about making some money out of it for Sammy. First let me confess that I have done what you told me not to. I went to Edison about my own affairs. The fact is that things had got so bad with me I could stand it no longer & if a change had not taken place I should have gone home. The other evening when Edison & myself had been working late together & were all alone I said "now Edison we have done nothing but talk about your finances & the Coys finances for some days if you do not mind ~~that~~ we will talk about my finances." I told him that I had not made any money in six months that in as much as ~~my~~ his[b] business occupied ~~a great d~~ all my time I could not go outside that business to make money as if I did my office affairs went wrong (which is so as I tried the experiment last summer)   I also told him that I had no money at all & that I could go along no further. I said I did not want a salary & would not take one, that I would leave the future to take care of itself but that he must show me some way to get enough to live on.[35] I told him I was sorry to go to him at such a time when everything was financially tight but that I was obliged to ask his advice   I told him I would not have thought of going to him if it were not that I was dead broke. He then proposed that out of any profits on these Village Plant Contracts I should have first $2400[c] a year & 20% of the profits remaining.[36] That is suppose we take a contract for Brockton for $30,000 (whole plant put up complete) and that our contracts for Boilers Engines Dynamos wiring & appliances &c came to $27 000.[37] Say that is all the work we did for the whole year we should have $3000 profit to divide—my $2400 would come out & then I should take 20% of the remaining $600. But if nothing is made I shall not get anything at all not even the $2400 so it cannot be said I am taking salary when it is entirely dependant on profits. Edison asked me if that was satisfactory & intimated that if it was not enough he wanted me to say what would be. I told him that whatever he did I was perfectly satisfied with; that it was very handsome but that I must still ask him to show me how I could get money to live until that business developes. He told me to draw $what money I needed from him & to debit it to my account & if this business turns out all right I shall pay Edison back just as early as I can. I may add that you also are to have 20% of the remaining profits after the ~~charge~~ amount[b] of $2500 to be paid me is deducted. Edison will take the other sixty per cent. Now I do not know what you will think of my going against your advice. I suppose you will call me a little fool &c &c   If so I cannot help it. It took me three weeks to speak to Edison & I felt pretty sore at being obliged to as I think he should have made the necessary arrangements at least to enable me to live without my asking him as he knew I was hard up. What do you think of it. I know you will not like my having spoken to him. I did not do it without thought. I was so dead broke that I did not care about staying longer & I came to the conclusion that I ought to speak to Edison. I think this is the first time I have gone against your advice since I came here.

Phila. Record. This Plant has been put all right & the Proprietor is now tickled to death with Electric Light[38]   Steringer[39] wished me to tell you this.

Central Stations. Pearl Street is working bang up.[40] Hastings gives a good deal of time down there & is adjusting everything   They have now between 5,000 & 6,000 connected & instructions have been given to go on connecting up to [–][a] 10,000 lights. This is with Edisons full approval & is perfectly safe in the light of present experience.[41] All the people now ~~getting~~ being connected are day consumers & I think before long I shall be able to ~~pledge~~ [st----][a] advise you that the daily average for 24 hours is 1000 lamps. As soon as we[b] get in that position our business will be a fine good paying concern. This business looks very good.[42]

Boston is having some "checks" in the formation of a Company but I think it is only because of very dear money both in New York & Boston[43]

Albany will doubtless be closed tomorrow night. Hastings & myself are going up there to see what we can do ~~during~~ up there. Edison is going to try & meet us there[44]

Then the Village Plants are going to be in great demand & we think there will be lots of this business

Telephones. The Bell Telephone Co[45] in their suits with Drawbaugh[46] find they cannot hold their ground at all on the Carbon Transmitter[47] unless Edison will agree to assist them in several interferences and unless he agrees to sign the papers for the division of one of his patents. The Bell Co sent Serrell[48] & Swan[49] of Boston to see Edison and he told them he would not do a thing till they gave him an agreement undertaking to put his name in large bold letters on every Transmitter. If Edison will only stick out he has got the Bell Co. He says he will insist that they put his name where Blakes is now & put Blakes in small letters.[50] He cannot get any money but he can certainly get some glory although it is rather late in the day for it.

Nothing is being done with your instruments at present.

It is just 3 a.m. & I think I will get to bed so au revoir   Yours as ever

Saml Insull[51]

Love to Edna & Maggie[52]

April 4th

I have just read over this letter & think parts of it look "blue." Of course such things as financial trouble is but temporary & cannot last many months. I wish you to understand that when we are all short of money Bergmann & Co get their share of what we have got just as much as the other Shops. Let me have a letter from you   I have had just one since you left but then I always have to blackguard you on this point

ALS (letterpress copy), NjWOE, Lbk. 3:120 (*TAED* LM003120). "Per SS Servia" written above dateline to indicate ship carrying letter. [a]Canceled. [b]Interlined above. [c]Obscured overwritten text. [d]Multiply underlined.

1. A former telegrapher, Edward Hibberd Johnson (1846–1917) was a longtime close associate of Edison and promoter of his inventions. Johnson had spent much of 1881 and 1882 in London working to commercialize the electric light. He went back in February 1883 and, despite Edison's suggestion in early March to return home, Johnson remained there into the summer. He was at this time vice president of the Edison Electric Light Co. Doc. 272 n. 13 and *TAEB* 1–6 passim; on Johnson's trip in early 1883, see Docs. 2047 and 2396.

2. Insull began writing in reply to Johnson's 18 March letter and, before closing on 2 April, acknowledged receiving two more from Johnson, none of which has been found. He wrote largely about Edison's efforts to improve dynamo efficiency and about the patent claims of John Hopkinson for a three-wire distribution system. Insull to Johnson, 1 Apr. 1883, LM 3:115 (*TAED* LM003115).

3. The brownstone building at 65 Fifth Ave. where Edison had offices (also the headquarters of the Edison Electric Light Co.) lay outside the distribution area of the Pearl St. station and was lighted by its own isolated plant. Insull had lived in a third-floor dormitory there and then with Edison at Menlo Park, but about this time was living a few blocks from the office at 38 West Eleventh St. *Trow's* 1884, 823.

4. Frank McGowan was a stenographer at the Edison Electric Light Co.'s office (Doc. 2261 n. 2). The editors have found only one lengthy letter written from Insull to Johnson since January, and it is in Insull's own hand.

5. John Hopkinson (1849–1898), the British electrical engineer and physicist, became a consultant to the Edison Electric Light Co., Ltd.,

in London about April 1882. Hopkinson had recently been trying to re-design the Edison dynamos. See Docs. 2180 n. 9, 2258, and 2419 (head-note).

6. The Edison Electric Light Co., Ltd. (organized in March 1882), owned Edison's electric light and power patents in Great Britain. Doc. 2221 n. 4.

7. See note 2 and Doc. 2419 (headnote).

8. Francis Upton (1853–1921), a university-trained mathematician and physicist, joined Edison's Menlo Park laboratory staff in 1878 and had a crucial role in designing and evaluating Edison's first dynamos. In late 1879 he also began working on incandescent lamp experiments, and at this time he was general superintendent and treasurer of the Edison Lamp Co. Docs. 1568 n. 1, 2260 esp. n. 2, *TAEB* 4 chap. 8 introduction, *TAEB* 5 passim.

9. Edison dynamo models were given alphabetic designations; for their general design characteristics see *TAEB* 6 App. 3.

10. Lamps designated "A" operated at about 110 volts; "B" lamps required half that voltage. Unless otherwise indicated, it was generally assumed that standard "A" lamps for the U.S. market were rated at 16 candlepower.

11. F.O.B. is the abbreviation for "free on board," a common element in sales contracts. The seller's obligations are discharged once the goods pass the ship's rail, at which point all costs and risks of transport are assumed by the buyer. Wharton and Lely 1883, 340; Gilles and Moens 1998, 134–37.

12. Edison cabled this offer on 2 April with the promise that he had doubled the output of his dynamos. Lbk. 16:68A (*TAED* LB016068A).

13. The C ("Jumbo") dynamos used at the Pearl St. central station in New York and the Holborn Viaduct demonstration plant in London were nominally rated for 1,200 16-candlepower lamps, though the New York machines could each run 1,400 lamps for a short time. Doc. 2238 (headnote), *TAEB* 6 App. 3.

14. Sparking between commutator bars and brushes was a serious problem, especially under conditions of heavy load, which reduced dynamo efficiency and could damage the machine. Although sparking was partly a problem of design of the armature, commutator, and brushes, the proper manufacture, operation, and maintenance of dynamos were also critical to its suppression. Edison had once prescribed coating commutator bars with mercury but discontinued the practice for health reasons. In general, sparking could be minimized by aligning brushes with the neutral (or non-sparking) point where current in the armature coil reverses direction as the rotating coils pass into a region of the magnetic field of opposite polarity. The neutral point advances in the direction of the armature as dynamo load increases, requiring a commensurate rotation of the brushes. Edison had already put considerable effort into the sparking problem, especially while testing his first large central station dynamo in 1881. See Docs. 1862, 1896, 2100, 2122 (headnote), 2126 (headnote), 2149 esp. n. 4, 2150, 2228, and 2484.

The editors have not positively determined the nature of the commutator brushes to which Insull referred. At Edison's direction, William Andrews had been testing brushes since late 1882. Made of copper, iron, or german silver wires separately insulated with mica or plaster

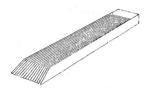

The Edison commutator brush, illustrated in 1884, consisted of a number of copper strips, so as to divide the spark when the contact was broken.

of paris, they were prone to excessive heating (see correspondence from Andrews to TAE in Edison Electric Light—Edison Machine Works—Testing Department; and Edison Electric Light—Edison Electric Light Co.—Testing Department; both DF [*TAED* D8235, D8330]). "Special resis Brushes sent by Mr Edison" were tried at the Machine Works several times beginning on 22 March (Charles Clarke notebook of "Electrical Experiments and Tests," pp. 17–20, 161–63, 170, acc. no. 1630, Box 30, Folder 2, EP&RI [*TAED* X001K3]). Thompson 1886 (47), republished from the second edition (1884, p. 48), described the Edison brush as "a number of copper strips placed edgeways to the conductor, and soldered flat against one another at the end furthest from the collector." One rationale of this design was "the subdividing of the spark at the contact." See also Doc. 2551.

15. Not identified.

16. Charles Batchelor (1845–1910) had been Edison's chief experimental assistant since 1873 (Doc. 264 n. 9). He went to Paris in 1881 to manage Edison's exhibit at the Exposition Internationale de l'Électricité. He remained there to establish and supervise Edison's factory for lamps and electrical equipment just outside the city, having previously helped set up Edison's first lamp factory at Menlo Park (see Docs. 1950 [headnote], 2086, 2111, 2112, and 2196). Edison's letters recalling Batchelor and Johnson are Docs. 2404 and 2407; the editors have not found that from Johnson. Batchelor visited New York from 30 April until 19 May (Batchelor journal, 30 Apr. and 19 May 1883; Cat. 1343:61, 71, Batchelor [*TAED* MBJ002061, MBJ002071]).

17. The editors have not found such letters from Insull. It is not clear what would have caused Edison's impatience with the Edison Electric Light Co.'s development of business at that time. The prototype generating station was still under construction in New York, and Edison had not yet conceived the idea of village plant systems for smaller cities and towns.

18. See Doc. 2424 (headnote).

19. The Edison Electric Light Co. owned Edison's U.S. patents for electric light and power under an 1878 agreement (Doc. 1576). Prospective sellers of equipment protected by those patents had to obtain a license from the company. Sherburne Eaton, who had largely managed the firm's affairs as vice president, continued to do so after he became president in October 1882; Edward Johnson replaced him as vice president. See Doc. 2356 esp. n. 9.

20. The Edison Co. for Isolated Lighting was formed in late 1881 to install individual lighting plants (such as for residences, factories, and mills) under license from the Edison Electric Light Co. It was a stock company in which the Edison Electric Light Co. held a controlling interest. Doc. 2189 n. 2; Sherburne Eaton memorandum, 15 June 1883, DF (*TAED* D8327ZAB).

21. That is, the main office in New York of the Edison Co. for Isolated Lighting. The company maintained regional agencies or departments in Boston and Chicago.

22. An experienced engineer and businessman, Miller Moore (1842–1930) was general manager of the Edison Co. for Isolated Lighting from 1882. Moore left in late 1883 to become an agent for Armington & Sims. He seems to have remained in that position less than a year before mov-

ing to the shipbuilding and engineering firm of Samuel F. Moore & Sons Co. as secretary and treasurer. Doc. 2221 n. 19; Gardiner Sims to TAE, 19 Nov. 1883; Moore to TAE, 5 Dec. 1883; both DF (*TAED* D8322ZDI, D8322ZEF); "Moore, Miller F.," Pioneers Bio.

23. Joseph Hutchinson began working as an accountant or clerk for the Edison Electric Light Co. before the end of 1881. At this time, he was the secretary of the Edison Co. for Isolated Lighting. Hutchinson remained associated with the Edison lighting business for many years. Hutchinson to TAE, 4 Apr. 1883, DF (*TAED* D8325V); Jehl 1937–41, 943, 946, 951.

24. Frank Seymour Hastings (1853–1924) worked with Egisto Fabbri in the shipping firm of Fabbri & Chauncey until 1882, when he became connected with the Edison lighting interests. At this time he was secretary of the Edison Co. for Isolated Lighting and treasurer of both the Edison Electric Light Co. and the Edison Electric Illuminating Co. of New York. Hastings held similar positions with successor companies. "Hastings, Frank Seymour," Pioneers Bio. *NCAB* 14:255; Insull to Johnson, 11 Apr. 1883, LM 3:135 (*TAED* LM003135).

25. Formerly an agent for the Edison Co. for Isolated Lighting, Massachusetts businessman Spencer Borden (1849–1921) became head of the company's nascent New England Department in the summer of 1882. He was also the agent in the region for the Edison Electric Light Co. The extended Borden family largely controlled the principal businesses of Fall River, Mass., though Spencer and his brother, Jefferson Borden, Jr., suffered a financial collapse in 1879. Doc. 2272 n. 5; "The Borden Family's Troubles," *NYT*, 6 Sept. 1879, 2.

26. The Edison Lamp Co. manufactured electric lamps in Harrison, N.J., under license from the Edison Electric Light Co. The firm was a partnership of Edison and several close associates, including its general superintendent, Francis Upton. It was formed under a variant name in 1880. See Docs. 2018 and 2343 (headnote).

27. Bergmann & Co. made sockets, fixtures, meters, and other small equipment for Edison lighting companies under license from the Edison Electric Light Co. It was established in New York in early 1881 as a partnership among Sigmund Bergmann, Edward Johnson, and Edison. See Docs. 1790 n. 8, 2091, and 2343 (headnote); see also note 33.

28. In partnership with Charles Batchelor, Edison organized the Edison Machine Works in 1881 to manufacture dynamos and other heavy electrical equipment. Its shops were on New York's Lower East Side. See Doc. 2343 (headnote).

29. The Electric Tube Co. manufactured underground electrical conductors for the Edison system. Minority stakes in this incorporated firm were held by Edison, Charles Batchelor, John Kruesi (its superintendent), and partners in Drexel, Morgan & Co. It was located in New York City until late April 1884, when it moved to Brooklyn. Doc. 2343 (headnote); Electric Tube Co. to TAE, 22 Apr. 1884, DF (*TAED* D8433O); map, p. 2.

30. The Edison Electric Illuminating Co. of New York was incorporated in December 1880 to provide electric light and power service in Manhattan. It operated the Pearl St. central station and the First District distribution system. See Docs. 2037 and 2243 (headnote).

31. The Edison Electric Light Co.

32. Sherburne Blake Eaton (1840–1914), a lawyer by training, was president and de facto manager of the Edison Electric Light Co. He held similar positions in the Edison Co. for Isolated Lighting, the Edison Electric Illuminating Co. of New York, and the Edison Electric Light Co. of Europe (Docs. 2120 n. 7 and 2356 esp. n. 9). Eaton resigned his offices and director positions in late 1884, though he remained active with the Edison Electric Light Co. as its counsel.

33. A skilled machinist, Sigmund Bergmann (1851–1927) began working for Edison in 1870. He opened his own shop in New York City in 1876 and maintained close ties with Edison, for whom he manufactured phonographs, telephones, and electric lighting equipment. In 1881, he joined Edison and Edward Johnson in forming Bergmann & Co. Bergmann was a native of Thuringia (later incorporated into Germany), and Johnson often referred to him as "Dutch," meaning German or Teutonic. See also note 27; Docs. 313 n. 1, 1680 n. 2, and 1790 n. 8; *OED*, s.v. "Dutch."

34. The New York firm of Drexel, Morgan & Co. had acted as Edison's bankers since 1878, when it took a dominant role in arranging funds for Edison's electric lighting research and development. The firm controlled Edison's patents in Britain for electric light and power. See *TAEB* 5–6 passim, esp. Docs. 1612, 1648, and 1649.

35. Fifteen months earlier, Insull pleaded to Johnson about improving his straitened finances, a subject on which he had "never said a word" to Edison: "I do not know what to do. Help me out. . . . Of course if I did not make some money outside (and lately that has been next to nothing) I should not be able to live." He worried that he was "wearing myself out with work & that if it goes on I must knock under & I naturally ask what is to be the compensation for such an extreme expenditure of my time and health. Just advise me & oblige." He reiterated these thoughts in a recent letter to Johnson (Insull to Johnson, 17 Jan. 1882 [pp. 10–11] and 6 Mar. 1883; LM 3:20, 109 [*TAED* LM003020, LM003109]; also cf. Doc. 2244). Insull initially received a salary from Edison and also some compensation from the manufacturing shops, for which he acted as secretary. He later recalled that almost immediately, he asked for a financial interest in Edison's inventions instead of a salary, a suggestion that pleased Edison, "not that he cared anything about the money . . . but he liked his immediate assistants to take a chance with him on their remuneration." He also remembered that Edison had given him during his first year as much as $15,000 in various securities (Insull 1992, 32–36).

36. Cf. Doc. 2417.

37. Spencer Borden had arranged in February to canvass the city of Brockton, Mass., for a village plant system. Three estimates were prepared at Edison's direction: one for 1,600 lights at about $46,500, one for 4,800 lights at about $70,000, and a third for 3,200 lights at an unknown cost. An operating company was formed sometime in March and a contract with Edison (for 1,600 lamps) concluded in May. Edison Electric Light Co. Bulletin 17:13, 6 Apr. 1883, CR (*TAED* CB017); Borden to TAE, 17 Feb. 1883, DF (*TAED* D83251); Insull to William Lloyd Garrison, Jr., 25 Apr. 1883; Insull to Johnson, 27 Apr. 1883; Lbk. 16:192, 201 (*TAED* LB016192, LB016201); see also Doc. 2436.

38. The *Philadelphia Record* contracted for a 250-light isolated plant

at its new building in central Philadelphia in mid-1882. The order pleased Sherburne Eaton, who believed the publishers to be personal friends of the local Maxim lighting company. Edison Electric Light Co. Bulletin 12:11, 27 July 1882, CR (*TAED* CB012); Eaton to TAE, 20 June 1882, DF (*TAED* D8226ZAB).

39. Formerly a gas engineer, Luther Stieringer began designing interior electric illumination for Edison in 1881, starting with the headquarters of the Edison Electric Light Co. in New York. He went on to a distinguished career as an illuminating engineer and was responsible for the artistic lighting of Edison exhibits at several major expositions. Doc. 2298 n. 11.

40. Edison's Pearl St. generating plant, the first permanent commercial electric central station, began supplying current throughout a district in lower Manhattan on 4 September 1882. See Doc. 2338.

41. This number of lights in simultaneous operation would have substantially exceeded the capacity of the six dynamos at the Pearl St. central station. The First District crossed the 10,000 lamp threshold by the end of May; in August the Edison Electric Illuminating Co. suspended new connections pending expansion of the generating plant. See Docs. 2243 (headnote) and 2338 n. 5.

42. Joseph Casho, superintendent of the Pearl St. station, had just prepared a report of expenses, cash receipts, electrical load, and other operating information; a copy was sent to Edison. Insull confided to Johnson that the station's collections of customer bills had been "poor" up to this time, but that he expected that Casho would improve them. Casho to Charles Clarke, 2 Apr. 1883, DF (*TAED* D8326F); Insull to Johnson, 11 Apr. 1883, LM 3:135 (*TAED* LM003135).

43. Formation of the Edison Electric Illuminating Co. of Boston was announced in late March; the editors have not determined the nature of the present "checks," and neither Edison nor Insull seems to have been apprised of them further. Insull scheduled a canvass of the city to gather statistical information in early 1884, but a central station was not built at that time. "The Edison Light in Boston," *NYT*, 29 Mar. 1883, 2; Insull to Sherburne Eaton, 7 Dec. 1883; Alfred Tate to William Lloyd Garrison, Jr., 14 Mar. 1884; LM 17:40, 18:356 (*TAED* LBCD4040, LBCD5356).

44. Edison, Insull, and Hastings visited Albany to resolve a dispute between an agent of the Isolated Co. and an Insull associate over the right to form an Edison central station company there (Insull to Johnson, 11 Apr. 1883, LM 3:135 [*TAED* LM003135]; see also Doc. 2409 n. 4). The parties reached a settlement but, despite a canvass of the city in January and some discussion about a central station location, it is not clear that they organized an illuminating company there (Alfred Tate to Hastings, 18 Jan. 1884, DF [*TAED* D8416AFL]; TAE to W. A. Graves, 18 Mar. 1884, LM 18:382 [*TAED* LBCD5382]).

45. The American Bell Telephone Co., located in Boston, owned the principal telephone patents in the United States. Originally incorporated in early 1879 as the National Bell Telephone Co. to exploit the inventions of Alexander Bell, it achieved an effective monopoly under a November 1879 settlement with Western Union that gave it control of the patents of Edison and Elisha Gray. Garnet 1985, 44–54; Bruce 1973, 260–71.

46. Daniel Drawbaugh (1827?–1911), a Pennsylvania mechanic and inventor, surfaced in 1880 as a claimant to have anticipated the Edison carbon telephone transmitter. His claim was pressed by the People's Telephone Co. against the American Bell Telephone Co. in a protracted but ultimately unsuccessful legal challenge to the Bell Co.'s patents (Obituary, *NYT*, 4 Nov. 1911, 13; Bruce 1973, 272–75; James Storrow to William Forbes, 23 Mar. 1881, Box 1167, NjWAT [*TAED* X012IBF]). Houston 1885 (78–103) describes the transmitter and presents Drawbaugh's priority claims.

47. The carbon telephone transmitter, which Edison invented and developed in 1876–78, altered the strength of an electric current from a battery through a wire in response to modulations of the speaker's voice. It was based on the principle that fine particles of carbon vary the resistance of an electric circuit according to the pressure exerted on them, in this case by a diaphragm acted on by sound waves (see Doc. 759 [headnote] and *TAEB* 3 and 4 passim). While simple in conception, such an instrument could take many specific forms; other inventors produced their own variations, leading to complex disputes over the breadth of patents. Patent Office interference proceedings involving several of Edison's patents or applications were still pending, and variant instruments competed with those manufactured for the Bell company. See Docs. 1270 and 1792.

48. Lemuel Wright Serrell (1829–1899), Edison's patent attorney from 1870 to 1880, had been involved closely with drafting many of Edison's telephone transmitter patent applications. Serrell continued to act on Edison's behalf regarding applications filed during his time of service. Docs. 110 (headnote) n. 2 and 2120 n. 5.

49. William Willard Swan (1837?–1911) was a Boston attorney largely engaged in patent matters. Davis 1895, 554; U.S. Census Bureau 1970 (1880), roll T9_561, p. 23.2000, image 0377 (Boston, Suffolk, Mass.); Holliday 1935, 70, 324.

50. Francis Blake devised a more reliable and accurate carbon transmitter in 1878. After some further modification, the instrument, commonly known as the Blake transmitter, was widely used by the American Bell Telephone Co. Docs. 1740 n. 5.

When Lemuel Serrell advised in mid-April that the U.S. Patent Office had declared a new interference involving patents of Edison and Blake and an application of Drawbaugh, Edison scrawled on the letter: "You must get that contract signed by the Bell Co." Some agreement was apparently reached, for the Bell Co., at Edison's request in November, sent one of their standard telephones "having the new form of lettering with 'Edison Carbon Telephone' " printed on it. Edison indicated he wished to use it in trying to persuade the United Telephone Co. in London to "follow the same course as that adopted by your Company." Serrell to TAE, 16 Apr. 1883; TAE to Theodore Vail, 28 Nov. 1883; Vail to TAE, 30 Nov. 1883; all DF (*TAED* D8370ZAQ, D8316BKX, D8374ZAL).

51. Samuel Insull (1859–1938) became Edison's personal secretary in February 1881. Having steadily gained Edison's trust since then, he had wide responsibilities beyond those suggested by his title. Insull had general oversight of Edison's financial affairs, advised him on business,

and coordinated the work of the manufacturing shops. Docs. 1947 n. 2, 2092, and 2343 (headnote).

52. Edna and Maggie Johnson were Edward Johnson's minor daughter and wife, respectively. Docs. 1790 n. 14 and 2258 n. 3.

−2421−

Draft to Porfirio Díaz[1]

[New York,] 27th Apl 83

Dear Sir[2]

~~Understanding that you are interested in~~ Thinking perhaps that you might be interested in the progress of[a] electric lighting it would afford me great pleasure to ~~explain~~ show you and your friends[b] the working of ~~our~~ my Central Station system ~~in~~ downtown and give you such explanation as you may desire in connection therewith.

The evening would be the best time to visit our works but I shall be glad to keep any appointment you may make[3]   Yours respectfully

DfL, NjWOE, DF (*TAED* D8303ZCB). Written by Alfred Tate. [a]"Thinking . . . progress of" interlined above. [b]"and your friends" interlined above.

1. José de la Cruz Porfirio Díaz (1830–1915), a Mexican general and political leader, dominated Mexican politics in the last half of the nineteenth century. He first served as president from 1876 to 1880 and again from 1884 to 1911. During his long administration, he sought to industrialize and modernize the country by increasing both foreign investment and the role of foreigners in the Mexican economy (*Ency. World Bio.* 4:534–36; Wasserman 2000, 161–66, 209–28). While Díaz was in New York, Jorge Hammeken y Mexia, a lawyer identified by Edison as Díaz's secretary, became interested in promoting Edison's inventions in Mexico (Zaremba 1883, 33; Edison to Samuel Taylor, 23 May 1883, Lbk. 16:441 [*TAED* LB016441]). On 12 May, he and Edison reached an agreement that gave Hammeken a year to form a syndicate for this purpose (Miller [*TAED* HM830180]). In 1889, Edison would send Díaz a phonograph as part of his efforts to promote the phonograph in Mexico (TAE to Díaz, 9 Oct. 1889, Lbk. 33:46 [*TAED* LB033046]).

2. This draft was addressed to "General Díaz   Windsor Hotel City." The editors have not found a finished version of this letter, but see note 3.

3. On 16 April, Porfirio Díaz and his wife, with other members of his party, were conducted on a tour by Edison and his wife, Edison Electric Light Co. secretary Frank Hastings, and Col. George W. Sherman, a friend of Díaz's who was then in New York negotiating with Edison and the Edison Electric Light Co. in regard to electric lighting in Argentina. They began their tour at a fire station outfitted with the Edison light and with automatic devices arranged by Edison to turn on the lights when the alarm bell rang. Several demonstrations were made of this system. The party then toured the Pearl St. station, the Edison Ma-

chine Works, and Edison's laboratory. Edison later recalled that Díaz's visit lasted two days and that he "took them around to railroad buildings, electric light plants, fire departments, and showed them a great variety of things." "Electricity in Engine-Houses," *NYT*, 17 Apr. 1883, 8; App. 1.C.14.

–2422–

To Henry Morton

[New York,] April 10th [188]3

My dear Mr Morton[1]

Referring to your favor of 10th inst[2] I am very sorry that I have not a Motor to let you have, but I am making a $\frac{1}{10}$, $\frac{1}{4}$, 1 and 2 horse power motors for Central Station work. I have a $\frac{1}{10}$ H.P. Motor finished and am testing it.[3]

It is not very economical being designed for cheapness. I will probaby be able to let you have one in about Two or Three weeks and I will try and come over to see you the first opportunity I have[4]   Yours truly,

Thos. A Edison   I[nsull]

L (letterpress copy), NjWOE, Lbk. 16:119 (*TAED* LB016119). Signed for Edison by Samuel Insull.

1. Henry Morton (1836–1902), a professor of chemistry and, since 1870, president of the Stevens Institute of Technology in Hoboken, N.J., first met Edison in January 1878 in connection with the phonograph. In 1880, he participated in tests of incandescent lamps, including Edison's, that led him to conclude that Maxim's were more efficient. *ANB*, s.v. "Morton, Henry"; see Docs. 1667 n. 4, 1927, 2010, 2107 n. 1, 2022, and 2033.

2. In his letter, Morton indicated that he was conducting tests of storage batteries that included running small electric motors in order to measure the current and power they produced. While noting Edison's opposition to storage batteries, he stated that "what we have here is decidedly better than any thing you have seen yet in that line" and invited him to visit the Stevens Institute. Morton read a paper "On the Storage of Electricity" and gave a demonstration to the New York Electrical Society at the Stevens Institute on 25 April, using a motor from the Weston Electric Co. and Edison incandescent lamps. Morton to TAE, 10 Apr. 1883, DF (*TAED* D8303ZCC); "Storage of Electricity," *NYT*, 21 Apr. 1883, 3.

3. For tests of these motors see John Ott to TAE, 10 and 20 Apr. 1883, both DF (*TAED* D8369D, D8369E); see also Doc. 2459.

4. The editors have not determined if Edison supplied the motor or visited Morton.

To Nathaniel Pratt[1]

Dear Sir,

Mr Insull has explained to me the terms you offered at your interview yesterday which I understand to be as follow:—

I am to give you all the orders I have for Boilers & you will allow me a drawback of fifty cents per horse power & quote prices to me exactly on the same basis of those given to the Edison Electric Light Co.[2] The terms of payment are to be as follows:—

25% with the order

50% 30 days after shipment

I to notify you the date ɫ I want the Boiler shipped   The balance (25%) to be paid sixty days from shipment.[3]

If this is your understanding please write[a] me a line to that effect[4] and let me have the necessary Price Lists[5]   Yours truly

Thomas A Edison

LS (letterpress copy), NjWOE, Lbk. 13:14 (*TAED* LB013014). Written by Samuel Insull. [a]Obscured overwritten text.

1. Nathaniel W. Pratt (1852–1896) was an engineer who at this time served as treasurer of the New York firm of Babcock & Wilcox. He would later become the company's president (Obituary, *NYT,* 13 Mar. 1896). Babcock & Wilcox was a noted manufacturer of stationary boilers, especially its patented "non-explosive" water-tube design, which carried a higher pressure than other types. The company had given Edison estimates for large central station steam plants in 1880 and provided the boilers for the Pearl St. station (Hunter 1985, 336–39; Docs. 1897, 2008, and 2125).

2. On 11 April, Pratt replied that the arrangement described by Insull was correct but clarified the terms of the drawback as "solely for boiler for which we have your personal order" and not boilers ordered by any other Edison companies. This letter also repeated the terms of payment stated by Edison in this document. DF (*TAED* D8323A).

3. In a letter to Armington & Sims, requesting these same terms for the purchase of steam engines, Edison noted that they were accepted by Babcock & Wilcox and his other suppliers for central station work. He gave as a reason for the arrangement that "in my contract for this kind of work I receive twenty five per cent cash on the signing of the contract and fifty per cent when the station is finished and twenty five per cent thirty days afterwards." TAE to Armington & Sims, 2 July 1883, DF (*TAED* D8316AEA).

4. On 24 April, Pratt wrote Insull to ask if Edison's letter and his reply should be understood as a contract. Edison's marginal note indicates "not a contract but an understanding," but in the formal reply that Insull sent on 7 May, he told Pratt "that the correspondence which has passed between yourself and Mr. Edison as to boilers forms a contract between you and him in connection with any orders for boilers that he may be able to place with you." Pratt then confirmed this understand-

ing in his letter to Insull of 7 May. Pratt to TAE, 24 Apr. and 7 May 1883, both DF (*TAED* D8323C; D8323D); Insull to Pratt, 7 May 1883, Lbk. 16:275 (*TAED* LBo16275).

5. The price list was enclosed with Babcock & Wilcox's letter of 11 April (see note 2). DF (*TAED* D8323B).

VILLAGE PLANT CONSTRUCTION   Doc. 2424

Local agents working on behalf of the Edison Electric Light Company had begun making plans in late 1882 to set up Edison central stations in Pennsylvania and Massachusetts. When Edison formed his Construction Department in May 1883, he took over plans already afoot in at least four places: Sunbury and Shamokin, in east central Pennsylvania,[1] and Brockton and Lawrence, in eastern Massachusetts. These were the first plants set up by the Construction Department, and all opened before the end of the year.[2]

The Massachusetts locales were substantial cities, but the towns of Sunbury and Shamokin were orders of magnitude smaller and less dense than New York, where Edison had built his first electric lighting system.[3] It was for such locations that he had designed the Village Plant distribution system. It used a higher voltage (330 volts) than the New York network, which substantially reduced the size of the copper conductors to carry the current. Edison set up a demonstration village system in Roselle, New Jersey, that began operating in January 1883.[4]

Edison had also devised a 220-volt three-wire system in the latter part of 1882, apparently independently of John Hopkinson and William Siemens.[5] Unlike the Roselle plan, in which each set of three lamps could be turned on or off only as a unit, the three-wire system provided individual control of each lamp. This control was possible because a neutral wire interposed in the circuit between each lamp in the pair permitted current to flow to (or from) either side as needed. Referring to the drawing from his United States patent, Edison explained that "when all the devices [lamps or motors] in any multiple-arc circuit are in use current will pass through all such devices, the current passing across from the positive to the negative main conductor; but if one or more translating devices are removed from any series circuit the excess of current which would otherwise affect the other lamps in the circuit is taken by the compensating central conductor, so that the other lamps remain unchanged."[6] A third wire, or compensating conductor, was placed between the main wires in each branch of the circuit all the way to the dynamos. The main conductors were sized to allow a certain electrical loss according to the resistance multiplied by the square of the current (I^2R). Because of the direct relationship between voltage and resistance, doubling the voltage meant that positive and negative conductors just one-fourth the standard size could be used. The third or neutral wire, through which little current would flow in a properly balanced system, could be smaller still. In his instructions to a lieutenant for determining the amount of copper required, Edison indicated that there would be a savings of 62.5 percent from a comparable two-wire network.[7] To reduce the cost still further, Edison was forced to accept high electrical losses resulting in a voltage drop of at least 10 percent along the feeder lines.[8] Even so, the metal remained a major part of the construction bill for each station.[9] The three-wire plan also introduced problems of properly balancing the electrical load on each side of the system, and of regulating voltage throughout (see Docs. 2505 and 2538 [headnotes]).

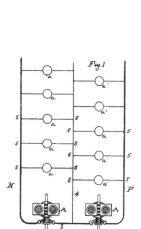

Patent drawing of Edison's basic three-wire system.

Plans for Edison central stations in Pennsylvania were pushed by Phillips Shaw, a merchant and manufacturer in Williamsport. A thriving city boasting concentrated personal wealth from the lumber industry, Williamsport had a nascent Edison illuminating company, the first licensed by the Edison Electric Light Company of New York, which made a public demonstration in mid-March.[10] Shaw became the Edison Electric Light Company's agent for Pennsylvania about that time.[11] Exercising considerable independence, he had solicited interest in Sunbury, a busy county seat on the Susquehanna River, and Shamokin, a major railroad junction about a dozen miles east of there. Both were a few score miles from Williamsport, and both were near the active western edge of the anthracite coal region where fuel for steam power was plentiful and illuminating gas expensive in some cases.[12] The region's mineral wealth flowed to Philadelphia or New York, but there was enough local interest to form a Shamokin company in November 1882.[13] When prospective local consumers, instead of outside investors, capitalized the company in May, Edison praised the Shamokin plan as a "new and successful idea" and reported that Charles Batchelor, his man in Paris, thought it would work there, too.[14] A Sunbury company was organized the following April; despite a reported visit by Edison about that time, there was little enthusiasm and most of the capital eventually was raised in Williamsport, where the board met. The Edison Electric Light Company's published bulletins announced the formation of these and some subsequent local illuminating companies, in several cases identifying the officers.[15]

Edison enclosed with Doc. 2424 an estimate of about $25,000[16] to set up distribution wires in Shamokin, equip the plant, and get it ready for the company to take control in thirty days. In early May, he went there at Shaw's insistence and signed a contract (Doc. 2438) with the Shamokin illuminating company for a system of 1,600 lights and one with the Sunbury company for 500 lights.[17] The plans for those towns (and probably most others) called for using 10-candlepower lamps, presumably to decrease further the amount of copper in the conductors.[18] The Shamokin group wished to set up the station building, boilers, and other equipment (see Doc. 2424). When they did not complete these tasks promptly, however, Edison resolved for the future (according to Samuel Insull) to "undertake to do the work ourselves and not leave any to the local people."[19] He appointed a superintendent of construction, William Rich, and dispatched him to Sunbury instead.[20] Electricians from Bergmann & Company's Wiring Department followed to wire the buildings.[21] As would become generally true of the village plants, the initial subscribers were largely commercial establishments.[22]

After Edison personally inaugurated service on 4 July at Sunbury—his first commercial three-wire distribution system—he stayed to oversee the plant's first days of operation and draw up a rudimentary set of operating instructions.[23] William Andrews, who became his chief electrician for village plants, remained longer, as did acrimony over the nature of the wiring, with the company at one point publicly threatening to prohibit any further wiring work.[24] Edison gave full control of the plant to the local company in early August despite misgivings about its high coal consumption and the quality of its fireman. Concerns about the capability of local skilled and semiskilled labor to operate the machinery with only a few weeks of training became characteristic of Edison's experience with the village installations. The steam plant later developed severe problems that the engineer apparently could neither recognize nor correct.[25] This problem became common in the village plants, and it bore out the truth of a statement Edison reportedly made in 1880, while planning his first New York station, that "steam engineering forms 75 percent. of the electric light."[26] He attributed reports of short lamp life to insufficient attention to manual voltage regulation; partly as a response, he devoted considerable time to developing a more reliable voltage indicator.[27] He also learned that even his own men did not necessarily operate or understand the plant as he would. With the feedback loop for innovation enlarged beyond his personal sight, he gruffly weighed comments and suggestions by mail.[28]

Sunbury foreshadowed other problems that dogged Edison's enterprise. The roof quickly began to leak, an indication of scrimping on generic construction to make the whole project more affordable.[29] The workmanship of the wiring made promoter Phillips Shaw feel "sick" and "ashamed."[30] Most crucially for Edison, the company ran short of money and deferred part of its payment to him. On the basis of substantial early profits, he accepted some of the outstanding balance in stock instead of cash, giving him a degree of influence over its operation. After the Edison Electric Light Company, which took stock in some of the local concerns as part payment of the license fee, Edison was the second-largest shareholder in the Sunbury company.[31] These problems coincided with a crush of demands for canvasses, estimates, and construction elsewhere. By working through them, however, he devised comprehensive policies for managing the business and dealing with local illuminating companies.[32] It was about this time that Edison declared he was taking a break from inventing to be "simply a business man for a year."[33]

Edison moved quickly to replicate the success in Sunbury by dispatching men to set up plants in Shamokin and Brockton.[34] With the Shamokin station nearly complete, he went there himself on 18 September to oversee final details and select a carpenter to assist with house wiring, the Wiring Department having been disbanded. He started the station (with its new short-core H dynamos) four days later.[35]

Shaw also tried to build on the Sunbury experience by recruiting investors for local companies in several other communities in the region. Among them was Williamsport, where he became the station manager.[36] Another was Hazleton, a town of about 7,000 to the east in Luzerne County.[37] Though it was a railroad center in the most intensely productive anthracite district in the state, Hazleton was economically diverse and had capital for local development.[38] An illuminating company was formed there late in 1883 that involved George Markle, Jr., son of a powerful coal operator and banker, who had ties to Wall Street.[39] Advised that "Markle means J[ames]. Hood Wright," a partner of Pierpont Morgan, and that the principals hoped to invest substantially in electric lighting, Edison paid special attention to this station. Like many of the others, it worked well electrically but suffered from physical and mechanical problems after it started in February 1884.[40]

In Massachusetts, where the Edison Company for Isolated Lighting had done much of its business, Spencer Borden found fertile ground for cultivating interest in village systems.[41] He started a canvass of Brockton, a major manufacturing town of about 14,000 residents, in February 1883 and, with William Lloyd Garrison, Jr., organized an illuminating company a month later.[42] Edison planned to put some conductors underground, perhaps as a concession to the gas companies, which opposed overhead wires for arc lighting.[43] In September, Edison was forced to put all the wires underground by what one of his electricians called "local pride in the fact that the streets are not defaced by many poles" and the fact that "the company has not a single dollar of Brockton money in its name."[44] The fact that the company's Boston backers were willing to pay for underground wires—several times what it would have cost to run them on poles, despite relatively cheap illuminating gas—suggests the differences of esthetics, local economies, and expectations for electric lighting between the New England manufacturing city and the Pennsylvania mining communities.[45]

A shortage of copper conductors delayed work in Brockton and Lawrence. Men from the Electric Tube Works, which fashioned underground conduits for Edison, laid out the network in trenches.[46] At nearly the last minute, Frank Sprague tried to requisition conductors for both two- and three-wire service in Brockton. Confused, Edison instructed him to use two-wire tubes already ordered. Although in retrospect Sprague claimed Brockton as the first underground three-wire system, it was apparently not entirely so.[47]

There was also some resistance by prospective subscribers to the high cost of wiring buildings. When it was suggested that Edison promise to pay for gas fixtures if electric customers were not satisfied, he scrawled on the letter (and later crossed out): "We are not so hard up for a market for our wares as to guarantee anything & make promises."[48] Leaving behind all the changes and uncertainties, Edison left for Brockton late on 29 September and, after a stopover in Fall River, started the station two days later.[49]

Construction in Lawrence and Fall River (where a "miniature central station" had opened in early 1882) continued, similarly subject to modifications, equipment shortages and other delays, bursts of impatience, and heavy cash outlays.[50] The question of future expansion arose in a number of towns. In the booming textile city of Fall River, the company decided at the outset to install conductors for an electrical load three times larger than their initial 1,600 light station could provide, though none of this capacity was intended for the city's thousands of mill hands.[51] When Edison personally supervised the start of the Fall River station (the last time he would do so) on 18 December, his template for designing and building small central station systems was essentially complete.

1. Plans for nearby Danville, Pa., were also moving along, and Samuel Insull considered going there on Edison's behalf in May. See Docs. 2441 and 2464; Insull to Philips Shaw, 24 May 1883, Lbk. 16:455 (*TAED* LB016455).

2. See Doc. 2437 (headnote). Fall River, Mass., and Tiffin, Ohio, also opened before the end of the year; see App. 2A. Documents pertaining to village plants comprise the largest group of Edison's extant records from this period. Incoming correspondence concerning specific plants is in the 1883 and 1884 folders of General, Canvassing, and Engineering folders of Electric Light—TAE Construction Department, DF (*TAED* D8340–D8343, D8439–D8442) and in 1883 and 1884 folders arranged by geographic location under Electric Light—TAE Construction Dept.—Stations, all DF (*TAED* D8344–D8363, D8443–D8461). Pertinent letters from Edison or Insull are in 1883 and 1884 folders of Outgoing Correspondence, both DF (*TAED* D8316 and D8416), the Construction Department letterbooks LM14–LM21 (*TAED* LBCD1–LBCD8), as well as General Letterbooks 17–18 (*TAED* LB017–LB018).

3. Population figures from the 1880 census are: Sunbury, 4,077; Shamokin, 8,184; Lawrence, 39,151; Brockton, 13,608; and Fall River, Mass., 48,961. In a decade of rapid growth, Brockton reached 20,783 in 1885. 1883–88, Population (vol. 1), Table 3, Northumberland (Pa.), Essex, Plymouth, and Bristol (Mass.) (316, 209, 210, 208); Kingman 1895, 588.

4. See Docs. 2350, 2336, 2353, 2364 n. 2, and 2417.

5. Docs. 2308 n. 1 and 2407 esp. n. 4 pertain to the earliest surviving direct evidence of the origins of Edison's three-wire system. He believed that his antedated Hopkinson's, which dated to earlier in 1882, if not before. Insull to Edward Johnson, 1 Apr. 1883, LM 3:115 (*TAED* LM003115); Israel 1998, 219.

6. Edison's U.S. Pat. 274,290 also included the idea that, at least in principle, additional compensating wires and proportionally higher voltage could be used. When a member of his Construction Dept. staff suggested something similar in July 1883, Edison responded, "You are behind with that idea   the greatest saving is on 3 wires   with 4 you

dont make proportionate saving." TAE marginalia on Harry Mather Doubleday to TAE, 21 July 1883, DF (*TAED* D8305J).

7. In an 1884 explanation and overview of the system, a top assistant in the Construction Dept. calculated the savings at 69 percent. TAE to William Andrews, 10 Aug. 1883, DF (*TAED* D8316ANI); Henry Guimaraes report, 29 Aug. 1884, Batchelor (*TAED* MB141).

8. Insull to Johnson, 25 Sept. 1884, Lbk. 18:419 (*TAED* LB018419).

9. Edison later blamed the tradeoff between electrical efficiency and the cost of materials for part of the economic difficulty of his central station business; see Doc. 2615. The editors have found few itemized particulars of materials expenses. Fall River, Mass., for which Edison was billed $4,000 for bare copper rods, may be representative. This cost was based on a list price of twenty-six cents per pound, on which he received a rebate of two cents (see Doc. 2515). Installed underground, the conductors were estimated to the local company at about $17,000. Limited information from other plants indicates that the cost of finished and installed conductors ranged from about seven dollars per light for a large station modeled on the Pearl St. plant to about twelve dollars for a much smaller one in Orange, N.J. TAE estimate for Fall River, 4 Oct. 1883, DF (*TAED* D8347K); Insull to Ansonia Brass & Copper Co., 27 Nov. 1883, Lbk. 13:25 (*TAED* LB013025); Charles Clarke estimate for 8,000-light central station, 18 Aug. 1883, Cat. 1237:89, Batchelor (*TAED* MBN006089); TAE to Francis Upton, 26 May 1884, LM 19:277 (*TAED* LBCD6277).

U.S. copper prices generally declined moderately for several years after 1882. The increasing production of western mines, improved transportation, and the end of effective collusion among Michigan producers more than offset the inflationary pressure of increased demand. The New York producer price at this time was around twenty-four cents per pound. Gates 1951, 61–63; Herfindahl 1959, 70–73, tables 1 and 21–22.

10. The Williamsport company was incorporated in May 1882, but construction there did not begin until the end of 1883. Lloyd 1929, chaps. 26–30; Edison Electric Light Co. Bulletins 17:19 and 18:36, 6 Apr. and 31 May 1883, both CR (*TAED* CB017, CB018); Nash, Rumm, and Orr 1985, 47; Alfred Tate to William Rich, 27 Dec. 1883, LM 17:252 (*TAED* LBCD4252).

11. Insull to Johnson, 11 Apr. 1883, LM 3:135 (*TAED* LM003135).

12. Gas in Shamokin was about four dollars per thousand cubic feet and, according to one report, ten dollars in Sunbury (Dublin and Licht 2005, 18–19; Jehl 1937–41, 1096; William Hammer notebook as Chief Engineer of Edison Electric Light Co., 1885–86, Ser. 1, Box 13, Folder 1, WJH; on Sunbury see Bell 1891, 480–500). In towns without gas service, such as nearby Mt. Carmel, prospective companies had to apply for a license from the Edison Co. for Isolated Lighting, which controlled rights to the Edison system in non-gas territory (see Doc. 2299 n. 4). The Edison Electric Light Co. later published some limited retrospective information on the price of gas in fourteen towns and cities (Edison Electric Light Co. circular p. 24, n.d. [1886?], PPC [*TAED* CA001D]).

13. Shaw was among the Shamokin company's directors. Most of

the early investors were Shamokin residents; a notable exception was Francis Upton. The company also sold bonds, largely to its stockholders, to help meet its first expenses. Incorporation certificate, 29 Nov. 1882, Guth; Jones 1995; Bell 1891, 627–28; Edison Electric Light Co. Bulletin 18:11, 31 May 1883, CR (*TAED* CB018); Edison Electric Illuminating Co. of Shamokin ledger (1883–99), pp. 1–3, PSuNCH.

14. TAE to Joshua Bailey, 6 May 1883, LM (*TAED* LM001310B).

15. Both companies were capitalized in May 1883, after a Shamokin delegation visited New York. Among the Sunbury directors were two prominent Williamsport attorneys (Seth T. and Frank McCormick) and a young physician (Thomas Detwiler); another (Charles Story) was from New York. Jones 1995, 70; Edison Electric Light Co. Bulletin 18:11, 31 May 1883, CR (*TAED* CB018); Collins and Jordan 1906, 293–96; Beck 1995, 51; U.S. Census Bureau 1970 (1880), roll T9_1153, p. 502.3000, image 0219 (Williamsport, Lycoming, Pa.).

16. The Shamokin estimate, like those for many towns, has not been found. An unsigned and undated statement of almost $18,000 in actual outlays through early November is in DF (*TAED* D8360ZDG). A surviving estimate for Fall River, Mass., may represent some typical costs. TAE estimate, 4 Oct. 1883, DF (*TAED* D8347K).

17. Much later, while trying to collect the balance due from the Sunbury company, Edison stated that he had never made an estimate nor promised to set up the plant for a specific price. Instead, because he was anxious to get the first plant installed, he agreed not to "charge more than any other customer." Edison billed $11,968, or $1,368 less than he claimed the project cost him. Shaw to TAE, both 3 May 1883, DF (*TAED* D8360B, D8360C); Insull to Spencer Borden, 6 May 1883, Lbk. 16:233 (*TAED* LB016233); see Docs. 2737 and 2464.

18. Edison considered the 10-candlepower lamp equivalent to a standard gas lamp burning coal gas. Insull to Spencer Borden, 11 July 1883, DF (*TAED* D8316AFT).

19. Edison subcontracted construction work in at least one and possibly two later instances. See Doc. 2636; Insull to Shaw, 22 May 1883, Lbk. 16:403 (*TAED* LB016403).

20. See Docs. 2444, 2447, and 2457; TAE to Henry Clark, 27 Sept. 1883, DF (*TAED* D8316AXF).

Postponement of the Shamokin plant's completion until September gave the station benefit of the new H dynamos manufactured over the summer. Sunbury got old style L machines. Although Edison promised to replace them with new models of the same capacity, they remained in service for some twenty years. Statement of Shamokin expenses, undated [November 1883?]; Edison Electric Illuminating Co. of Shamokin report, undated [June 1883?]; both DF (*TAED* D8360ZDG, D8457H3); *Edisonia* 1904, 141–42; see Doc. 2498.

21. See, e.g., Doc. 2464 n. 12.

22. See, e.g., Edison Illuminating Co. of Brockton customer list, 17 Nov. 1883, DF (*TAED* D8346ZBN1). Two hotels were the first Sunbury buildings lighted (Beck 1995, 53. On the wiring department see Insull to Johnson, 21 May 1883, LM 3:150 (*TAED* LM003150); concerning other early customers see Andrews to Insull, 8 Aug. 1883; and Frank McCormick to Insull, 8 Dec. 1883; both DF (*TAED* D8361ZCB, D8361ZDZ).

23. See Docs. 2477 and 2484. Among the "Fourth of July Notes" in the *Sunbury News* of 6 July was this brief paragraph: "The Electric light appeared, for the first time in Sunbury, at the Central Hotel in the evening. It was a complete success, and was inspected and admired by large numbers of citizens. Thomas A. Edison, the inventor, was here." Jehl 1937–41, chap. 103 provides a retrospective account and several anecdotes concerning construction and early operation of the Sunbury plant, and *Edisonia* 1904 (141–42) discusses its technical specifications.

24. Andrews to TAE, 11 July 1883; TAE to Shaw, 2 Aug. 1883; both DF (*TAED* D8361ZAV, D8316ALD); see Doc. 2496 esp. n. 2.

25. Long night shifts, to which most engineers or machine tenders would not have been accustomed, exacerbated some of the problems. It is not clear what type of workers local Edison illuminating companies wanted to or could hire, but it is likely that stationary steam engineers would have been looked upon favorably. In Sunbury, the Pennsylvania Railroad's large shops for locomotive and car repair were probably the major employers of the type of labor that Edison's company required. See Docs. 2484, 2500, 2563, and 2603; Bell 1891, 484–92.

26. *TAEB* 5 chap. 6 introduction, esp. n. 4.

27. McCormick to TAE, 5 Nov. 1883, DF (*TAED* D8361ZDO); see Docs. 2505 (and headnote), 2538 (and headnote), 2584, 2592, and 2669.

28. See, e.g., Docs. 2490, 2505, 2582, and 2615.

29. Problems with the roof were among a list of grievances cited by the company in refusing to pay Edison in full; to the president, the entire Sunbury project looked by 1884 "very much like a swindle." Roofing failures appeared at other plants before the final payments were made. In March 1884, Insull demanded of Rich: "why cannot roofs in Pennsylvania be fixed immediately   this matter is causing us great trouble and keeping us out of considerable money." Frank McCormick to TAE, 18 July 1884, DF (*TAED* D8458ZAE); Insull to Rich, 11 Mar. 1884, LM 18:336A (*TAED* LBCD5336A); see also Doc. 2677.

30. Shaw to TAE, 23 July 1883, DF (*TAED* D8361ZBH1).

31. TAE to McCormick, 10 Sept. 1883; McCormick to TAE, 19 Sept. 1883; Insull to Bergmann & Co., 24 Mar. 1884; all DF (*TAED* D8316AUI, D8361ZDE, D8416AZA); Insull to Frank Sprague, 10 Sept. 1883, LM 15:41 (*TAED* LBCD2041); see also Doc. 2603.

32. See, e.g., Docs. 2472, 2478, 2486, 2497, 2578, 2532, 2559, and 2597. See Doc. 2575 on Frank Sprague's substitution of mathematical for physical models to determine physical dimensions of the conductor networks.

33. See Doc. 2503.

34. Whether through oversight or a shortage of cash, Edison's men at Sunbury were not paid promptly and had trouble getting to Shamokin. When William Andrews complained, Edison noted pointedly for Insull: "This will never do to discourage our men in this manner   it takes that rum out of them that I require." TAE marginalia on Andrews to Insull, 18 Aug. 1883; TAE to Spencer Borden, 24 July 1883; both DF (*TAED* D8361ZCX, D8316AIQ); see also Beck 1995, 61–62.

35. TAE to Andrews, 18 Sept. 1883, LM 15:96A (*TAED* LBCD2096A); Bergmann & Co. to TAE, 5 Sept. 1883; Insull to Andrews, 3 Sept. 1883; both DF (*TAED* D8324ZAE, D8316ASC).

36. See App. 2.

37. U.S. Census 1883–88, Pop. (vol. 1), Table 3, Luzerne (316).

38. U.S. Census 1883–88, Manufacturers (vol. 15), Table 27; Dublin and Licht 2005, 22–23, 28.

39. Markle's brothers Alvan and John were also among the stockholders. Other investors included W. R. Longshore, a local physician, and Peter Kellmer, who established the town's first photographic gallery and at this time was building the Kellmer Piano & Organ Works. *NCAB* 24:138–39, 18:153–54, C:525; Sherburne Eaton to Insull, 4 Mar. 1884, DF (*TAED* D8439ZAC); Edison Electric Lighting Co. of Hazleton minutes, 15 and 16 October 1883, pp. 2–3, PP&L; Bradsby 1893, 308, 1054, 1115, 1159; Beck 1995, 61.

40. Eaton to Insull, 20 Feb. 1884, DF (*TAED* D8455ZAL); see Doc. 2617.

41. *TAEB* 6 App. 2 and Doc. 2272 n. 5.

42. Spencer Borden to TAE, 7 Feb. 1883, DF (*TAED* D8325I); Insull to W. A. Graves, 6 Mar. 1883, Lbk. 15:422A (*TAED* LB015422A).

43. Insull to Johnson, 27 Apr. 1883; TAE to Garrison, Jr., 31 May 1883; Lbk. 16:201, 17:7 (*TAED* LB016201, LB017007); Harris & Robinson to TAE, 22 May 1883; William Jenks to TAE, 8 June 1883; Garrison, Jr., to Insull, 9 June 1883; all DF (*TAED* D8344B, D8346J, D8346K).

44. Sprague to TAE, 16 Sept. 1883 (pp. 4–5), DF (*TAED* D8346ZAG). The president, treasurer, and directors of the Edison Electric Illuminating Co. of Brockton resided in Boston, and Henry Villard agreed to subscribe for $10,000 of its stock. Kingman 1895, 473–74; Villard to Garrison, Jr., 24 July 1883, letterbook 51:262, Villard.

45. In a newspaper interview about the dangers of urban overhead electric wires, Edison stated that it cost about $500 per mile to place wires on poles and 50 cents per foot ($2,640 per mile) to put them underground ("What Edison Says," *New York Herald*, 20 Apr. 1883, DF [*TAED* D8320H2]; Jenks 1885, 7, 15; cf. Doc. 1707 n. 9). Illuminating gas in Brockton was only $2.25 per thousand feet in October 1883. This was similar to charges recorded for Lawrence and Lowell, Mass., in an undated book of "Gas Statistics" in which Edison's staff compiled information about the ownership, capitalization, and fees of scores of gas companies. They are arranged alphabetically by state, with most states from Alabama to New York represented (Edison Illuminating Co. of Brockton "Description of Edison Electric Light Plant," 24 Mar. 1885 [p. 7], PPC [*TAED* CA003A]; Gas Statistics Book No. 1 [n.d.], Electric Light Companies—Domestic: Edison Construction Department, CR, NjWOE).

46. TAE to Rich, 10 Aug. 1883, DF (*TAED* D8316AMY); Insull to Frank Sprague, 10 Sept. 1883, LM 15:41 (*TAED* LBCD2041); *Lawrence Daily Eagle*, 25 Aug. 1883, 4; ibid., 15 Sept. 1883, 4.

47. Jenks 1885, a republication of an *Electrical World* article, provides an overview of the Brockton station's history, specifications, and finances but does not address the type of conductors placed in the ground. Sprague to TAE, 16 Sept. 1883 (pp. 11–14); TAE to Sprague, 21 Sept. 1883; both DF (*TAED* D8346ZAG, D8316AWC); see Doc. 2575.

48. Some of the subsequent wiring was done by a separate company started by Walter Paine and William Jenks, the Brockton Wiring Co.

(renamed the New England Edison Wiring Co. at the end of 1883 and the New England Wiring Co. in early 1884). Edison had his own men do it in other places, urging them to hire the cheapest help available. Garrison, Jr., to TAE, 23 July 1883; TAE marginalia on William Jenks to TAE, 6 June 1883; both DF (*TAED* D8346T, D8346I); Jenks 1885, 6; Paine typescript, n.d.; see Doc. 2526.

49. Insull to Garrison, Jr., 27 Sept. 1883, DF (*TAED* D8316AXE).

50. The first Fall River station was essentially a demonstration organized by Spencer Borden. It powered 126 lamps in several stores and a telegraph office over two blocks, about four hundred feet from the generating plant. Edison Electric Light Co. Bulletin 7:5, 17 Apr. 1882, CR (*TAED* CB007).

51. See Doc. 2436; the Lawrence company evidently adopted the same course. W. H. Dwelly, Jr., to TAE, 7 Sept. 1883; TAE to Electric Tube Co., 10 Oct. 1883; E. H. Lord to TAE, 16 Oct. 1883; all DF (*TAED* D8347G, D8316AZT, D8348ZAC); on the growth and economic conditions of Fall River, see Cumbler 1979, 103–14.

–2424–

To Phillips Shaw[1]

[New York,] April 10th [188]3

Dear Sir

I beg to hand you herewith estimate for the Installation of the Shamokin Plant amounting to $24 920.[2]

If the gentleman forming the Shamokin Company[3] thinks they can purchase and erect any part of the plant for a less price than that named in the estimate, I shall be glad to cut out any such items if they so desire provided that they will do the work with sufficient rapidity so as not to delay my part of the construction and as good and cheap as I should do it,[4] and according to the requirements of the Engineering Department of the Edison Electric Light Co[5] to which all my contracts conforms.

You will notice that the House connections and wiring inside of Houses are not included in this estimate. It was impossible for me to quote a set price on this work as it is of course uncertain how much of it there will be.

I will however undertake to connect consumers with the Mains do the inside wiring in houses, supply all cut outs and bring the wires to the fixtures for $2.75 per light providing you give me an order to wire not less than 150 lights. The sockets Temporary fixtures and shades to the number of 250 lights are provided for in the estimate   I enclose form of contract which I should like executed by the gentlemen forming the Shamokin Company and immediately you telegraph me that the contract is signed[6] I will give instructions for the immediate preparation of the Plant which I hope to have running inside of Sixty days   Yours truly

Thomas A Edison

LS (letterpress copy), NjWOE, Lbk. 16:177 (*TAED* LB016117).

1. Phillips B. Shaw, a merchant and manufacturer in Williamsport, Pa., became the Edison Electric Light Co.'s agent for Pennsylvania about this time; he was also involved in forming the Ohio Edison Installation Co. He had tried unsuccessfully in 1882 to broker the commercial use of Edison's electric railroad technology; by July of that year, he was inquiring about estimates for "Village plant" systems. Doc. 2350 n. 1; Obituary, *Williamsport (Pa.) Gazette and Bulletin*, 2 Feb. 1937, 2; Insull to Edward Johnson, 11 Apr. 1883, Lbk. 3:135 (*TAED* LM003135); Shaw to TAE, 11 and 22 May, 15 and 28 June, 24 and 31 July 1882; TAE memorandum on Ohio Edison Installation Co., undated 1883; all DF (*TAED* D8249M, D8249R, D8249X, D8249ZAH, D8220W, D8220ZAC, D8353ZBQ); Beck 1995, 55–56; see also headnote above.

2. See headnote above and Doc. 2438. The estimate has not been found, but Samuel Insull sent copies to George Foster Peabody, partner in the investment bank of Spencer Trask & Co., and George Solon Ladd, Edison's franchisee in San Francisco, as examples of the type of business Edison expected to do. Edison contracted for a lesser amount, presumably after the Shamokin company decided to do some structural and mechanical work itself. Insull to Peabody, 24 Apr. 1883; Insull to Ladd, 7 May 1883; Lbk. 16:188, 261 (*TAED* LB016188, LB016261); *DAB*, s.v. "Peabody, George Foster."

3. The Edison Electric Light Co. of Shamokin was incorporated in November 1882, though not capitalized until May 1883. Beck 1995, 51; see also headnote above.

4. The local company planned to erect the building and stack, set up steam piping, and place the foundation for the engine. Edison rejected the first blueprints because the building could not accommodate any expansion of the plant's capacity. Edison complained that construction delays, which occurred almost immediately, would "seriously inconvenience" him on account of material already ordered. Samuel Insull to Charles Campbell, 7 May 1883, DF (*TAED* D8360F); TAE to William Douty, 19 and 28 May 1883, Lbk. 16:376, 487 (*TAED* LB016376, LB016487).

5. This department was formed sometime in 1881 or early 1882 to manage the technical and cost specifications of central system components. Edison reorganized it in July 1883. Edison Electric Light Co.— Engineering Department, DF (*TAED* D8227); see Doc. 2488.

6. The contract is Doc. 2438. Shaw telegraphed that it was signed on 3 May, one day before the official date of the agreement. Shaw to TAE, DF (*TAED* D8360B).

Edison spent most of his working hours at his offices in the Edison Electric Light Company building on Fifth Avenue during the spring and summer of 1883. Largely unable to direct laboratory research in person, he depended instead on written reports from longtime associate John Ott, a skilled machinist and instrument maker.[1] Ott oversaw experiments on the top floor of Bergmann & Company's factory on Avenue B, more than a mile from the Fifth Avenue office, where Edison had set up a laboratory in October 1882 after locking his Menlo Park complex.[2] Edison also traveled several times during this period, including a brief trip in the first week of April. Doc. 2425 is Ott's first extant report; there were eventually more than a dozen between April and August. They dealt largely with the subjects of this one: lamps, motors, and electric meters. There are only a few surviving directives from Edison in this period, neither in immediate reply to a communication from Ott.[3]

Edison was also in communication during the year with Francis Upton and William Holzer at the Edison Lamp Company and with William K. L. Dickson at the Testing Room of the Edison Machine Works regarding ongoing experiments at the manufacturing shops. Much of this correspondence has been lost.

1. When he first moved to New York in 1881, Edison entrusted his Menlo Park machine shop to Thomas Logan, who kept him similarly apprised. See Doc. 2074 (headnote).
2. See Docs. 2343 and 2356.
3. Most of Ott's reports are in New York Laboratory, DF (*TAED* D8369), though a few are in Electric Light—Edison Lamp Co.—Test Reports; and Electric Light—Edison Lamp Co.—General; both DF (*TAED* D8333, D8332). For Edison's directions, see Doc. 2676 and TAE to Ott, 4 and 14 Aug. 1883, both DF (*TAED* D8316ALH, D8316ANZ9).

–2425–

From John Ott[1]

NEW YORK April 10[–]11 1883[2a]

Dear Sir

I hade Force[3] make glass vessil to fill with City gass and fasten carbon, with an Arc, by depossiting hydro carbon[4]

Had Lew[5] make clamping device to tie ~~fib~~ drawn fibers on.[6]

Had Bergman repair small rolls, to roll out platinum wire for clamps.

Made small device to cut of sleaves for clamps.[7]

Fastened fiber on inside part by depositing hydrocarbon with an arc under petrolaum. this worked to a sirten extent, it being impossible to deposit suficient quanity around the clamp and fiber, as it has a tendency to build up in threads and not in lumps

The small armiture is finashed and tested. I ran it one hour without raising the heat the least. I alsos ran a belt over the pulley and put $14\frac{1}{2}$ pounds on [------][b] without diminishing the speed much.

I then made the following test

Ran Moter without load

speed 3820 rev,
resist in Field $6.^{7}/_{10}$ ohms
resist in Armature $10.^{56}/_{100}$
length of Armature   6 inch
diam of Armature $2^{7}/_{8}$
div[isions]   26
Con[nections]   13

I then loded the machine in the following manner[8]

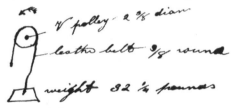

This almost stoped the moter

Had Bergman put on two more lairs on field magnets, witch I think will intencify the field, and much improve the moter.

Hamilton[9] is working on meter experiment.

Kelog[10] made two models for Bergman showing the winding of armature, to avoid mistakes.

I shall have more experiments tested tomorrow[11]   Yours truly

J. F. Ott[12]

ALS, NjWOE, DF (*TAED* D8369D). Letterhead of Thomas A. Edison. [a]"NEW YORK" and "188" preprinted. [b]Canceled.

1. See headnote above.

2. Ott dated each page of his report: the first "April 10" and the latter two "April 11."

3. Martin Force (b. 1848?) began working for Edison in 1876, first as a carpenter and then, about 1877 or 1878, as an experimental assistant (Doc. 1039 n. 3; *TAEB* 5 App. 2; U.S. Census Bureau 1970 (1880), roll T9_790, p. 275.3000, image 0391 [East New Brunswick, Middlesex,

N.J.]). His responsibilities having gradually increased, he helped set up Edison's exhibit at the International Exposition in Paris in 1881 and remained in Europe on behalf of Edison's electric lighting interests. He returned to the United States on 28 April 1882 and was soon working again as an experimental assistant at the Menlo Park laboratory (see Docs. 2128 n. 1, 2196, 2214, 2384; Force's testimony, 20, *Mather v. Edison v. Scribner*, Lit. [*TAED* QD003020]).

4. This process was evidently an attempt to connect the carbon lamp filament and metallic lead-in wire by depositing hydrocarbons from municipal illuminating gas. On 2 April, Ott sketched a container that could be filled with "Oils of diff kind," likely for the same purpose. N-82-12-21:77, Lab. (*TAED* N150077).

Edison's standard manufacturing process was to electroplate the connection. He had been experimenting for some time with applying hydrocarbon coatings to filaments. In the first week of April, he executed a patent application for manufacturing carbon filaments by depositing hydrocarbon material between the electrodes of an electric arc. Earlier in the year, he had also directed research on carbonizing various sugary syrups to fuse the ends of multiple-fiber filaments (see, e.g., Docs. 2081, 2098 [headnote], 2346, 2363, 2411; U.S. Pat. 354,310). In mid-April, Ott reported that he could not clamp six fibers to a single point on a wire because they would "shift and break or crumble" under pressure; his efforts to attach individual carbons were more successful (Ott to TAE, 20 Apr. 1883, DF [*TAED* D8369E]).

5. Not identified.

6. This device may have been like one that Ott sketched at the end of March to hold fibers for "Carbonizing or Hydrocarbon deposit," which is similar to that drawn by Edison in Doc. 2415. Unbound Notes and Drawings (1883), Lab. (*TAED* NS83ABX); N-82-12-21:71–73, Lab. (*TAED* N150071C).

7. On 20 March, Ott sketched a tool for holding a filament with a strip of platinum while carbonizing in the presence of illuminating gas.

John Ott's 20 March sketch of an arrangement for clamping a lamp filament while depositing hydrocarbon.

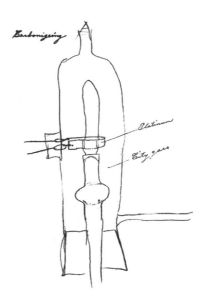

He made a clamp on 12 April; after testing, he seemed pleased that it was "large enough to radiate heat of[f] preventing tempiture raising to red heat." N-82-12-21:67, 87, Lab. (*TAED* N150067, N150087).

8. Figure labels are "V polley   2⅜ diam," "leathr belt ⅜ round," and "weight 32¼ pounds."

9. Edison hired H[ugh?]. de Coursey Hamilton, a native of Suffolk, England, as a laboratory assistant in late 1882 at the request of the senior Hugh Hamilton, and upon the recommendations of George Gouraud and Dillwyn Parrish. (Parrish, the largest investor in Edison's British telephone enterprise, was a Philadelphia-born civil engineer, not a London merchant as stated in Doc. 1742 n. 1). The young Hamilton may have made experiments at the lamp factory in Harrison; he later worked for at least several years at Edison's new laboratory in West Orange. Israel 1998, 270; Hugh Hamilton to TAE, 14 and 17 Nov. 1882; Gouraud to TAE, 22 Aug. 1883; Parrish to TAE, 24 Aug. 1884; all DF (*TAED* D8213ZAA, D8213ZAB, D8213P, D8213R); "Edison's Phonograph in Glasgow," *Bruce Herald [New Zealand]*, 6 Hakihea 1889, 5.

10. Edward Kellogg worked on a variety of experiments for Edison in 1883. He was working on lamps the following year, when he reportedly suffered mercury poisoning. *TAED* s.v. "Kellogg, Edward"; see Docs. 2696 and 2702.

11. The editors have found no additional reports from the next week. On 16 April, Ott was in Boston, evidently trying to set up a motor at the Bijou Theater. A few days later he was back in New York, where he summarized for Edison a motor test like the one described in this document. The armature attained 4,466 revolutions per minute but slowed to sixty-five with eighty-two pounds applied to the pulley. Spencer Borden to TAE, 14 Apr. 1883; Ott to Samuel Insull, 16 Apr. 1883; Ott to TAE, 20 Apr. 1883; all DF (*TAED* D8325S, D8325T, D8369E).

12. John Ott (1850–1931), an expert machinist, had worked for Edison in Newark and Menlo Park. He was Edison's principal experimental instrument maker and a trusted laboratory assistant. See Docs. 623 n. 1 and 1321 n. 1; *TAEB* 5 App. 2.

–2426–

To George Bliss[1]

[New York,] 12th Apl [188]3

Dear Sir

Your favor of 9th came duly to hand.[2]

I think the Western Coy[3] should bear the expense of an Exhibit at the Railway Exhibition[4] as it will most decidedly benefit them more than anyone else.[5]

The Isolated Co bore the expense of the exhibition at Boston[6] from which every one of our Companies derive benefit & they are going to exhibit at the Louisville Exhibition[7] & you will derive benefit from that as much as they will.

If you decide to exhibit you should make a good display at least 350 lights.

As to an Electric Rail Road Exhibit I cannot give the time

necessary to do it properly, & I therefore think we shall have to give up the idea[8]   Yours truly

Thos. A Edison   I[nsull]

L (letterpress copy), NjWOE, Lbk. 16:137 (*TAED* LB016137). Written by Samuel Insull.

1. A longtime Edison associate in Chicago, George Harrison Bliss (1840–1900) had been involved previously in promoting and commercializing Edison's electric pen, phonograph, and telephone. With Edison's intercession, he had established the Chicago office of the Edison Co. for Isolated Lighting by early 1882; at this time he was also general superintendent of the Western Edison Light Co. Docs. 861 n. 8, 2278 n. 8; Andreas 1886, 3:598.

2. Bliss recounted in this letter his conversation with the manager of the forthcoming Railway Exposition in Chicago (see below), and attempted to persuade Edison to mount exhibitions of his electric light and railway. Bliss to TAE 9 Apr. 1883, DF (*TAED* D8372B).

3. The Western Edison Light Co. was formed in May 1882 to sell and install electric lighting plants in Illinois, Iowa, Wisconsin, and Minnesota. Doc. 2299 n. 1.

4. Plans were announced in December 1882 for a National Exposition of Railway Appliances at the Inter-State Exposition buildings in Chicago from late May to late June 1883. Organizers promised a "series of scientific and practical tests, to be made by well-known scientists and carefully selected committees, extending to every article and every description of material susceptible of reliable test." Members of the exhibition's board of commissioners included Lucius Fairchild (former Wisconsin governor), president; George Pullman, vice president, and representatives of other railroad manufacturing and supply firms. "National Exposition of Railway Appliances," *Sci. Am.* 47 (23 Dec. 1882): 400; "Railway Machinery," *Washington Post*, 25 Mar. 1883, 5.

5. After enthusiastically describing the proposed exhibition (see note 2), Bliss concluded that Western Edison likely would not put up a lighting display without financial participation by the Edison Electric Light Co. because "Our territory is so limited and the benefits to be derived from the exhibit so wide spread that we do not feel like incurring the expense without some co-operation from New York."

6. Spencer Borden had managed the Edison exhibit at the second annual New England Manufacturers' and Mechanics' Institute Fair, which opened in September 1882. Doc. 2318 n. 9; "The Institute Fair," *Boston Globe*, 1 Nov. 1882, 4.

7. See Doc. 2465 n. 3. The Southern Exposition ran in Louisville from 1 August until November. Originally conceived as a display of Southern agriculture and manufacturing (following the commercial success of Atlanta's International Cotton Exposition in 1881), the exhibition took on a broader character and included a large and well-received collection of fine art (including works loaned by Edison backer Egisto Fabbri). Findling and Pelle 1990, 84–85; Findling 1996, 29–31; "The Southern Exposition," *NYT*, 30 May 1883, 5; "A Great Southern Show," ibid., 2 Aug. 1883, 1; "Edison's Lights for the Louisville Exposition," *Baltimore Sun*, 5 Jul. 1883, 1; "The Louisville Art Exhibition," *Art Amateur* 9 (Nov. 1883): 122.

8. Edison noted on Bliss's letter (see note 2) only: "They ought to put in a new K [dynamo] but havent time to do anything on RR now—"

-2427-

To Charles Bruch[1]

[New York,] April 12— [188]3

Dear Sir

Referring to yours of 11th inst.[2]

I have contemplated establishing a department at my shops for the instruction and preparation of men to take positions in our Companies for the establishment and operation of our Light Plants.[3]

The duration of the instruction will depend entirely upon the ability and application of the parties themselves and no charges will be made for the instructions which will be entirely of a practical nature   Yours Truly

Thos. A Edison   I[nsull]

L (letterpress copy), NjWOE, Lbk. 16:130 (*TAED* LB016130). Signed for Edison by Samuel Insull.

1. Charles Patterson Bruch (1860–1927) was a Western Union telegraph operator who at this time headed the complaint bureau of the operating room at the New York headquarters building. For a period in 1888, he acted as Ezra Gilliland's secretary at the Edison Phonograph Co.; he later became a vice president and general manager of the Postal Telegraph Co. Taltavall 1893, 168–69; Obituary, *NYT*, 28 Dec. 1927, 23; Bruch correspondence in Gilliland letterbook, LM 22 (*TAED* LM022).

2. Bruch's letter to Edison has not been found.

3. Edison received sporadic inquiries, including several around this time, about opportunities for practical education in electrical engineering, for which no formal training existed at this time. Edison made replies generally similar to this one, though on an earlier occasion he advised that the "only way" to gain expertise was to get "a position in some first class Electrical concern" such as the Edison Electric Light Co. (Samuel Insull to John Golding, 4 Apr. 1883; TAE to George McKibbon, 4 Apr. 1883; TAE to Edwin Miller, 24 May 1883; TAE to F. A. Douglas, 29 May 1883; TAE to J. McDonough, 6 June 1883; Lbk. 16:99, 99A, 459; 17:1, 77 [*TAED* LB016099, LB016099A, LB016459, LB017001, LB017077]; William Elasker to TAE [with TAE marginalia], 17 Aug. 1882, DF [*TAED* D8204ZFR]). At least one of the queries came from a prospective employer for whom, like Edison, the dearth of trained electricians was an impediment (Stephen Hoe to TAE, 26 March 1883, DF [*TAED* D8303ZBT]; TAE to Hoe, 29 Mar. 1883, Lbk. 16:44A [*TAED* LB016044A]; cf. Doc. 2643). The Testing Room at the Edison Machine Works had carried out a limited educational function since 1882 (Doc. 2343 (headnote); Jehl 1937–41, 962–63; Rosenberg 1990, 34; see also Docs. 2428 n. 2, 2434, 2484, and 2530; and Jehl 1937–41, 962–65).

To Abram Hewitt[1]

[New York,] 12th Apl [188]3

Dear Sir,

Your favor of 2nd inst came to hand whilst I was away in Canada—hence the delay in replying thereto[2]

If Columbia College does not accept the proposal I made them, the Stevens Institute Hoboken have the next claim on my collection of Instruments.[3] If they should refuse to take the matter up in the proper manner then I shall be able to offer the collection to you.[4]

I should be glad to see a class for the study of Practical Electrical Engineering started at Cooper Institute[5] as I think it would be a great advantage to the class of men you get there besides being advantageous to such business enterprises as my own   Yours truly

Thos. A. Edison   I[nsull]

L (letterpress copy), NjWOE, Lbk. 16:131 (*TAED* LB016131). Written by Samuel Insull.

1. Abram Hewitt (1822–1903) was an iron manufacturer, congressman, and educational philanthropist who later became mayor of New York. A son-in-law of Peter Cooper, he was active in the management of the Cooper Institute (see note 5). *ANB*, s.v. "Hewitt, Abram Stevens."

2. Edison attended an electrical exhibition at Cornwall, Ont., in early April (see Doc. 2418 n. 10). Prompted by a newspaper report "to the effect that if Columbia College does not undertake the duty, you will establish a school yourself," Hewitt had reminded Edison of their earlier conversation about founding "a school of electricity in the Cooper Institute" (Hewitt to TAE, 2 Apr. 1883, DF [*TAED* D8303ZBY]). That article was probably "Questions Affecting Columbia" (*New York Tribune*, 30 Mar. 1883, [8]) about Columbia's effort to establish new programs of study, including a school of electrical engineering. In response to rumors that he would set up his own engineering program if Columbia did not do so, Edison reportedly told the newspaper: "I am going to do so anyway. . . . I can't wait for the trustees, they are too slow. I need the men. My work is seriously retarded by my inability to obtain competent engineers." Other electrical manufacturers also found it difficult at this time to hire young men with education or practical experience in electrical engineering; several instituted in-house training programs as a result (Rosenberg 1990, chap. 2).

3. See Docs. 2173 esp. n. 12, 2360, 2408, and 2434 regarding Edison's wish to donate instruments from his exhibit at the 1881 Paris International Electrical Exposition to help set up an electrical engineering program in or near New York City. Edison later disavowed having made any promise to the Stevens Institute. The University of the City of New York (now New York University) and Lehigh University also expressed interest. See Doc. 2479; TAE to George Barker, 28 June 1883; Albert Gallatin to TAE, 28 May 1883; both DF (*TAED* D8316ADR, D8303ZDO); TAE to Edwin Miller, 24 May 1883, Lbk. 16:459 (*TAED* LB016459).

4. Hewitt replied on 13 April that the Cooper Union would "establish a school in practical electrical engineering, without reference to the action of Columbia College or the Stevens Institute," and he asked for Edison's help "in procuring such apparatus as may be necessary." Edison promised to furnish instruments "at cost price and free of all royalty." Hewitt to TAE, DF (*TAED* D8303ZCD); TAE to Hewitt, 15 May 1883, Lbk. 16:336 (*TAED* LB016336).

5. The Cooper Union for the Advancement of Science and Art was founded by in 1859 by Peter Cooper (1791–1883) to provide free training to working-class men and women in engineering, architecture, and art. Under Hewitt's direction, the school established a degree-granting engineering program in 1886. Cooper had just become ill when Hewitt wrote his 2 April letter to Edison; he died on 4 April. *Ency. NYC*, s.v. "Cooper Union (for the Advancement of Science and Art)"; *ANB*, s.v. "Cooper, Peter"; "Death of Peter Cooper," *NYT*, 5 Apr. 1883, 1.

–2429–

Technical Note: Electric Light and Power

[New York, April 12, 1883?[1]]

New System[2]

its to take power*fer* from waterfall & take 10 miles, then distribute

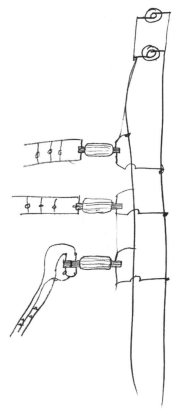

3 wire system   3 windings on core and 3 commutators

3 separate central stations independent of each other, mult
arc   very high resistance in primary

[Witness:] H. W. Seely[3]

X, NjWOE, Lab., Cat. 1149 (*TAED* NMo18AAF).

1. Henry Seely signed and dated the document on this day when he received it from Edison.

2. Edison briefly considered similar electro-mechanical means of reducing the voltage of direct current in the first part of 1882 (see Docs. 2242 and 2284). He copied the sketch in this document from an undated pocket notebook entry (PN-82-04-01, Lab. [*TAED* NPo16N]) and it became the basis for the first figure in a patent application he executed on 8 May 1883. The resulting specification described a system in which high voltage "currents can be economically transmitted from a distant source of energy . . . to the town, village, or other locality which is to be supplied, and there distributed from different independent central stations, situated at convenient points." The system consisted of high-voltage generators, conductors, and voltage-reduction devices. Each such reducing device "consists of a field-magnet and an armature-core, on which are wound two sets of coils—one of fine wire, connected with a commutator whose brushes are placed in the multiple-arc circuit from the main conductors, the other of coarser wire, connected with a commutator from whose brushes the [local distribution] circuit or circuits extend." The field magnets are not shown. The independent windings allowed the armature to function both as a motor (driven by the incoming high-voltage current) and a generator (inducing lower-voltage current for local consumption). Edison specifically pointed out that adding a third winding to the armature and modifying the commutator accordingly, as shown at the lower left of the drawing, would adapt it to his three-wire (or "compensating") distribution system (U.S. Pat. 287,516).

3. Henry W. Seely was associated with Richard Dyer's patent law practice and often witnessed or acknowledged receipt of documents from Edison. He went into partnership with Dyer in mid-1884, quite likely around the time he was admitted to the bar, and thereafter acted as a patent attorney for Edison until about 1897. Seely is widely credited with holding the first U.S. patent for an electric flatiron. See Docs. 2298, 2346 n. 15, and 2681; *TAEB* 6 App. 5; on the changing role of patent practitioners without bar accreditation, see Swanson 2009 esp. pp. 537–41.

–2430–

From Henry Howard

Providence, R.I., April 17th. 1883[a]

Dear Sir:—

Mr. Sims,[1] who is now sitting by my side, has related some of the conversation which he had with you in New York a few days since in regard to your having an interest in the engine business. Confidentially, the facts are these: we are about enlarging our capital, to what we think more nearly comports with the actual value of our franchises and property,—to wit, five hundred thousand dollars. I do not mean to say that the value is fully five hundred thousand dollars, but certainly three hundred. None of us would sell any of our stock any less than at the rate of three hundred thousand dollars. But so far as you are concerned, the case is entirely different. The relationship between us is something more than a pecuniary one. You have, as we all feel, treated us fairly and handsomely. We have endeavored to treat you in the same way. Your connection in every respect is a pleasant one. I cannot speak of course for anybody but Mr. Sims and myself; but, so far as we are concerned, we would be in favor of each one of us contributing a portion of our stock to make up an amount for you at the original price, to wit—one hundred thousand dollars. Of course we do not mean by this that an offer of that kind should be extended to others, as all of us would rather buy than sell at the rate of three hundred thousand. If you would make up your mind how much you would like to have for yourself personally, and would write Mr. Sims about it, we will see what we can do.[2] Truly Yours[b]

H. Howard[3]

TLS, NjWOE, DF (*TAED* D8322ZAG1). Letterhead of Providence Telephone Co. [a]"Providence, R.I." and "188" preprinted. [b]"Truly Yours" written by Howard.

1. Gardiner Sims (1845–1910) formed a partnership with Pardon Armington to manufacture steam engines in the late 1870s in Lawrence, Mass., where he had been superintendent of the J. C. Hoadley Engine Works. In 1881, they developed a high-speed engine for the Edison electric system, for which they became a major supplier. The following year they relocated to Rhode Island and incorporated as Armington & Sims Co. See Docs. 2078 n. 1 and 2131 n. 7; Freeman 1901, 787.

2. Sims wrote Samuel Insull the same day to provide information about the existing stockholders. He noted that "we have got to have new capital as our business is increasing" and indicated that several stockholders would be willing to sell stock to Edison at par value. He informed Insull that when the company increased its capital stock the existing stockholders would receive new stock in proportion to that they already held. The newly enlarged company was incorporated as the Armington & Sims Engine Co. in May 1883 with its capital stock not to exceed $500,000. No evidence has been found regarding Edison's acquisition of stock in either the old or new company. Sims to Samuel Insull, 17 Apr. 1883, DF (*TAEB* D8322ZAG); "An Act to Incorporate the Armington & Sims Engine Co.," *Acts and Resolves Passed by the General Assembly of the State of Rhode Island and Providence Plantation at the May Session, 1883*, 12–31.

3. Henry Howard (1826–1905), who served two terms as Republican governor of Rhode Island in 1873–75, became president of Armington & Sims after the company moved to Rhode Island and incorporated in 1882. *NCAB,* 9:90; Sobel 1978, 1353; letterhead, Armington & Sims to Edison Machine Works, 15 Aug. 1882, DF (*TAED* D8233ZDK).

–2431–

Agreement with Stephen Field, Simeon Reed, Sherburne Eaton, and the Edison Electric Light Co.

[New York,] April 20, 1883[1a]

MEMORANDUM of an AGREEMENT between Stephen D. Field[2] and Simeon G. Reed,[3] on behalf of themselves and associates, of the First part, and Thomas A. Edison and S. B. Eaton on behalf of themselves and their associates, (including the Edison Electric Light Co.), of the Second part.

WHEREAS, Mr. Field and Mr. Edison have each been engaged in experiments and inventions having for their object electrical propulsion on Railways and have each obtained Patents and applied for other Patents, and it has been proposed to unite the two interests,[4] and

WHEREAS, Agreements have been entered into between the parties hereto of even date herewith, for the formation of another Company uniting all of the interests excepting the rights to use upon elevated railroads in the City of New York,[5] and

WHEREAS, it is proposed to unite the two interests for use upon elevated railroads in the City of New York

It is, therefore AGREED as follows:

I. A CORPORATION, under the laws of the State of New York, shall be formed within six months from the execution of this Agreement,[6] to which shall be transferred all the inventions now owned or controlled by the parties hereto, being exclusively applicable to electrical propulsion on Railways, but not including lighting and heating by electricity, to be used only upon the Elevated Railroads in the City of New York now existing, or that may hereafter be extended from the present Elevated Roads, or independently built.

All future inventions in Electric RRs[7b] of the said Edison made prior to January 12th. 1886,[8] and all future inventions of said Field made at any time hereafter which may be exclusively applicable to electrical propulsion on Railways (but not including lighting and heating by electricity) shall also be transferred to, the said Corporation for use upon the said Elevated Railroads. And the said Corporation shall also receive exclusive licenses to use on the said Elevated Railroads all inventions in E RR[c] which have been made or may be made by the said Edison before January 12th. 1886, and which have been made or may at any time be made by the said Field, incidental to such propulsion, exclusive of lighting and heating by electricity.

The stock of the said Corporation shall be used for the purchase of the rights or license to use upon the said Elevated Railroads, and shall be divided as follows:

(A)   Sufficient thereof shall be sold to pay into the Treasury of the parent Company above referred to, a sum equal to the proportion that the Capital of the two Corporations bears to the Capital of this Corporation in the amount paid by the parent Company to the said Field and Edison, to reimburse them for their cash outlay in experiments, or in other words, this Corporation shall contribute pro rata, with the Capital of the Companies towards such reimbursement.

(B)   And after such reimbursement the remaining stock shall be divided; fifty per cent (50%) to the parties of the First part, and fifty per cent (50%) to the parties of the Second part.

The Stock to be divided between the Field and Edison Interest, shall be deposited with a Banking Company, and placed under the charge of a Special Committee of three, one named by the Field interest, one by the Edison interest, and a third to be named by these two; and none of the said stock shall be sold, pledged or used, except under the direction of this Special Committee. This Committee shall give to each owner of the stock a receipt specifying the number of his shares, and whenever he wishes to sell the same or any part thereof, he shall give notice to the Committee and they shall sell it and give him the proceeds if the price to be obtained be such as they think it right to take, having regard to the interests concerned, but not below the price at which he may limit it. When either interest desires to sell, the other shall be notified, and shall be at liberty to contribute one half or less of the amount to be sold at the same time and price, and the proceeds shall be divided pro rata.

At such time as the Field and Edison interest may dissolve the Committee, which must be within two years of the date hereof, the stock not sold and belonging to each owner shall be returned to him.

It is further agreed, that should the parties hereto, consent to forego the formation of a Corporation as herein provided a sale to the Elevated Railroad Companies of the rights to use the Patents and inventions may be made outright, and the proceeds apply as the stock of a Company is to be applied.

IN WITNESS WHEREOF the parties hereto have hereunto set their hands and seals this

Twentieth day of April, 1883.   (Signed)[9]
Signed, Sealed and Delivered in
the presence of   Stephen D. Field.
as to S. D. Field and S. G. Reed.   Simeon G. Reed.
Ed. P. Howell[10]
Chas. Edgar Mills.[11]

as to T. A. Edison and S. B. Eaton and
S. B Eaton P[residen]t.   T.A.E.
                                                                  S.B.E.
                                        E.E.L. Co. by S.B.E. P[residen]t.

D (typed copy), NjWOE, Cat. 2174, Scraps. (*TAED* SB012BCG). [a]Date from document, form altered. [b]"in Electric RRs" interlined above by Edison; insertion point after "All" altered for clarity. [c]"in E RR" interlined above by Edison; insertion point after "all" altered for clarity.

1. The agreement of this date was superseded by a similar one among the same parties signed on 26 April, to which subsequent ancillary contracts referred; see notes 5 and 7 below. TAE agreement with Edison Electric Light Co. and Electric Railway Co. of the United States, 18 May 1883; Electric Railway Co. of the United States agreement with Stephen Field, 18 May 1883; both Miller (*TAED* HM830184, HM830183).

2. Electrical inventor, engineer, and entrepreneur Stephen Dudley Field (1846–1913) developed an electric railroad locomotive in 1880, which became part of an 1881 patent interference proceeding with respect to applications of Edison and Werner von Siemens. Edison, who had prior business contacts with Field, initially dismissed the invention but did not contest the interference after formation of the Electric Railway Co. Field had recently arranged to test electric motors on an elevated rail line in Chicago. Field had several prominent uncles, including Atlantic cable telegraph promoter and rail entrepreneur Cyrus Field (see note 5) and sitting Supreme Court justice Stephen Johnson Field; he was also an associate of George Solon Ladd of California in telegraphy and electrical work long before the latter's own acquaintance with Edison. See Docs. 541 n. 2, 1056, 1968 n. 1, 1987 n. 3, 2237 n. 8; "Stephen D. Field, Car Pioneer, Dead," *NYT*, 19 May 1913, 9; Patent Application Casebook E-25-36:90, PS (*TAED* PT020090); TAE to Henry Spang, 15 Aug. 1884, DF (*TAED* D8416BUM1); "Electric Motors," *Chicago Daily Tribune*, 9 Mar. 1883, 6; Kettenburg 1978, 4.

3. Simeon Reed (1830–1895) was a steamboat operator and financier who helped start the highly profitable Oregon Steam Navigation Co. In 1879, Henry Villard merged the company into his Oregon Railway

& Navigation Co., of which Reed became vice president and manager. When the Electric Railway Co. of the United States was organized, Reed became president; he was succeeded at the end of 1884 by Edward Johnson. He was also involved with a number of other transportation and mining companies, including construction of the Canadian Pacific Railway through British Columbia. *DAB*, s.v. "Reed, Simeon Gannett"; Johansen 1936.

4. In December 1882, Edison had rejected a consolidation of his electric railroad patents in the United States and abroad with the Siemens interests (see Doc. 2379). He reportedly also rebuffed an initiative from Field's representatives early in the year. In view of his anticipated central station business and the Edison Electric Light Co.'s fear of losing control of a valuable patent, he reconsidered and entered negotiations about the beginning of April. Edison gave Field access to his experimental rail line at Menlo Park. Sherburne Eaton's testimony, *Electric Railway Co. v. Jamaica and Brooklyn Road Co.*, p. 640, Lit. (*TAED* QE001638); "Edison and Field Combine," *NYT*, 29 Apr. 1883, 7; "Electricity for Street Cars," ibid., 16 May 1883, 2; "Edison's Electric Railroad," ibid., 13 Sept. 1883, 8.

5. Sherburne Eaton testified a decade later that this distinction was made in deference to Cyrus Field's controlling interest in the New York elevated lines, which, it was hoped, would lead to rapid adoption of electric traction there. The similar agreement signed on 26 April maintains this distinction but, no company having been formed within about a year, the signatories annulled the distinction by mutual consent in March 1884. The resulting combined agreement remained in force until 1890. See also note 7; Eaton's testimony, *Electric Railway Co. v. Jamaica and Brooklyn Road Co.*, pp. 639–41, Lit. (*TAED* QE001638); TAE agreements with Reed, Field, Eaton, and the Edison Electric Light Co., 26 Apr. 1883 and 6 Mar. 1884, Miller (*TAED* HM830174, HM840212).

6. Articles of incorporation of the Electric Railway Co. of the United States to operate within the United States and Canada, were signed on 20 April. Another set of papers, the same except for the list of trustees, was executed and notarized on 5 May. The nine trustees included Eaton, Reed, Field, Grosvenor Lowrey, Robert Gallaway, and Charles Rogers; Henry Villard appeared on the 20 April list but not on the second one. The company was capitalized at $2 million. By consent of the Edison parties, the Field interests had authority to run the business. Electric Railway Co. of the United States articles of incorporation, 20 Apr. 1883, Cat. 2174, Scraps. (*TAED* SB012BCF); Electric Railway Co. of the United States articles of incorporation, 5 May 1883, Miller (*TAED* HM830176); Edison Electric Light Co. annual report, 23 Oct. 1883, in company Bulletin 20:41, 31 Oct. 1883, CR (*TAED* CB020442).

Retreating from his prior optimism that electricity would be an economical form of power for long-distance rail lines through lightly populated agricultural regions, Edison no longer expected it to power either on heavy freight trains or long distance passenger service. According to a newspaper account from this time, however, he believed that "for street cars, elevated railroads and other short lines it will be found invaluable." "Electric Railway Trains," *Washington Post*, 13 May 1883, 7; cf. Docs. 1745, 2152.

7. This restriction and the related one in the same paragraph, both inserted by Edison, were omitted from a second agreement signed by the same parties on 26 April. That document, however, included a wholly new penultimate paragraph stipulating that: "The term Electric propulsion on Railways as used herein, includes all and only such electrical means, agencies, and devices, as are employed for the purpose of moving railway cars or engines upon railway tracks." The new contract was identical in all other substantive respects, although a witness other than Charles Mills attested to the signatures of Reed and Field. The same stipulation was made in the 26 April instrument covering areas outside New York (see note 6). TAE agreement with Reed, Field, Eaton, and the Edison Electric Light Co., 26 Apr. 1883, Miller (*TAED* HM830173); Eaton's testimony, *Electric Railway Co. v. Jamaica and Brooklyn Road Co.*, pp. 638–39, Lit. (*TAED* QE001638); Eaton to TAE, 23 Apr. 1883, Cat. 2174, Scraps. (*TAED* SB012BCH).

8. Under terms of an 1881 contract, the Edison Electric Light Co. agreed to license to a new company, called the Edison Electric Railway Co., any Edison patents issued before this date pertaining to electric railroads. Edison later claimed that this entity continued to hold property connected with his railroad experiments at Menlo Park. Some correspondence was conducted under that name, but the editors have not determined the company's legal standing. TAE agreement with Edison Electric Light Co., 12 Jan. 1881, Miller (*TAED* HM810140); see, e.g., Benjamin Rhodes to Edison Electric Railway Co., 22 Jan. 1883, Cat. 2174, Scraps. (*TAED* SB012BBX); Edison Electric Railway Co. to Middlesex County (N.J.) Sheriff, undated May 1884, DF (*TAED* D8403ZDD).

9. The position of "(Signed)" is ambiguous, but the word most likely was intended to indicate the signatures of all the parties.

10. Not identified.

11. Charles Edgar Mills was identified as a lawyer, notary, and "commissioner for the States and Territories" at 115 Broadway in New York. *Trow's* 1884, 1173.

–2432–

From William Andrews

New York, April 21st *1883*[a]

Dear Sir,

I will attend to the test on E Dynamo[1] as per your letter of yesterday as soon as possible—[2]

What am I to do with the pair of large insulated copper brushes received yesterday? They are evidently intended for a "C" Dynamo, and I cannot test them here on a smaller one—

I have made a test today on the 450 Light Dynamo—[3] The field measures about 8 ohms connected like an ordinary 110 Volt K Dynamo—[4] The armature measures .015 ohm—

With a load of 450 "A" Lamps (8½ to H.P.) I took the following readings:—

Speed	E.M.F. at Dynamo
1120	105 Volts
1140	107 "
1160	109 "

Drop of E.M.F from Dynamo to end of Cable= 3.5 Volts—

Ran for about 4 Hours with above Load,— Armature was quite hot commutator & bearings cool, Field Cores just warm— I wish we Could have got one more wire round armature, which would have reduced its resis. to about .012 ohm. It would then carry 450 lamps easily— As it is I think that number makes rather a heavy load for it—Especially when used for Small Central Station purposes, which means I suppose 24 Hours a day   For this work I think 400 Lamps are all that can be safely Counted on with present armature[b]

I expect to continue the test all day Monday—and will then send you another report[5]   Respectfully Yours

W. S. Andrews

⟨Cannot another wire be put on= In Village plant stations it will only have to carry 450 for 3½ hours   before & after that the Load aint ½ 450=⟩[6]

ALS, NjWOE, DF (*TAED* D8330W). Letterhead of Edison Electric Light Co. [a]"*New York*," and "*188*" preprinted. [b]"Especially when . . . present armature" written below signature with an "x" to indicate placement in text.

1. This dynamo was a small model, which Edison considered "experimental" and "unsatisfactory" at the end of 1882. By the third week of April, however, William Andrews was testing experimental models of much larger capacity, and Edison was evidently hoping to supply one to Boston's Bijou Theatre. See *TAEB* 6 App. 3; TAE to George Bliss, 29 Nov. 1882, Lbk. 14:481 (*TAED* LB014481); Spencer Borden to TAE, 13 and 14 Apr. 1883; Andrews to TAE, 9, 13, 16, and 18 Apr. 1883; all DF (*TAED* D8325R, D8325S, D8330R, D8330R1, D8330S, D8330U, D8330V).

2. Edison's letter has not been found but may have concerned tests of an E dynamo with the "field blocks raised higher off iron base," which Andrews expected would "undoubtedly increase power of machine." Andrews to TAE, 16 and 18 Apr. 1883, with TAE marginalia, both DF (*TAED* D8330U, D8330V); TAE to Andrews, 17 Apr. 1883, Lbk. 16:158 (*TAED* LB016158).

3. See Doc. 2419 (headnote).

4. That is, each pair of field coils connected in parallel with the others. The standard K dynamo had a total resistance in the field of 13.5 ohms. Doc. 2126 (headnote); Andrews 1924, 166.

5. Andrews reported on Monday, 23 April, that the results given here were incorrect because only about 420 lamps had actually been on the test circuit. When he tried again with the full load, he unsurprisingly found the voltage too low and concluded that the field magnets had not

reached their saturation point. The next day, he connected the field cores in parallel to reduce resistance in the field circuit and increase the current flowing through the magnets. This change raised the voltage but reduced the amount of current to the outside line. Andrews then re-wound the magnets and armature, and he connected the field coil pairs in series. Andrews to TAE, 23 and 24 Apr., 12 May 1883; Andrews test reports, 24 Apr. and 12 May 1883; all DF (*TAED* D8330X, D8330Z, D8330ZAH, D8330ZAA1, D8330ZAI).

6. Edison's marginalia is essentially the same as his full reply to Andrews on 23 April. Lbk. 16:181 (*TAED* LB016181).

–2433–

To Edward Johnson

[New York,] 23rd Apl [188]3

My Dear Johnson

I received your favor of 11th inst[1] this morning and at once cabled you "Send Sprague."[2] I propose using him in connection with the establishment of the "small Town Plants.

You ask "How is my other protege panning out? (Benton)."[3] At present I think I must say "N.[o]G.[ood]" Benton does not seem to know any more about our business now than the first day he came here & so far he has got no orders except one—Port Jervis Small Town Plant.[4] I am afraid he will not pan out as you thought Yours Sincerely

T A Edison

Other agents come & study the thing out Exhaustively Cant get him to sit down 2 minutes. He is entirely unable to present the subject at all[a]

LS (letterpress copy), NjWOE, Lbk. 13:15 (*TAED* LB013015). Written by Samuel Insull. [a]Signature and postscript written by Edison.

1. Johnson had written that Frank Sprague was hoping to work for Edison despite having other job offers. Sprague was

very Anxious to get to work— He was telling me a few days ago about some excellent Ideas he has in re—to Electric Motors—Railways—& was asking me to advise him in the matter— The Idea then took possession of me that He is of all men the very one to take charge of your Railway Experiments He understands the matter theoretically & practically & would fill that position I am sure to your entire satisfaction— It certainly needs some one to take it in hand for you.

Johnson suggested that Sprague be assigned to obtain a contract for a new London Underground line that he feared would otherwise go to the Siemens interests. Johnson to TAE, 11 Apr. 1883, Cat. 2174, Scraps (*TAED* SB012BCC).

2. This is the full text of Edison's 23 April cable to Johnson (Lbk. 16:172B [*TAED* LB016172B]). Frank Julian Sprague (1857–1934)

studied engineering and electricity at the U.S. Naval Academy, graduating in 1878. With permission from the Navy, he had served as secretary to the awards jury of the Crystal Palace exhibition in London in 1882 (earning Edward Johnson's approbation as "an enthusiastic advocate of everything that is Edisonian"), but by this time he had substantially overstayed his leave. Lacking the financial resources to employ him with the British Edison company, and knowing Sprague did not "have enough surplus pocket money to allow him to loaf long—beside he is not one who can endure it long," Johnson secured his engagement by Edison (Doc. 2258 n. 13; Johnson to TAE, 21 May 1882 (p. 6), DF [*TAED* D8239ZBU1]; Johnson to TAE, 11 Apr. 1883, Cat. 2174, Scraps. [*TAED* SB012BCC]). After working for Edison for about a year, Sprague left to follow his own interests as an inventor and entrepreneur (see Docs. 2656 and 2657), forming the Sprague Electric Railway and Motor Co. in November 1884 to advance his own designs for motors and railway systems. That firm benefitted from the connections and capital of Edward Johnson (who served as its president) and other Edison associates, including Charles Batchelor and Sigmund Bergmann; it merged with the Edison Electric Light Co. into General Electric in 1890. Sprague made a pioneering street railway installation in Richmond in 1887 and devised a system, installed in Chicago in 1897, for controlling trains with independently powered motors (Middleton and Middleton 2009, 3–60; Dalzell 2009, 54–55, 58–90, 103–104, 115–24).

3. Charles Abner Benton (1847–1939) began working for Edison at Johnson's recommendation in about 1881. Despite the opinions expressed in this document, Benton remained connected with Edison's electric lighting businesses for several years. He had a private understanding at this time to receive $250 per month through Johnson, for whose house he was caring while Johnson was in England. Doc. 2409 n. 9; Insull to Johnson, 11 Apr. 1883, LM 3:135 (*TAED* LM003135); "Benton, Charles Abner," Pioneers Bio.

4. The editors have found no evidence concerning a central station in Port Jervis, N.Y., a town on the Delaware River about fifty miles northwest of New York City.

–2434–

To Andrew White[1]

[New York,] April 23d [188]3

Dear Sir,

Referring to your favor of 16th inst.[2] I have already committed myself to Columbia College and must give them time to decide but I have the impression that they will do nothing—[3] after that the Stevens Institute desire to establish a course. I have several men who were examined at the Stevens institute and they are the best I have.[4] The Cooper Institute are about starting a Night School to learn Engineers and others who already have positions, but of course this is a different matter.[5] There[a] is a[b] great advantage in having the School near New York, but if neither Columbia College or the Stevens Institute will start a course in the proper manner, I will be very happy to cooperate with you.

The tendency at both places ~~are~~ is[c] towards too much theory and impracticable Professors[6]   Yours truly,

Thos. A Edison   I[nsull].

L (letterpress copy), NjWOE, Lbk. 16:173 (*TAED* LB016173A). Signed for Edison by Samuel Insull. [a]Obscured overwritten text. [b]"is a" interlined above. [c]Interlined above.

1. Andrew Dickson White (1832–1918) was a university president and diplomat. He was the first president of Cornell University, serving from its opening in 1868 until 1885, and was an early advocate of co-education and the establishment of curricula in science, engineering, and agriculture. *ANB*, s.v. "White, Andrew Dickson."

2. Cornell's trustees approved the inauguration of a program in electrical engineering on 26 March 1883, making it the nation's second program after the one established at the Massachusetts Institute of Technology in 1882. Fourteen undergraduates and two graduate students entered Cornell's program in its first academic year, 1883–84. Aware that Edison wanted to start a college program in electrical engineering and to donate his collection of equipment from the Paris Electrical Exposition of 1881, White made a forceful case in favor of Cornell. He emphasized the practical nature of Cornell's educational philosophy, its well-equipped laboratories and physical plant, its just-established program in electrical engineering, and key support from industrialists like telegraph entrepreneurs Hiram Sibley and Ezra Cornell. In contrast to leaders at other institutions, White promised immediate action to start an "Edison Course of Electrical Engineering," which could "begin graduating well trained electrical engineers" within a year. Physics professor William Arnold Anthony also strongly supported Cornell's case, imploring White to make "every effort" to get Edison's exhibition equipment and support. Anthony to White; White to TAE; both 16 Apr. 1883, both DF (*TAED* D8303ZCE1, D8303ZCE2). For a general account of early electrical engineering education at Cornell, see Rosenberg 1990, chap. 4.

3. See Docs. 2427 n. 3 and 2428 n. 2.

4. At least two of Edison's assistants, John Howell and Albert Herrick, had attended Stevens Institute (Doc. 2129 n. 4). In response to a recent inquiry, Edison stated that the Institute was "probably as good if not the best place at present to study electrical science and engineering." Edison was markedly less enthusiastic about Stevens in July, when the institute announced a program of applied electricity. Doc. 2129 n. 4; "Herrick, Albert B.," Pioneers Bio.; TAE to G. W. Baker, 24 Mar. 1883, Lbk. 16:17 (*TAED* LB016017); cf. Doc. 2479.

5. See Doc. 2428.

6. On 10 May, Cornell's William Anthony wrote Edison directly to claim that electrical theory and practice went hand-in-hand, and that Cornell excelled in teaching both. As an example, Anthony told Edison of a student who had designed and built a small dynamo with little help from his professors. Anthony also emphasized that Cornell already had

an established electrical engineering curriculum and that it had spent about $20,000 in the past two years on state-of-the-art equipment. He invited Edison to visit Cornell to see firsthand how the school sought "to Educate Electrical Engineers who shall know how to apply theory to practice. I know we should be able to do more and better work with your cooperation." Edison replied on 15 May to say that he did not have the time at present to visit the university. Anthony to TAE, 10 May 1883, DF (*TAED* D8303ZDD); TAE to Anthony, 15 May 1883, Lbk. 16:328 (*TAED* LB016328A).

–2435–

From Francis Upton

EAST NEWARK, April 30, 1883[a]

Dear Mr. Edison:

The money question is still exceedingly important and our financial situation in growing constantly worse.[1]

We have drawn down our expenses as far as we can practically and run at all. In March when running perfectly full for over[b] 15 000 lamps a week we had on 324 hands, quite a number of them learners.[2]

We now have [on?][c] 127 hands on our pay roll and run only day times.

During the month of April we have had orders for 7500 lamps only, which is the smallest month for a year, and less than one quarter of the average for the past six months.[3]

Our books show that you have given us as agent about $1046.00 during the month, and no other help. I have been compelled to put in $1775.00 more as a loan to pay ~~one~~ pay roll and help out on another.

Dyer[4] informs me that over six Newark parties called[b] here in one day to collect the money owed to them last week.[5]

You should make an effort to help us on this ground.

We expect to handle a large business next year and may want credit in doing so. Unless we pay the money that we now owe[d] we shall have great trouble in getting the credit we desire. Personally I have a great deal of financial pride. I do not owe any money and never have owed any, and hope never to. I am aware that you laugh at such a pride, yet it is born in me. I have loaned money to the Lamp Co. to help their credit, as I feel a personal pride in this company.

Now I ask you as earnestly as I can, that you procure for us $5,000 outside of the money due us for lamps, which we need for pay-rolls and running expenses, so that we can place it into bills by May 2 of this week.[6] Yours Truly

Francis R. Upton.

ALS, NjWOE, DF (*TAED* D8332ZAL). Letterhead of Edison Lamp Co. a"EAST NEWARK," and "188" preprinted. bInterlined above. cCanceled. dObscured overwritten text.

1. Unable to show a consistent profit, the Edison Lamp Co. was chronically short of cash. It met payroll and paid bills by a series of assessments on its partners' shares and by outright loans from Upton and Edison (see Docs. 2259, 2297, 2343 [and headnote], 2368, and 2394). Two weeks earlier, Upton suggested converting his loan to an increased ownership stake in the partnership. Upton blamed the current crisis on "extraordinarily light" orders (especially from abroad), a growing inventory of lamps rated for the wrong voltage, and the difficulty of collecting nearly $10,000 in accounts receivable. He sent "a list of pressing bills" totaling about $3,600 on 1 May; the factory owed about $32,000 in all (Upton to TAE, 18 Apr. and 1 May 1883, both DF [*TAED* D8332ZAJ, D8332ZAO]).

2. Upton had tried to make March "a sample month of what we can do if we had orders enough. . . . We are now in complete order to make 15 000 lamps a week, but have cut down our working force so as to make 10,000 a week or less" by running only five days per week. He reported in early May that the factory was "down to fighting weight, and can keep from running behind in cost of lamp on 6000 a week." Upton to TAE, 6 Apr. and 7 May 1883, both DF (*TAED* D8332ZAI, D8332ZAU).

3. The factory took orders for 7,889 lamps in April, down from 24,686 in March. Orders rebounded to 21,284 in May. Edison Lamp Co. to TAE, 2 Apr., 1 May, and 1 June 1883, all DF (*TAED* D8332ZAG, D8332ZAN, D8332ZBD).

4. Philip Sidney Dyer (1857–1919) had been the bookkeeper for the Edison Lamp Co. since January 1881. He was the son of Edison's former patent attorney, George W. Dyer, and the brother of his present one, Richard N. Dyer. Doc. 2076 n. 7; "Dyer, Philip S.," Pioneers Bio.; U.S. Census Bureau 1970 (1880), roll T9_122, p. 73.3000, image 0149 (Washington, D.C.).

5. Dyer had recently sent Upton a list of eighteen Newark suppliers who "<u>must</u> be paid very soon   We have staved them off about as long as we can"; their bills totaled about $1,400. He compiled a similar list of Newark and New York vendors on 1 May. Dyer to Upton, 23 Apr. and 1 May 1883, both DF (*TAED* D8332ZAK, D8332ZAM).

6. After Upton warned on 7 May that "We must have some money or bust," Samuel Insull provided $1,000 in cash and agreed to pay a $1,400 note to Bergmann & Co. He asked for an additional $3,800 to carry the factory through his absence in early July. In mid-summer, Upton notified Edison that despite significant investments in plant and equipment, the factory carried essentially the same debt load it had at the start of the year. He cautioned, however, that "we have no cash and are struggling from hand to mouth to meet every little payment that we make." Upton to Insull, 4 and 7 May and 29 June 1883; Upton to TAE, 23 July 1883 enclosing 2 July statement; all DF (*TAED* D8332ZAS, D8332ZAT, D8332ZBO, D8332ZCB, D8332ZCB1); Insull to Upton, 7 May, Lbk. 16:239A (*TAED* LB016239A).

To William Lloyd Garrison, Jr.[1]

[New York,] 3rd May [188]3

Dear Sir,

I enclose you herewith statement of probable Running expenses & of profits on the several estimates already sent you. As soon as I have copies printed I will send you a number of them.[2]

If you desire it we can reduce the amount of the item for underground conductors by only putting down enough conductors to carry 1600 or 3200 lights (according to which Plant you may elect to install) & leave the remainder of the conductors to be put in when you desire to extend the business. I hope to send you some information on this subject either to-morrow or next day[3]

I understand from Mr Borden that you are contemplating wiring the houses at Brockton free of charge to the Consumers   I think this is entirely the wrong course to pursue. A Gas Company does not do House Piping free of charge to the Consumers & I see no reason why we should. I trust that you will give up any such idea as I feel confident that you can get the Customers to do the work

I enclose you Copy of Balance Sheet of the Waltham Gas Co[4] from which you will see that a Gas Plant for a Town of 11 700 inhabitants   Yours truly

Thos. A. Edison   I[nsull]

L (letterpress copy), NjWOE, Lbk. 16:227 (*TAED* LB016227). Written by Samuel Insull.

1. Son of the renowned abolitionist and reformer, William Lloyd Garrison, Jr. (1838–1909) headed a wool trading firm in Boston from approximately 1866 to 1883, when he took up the promotion of Edison's electric lighting system. In 1883, he was the founding treasurer of Edison Electric Illuminating Cos. in both Boston and Brockton, Mass. Already prominent in the woman's suffrage and racial equality movements, Garrison was later associated with single-tax and anti-imperialist causes; he was also a brother-in-law of Henry Villard. Obituary, *Boston Daily Globe,* 13 Sept. 1909, 1; "The Edison Light in Boston," *NYT,* 29 March 1883, 2; "Electrical Interests," *Boston Daily Advertiser,* 5 Oct. 1883, 8.

2. The editors have not found the enclosure, but Insull sent multiple printed copies of expected profits and expenses on 6 May. Estimates for Brockton central stations of 1,600, 3,200, and 4,800 lights were finished in late April and corrected soon afterward. Insull to Spencer Borden, 6 May 1883; Insull to George Peabody, 24 Apr. 1883; Insull to Garrison, both 25 Apr. 1883; Insull to Edward Johnson, 27 Apr. 1883; Lbk. 16:233, 188–89, 192, 201 (*TAED* LB016233, LB016188, LB016189, LB016192, LB016201).

3. Edison shortly sent estimates for plants of 1,600 lights ($28,784)

and 3,200 lights ($46,924). These figures were modest revisions of an earlier estimate, which had included a plant for 4,800 lights ($70,267). The Brockton company contracted for a plant of 1,600 lamps, though it evidently expanded the system by October 1884, when it had 2,000 in operation. TAE to Garrison, Jr., 8 May 1883, Lbk. 16:251 (*TAED* LB016251); Samuel Insull to Garrison, Jr., 25 Apr. 1883, LM 16:192 (*TAED* LB016192); Edison Construction Dept. inventory, 19 Dec. 1883, Lbk. 4:179 (*TAED* LBCD4179); Edison Electric Illuminating Co. of Brockton brochure, 24 Mar. 1885, PPC (*TAED* CA003A); see also Doc. 2420 n. 37 and App. 2A.

4. The Waltham (Mass.) Gas Light Co. was incorporated in 1853; it added electric lighting to its business in 1887. The enclosure has not been found. "Massachusetts Legislature Session of 1853," *Boston Daily Atlas*, 31 Mar. 1853, 2; Gas Commissioners 1887, 57.

THOMAS A. EDISON CONSTRUCTION DEPARTMENT   Doc. 2437

An impatient Edison saw his "Edison system" of central station electric lighting at a crossroads in the late winter and early spring of 1883. The system was a reality in New York's Financial District and in Roselle, New Jersey, on a small scale, but neither Edison nor the Edison Electric Light Company had a suitable system—organizational, financial, and technical—for selling and constructing central stations elsewhere. Edison foresaw a large market in small cities and towns but feared the business "would go to ruin" in the Edison Electric Light Company's hands.[1] He came to believe "that if the business is to be made a success it must be by our personal efforts and not by depending upon the officials of our Companies."[2] He saw the need for new sources of outside capital, and the year had started auspiciously when financier Henry Villard proposed contracting with the Edison Electric Light Company for "lighting all the cities & towns along the main line & branches of the Northern Pacific" railroad. A recent conversation with banker George Ballou had also lifted his hopes for fresh investment, but neither Villard nor Ballou brought new funds in the short or long run.[3] Searching for another way to proceed, in late March, Edison sketched a partnership arrangement with Edward Johnson, Samuel Insull, and Charles Batchelor.[4] He already had comparable arrangements with trusted associates as partners in his manufacturing shops, but this plan, too, failed to materialize.

Then in mid-April, Edison hit upon the notion of forming the Thomas A. Edison Construction Department to accomplish his aims. His first apparent reference to "my Central Station Construction Department" came in a power of attorney[5] authorizing Samuel Insull to carry on the business in Edison's name. Edison executed the document immediately upon receiving an urgent summons to Pennsylvania to sign construction contracts with two nascent electric illuminating companies seeking to bring the Edison central station system to their respective towns.[6] He left no clues as to the origin of the idea. His sketchy plan for the new entity created enduring ambiguities over its specific functions and its relations with existing organizations.

The Construction Department provided an informal financial and administrative framework in which Edison and Insull could manage a variety of transactions over a wide geographic area. The idea was not entirely novel. The organization of specialized construction companies had precedents in capital-intensive projects like submarine telegraphy and telephone exchanges, and there were by this time also numerous examples of independent contractors and suppliers in electric lighting. The tradition of referring to the Construction Department as a company goes back at least to 1894, but it functioned as a contractual surrogate for Edison, who was personally liable for its obligations.[7] The Department had no independent legal standing, nor was it formally a branch of another entity, like the Edison Electric Light Company. Within the unwritten understandings among the principals, however, the Company used the Construction Department's services and exercised some oversight of its operations.

The Construction Department promised to benefit both Edison and the Edison Electric Light Company. The firm could trade on Edison's name, now intimately connected with electric lighting, to sell village plants. Those stations promised to be far cheaper to build than the flagship underground system in New York, whose construction had cost something on the order of $50 per lamp of generating capacity. Owing to the three-wire design, relatively small service areas, and overhead wires in most systems, the cost of Edison's village plants typically fell somewhere between $10 and $30 per lamp. Even these figures represented a much larger investment than either isolated plants or the arc lighting systems of Edison's principal competitors. Faced with an uncertain investment climate (one that would notably worsen in 1884), the company's backers were unwilling to put more of their own money into the enterprise, but they could attract local investors with Edison's personal participation. For the company, which owned Edison's patents (and little else), village plants would preserve the value of the patents and generate income from the license fees of local illuminating companies. Edison, for his part, recognized the proliferation of central stations as the best means to increase demand for his manufactured products, especially lamps, for which he was entitled to royalties from the company and from which he also expected to obtain substantial direct profits.

It was not necessarily in anyone's interest to specify too closely the relationships among Edison, the Electric Light Company, or the Edison Company for Isolated Lighting. The latter firm, by prior contract, controlled Edison's patents in areas without municipal gas service.[8] Adding to the confusion, the two Edison companies had overlapping officers and investors. Sherburne Eaton served both as president and often failed to differentiate these roles in his prolific correspondence.

Aware of preliminary interest from hundreds of towns, Edison and Insull had the Construction Department's essential elements for canvassing, mapping, estimating and contracting for village plants in place by the end of May.[9] During the summer, Edison advertised his intentions and expectations. Supremely confident of his ability to carry out a large amount of work in this way, he reportedly boasted in June to the *Chicago Daily Tribune* that "we have the means, shops, and a system so that I am able to take a contract to light a whole town, and do so within sixty days." Elsewhere, he explained: "I could take hold and push the system better than any one else. It is so complicated that I do not feel like trusting it to new and untried hands, because science and dollars are so mixed up in it."[10] However, the scale and expense of this work prompted him to make more or less exclusive agreements with manufacturers of three crucial components outside of his control: steam boilers, engines, and copper wire. He told a New York newspaper in July that he was "a regular contractor for electric light plants," evidently referring to his contracts with local illuminating companies and perhaps also to his evolving intermediary role in the Edison Electric Light Company's business.[11]

Edison's method was to contract with a local illuminating company, one formed in advance with the assistance of agents tied to the Edison Electric Light Company, the Edison Company for Isolated Lighting, or both.[12] He agreed to install a complete electric lighting system of a certain size, with his employees generally performing or supervising the work (except for the wiring of several initial towns, which he subcontracted to the Wiring Department of Bergmann & Company). After a trial period, he would turn the plant over to the local company, which would own and operate it.[13]

Before offering a contract, Edison made a cost estimate based on a careful canvass of the district to be illuminated and a "determination" of the copper conductors required.[14] There was some ambiguity over coordinating the preliminary steps and responsibility for their costs, though by early summer he generally had made more systematic arrangements with regional agents of the two controlling companies.[15] He stipulated cash payments for the plants at intervals (before and after construction and following the trial period) but found himself accepting stock shares in undercapitalized illuminating companies.[16] Edison had hoped to make a modest profit from the construction work but soon found he was losing about $1,500 dollars on each of the first few stations.[17] The high initial costs of the plants in some cases exceeded the enthusiasm of local investors. In addition to paying for the purchase and installation of equipment, each company had to buy patent licenses from the Edison Electric Light Company (often paid in stock) equal to one quarter of its capitalization.[18]

Edison delegated a large amount of operational responsibility to Samuel Insull. Insull routinely drafted and signed letters for Edison, and he coordinated the movement of people and equipment through several states at each stage of the canvassing and construction process. He also conducted extensive correspondence under his own name and often conflated his voice with Edison's. Insull was a dozen years younger than Edison and had worked for him for just over two years, but he clearly relished the authority.

Edison laid out large sums of money to get the business started. A Construction Department ledger book shows that by late September, he had advanced about $43,000 on behalf of five central stations to pay for everything from wages to switches to boilers.[19] Slowly the debts were cleared off the books, though not always in cash, and often only after strident reminders from Insull.[20] Subsequent payments from these and other stations often came only after unpleasant haggling over the quality of the workmanship.

Edison's evolving relationships with the two patent-controlling companies complicated his personal financial outlook. Sherburne Eaton initially had helped coordinate canvasses and estimates for large cities like New Orleans and St. Louis.[21] From the fall 1883, however, the Edison Electric Light Company played a more active role in soliciting business for the Construction Department by deciding where to canvass, make estimates, and tender contracts.[22] The company was to reimburse the expense of such preliminary work, but with that promise came greater control and a corresponding erosion of Edison's authority and his own expectations for the business.[23] Edison had indeed become a "regular contractor."

Edison eventually became frustrated that "the Agents of the Light Company are not forming any new Companies" to operate central stations.[24] In April and May 1884, with tens of thousands of dollars owed him by the Edison Electric Light Company and local illuminating companies, in addition to the ongoing expenses of his engineering and electrical staffs, Edison declared his wish to leave the construction business. His effort to hand the whole business to Edison Electric Light was complicated by the company's attempt to acquire at least part ownership of the manufacturing shops, which, now that the Edison Lamp Company had passed the break-even point, were attractive acquisitions. His decision may have been aided by the recent memory of his Florida vacation with Mary in February and March, when he again found time to fill notebooks with inventive ideas.[25] For almost a year, Edison had laid aside the pleasurable labor of inventing to be "simply a business man," running what was sometimes derisively called the "Destruction Department."[26] Ultimately, however, Edison understood that his stations were too expensive. Despite design compromises, "the 1st investment is the trouble in

Edison central stations constructed 1883–84.

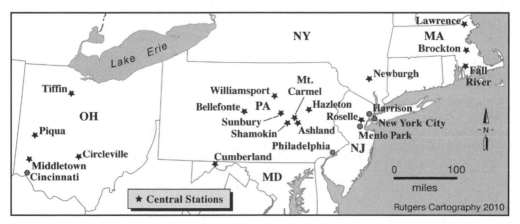

pushing our biz," he realized, and this fact would remain a major hurdle in developing the central station model of electric light and power that he envisioned.[27] The solutions to this problem required nearly a decade, the formation of industrial giants General Electric Company and Westinghouse Electric, and creative new ideas about financing capital construction.[28] In May 1884, Edison negotiated the temporary assignment of essential Construction Department employees to the Edison Electric Light Company. The Edison Company for Isolated Lighting officially took over the central station business on the first of September.[29]

1. Doc. 2417 n. 1.

2. Doc. 2407.

3. Villard to Sherburne Eaton, 2 Jan. 1883, letterbook 47:6, box 122, Villard (MH-BA); see Doc. 2413.

4. See Docs. 2417 and 2420; see also Israel 1998, 219–25.

5. Doc. 2437.

6. One of those contracts is Doc. 2438. Shaw to TAE, 3 May 1883, both DF (*TAED* D8360B, D8360C).

7. Luther Stieringer testified in an 1894 lawsuit in federal court: "In order to advance central-station lighting, Mr. Edison had organized a construction department known as The Edison Central Station Construction Company." *Edison Electric Light Co. v. F. P. Little Electrical Construction and Supply Co., et al.*, Pleadings and Complainants Prima Facie Proofs, p. 40 (New York: C. G. Burgoyne); see also Doc. 2709.

8. Doc. 2299 n. 4.

9. Insull to Edward Johnson, 11 Apr. 1883, LM 3:135 (*TAED* LM003135); see also Docs. 2466 and 2476.

10. "Promoting the Electric Light," *Electrical World* 1 (4 Aug. 1883): 489. By July, Edison had closed contracts for ten plants and had another three dozen pending ("Edison," *Chicago Daily Tribune*, 19 June 1883, 8).

11. Edison's principal suppliers were, respectively, Babcock & Wilcox, Armington & Sims, and the Ansonia Brass & Copper Co. (see Docs. 2423, 2458, and 2515). The "regular contractor" quotation is from Doc. 2503.

12. Other agents included Harris & Robinson in Connecticut, Spencer Borden in Massachusetts, Henry Clark in Maryland, Washington, D.C., and western Pennsylvania, and, soon after, Archibald Stuart and the Ohio Edison Electric Installation Co. in Ohio. The individuals in this list were formally affiliated with the Edison Co. for Isolated Lighting (Letterhead, Henry Clark to TAE, 27 June 1883, DF [*TAED* D8325ZBA]). A list of agents and their affiliations and commission rates at the end of 1883 is in Eaton to Edison Electric Light Co. (pp. 10–11), 27 Dec. 1883, Batchelor (*TAED* MB109); see also Doc. 2725 n. 3.

13. The Ohio Edison Electric Installation Co. contracted with Edison for plants in that state and took possession of them upon completion. In at least one case (Circleville), Edison subcontracted actual construction work to the company. He also expressed a willingness to

make similar arrangements with the Western Edison Light Co. in Chicago, but apparently never did so. TAE agreement with Ohio Edison Electric Installation Co., 16 May 1884, Miller (*TAED* HM840216); see Doc. 2478.

14. Edison took over the canvassing process from regional agents and centralized it in New York by summertime, about the time he developed an "entirely novel process" for making "a perfect canvas in two days" (see Docs. 2447, 2459, and 2478). The canvass was the basis for drawing a map of the conductors, which also occurred in New York (Insull to TAE, 10 July 1883, DF [*TAED* D8316AFF]). Then Hermann Claudius and a small staff at the Menlo Park laboratory made the determination by building a physical model of the proposed illumination district, as they had done for the New York central station. Some estimates were made without benefit of a survey, apparently from a careful examination of local maps (see App. 2B). The physical model became the basis for making blueprints showing the locations of feeders and mains. Claudius was dismissed after Frank Sprague (and others) devised more accurate and much faster mathematical methods of determining the conductors. Copies of the blueprints went to the local company with a request for information about a possible central station; Edison (or Insull) developed a form letter to solicit this information (Edison Construction Dept. form letter, n.d. [1883?], Undated Notes and Drawings [c. 1882–1886], Lab. [*TAED* NSUN08:70]).

15. See Docs. 2447, 2468, 2472, and 2476.

16. See, e.g., Doc. 2498.

17. See Doc. 2498; TAE to Phillips Shaw, 18 July 1883, DF (*TAED* D8316AHU).

18. The Edison Electric Light Co. passed a small portion of the license fees to its regional agents; see Docs. 2442 and 2454.

19. Construction Dept. "Trial Balance[s]" prepared by John Randolph also show in detail Edison's running expenses as of 1 September and 1 October (Edison Construction Dept. statements, 1 Sept. 1883 and 1 Oct. 1883, Miller [*TAED* HM830186E, HM830186F]). Summaries of expenses for individual central stations are in Edison Construction Dept. Ledger (1883–1886), esp. pp. 2–17, Accts. (*TAED* AB033); routine financial and bookkeeping records are in Electric Light—TAE Construction Dept.—Accounts, DF (*TAED* D8440). Probably the largest single expense was for copper conductors, for which Edison paid more than $20,000 between October 1883 and January 1884 (Statement of TAE account with Ansonia Brass & Copper Co., n.d. [Oct. 1884], DF [*TAED* D8421ZAB1]).

20. The Sunbury, Pa., company paid part of its obligation in stock; at least one other company sought to do the same. See Doc. 2424 (headnote); Insull to William Lloyd Garrison, Jr., 7 Mar. 1884; TAE to Garrison, Jr., 3 Apr. 1884; both DF (*TAED* D8416ATX, D8416BCE).

21. See Doc. 2482.

22. See, e.g., Doc. 2604.

23. Regarding reimbursement of Edison's outlays, see Docs. 2704 and 2736. With a key change of officers at the end of 1883 and a worsening business climate in the new year, the Edison Electric Light Co. was increasingly interested in managing its own costs. See Doc. 2569 and Eaton to TAE, 5 Nov. 1883, DF (*TAED* D8327ZBK).

24. Doc. 2655.

25. See Docs. 2607 and 2609 (headnotes).

26. Edison's declaration in the interview enclosed with Doc. 2503 came about two and a half months after he threw himself into the central station business in May 1883. Insull recalled the "Destruction Department" nickname in a 1909 set of reminiscences incorporated into Dyer and Martin 1910 (428); Insull reminiscences, Feb. 1909, p. 164, Meadowcroft (*TAED* MM010DAH).

27. See Doc. 2615.

28. Chandler 1990, 214; Kobrak 2007, 52.

29. See Docs. 2658, 2661, 2672, 2677, and 2725.

–2437–

Power of Attorney to
Samuel Insull

[New York,] May 3, 1883[a]

Know all Men by these Presents, That I Thomas A Edison of the City, County, and State of New York[b] have made, constituted and appointed, and by these presents do make, constitute and appoint Samuel Insull[c] my true and lawful attorney for me[d] and in my[d] name, place and stead to sign contracts for the erection of Edison Electric Light Installations to receive payments and give receipts for same to make settlements and transact any other business whatsoever [----][e] appertaining to my Central Station Construction Department[1][f] giving and granting unto my[d] said attorney full power and authority to do and perform all and every act and thing whatsoever requisite and necessary to be done in and about the premises, as fully to all intents and purposes, as I[d] might or could do if personally present, with full power of substitution and revocation, hereby ratifying and confirming all that my[d] said attorney or his[d] substitute shall lawfully do or cause to be done by virtue hereof.

In Witness whereof, I[d] have hereunto set my[d] hand and seal the 3rd[d] day of May[d] in the year one thousand eight hundred and eighty three[g]

Thomas A Edison [Seal]

Sealed and delivered in the presence of Wm H Meadowcroft[2]   State of New-York, City and[h] COUNTY OF New York[i]

PDS, NjWOE, Miller (*TAED* HM830175). Preprinted power of attorney form of W. Reid Gould, law blank publisher; notarization omitted. [a]Date from document; form altered. [b]"I Thomas . . . New York" written by Samuel Insull. [c]"Samuel Insull" written by Insull. [d]Written by Insull. [e]Canceled. [f]"to sign contracts . . . Construction Department" written by Insull and followed by diagonal line through remaining blank space. [g]"eighty three" written by Insull. [h]"City and" written by William Meadowcroft. [i]"New York" written by Meadowcroft.

1. See headnote above.

2. British-born William Meadowcroft (1853–1937) had been an assistant chief clerk in the law partnership of Sherburne Eaton, whom he followed to the Edison Electric Light Co. in 1881 as an assistant. He became involved in a variety of Edison enterprises and helped manage the affairs of the Edison Ore Milling Co. and the Edison Electric Light Co. of Europe. Meadowcroft remained closely associated with Edison; in 1910, he became an "assistant and confidential secretary," a position he held until Edison's death. "Meadowcroft, William Henry," Pioneers Bio.; Obituary, *NYT,* 16 Oct. 1937, 19.

–2438–

Agreement with Edison Electric Illuminating Co. of Shamokin

[Shamokin, Pa.,[1]] May 4, 1883[a]

Memorandum of agreement[2] made this 4th day of May 1883 Between The Edison Electric Illuminating Co of Shamokin Pa hereinafter called party of the first part and Thomas Alva Edison of the City, County, and state of New York hereinafter called party of the second part.

The party of the second part upon receiving permission from the Edison Electric Light Company of New York agrees to furnish and install for the parties of the first part an Edison Electric Light central station plant necessary to furnish a maximum of one thousand and six hundred—ten candle lights over a system of pole lines as per annexed map (marked exhibit A) for the sum of Nineteen Thousand Two Hundred & nine dollars, according to statement of labor and material hereunto annexed. (Marked Exhibit B)[3]

In consideration of the party of the second part undertaking to install and furnish said plant the party of the first part hereby agrees to pay to the party of the second part the following sums. Four Thousand Eight Hundred and Two $^{25}/_{100}$ dollars within thirty days of the signing of this agreement, Nine Thousand Six Hundred and Four $^{50}/_{100}$ dollars when the installation is ready to supply light and the balance Four Thousand Eight Hundred and Two $^{25}/_{100}$ dollars after the installation has been in condition to supply light for thirty days.— The party of the second part further agrees to use due diligence in putting into practical operation the said installation and to do all work specified in Exhibit B and in connection therewith in the best and most workmanlike manner, and in compliance with the requirements of the Engineering department of the Edison Electric Light Company of New York. It is further understood and agreed by and between the parties hereto that if from any cause P. B. Shaw fails to furnish the cash payments on the bonds of this company as agreed, that the said party of the second part agrees to take and accept said bonds at par in such sum or sums as may be due him in cash.

It is agreed by the party of the first part that the work in connection with the installation of the Plant which is not included in statement annexed shall be done[b] in such manner and at such times as not to interfere or delay the work of the party of the second part, and in case of a failure so to do on[c] the part of the party of the first part the party of the second part shall be reimbursed the amount of any extra expense in consequence thereof by the party of the first part—

In witness whereof the party of the first part has caused these presents to be subscribed by its proper officers thereunto duly authorized by the board of Directors and the party of the second part has hereunto set his hand and seal the day and year first above written

Witness present   Wm. H. Douty[4]   President
A Robertson[5]   John Mullen[6]   Treasurer
                                          Wm. Beury[7]   Secretary
                    of the Edison Electric Illuminating Co of Shamokin
Sam'l Insull   Thomas A Edison

DS, ICL, Insull (*TAED* X077AB). [a]Date from document, form altered. [b]"shall be done" interlined above. [c]Added later.

1. Phillips Shaw telegraphed from Shamokin on 3 May that this agreement had been signed by the local parties and asked that Edison "come by penna [rail]road to sunbury tonight   I will meet you." Edison evidently declined, because Shaw wired again: "message recd   I regard it absolutely necessary for you to come   cannot think of negative reply   you can return in six hours." Edison and Insull left that evening and did not return until late on 5 May. Shaw to TAE, 3 May 1883, both DF (*TAED* D8360B, D8360C); Insull to Spencer Borden, 6 May 1883, Lbk. 16:233 (*TAED* LB016233).

2. The general terms of this agreement closely resemble those Edison made with other local illuminating companies. About a half dozen other such agreements have been found; some required the first payment to be made upon execution of the contract. See TAE agreements with Edison Illuminating Cos. of Fall River, 4 Oct. 1883; Newburgh, 29 Oct. 1883; and Hazleton, 15 Nov. 1883; all DF (*TAED* D8347J, D8351V, D8340ZGJ) and TAE agreements with Edison Illuminating Cos. of Tiffin, 24 Oct. 1883; Williamsport, 26 Nov. 1883; and Middletown, 28 Nov. 1883; all Miller (*TAED* HM830197, HM830201, HM830202).

3. Neither exhibit has been found. Edison's April estimate was higher; see Doc. 2424 (and headnote).

4. William H. Douty (b. 1837) owned W. H. Douty Dry Goods. He had operated a mining tract since 1882 and was later a director of the Shamokin Board of Trade. Bell 1891, 627–28, 385; Douty to TAE, 5 Jan. 1884, DF (*TAED* D8457C).

5. Scottish-born Andrew Robertson (b. about 1831) lived in Potts-

town, Pa., but was active in Shamokin commercial enterprises, particularly as a colliery operator and manager. He also helped introduce water and gas service (and later electric arc lights) to Shamokin. Bell 1891, 627–28, 892–94.

6. John Mullen (b. 1838) owned John Mullen & Co., which manufactured mining machinery, and the Anthracite Foundry and Machine Works. He was also a director of the Shamokin Gas Light Co. and president of both the First National Bank in Shamokin and the Shamokin Coal and Coke Co. of May-Beury, W.Va., among other business ties. Bell 1891, 618, 906–7.

7. William Beury (b. about 1844) was a Shamokin gunpowder manufacturer. He was later founding treasurer of the Shamokin Arc Light Co.; by the end of the decade he (or a namesake) was associated with John Mullen in the Shamokin Coal and Coke Co. U.S. Census Bureau 1970 (1880), roll T9_1164, p. 133.3000, image 0011 (Shamokin, Northumberland, Pa.); Bell 1891, 628; "Black Diamonds," *Washington Post*, 2 Aug. 1889, 2.

–2439–

To Henry Villard[1]

[New York,] May 7th. [1883]

Dear Sir:—

I have noticed that the Evening Post has been very careful to publish of late anything derogatory to our system of electric lighting, the latest item being on April the 28th. when the published the enclosed paragraph from the London Truth.[2] It happens that the London Truth is entirely in error, that the system in use in the House of Commons at the date mentioned was the Swan light which was ordered out in consequence of such failures as spoken of and since then our light has been installed in several of the rooms of the House of Commons and our information is that there has not been a single failure on the part of our lights since they were first started.[3]

Cannot you do me the very great favor to take steps to prevent the Post from publishing such items as they must certainly cause us considerable injury?[4] If the managers of the Post would take the trouble to make inquiries I am sure they would find that we give general satisfaction wherever we install plants.

If the Post would take the trouble to publish all the items they get in our favor as well as all they get against us, I am sure we would be perfectly satisfied. Apologizing for troubling you in this matter, believe me, Very truly yours,

TL (carbon copy), NjWOE, Lbk. 16:291 (*TAED* LB016291).

1. A former newspaper reporter, transportation capitalist Henry Villard (1835–1900) acquired a controlling interest in the New York

Evening Post in 1881, though he reportedly exercised no influence over its editorial department (*DAB*, s.v. "Villard, Henry"). A director of the Edison Electric Light Co., he had previously underwritten some of Edison's electric railway experiments, and he became a founding trustee of the Electric Railway Co. of the United States about this time. His new mansion in Manhattan had recently been wired for Edison's electric light, and he was among the organizers of a prospective Edison illuminating company in Boston (see Docs. 2152 n. 2, 2195, and 2431 n. 5; Edison Electric Light Co. Bulletin 17:9, 6 Apr. 1883, CR [*TAED* CB017]; "The Electric Light," *Boston Daily Advertiser,* 17 Feb. 1883, 8).

2. The editors have not found the enclosure from *Truth* nor identified the *Evening Post* article to which Edison referred.

Truth was a weekly newspaper known for its satire and gossip as well as political and financial journalism. It was owned and written by Henry Du Pré Labouchere, scion of a Dutch banking house and a member of Parliament for Northampton and the radicals (*Oxford DNB*, s.v. "Labouchere, Henry Du Pré [1831–1912]"). Generally skeptical of the 1882 mania for electrical shares and financial speculation in patents, the paper particularly criticized the Anglo-American Brush Electric Light Corp., Ltd., whose troubles were emblematic of the financial disorder soon known as the Brush Bubble (see Doc. 2317 n. 8). The Edison Electric Light Co., Ltd., had advertised in the 1 March issue ("Bubble Companies," *Truth* 8 [18 Jan. 1883]: 87; "The Edison Electric Light Co., Ltd.," ibid., 8 [1 Mar. 1883]: 303; Hughes 1962, 29; Wilson 1988, 14–15).

3. An exhibition of Joseph Swan's incandescent lighting system had been installed at the House of Commons in June 1881. It was taken down a week or two later in deference to members' unfavorable opinions ("Electric Light in the House of Commons," *Electrician* 7 [9 July 1881]: 99, 112–13). The Edison Electric Lighting Co., Ltd., began installing equipment for lighting the House of Commons dining rooms and library in March 1883; service began on 2 April. Different sources give conflicting accounts of the number of dynamos (1 or 2) and lamps (150 to 260); the plant was expanded at the end of the year or early 1884 (Edison Electric Light Co. Bulletin, 17:27, CR [*TAED* CB017]; "Home," *Graphic* [London], 7 Apr. 1883, 10; "The Edison System in the House of Commons," *Electrician* 10 [7 Apr. 1883]: 482; "Notes. Electric Lighting," *Teleg. J. and Elec. Rev.* 14 [16 Feb. 1884]: 134; see also Docs. 2679 and 2451). The 12 April issue of *Truth* reported that Edison lamps in the library had suddenly stopped working twice during their first ten days of operation and that Labouchere himself had counted four dark lamps in one chandelier ("Notes," *Truth* 8 [1883]: 497). A different problem occurred in August, when the automatic regulator "split and cracked from the great heat" (Arnold White to Johnson, 9 Aug. 1883, DF [*TAED* D8338ZBF]).

4. The editors have not found a reply from Villard, who was traveling through Montana on 8 May. The *Electrician* (New York) referred in a brief published note to Edison's protest to the *Post.* "Henry Villard Given a Reception at Helena, M.T.," *Chicago Daily Tribune,* 8 May 1883, 6; "Note," *Electrician* (N.Y.) 2 (June 1883): 170.

Technical Note:
Electric Light and
Power

[New York, May 7, 1883?[1]]

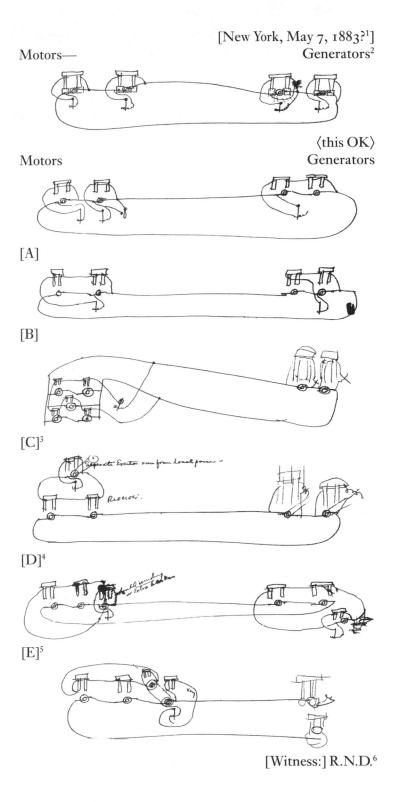

Motors—   Generators[2]

⟨this OK⟩

Motors   Generators

[A]

[B]

[C][3]

[D][4]

[E][5]

[Witness:] R.N.D.[6]

X, NjWOE, Lab., Cat. 1149 (*TAED* NM018AAI). Document multiply witnessed and dated.

1. Richard Dyer signed and dated the document on this date when he received it from Edison.

2. The drawings in this document represent variations for regulating, by electrical means, each of several dynamos and motors. Edison executed seven closely related patent applications on 1 June. Like these drawings, the applications applied to switches and resistance coils in the circuit to control the field coils of each generator or motor. The armatures are connected in series, but in some of the arrangements Edison provided a shunt circuit for each pair of field magnets in order to permit individual machines to operate independently or to be removed entirely from the circuit without affecting the remaining ones. In other arrangements, several field coil pairs are in the same circuit, effectively controlling a group of motors or dynamos; he obtained additional permutations by stipulating an outside electrical source to excite the field coils in some cases. U.S. Pats. 370,125; 370,126; 370,127; 370,128; 370,129; 370,130; 370,131.

These arrangements were generally applicable to the problem of generating, transmitting, and applying high voltage currents (see Docs. 2242, 2276, and 2284). They were also relevant to using electric power under conditions of varying load, such as for railroads. Several of the June applications referred to a patent Edison had obtained in 1881 for using electric power (U.S. Pat. 248,435).

3. Figure labels are "Seperate Exciter run from local power—" and "Receiver."

4. Figure labels are "double winding or extra brushes."

5. Figure label is "Vary."

6. Richard Nott Dyer (1858–1914, brother of Philip) became Edison's principal patent attorney in early 1882, when his father and Zenas Wilber, who had filled that role, dissolved their partnership. An 1879 law graduate of Georgetown University, Dyer practiced in Washington, D.C., and Chicago before moving to New York in 1881. Doc. 2203 n. 25; Obituary, *NYT*, 15 Jan. 1914, 9; Association of the Bar 1915, 181–82.

–2441–

Draft to J. Ferro[1]

[New York, May 11, 1883][2]

Dear Sir

The letter from J V Magallon[3] just at hand— Its nature is such that it would be an injustice to[a] all concerned to attempt quote a price for apparatus to ~~prefer~~ to give an ~~uncertain~~ unknown result the letter is to vague. The party writing ~~has no~~ does not give sufficient explanation as to the amount of light each Lamp should give or if they are arc or incandescent. I should not be willing to sell the[b] machinery without a certainty that it would be successfull in every respect & give satisfaction to the purchaser both as to working, economy, & results required, hence, I ask that you write your correspondent to obtain the following[4]

A Map of Baranquella,[5] shewing the squares streets, etc. such map should be drawn to scale so measurements can be taken.

2nd   Let the position of the street Lamps be marked by[a] a cross X on the map. also the position of the houses that are to use the light with the number of lights marked & the average hours that the parties will burn these lights or in other words the average number of ~~lights~~ hours[b] the party will burn each light every night—

Also mark on the map the position of the ground or the building with its width length of ground[a] or building also the heighth of the ceiling if there is already a building—

3   The price of steam coal. If the water is good or bad for steam boilers. The price of & method of obtaining water. The price of wood

If the streets are paved or unpaved or any portion paved if any[b] paved mark on the map—

If there are carts in Baranquella capable of handling a large boiler.

The duties on machinery such as will be used in Electric Lighting—

The Cost of Lumber Common per 1000 feet

The Cost of Brick per thousand feet—

The average closing time at night of the stores.

The average time people go to bed in private houses—

The population of Baranquella.

State that ~~our light~~ we make several sizes of lamps, all of which can be worked on the same wire. No 1 equal to best gas jet. No 2 Equals $1^{3}/_{4}$ of gas jet [-][c] or 3 Kerosene jets. No 3 Equals 3 gas jets. or 5 Kerosene burners   No 4. equals 5 gas jets of 8 Kerosene lamps— No 5. Equals 10 gas jets or 16 Kerosene jets—

We obtain 10 jets each equal to a gas jet for every indicated horsepower of the Engine, & the other Lamps in proportion, that is to say 5 Lamps of 2 jets each, 1 lamp ~~of~~ giving a light equal to 10 gas jets—

I herewith send you as a sample The town of Danville Pennsylva[6] shewing that portion of the Town where the lights are most used, that is to say the best portion of the town that would pay the greatest profit on the smallest investment. The lines running in the blocks with figures at the end indicate the number of gas jets. The lines in the Streets show our wires. The Star within red circle shew the Central station, & red lines the wires from the street wires to the station= There are ove about 1800 gas jets in this area. The Cost of the whole thing errected complete is about 29,000.[a] including the building wires & all apparatus at the stations & meters= The Extra cost of Errecting at Baranquella would be the fares of Experts freight Custom duties & increased cost of material which must be obtained at Baraquella—

please Explain to your correspondent that the wires are on poles, that a wires is run from the main street wires where they enter a meter & from that to different parts of the building where the lamps are situated. The cost of wires & fixtures in the building are paid by the person using the light— The meter measures accurately the amount of light used in the house the unit of light is the amount of light obtained in one hour from a candle burning 120 grams per hour of tallow. 10 such candles are Equal to the average coal[b] gas jet in practice. at Danville we sell each gas jet for one cent per hour or rather a light equal to[d] 10 candles for 1 hour for 1 cent. Two thousand candles of light measured on the meter equal the light given by[e] 1000 Cu f Cubic feet of common coal gas   for this we charge $2.

The dividends will be about 28 per cent at Danville on the cost of the complete plant=

Mr Ferro   P.S. These are cost prices at Danville we shall have to all your & our profit on   Edison[7f]

ADf, NjWOE, DF (*TAED* D8336L). [a]Obscured overwritten text. [b]Interlined above. [c]Canceled. [d]"a light equal to" interlined above. [e]"the light given by" interlined above. [f]Postscript written in an unknown hand.

1. Joaquin Ferro (1837?–1907) was a Colombian-born shipping and commission merchant in New York City. The editors have not found any correspondence from him to Edison. U.S. Census Bureau 1970 (1880), roll T9_895, p. 456.4000, image 0314 (New York City, New York County, N.Y.); "Surrogate Notices," *NYT*, 2 Aug. 1907, 11.

2. This date is taken from a letterpress copy of the finished letter to Ferro prepared by a secretary (probably William Meadowcroft). Lbk. 16:297 (*TAED* LB016297).

3. This letter has not been found, and the editors have not identified Ferro's correspondent.

4. Ferro had recently received an estimate from the United States Electric Lighting Co. for the installation of street lighting in Barranquilla, Colombia. Despite Edison's professed reluctance, sometime in April or May he personally wrote and signed a summary estimate of the equipment needed for a mixed incandescent and arc light installation. He recalculated the costs on 12 May and sent two versions of the figures to Ferro soon afterward. United States Electric Lighting Co.

to Ferro, 4 May 1883; TAE memorandum, c. 19 Apr. 1883; TAE to Ferro, 12 and undated May 1883; all DF (*TAED* D8336K, D8336H1, D8336M, D8336T); TAE to Ferro, both 14 May 1883, Lbk. 16:319, 321 (*TAED* LB016319, LB016321).

5. The Caribbean port of Barranquilla, on the Magdalena River in northern Colombia, was a commercial center but not a leader in electrification at this time. Nichols 1954, 171; Rippy 1945, 134–35; Posada Carbó 1996, xii, 116 (fig. 3.1).

6. With Samuel Insull and Phillips Shaw, Edison was assembling estimates, ground plans, and a hint of local capital for a 1,600-lamp village plant in Danville, a borough on the Susquehanna River in east central Pennsylvania (Insull to Phillips B. Shaw, 22 May 1883; Insull to D. Bright, 25 May 1883; Lbk. 16:403, 464 [*TAED* LB016403, LB016464]; Shaw to Insull, 24 May 1883, DF [*TAED* D8340U]; *Columbia Gazetteer*, s.v. "Danville"). In July, however, after realizing he was underestimating the cost of central stations by about $1,500, he reluctantly went ahead with the "losing business" in Danville but told Shaw he would withdraw the estimate if the contract were not closed in thirty days. The plant was not completed (TAE to Shaw, 18 July 1883, both DF [*TAED* D8316AHT, D8316AHU]; cf. Doc. 2456 and see also App. 2).

7. Edison appended a nearly identical postscript below his signature on the formal letter to Ferro (see note 2).

–2442–

Samuel Insull to John Culbertson[1]

[New York,] May 13th. [1883]

Dear Sir:—

I am in receipt of your favor of the 8th. inst.,[2] and in reply would beg to state that we are not at present in need of any assistants in the purely electrical department of our business. We are, however, just now developing the lighting of small towns throughout the United States by means of Mr. Edison's system and I think that in this connection we might find something for you that would eventually turn out very profitable. We have got a number of agents out in different states who form local companies for us and influence local capital to go into our business in competition with their local gas companies. As a rule we do not pay any salary for this work but allow our agents a liberal share of such stock and cash as our parent company receives for the license they give to the local company. If you were disposed to go into this business I should be happy to supply you with further information. Any assistance that I can render you in this connection I shall have the greater pleasure in doing so as I have a lively recollection of the kind manner in which you treated a number of our telephone men whom Mr. Edison sent to Antwerp in 1880 under arrangements with Mr. Hubbard.[3] You may possibly remember my communicating with you at that time. Very truly yours,

TL (carbon copy), NjWOE, Lbk. 16:313 (*TAED* LB016313).

1. John Norcross Culbertson (1843–1927) worked for the A & P Telegraph Co. and the Pennsylvania Railroad telegraph department before completing a medical degree in Buffalo, N.Y., in 1878. Gardiner Hubbard recruited him about that time to establish Bell telephone exchanges in the state and, in 1880, designated him for similar work in Europe for the International Bell Telephone Co. Culbertson was based initially in Antwerp but also spent considerable time in Russia before returning to the United States in 1882. He operated a ranch in Montana for several years before going back to Buffalo, where he became superintendent of the Bell exchange in 1885. He remained at that post until 1903. Culbertson 1926, 2–11, 22–23; Obituary, *NYT*, 2 Mar. 1927, 25.

2. Not found.

3. Gardiner Hubbard (1822–1897, father-in-law of Alexander Bell, was a patent lawyer and chief organizer of the Bell telephone interests (*ANB*, s.v. "Hubbard, Gardiner Greene"; Doc. 1011 n. 5). Edison had sent about a half dozen "telephone inspectors" to Britain in 1879 and early 1880 to set up local exchanges, but by March 1880 there was not enough work for them. Hubbard was negotiating with George Gouraud, Edison's representative in London, to consolidate the Bell and Edison telephone interests in Europe, and at least four of the Edison inspectors evidently joined the Bell company in Antwerp (Docs. 1808 n. 2, 1845 n.2; Gouraud to Edison, 21 May 1880; Edward Bouverie memorandum, 2 July 1880; both DF [*TAED* D8049ZEK, D8049ZFS]).

–2443–

To George Bliss

[New York,] May 16th [188]3

Dear Sir

Will you please inform me how many Maxim Incandescent Light Plants are now being operated in the Territory worked by you and the names and addresses of the parties working them.[1]

Also full particulars of any Maxim Incandescent plant that have been thrown out and the addresses of the parties who used the Light. I should further like to have the names and addresses of any parties in your Section who are using or who have used Maxim Incandescent Light, the current for which is supplied from a Central Station.[2]

Thanking you in advance for your courtesy and hoping you will favor me with an early reply[3] I remain   Yours truly

Thomas A Edison

LS (letterpress copy), NjWOE, Lbk. 16:342 (*TAED* LB016342).

1. Edison sent virtually identical letters dated 15 or 16 May to at least ten other agents of Edison companies. In a circular distributed in November, Spencer Borden attributed them to the president of the Edison Co. for Isolated Lighting (Sherburne Eaton) as a way of gauging whether the company's losses justified the expense of suing the Maxim interests for patent infringement (Borden circular, Nov. 1883, *Edison Electric Light Co. v. U.S. Electric Light Co.*, Complainant's Rebuttal—Exhibits [Vol. 6], pp. 4429–32, Lit. [*TAED* QD012G4429]). Besides Bliss, other recipients of Edison's inquiry included: George Bancroft (Springfield, Mass.), Spencer Borden (Boston), Benjamin Card (New York City), Henry Clark (Baltimore), George Fairbanks (Cleveland), G. F. Haughey (Cincinnati), Preston Hix (Louisville), John Hoskin (Philadelphia), George Ladd (San Francisco), and John R. Markle (Detroit).

Although electrical inventor Hiram Stevens Maxim (1840–1916) was no longer directly involved with the United States Electric Lighting Co., Edison felt a special antipathy toward him and regarded the Maxim incandescent lamp as an ill-gotten copy of his own (see Docs. 1617 n. 4, 2002 nn. 2–3, 2021, 2033, and 2092; Friedel and Israel 1986, 193–94). The United States Electric Lighting Co. was receiving ample publicity just at this time for its arc lights on the new Brooklyn Bridge (see Doc. 2455 esp. n. 8).

2. The inquiry was apparently of special interest to Phillips Shaw, to whom Insull forwarded the replies (Samuel Insull to Shaw, 22 and 23 May 1883, Lbk. 16:403, 446 [*TAED* LB016403, LB016446]). The Edison Co. for Isolated Lighting compiled a fifteen-page memorandum in June "giving authentic information about Maxim Lights," including establishments from which plants had been removed (Edison Co. for Isolated Lighting memorandum, 9 June 1883, DF [*TAED* D8325ZAT]). This "strictly confidential report for the use of agents" was subsequently publicized by the New England Weston Co., an ally of the United States Electric Lighting Co., for its own purposes. According to Spencer Borden (see note 1), there were only thirty active Maxim installations; another two dozen had been rejected by their users. Edison thought the report was "doing us so much good" that he asked Borden to enlist a Massachusetts textile mill to gather additional information in its own name (Insull to Shaw, 27 July 1883, DF [*TAED* D8316AJZ]; TAE to Borden, 8 Aug. 1883, LM 14:326 [*TAED* LBCD1326]). In 1882, Edison had assembled a short list of Maxim isolated lighting installations that had been removed (see Doc. 2216).

3. Bliss supplied a short list of Maxim installations in the vicinity of Chicago. He cautioned that "The United States people are very active in this region. Where we go either to sell a plant or organize a company they are on our heels immediately and offer all sorts of inducements to parties" (Bliss to TAE, 22 May 1883, DF [*TAED* D8364O]; see also Doc. 2468). At least eight other recipients responded to Edison's inquiry (Bancroft to TAE, 17 May 1883; Card to TAE, 23 May 1883; Clark to TAE, 25 and 28 May 1883; Fairbanks to TAE, 17 May 1883; Haughey to TAE, 22 May 1883; Hix to TAE, 22 May 1883; Hoskin to TAE, 17 May 1883; Ladd to TAE, 23 May 1883; all DF [*TAED* D8325ZAB, D8325ZAH, D8325ZAL, D8325ZAQ, D8325ZAA, D8325ZAF, D8325ZAE, D8325Z, D8325ZAI]).

To William Rich[1]

[New York,] 18th May [188]3

Dear Sir,

Confi[r]ming[a] the verbal arrangement made wi[th][a] you by Mr Insull I beg to state th[at][a] in consideration of your taking the [su]perintendence[a] of the work (such as I [ma]y[a] from time to time indicate to [y]ou)[a] in connection with the Constr[uc]-tion[a] of Electric Light Central Station Plants[2] which I may contract to put up I will pay you monthly a salary of one hundred dollars ($100.00 your travelling expenses and will allow you two dollars per day for boarding and other[b] expenses during such times as you are absent from New York.

It is my intention to present you with a certain sum on each job providing your work is performed satisfactorily to myself the amount of said sum however to be settled entirely by me   Yours truly

Thomas A Edison

LS (letterpress copy), NjWOE, Lbk. 13:018 (*TAED* LB013018). Written by Samuel Insull. [a]Copy damaged. [b]"and other" interlined above.

1.William D. Rich (b. 1830) had been working as a miner in North Carolina in 1880. In 1881, Edison engaged him for a prospecting trip to that state, but the editors have found no information about the circumstances leading to his employment by Edison at this time. United States Passport Applications, 1795–1925, online database accessed through Ancestry.com, 23 Mar. 2009; U.S. Census Bureau 1970 (1880), roll T9_973, p. 389.4000, image 0771 (Griffin, Nash, N.C.); see documents to and from Rich in Mining—General, DF (*TAED* D8138).

2.Edison created this position in frustration with the slow pace of construction work performed by the local illuminating company in nearby Shamokin, Pa. (see Doc. 2424 [headnote]). He personally wrote a general letter of introduction for Rich as his "Supt of Construction" on the same date as this document. The next day, Samuel Insull wrote two letters on his behalf to principals of the Edison company in Sunbury, Pa., where Rich went on 20 May. Insull promptly asked him to provide cost estimates for proposed changes in the station building there and also for the entire station at Brockton, Mass. Edison letter of introduction, 18 May 1883; Insull to C. B. Story, 19 May 1883; Insull to Phillips Shaw, 19 May 1883; Insull to Sherburne Eaton, 20 May 1883; Insull to Rich, both 22 May 1883; Lbk. 13:19; 16:361, 367, 379, 414, 418 (*TAED* LB013019, LB016361, LB016367, LB016379, LB016414, LB016418).

To George Warren[1]

Dear Sir:[2]

I suppose the station scheme has fallen through[3] hence I have got our folks to agree to put in complete plant (wiring and lamps excepted) for two thousand (2000) lamps burning at one time, and run the same themselves providing [they are?][a] paid the same as gas i.e. 2500 Standard Candles for one hour for \$2.25,[4] providing Some of you gentlemen will loan the price of the machinery at five (5) per cent. The amount is \$38 865. We will give you the privilege to take the machinery after one year for the loan and interest, interest not to commence until regular Opera season[5] but loan to be made when machinery installed and tested, time to be three (3) years.[6] Yours truly

(Signed) Thomas A. Edison

L (transcript), NNMetOp, Board of Directors minutes, 23 May 1883, 2:8 (*TAED* X708A). Decimal inserted in monetary expression for clarity. [a]Illegible.

1. A cofounder and original director of the Metropolitan Opera House Co. of New York, Ltd., George Henry Warren (1856–1943) was a member of the building committee; he became vice president of the company on 23 May 1883. Warren had also served recently on a special committee for the completion and improvement of the house, presently under construction on Broadway between Seventh Ave. and Thirty-ninth and Fortieth Sts. "The Next Opera Season," *NYT*, 3 Apr. 1880, 5; "The New Opera-House," ibid., 7 Apr. 1880, 5; Obituary, ibid., 4 June 1943, 21; Homberger 2000, 232–33; NNMetOp, Board Minutes 1:15, 176; 2: 4–7, 12–13.

2. Edison's letter was read to the board of directors by William Vanderbilt, a member of the building committee. Vanderbilt became committee chairman on 23 May 1883, succeeding Egisto Fabbri, the Opera's founding treasurer, who had quit the board in January 1883. In addition to Fabbri, there were a number of connections between the Metropolitan Opera Co. and the Edison Electric Light Co. Between May 1881 and January 1883, if not longer, each maintained offices in the same building at 65 Fifth Avenue. Calvin Goddard was secretary of both companies during this period until he resigned from each in February 1883. NNMetOp, Minutes 1:100, 116, 204, 223; 2:13); Sherbune Eaton memorandum to Edison Electric Light Co. executive committee, 2 Feb. 1883, DF (*TAED* D8327N).

When asked in April 1882 "whether the opera-house would be lighted with the Edison electric light," Goddard reportedly said that "he did not know, but he thought it was likely." In December of that year, the Metropolitan Co. projected its lighting costs at \$10,000 but did not specify gas or electric illumination. After a motion by director William Whitney on 16 April 1883, the Opera's building committee "were requested to have the Opera House wired for electric lights if the cost should not much exceed \$4300, the price named by the Edison Electric

Lighting Co." "The New Opera-House: Work Now Rapidly Pushed Forward," *NYT*, 23 Apr. 1882, 9; NNMetOp, Minutes 1:183, 237.

3. The Edison Electric Illuminating Co. of New York originally projected its second service district to encompass the area from Thirty-fourth to Forty-fourth Sts. between Madison and Eighth Aves. (see *TAEB* 6 chap. 4 introduction and Doc. 2223 n. 3). What Sherburne Eaton called a "boom in real estate . . . owing to the theatres, hotels and apartment houses" in those blocks prompted the company to consider siting the generating plant outside the service district. Eaton specifically asked Edison's opinion in early March about running conductors to the Metropolitan from one of the station locations being considered. Within a few days, however, the Edison Illuminating Co. had redrawn the proposed service boundaries along Twenty-third and Thirty-fourth Sts. and formed a subcommittee (including Fabbri) to consider how to finance construction (Eaton to TAE, 7 Mar. 1883; Edison Electric Illuminating Co. [N.Y.] report, 12 Mar. 1883; Edison Electric Illuminating Co. [N.Y.] memorandum, 12 Apr. 1883; all DF [*TAED* D8326D, D8326E, D8326F1]; see also Doc. 2669).

4. This was the approximate retail price for 1,000 cubic feet of gas, the nominal amount consumed in one hour by 200 standard gas jets. Each such jet would be rated for about 16 candlepower but would provide less light in practice. The total given here of 2,500 aggregate candlepower was consistent with Edison's direct experience with gas (see Doc. 1990).

5. The first season began on 22 October 1883 with Gounod's *Faust*. "The New Opera House: First Performance," *NYT*, 23 Oct. 1883, 1.

6. Edison's proposal was referred to the Finance Committee. On 22 August, "The question of electric light for the dome was referred to the President [James A. Roosevelt] with Power," but a few days later the board authorized secretary Edmund Stanton to contract with the United States Electric Lighting Co. "to furnish Electric light in the boxes." Metropolitan Opera to TAE, 24 May 1883, DF (*TAED* D8325ZAJ); NNMetOp, Minutes 2:23, 81.

–2446–

To A. J. Lawson[1]

[New York,] May 21/[188]3

Dear Sir:

In reply to your favor of the 18th inst.[2] I would state that our overhead system is in no wise effected by lightning. We use a metallic circuit throughout and it is only where the ground circuit is used that lightning can do any damage at all.[3] In telegraphing, when it is impossible to do anything owing to lightning it is the practice to connect up the wires in metallic circuit where the operators experience no trouble whatever. We have had a number of severe storms since we started the village plant at Roselle[4] but at no time has our system been effected by lightning. Mr Flemming[5] is entirely in the wrong if he thinks it impossible to run electric light wires overhead.

Of course it is better to use the underground system but as this is considerably more costly than the overhead there are many places that would be absolutely debarred from the advantages of Electric light if we could not use overhead wires. Yours truly

Thos. A Edison   I[nsull]

L (letterpress copy), NjWOE, Lbk. 16:386 (*TAED* LB016386). Written by Samuel Insull.

1. A. J. Lawson visited New York in August 1882 to gather information, especially concerning the costs of dynamo manufacture, for the promoters of Edison's electric light interests in Canada. After a brief stint in Paris with the Société Électrique Edison, he returned to Canada. In April 1883, he participated in an electrical exhibition at Cornwall, Ontario, which Edison attended. In 1884, Lawson was assistant manager of the Canadian Dept. of the Edison Co. for Isolated Lighting; he eventually took on major responsibilities for the Edison Electric Light Co.'s manufacturing and isolated lighting businesses in Canada (Calvin Goddard to TAE, 25 Aug. 1882; Lawson to Samuel Insull, 6 Apr. 1883 and 11 Apr. 1884; Edison Co. for Isolated Lighting memorandum of employees, 1 May 1884; Edison Electric Light Co. minutes, 10 Jan. 1889; all DF [*TAED* D8237ZAB, D8367J, D8465V, D8425C, D8935AAB]). After ending his ties with Edison, he evidently became engineer and general manager for the County of London and the Brush Provincial Electric Lighting Co. ("Miscellaneous Companies," *Times* [London], 31 Jan. 1899, 15).

2. Lawson's letter concerned a proposed village-plant central-station system in the port town of Sarnia, Ontario. Local anxiety about the safety of overhead wiring during lightning storms had prompted him to request Edison's "opinion as to the apprehended danger." Edison's marginal comments on the letter formed the basis for his reply. Lawson to TAE, 18 May 1883, DF (*TAED* D8340P); see also Doc. 2496.

3. In telegraph or telephone practice, a metallic circuit was one that used a wire or other metallic conductor instead of the earth for the return path (*OED*, s.v. "metallic" 7). In his marginal note on Lawson's letter (see note 2), Edison claimed that "Lightning or Aurora never interfere with Metallic circuits. as far as the lightning concerned our circuit is open both Ends is in fact on straight wire to the Aurora or Lightning." A set of "Instructions for Running Wires for the Edison Incandescent Electric Light" issued by the Edison Electric Light Co. in 1882 formally prohibited ground circuits (Jenks 1893, 19–20).

4. Edison built a demonstration of his village plant system for small towns in Roselle, N.J., that began operating in January 1883 through an area of one-half mile radius. To further reduce construction costs, he had the conductors placed on poles instead of underground. Doc. 2336 nn. 2 3; Prescott 1884a, 228–29; Hix 1979, chap. 5; Edison Electric Light Co. Bulletin 17:5, 6 Apr. 1883, CR (*TAED* CB017).

5. Michael Fleming (1841–1892) was born in Ireland and raised in New York City, where he started a career in telegraphy as a messenger boy. A proficient operator by age sixteen, he moved to Ontario and

eventually settled in the town of Sarnia. There he became a chief rail-road telegraph operator and a bank manager, longtime member of the town council, and mayor of Sarnia (Beers 1906, 30–31). Fleming was already known to Edison, having corresponded with him in 1876 about the Sarnia Street Railway Co. (see Doc. 794).

–2447–

To Anson Stager[1]

[New York,] May 21st [188]3

Strongly advise that canvass of Village plants be made under directions this Office and that we bid on doing whole job complete. I have large gang experienced men and work quick guaranteeing work. can save pile money and so dispose Street Mains get greatest dividend smallest investment perfectly sure. I can put up plant complete working order for less money than even our own Company here[2]

Edison

L (telegram, letterpress copy), NjWOE, Lbk. 16:381 (*TAED* LB016381). Written by Alfred Tate.

1. Anson Stager (1825–1885) was president of the Western Edison Light Co., which he had helped to organize. He was a longtime Western Union Telegraph official and founding president of Western Electric. Docs. 817 n. 13 and 2290 n. 8.

2. Edison sent this telegram immediately upon taking control of the village plant construction business himself (see Docs. 2424 [headnote] and 2444). When George Bliss acknowledged receipt the next day, he suggested meeting Edison in New York before making further arrangements. In early June, Stager visited Edison, who then decided to go to Chicago instead. Bliss to TAE, 22 May 1883; TAE to Bliss, 11 June 1883; both DF (*TAED* D8364N, D8316AAA); TAE to Stager, 5 June 1883, Lbk. 17:56 (*TAED* LB017056).

–2448–

To Emil Rathenau[1]

[New York,] May 22 [188]3

Dear Sir:

I have received from the Secretary of the Committee of the Elektrotechnischer, Verein, A notice that my application for membership to that Society has been received.[2]

I am very much obliged to you for transmitting my application to the Society through Dr Siemens.[3]

Mr Batchelor during his recent visit here told me that he thought that you would be willing to write me occasionally on the progress of Electric lighting in Germany; if not troubling you to much to do so I shall esteem it a very great favor.[4]

Our business here is progressing wonderfully. In New York

our Central Station has been working very successfully ever since September 4th of last year from which date current has been in our mains continously.

We are doing a very large isolated business all throughout the United States[5] and just recently we have started a new Department, viz: the lighting of small towns and villages using from 1000 to 3000 lights. This department of our work also promises to be a perfect success. Yours truly

Thomas A Edison.

LS (letterpress copy), NjWOE, Lbk. 16:405 (*TAED* LB016405). Written by Samuel Insull.

1. Electrical entrepreneur Emil Rathenau (1838–1915), a native of Berlin, attended the Polytechnical colleges of Hanover and Zurich. He gained experience as a draftsman and mechanical designer in Greenwich, England, before acquiring his own foundry in Germany. After seeing the telephone at the 1876 Centennial Exhibition in Philadelphia, Rathenau set up a telephone exchange in Berlin. While visiting the 1881 Paris International Exhibition, he became similarly enthusiastic about Edison's electric lighting system and committed himself to commercializing the new technology in Germany. Backed by three German banks, Rathenau negotiated the formation of a temporary syndicate for central station and isolated lighting in Germany in 1882. With similar support in early 1883, he organized the Deutsche Edison Gesellschaft für angewandte Elektrizät (DEG) on a permanent basis to exploit the incandescent lighting patents of Edison and arc lighting patents of other inventors. Wile 1914, 71–78; Kocka 1999, 55; Hughes 1983, 67; see also Docs. 2298 n. 9, 2392, and 2555 n. 1.

2. The Berlin Elektrotechnischer Verein (Electrotechnical Association) was established in 1879 by Werner Von Siemens and postmaster general Heinrich von Stephan. The society further prompted the establishment of the Verband Deutscher Electrotechniker in 1894, which became the leading professional organization for the electrical fields in Germany (Feldenkirchen 1994, xxii; Hughes 1983, 177). The notice of 26 April, signed by the secretary Geh. Unger (also a postal official), stated that Edison's application would be considered at the next meeting. Edison became an honorary member in 1890, only the second person so designated to that time (Elektrotechnische Verein to TAE, 26 Apr. 1883, DF [*TAED* D8311C]; Görges 1929, 10, 19).

3. Werner von Siemens (1816–1892), preeminent German electrical engineer and inventor, was a cofounder and principal of the manufacturing firm of Siemens & Halske. Doc. 2173 n. 19; Feldenkirchen 1994, chaps. 2–4.

4. See Doc. 2487.

5. See *TAEB* 6 App. 2.

−2449−

To William Anderson[1]

[New York,] May 23/[188]3

Dear Sir:

I shall be glad if you can write me a letter stating that there is no recorded instance within the knowledge of the American Board of Fire Underwriters[2] where a fire has been caused by the Edison Electric Light System.[3]

I shall be glad to know if I can make use of such a letter in our catalogue or pamphlets that we may issue[4]   Yours truly

Thos. A. Edison

LS (letterpress copy), NjWOE, Lbk.16:445 (*TAED* LB016445). Written by Alfred Tate.

1. William A. Anderson was chairman of the Committee on Police and Origin of Fires of the New York Board of Fire Underwriters. The committee was established in 1881 to propose standards for electric lighting and wiring. Doc. 2298 n. 5; "Fire Underwriters' Election," *NYT*, 22 May 1883, 2; Jenks 1893, 16–18.

2. Edison probably meant the New York Board of Fire Underwriters (with which Anderson was connected), chartered by the state legislature in 1867, rather than the National Board, organized in 1866. Brearley 1916, 14–19; "Public Notices," *NYT*, 20 June 1867, 6.

3. By coincidence, an explosion occurred a few weeks later at a building in which the first floor was lighted by current from Edison's Pearl St. central station. The *New York Times* declared the event "the first of the kind in the history of electric lighting in which the electricity seemed to be a factor." According to the newspaper, plant superintendent Charles Chinnock determined that one of the wires entering the basement had become grounded against a gas service line and heated the pipe until it burst. "Events in the Metropolis   Two Singular Explosions," *NYT*, 12 June 1883, 8; see also Doc. 2533.

4. Anderson replied the next day that although he would like to comply, "it is a rule of our N.Y. Board of Fire Underwriters to refuse to endorse any particular patent, by name and as Chairman of one of the Standing committees of the Board I must obey the rules. I would add, that personally I think the Rule should be rescinded." Edison responded that he, too, would like the rule withdrawn, "as it is natural that we should like to have a little encouragement in the shape of a letter endorsing what we have done towards preventing fire in connection with electric light apparatus." Anderson to TAE, 24 May 1883, DF (*TAED* D8320N); TAE to Anderson, 25 May 1883, Lbk. 16:466 (*TAED* LB016466).

−2450−

To Cyrus Field[1]

[New York,] May 23/ [188]3

Dear Sir:

I learn that a concern called the Brush Swan Co.[2] are bidding to wire your building.[3] These people have never had any experience in wiring and have'nt a single installation in the country. What they would do would be utterly worthless. Mr

Hannington[4] who presents this letter is one of the firm of Bergmann and Co. who have done a large amount of work for our Co'y, which has been perfectly satisfactory to our Company and the Board of Fire underwriters. They have wired over nine hundred buildings in New York alone, of which over 500 are using the light. Yours truly

Thomas. A. Edison

LS (letterpress copy), NjWOE, Lbk. 16:422 (*TAED* LB016422). Written by Alfred Tate.

1. American merchant and capitalist Cyrus Field (1819–1892) became a noted public figure in the United States and Britain after promoting the first successful Atlantic cable in 1866. He also established the American Telegraph Co. More recently, he had acquired a controlling interest in the New York Elevated Railroad Co. and, since 1882, had published the *New York Mail and Express. ANB*, s.v. "Field, Cyrus West"; Carter 1968, 306–24.

2. The Brush lighting interests in the United States had previously licensed the Lane-Fox incandescent lamp patents. When this system did not work as well as expected, a convention of Brush illuminating companies agreed in April 1883 to adopt the Swan lamp. At the same time, a permanent organization of the independent Brush subcompanies was formed. About this time, the Brush Electric Light Co. began manufacturing Swan lamps with filaments of treated cotton thread, and it planned to use them in conjunction with Brush storage batteries in a distribution system for arc lights. "Brush Electric Light Companies in Convention," *Electrical World* 1 (21 Apr. 1883): 245; Bright 1972, 72–73; Passer 1972, 20–21; "The Brush-Swan Electric Light," *Sci. Am.* 47 (30 Dec. 1882): 423; "Electrical Notes," *Electrical Review* 2 (28 June 1883): 9; see also Doc 2239 n. 4.

3. Field was the principal builder of the Washington Building (often called the Field Building) at 1 Broadway, at Battery Place. Begun in 1882, it was partially occupied in 1884 by the Hay Exchange, among other commercial tenants. Ten stories tall, it was one of New York's highest buildings. Landau and Condit 1996, 125–27; "New Quarters for the Hay Exchange," *NYT*, 27 Mar. 1884, 8.

4. Charles F. Hanington (b. 1842) had been a janitor until about the time he became an office assistant to Insull, in the latter part of 1881. By 1883, he was employed by Bergmann & Co. to install wiring in various buildings in New York. In early summer and again in the winter of that year, he was sent to Pennsylvania to oversee the wiring of the new central stations in Sunbury and Shamokin. He remained in Edison's employ for several years; he identified himself decades later in the federal census as an inventor. Often misspelled in correspondence, including this document, Hanington's name was listed incorrectly as "Hannington" in the *TAEB* 6 index. Doc 2171 n. 5; U.S. Census Bureau 1970 (1880), roll T9_874, p. 62.2000, image 0562 (New York City, New York, N.Y.); ibid. 1992? (1920), roll T625_1048, p. 21B, image 864 (Kearny Ward 2, Hudson, N.J.); U.S. Dept. of State n.d., roll 298, image 607; Andrews to TAE, 7 Aug. 1883, DF (*TAED* D8361ZBX); Bergmann & Co. to Edi-

son Construction Dept. 18 Dec. 1883, DF (*TAED* D8324ZBH); TAE to Hanington, 31 May 1883, Lbk. 17:5 (*TAED* LB017005).

Edison personally wrote Hanington that the Edison Electric Illuminating Co. "would not run mains so far from station [on Pearl St.] as Field Bldg— If they want the light right off ie as soon as the bldg is finished they should buy an Isolated plant." He advised putting in an engine and boiler for at least 500 lights but a dynamo only for 300 in order to reduce the expense. The editors have found no evidence of further action regarding this building.

–2451–

To George Shaw-Lefevre[1]

[New York,] May 23 [188]3

Sir

I notice in the Parliamentary Report in the London Daily News May 11th[2] that some discussion took place in the House of Commons with reference to my system of electric lighting[3]

I write to thank you for what you said in my behalf on reply to Lord Randolph Churchills[4] remarks reflecting on myself   Respectfully

Thomas A Edison.

ALS (letterpress copy), NjWOE, Lbk. 16:428 (*TAED* LB016428).

1. George Shaw-Lefevre (1831–1928), later Baron Eversley, was a member of Parliament for Reading from 1863 until 1885. As the First Commissioner of Works since November 1880, he had invited the Swan company to install electric lighting in the new Royal Courts of Justice in the Strand; the job was designed by R. E. B. Crompton at a cost of at least £7,000 and demonstrated at the Royal opening on 4 December 1882. Lefevre subsequently accepted a proposal from the Edison Electric Light Co., Ltd., to install experimentally its system in portions of the House of Commons for the nominal fee of £100. *Oxford DNB*, s.v. "Lefevre, George John Shaw-"; "Town And Country Talk," *Lloyd's Weekly Newspaper* (London), 4 Mar. 1883, 5; "Imperial Parliament," *Leeds Mercury*, 11 May 1883, 8.

2. Samuel Insull noticed the article and "suggested to Edison that he should write to Mr Shaw Lefevre." Insull to Edward Johnson, 23 May 1883, Lbk. 16:432 (*TAED* LB016432).

3. This discussion dated back to April, when Lord Randolph Churchill began to question the expense of the House of Commons demonstration and lack of competition in awarding the job. Lefevre explained that he considered competitive proposals unnecessary because the experiment would cost no more than gas lighting, and the Government was under no further obligation to the Edison company, whose work had previously proven reliable in lighting the Post Office ("Imperial Parliament," *Daily News* [London], 18 Apr. 1883, 2). Sir John Eldon Gorst reintroduced the matter on 10 May after complaining about the "varying brilliance of the electric light in the reading room." In the ensuing exchange, Churchill twice described Edison as a "Yankee Adventurer" who not only had stolen the patent rights of Joseph

Swan, but whose company now had been favored with a contract for "greatest advertisement they could possibly give." Lefevre answered that "Mr. Edison was one of the most distinguished scientific men of the day, and was beyond all suspicion of the character mentioned by his noble friend." Lefevre also suggested that Churchill was arguing "in the interests of some rival company to that which had been lighting the House," in which Crompton was involved. Churchill denied any personal interest in the electric companies and pointed out a potential conflict for two of Edison's backers with political portfolios: John Lubbock and Edward Plydell Bouverie. In his memoirs, Crompton claimed to have had unspecified political conversations with Churchill, Gorst, and others while working on the Churchill house ("Imperial Parliament," *Daily News* [London], 11 May 1883, 2; Crompton 1928, 107–8; "Electric Lighting in the House of Commons," *Electrician* 10 [21 Apr. 1883]: 532; "Electric Lighting in the House of Commons," ibid., 11 [19 May 1883]: 5).

4. Lord Randolph Henry Spencer Churchill (1849–1895) was a conservative member of Parliament for the borough of Woodstock and the leader of an independent-minded group of Tories known as the "Fourth Party." He also was the father of Winston Churchill, the future prime minister. *Oxford DNB*, s.v. "Churchill, Lord Randolph Henry Spencer."

–2452–

From William K. L. Dickson

Edison Machine Works Testing Room [New York,]
May 23d/83

Dear Sir

During my illness of the last few days I roughly outlined in ink several designs for lamp brackets & such like which I now send for your inspection   I do not send the others as I find from comparing catalogue ~~the~~ of Bergman that several are <u>somewhat</u> similar—

I shall soon send you others & shd you like any of the designs let me know so as to work them up for the workmen—

If you only knew how I am heart in soul in all your inventions & all you do[a] you would now & then stoop to assist & better my prospects in life

Hoping to hear from you personally or other wise—[1] I beg to remain   Yours devotedly

W Kennedy Laurie Dickson[2]

ALS, NjWOE, DF (*TAED* D8330ZAN). [a]"& all you do" interlined above.

1. Edison replied on 26 May, in a letter written and signed on his behalf by Samuel Insull, that all designs for fixtures should be submitted to Sigmund Bergmann, "who has complete control of that character of work." Lbk. 16:480 (*TAED* LB016480).

2. William Kennedy Laurie Dickson (1860–1935) was born in France to parents of Scottish descent who themselves were apparently born in the United States. Dickson first applied to work with Edison in 1879, when he was living in London. Dickson stated then that he "had a good English education, can speak French and German . . . have a fair knowledge of accounts, and draw well. For all these things, I have certificates from the Cambridge Examiner." According to Dickson's biographer, there is no record of his enrollment at Cambridge, but he may have been the student identified as "Dixon" who was examined in 1877. Edison did not hire him at that time, and Dickson and his family moved to Richmond, Va., later in that year. Dickson applied again at the end of March 1883 and was working in the Testing Room at the Edison Machine Works at the time he wrote this letter. By the beginning of 1884, Dickson appears to have been placed in charge of the Testing Room. In late 1884 or early 1885, he became chief electrician of the Electric Tube Co. Sometime in late 1885, he appears to have become an experimenter at Edison's New York laboratory, and he moved with Edison to the laboratory at the Edison Lamp Works in Harrison, N.J., in 1886. Dickson then accompanied Edison to the new West Orange, N.J., laboratory when it opened in late 1887, where he worked not only as an experimenter but as Edison's photographer. Dickson was especially important in the work on ore milling and motion pictures and is often credited as co-inventor of the latter technology. Dickson made many of the early Edison films but left Edison's employ in 1895 in a dispute over his role in the Edison film business and his association with competing film interests. In 1892, he and his sister Antonia published *The Life and Inventions of Thomas Alva Edision,* which included many of his photographs, and in 1895 they published *History of the Kinetograph, Kinetoscope & Kinetophonograph.* Dickson to TAE, 17 Feb. 1879; Dickson to Insull, 28 Mar. 1883; both DF (*TAED* D7913K, D8313H); Spehr 2008.

–2453–

To Calvin Goddard

[New York,] 26th May [188]3

Calvin Goddard[1]   care Drexel Harjes[2]   Paris

Private   Would you return take presidency Isolated with absolute control. Complete separation from Parent[3] and when answer[4]

Edison

L (telegram, letterpress copy), NjWOE, Lbk. 13:20 (*TAED* LB01302); a variant copy is in DF (*TAED* D8325ZAN). Written by Samuel Insull.

1.Calvin Goddard (c. 1837–1892), founding secretary of the Edison Electric Light Co. and the Edison Electric Illuminating Co. of New York, became a director of the Edison Co. for Isolated Lighting in 1882. He resigned the first of these positions, and probably all of them, in early 1883, when he also vacated a trustee position in the Gramme Electrical Co., a consortium of electrical interests, including Edison's. He returned to a position of financial authority in the Edison Electric Light Co. (probably chairman of the Finance Committee) by May 1884. Goddard was involved in finance throughout a career that began and ended in Chicago, with an interruption for service in the Union Army. He joined the Edison interests after having been treasurer of Wells, Fargo & Co. and the New York and New England Railroad. Goddard later became secretary of the New York Underground Railway Co., which proposed to build an electric subway. He returned to Chicago about 1890, having already helped to organize the Chicago Rapid Transit railroad, of which he was president until his death. He died in San Francisco while reportedly seeking respite from overwork. Docs. 2106 n. 3, 2189 n. 2; Sherburne Eaton memorandum, 2 Feb. 1883; Samuel Insull to Frank Hastings, 15 May 1884; both DF (*TAED* D8327N, D8416BPE); "The Gramme Electrical Company," *NYT*, 1 June 1883, 5; "Underground Railroad Plans," ibid., 27 July 1887, 8; Obituary, *Chicago Daily Tribune*, 6 Apr. 1892, 3.

2. Drexel, Harjes & Co., the Paris affiliate of Drexel, Morgan & Co., was one of the leading private banks in Europe. Doc. 2155 n. 2.

3. That is, the Edison Electric Light Co.

4. Goddard went to Europe soon after leaving his positions in the Edison companies, taking with him Edison's power of attorney to make contracts for electric lighting and railroads in Switzerland (for which he was to receive a 10 percent commission). He cabled on 28 May from Carlsbad, a popular tourist destination in Bohemia known for its sulphur springs: "If all interests desire should be favorably disposed I write fully." In a letter the same day, Goddard explained that he was "wholly unprepared" for the offer of the Edison Co. for Isolated Lighting and would consider it only if "assured of the support of those representing what I may call the commercial interest. . . . whoever is at the head of the Company should have the confidence and support of all parties." He referred to his opposition to the segregation of manufacturing from other Edison interests and to the establishment of an installation company (such as the Construction Department), for which he feared the Isolated Co. would merely solicit orders while taking the risk upon itself. Though he also hesitated on account of a desire for work "less wearing, where if there were not so much glory to be won there would be less anxiety," Goddard promised to "come and devote myself heartily" to it in September if Edison insisted. TAE power of attorney to Goddard, 13 Feb. 1883, Lbk. 13:10 (*TAED* LB013010); Goddard to TAE, 28 May 1883, both DF (*TAED* D8325ZAO, D8327Y); *WGD*, s.v. "Karlovy Vary."

Goddard ultimately did not take up the Isolated Co. office. Shortly after returning to New York, he became connected with the Time Telegraph Co., but assisted with estimates for foreign lighting plants. He also offered to help Samuel Insull with the Construction Department's bookkeeping. Goddard to Alfred Tate, 9 Nov. 1883; Samuel Insull to Goddard, 21 Nov. 1883; both DF (*TAED* D8336ZAI, D8316BID).

Samuel Insull to Charles Brown[1]

[New York,] May 26/ [188]3

Dear Sir:

I have your favor of the 25th inst. and in reply beg to state that the terms we make with promoters of Companies formed for lighting of small towns are as follows:—[2] Our parent Company requires 25% (20% in stock and 5% in Cash) of the capital stock of each local Comp'y formed, for the right of using our patents and they pay their Agents for forming the Company 15% of the amount of stock and cash which they receive from the local Company— In a town of 10,000 inhabitants the plant would cost you $30,000 to $35,000. At the same time that the Company signs the contract giving the license Mr Edison stands ready to sign a contract undertaking to put the plant up for a given sum so whenever people go ahead they know exactly what their expenses are going to be. If you could give me a call I could explain this matter to you in a few minutes and give you such information as to enable you to discuss the matter intelligently   Yours truly

Saml Insull   Private Secretary

ALS (letterpress copy), NjWOE, DF (*TAED* LB016476).

1. Charles Albert Brown (1858–1938) was a manager in the New York office of the Western Electric Mfg. Co. He later worked in a similar capacity in Chicago before starting a successful career as a patent attorney in that city. Insull incorrectly addressed this letter to him as "Charles R. Brown." *NCAB* 28:30.

2. Brown asked Insull about the "terms you make to companies in small towns, in competition with gas," particularly capitalization and licensing requirements. Insull based his reply on the notes Edison made on Brown's inquiry. Brown to Insull (with TAE marginalia), 25 May 1883, DF (*TAED* D8340V).

To George Bliss

[New York,] June 4/[188]3

Dear Sir:

I am in receipt of your favor of the 30th ulto. and I am much obliged for the information contained therein.[1]

It is my strong impression that the Electric Rail Road Exhibition will be a failure as the gentlemen who have charge of it do not seem at all anxious to profit by our experience.[2]

I certainly think that you should insist on proper tests being made as to the relative merits of the various incandescent lights.[3] When you are ready we will send Mr Howell[4] of the Lamp Factory with all apparatus; he thoroughly understands testing and my impression is that the Maxim people will be afraid to compete against us.

I think you should do all you possibly can to dwarf the exhibit of the United States Company[5] by putting more lights in than they have.[6] It would be a good thing for you to engage a boy and put him in the Exhibition to watch the Maxim Exhibit and so get a Statement of breakages and extinctions during the whole Exhibition.[7]

The lighting of the Brooklyn Bridge by the United States Company is a hopeless failure.[8] I am having a man take exact record of extinctions there and will send you copy of the results; they seem to be as incompetent to deal with the arc light as they are with incandescent lighting. They are going to take out all their wires, which have been laid down at enormous expense on the bridge, and re wire the whole structure; they keep from eighteen to twenty men on the bridge every night to keep lamps going.[9]

With reference to the Central Station business I am altogether too busy to go to Chicago just now;[10] we have four Central Stations already being erected in different parts of the Country and are negotiating for at least thirteen more.[11] I think that you should come on here and see me about this business. I shall be very glad to take contracts for doing this work and am sure I can do it for less money than your Company can as I have men and organization specially for this Department of our business.

Let me know when you are coming   Yours truly.

Thos. A Edison   I[nsull]

L (letterpress copy), NjWOE, Lbk. 17:48 (*TAED* LB017048). Written by Samuel Insull.

1. George Bliss reported the successful start of the as yet incomplete Edison lighting display at the Chicago railway exhibition. He also provided information about the United States Electric Lighting Co.'s exhibit, promising that "Although only a part of our lights are in operation it is evident that we will make the United States Company sick before this show is over." Edison's extensive notes on Bliss's letter formed the basis for this reply. Bliss to TAE, 30 May 1883, DF (*TAED* D8364Q).

2. Immediately after its founding in May, the Electric Railway Co. of the United States (see Doc. 2431) began putting together an exhibit for Chicago. Managed for the company by veteran telegraph electrician Frank Rae and his assistant Clarence Healy, the exhibit evidently made little use of Edison's patents. It used Field's third-rail system and Weston dynamos from the United States Electric Lighting Co. for its generator and motor. A single locomotive (named *The Judge* in honor of Field's uncle and U.S. Supreme Court justice Stephen Johnson Field)

and attached car ran through a 1,500-foot oval track, starting in early June. Though Bliss complained that "The whole arrangement seemed crude to me compared to your machinery at Menlo Park," it was dismantled and sent to Louisville for the Southern Exposition there over the summer. "The Electric Railway at the Chicago Exposition," *Electrician* 11 (25 Aug. 1883): 349–54; Bliss to TAE, 7 June 1883, DF (*TAED* D8364S); Martin 1889, 62–69.

3. Bliss wrote (see note 1) that "Awards are to be given and the Edison [lighting] exhibit has been entered for competition. We want to insist on the proper tests being made and I wish you would advise what we shall have done to thoroughly defeat the United States people. The Electrical exhibitors are all down on the United States people for the tricks which they have played and will unite with us in demanding anything which is right."

4. John Howell, an 1881 graduate of the Stevens Institute, headed the testing department at the Edison lamp factory (Doc. 2129 n. 4). Edison ordered that Howell be held "in readiness to go to Chicago on receipt of telegraph advices by me" but later decided that he would go only when a test was sure to be made. A jury was designated, but it does not appear that Howell went (TAE to Francis Upton, 11 June 1883; TAE to Bliss, 11 June 1883; Bliss to TAE, 7, 12, and 13 June 1883; all DF [*TAED* D8316AAG, D8316AAA, D8364S, D8364T, D8364V]).

5. The United States Electric Lighting Co., formed by the New York Equitable Life Assurance Co. in 1878, controlled the lighting patents of Hiram Maxim and the dynamo and arc lighting patents of Edward Weston. Doc. 1617 n. 4; Passer 1953, 32–33; Carlson 1991, 185–86.

6. The Western Edison Co. put up a display of several hundred lamps, including colored globes and lamps of different intensity in the same circuit; they also illuminated press rooms and lecture rooms. According to a news account, "The fountain in the centre of the main building was lighted by 150 Edison lamps placed under the spray and falling water. This formed the most attractive feature of the entire exposition, and hundreds of people were constantly seated in the adjacent chairs to observe the effect." "Edison," *Chicago Daily Tribune,* 24 June 1883, 8.

7. Bliss replied that the company was "up to all sort of tricks and to attempt to refute their many falsehoods is out of the question. They are doing an immense amount of advertising and I will send all the papers to Maj. Eaton so their statements may be understood" (Bliss to TAE, 7 June 1883, DF [*TAED* D8364S]; see also Doc. 2468). After Charles Batchelor suspected an exhibitor at the 1881 Paris International Electrical Exhibition of surreptitiously replacing broken lamps, he planned to monitor other displays (including that of Hiram Maxim). Edison suggested doing the same thing at the Crystal Palace exhibition in 1882 (see Docs. 2148, 2155, and 2216).

8. The United States Illuminating Co., a subsidiary of the United States Electric Lighting Co., won a contract for about $18,000 to illuminate the new Brooklyn Bridge with seventy Weston arc lamps. (The Edison Electric Light Co., also among the half dozen potential contractors, had bid $21,500; the committee generally favored the more intense arc lamps.) Problems immediately appeared when the bridge opened to traffic in late May, and Edison warned Seth Low, mayor of Brooklyn, of a possible panic on the crowded structure should the lights fail. (A

daytime panic did occur a week later, killing twelve people outright and injuring dozens). The *New York Times* editorialized that the trouble was "not because of insufficiency of lamps, but because of irregularities in the current or in the adjustment of the carbons, which has resulted in the frequent extinguishing of different lights. It has been a very rare circumstance indeed for all the lamps to be lighted at one and the same time." The paper stated on 30 May that workers were installing new lamps that would "be placed on springs, to act with the vibrations of the bridge." In response to Bliss's request for information on the subject, Edison instructed Samuel Insull to send copies of "NY Times & Sun where they mentioned lighting Brooklyn bridge failure." The "confidential report" on Maxim lamps sent to Edison agents in June (see Doc. 2443 n. 2) included (p. 11) a tabulation of lamps that failed on the bridge between 30 May and 5 June. "Estimates of Lighting the Brooklyn Bridge," *Sci. Am.* 48 (3 Mar. 1883): 129; Woodbury 1949, 110; "Letter to the Editor," *Electrical Review* 2 (24 May 1883): 3; TAE to Low, 23 May 1883, Lbk. 16:423 (*TAED* LB016423); "The Brooklyn Bridge," *NYT*, 13 Mar. 1883, 8; "Dead on the New Bridge," ibid., 31 May 1883, 1; editorial, ibid., 28 May 1883, 4; "The Bridge Illuminated," ibid., 20 May 1883, 7; "Regulating Bridge Travel," ibid., 30 May 1883, 8; McCullough 1972, 543–44; Bliss to TAE (with TAE marginalia), 7 June 1883, DF (*TAED* D8364S); cf. "Illuminating the Bridge," *New York Star*, 20 May 1883, Cat. 1018:11H (*TAED* SM018032b); see also Sherburne Eaton to TAE and Bliss, 18 June 1883, DF (*TAED* D8327ZAC).

9. According to a private report by a law firm engaged by Edison, the bridge lighting was so unsatisfactory by the end of June that the parent United States Electric Lighting Co. took over its subsidiary's contract and planned to make a new installation itself. The report also credited the original contract to the company's political influences, a possibility consistent with the bridge project's history of corruption and cronyism. Ecclesine & Tomlinson to James Russell, 27 June 1883, DF (*TAED* D8327ZAJ); Trachtenberg 1965, 99–113.

10. Edison changed his mind a week later and was in Chicago on 18 and 19 June. He met with several prominent citizens, including railroad innovator and manufacturer George Pullman, and gave an interview to the *Chicago Daily Tribune* on the state of electric lighting and traction. TAE to Bliss, 11 June 1883; Samuel Insull to John Hoskin, 18 June 1883; Bliss to Insull, 20 June 1883; all DF (*TAED* D8316AAA, D8320S, D8364W); "Edison," *Chicago Daily Tribune*, 19 June 1883, 8.

11. See Doc. 2424 (headnote) and App. 2.

–2456–

Samuel Insull to
Edward Johnson

New York, June 4/ *1883*[a]

My Dear Johnson:

Sprague has been here and presented his letter of introduction to Edison—[1] In that letter you refer to various matters in connection with the reorganization of the London Company,[2] and Edison thought that he read "between the lines" that your inclination is to favor an amalgamation with the Swan Company—[3] I told him that I did not think it was so but promised to write you and state that he is very much opposed to any such idea— He thinks it would be fatal to the future developments of our English business if any amalgamation with Mr Swan took place— Edison says that it is simply the amalgamation of a system of Electric Lighting with a manufacturer of lamps—

I state this simply because he asked me to and not because I think you have any such ideas as Edison is half inclined to attribute to you—

We got your brief cable stating that you had obtained full control without contest and we are now awaiting the receipt of letters from you giving details.[4] If the control is really full and absolute, I congratulate you, and I presume it is or else you would not have telegraphed as you did— I certainly think it is time that the Arnold White regime was brought to an end—[5]

I heard the other day that Borden and Benton proposed telegraphing you to come here and settle some disputes of Borden's with Moore— They were going to telegraph you because Edison was half inclined to side with Moore— Benton is very fond of using your name in this connection and stating that he will get Mr Johnson to fix matters—

I heard him do so once to Eaton with relation to some matters in connection with the Albany Co'y— I think it does you a great deal of harm to have your name handed around in such a manner—

Bergmann came to me about the subject and asked me if I wrote you to tell you so. I have no concern it it myself beyond not liking to see your name used in a manner that it should not be, but as Bergmann wanted me to write you I thought I would.

Illuminating Company:— You will have heard ere this that Chinnock[6] has at last come in our business— He has been appointed Supt. of the First District with very full power; he has business charge of it as well as technical charge and although he has only been there since the first he grasps the difficulty amazingly well and will I think make a great success of it. Edison has made arrangements with him by which Chinnock gets quite a block of stock as soon as the station pays 5% on $600,000. and he is going to work very hard to get this stock. In fact every part of this illuminating business is now assuming a right position. Cassho[7] seems to have thoroughly bitched matters and it will take all Chinnocks time to get them straightened—

Our small town plant business looks very well indeed—
We have Sunbury half finished; Shamokin is well under way;
work will be started at Brockton in about two days; Lawrence
will be commenced as soon as we have the underground conductors ready—[8] St Louis promises to go through; our contract for this place alone will be $150,000. and prospects in this
connection look very brilliant.[9] Of course this is the best kind
of work that we can deal with; it means plenty of work for our
shops.

Isolated:— The Isolated business also promises very well
indeed— Moore got an order for 3000 lights somewhere in
Kentucky[10] today and this you might call "the opening of the
ball": our season is going to be a very heavy one I am sure.

Financial: Notwithstanding the above and the brilliant
prospect before [us?][b] we are all of us just now on our backs
financially— Bergmann seems about the worst off—or rather
he howls the loudest— We manage however to give him relief
all the time; sometimes it is very difficult to see where we can
raise money but we do it somehow—

I should like to Know how we are going to get paid from
London. They owe us at the present moment $4000. and some
odd dollars; about $500. they have against us which leaves un-
disputed $3500. and when we cabled you for remittance the
other day they only sent half of this— It is all very well for you
to write nicely turned phrases and complain that the London
Company have to pay up their bills in order to support the
home companies but such is not at all the case— The London
Company are the worst payers we have today— They have
just sent us an order for $2000. worth of lamps and I suppose
we shall have to pay the Lamp Company and carry that item
on our books for a couple of months before we get the money
from them—[11] If we had plenty of money such matters would
not be noticed but as we have not and shall not have for a year
or eighteen months to come I think you should make arrangements for the London Company to do the same as everyone
else does—pay for things on receipt— If we can only get rid
of some of their machines we will pay ourselves.[12]

Household Affairs:— Your house paper bill, of which I
wrote you the other day, will have to be paid—[13] The people
have been in to see me about it three or four times— I have
never seen the bill. Mrs Harrison had it and handed it to Benton to take it and have it fixed as it is about $100. too much—;
she says that Benton has never attended to it; I have written
him today, at Mrs Harrison's request, asking for it— The bill

as it stands now is $630. This is the only bill of consequence pressing now— In fact it is the only bill, I think, against you that has not been paid with the exception of Bauman's[14] bill for $800 & some odd dollars— The house a/cs I will take care of all right, until you come home some way or another.

Why the Hell don't you write me a letter:—if it is only to say Dear Sam and sign your name to it— It is about the most satisfactory business to keep on writing to you and get no replies and about once in three or four weeks I have to blow off the safety valve by swearing at you, but I find it makes not the slightest difference so will not inflict on you any elaborate scolding—

Sprague has gone home for a week— When he comes back here he will look into Mr Edisons experiments in dynamos and he will in all probability cable you by the time you get this that Edison's results with the Z dynamo are considerably better than Mr Hopkinsons   This will be news to you I presume. Edison says that he and Sprague figured it out and he (Edison) got more lamps than did Mr Hopkinson, ɪif you figure out the relative economy of the lamps used by Edison in his test and those used by Hopkinson in his—

Mrs Edison & family have gone to Menlo Park today for the summer; my impression is she will stay there about a month—[15]

Mr Edison will remain in the City [by?][c] all[d] the week and go out every Saturday night.

Hannington has gone around to Grammercy Park[16] to live in the house till the family come back in the autumn— When are you coming home—? Sincerely Yours

Saml Insull   T[ate][17]

L (letterpress copy), NjWOE, LM 3:160 (*TAED* LM003160). Stamped Thomas A. Edison to Edward H. Johnson. Written by Alfred Tate; "Per S.S. Abyss" written above handstamp to indicate ship carrying letter. [a]"*New York*," and "*188*" preprinted. [b]Illegible. [c]Canceled. [d]Interlined above.

1. Johnson had Frank Sprague carry a letter to Edison that began: "This is Sprague— He would probably tell you that much—but he wouldn't tell you that the pay of an Ensign in the U.S. Navy isn't sufficient to keep a man in Idleness long on shore—and that I have kept him here so long to have him try the Hopkinson dynamo for me—that he is Dead Broke." Johnson to TAE, 18 May 1883, DF (*TAED* D8320J).

2. One item on the agenda for the 30 May meeting of the directors of the Edison Electric Light Co., Ltd., was to fill the seats of the the the retiring Shelford Bidwell and Sir John Lubbock. Johnson had written (see note 1) that he planned to gain control of the board, and he ex-

pected that "All the Big Questions of Patent Fight—Reorganization into a Public Co—Amalgamation with Swan—Etc Etc are to come up Immediately our New Board is in office." Edison Electric Light Co., Ltd., to TAE, 15 May 1883; Edison Electric Light Co., Ltd. statement, 25 May 1883; both DF (*TAED* D8338ZAC, D8338ZAF).

3. The Swan United Electric Light Co., Ltd. (organized in May 1883 as the successor to Swan's Electric Light Co., Ltd.), controlled the patents of British chemist and inventor Joseph Swan (1828–1914), who had created an incandescent carbon lamp. It had a broad charter for the production and distribution of electricity for lighting, power, and transportation, as well as for related manufacturing (see Doc. 2292 n. 8). Though Edison had long resisted any suggestion of an alliance with Swan (see, e.g., Doc. 2178), Johnson had told him (see note 1) that after the board takeover, "We shall have to make some important modifications of our View of Matters & things ere we Can get ourselves into a strong financial position—so be prepared for surprises."

4. Johnson cabled on June 1 that he had "obtained full control without contest." Edison subsequently acknowledged two letters "giving me an account of the struggle you have had in London"; these have not been found, but see Doc. 2467. Johnson to TAE, 1 June 1883, LM 1:314D (*TAED* LM001314D); TAE to Johnson, 13 June 1883, DF (*TAED* D8316AAQ).

5. As secretary of the Edison Electric Light Co., Ltd., Arnold White (1848–1925) ran the company's daily affairs and finances; he had also recently been elected to fill an empty seat on the board (Doc. 2221 n. 5; Edison Electric Light Co., Ltd., statement, 25 May 1883, DF [*TAED* D8338ZAF]). White's management had led to sharp disagreements with Edison; Insull and Johnson held him in low regard (see Docs. 2339, 2347, 2356–57, 2374, 2375, 2403, and 2467).

6. Charles Edward Chinnock (1845–1915), superintendent of the Metropolitan Telephone Co. of New York, replaced Joseph Casho as superintendent of the Pearl Street station. He received from Edison an option for seventeen shares of Edison Electric Light Co. stock or $5,000 cash provided the station's earnings yielded a 5 percent dividend by June 1884. (Chinnock received ten shares in February, 1885 [Voucher no. 45 (1885)].) A native of London, Chinnock had worked in Brooklyn as an inventor and instrument maker and was a party to one of the Telephone Interference Cases. After his service at Pearl St., he was key in organizing the Edison Electric Illuminating Co. of Brooklyn and also became vice president of the Edison United Manufacturing Co. TAE agreement with Chinnock, 22 May 1883, Miller (*TAED* HM830185); Martin 1922, 60; Obituary, *NYT,* 12 June 1915, 11; *Edison v. Blake v. Chinnock* Litigation Records, i–ii, TI 6 (*TAED* TI6).

7. Joseph Casho (1840–1924) was the superintendent at the Pearl St. central station from its opening in September 1882. He left about this time because, according to one later account, having had extensive prior experience with slow Corliss engines, he was uncomfortable with the high speed engines used at the station. Doc. 2403 n. 6.

8. See Doc. 2424 (headnote) and App. 2A.

9. Local backers of the prospective illuminating company in St. Louis sought municipal approval in May. After agent William Brewster had the city canvassed, Edison felt confident that "St. Louis is by far the

cheapest City for lighting that has come under my notice." He drafted a contract to build a central station and distribution system of 5,400 (16 candlepower) lamps for $115,625 and also prepared a statement of probable operating expenses and profits (Brewster to TAE, 23 and 26 May 1883; TAE to Brewster, 29 June 1883; Insull to Brewster, 3 July 1883; all DF [*TAED* D8340T1, D8350A, D8316ADU, D8316AEG]; TAE to Brewster, 23 May 1883, Lbk. 16:426A [*TAED* LB016426A]). However, the project languished. Blueprints for the twelve-dynamo station were prepared in the fall (several were published in Hagen 1885 [leaf following 104]), but the pending contract blocked other business after Edison decided in October that he did not want "New Orleans nor any other Southern towns canvassed till St Louis closed." At least one key investor remained indecisive until Edison left the central station construction business the following spring (TAE to Brewster, 1 Oct. 1883; Alfred Tate to Insull, 15 Oct. 1883; LM 15:278, 392 [*TAED* LBCD2278, LBCD2392]; Brewster to Pierre Chauteau, 30 May 1884, DF [*TAED* D8439ZBI]).

10. Presumably the Southern Exposition in Louisville (see Doc. 2465).

11. Edison had been anxious for about $9,600 due from the Edison Electric Light Co., Ltd., in March. He cabled Johnson on 16 May and again on 21 May that he was depending on settlement of this "long overdue" account to pay a note of his own. Insull complained in mid-June that the company had been making "scandalous" reductions in the amounts owed "for which they absolutely refused to give any explanation whatever." He asked Johnson to "fix matters up, now you have control," noting that Edison had instructed him in the meantime to suspend shipments of all equipment (including the $2,000 lamp order) to London (TAE to Edison Electric Light Co., Ltd., 29 Mar. 1883; TAE to Johnson, 21 May 1883; LM 1:303A, 312A [*TAED* LM001303A, LM001312A]; TAE to Johnson, 3 Apr. and 16 May 1883, Lbk. 16:84, 339A [*TAED* LB016084, LB016339A]; Insull to Johnson, 11 June 1883, DF [*TAED* D8316AAB]). Arnold White gave Edison a bleak assessment of the company's financial condition in mid-June. A few weeks later, however, the firm advanced $4,000 on two engines for the Criterion Theater, which Insull arranged to have shipped promptly (White to TAE, 15 June 1883; Insull to TAE, 10 July 1883; both DF [*TAED* D8338ZAL, D8367Y3]; see Docs. 2467 n. 14 and 2485).

12. Edison had promised in late 1882 to try to sell some of the dynamos charged to the British company, including six C central station machines still in New York (see Doc. 2374 esp. nn. 4–5). In reply to a recent inquiry about the matter, Edison expressed hope that, despite his own large inventory, he could dispose of some of the big machines in a forthcoming order. Edison Electric Light Co., Ltd., to TAE, 24 Apr. 1883, DF (*TAED* D8338U); TAE to Edison Electric Light Co., Ltd., 7 May 1883, Lbk. 16:268 (*TAED* LB016268).

13. Insull referred to bills for furnishing Johnson's house. Insull to Johnson, 21 May 1883, LM 3:150 (*TAED* LM003150).

14. Baumann Bros. was a prominent dealer in "Elegant Home Furnishings" at 22–26 E. Fourteenth St. in New York. They claimed to have the "finest, largest, and most varied stock of furniture, carpetings, and interior decorations" in the United States. Baumann Bros. adver-

tisement, *New York Herald,* 9 Sept. 1883, 22; "Enlarging Their Facilities," *NYT,* 6 Jan. 1884, 7.

15. Edison gave orders to discharge all domestic servants from the family's New York home as of 1 June. TAE to Marie Thomas, 15 May 1883, Lbk. 16:339 (*TAED* LB016339).

16. The Edison family moved to a rented home at 25 Gramercy Park, in a fashionable New York neighborhood, in October 1882. See *TAEB* 6:2; Doc. 2341.

17. Canadian-born Alfred Ord Tate (1863–1945), a telegraph operator who had worked in other departments of various railroads, became Insull's assistant in early May. This position was arranged by an acquaintance of James MacKenzie, the man who had taught telegraphy to Edison. Insull announced Tate's arrival to Johnson: "You see at last I have a short hand man," part of an "organized establishment to do the heavy work" of the Edison Construction Dept. Tate succeeded Insull as Edison's private secretary in 1887 and held that position until 1894, after which he was involved in battery development and manufacture. Tate 1938, 15–50; Insull to Johnson, 21 May 1883, LM 3:150 (*TAED* LM003150).

–2457–

To Bergmann & Co.

[New York,] June 5/[188]3

Dear Sirs:

I have undertaken to have light running at Sunbury Pa., on the 4th of July[1] and the only possible cause of delay will be your failure to get poles for the work there. I understand that nothing has been done so far in relation to this matter and I write this to remind you that poles dont grow right on the exact spot where they will be needed.[2] Yours truly

Thos A Edison

I would also remind you that the Almighty has'nt yet grown any trees which attain the necessary height and diameter within a week— TAE

LS (letterpress copy), NjWOE, Lbk. 17:68 (*TAED* LB017068). Written by Samuel Insull.

1. On 1 June, a local newspaper had published 4 July as the starting date for what would be Edison's first commercial village plant. Frank McCormick, president of the Edison Electric Illuminating Company of Sunbury, had requested Edison to have the station operating by this day, ahead of the contracted schedule. The town was planning for "a boating regatta, the opening of a new rail road and a general celebration," which would provide "an excellent opportunity to exhibit our light." Edison replied on 4 June that he had "gone as far ahead with the work as possible until our machinery is ready. This has all been ordered and will be at Sunbury within a week. There will then be very little time lost in getting the whole plant started." *Sunbury News,* 1 June 1883, 182;

McCormick to TAE, 1 June 1883, DF (*TAED* D8361D); TAE to Mc-Cormick, 4 June 1883, Lbk. 17:52 (*TAED* LB017052).

2. On 18 June, Edison again pressed Bergmann & Co. to hurry the Sunbury work. The town issued a permit to erect the poles by 19 June. At the end of the month, Charles Hanington, whom Bergmann & Co. had dispatched to help with the installation, reported that although "the entire change of the plan of running the Pole line" and heavy rains had delayed the work, his crew had "been using every effort" and would be finished by 4 July. TAE to Bergmann & Co., 18 June 1883; McCormick to TAE, 19 June 1883; Hanington to Insull, 29 June 1883; all DF (*TAED* D8316ABH2, D8361U, D8340ZAV).

[New York,] June 5/[188]3

Gardner C. Sims

Probably want all Engines you can make. Be careful about taking outside orders. Send on exact list Engines building.[1] Moore will send order tomorrow for three 3ᵃ fourteen half Engines for Cincinnati[2] and Louisville Ffairs; three must be shipped before tenth July; one tenth August. Borden will require one for Boston theatre.[3] Want for my Central Station work ten fourteen half by thirteen,[4] one in two weeks, others following within about two months; also nine Eight half by ten mostly right off. Negotiating many other contracts additional. All this besides regular isolated business which will be very heavy. If necessary work nights; will wire you up and put in dynamos. Letter tomorrow[5]

Thos. A. Edison   T[ate]

L (telegram, letterpress copy), NjWOE, Lbk. 17:069 (*TAED* LB017069). Written by Alfred Tate. ᵃCircled.

1. Sims did so the next day. DF (*TAED* D8322ZAO).

2. The annual Cincinnati Exposition, scheduled from 5 September to 7 October, featured manufacturing industries, machinery, and decorative arts. In competitive trials, principally against the Maxim lamp of the United States Electric Lighting Co., the Edison company received first prizes for its incandescent lamp, dynamo for an incandescent system, and overall incandescent system. "The Cincinnati Exposition," *NYT,* 27 Aug. 1883, 1; "Mr. Edison Ahead," ibid., 9 Oct. 1883, 5; Edison Electric Light Co. Bulletin 20:8, 31 Oct. 1883, CR (*TAED* CB020).

3. This reference was probably to the Park Theater, for which Spencer Borden was arranging two dynamos and up to 800 lights. However, the theater does not appear on a list of isolated installations sold by April 1884. Borden to TAE, 15 May 1883; Borden to Luther Stieringer; both DF (*TAED* D8340M, D8340N); TAE to Borden, 16 May 1883, Lbk. 16:354 (*TAED* LB016354); List of isolated plants, Edison Electric Light Co. Bulletin 22:5, 9 Apr. 1884, CR (*TAED* CB022520).

4. That is, ten engines having a 14½-inch diameter cylinder bore and

13-inch piston stroke. By convention, steam engines were categorized by cylinder bore diameter and piston stroke length. Until about this time period, most engines had a stroke considerably larger than the cylinder bore and operated relatively slowly. Engines with shorter strokes were suited to high-speed operation because they required less piston travel.

5. Acknowledging the 6 June response from Sims (see note 1), Edison reiterated his "great concern . . . that you should not make contracts which will interfere with our getting Engines; our business promises to be very large indeed as does also that of the Isolated Company." He emphasized the importance of prompt delivery but concluded that "unless you take some large outside contracts I feel certain you will be able to do what we require." TAE to Sims, 7 June 1883, Lbk. 17:108 (*TAED* LB017108).

-2459-

To Spencer Borden

[New York,] June 7th [188]3

Dear Sir:

I have your letter of the 5th inst and in reply would beg to state that I now have a canvasser who can make a perfect canvas in two days by an "entirely novel process."[1]

I am going to attend to this canvassing business myself and have my own man do it so for the future do not trouble about starting new canvasses without notifying me of what there is to do and I will arrange for my man to do the work.

There is no trouble about heating wax in the shoe business; we can do it to perfection.[2]

We have a motor now, the first size which will run from one to four sewing machines—probably run about two leather sewing machines: the cost in large lots will be about $35.00; the local Company will rent them for say $1.75 per month and charge for current on the meter. We will have another size ready—3/4 h.p.—in about six weeks.[3]

If enough power can be sold at Haverhill during the day there is no reason whatever why we should not run; all I object to is running in a place where there is no money; if we can make money I am prepared to run at any time. Yours truly

Thos. A Edison   I[nsull]

L (letterpress copy), NjWOE, Lbk. 17:115 (*TAED* LB017115). Written by Alfred Tate; signed for Edison by Samuel Insull.

1. Borden told Edison he would "write our canvasser to get a careful acct. of machines" used for mechanical power in small shoe-making shops of Haverhill, Mass. Edison's remark about his canvasser may have referred to Charles Cooke. The "entirely novel process" is probably that described in Doc. 2499. This reply is based on a draft written by

Edison on the back of Borden's letter. Borden to TAE, 5 June 1883, DF (*TAED* D8340Z).

2. Borden described in his letter (see note 1) small lamps to warm tools used to work wax in Haverhill's shops: "An excessive heat, such as for soldering, is not required, merely the temperature necessary to keep the wax soft. These lamps are so dangerous that insurance companies demand enormous rates for these shoe factories, the lamps being in every instance the cause of the fires. Now, they want to Know if the electric current can be made to warm these tools."

3. Borden also told Edison (see note 1) about the numerous small Haverhill shops that hired

> power for a few sewing machines each, and for other light machinery. This power is conveyed by tremendously long lines of shafting, several hundred feet under all the building on one side of a street from building to building. This can all be cut out if your motor business is in shape to take this, and they want to Know if your motors for single machines, or a single room are in such condition that they can be put on the circuit, so as to give business to the station all day. There is lots of this business in Haverhill.

A month later, Edison reported he was making ⅕ horsepower motors and preparing drawings for larger sizes. TAE to E. H. Lord, 17 July 1883, DF (*TAED* D8316AHO); see also Docs. 2459, 2567, and 2579.

Self-contained so-called "fractional horsepower" steam engines were generally not suited for commercial work, though small manufacturers sometimes rented steam power from a large central boiler. The shafting arrangement described by Borden was apparently not a typical alternative. Transmission of power within factories by complex series of shafts and belts, however, was as common as it was wasteful, frequently consuming one-fifth to one-half of the supplied power. Even with somewhat later metallurgical improvements, one authority calculated that a shaft for conveying 170 horsepower through 200 feet would require a five-inch diameter and weigh about ten tons (Hunter and Bryant 1991, 53–62, 115–20). The use of steam power in shoemaking was distinctive among major industries in Massachusetts. That trade's 294 engines were the most of any type of business in 1880. They were generally also much smaller than in other applications, averaging only about 23 horsepower apiece (Hollerith 1883, Table III, p. 23).

–2460–

From Edward Hastings

Boston Mass June 7 1883. 3:20[a]

Thomas A Edison

Where Are Our reflector lamps[1]   We are in total Darkness so to Speak   please answer

E Hastings[2]

⟨Ask Upton if he is making any more coiled 100 cp Lamps[3] & if he can accomplish the object that the theatre people are getting impatient= Answer Hastings to have patience.⟩[4]

L (telegram), NjWOE, DF (*TAED* D8332ZBF). Message form of Western Union Telegraph Co. ᵃ"1883." preprinted.

1. Hastings was referring to lights in the Boston Bijou Theatre (see Doc. 2768 n. 4). In a testimonial letter of 18 May, he had stated that the Edison system had been operating without fault or need for alterations for nearly six months (Edison Electric Light Co. Bulletin, 18:8–9, 31 May 1883 CR [*TAED* CB018]). His present request seems to seek a supplement, if not a replacement, for the theater's stage lighting, where lamps were installed in a concave arch of tin behind the proscenium ("The Bijou," *New York Graphic,* 9 Mar. 1883, in Boston Bijou, MH-HT). Francis Upton had written from the Lamp Co. on 3 May: "We are going to make a spiral 100 candle power lamp and treat it so as to give 150 candles and try it in place of calcium light." A few weeks later he reported, "We have tried a spiral 100 c.p. in a reflector and think that it works well. The ones we made had too high volts for the Bijou Theatre. We are dosing them and hope to have [~~them~~?] one or two ready by tomorrow" (Upton to TAE, 3 and 21 May 1883, both DF [*TAED* D8332ZAQ, D8332ZAY]).

2. Edward Holland Hastings (1851–1889) had attended Harvard and had a brief stint in banking. Hastings became treasurer of the Bijou Theatre in late 1882 (*Third Report of the Secretary* 1887, 97; *Secretary's Report* 1905, 77). In November of that year, he and his brother (T. Nelson Hastings) formed a partnership with veteran actor-manager George H. Tyler to complete the remodeling of Boston's old Gaiety Theatre as the Bijou Theatre Co. (Articles of Association, 9 Nov. 1882; Edward Johnson and Edward Hastings, memorandum of alterations, 1 Dec. 1884; both MH-HT; Bacon 1886, 43–45). Hastings became president of the theater in January 1883.

3. Edison and the lamp factory had devised 100-candlepower carbon lamps in 1881 and 1882, intending them mainly for streetlights in Great Britain, and the factory began to produce them at the end of 1882 (see Docs. 2202, 2238, 2275, 2292, 2334, and 2339). When Upton returned to this subject in a 13 June letter, he told Edison: "We think that our 100 candle lamps will give perfect satisfaction in a short time. We have made some and found them right. The last shipment we think will be found better than the first as we have largely gotten over trouble with clamping. The spiral 100 candle power lamp will be found adapted for side and head lights of ships" (Upton to TAE, 13 June 1883, DF [*TAED* D8332ZBG]). At the end of the month, Edison filed a patent application (Case 584) on a light fixture in which several lamps were arranged radially within a reflector (Patent Application Casebook E-2538:232, PS [*TAED* PT022232]). In Paris at about the same time, Charles Batchelor was working on regulators for theater lights (Batchelor to TAE, 30 Aug. 1883, DF [*TAED* D8337ZDR]).

4. Alfred Tate replied in Edison's name for Hastings to "Have patience for a little while." TAE to Hastings, 8 June 1883, Lbk. 17:104A (*TAED* LB017104A).

Memorandum:
Electric Light and
Power Patent

[A]

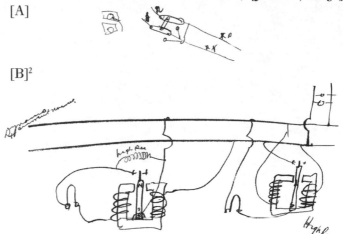

[B][2]

Mention that apparatus might be used for working other things & claim broadly polarized app. in combn with electric lighting system wherein lamps are not affected by changes in current.[3]

Electrolytic[b] Meters[c] cant be used but there is no need of making mention of this—[4]

[Witness:] R N Dyer

ADDENDUM[d]

[New York,] June 12, '83[e]

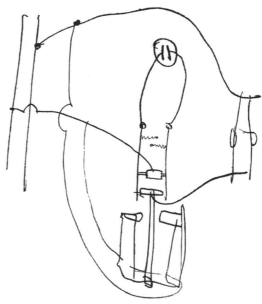

⟨579[5f]   Add this to diag⟩[6g]

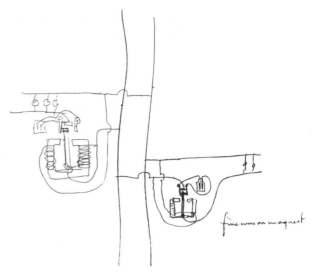

[Witness:] H W Seely

AD, NjWOE, Lab., Cat. 1149 (*TAED* NM018AAO, NM018AAP). Letterhead of Edison Electric Illuminating Co. of New York. ᵃ"*New York*," preprinted. ᵇInterlined above. ᶜObscured overwritten text. ᵈAddendum is an AD; its reverse contains a faint, incomplete drawing which has not been reproduced. ᵉDate written by Henry Seely. ᶠFollowed by dividing mark. ᵍMarginalia probably written by Richard Dyer.

1. Richard Dyer signed and dated the document on this day when he received it from Edison.

2. Figure labels are "Dynamo current [mains?]," "high res," and "High Res." The two apparatus shown in this sketch were incorporated into the single drawing of the subsequent patent (see note 3).

3. Edison executed a patent application embodying this idea on 25 June. He intended the device to enable the operation of

> apparatus located at different points in an electric-light circuit from
> the source of supply without the employment for that purpose
> of a circuit separate from the light-circuit, and more especially
> to provide means . . . for turning on and off from the source of
> current-supply certain lamps—as, for instance, street-lamps, . . .
> thus avoiding the expense and inconvenience of operating them
> separately by hand. This I accomplish by . . . polarized apparatus
> located in connection with the lamp or lamps or other devices to be
> operated, which polarized apparatus is actuated by a reversal in the
> polarity of the current at the source of supply, a suitable switch be-
> ing provided at or near the generator or generators for this purpose.
> [U.S. Pat. 430,934]

Edison also sought patent protection around this time for a device to keep lamps wired in series (such as in the two-wire 330 volt system at Roselle, N.J.) operating when one failed. He also explored arrangements for instantly switching a substitute for a failed lamp into the circuit. Cat. 1149, Lab. (*TAED* NM018AAK, NM018AAQ); U.S. Pat. 438,301.

Edison had previously used polarized relays in a number of telegraphic inventions, particularly for conditions, such as multiple telegraphy, in which devices had to discriminate among signals intended to affect only some of the apparatus in a circuit. See *TAEB* 1 passim., esp. Docs. 11–12, 275–79, and 300 (and headnotes).

4. Edison devised a way around this limitation by the next day, when he wrote the addendum to this document. The lower figure of that sketch was incorporated into the patent drawing as the third (of three) devices represented. An electrolytic cell is in circuit at right, connected with a device for changing the polarity of its connections, essentially as represented in the patent specification. Edison's application stipulated that the invention "is preferably employed in a system wherein the regular lamps or other translating devices are not affected by changes in the polarity of the current." It included, however, provisions for "translating devices . . . which are affected by changes in the polarity of the current—such as my electrolytic-cell meter or arc lamps—and the current in such translating devices will be maintained of one polarity by similar polarized apparatus" which would reverse the connections of the meter or other device. Four of the twelve draft claims pertained to such arrangements. In the face of the Patent Office's objections, these were consolidated into a single claim; the body of the specification was essentially unchanged. Pat. App. 430,934.

5. Edison's patent attorneys assigned this case number to the resulting application.

6. The addition referred to is the relay on the left side of the circuit. The text in the lower right is "fine wire on magnet."

–2462–

To Guiseppe Colombo[1]

[New York,] June 12. [1883]

Dear Sir:—

I would suggest that you should put safety catches on BOTH POLES in the street boxes of all kinds and in the station on the feeders and dynamos. We have had trouble here in New York from not having this done, there having been a combination of circumstances whereby a cross occurred beyond the influence of the safety catches, which heretofore have only been on one pole.[2] Very truly yours,

TL (carbon copy), NjWOE, DF (*TAED* D8316AAL).

1. Guiseppe Colombo (1836–1921) was an engineering educator, entrepreneur, and statesman in Milan. He formed an Edison lighting syndicate and began constructing a Milan central station in 1882. Docs. 2332 nn. 2–3, 2343 n. 14.

2. According to later accounts, the fuses or "safety catches" used throughout the New York system were not placed exclusively on either the positive or negative branch. Edison sent a letter identical to this one

the same day to Willis Stewart, who was overseeing Edison installations in Chile. He likely was prompted by several recent incidents. On 9 June, horses reportedly received shocks through the ground on Maiden Lane from a broken Edison conduit. A large feeder line suffered a short circuit the next day, described by Samuel Insull in Doc. 2464. A gas explosion that same day in a Front St. building was determined to have been caused by an Edison service line becoming grounded against a gas pipe (TAE to Stewart, 12 June 1883, DF [*TAED* D8316AAP]; Martin 1922, 65–66; "Horses Attacked by Electricity," *NYT*, 10 June 1883, 9; "Events in the Metropolis Two Singular Explosions," ibid., 12 June 1883, 8; see also Doc. 2449 n. 3 and cf. Jehl 1937–41, 1063–64, 1083–85). When Charles Batchelor received word in Paris of a "cross in Central Station mains, the first thing I thought of . . . was: 'Where were your safety catches?' " William Hammer later claimed to have used fuse blocks on both poles of the Holborn Viaduct system he helped construct in London in 1882 (Batchelor to John Kruesi, 23 July 1883, Cat. 1331, Batchelor [*TAED* MBLB3105]; Jehl 1937–41, 1064).

–2463–

*From Charles
Batchelor*

Ivry, le June 12th 1883[a]

My dear Edison,

In testing lamps at 64 c.p. do you make them give 64 candles all the time by keeping them in Photometer & regulating— The way I do is as follows:— Test lamp in Photometer for 64 candles & see how many volts & then keep this volts on it until it gives out regardless of candlepower— Am I right?[1] Yours

Batchelor

⟨We do the same—but get Economy[2] at same time, & also sometimes put lamps back after burning some time to get fall in Candle power[3] Does Upton send you curves[4] we are making better curves than ever before⟩[5]

ALS, NjWOE, DF (*TAED* D8337ZBX). Letterhead of Société Industrielle & Commerciale Edison. [a]"Ivry, le" preprinted.

1. Batchelor had begun commercial production of Edison lamps at the factory of the Société Industrielle & Commerciale Edison in Ivry-sur-Seine, a southeastern suburb of Paris, in the latter half of 1882; by this time the factory reportedly turned out 400 lamps per day (Edison Electric Light Co. Bulletin 17:28, 6 Apr. 1883, CR [*TAED* CB017]). As manager, Batchelor had quality control challenges similar to those faced by Edison and Francis Upton in domestic manufacture. The test he describes here was designed to measure the life of selected bulbs at above-normal voltage, from which their durability under normal conditions could be extrapolated (see Doc. 2177 [headnote]). Batchelor's own notes and records of lifetime tests, made sporadically from late May through the end of the summer, are in Cat. 1235:1–24, Batchelor (*TAED* MBN012001–MBN012024).

2. That is, the number of lamps that one horsepower could run simultaneously at a given intensity.

3. A lamp's illuminating power decreased with use because of a gradual darkening of the glass globe. See Doc. 1898 (headnote).

4. Lamp "curves" graphically represented the range of lifetimes in a sample of test lamps. See Doc. 2177 (headnote).

5. Edison's marginalia was the basis for a typed reply sent to Batchelor on 25 June. DF (*TAED* D8316ACP).

–2464–

*Samuel Insull to
Charles Batchelor and
Edward Johnson*

[New York,] June 12th. [188]3

Batchelor & Johnson[a]

Things are progressing here at quite a pace.

CENTRAL STATION PLANTS. The contracts absolutely closed are as follows:

Sunbury, 500 lights, $11,000
Shamokin, 1,600 lights, $25,000
Lawrence, 3,200   " $32,000
Brockton, 1,600   " $31,000
Danville,[1] 1,600   " $23,000

All following are being negotiated and promise to close very shortly:

St. Louis, $150,000
Lowell, 3,200 lights, about $50,000[2]
Hamilton, Ohio, about $35,000
Utica, $85,000[3]

and a whole host of others too numerous to mention. A syndicate has been formed to work Ohio[4] on the following terms: They are to guarantee 24 plants a year of an average cost of $20,000 each. This guarantee lasts for 5 years. After the 5th. year, I think double the number, namely, 46 plants a year of $20,000 each. This contract is made with Cincinnati capitalists, all of whom are extremely wealthy and we expect great results from it. We have now got our men working at Sunbury, Brockton and Shamokin. We expect to start them at Danville next week and Lawrence about the week after. Business looks as if it is going to be larger than our most sanguine expectations could anticipate.

Our isolated business too is getting to be tremendous. We have received an order for 4,400 lights for Louisville[5] and about 8 other orders within the last week, the details of which I am not familiar with. Down in New England we have got a large number of very large factories and mills to light. We are also getting quite a considerable number of theatres, in fact Edison light is commencing to boom for the season, and I think now the boom has commenced it will never stop.[6]

ILLUMINATING COMPANY. The illuminating business is running very well in the City. Chinnock, the new superintendent, has already instituted considerable improvements down there, and after looking into the matter very carefully he says that he knows he can make it an entire success. On Sunday they had a peculiar accident by a combination of circumstances, which I do not exactly understand. The largest feeder got crossed. The lights went down immediately of course. Chinnock was home sick, Casho and Bradley[7] were at the office, Casho wanted to shut down but Bradley insisted upon putting on more engines. They put on engine after engine until they had 5 on, then they slowed down the engines, opened the throttle suddenly and after doing this a number of times they bursted the cross. When discovered it was found that over a foot of the largest feeder we use was melted off.[8] Of course the above is the only way of preventing the stoppage of the Station. This is the first narrow escape, and a very serious one to, that we have had. Arrangements are now being made to extend the present District to Broadway from the Times Building to Wall Street. First of all we are only going to run on one side of Broadway, then we are going to run on both sides. This will take in the streets between Nassau and Broadway as well. This matter is now being figured out.[9] We have got in the Station an ampere indicator which indicates exactly the number of lamps on at any one time.[10] They are now getting up a recorder so as to keep an exact curve of the indications of this instrument.[11]

The new wiring department[12] is beginning to get quite a good deal of work. Very little of outside wiring but still a great deal of work from the Illuminating Company and the Isolated Company, both from the home office and the New England Department.

FINANCIAL. We are all of us still very closely pressed for money. The main cause is the shortness of money of the Illuminating and Parent Companies. If these two companies were not so hard up and could pay up their bills we should be in very good condition to-day. Of course the endeavor of Edison to swing this new construction business will keep us somewhat short of funds for some time to come notwithstanding we shall have plenty of work, however this is only temporary.

Edison goes to Chicago either the end of this week or the commencement of next, to wake up the Western Company and start them on central station work.[13] Very truly yours,

TL (carbon copy), NjWOE, DF (*TAED* D8316AAM). [a]"Batchelor & Johnson" handwritten.

1. In a letter to Francis Upton the same day, Insull listed Danville, Pa., Lawrence, and Hamilton as plants that "will be closed this week." Insull to Upton, 12 June 1883, DF (*TAED* D8316AAO); cf. Doc. 2456 and see also App. 2.

2. Edison sent an estimate of $52,240 for a Lowell plant of about 4,000 lamps two days later. He thought it "best to start a plant in the first place of 3,200, although laying down mains for the total lamps in the district." Though initially expecting to close a contract before 23 June, in mid-July he provided a statement of anticipated "Profits and Expenses," including depreciation and the quantity of light to be sold. This information apparently did not persuade prospective investors, and, with no contract executed by late September, Edison withdrew his tender. TAE to Spencer Borden, 14 June and 13 July 1883; TAE to William Rich, 18 June 1883; all DF (*TAED* D8316AAU, D8316AGL, D8316ABD); TAE to Borden, 25 Sept. 1883, LM 15:206 (*TAED* LBCD2206); see also Docs. 2469 and 2472.

3. See Doc. 2519.

4. The Ohio Edison Electric Installation Co. was incorporated on 1 June 1883 in Covington, Ky., though its business offices were across the river in Cincinnati. Phillips Shaw's name appears on a handwritten list of eight incorporators. Its authorized capital was $250,000. According to a report on the firm's creditworthiness requested by Insull, "its stockholders without one or two exceptions are wealthy influential business men of this city and Covington KY and their reputation is said to be first class in all respects." The company paid the Construction Dept. to build village plants on behalf of local Edison illuminating companies. Ohio Edison Electric Installation Co. memorandum, undated 1883; Insull to Bradstreet Co., 16 July 1883; Bradstreet Co. report, 21 July 1883; Stuart to TAE, 10 July 1884; all DF (*TAED* D8353ZBQ, D8316AGW, D8353E, D8449ZBI); see also Doc. 2497.

5. That is, the Southern Exposition in Louisville; see Doc. 2465.

6. The editors have not itemized the isolated plants constructed or contracted for at this time. However, the Edison Electric Light Co. printed sometime in June 1883 a comprehensive list of approximately 420 isolated plants itemized by type of establishment. This list may be compared with a similar compilation (arranged in broader categories) of 334 plants in the United States and abroad through the month of May, which was the basis for the analysis of the isolated business in *TAEB* 6 App. 2. The company's list of 307 plants sold (not necessarily completed) in the United States and Canada from May 1883 to early April 1884 also provides an indication of its business over the next ten months. Edison Electric Light Co. list of central stations and isolated plants, undated June 1883, DF (*TAED* D8320Y); Edison Electric Light Co. Bulletins 18:30 and 22:5, 31 May 1883 and 9 Apr. 1884, CR (*TAED* CB018, CB022).

7. Charles Schenck Bradley (1853–1929), an employee of the Edison Electric Illuminating Co. of New York since about 1881, was responsible for testing and inspecting the underground conductors at the Pearl St. station. He created a voltmeter that was used on the distribution system's feeder lines. Bradley helped set up the central station at Brockton, Mass., before leaving the Edison interests later in 1883. He became a successful inventor and engineer in his own right, devising a polyphase motor, synchronous motor, and rotary converter for alternating current systems. He also invented a commercial process for fixing nitrogen with atmospheric oxygen. Obituary, *NYT*, 5 Mar. 1929, 30; Jehl 1937–41, 949; "Bradley, Charles Schenck," Pioneers Bio.; Hughes 1983, 118–19, 121; Jacobs 1902, 306; see also Doc. 2538 (headnote).

8. See Doc. 2462 esp. n. 2. This incident is recounted in "Notes on Work Done by Charles S. Bradley" in Bradley's Pioneers Bio.

9. This plan would have involved expanding the First District several short blocks to the west (see map, p. 2). An additional 15,000 feet of underground conductors (approximately 18 percent of the original total) were laid during the latter half of 1883, although the editors have not ascertained where this occurred. Clarke 1904, 46, 49.

10. According to a description given by Insull at the end of the summer, Charles Bradley, using the ampere indicator in the main circuit at the Pearl St. station (see note 11),

> carefully recorded the movements of the needle and compared these with the steam cards taken at the same time and then arrived at the exact number of lamps indicated as being on by the movements of the needle. He then prepared a scale, and attached it to the recorder, so that when a deflection of the needle takes place it will point out exactly the number of lamps burning on the system at that moment. The thing works very well indeed. . . . If it happens to be a showery day, and a cloud suddenly comes up, the number of lights put in within a few moments is very large, as is shown by the deflection of the needle. [Insull to Charles Batchelor, 3 Sept. 1883, DF (*TAED* D8316ARP)]

11. John Ott had sketched a recording ampere meter about a week before. It consisted of a horseshoe magnet in the main circuit, which apparently acted on a pivoted armature and stylus to inscribe a record on a moving roll of chemical paper. Edison executed a patent application in October for a similar recording device actuated by a galvanom-

Ampere meter with recording device, perhaps like that adopted at the Pearl St. station about this time, sketched by John Ott on 4 June.

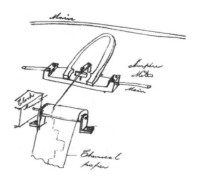

eter needle. During much of July, Ott and Edwin Kellogg tried scores of experiments "with chemical paper with the spark of an induction coil, for ampier meter to be used at central station." N-82-12-21:97, N-82-12-04:39–101; Lab. (*TAED* N150097, N145039–N145099); U.S. Pat. 307,030.

12. The Wiring Department of Bergmann & Co. was formed sometime after a February outline of its proposed participants (Luther Stieringer and Charles Hanington), purpose (pole lines and building wiring), and prices, but before late May. It was headed by Edwin Greenfield, who had been in charge of wiring buildings in the New York First District. Bergmann & Co. memorandum, 8 Feb. 1883, DF (*TAED* D8324E); Insull to Johnson, 21 May 1883, LM 3:150 (*TAED* LM003150).

13. Edison left on Monday, 18 June, for a brief Chicago trip. See Doc. 2455 n. 10.

–2465–

To Edward Johnson

[New York,] June 14th 1883

Fiftyseven London[1]

Things taken tremendous change five Central Stations building eighteen others ready to close.[2] Contract signed Light Louisville Exposition five thousand lights, forty five days from today good profit.[3] hasten London scheme return soon.[4]

L (telegram, letterpress copy), NjWOE, Lbk. 17:141 (*TAED* LB017141); a copy is in LM 1:316C (*TAED* LM001316C). Stamped "From the Office of Thomas A. Edison"; "Cable" written at top.

1. This was Edward Johnson's registered cable address in London. *TAEB* 6 App. 4.

2. In a letter that Johnson would not yet have received, Insull advised him of a $30,000 contract for Lawrence, Mass.: "This makes $100,000 worth of contracts absolutely closed. We have got several hundred thousand dollars more under way, and before you get back here I think we shall have half a million dollars worth of work in the course of construction." Insull to Johnson 11 June 1883, DF (*TAED* D8316AAB).

3. The Edison Co. for Isolated Lighting executed a formal agreement on 3 July on behalf of the Edison Electric Light Co. to illuminate the Southern Exposition in Louisville, the terms of which made the company a contractor rather than an exhibitor. The plant reportedly cost more than $75,000 (and perhaps as much as $100,000), with the Exposition paying "five cents a night for each lamp, or $23,000 for 100 nights." Four thousand six hundred 16-candlepower lamps were to illuminate the gatehouses and more than half a million square feet of exhibition space. Three Armington & Sims engines (plus one for night use) and additional power from the Exposition's large Corliss engine ran fourteen or fifteen new "H" dynamos. Work began on 9 July, and while some lights were running by the end of the month, the entire installation was not completed before mid-August. Illuminating engineer Luther Stieringer, who designed and oversaw construction of the dis-

play, recalled it as "the first to bring the light up gradually from nothing to full candlepower"; it eventually won four first-place jury awards and was reinstalled for a similar exhibition in 1884. Reportedly the largest single installation of incandescent lighting to that time, in the words of historian David Nye, it "briefly made Louisville the brightest spot on the planet." The Louisville exposition was the first of its size to be open at night, and visitors came for the express purpose of seeing the lights come on. "Letter to Editor: The Southern Exposition at Louisville," *Electrician* (N.Y.) 2 (Sept. 1883): 281; "Edison's Lights for the Louisville Exposition," *Baltimore Sun,* 5 July 1883, 1; *Edisonia* 1904, 143; TAE marginalia on Sidney Paine to TAE, 14 Sept. 1883; Edward Johnson to Arnold White, 24 Aug. 1883; Harry Doubleday to Samuel Insull, 11 Aug. 1883; all DF (*TAED* D8325ZCM, D8338ZBJ, D8366X); Edison Electric Light Co. Bulletins 19:23 (15 Aug. 1883), 20:14 (31 Oct. 1883), and 22:8 (9 Apr. 1884), CR (*TAED* CB019, CB020, CB022); Stieringer 1901a, 187; Nye 1990, 3; *Great Southern Exposition* 1886, 15.

4. It is uncertain what Insull specifically meant by the "London scheme," as Johnson had been writing about several major issues confronting the new board of directors of the Edison Electric Light Co., Ltd. Among possible readings of the phrase is the effort to obtain provisional orders authorizing a London central station. It could also refer to resolving the pending question of amalgamation with the Swan United Electric Light Co., Ltd. Johnson to TAE, 18 May 1883, DF (*TAED* D8320J); see Docs. 2456 and 2467.

–2466–

Samuel Insull to Edward Johnson

[New York,] June 19th. [188]3

My Dear Johnson:—

Sprague is here, but Edison has no letter from you stating what arrangements you have made with him.[1] The other day Sprague asked me for a check for $200. on account and I gave it to him, but I do not know whether to charge it against his travelling expenses or against his salary, and I do not wish to ask him whether you arranged for Edison to pay his expenses or whether he was to pay them himself. I wish you would advise me on this point as if the check was on account of salary. The young man is going it rather steep for a start off. Edison is going to use him in starting village plants, and although he thinks Sprague quite a smart fellow, his impression is that he is a pretty expensive man, as we do not have to pay Andrews any such salary as that paid to Sprague, and Andrews is a man of wider experience, however, if he does good work his salary will not be of much account.

I think it is about time you wrote me. I have come to the conclusion that your wife is the correspondent of the family, and I am going address my letters to her in the future, you do not write a fellow worth a cent and if there was not a lady in the room while I am dictating this, I would swear at you like a trooper, however I will give you a little information. Edison telegraphed for you to come home[2] because he thinks we are getting in the position of Bill Nye's dog, who swallowed some soft plaster of paris and as it began to harden came to the conclusion that he had bitten off more than he could chew.[3] We want you to help us do some of the chewing. Our mastication is pretty good but it not infallible. Edison went to Chicago yesterday to wake up the people there. This morning Eaton, Hastings, Borden and myself had a conference to talk over the question of instructing our agents to get no more towns, until we had dealt with those already on hand. We have about six at Menlo Park waiting to be determined and we have about fifteen others waiting to be canvassed. Eaton's inclination was to stop. I expressed what I knew were Edison's views and suggested that the agents should pile in all the towns that they could and we would find some way of taking care of them. We are in a position to deal with this central station work very rapidly, the only weak point being Claudius' determining.[4] As soon as Edison gets back from Chicago he will put this right I think. I think that we can erect fifteen or twenty stations simultaneously which means to carry about $500,000 to $750,000 worth of work at once. The building of the station is a small question, the great point is to get them started and leave experts in charge of them for a month to work them and to teach local employees how to do our business. Now Edison wants you here to take charge of these experts, in fact to start the stations, and I think by the time you get home there will be piles of this kind of work for you to do. I cannot write any more just now but will do so to-morrow. Very truly yours,

TL (carbon copy), NjWOE, DF (*TAED* D8316ABK).

1. Johnson had alerted Edison to Sprague's financial predicament (see Docs. 2433 n. 2 and 2456 n. 1), but the editors have not identified the terms of his employment.

2. Edison's cable is Doc. 2465. Johnson sailed on the *Alaska* on 7 July. See Doc. 2467; Samuel Insull to Charles Batchelor, 16 July 1883, DF (*TAED* D8316AGT).

3. The eponymously titled short story about his dog, "Entomologist," appeared in an 1882 edition of works by American humorist and journalist Bill Nye. Nye 1882, 198–99; *ANB*, s.v. "Nye, Bill."

4. Electrical engineer Hermann Claudius, hired by Edison in December 1880 to create a scale working model of the distribution system in the New York First District, now supervised a small staff at Menlo Park making similar determinations for village plants under conditions of strict secrecy. With a reputation for "meticulous fidelity and math-

ematical precision," Claudius was considered to be painstakingly slow. Edison heard at about this time that the men of the Determination Department tended "to loaf during the day and work at night, and so get double pay"; they reportedly had spent one recent afternoon out "gathering wild strawberries." An unspecified "conference" at Menlo Park on 18 June led, according to Insull, to "new methods of determining the conductors which will very materially expedite matters." Docs. 2028 n. 2 and 2337; TAE to Charles Clarke, 18 June 1883; Insull to Spencer Borden, 23 June 1883; both DF (*TAED* D8316ABC, D8316ACJ).

–2467–

From Edward Johnson

[London,] June 20/1883

My Dr Edison[1]

I have not been writing you lately as I have been in daily expectation of getting things into such shape that I could write and say everything was definitely arranged—but since that point has not yet been attained and you appear to be anxious to know how matters stand I'll postpone it no longer—

A few hours before the annual meeting[2] was to take place Bouverie[3] & Wade[4] met and arranged a compromise— The meeting was then held and the report read and endorsed   The Old Board of Directors re-elected and so far as surface appearances were concerned everybody was unanimous that everybody else was all that could be desired— This for the benefit of the Public— The terms of the compromise were to be subsequently carried out in the following manner— The Board were to meet. Bouverie and Moore[5] (lately made a Director) were to resign— 2 of our men were then to be elected to fill the vacancy and <u>One</u> new Director was to be added (by us) to make up a Board of 7. This constituted the Board thus—4 Old Directors 3 New—Sir John Lubbock[6] was then to take the chair for 3 months—when he insists upon retiring— <u>We</u> then elect his successor—thus giving us 4 and they retaining 3— As Sir John Lubbock was chosen by us in any event to remain a member of our New Board— his taking the chair[7] was practically the same as our putting our own man in at once— Especially in view of his doing so upon the understanding that the Post had been conceded to us by the terms of the compromise and our taking absolute possession was only deferred for 3 months to keep up the appearance of continuity in the eyes of the Public— (This fear of the Public is at the bottom of the weakness of our English friend)—

In addition to the above— We were to have the Appointment of the General Manager—White to be retired from that position= This then was practically a Complete surrender to us & we were bound to so consider it and express our satisfaction by our action of the General Meeting—which we did— thus fulfilling Our part of the understanding— Now we are awaiting their fulfillment[a] of theirs— This is how it stands at present—

Bouverie and Moore have resigned   Leyland[8] and Stewart[9] have been elected in their stead— The other Director is chosen and will take his seat as soon as the Qualification clause of our By Laws can be amended by reducing the required holding from 500 shares to 100— Our man will not buy 500—and as all concede it is too high—no one objects to his action, but agree to reduce it— This is a work of time—42 Days— Meantime he sits at the Board and takes part in the discussions—but is not allowed to vote—

Sir John has taken the Chair—so you see the present Constitution is as follows—

Chairman Sir John Lubbock
Director Shelford Bidwell[10]
"   Arnold White
"   Lord Anson[11]
"   Fredk R. Leyland
"   John Stewart
"   W. C. Watson—[12] (not yet voting)

Mr Stewart only occupies the position for a Mr Scoble[13] who, like Mr Watson is in attendance but is not qualified to vote— Mr Stewart will give way to him when the qualification matter is adjusted— Thus you see that my new forces are there— but as yet can only exercise their influence—not their voting power—at least not in full— However the arrangement works to the extent of my getting a proper hearing upon all the matters I bring up—which I could not get[a] before= Right here however comes in the difficulty which I am waiting to have removed—

White has not yet been replaced and until he is—He holds the reins of office and as you can easily understand uses them for the purpose of frustrating my every attempt to incorporate improvements or new methods of doing business— This friction has become so great that I have finally put it to the Board that they are not quite on the lines of good faith with us in procrastinating the question of substituting a new G.M. For Mr Arnold White— This has developed the fact which I began to suspect—that A.W. is playing his cards to hold the office in spite of the Agreement— As he has said to more than one was his intention—you can imagine the game he is playing and the cards that are in his hand to play it with— He misrepresents everything I do or say—and doing it behind my back I have no defence— This puts an effectual check to all my proceedings—of course I am not content to depart these shores without at least giving them a foretaste of what a little well directed effort can do—so I have asked that a committee be appointed to confer with me as to the methods of business procedures to be hereafter adopted— This Com. will be appointed tomorrow

I have further asked through my own Directors that the Board carry out at once the Gen'l Manager feature of the agreement   This they say they will ask the Board to do at once or explain their tardiness— There is no doubt this pressure will bring about the desired action   Or if not to put a new manager in at once—to confer upon me temporarily the power of the office—in which case I will put some life into the concern in short order— I have a lot of inquiries for Isolated Plants. I carry them on to a certain point then have to turn them over to White—who of course is only too glad to see them die, thus to demonstrate that his contention that there is no real demand for Isolated plants is correct—[14] I hear & know enough of his methods however to know that he makes no effort to secure them and further I know that among them there are a number who would close with me if I had the power— I spent 3 hours today in talking Spiers & Pond[15] into a 600 Light Plant for the Criterion Theatre[16] and although the cost is far in excess of what others offer to do it for—Spiers[17] told me today If I would stay & do the work myself He would sign the contract— I told him that was impossible & showed him your Telegram[18] about the 5 Central Stations Building & 18 others ready to sign contracts then showed him my "V.P." Card[19] & asked him If it was a reasonable request— He thought not— but said he had been so damnably humbugged by the 'E.L. Companies' that he wouldn't trust any of them—Our own among the number   However I finally persuaded him to accept Hopkinson in^a lieu of me   Now the matter only remains for final action upon the detailed Estimate I am to obtain from the office tomorrow— White will of course try his old game of coming in between me & S&P but this time I have made up my mind to make the matter an issue before the Board and shall probably bring on a crisis by insisting upon being vested with authority to complete the transaction—

Now as to the Other side of the Ledger for of course there is another side in this as in all transactions

Early in the canvass of affairs I replied to the criticism upon you—that you had no real care for the prosperity of the concern—by saying that when they showed a little more disposition to proceed to help themselves they would find you ready to do your part— This cropped up again later in the day when I was trying to secure new men they one and all said as at present Constituted our Company must fail and unless a radical change was made in the agreement with you—they would not join the concern or put a penny in it—[20] This was also the Opinion of Wade—and in fact of every one—even outsiders who have no interest at all in the Company   Finding this view so omnipotent I yielded to it and expressed the opinion that you would do the right & fair thing If they would join the Board and vigorously set about putting the Company upon its legs.[21] Fearing I was taking too much upon myself I consulted several Bankers upon the financial question explaining to them the terms of the contract with you   They were all unanimous to the effect that no money could be obtained from the Public upon terms permissable under it— J S Morgan & Co[22] concurred in this. I thereupon put the question to Pierpont Morgan[23] who happened to be here at the time and he said I had done quite right or something to that effect— The negotiations have therefore gone on upon that understanding— Now that we are beginning to get into working shape— my new people have shown that they grasp the business by the right end, in demanding that the financial question be dealt with first and all minor ones follow—since the minor ones can only be dealt with upon the basis of cash in the Till—and we are at present in debt to our Bankers some 10,000£

Everytime the discussion turns this way up comes the Query How. The answer is always the same—"First get Edison to relieve us of the Burdensome Character of our Contract"— Leyland—the longest headed[24] of the lot, is foremost in pushing this matter & insists that, as upon it everything else depends—it shall be at once adjusted— In this I am bound to concur—but am insisting now that they give me formal assurance that my recommendations be carried out otherwise I will not ask you to make the required concessions— To this I am to have my final answer at the next Board—in a day or two— As soon as we thus exchange undertakings I am to sail for New York to explain matters fully to you and induce you to give them what they ask— This will probably be all arranged in a few days and I go out by the Alaska which sails July 7th[25]   However you will learn by Telegraph of my coming before you get this= I am not now going to broach to you the proposed alterations in the contract   I do not care to knock you down till I am by to pick you up— Beside I want to get everything in hand before I have any indication from you—or rather before <u>they</u> have—that you will or will not meet them— I have therefore informed them that you cannot be dealt with in the matter by mail or Telegraph that your understanding of the situation is too meagre to enable you to form a proper judgment— All of which is of course perfectly true— Nor am I going into the question of Amalgamation as invited by a remark in[a] Insulls last letter— recd today.[26] All that will be better done when I see you—

The Central Station business[27] will now receive such attention that it is very probable a site will be selected within a few days or weeks and work begun at once—The intention being to have at least a portion of it in operation this fall—

The provisional order bill will probably be passed by Parliament in a few days—it is now before them— This will be the signal to begin—and the men I have selected are all pledged to give this branch of the business their principal attention— The money will be raised for it—and if my recommendation concerning it is carried out as I am sure it will be—Hopkinson will come out to see you upon the whole subject ere beginning work— He has some radical changes to propose—but wishes your opinion on them first— Hopkinson has lent me Strong support in my fight against White and Sir John Lubbock is largely guided by his opinions—so we owe him our cordial good will—

This Central Station will be the largest yet undertaken or proposed— It includes the whole of the Parishes of St James St Martins in the Fields and a large part of the Strand—.[28] It is considered to be the Plum of E.L. in the World— It includes about a Dozen Theatres—more than that no. of Large Hotels—All the Pall Mall Clubs—and in brief covers what is known as "the dissipated District of London that is— where the people never go to bed & consequently use Light all night— Trafalgar Square is its centre— The Strand Regent St. Piccadilly & such like streets are within its boundaries   The fight for this District has been great— but we have won and much credit is due Waterhouse[29] as it has been he who piloted us through the nastiness of the Parish Local Boards— The Dogmatic decrees of the Board of Trade[30] and the Ignorance of Parliament   A gamat a thousand fold worse than the New York Board of Aldermen[31]

But All this will keep— What is wrong with Mr Fabbri, has he taken affront at anything I have said or done?[32] I have never heard a word from him since I left New York—not even an acknowledgment of my Letters

This job has been the toughest and the most unsatisfactory of any I have yet undertaken, I shall finish it with as much credit as I can   Then If you or any one else ever get me to come this side of the Atlantic again it will be simply because you discount my great expectations & put the hard cash in Bank to my credit. No more 5% for me—no nor 25— Nothing short of the hard stuff— I've had plenty of Glory—but where is the sums over & above 100,000$ which my contract so carefully provided for a proper division of?[33]

Echo answers where:[34] Yes and will answer until my children are old enough to hear and understand the cry—

I am glad to have so good & cheering news from you as your last cable and am only too anxious to comply with your request to hurry home—but a finish is essential to all work—and I must put one on this last transatlantic job of mine. From what I hear I imagine some injury is being done me by a too free and unauthorized use of my name by some of the home malcontents[35]   You must guard me against this   I cannot be in two places at once— However I'm not much concerned about it—

One more thing= If Insull should badly need for me[b] a few hundred Dollars ere I get home and Bergmann is unable to accommodate him—do what you can to help him and I will recoup you somehow on my[a] return—

I've got a Dozen of Swans new High Resistance Lamps[36] which I am bringing you for Testing— If there is anything else you particularly want cable me on receipt of this and I will bring it— Very Truly Yours

Edward H. Johnson

My pious regards to everybody including our "Double Headed Dutchman"[37] and the new Illuminant Chinnock

ALS, NjWOE, DF (*TAED* D8338ZAO). [a]Obscured overwritten text. [b]"for me" interlined above.

1. Johnson addressed this letter to "Thomas A. Edison Esquire Electrical Prestidigitare."

2. Soon after the Edison Electric Light Co.'s second general stockholders' meeting on 30 May, an agreement was drafted to consolidate with the Swan United Electric Light Co. Edison Electric Light Co.,

Ltd., to TAE, 15 May 1883, DF (*TAED* D8338ZAC); see Docs. 2456, 2483, 2431, and 2514.

3. Edward Pleydell Bouverie (1818–1889), chairman of the Edison Electric Light Co., Ltd., remained invested in the Edison & Swan United Electric Light Co., Ltd. He was also chairman of the Corporation of Foreign Bondholders, a Privy Councillor, and former member of Parliament for Kilmarnock. Docs. 1765 n. 2, 2228 n. 23; "Electric Lighting," *Times* (London) 23 Apr. 1883, 6; "This Morning's News," *Leeds Mercury,* 14 Jan. 1881, 5; lists of stockholders are Summary of Share Capital & Shares in the Edison Electric Light Co., Ltd., and Edison & Swan United Electric Light Co., Ltd., 13 June 1883, 11 Nov. 1884, and 14 Aug. 1885 [1926], CR (*TAED* CF001AAO, CF001AAP, CF001AAQ).

Bouverie, with Arnold White, Theodore Waterhouse, and representatives of the Swan company, had met in April with Joseph Chamberlain, who as president of the Board of Trade had authority to administer the Electric Lighting Act of 1882. The group pleaded unsuccessfully for a reduction in the amount of capital that the Board required lighting companies to pledge as a bond. "Electric Lighting," *Times* (London) 23 Apr. 1883, 6.

4. Richard Blaney Wade (1820–1897) was chairman of National Provincial Bank of England, Ltd., bankers in the £500,000 first issue of 100,000 shares for the Swan United Electric Light Co., Ltd., in 1882. He was also a founding director of the Edison Electric Light Co., Ltd., and had resigned as a director some months earlier, though he continued to hold stock in the combined company. Obituary, *Times* (London), 31 July 1897, 11; Cassis 1994, 56, 57; "Advertisements and Notices," *Pall Mall Gazette* (London), 23 May 1882, 13; Edison Electric Light Co., Ltd., statement, 25 May 1883; Waterhouse to TAE, 7 Dec. 1883; both DF (*TAED* D8338ZAF, D8338ZCA); see note 3.

5. Michael Miller Moore was recently elected to the board of the Edison Electric Light Co., Ltd., and he remained invested in the Edison and Swan United Co. (Edison Electric Light Co., Ltd., statement 25 May 1883, DF [*TAED* D8338ZAF]; see note 3). An American, Moore had been a director of the Edison Telephone Co. of Glasgow, Ltd., and an investor in the Edison Telephone Co. of London, Ltd. His business interests also included compressed air traction, mining companies, and later, mortgage banking. Samuel Insull later characterized him as "a gentleman of discretion" (Doc.1854 n. 5; Insull to Upton, 9 Nov. 1886, Lbk. 23:82 [*TAED* LB023082]; "Meetings of Public Companies," *Daily News* [London], 26 Feb. 1881, 7; "Public Companies," *Times* [London], 12 Nov. 1886, 14; "Public Notices," *Freeman's Journal and Daily Commercial Advertiser* [Dublin], 18 July 1891, 2).

6. Sir John Lubbock (1834–1913), first Baron Avebury, was head of the banking house of Robarts, Lubbock & Co. and a member of Parliament. He was also as a founding member of the X Club, an influential dining club of Victorian gentlemen and scientists who met monthly. An original investor in the Edison Telephone Co. of London, Lubbock had become actively involved with Edison's lighting interests in Britain after an 1881 conversation with Egisto Fabbri. A founding director of the Edison Electric Light Co., Ltd., Lubbock was due to retire from the board by rotation at this meeting. Doc. 2228 n. 21; Barton 1990, 54–55;

Lubbock to TAE, 8 Dec. 1881; Edison Electric Light Co., Ltd., statement, 25 May 1883, DF (*TAED* D8133ZBC, D8338ZAF).

7. Lubbock ultimately declined to remain chairman because it would "require more time & thought than I could possibly give." Lubbock to TAE, 26 Oct. 1883, DF (*TAED* D8338ZBV); see note 3.

8. An important patron of Aesthetic Pre-Raphaelitism, Frederick Richards Leyland (1831–1892) was managing owner of the Leyland Line, a Liverpool steamship navigation company that served Boston, the Mediterranean, and the Black Sea. He was a founding director and deputy chairman of the Edison & Swan United Electric Light Co., Ltd.; in 1888, he became a director of the United Telephone Co., Ltd. *Oxford DNB*, s.v. "Leyland, Frederick Richards"; Obituary, *Liverpool Mercury*, 6 Jan. 1892, 6; "The Merchant Shipping (Grain Cargoes) Bill," ibid., 6 July 1880, 6; "Trade and Finance," *Leeds Mercury*, 26 Apr. 1888, 4; Edison & Swan United Electric Light Co., Ltd., annual report, 21 July 1885, DF (*TAED* D8534M); see also n. 2.

9. John Stewart of the National Provincial Bank of England, Ltd., also held shares in Edison and Swan United Electric Light Co., Ltd. (see note 3).

10. Shelford Bidwell (1848–1909) was a barrister with extensive scientific interests who had been a juror at the 1881 Paris Exposition, where he exhibited his telephotographic instrument for the electrical transmission of still pictures. He had more recently been investigating microphonic contacts. Bidwell had made a thorough review of Edison's British electric lighting patents at the end of 1881. He was due to retire from the Edison company at this meeting, but he remained financially interested in it and became a director of the Edison and Swan United Electric Light Co., Ltd.; Doc. 2211 n. 11; *Oxford DNB*, s.v. "Bidwell, Shelford"; Wood, 1882, 756–59; Benjamin, 1886, 264; "Recent Telephonic and Microphonic Researches," *Electrician* (N.Y.) 2 (May 1883): 126; Edison Electric Light Co., Ltd., statement, 25 May 1883; Waterhouse to TAE, 7 Dec. 1883; both DF (*TAED* D8338ZAF, D8338ZCA); see also note 3.

11. Lord Thomas Francis Anson (1856–1918), the third Earl of Lichfield, was a founding director of the Edison Company; he became a director of the Edison & Swan United Electric Light Co., Ltd.; he retained that position until his death. Also an original investor in the Edison Telephone Co. of London, Anson remained closely involved with its successor for many years. Docs. 1799 n. 16 and 2231 n. 3; Edison & Swan United Electric Light Co., Ltd., annual report, 21 July 1885; Waterhouse to TAE, 7 Dec. 1883; both DF (*TAED* D8534M, D8338ZCA); Bassett 1900, 139.

12. William Clarence Watson (1844–1906) was a London merchant and Ottoman vice-consul who remained invested in the Edison & Swan United Electric Light Co., Ltd. Obituary, *Times* (London), 10 Feb. 1926, 11; "Partnerships Dissolved," ibid., 10 January 1883, 7; "The Famine in Asia Minor Relief Fund," ibid., 26 Apr. 1875, 10; see note 2.

13. Andrew Richard Scoble (1831–1916) remained invested in Edison & Swan United Electric Light Co., Ltd. A member of the Queen's Council since 1876, Scoble was advocate general for Bombay (1872–1877) and became a Privy Councilor in 1901. His other directorships included the Oriental Bank Co. and the Commercial Union Assurance

Co. See note 3; "Biographies of Candidates," *Times* (London), 24 Nov. 1885, 3; "The Judicial Committee of the Privy Council," ibid., 4 Dec. 1901, 9; "Death Of Sir A. Scoble," ibid., 19 Jan. 1916, 4; "Railway and Other Companies," ibid., 5 June 1879, 7; "Advertisements & Notices," *Pall Mall Gazette* (London), 30 Nov. 1883, 13.

14. Arnold White had recently sent Edison his own analysis of the British company's situation. Much of it mirrored Johnson's, particularly regarding the terms of Edison's contract, the difficulty of raising capital, and the deep distrust between the two men. White, however, expressed himself as "dead against isolated business in England, except ships." He described the intense competition for this type of business, even at a loss to the companies involved, and argued that:

> The possession by us of £30,000 worth of machinery is another element in our case. I am quite sure you would help us if you could. We canot sell, because folks won't buy. If we rent we must first spend four sovereigns for every thirty shillings we have invested in dynamos, & that means raising £100,000 for isolated business. Moreover folks won't rent plant for more than a three year term at most so that after spending the money necessary to put up an installation, we either have to charge a prohibitive rent or run the risk of heavy loss at the expiry of the term, by having unsaleable plant thrown back on us.

Pointing out that Johnson had failed to sell any isolated plants himself, he ridiculed Johnson's breezy

> American teaching "Sell light & nothing but light." I answer, "So we will where we can—that is in those areas we have acquired by Provisional Order, but in those many cases where other folk have the area we must either sell machinery & lamps or keep out altogether." Now this is no matter of opinion, it is a plain matter of fact—that where we have no Parliamentary rights we cannot exact royalties on light or horsepower. If we attempted to do so, there are plenty of folks springing up, who have learned enough from you, Mr Edison, to undercut us, & yet patch along somehow. [White to TAE, 15 June 1883, DF (*TAED* D8338ZAL)]

15. Spiers & Pond, Ltd., restaurant and catering entrepreneurs, operated railway refreshment rooms throughout the United Kingdom and numerous dining establishments around London. The original partnership ceased to exist in 1881, with the death of Christopher Pond, but the business continued as a privately capitalized, limited partnership under the direction of its surviving cofounder, Felix William Spiers (1832–1911). Dickens 1998, 108 n. 1; Thorne 1984, 240–241, "Advertisements & Notices," *Pall Mall Gazette* (London), 1 Apr. 1882, 16.

16. The Criterion Theatre in Piccadilly, designed by architect Thomas Verity, was built beneath the Criterion Restaurant and opened in 1874; it was owned by Spiers & Pond, Ltd., the proprietors of the restaurant as well. The Metropolitan Board of Works condemned the theater in November 1883 on grounds of fire safety, and Edison lighting was installed throughout during the subsequent reconstruction. ("Occasional Notes," *Pall Mall Gazette* [London], 4 Nov. 1882, 3; "Criterion

Theatre," *Teleg. J. and Elec. Rev.* 14 [1884]: 337; Tidd 1889, 460–61; Sheppard 1960, 29:254–56). The proprietors' insistence that the lighting plant "should be <u>noiseless</u>," or at least produce "less noise than the Houses of Parliament Installation" led Arnold White to wonder how two engines could operate more quietly than one. Charles Batchelor was expected to design a regulator for the theater. The plant began operating in April 1884 (Spiers & Pond to Edison Electric Light Co., Ltd., 11 July 1883; White to Johnson, 12 July 1883; H. S. Trehearne to TAE, 14 July 1883; all DF [*TAED* D8338ZAT, D8338ZAV, D8338ZAW]); see Doc. 2679 n. 1.

17. Felix William Spiers; see note 15.

18. Doc. 2465.

19. Johnson was elected vice president of the Edison Electric Light Co. in October 1882. See Doc. 2356 esp. n. 9.

20. In addition to an initial cash payment, Edison was to receive one £10 B share for each £10 A share (which conferred greater voting power and preferential dividends) actually paid in to the company (see Doc. 2231 esp. n. 2; TAE agreement with Edison Electric Light Co., Ltd., 31 Mar. 1882, CR [*TAED* CF001AAF]). In early 1880, he had renegotiated a similarly favorable contract with Bouverie and the Edison Telephone Co. of London when that firm could not raise capital to head off a crisis (see Docs. 1903, 1908, 1918 esp. n. 3, and 1928).

21. See Docs. 2483 and 2493.

22. J. S. Morgan & Co., an American private bank at 22 Old Broad Street in London, was founded in 1864 by Junius Spencer Morgan. It had provided numerous banking services to the Edison Electric Light Co., Ltd., and was negotiating on behalf of the Edison company's interests in the amalgamation with Swan. Burk 1989, 16, 24, 50–51; see also Doc. 2493.

23. John Pierpont Morgan (son of Junius; 1837–1913) was the senior partner of Drexel, Morgan & Co., which controlled Edison's lighting patents in Great Britain (see Doc. 1649) and whose principals, under Morgan's direction, had given crucial financial support to Edison's research. Morgan apparently had sailed from New York on 5 April and began his return voyage from Liverpool on 24 May. See e.g. Docs. 1494 n. 4 and 1671; J. P. Morgan to Walter Burns, 27 Mar. 1883; George Knight to J. Duncan & Sons, 30 Apr. 1883; both J. Pierpont Morgan Letterbook 2:329, 340; NNPM.

24. That is, having great foresight or discernment. *OED*, s.v. "long-headed" 2.

25. Johnson did so, reaching New York late on 15 July. Samuel Insull to Charles Batchelor, 16 July 1883, DF (*TAED* D8316AGT).

26. See Doc. 2456.

27. Edison had long hoped to establish a permanent central station in London; see Docs. 2197, 2292, and 2374.

28. In its Electric Lighting Orders Confirmation Act of July 1883, the Select Committee of the House of Commons confirmed the St. James' and St. Martin's (London) Electric Lighting Provisional Order, which authorized the Edison Electric Light Co., Ltd., to erect and maintain electric lines and works, and to supply electricity within the Parish of St. James, Westminster, the Parish of St. Martin in the Fields, and portions of the district of the Strand Board of Works (United King-

dom. Parliament. House of Commons. 1883a [224], 233). These areas were included in a long area of proposed Edison service, along the arc of the Thames roughly between the Westminster and London Bridges, outlined on a large London map prepared for the Edison company's application. Acc. no. CLA/047/LC/04/264, CPF, LMA.

29. Theodore Waterhouse, solicitor of Waterhouse, Winterbotham & Harrison, was a founding director of the Edison Electric Co., Ltd., as well as of the Edison Telephone Co. of London, Ltd. He had helped carry out reviews of Edison patents owned by both companies and guided modifications through the legal process. The Waterhouse family, including Theodore and his notable brothers, accountant Edwin Waterhouse and architect Alfred Waterhouse, carried substantive investments in the Edison Electric Light Co., Ltd., as well as the Edison & Swan United Electric Light Co., Ltd. See Docs. 1799 n. 5, 1825, 1829, 1870, 2187 n. 20, 2231, 2270; *Oxford DNB*, s.vv. "Waterhouse, Edwin" and "Waterhouse, Alfred"; Waterhouse 1894, 86–87; see also nn. 2–3.

30. The Board of Trade had authority under the Electric Lighting Act of 1882 to grant licenses and provisional orders related to the supply of electricity for lighting throughout Great Britain and Ireland. With Parliamentary oversight, the Board could alter or repeal any rules pertaining to applications or related matters. Higgins and Edwards 1883, 1, 12–13, 17.

31. Edison had personally joined in the effort in 1880 and early 1881 to persuade New York City's Board of Aldermen to grant the Edison Electric Illuminating Co. of New York authority to lay its conductors under the streets. See Docs. 2035, 2037–38, 2039 esp. n. 2.

32. A partner of Drexel, Morgan & Co. since 1876, Egisto P. Fabbri (1828–1894) had extensive involvement with Edison's global electric lighting ventures. The founding treasurer of the Edison Electric Light Co., Fabbri was a trustee for the sale of Edison's patents in Sweden, Norway, and Britain, and had worked with Johnson to set up the Holborn Viaduct station and help organize the Edison company in London. He also had helped manage Drexel, Morgan & Co.'s interest in the Edison Electric Light Co. of Europe and had arranged for control of Edison's patents in much of South America by Fabbri & Chauncey, the shipping and commission house he led until 1876. Letterhead, Edison Electric Light Co. to TAE, 19 Nov. 1878, DF (*TAED* D7820ZBV); Docs. 1566, 2135, 2144 n. 3, 2161 n. 4, 2178, 2180, 2190, 2197, 2211, 2339 n. 7; Hall 1895–96, s.v. "Fabbri, Egisto Paulo."

Johnson may have recalled the opinion expressed by Edison in Doc. 2187 that Fabbri "is entirely too sensitive. He takes as slights things that are never intended as such draw wrong infrences and misunderstands motives. I think he is one of the nicest men to get along with I ever knew but is so fearfully sensitive." Fabbri may well have been preoccupied with other business, such as financing the Northern Pacific railroad system. Also, Fabbri & Chauncey, headed by his brother Ernesto G. Fabbri from 1876 until his death in July 1883, was at this time buckling under competitive pressure from W. R. Grace; the firm was dissolved in 1884. Drexel, Morgan & Co. to J. P. Morgan (in London), 1 May 1883, Letterbooks, PML; "The Annual Meetings," *NYT*, 8 June 1883, 2; James 1993, 187; Lemann 1999, 12.

33. The editors have not determined the provisions of this agreement.

34. Johnson was probably referring to the penultimate stanza of "The Bride of Abydos" by George Gordon (Lord) Byron, which concludes (27, ll. 662–63) with the lamentation for a dead daughter:

Hark—to the hurried question of Despair!
"Where is my child?"—an Echo answers—"Where?"

35. See Doc. 2456.

36. Charles Batchelor tested "Type Swan new model" lamps in Paris at the end of June, when he measured the resistance of 16-, 20-, and 25-candlepower lamps at 96 to 99.7 ohms; they required roughly 100 volts. These characteristics may date to the end of 1882, when the Swan United Co. began manufacturing a 100-volt 100-candlepower lamp. The editors have not determined the resistance of previous Swan commercial lamps, but a report from the 1881 International Exposition at Paris indicated that they fell within the wide range of 30 to 100 ohms. Until at least that time, Joseph Swan strove to increase the brilliance, not the economy, of his lamps. Cat. 1235:7 (item 11), Batchelor (*TAED* MBN0120007); Batchelor to TAE, 2 July 1883, DF (*TAED* D8337ZCP); "Swan Lamps," *Electrician* 10 (2 Dec. 1882): 49; Heap 1884, 194; Swan 1881, 284.

37. Johnson often referred to Sigmund Bergmann, a native of Thuringia (incorporated into Germany in 1871), as "Dutch," meaning German or Teutonic. *OED*, s.v. "Dutch."

–2468–

To George Bliss

[New York,] June 23rd. [188]3

Dear Sir:—

Will you please ascertain the exact truth concerning the removing of Maxim lamps in Chicago every morning and the replacing of them with new ones.[1]

I have instructed my canvasser,[2] who is now working in St. Paul and Minneapolis, when he gets through to go first to Davenport, then to Appleton and then to Milwaukee.[3] All these canvasses should be finished inside of a month. The cost of canvassing and determining the conductors at Menlo Park is charged to the local people through our estimate if the scheme goes through, but if the company is not[a] formed, the actual cost will be charged to your Company.[4] Exactly what it would be I cannot say at the moment as we are making considerable changes at Menlo Park, in order to cheapen and expedite the work.[5] I should imagine, however, that in a place like Davenport the completed canvass map & tender[b] would cost about $200.

Will you please have diligent inquires made all through your territory as to the number of Maxim plants, both arc and incandescent, how the owners of them like them, also what plants have been thrown out and the cause of same? I desire to get this information, as we are compiling a history giving every detail with relation to the lights of our opponents.[6] Very truly yours,

TL (carbon copy), NjWOE, DF (*TAED* D8316ACI). [a]Interlined above by hand. [b]"map & tender" interlined above by hand; insertion point after "would" altered for clarity.

1. Bliss answered that he could not "give you the exact truth in regard to the removal of the Maxim Lamps, every morning. . . . Our men were not placed as spies upon that Co., but they understood it to be a fact during the latter part of the [Railway] Exposition, that the Maxim lamps were changed daily." He reported that the Maxim company had replaced thirty-two lamps on the last day of the exhibition, of which twenty-seven failed by day's end. Edison asked for an affidavit to this effect and urged Bliss "not be so pious about getting at the truth. . . . We want every scrap of information about them. It must be simply the naked truth, as we only propose to make statements that we can stand by." Bliss sent sworn statements in July, but these have not been found. Bliss to TAE, 25 June and 2 July 1883; TAE to Bliss, 28 June 1883; all DF (*TAED* D8364X, D8364ZAD, D8316ADQ).

2. Charles H. Cooke (1847?–1914) started canvassing for Edison in June 1883 in St. Paul. He expected to finish there on 9 July and three days later was in Davenport, after which he had instructions to go to Appleton and Milwaukee. A civil engineer, Cooke had worked on construction of the Union Pacific Railroad; at the time of his death, he was involved in a long-term project to build a hydroelectric dam on the Delaware River. Cooke to Samuel Insull, 18 June and 7 July 1883; Insull to Cooke, 23 June 1883; all DF (*TAED* D8342A, D8342P, D8316ACY); Insull to Cooke, 12 July 1883, LM 14:141B (*TAED* LBCD1141B); Obituary, *NYT,* 3 Aug. 1914, 11.

3. According to the *Davenport Gazette*'s interview of Bliss, the local gas company, prompted by the prospective formation of an Edison light company, proposed to build its own electric light central station. Bliss called its plan to generate steam from coal gas a "queer proposition." "Edison's System," *Davenport Gazette,* 22 May 1883, in *Edison Electric Light Co. v. U.S. Electric Lighting Co.* (1885–1892), Lit. (*TAED* QD012G4421).

4. Bliss represented the Western Edison Light Co. The expense of making the canvass, map, and electrical determination for each plant was often several hundred dollars (see also Doc. 2478). The sum of this preliminary work on approximately eighty prospective stations exceeded eleven thousand dollars. Edison Construction Dept. statement, 25 Mar. 1884, DF (*TAED* D8441I1).

5. Edison may have been thinking of the mathematical methods of calculating the feeders and mains, which he adopted in place of the time-consuming construction of physical models sometime after the middle of August. Samuel Insull to Charles Batchelor, 14 Aug. 1883, DF (*TAED* D8316AOA5); see also Docs. 2466 and 2575.

6. See Doc. 2443 esp. n. 2.

From Spencer Borden

Boston, June 26 [24?]th[1] 1883.[a]

Dear Sir:

Your favor of yesterday duly at hand, also telegram from Insull this A.M. about Canvasser.[2] The only reason I have not Companies started in the various towns already worked up, is that I have no actual figures from which to work, the estimates are still in your hands. Brockton, Lawrence and Lowell are the only estimates I yet have, and we have organizations in all of those places.

As to your putting a man in some of the N.E. Dept. territory, my only objection is that I have to pay him, either with salary or division of Commissions, and I believe my Commission of 5% is the lowest of any arrangement the Company has with any of its Agents, others being from 15% to 50% of the Company's allotment of stock. My contract is made so I make no point of this, only it is too small to be divided.

Further, we are working this territory for all any one Can, until one or two stations start, then we can give you all the business of that kind that You Can attend to. We are working this whole business in all directions. We do not now need any more force on the ground, and it would be no advantage to incur the additional Expense. Give us the figures as fast as we put in Canvasses, and we can keep you busy. I was sorry not to see you Monday last, but may be down again in a few days.

When Can you give me the figures for Fall [River], Portland, & Haverhill?[3] Yours very truly

Spencer Borden.

⟨we do not have an agent that gets over [-][b] ~~20 p~~ 20 percent & 90 per cent of the agents get only 10 percent[c]

why not go into a town & get parties together [~~stating that if?~~][b] & ask them if they will put up the money If you make a canvass & tender & present the same & can prove it to be a good investment— If they agree to this then send where & we will make the Canvass=

When[d] I spoke about agents going into your territory I of course thought you would get your 5 pct & not have to pay anything out but merely do all you could to help the agent— Fall River is being determined= Portland & Haverhill is being mapped— E⟩[4]

ALS, NjWOE, DF (*TAED* D8340ZAM11). Letterhead of Edison Electric Light Co. and Edison Co. for Isolated Lighting, New England Dept.; Spencer Borden, manager. [a]"Boston," and "188" preprinted. [b]Canceled. [c]Followed by "over" to indicate page turn. [d]Obscured overwritten text.

1. Although Borden dated this letter 26 June, Edison replied to it on 25 June, to which Borden responded the following day in Doc. 2472. TAE to Borden, 25 June 1883, DF (*TAED* D8316ACR).

2. Insull's telegram has not been found. Edison's letter may have been the one about his decision to hire George Wilber, who until recently had been canvassing in Massachusetts, to continue doing so in New England, including areas under Borden's agency. Borden to Samuel Insull, 19 June 1883; TAE to Borden, 23 June 1883; Insull to Borden, 23 June 1883; all DF (*TAED* D8340ZAK, D8316ACL, D8316ACJ).

3. See App. 2.

4. Edison's notes were the basis for a typed reply (see note 1).

–2470–

To Preston Hix

[New York, June 26, 1883[1]]

~~Preston Hix~~ W. Preston Hix[2]

Endeavor to prevent Arc Lights of any kind being used inside   it will spoil the effect which will be beautiful. Arc Lighting there[a] is too much like a Bowery Show[3] entirely in bad taste. Maxim Co run[b] their lamps up[b] above candle power to mislead the public & then Every morning before opening changed the blackened lamps[4]   We run our lamps at [~~their?~~][c] the candle power we guarantee purchasers.[5]

Edison

ALS (telegram), NjWOE, DF (*TAED* D8366C). Message form of Western Union Telegraph Co. [a]Interlined above. [b]Obscured overwritten text. [c]Canceled.

1. Edison evidently sent this undated message in response to Preston Hix's telegram this day stating that the contract for the Louisville electrical exhibition was "settled as we want it except arc companies in building   [James?] brown will call to see you or [Miller] moore & settle that point." Edison elaborated on his reply in a 26 June letter (see note 5). Hix to TAE, 26 June 1883, DF (*TAED* D8366D); see also Doc. 2465.

2. A native of South Carolina and a former Confederate Army captain, William Preston Hix (1836–1911) made a career of photography and portrait painting before moving to New York City in 1882 and becoming a general agent and electrical contractor. Hix represented Edison's interests in Kentucky and Virginia at this time; he remained associated with Edison lighting for many years. Obituary, *NYT,* 24 Oct. 1911, 13; Chappell and Chappell 1983, 278; "Personal Notes," *The Art Amateur* 6 (Feb. 1882): 68; Teal 2001, 149; Hix to TAE, 22 May 1883, DF (*TAED* D8325ZAE).

3. A mile-long street in lower Manhattan, the part of the Bowery to which Edison refers was at this time a center of cheap entertainment peppered with popular theaters, music halls, and beer gardens where patrons were "dazzled by the lights." *Ency. NYC,* s.v. "Bowery"; "Glance at the Bowery," *NYT,* 28 Mar. 1880, 10.

4. See Doc. 2468.

5. Edison sent Hix a letter on 26 June specifically to explain that "We understand business of lighting better than anyone else and we know if arc lights are put in building the general effect of the main lighting will be spoiled. It will be a great mistake to permit Arc lights as they so contract the pupil of the eye as to make everything abnormal and unnatural" (Lbk. 17:151 [*TAED* LB017151]). The Edison company's contract with the Exposition reportedly specifically excluded arc lighting from the main building, as a consequence of which the United States Electric Lighting Co. was "confined to one of the annexes and to general illumination about the main building" ("Letter to Editor: The Southern Exposition at Louisville," *Electrician* [N.Y.] 2 [1883]: 281).

–2471–

To Francis Upton

[New York,] June 26th. [188]3

My Dear Mr. Upton:—

I have your letter of the 25th. inst.[1] I am delighted that your business should be in such condition as to allow of your taking a vacation, in fact I really envy you, it certainly shows that the Lamp Company's finances must be in a very flourishing condition, and the cause for the anxiety which you experienced some weeks back as to shortness of money and the necessity for shutting down the factory must have passed away.[2] Very truly yours,

TL, NjWOE, DF (*TAED* D8316ACW).

1. Upton had informed Edison that he would "take my vacation during the first two weeks in July. This will be the first vacation of any length I have had since I have been with you." Replying to this document, he stated that he had "found the present always so very tight for money and the future so very flush, that I thought possibly by dropping two weeks out of the present I could by some chance strike the bountiful future." Upton did leave the lamp factory for the first part of July, leaving Samuel Insull in charge of its finances. Upton to TAE, 25, 28, and 29 June 1883; all DF (*TAED* D8332ZBJ, D8332ZBL, D8332ZBM).
2. See Doc. 2435.

–2472–

From Spencer Borden

Boston, June 26th, 1883.[a]

Dear Sir:

Your favor June 25th duly at hand.[1] The policy you suggest in getting capital for local companies is that which I try in every instance. In only one case, that of Lowell, has it been assented to by local capitalists. Wherever else I have tried it on they have demurred and asked to see the figures before committing themselves. In Portland they agreed to pay $150. for the canvass, in other places we could not secure even that. I, however, try, and shall continue to try the method you suggest, in every instance, which is a sure and corroborative proof that "great minds think alike."

Your remarks concerning the coming of other agents into the territory of this department to seek to organize companies quite changes the conditions as indicated in your former letters,[2] or my understanding of them. As explained at present I am willing to make an arrangement such as suggested, and will settle it with you, the next time I come to New York. My coöperation in any scheme to forward the business of the company can always be depended on. Yours very truly

Spencer Borden

⟨You[b]   You entirely misunderstand my letter   please read it over again, & you will learn that nothing is asked of the local people except that to agree to put money into the scheme if you present a location & can show them that it is a profitable scheme= The broad idea is that a canvass will be made & tender drawn up in every case where there is reasonable assurance that if it is favorable the local parties will go in— Until they have seen the tender & organize they pay no money whatever— Read the letter again= E.⟩[3]

ALS, NjWOE, DF (*TAED* D8340ZAO). Letterhead of Edison Electric Light Co. and Edison Co. for Isolated Lighting, New England Dept.; Spencer Borden, manager. [a]"Boston" and "188" preprinted. [b]Followed by "over" as page turn.

1. Edison's draft of his 25 June letter is contained in Doc. 2469. TAE to Borden, 25 June 1883, DF (*TAED* D8316ACR).

2. Borden exchanged a number of letters with Edison about this time, but his reference here is unclear. Edison had recently told Borden of his offer to W. H. Allen, an Ohio acquaintance, of the village plant agency for New Hampshire, an offer he quickly rescinded. TAE to Borden, 7 June 1883, Lbk. 17:95 (*TAED* LB017095); Allen to TAE, 11 June 1883, DF (*TAED* D8325ZAU).

3. The editors have not found a formal reply based on Edison's notes.

–2473–

To the New York World

[New York,] June 27th. [188]3

Dear Sirs:—[1]

I require a colored man as coachman, to live in the country. If you have any applying for positions, I shall be glad if you will send them to see me between three and four this afternoon or from nine to ten to-morrow morning.[2] Yours truly

TL (carbon copy), NjWOE, DF (*TAED* D8316ADJ2).

The Edisons' Menlo Park household staff included several African-Americans, seen in this undated photograph.

1. This letter was addressed to the "Employment Bureau, World Office."

2. The Edison family's Menlo Park staff included several African-Americans, visible in an undated photograph at the home (with a horse), but the editors have found no other correspondence on this subject and have not identified whom, if anyone, Edison hired. In an oral history interview many years later, Edison's daughter Marion recalled that an African-American coachman, his wife, and at least one child lived in an apartment over the Menlo Park stable. Edison possibly sought a coachman for the summertime enjoyment of his children. In October, Samuel Insull asked Mary Edison's approval to "sell the gray pony and the children's wagon and brown harness for $100. . . . the wagon is very much dilapidated and the pony is worth comparatively little." Oser 1956, 1; Insull to Mary Edison, 8 Oct. 1883, DF (*TAED* D8316AYZ).

–2474–

From Henry Clark

Erie, Pa. June 27th 1883.[a]

Dear Sir:

Your letter in re Maxim light this moment received.[1] I was intending to take the train this A.M. and meander to Washington slowly. I will now hasten somewhat. Swazey[2] told me that if your light could do one-half what those who had isolated plants said it would it would abundantly satisfy the gGov. He spoke to me about my proposition for the Bureau of Printing and Engraving in such a way that I felt sure he would report favorably. I will call the friend whom you met to Washington and we will see to this matter.[3] All that my friend asks for services now is his "travelling expenses," if successful a fair compensation. Please notify me here by wire to go ahead on that basis if you so desire.

A successful way of introducing the light into "Public Buildings" would be to put in the experimental plant into the Capitol.[4] I obtained the consent of Architect Clark[5] to do this. Only a small plant a Z dynamo plenty large enough. Direct here and telegrams or letters will reach me quickest

Mr. Dyer[6] leaves this A.M. for Renovo.[7] Mr. Selden[8] will be in N.Y. the last of this or first of next week and will probably call upon you. If I change my mind and conclude to start direct for W. I will wire you. I think I will ask my friend, Mr. Richards,[9] to call upon you before he goes to W. Very truly yours,

Henry A. Clark[10]

⟨Insull— Write private note to Clarke that, if we can work the govt racket we can pony up a private commission to the proper party⟩[11]

ALS, NjWOE, DF (*TAED* D8325ZBA). Letterhead of Edison Co. for Isolated Lighting, agency for Maryland, Western Pennsylvania, and District of Columbia. [a]"1883" preprinted.

1. In a 23 June letter addressed to Clark in Baltimore, Edison reported that

> We have had a big fight with the Maxim Company with relation to lighting Public Buildings in the U.S.   Secretary [of the Treasury Charles] Folger appointed Mr. Swazey, whose business relates to public buildings, to investigate ours and other incandescent lights, and Mr. Swazey made a very one sided report in favor of the Maxim Company. He did not investigate our light at all, he merely came here and talked so very favorably of everything he saw here, that I was impressed with the fact that he would make a report in favor of the Maxim Company, as he was entirely too effusive. [DF (*TAED* D8316ACE)]

Edison had previously sent Clark a general inquiry about Maxim installations similar to Doc. 2443 (TAE to Clark, 16 May 1883, Lbk. 16:345 [*TAED* LBo16345]).

2. Theodore F. Swayze (1845?–1926) was appointed assistant superintendent of the Treasury Department building in 1882. He became chief clerk of the Treasury in 1884, a position to which he returned in 1897 after a hiatus as secretary to the president of the United States Express Co. U.S. Census Bureau 1970 (1880), roll T9_123, p. 13.2000, image 0029 (Washington, District of Columbia); "Current Capital News," *Washington Post*, 8 Aug. 1882, 1; "Government Gossip," ibid., 27 Apr. 1884, 5; "Changes at the Treasury," ibid., 24 Mar. 1897, 2; Obituary, ibid., 25 Feb. 1926, 3.

3. The editors have found no information about this individual or the encounter.

4. Beginning in 1880, various electric light apparatus were tried in the Capitol, none of them satisfactory. By this time, the House had an appointed electrician. Arc lamps for lighting the grounds were installed in 1884 by the United States Electric Lighting Co. and the Brush-Swan Co. Edison incandescent lights installed in cloak rooms and lobbies in 1885 met with general approval. Brown 1970 [1903], 2:159–60; "Architect Clark's Report," *NYT,* 14 Sept. 1883, 2; "An Electric Light Illumination," ibid., 31 Mar. 1884, 4.

The Edison and United States Electric Lighting Cos. collaborated in September 1883 to persuade the Treasury Secretary to permit electric lighting of that building, with the expectation that the two companies would then bid against each other to carry out the work. Edison may have been involved somewhat later in a similar effort with the Brush-Swan Co. to light the Bureau of Engraving ("To Light Up the Treasury," *NYT,* 12 Sept. 1883, 2; A. A. Hayes to TAE, 17 Nov. 1884, DF [*TAED* D8403ZIM]). The Edison Co. for Isolated Lighting arranged in 1885 to light the Senate chamber and cloakrooms and installed a plant early the next year, while the United States Electric Lighting Co. was to light the House side of the Capitol. The Edison company, which had illuminated the Congressional Record composing room since 1882, also bid for the Government Printing Office and the Washington Monument (see 1885 and 1886 correspondence between Uriah Painter and Edward Johnson in UHP (*TAED* X0154), esp. Painter to Johnson, 28 Feb. and 10 Dec. 1885 and 9 Apr. 1886; Painter to U.S. Senate Sergeant at Arms, 1 Feb. 1886; Painter to Bergmann & Co., 1 Feb. 1886; all UHP [*TAED* X154A4CA, X154A4FJ, X154A5CG, X154A5AT, X154A5AQ]).

5. Edward Clark (1824–1902) became assistant architect of the Capitol extension in 1851 under his teacher and mentor, Thomas U. Walter. Starting in 1857, he also supervised expansion of the Patent Office and Post Office buildings in Washington. He succeeded Walter as Architect of the Capitol in 1865 and held that position until one month before his death. *NCAB* 11:223; Frary 1969 [1940], 237–38.

6. Paul D. Dyer did central station canvassing for Edison. The first evidence of this work is from Utica, N.Y., in early June. He went from there to Erie and then, at Edison's instruction, to Renovo, Pa. He likely was the same Paul Dyer born about 1862 to George W. Dyer, of Washington, D.C., who had retired as Edison's patent attorney by this time, and whose sons George L., Richard, Philip, and Frank all had professional associations with Edison. In 1886, Paul D. was working unhappily for the Central Edison Light Co. in Cincinnati; less than two years later he was managing the Western Edison Electric Co. in Denver. Insull to Paul Dyer, 8 June 1883, Lbk.17:118 (*TAED* LB017118); TAE to Dyer, 11 June 1883; Dyer to Insull, 17 Oct. 1886 and 17 May 1888; both DF (*TAED* D8316AAD, D8603ZCF, D8814ACG); U.S. Census Bureau 1970 (1880), roll T9_122, p. 73.3000, image 0149 (Washington, D.C.); see also Doc. 2562.

7. A small railroad town on the Susquehanna River in central Pennsylvania, about 80 miles northwest of Sunbury. *WGD,* s.v. "Renovo."

8. Clark may have meant Charles L. Selden (1849–1930) of Baltimore, superintendent of the B&O Railroad's telegraph department. He and Edison had a personal acquaintance from their days as telegraph operators in Cincinnati. Selden was appointed to the first board of di-

rectors of the B&O's Technological School for employees in 1885. U.S. Census Bureau 1980? (1900), roll T623_613, p. 1B (Baltimore City, Md., enumeration district 152); "Miscellaneous," *Van Nostrand's Engineering Magazine* 32 (1885): 352; Obituary, *NYT,* 3 May 1930, 13.

9. Not identified.

10. Henry Alden Clark (1850–1944) was an agent of the Edison Co. for Isolated Lighting for Maryland, Washington, D.C., and western Pennsylvania. A native of Erie County, Pa., Clark attended Harvard University and then Harvard Law School, from which he graduated in 1877. He practiced in Fall River, Mass., before moving to Erie in 1882, though he maintained an office in Baltimore for his Edison agency. Clark was involved with Edison lighting until about 1887, after which he became active in Erie politics as a Republican and eventually served a single term (1917–1919) in the U.S. House of Representatives. *BDUCS,* s.v. "Clark, Henry Alden"; Woods 1885, 400.

11. The editors have not found a reply from Insull.

-2475-

To William Jenks[1]

[New York,] June 30th. [188]3

Dear Sir:—

Referring to your favor of the 26th,[2] I should be glad to have a plan of the location obtained for pole lines at your earliest convenience. With reference to employees for the station, I do not think it would do to engage people ahead of the time that we want them. I have always got a curiosity to gaze upon the physiogomy if possible of the men who are going to be engaged in our work.[3] Very truly yours,

TL (carbon copy), NjWOE, DF (*TAED* D8316ADW).

1. William J. Jenks (1852–1918) was manager for the Edison illuminating company in Brockton, Mass. A former telegraph operator, Jenks had managed the Brockton Telephone Co. and was also involved with Brush lighting interests there before he joined the Edison organization about February 1883. He later helped organize the New England Wiring and Construction Co. He worked for the Edison Co. for Isolated Lighting and other Edison companies; he also became secretary of the Association of Edison Illuminating Companies. "Jenks, William J.," Pioneers Bio.; U.S. Census Bureau 1970 (1880), roll T9_551, p. 293.3000, image 0095 (Brockton, Plymouth, Mass.); Jenks testimony, *Edison Electric Light Co. v. F. P. Little Electrical Construction and Supply Co.,* 26–28.

2. Jenks asked Edison's opinion about hiring a local fireman and a "meter boy" who had worked briefly at the Pearl St. generating station "for the express purpose of fitting himself for the work here." Edison's marginalia on that letter is the basis for this typed reply. William Jenks to TAE, 26 June 1883, DF (*TAED* D8346Q).

3. Edison expressed at least occasional interest in the physiognomy of prospective employees, even those from prominent families. Around this time he asked for photographs of the sons of Chicago cartographic

publisher W. H. Rand and Boston postmaster Edward Tobey before considering their employment. Years later, he articulated some very broad principles according to which he judged the photographs he sometimes required of prospective employees. See Doc. 2485 n. 1; George Bliss to TAE, 10 Apr. and 7 May 1883; TAE to Spencer Borden, 3 July 1883; all DF (*TAED* D8313L, D8364J, D8316AEM); Israel 1998, 445.

–2476–

From George Bliss

Chicago, June 30 1883[a]

Dear Sir:

Our operations are at a stand still until your men arrive to prepare data for Village Plants.

We have several towns ready in addition to the [those?][b] named

Neenah Wis and Racine Wis are in special haste.[1]

Don't fail to send us a man without delay. Sin. Yours

Geo. H. Bliss   Gen'l Sup't

⟨We[c]   Our man will reach your place in good time. I dont see why you should be at a standstill when you have about 225 Towns left. go ahead & get the reasonable assurance from local people then order the canvass. I dont want to start another man in your territory unless you order more canvasses but will do so immediately you order more canvasses   Our man can do 2 a week like Davenport[2]   E⟩

ALS, NjWOE, DF (*TAED* D8364ZAA). Letterhead of Western Edison Light Co., George Bliss general superintendent. [a]"Chicago," and "188" preprinted. [b]Illegible. [c]Followed by "Over" as page turn; remainder of marginalia written on back of letter.

1. These stations were not completed, but see App. 2.B.

2. Edison's draft formed the basis of a typed reply on 3 July. The exchange with Bliss continues in Doc. 2486. TAE to Bliss, 3 July 1883, DF (*TAED* D8316AEQ).

July–September 1883

Edison traveled to Sunbury, Pennsylvania, on 3 July to supervise the opening of the first village plant installed by the Edison Construction Department. When the lights went on at the City Hotel on the Fourth of July, Sunbury became the first one lighted by Edison's three-wire system with overhead conductors. He remained in the city for a week to further study the system and its numerous trouble spots, during which time he prepared the outline of a memorandum for future managers of three-wire stations. Later that summer, Edison and his engineering staff drew up several sets of examination questions for prospective station engineers. In a further attempt to codify the operational knowledge for running and maintaining central stations from a distance, Edison exchanged numerous letters with William Andrews, who had overseen construction in Sunbury and then stayed behind. Andrews subsequently reported a number of unexpected problems that prompted Edison to improve several elements of the system, including regulation of the distribution network, lightning protection, and wiring connections for a spare dynamo.

During the summer, the Construction Department expanded its efforts to canvass towns throughout New England, the Midwest, and the Upper South. The engineering force had trouble keeping up with the volume of work, particularly building physical models of central station systems to determine the conductors needed for each plant, leading Edison to adopt Frank Sprague's mathematical methods. As a consequence of the boom in the central station business, Edison told a reporter for the *New York Evening Post* that he planned to be "simply a business man for a year. I am now a regular contractor for electric light plants, and I am going to take a long vacation in the matter of inventions."[1] However, Edison continued to oversee experimental work at his laboratory and shops. He also drafted several patent applications, including one for the direct conversion of coal into electricity.

By the end of the summer, stations were being built in Shamokin, Pennsylvania, and in Brockton and Lawrence, Massachusetts. Edison traveled to Shamokin around 20 September to oversee the completion and start of the station. He then went to Massachusetts in connection with Brockton and a proposed plant in Fall River.

To supply the growing business, Edison sought to lock in his principal suppliers. Chief among these firms were the Ansonia Brass and Copper Company, which provided copper conductors, Armington & Sims, which built steam engines, and boiler makers Babcock & Wilcox. He also replaced the management of his Machine Works in the face of allegations of financial malfeasance. Despite the promising outlook for the central station business, the financial condition of the Edison Electric Light Company was emerging as concern for its directors. They began to address its indebtedness to the Isolated Company by tightening the vice on expenses, beginning with the engineering department and Edison's laboratory.

Edison's foreign companies also anticipated increased trade, and they sought permission to set up their own manufacturing operations. Edison refused to provide men to start factories in Germany and England. However, under pressure from his British capitalists, he reluctantly approved a proposed merger with the Swan interests. In order to satisfy Canadian patent law, Edison did establish a small factory in Hamilton, Ontario, under the direction of his brother-in-law, Charles Stilwell. Regardless, the Harrison, New Jersey, lamp works supplied all the carbons for that plant and the factory in Ivry, France.

Edison took a personal interest in setting up the isolated plant and stage effects used in the New York production of *Excelsior,* an allegorical ballet of Progress, which he and Mary attended soon after it opened in late August.[2] Edison's family likely spent part of June and July in Menlo Park, as they had done the previous year, but it is unclear how much time he himself was there.[3] Although he ordered $25 worth of fireworks for Menlo Park on 2 July, he spent the Fourth of July holiday in Sunbury. Edison, along with William Holzer, briefly joined Mary while she, daughter Marion, and sister Alice (Holzer's wife) vacationed at the Long Beach Hotel on Long Island during the first half of August. By September, the Edison family was back in New York City. They used Mary's health and purported medical orders that she give up housekeeping to break their lease on the Gramercy Park house; by the beginning of October, they had moved to the Clarendon Hotel.

1. See Doc. 2503.
2. Doc. 2502 n. 2.
3. TAE to Stirn & Lyons, 2 July 1883, Lbk. 17:154 (*TAED* LB017154A). Regarding the family's stay in Menlo Park, see Doc. 2489 n. 2.

–2477–

Samuel Insull to Phillips Shaw

[New York,] July 2nd [188]3

Edison will be at Sunbury on the fourth.[1]

Sam'l. Insull.

ALS (telegram, letterpress copy), NjWOE, LM 14:76A (*TAED* LBCD1076A).

1. The Sunbury village station was the first fully commercial 330-volt plant using Edison's three-wire system (see Doc. 2424 [headnote]; Jehl 1937–41, 1089–1101; *Edisonia* 1904, 141). Edison's personal attendance was therefore important for technical, business, and publicity reasons. The opening of the plant on the Fourth of July holiday, which coincided with a local sailing regatta, was a goal of the local company and Edison alike (see Doc. 2457 n. 1). The opening was not Edison's first visit to Sunbury, as he had supervised equipment installation earlier in June and periodically boarded at the City Hotel in town in the weeks prior to startup—visits that led to erroneous local beliefs that he actually lived in Sunbury for a year or more. Edison arrived on this trip by 3 July and apparently was present for a test run that night (Gearhart 1936, 177–78; Neff 1946, 186; Jones 1995 [1984]).
   The actual lighting of the plant reportedly went smoothly. About 100 lamps of the 400-lamp capacity were lit on the first evening, and Edison remained at Sunbury for several days afterward. A Williamsport newspaper declared that "Mr. Edison has made a great success, as the lighting at Sunbury fully proves." Edison also used the occasion to give a number of interviews to local papers in the following days, which allowed him to promote his system's cost, reliability, and benefits to the individual consumer. *Edisonia* 1904, 141–42; "The Electric Light," *Northumberland County Democrat*, 13 July 1883; "The Edison Light," *Williamsport (Pa.) Gazette and Bulletin*, 6 July 1883, 1; see also Docs. 2424 (headnote n. 23) and 2437 (headnote).

-2478-

To George Bliss[1]

Dear Sir:—

I am in receipt of your favor of the 25th. June, with relation to the central station plant business.[2]

My canvasser at the present time is in Minneapolis, and I have written him to advise you how soon he will leave for Davenport.[3] I would much prefer that you would not send anybody too co-operate with him in his work, as he has his full instructions from me, and knows exactly what I want, and if he is interfered with and proffered advice by other people, I can hardly hold him responsible for not following my instructions.

Mr. Clark[4] and yourself are entirely of the wrong impression with relation to what I said about keeping a canvasser in your territory, at my own expense, to prepare data for estimates. If the Edison Electric Light Company want an estimate on a town, it is their business to supply me the with the data to estimate on it. That is they ought to tell me exactly what they want, and then I should give them an estimate for the cost of supplying them with what they want. However, we found that this plan would not work very well. Our canvassers were all very unsatisfactory, and I have therefore taken this canvassing work on my shoulders for the benefit of our Companies, that they should have reliable data on which to base figures for a central station. I fail to see why I should pay for the making of a canvass, which forms part of the work of the Parent Company or its licensees in getting up their local companies. So far as the contractors profits are concerned, as I told you when I was in Chicago, I know that I can put these stations up for less than your Company can. If, however, they want to put them up themselves, I am quite satisfied that they should do so. But in such a case they must not hold me responsible for failure in the business. I would point out to you, that if I agreed to what you propose, namely, to bear all expenses in connection with a canvass, I would be assuming the expenses which is legitimately chargeable to the Western Company, as part of their exploitation expenses, in case of a failure [--][a] an installation. You seem to lose sight entirely of the fact that the Western Company would certainly gain some advantage if a central station was installed in their territory. Of course if the plant is ordered for any town we canvass, you will not have to bear the expense of the canvass. I think I will take care of this in my General Expenses in connection with the estimate.[5] Yours truly,

TL (carbon copy), NjWOE, DF (*TAED* D8316AEO). [a]Illegible.

1. In his capacity as de facto manager of the Edison Construction Dept., Samuel Insull conducted routine correspondence based on Edison's notes and instructions. This document is in direct reply to an inquiry from Bliss on which Edison instructed: "Insull write a plain biz letter." TAE marginalia on Bliss to TAE, 25 June 1883, DF (*TAED* D8364Y).

2. Bliss responded to Doc. 2468 that Edison's position "puts a new aspect in the case. Of course we expect to act with prudence and not call for estimates except when responsible parties are ready to make an investment." He protested, however, that "as you make all the Contractor's profits from actual installation, is it not equitable for you to pay the expense of estimates in case of failure to obtain contracts." Bliss to TAE, 25 June 1883, DF (*TAED* D8364Y).

3. No such instruction has been found, but see Doc. 2486 n. 1.

4. Chicago businessman and politician John Marshall Clark (1836–1918), vice president and treasurer (later president) of the Western Edison Light Co., exhibited the company's Edison isolated lighting installation for the local press in October 1882. Trained in civil engineering at the Rensselaer Polytechnic Institute, Clark served in the Union Army until 1864, when he returned to Chicago. He ran unsuccessfully for mayor in 1881 as a member of the Common Council. He later held several other political positions, including customs collector. At various times he also served as president and director of the Chicago Telephone Co. "The Edison Light," *Chicago Daily Tribune*, 29 Oct. 1882, 22; Obituary, ibid., 7 Aug. 1918, 11; letterhead, Clark to TAE, 2 June 1886, DF (*TAED* D8631C); *WWW-1*, s.v., "Clark, John Marshall."

5. See Doc. 2486.

–2479–

From Henry Morton

HOBOKEN, NEW JERSEY. July 5th/83[a]

My dear Mr. Edison

The inclosed slip will tell you what we are doing in the line of instructing men with a view to making them useful in the subject which has received so large a development by reason of your work.[1]

One of the things we want is a collection of instruments of measurement and specimens of what may be called standard electrical fittings. I have supplied myself the means to get a number of such things as resistance coils, galvanometers and the like but am unable to get all that it would be desirable to have in this direction, where a large number of duplicates would greatly facilitate the work of instruction.

Now if it happens to be easy for you to give us any aid in this matter in the way of specimens of your standard applyances such as cut-outs illustrations of conductors with branch connections or any thing else that you would like to have a young man made familiar with before you employed him in your business, we should be highly appreciative of your kindness, or if among your museum of instruments you find any standard instruments of electric measurement good enough to use in instruction but not too good to put into the hands of learners, we should be obliged for, and aided by the loan of such as could spare.

While I do not propose to wait for any help in starting our new Department, I am fully aware of the great advantages that would could come from the aid of others, and am prepared to accept and acknowledge all assistance great or small which comes from those who appreciate the importance of that which are endeavoring to accomplish.

I have no wish to importune you, but only to draw your attention to the matter and so leave it to your own judgement. Very truly yours

Henry Morton.

⟨I have been so badly used and so many pirates like Maxim & Weston[2] are constantly copying what I do even to the design without labor to them that I have turned a cynical savage and shall continue hereafter to keep everything secret & silent, in other words paddle my own canoe   Edison⟩[3]

ALS, NjWOE, DF (*TAED* D8303ZEH). Letterhead of Stevens Institute of Technology. [a]"Hoboken, New Jersey." preprinted.

1. According to the enclosed *New York Times* article, Morton had personally given money to establish a program of applied electricity (largely aimed at mechanical engineering students) at the Stevens Institute, and William Geyer had been named to its chair. Morton reportedly did not believe the Institute should organize a separate department of electrical engineering at this time ("Stevens Institute Graduates," *NYT*, 15 June 1883, 8). Beginning with the 1883–84 academic year, the Stevens Institute offered instruction in applied electricity within its department of Applied Chemistry. The curriculum included classroom and laboratory work in electrical theory and measurement, battery construction and management, arc and incandescent lighting, and dynamo design and operation. "Electrical Engineering at the Stevens Institute of Technology," *Electrical World* 1 (14 July 1883): 436; "Electrical Studies," ibid., 2 (20 Oct. 1883): 120.

2. Dynamo designer and manufacturer Edward Weston (1850–1936) was (since 1882) chief electrical engineer and manufacturing superintendent of the United States Electric Lighting Co., which had absorbed the Weston Electric Light Co. He had already made important improvements to the Maxim incandescent lamp, particularly the technique of treating lamp filaments with a hydrocarbon vapor. Even more importantly, Weston had patented a process for making filaments from a celluloid material he called tamidine, which the United States Electric Light Co. adopted about the end of 1882. Edison had had an acrimonious dispute with Weston in 1879 regarding the efficiency of his dynamos. See

Docs. 1832 esp. n. 1, and 2218 n. 2; Passer 1953, 32–33; "The International Electrical Exposition, Philadelphia [third paper]," *Sci. Am.* 51 (27 Sept. 1884): 192; "The International Electrical Exposition, Philadelphia [sixth paper]," ibid. 51 (25 Oct. 1884): 261; Woodbury 1949, 109–27; U.S. Pat. 264,987; see also U.S. Pats. 264,985 and 292,720.

3. The editors have not found a formal reply to Morton. Edison's reluctance to participate was a departure from his broad enthusiasm for developing electrical engineering education. He had in the past spoken well of Stevens and specifically expressed a willingness to help (see Docs. 2408 n. 2, 2428, and 2434; also cf. Doc. 2643). However, when George Barker queried him in June about a rumored donation of his instruments from the 1881 Paris Exposition, Edison replied that he had "made no firm promises to any except Columbia" and would not "give it to any of them much less Stevens" (marginalia on Barker to TAE, 26 June 1883, DF [*TAED* D8303ZEC]). Morton had helped make an independent test of the efficiency of the Edison lamp. Although Edison had had some sharp disagreements with him (see Docs. 1927, 2010 esp. n. 3, 2017, and 2033), he was willing to trust Morton to "do justice" again as a jury member for lighting tests at the Chicago railway exposition (TAE to George Bliss, 13 June 1883, Lbk. 17:138 [*TAED* LB017138]).

–2480–

To Joshua Bailey

[New York,] July 6th [188]3

My Dear Mr. Bailey:—[1]

We have received a great many telegrams from Paris and from you with reference to the models for Siemens.[2] You seem to have lost sight of the fact, or Batchelor must have omitted to inform you, that they were not even designed at the time you originally cabled for them. We shipped a model H 450 light machine immediate it was tested,[3] and so soon as other machines come along we will ship them also. Everything is being done in order to push them forward as quickly as possible, but where it is a case of designing a machine, rather than turning them out in the ordinary course of manufacture, a great deal of time is consumed in experimenting[a]   It is almost impossible to promise a date for shipments for these various machines, as it is very difficult to foretell the result of the experiments. The main features of the various machines have all been decided upon, but there is a great deal of details which require very careful experimenting. What I wish you to understand is that when you ask for machines, we do not forget all about it and overlook the matter, until your next cable arrives. It would appear, however, that you must have the impression that we are very forgetful in this matter, if you cable us and we find it impossible to name a date, and forthwith cable to somebody else to jog our memories.

I think the 100 light machine will be shipped in the course of a week, but it is rather difficult to promise absolutely.[4]

I received your letter of introduction of Mr Von Muller,[5] and I have done all I possibly could to make his trip here pleasant. He seemed to be very much delighted with everything he saw, and I understand that he is now making a trip West. When he comes back to New York again, he will doubtless call and see us, before he returns to Europe.[6]

Our business here promises to be very brisk indeed. We are erecting four small central stations, two in Pennsylvania and two in Massachusetts, and are negotiating for about half a dozen more. We have no less than forty towns being canvassed where companies will be formed, provided we can make a good showing in our estimates.[7]

Our Isolated Company's business also promises to be very large this Summer. We are putting 4,500 ~~sixteen~~ candle lights into the Louisville Exposition, and the Isolated Company contracted to do the whole of the work within forty days of the signing of the contract. We have got about twenty more days to finish in. There is no reason for thinking that the Isolated Company will break their part of the agreement. This is pretty quick work we consider.

Can you not write me now and then and give me an idea of the business in Europe?[8] Very sincerely yours,

TL (carbon copy), NjWOE, DF (*TAED* D8316AEV). ᵃObscured overwritten text.

1. Joshua Bailey (probably American), a tireless promoter with complex allegiances, was based in Paris, where he had entered the Edison orbit in 1878 in connection with the telephone business in Europe. He became involved with the Edison electric light enterprises in 1881, when he helped make arrangements for the Paris Exposition. As an agent for the Edison Electric Light Co. of Europe, he served more recently as intermediary between New York and the Edison company in Milan (see Docs. 1213 n. 1, 2112 n. 8, 2114, 2120 [headnote], 2235 n. 5, 2332 n. 3, 2369 n. 3, 2399). He helped negotiate a patent license in March 1883 under which Siemens & Halske would manufacture Edison electric lighting equipment, including dynamos, for the German Empire (see Doc. 2392; Bailey to TAE, 24 Feb. 1883, LM 1:293B [*TAED* LM001293B]).

2. Since the end of March, Bailey and the Société Électrique Edison had together sent more than a half dozen cables pressing Edison to ship various dynamos for Siemens & Halske to use as manufacturing templates or "models." He warned in early June that the German firm was threatening to break its contract with Edison over the issue, a possibility reiterated by Charles Batchelor. Bailey to TAE, 29 Mar., 18 Apr., 20 and 25 June 1883; Société Électrique Edison to TAE, 21 May, 7 and

16 June 1883; Batchelor to TAE, 8 June 1883; LM 1:302A, 307C, 316D, 317C, 312B, 317A, 315B, 315C (*TAED* LM001302A, LM001307C, LM001316D, LM001317C, LM001312B, LM001317A, LM001315B, LM001315C).

3. Edison cabled Bailey "Models being built" in late April (about the time of a significant redesign of the K dynamo; see Doc. 2419 [head-note]). Although Edison understood the German parties to want samples of each dynamo, the only model ready for production at this time was the 450-light H design. He and his associates gave several assurances about the forthcoming dynamos before shipping an H machine on or before 16 June. TAE to Bailey, 20 Apr. and 5 May 1883; TAE to Société Électrique Edison, 7 June 1883; TAE to Deutsche Edison Gesellshaft, 16 June 1883; LM 1:308A, 310A, 315A, 317B (*TAED* LM001308A, LM001310A, LM001315A, LM001317B); TAE to Charles Batchelor, 11 June 1883, DF (*TAED* D8316AAC); TAE to Société Électrique Edison, 8 May 1883, citing Batchelor to Bailey, 5 May 1883; Samuel Insull to Société Électrique Edison, 7 May 1883, Lbk. 16:249, 267 (*TAED* LB016249, LB016267).

4. The editors have not confirmed the shipment of this machine.

5. Bailey addressed a letter to Insull introducing "Mr. von Müller of Munich, who comes to New York to study the subject of incandescent lighting." Charles Batchelor wrote a similar letter to Edison (on behalf of the "Engineer of the Government of Bavaria") and reportedly also to Sherburne Eaton and Charles Clarke. Bailey to Insull, 25 May 1883; Batchelor to TAE, 18 Apr. 1883; both DF (*TAED* D8367Q, D8337ZBI).

Oskar von Miller (1855–1934) was a principal organizer, with Emil Rathenau, of the Deutsche Edison Gesellschaft für angewandte Electricität (DEG). A recent engineering graduate of the Munich Polytechnic, Miller was working for the Bavarian government when he visited the Paris Exposition in 1881. Edison's exhibit convinced him of electricity's domestic practicality, its potential to harness the abundant water power of his native Bavaria, and its virtues as a public spectacle. Von Miller enthusiastically helped organize an international electrical exhibition in Munich the following year. It was there that von Miller met and became associated with Rathenau. A vigorous promoter of high voltage alternating current distribution systems, von Miller spent much of his subsequent career working toward the electrification of Bavaria and the development of an electrified railroad network. He also had prominent advisory capacities in major electrical displays in Frankfurt (1891) and Chicago (1893). Perennially interested in dramatic public presentations of technological advances, von Miller put forward the founding plan of the Deutsches Museum in 1903. *NDB*, s.v. "Miller, v., 2. Oskar"; Hughes 1983, 51, 66–67, 131, 334–50; Duffy 2007, 517–24; Füssl 2005, passim.

6. In New York at the start of an extended United States trip, von Miller made an excursion with Edison to the Menlo Park laboratory. He reportedly stopped in New York again before sailing in August, but the editors have found no evidence that he met with Edison that time. Von Miller wrote aboard the steamship *Gallia* expressing to Edison "his innermost thanks for all the kindness you showed me. I am particularly happy to have made your personal acquaintance, and very pleased

that I was able to learn so much in visiting your establishment and in listening to your explanations." Edison had given him an autographed photo and a "small dynamo," apparently a model of a short-core version of the L machine with a commemorative plaque, for which von Miller also thanked him. Von Miller to TAE, 9 Aug. 1883, DF (*TAED* D8303ZEW); Füssl 2005, 64–65.

7. Edison referred to construction in Sunbury and Shamokin, Pa., and Brockton and Lawrence, Mass. See Docs. 2424 (headnote) and 2456 and, regarding pending canvasses, also App. 2.

8. An inveterate conjurer of patent ownership schemes, Bailey was soon writing to Edison (as well as Charles Batchelor and reportedly also Sherburne Eaton) about a new "International Edison Company for Electric Light and Motive Power" ("Notes on the Formation of an International Edison Co." with draft articles of association, n.d. [1883], PPC [*TAED* CA006A]). At one point he seems to have hoped to include the Swan interests, but various notes and draft articles of association describe an Edison-only company for central stations, isolated plants, and manufacturing throughout continental Europe (excluding Portugal, Sweden, Norway, and the United Kingdom) and some colonies. Bailey suggested basing the firm in London or Paris, on which Edison wrote: "Dont believe in London   Paris might do." Bailey continued to revise the plan, but Edison remained noncommittal and nothing seems to have come of it (Bailey to TAE, 10 Aug. [with TAE marginalia] and 12 Sept. 1883; Eaton report on proposed International Co., 5 Oct. 1883; all DF [*TAED* D8337ZDN, D8337ZEF, D8331ZAD1]; Bailey to Batchelor, 8 and 14 Dec. 1883, Batchelor [*TAED* MB099, MB102]).

–2481–

To John Lieb[1]

[New York,] July 6th. [188]3

Dear Sir:—

I beg to confirm receipt of your cable dated June 29th, as follows:

"Started station success tonight, Manzonia Theatre lighted."[2]

I have very great pleasure in congratulating you on the successful starting of the Milan Station, and shall hope soon to hear of your having a great number of lights connected with it. Very truly yours,

TL (carbon copy), NjWOE, DF (*TAED* D8316AET).

1. After having worked for the Edison lighting business since early 1881, John William Lieb (1860–1929) became the principal electrician of the New York Pearl St. central station when it opened in September 1882. Giuseppe Colombo was so impressed with his ability that he asked Edison to assign Lieb to Milan to oversee installation of the central station there. Lieb arrived in Milan in December 1882. He was promoted to chief engineer of the Italian Edison company in late 1883 and remained abroad until 1894. Doc. 2369 n. 7.

2. This is essentially the text of Lieb's cable, which was received in

New York and entered into Edison's cable book on 29 June (DF [*TAED* D8337ZCI]; LM 1:318D [*TAED* LM001318D]). The Teatro Manzoni, built in 1872 as the Teatro della Commedia and renamed in 1873, specialized in productions of drama and music, especially opera buffa and light opera, though it had begun to take a more serious place in Milanese theatrical affairs. It was the site of a celebration of the inauguration, on 28 June, of service by the Edison central station. Lieb wrote a more detailed description of the evening and the station generally a few days later. Open to the general public, the event consisted of "two Comedys, arranged to show the effect of the light on the costumes, and a Mandolin Concert. Between the acts an exhibition was given to the audience of the regulation of the lights in the Main Hall, foot lights and upon the stage the public receiving the display with great applause. The press was very enthusiastic . . . and urged the immediate adoption of the system in all the Theaters." Regular service throughout the distribution area began on 2 July (Lieb to TAE, 1 July 1883, DF [*TAED* D8337ZCO]; *NGDO*, s.v. "Milan, 4").

The Milan plant was the first permanent Edison central station in Europe. Started with four Jumbo dynamos, it served a district centered around the famed Galleria Vittorio Emanuele, noted for its theaters, restaurants, cafes, and shops, which largely comprised the first customers. Giuseppe Colombo had made a point of proving the practicality of electric lighting for theaters, and the Teatro alla Scala was also within the service district. (An independent report completed in June and forwarded to Edison showed substantially lower maximum temperatures at various points throughout a Munich theater when it was lighted by electricity instead of gas. Philip Seubel to TAE, 20 June 1883, enclosing Max von Pettenkofer report, 13 June 1883; both DF [*TAED* D8337ZBZ, D8337ZCA]). Ironically, the Milan generating plant was in a building specially constructed on the site of the former Santa Radegonda theater, which the Italian company had acquired. That company had been formed in July 1882 and gave Edison its order for central station equipment in September of that year; it had protracted conflicts with Edison over the specifications and payment for the plant. Two Jumbo dynamos (probably those from the Holborn station in London) were added in the late summer or early fall, and in November the plant began operating around the clock. By the end of 1884, it supplied current to 5,500 10-candlepower lamps (Edison Electric Light Co. Bulletin 19:25, 15 Aug. 1883, CR [*TAED* CB019]; "The Edison Central Station at Milan," *Engineering*, 31 Aug. 1883, Scraps. [*TAED* SM018094a]; "The Edison Electric Light in Italy," *Electrician* 11 (28 July 1883): 242; "The Edison Central Station in Milan," ibid. 13 (26 July 1884): 248–50; "Electric Lighting of the Scala Theater at Milan," *Sci. Am. Supp.* 17 (31 May 1884): 1008–9; "The Electric Light in Theatres," *Electrical World* 3 (26 Apr. 1884): 135; Docs. 2235 n. 2, 2332 n. 3, 2343 n. 15, 2399, and *TAEB* 6 chap. 6 introduction; Clarke 1904, 53; Colombo to TAE, 4 Oct. 1883, DF [*TAED* D8337ZEO]).

From Sherburne Eaton

[New York,] July 7th. 1883.

Mr. Edison,

In my financial statement to the Directors of June 6th. I stated that the expenses of the engineering department had been going on thus far this year at the rate of $35,000 a year or nearly $3,000 a month.

Those expenses have now been cut down more than one half and are only about $1200 a month for wages. To day they are as follows, by the month:

Clarke[1]	$292
Solden[2]	150
Atkins (blue print)[3]	47
Dodge[4]	85
Burnham[5]	92.50
Total	666.50

Menlo Park expenses, most of which can be charged up to customers ultimately:

Cladius[6]	$125
Gmur[7]	65
Anderson (mechanic)[8]	65
Bengsten (mechanic)[9]	65
Burnett (electrician)[10]	63
Andrews (electrician)[11]	63
Van Clief (carpenter)[12]	52
Total	498
Grand Total	$1,164.50

Of course there are some expenses besides wages, but after allowing for this I think the expenses of the engineering department, are to day cut down over one half.

EDISON'S LABORATORY EXPENSES. Up to the date of my report, June 6th., these expenses were going on at the rate of about $40,000 this year; from January 1st. They have been greatly reduced. The laboratory expenses for the last 4 weeks, ending July 7th., have averaged only $103 a week, or $412.79 for the 4 weeks. To this should be added rent and certain miscellaneous charges. The expense even then, at the present rate, would probably not be much greater than $10,000 per year.[13a]

Now if we can keep down the patent fees, law expenses, and not give any new retainers to lawyers and experts, we stand a pretty good chance of keeping our annual expenditures inside of our receipts, which ought to be (if a 10 per cent dividend is paid) $37,500 from the Isolated Co., besides the various 5 per cent cash payments from subordinate companies.

But the great problem is how to wipe out the debt of the Light Co. to the Isolated Co., for borrowed money. This debt amounts to about $30,000.[14]

S. B. Eaton   per Mc.G[owan].

TL, NjWOE, DF (*TAED* D8327ZAN). Monetary expressions standardized for clarity. [a]Followed by dividing mark.

1. Charles Lorenzo Clarke (1853–1941) was chief engineer of the Edison Electric Light Co. Trained as an engineer (with two post-graduate degrees from Bowdoin) and draftsman, Clarke began working at Edison's Menlo Park laboratory in 1880 and played a major role in designing dynamos. Docs. 1921 n. 3 and 2125 n. 2; Little 1909, 3:1242–44.

2. Gustav Soldan (b. 1838?) emigrated from Hesse-Darmstadt in 1864. Soldan was a self-identified mechanical or civil engineer who supervised the engineering department's drafting room at the Edison Electric Light Co.'s headquarters. He moved to the Edison Machine Works in August (see Doc. 2512) and remained there until about June 1884; he was later associated with the pharmaceutical firm Merck & Co. Doc. 2269 n. 4; Jehl 1937–1941, 929–30, 987; U.S. Census Bureau 1970 (1880), roll T9_784, p. 398.3000, image 0277 (Jersey City, Hudson, N.J.); ibid., 1882? (1900), roll T623_979, p. 16B (Jersey City, Hudson, N.J.).

3. Joseph F. Atkins (b. 1849) worked at the Construction Dept.'s offices on Fifth Ave. from May 1883 to June 1884. Identified in a later census as a clerk, Atkins probably filed or copied blueprints rather than drew them. U.S. Census Bureau 1982? (1900), roll T623_1124, p. 6B (New York [Manhattan], N.Y.); Atkins to Arthur Williams, 16 Oct. 1919, EGF (*TAED* E1928BA).

4. Robert J. Dodge worked in the mapping department on prospective central plants. Previously, he had scouted locations for the generating station of the proposed Madison Square (Second District) plant in New York. Dodge to TAE, 9 Nov. 1882, 3 and 7 July 1883, all DF (*TAED* D8223O, D8362B, D8340ZBB).

5. Charles E. Burnham (1857?–1898), born in New York to Canadian parents, acted as Charles Clarke's secretary at the Edison Electric Light Co.'s Fifth Ave. offices. Jehl, 1937–1941, 930; U.S. Census Bureau 1970 (1880), roll T9_888, p. 630.1000, image 0355 (New York City, New York County, N.Y.); New York City death certificate 2563, 4 July 1898, online database of New York City Deaths (1892–1902), accessed through Ancestry.com on 22 Sept. 2008.

6. Hermann Claudius.

7. B. Gmur did central station mapping and electrical determining, but he is otherwise unidentified. TAE to Eaton, 21 May 1884, DF (*TAED* D8416BQG).

8. Hugo Andersson (1857–1940), a Swedish machinist, engineer, and instrument maker, was making Edison telephones at Sigmund Bergmann's shop in New York in May 1881 when Edison ordered him detailed to Menlo Park to construct instruments for the Paris Exposition. After a hiatus with Bergmann from late 1882 through early 1883, Andersson returned to Menlo Park, where he worked on the electric railway and, about this time, experimented on carbonizing molds for lamp

filaments. His employment there ended on 18 August 1883. Andersson left the United States in 1884 and eventually returned to Sweden, where he worked as a telephone engineer and government electrical inspector. Jehl 1937–41, 685–86; "Andersson, Hugo," Pioneers Bio.; Andersson to Francis Jehl, 18 Sept. and 9 Nov. 1930, Ser. 5, Edison Pioneers, EP&RI (*TAED* X001J2A, X001J2B).

9. Not identified.

10. Not identified.

11. Edison had appointed William Andrews as chief electrical engineer for his Central Station Construction Dept. on 1 June. "Andrews, William Symes," Pioneers Bio.

12. Cornelius Van Cleve (known as "Neal"; b. 1842?) was married to Harriet "Hattie" Stilwell, the half sister of Mary Stilwell Edison. He began working in the laboratory at Menlo Park about the middle of 1880 and had been closely involved in running Edison's experimental electric railroad. Samuel Insull arranged in August for Van Cleve to operate the line for Charles Rogers of the Electric Railway Co. of the United States and his unnamed guests. Doc. 2227 n. 6; Insull to Rogers, 28 Aug. 1883; TAE to Van Cleve, 28 Aug. 1883; both DF (*TAED* D8316AQK, D8316AQM); Rogers to Insull, 30 Aug. 1883, Cat. 2174, Scraps. (*TAED* SB012BCK).

13. The Edison Electric Light Co. had contracted in 1878 to support Edison's electric light research (see Doc. 1576). Records of Edison's laboratory account are incomplete, but those from November 1882 to May 1883 appear to show much smaller expenditures than those given by Eaton (Personal Ledger #5:259, Accts. [*TAED* AB003:141]). Experimental expenses, recorded in Edison's ledger books under a number of distinct headings, were summarized each week for the Edison Electric Light Co. until 1882 (Electric Light Co. Statement Book [1880–1882], Accts. [*TAED* AB032]; Doc. 1562 is an early example of the weekly reports). These reports also give a lower yearly total than that projected by Eaton. The company reimbursed Edison's $2,500 annual rent of the laboratory space from Bergmann & Co., which generated some contention (see Doc. 2356 n. 7).

14. Eaton put the debt at $67,000 only a month later. The total resulted in part from prior loans from the Edison Co. for Isolated Lighting (which also provided cash to other Edison enterprises) and the installation of the experimental Roselle, N.J., plant. The parent company owned a majority of stock in the Isolated Co., whose available capital had recently been increased by $250,000 through the issue of new shares. The amount dropped slightly below $20,000 in early 1884. In mid 1885, a special committee on the financial affairs of Edison lighting companies recommended that the debt be paid by transfer of the Edison Electric Light Co.'s full three-fourths interest in the Roselle, N.J., plant as well as its ownership of the lighting and office equipment at 65 Fifth Ave. The balance was to be paid by four unspecified promissory notes. Eaton to TAE, 9 Aug. 1883; Edison Co. for Isolated Lighting report, 4 June 1885; both DF (*TAED* D8325ZBT, D8526U); Eaton to Edison Co. for Isolated Lighting, 27 Dec. 1883, Batchelor (*TAED* MB109); Edison Electric Light Co. annual report, 23 Oct. 1883, in company Bulletin 20:41, 31 Oct. 1883, CR (*TAED* CB020442); Edison Electric Light Co.

report, 31 Mar. 1884, DF (*TAED* D8427ZBA); Edison Co. for Isolated Lighting annual report, 18 Nov. 1884, PPC (*TAED* CA002D).

–2483–

From Theodore
Waterhouse

[London,] July 7th 1883.

My dear Sir,

Before this reaches you Mr. Johnson will probably have informed you of the negotiations which have been entered into with the Swan Company for an Amalgamation of that Company with our own, and will have placed in your hands a short statement of the terms proposed.[1]

You may perhaps like to hear from me what are the considerations which have weighed with the Edison Board in coming however reluctantly to the conclusion that this Amalgamation is desirable, and which induce us to think that your interests as well as those of the A Shareholders will be best served by this step.

The only patent which at present seems likely to secure a monopoly of your System is the Carbon Filament Patent of Novr. 1879

The difficulties in maintaining that Patent are two,[2]

(a)   Alleged prior publication by Swan

(b)   the omission in your Patent to describe the process of getting rid of the atmospheric air from the filament, a process subsequently patented by Swan and by a later patent patented by yourself.

It is clear that amalgamation with Swan to a considerable extent gets rid of both these difficulties. The alleged prior publication was mainly if not exclusively in oral lectures and unless Mr. Swan insists on and gives personal evidence as to this publication it is not likely that it can be successfully maintained by others.

With regard to the 2nd. point although there is good ground to hope that your Patent would be sustained in spite of this defect there is, at the same time, some risk that it might on this point be held bad, and even if sustained it is possible (and this is a point to which some of the advisers of the Company are inclined to attach great importance) that Mr. Swan's Patent for getting rid of the air in the Conductor, might prevent us from Manufacturing Lamps in which this process is employed.

With regard to our financial position, the £100,000 paid up on the formation of the Company has been more than spent, a considerable portion being locked up in Machinery which is likely to be superceded by machines of a newer and better type— The Board feels it hopeless having regard to the present feeling of the investing Public in this Country on the subject of Electric Lighting to get more Capital subscribed through a further issue of A Shares,[3] while at the same time a Call would excite great dissatisfaction on the part of many of the existing Shareholders— This condition of things paralyzes the Company's operations and of course renders the B Shares for the time being valueless.

We cannot ignore the fact that our Company instead of finding an unoccupied field for its own operations has been subjected to severe Competition in almost every installation for which it has tendered.

Its most formidable competitor has been the Swan Company which had had a considerable start, had already established itself in the public estimation, and which no doubt possesses at the present time a valuable goodwill.

Moreover all Electric Lighting has enormous legal and other difficulties to overcome in this Country before it can establish itself on its true basis from Central Stations—difficulties which can only be successfully met by a Company with a large Capital and otherwise strong.

Under these circumstances it was felt that there would be a greatly improved prospect of remunerative business and of an ultimate monopoly of Electric Lighting if a union could be effected with the Swan Co. and there is no doubt that a union will greatly increase the value of the interests on both sides— The only difficulty was felt to be as to the terms on which such a Union should be carried out.— The Directors of the Edison Co. do not think that those which have been arranged are as much in our favor as the equity of the case demands but it was clear that no better ones could be obtained and there is little doubt that the B shares in the new Company will be of much larger and more immediate value than the B shares of the existing Company.— This larger value is in our judgment obtained,—(in spite of their being postponed in order to attract Capital to a Preferential Dividend of $7\frac{1}{2}\%$ and of their taking only $\frac{1}{4}$th. of the profits after that dividend.)—[4]

(1)   Through their being immediately entitled to dividends and not deferred till the £20,000 already paid to you has been repaid—that sum being now treated as a payment out and out.[5]

(2)   Through there being deferred shares of a Company which is likely at once to work on a scale many times larger

than either of the separate Companies and to a satisfactory profit;—whereas as matters at present stand with the Edison Company any prospect of dividend on the B Shares must be regarded as more or less problematical.

The name of the New Company will probably be "The United Electric Light Company" but we have stipulated that whenever the Inventors are referred to your name shall, as is right, precede Mr Swan's.[6]

I must confess that I had looked forward to more speedy success and a more independent existence for the Edison Company and I should be very loath to advise or take part in a step which is not desirable in your interests as well as in those of the A Shareholders—but after giving the best consideration to the matter of which I am capable, I believe that all these interests will be advanced by what is contemplated, and that in this way we have our best chance of securing what should be our great ultimate aim—a general adoption of your system of central station lighting— I am, my Dear Sir   Yours faithfully

Theodore Waterhouse.

LS, NjWOE, DF (*TAED* D8338ZAR1). Letterhead of Waterhouse & Winterbotham.

1. The statement has not been found, but see Edward Johnson's analysis of the British company's position in Doc. 2467 and also that of Arnold White summarized in Doc. 2467 n. 14. Also on 7 July, the firm of Waterhouse, Winterbotham & Harrison acknowledged receipt of Edison's voting proxies for his 3,000 B shares in the Edison Electric Light Co., Ltd. These were forwarded to John Douglas, in whose name Edison's shares were jointly registered on the shareholders list of June 1883. An additional 5,000 A shares were registered solely in Douglas's name, while another 5,000 B shares were held commonly by Douglas, Egisto Fabbri, and James Hood Wright of the Morgan firms, with which Douglas was associated at 22 Old Broad St. in London. Waterhouse, Winterbotham & Harrison to TAE, 7 July 1883, DF (*TAED* D8338ZAR); Summary of Share Capital & Shares in the Edison Electric Light Co., Ltd., as of 13 June 1883 [1926], CR (*TAED* CF001AAO).

2. Edison's British Patent 4,576 of 1879 (Cat. 1321, Batchelor [*TAED* MBP019]) was his first and most fundamental on the incandescent carbon electric lamp. His backers in Britain had long been troubled by the fact that it implied but did not specifically claim a crucial process for purging air from the filament during manufacture, which Edison had claimed in a previous specification for metallic conductors (Brit. Pat. 2,402 [1879], Cat. 1321, Batchelor [*TAED* MBP017]). This omission was corrected in Edison's British Patent 562 of 1881 on the "Manufacture of Carbon Burners, &c.," but Joseph Swan had in the meantime filed a specification (Brit. Pat. 18 [1880]) on essentially the same process in January 1880 (Docs. 2203 [and headnote] esp. nn. 14, 21–22; and 2187

esp. n. 20; Dredge 1882–85, 2:xcvii). Edison's 1881 patent was probably the one about which Johnson complained in December of that year, not the 1880 patent cited in Doc. 2203 n. 13.

Backers of Joseph Swan claimed that his demonstration of a carbon lamp in 1878 and early 1879 anticipated Edison's British patents for an incandescent carbon filament. Edison vehemently argued both that Swan's lamp was unlike his own, and that in any case, Swan had failed to make a legally sufficient claim to it at the time. See Docs. 2165 and 2187.

3. The capital stock of the Edison Electric Light Co., Ltd., was divided equally between A (sometimes called ordinary shares) and B shares, each having par value of £10. The A shares, reserved for investors, were solely entitled to a so-called preferential dividend of five percent annually of the company's paid-up capital. The B stock, given to Edison in payment for his patents, shared equally with A shares in profits remaining after payment of the preferential dividend and had one-half the voting power of the others. TAE agreement with Drexel, Morgan & Co., Grosvenor Lowrey, Fabbri, and Bouverie, 18 Feb. 1882, CR (*TAED* CF001AAE1); see also Doc. 2231 n. 2.

4. Under terms of the October 1883 final agreement between the Edison and Swan companies, Edison interests entitled to B stock would receive one share for every three A shares in the combined company. However, in deference to Edison's objections, the preferential dividend payable to A shares (before allocation of remaining profits among A and B shares) was reduced to seven percent of the company's paid-in capital (see Docs. 2493 and 2514). Each B share had only one-half a vote, while A shares had a full vote. The relative diminution of B stock also meant a corresponding reduction (to one-quarter) in its portion of the profits remaining after allocation of the preferential dividend to A shares. Edison and Swan United Electric Light Co. agreement with Edison Electric Light Co., Ltd., Swan United Electric Light Co., Ltd., and George Black, 1 Oct. 1883, DF (*TAED* D8338ZBS3).

5. Under article thirteen of the original contract, the company held half of the B shares allotted to Edison until the dividends, which otherwise would have been paid on their account, equaled the £20,000 advanced to him. TAE agreement with Drexel, Morgan & Co., Grosvenor Lowrey, Fabbri, and Bouverie, 18 Feb. 1882, CR (*TAED* CF001AAE1).

6. See Docs. 2493 and 2514.

–2484–

Memorandum: Village Plant Operation

[Sunbury, Pa.] July 8/83[1]

Instructions
Sparking of Dynamo[2]
Causes—[3]

    1[a]   Brushes not set on the neutral point
    2nd[a]   One or more of the commutator bars have raised up
    3rd[a]   Rough Commutator

4th[a] Bad ends on brushes whole of the brush not touching commutator

5th[a] Grooved commutator worn by brushes or sparking
continuously from one part of the brush

6th[a] Brushes not opposite each other.

7th[a] Brushes pressing against commutator too lighty

8th Brushes stick in holder

9th Little ring of fire running around end of Commutator[4]

To fix the

1st[a]= Move the arm carrying the brushes back or forward
until you reach the point where there is no sparking   this will
be the neutral point with the load then connected

2nd[a] File down the bar or bars with care so as not to file a
flat place on bars not high. The high bars will be easily found
by one edge being roughened by the sparking more than the
others, also you can see & feel the high bar or bars, one ~~sig~~
single high bar will knock & vibrate the brush making a spark
& owing to high speed of revolution the sparking wil[l][b] appear as if on all the bars.

3rd[a] Smooth the Commutator by the finest sand paper
polishing with well worn sand paper—do not use emery
paper

4th[a] The brushes if there has been any sparking should
be taken out every day and dressed up for the nights run, by
setting brushes properly & keeping them in the neutral point
so as to have scarcely any sparking. the commutator will polish nicely & brushes will run for weeks, but sparking cuts
& roughens the commutator as well as the brushes, hence if
it occurs it will require attention constantly   The neutral
or nonsparking point changes with the load of lamps with
scarcely any load on ~~machines~~ the brushes will down on the
side of the Commutator   as the load Increases[c] the neutral
point will shift in the direction of rotation until with full load
the brushes will be advanced way up near the top of the commutator.

5th[a] Every 2 or 3 days set the brushholder pieces along
the brushholding bar connected to the arm and thus change
the position of the ends of the brushes resting on the commutator   by thus shifting you will be enabled to prevent the
formation of grooves by allowing the brushes to run in one rut
too long & thus wear down the Commutator Evenly over the
entire surface.

6th[a] It is essential that there should be the same number
of commutator blocks between each brush   if there were 40

blocks for instance & the brushes were 2 blocks wide each you should so set the brushes that you could count 18 blocks on the Commutator from brush to brush on either side around the Commutator   If this is not attended to & one brush is nearer to the other one way around the commutator than around the other way, it will be difficult to get both brushes in the nonsparking or neutral point for when one brush is in the neutral point the other will be out.

7th[a]   The pressure of brushes against the Commutator should be as light as will permit the current being taken off with sparking any excess of pressure will on help wear the brushes & commutator out. Sometimes a high bar will knock the brush having a light pressure so badly that until the high bar can be fixed a heavy pressure will for the time being is necessary

8th[a]   Sometimes especially when the holder gets hot, ex-pansion & at other times burnt oil etc will cause brushholder to stick— when the brush sticks the brush itself will in little while ~~just~~ leave the commutator making no contact with it yet be so close to it that it appears to be in contact & working all right   if you should lift the other brush it would open the main circuit & make an ugly spark   alway[s][b] try the brushes before starting to see they work freely in the holder[d]

After a station is loaded up with nearly[e] all the lamps that was originally intended the speed of the Dynamo may be in-sufficient to give the requisite pressure to the Current to bring the lamps up to their proper candle power   when this occurs & it is found with the load on that the Engine is going at her regular speed and[c] the regular[5] is so turned to get the great-est pressure that the pressure falls short the speed of the En-gine can be increased just sufficient to obtain the right pres-sure   The pressure increases with the speed of the Dynamo while the Current depend on the number of lamps burning.

Lamps=

The test lamps are defective lamps of which there are a per-centage made at the lamp factory[6]   they are suitable for test-ing purposes and they should not be run to[e] as high candle power as regular lamps so as to make them last as long as possible; they cost 25 cents. The Regular 10 candle lamp cost 60 cents, the 16 candle, 340c

An account of lamps out Lamps in stock and broken Lamps returned from Customers must be kept ~~every~~ so that over any period you can tell how many lamps were rec[eive]d as broken at the Central station   people are liable to steal the lamps hence you should keep you lamp account well up— [d]

All service wires running from the mains into the Consumers house must be run with insulated wire and placed at least 2 feet apart so that in the most violent wind storm they cannot blow together[d]

with the 3 wire system perfection of working would be attained when the Customers were so placed on the A or B side that there was an equal number of lamps on both sides burning   In starting the consumers are so connected that it is expected that there will be an even number but after a while it is noticed that the B side has a greater number of lights than the A which is ascertained by the deflection of the amperemeter to the right or left   the next Customer can be connected on the A side & by adding customers to the A side as balance is soon attained   when there is an exactly the same number of lights on the A & B sides the ampere meter at the station will stand at Zero .o. Some of the heavist Customers should have 3 wire service lines in their premises & be provided with a switch so that when you are badly out of balance the customer can quickly be thrown over from one side or the other   4 or 5 such customer will be sufficient to effect a balance when badly out if the town is small[7]

Thos A Edison

ADS (facsimile), Jehl 1937–41, 1102–13; two typed transcripts are in EP&RI (*TAED* X001J3A, X001G2BD). [a]Multiply underlined. [b]Facsimile copied off edge of page. [c]Obscured overwritten text. [d]Followed by dividing mark. [e]Interlined above.

1. Appended to one of the transcripts of this document is a typed statement, attributed to William Andrews, that Edison wrote the text at about midnight on 8 July for the benefit of Andrews and "those who were to assume the management and operation of small three-wire stations" such as the one at Sunbury. EP&RI (*TAED* X001J3A).

2. See Doc. 2420 n. 14.

3. Some of the causes below are attributable largely to design problems or improper operation, but others clearly involve manufacturing deficiencies. Difficulties with commutators and brushes appear with some frequency in the "Abstracts of Defect Book Reports," a weekly compilation of trouble reports, most from isolated lighting plants, which Sherburne Eaton began forwarding to Edison in April 1882. Electric Light—Edison Co. for Isolated Lighting—Defect Reports, DF (*TAED* D8222).

4. Construction Dept. employee Harry Ward Leonard discussed at some length the occurrence of this problem at the Mt. Carmel, Pa., station several months later. He suggested that it was "probably caused by a cross between successive bars of the commutator." Edison wrote

above this and then crossed out "No its copper dust between bars on end—clean & shellac the ends [little?]." At the top of the letter he wrote "Send this to Soldan—this occurred at Fall River & must be looked into quickly & remedied." Leonard to TAE, 7 Feb. 1884, DF (*TAED* D8456ZAD).

5. Edison presumably meant voltage "regulator."

6. Sometimes called resistance lamps, a number of these could be used to test dynamos or other equipment off-line. The Pearl St. station in New York had a bank of 1,000 lamps for this purpose; see Doc. 2243 (headnote with figure).

7. Edison had outlined a patent application in March for placing electromagnetic relays for this purpose "in every 10th house   the moment any tendency of overloading one circuit occurs some of the automatic devices will throw Lamps from the weak to the strong side until equilibrium is obtained." Unnecessary in a large distribution system, such an arrangement was "valuable where there are but few consumers & occasions arise where the circuits are very much out of balance" (Cat. 1149, Scraps. [*TAED* NM018AAD]). In a patent application filed the same month, Edison sought to maintain equal electromotive force on each side of the circuit by having an adjustable rheostat on each side of the circuit. Noting that "the translating devices [lamps] of the system are preferably arranged in such manner that the number on one side of the compensating-conductor will constantly remain about the same as that of the other," he also provided a switch that would enable a set of lamps or a building to "be transferred from one side of the system to the other, should the numbers become so unequal as to render such a change desirable" (U.S. Pat. 283,984). Edison filed another application in April 1883 (U.S. Pat. 283,983) on various arrangements of electromagnets for automatically switching lamps from one side of the circuit; the engineer at the central station could also do this manually.

–2485–

From Samuel Insull

New York, July 10th. 1883.

My Dear Edison:—

I wrote you asking if I could engage another canvasser and you have not replied.[1] We have a number of towns that should be canvassed, and although this work is getting on considerably ahead of our ability to take care of it in the office, yet I think we should put as many canvassers in the field as are necessary to satisfy our agents. Three weeks ago I did not think the business justified it, recent developments lead me to the conclusion that it does. We shall also require another map man. I am sure that the one man now here cannot take care of all our map work.[2] He has just finished the map of Erie. He has now got St Paul, Renovo and Bellefonte to deal with. The Hartford canvass will be here to-morrow, Minneapolis canvass in a day or so after, and Pittsfield and Louisville soon after that. We have got Elysville, Utica and Erie maps waiting for you to red line, so that they can go to Menlo,[3] others will quickly follow.

I telegraphed you the other day asking you what I should do about regulators and switches for the Louisville Exposition. You have not yet answered [--][a] my telegram.[4] You see there are piles of matter requiring your attention, and I am very anxiously looking forward to your return, else I am afraid the plaster will grow so horribly thick that it will be set hard in our stomachs.[5]

I have got an order from the London Company for two 13 × 13 engines for the Criterion Theatre, London. Sims has undertaken to ship them before August first.[6]

I have cabled Bailey as you have suggested and hope that will result in an order.[7]

The Isolated Company sent an order to the Machine Works for one 100 light machine. Rocap[8] wrote asking me whether the Isolated Company thought we kept a toy shop.[9] I went to Hutchinson and Hastings, insisting on an order for ten, and got it, and they have undertaken to pay cash on delivery.[10]

The London Company sent a remittance for $4,000 to pay for the two engines they ordered. This greatly relieves the financial plaster, which has been getting horribly thick, since I got back.

Eaton has gone off for a holiday of two or three days with a Postmaster General and two or three other swells.[11]

Brewster[12] has got back from St. Louis. The people there have bought a lot for the central station, and we are dead sure of that contract sometime next Fall.

Moore, I think, is going to get left on his engines for Louisville, he is going down to Providence to see about them.

I wish you would go for Hanington bald headed.[13] He has rendered me a bill for $67.05 for work done at the central station. I suppose we will have to pay the bill, but I look upon it as a scandal. If I had my way I would charge Mr. Sprague with it, as I presume most of the expense was incurred owing to his little escapade with the engine.[14] Even taking that into consideration, the bill had no business to be any such amount as it is.[15] Hanington cannot know much about handling men. if iIt is seems to be his[b] his rule to get the least possible work out of the greatest number of men in the greatest possible length of time. Yours very sincerely,

Samuel Insull

TLS, NjWOE, DF (*TAED* D8367Y3); a carbon copy is in DF (*TAED* D8316AFF). [a]Canceled. [b]"seems to be his" interlined above.

1. In one of at least eight letters or telegrams Insull dispatched to Sunbury that week, he reported to Edison on 5 July that he believed it necessary to hire another canvasser and was "on the track of a man, who is coming to see me to-morrow. . . . If he turns out satisfactory, can I send him off before you return? Will you trust to my judgment of his physiognomy? Please wire reply." DF (*TAED* D8340ZAY); cf. Doc. 2475.

2. Louis F. Ruf (b. 1865?) apparently had primary responsibility for drawing maps until he left at the end of August for a job with the West Shore Rail Road. Some maps were also assigned to Charles F. Richardson, head of the canvassing and mapping department since early July. Mapping had been done at the Construction Dept.'s offices in New York and was soon put within the purview of the Edison Electric Light Co.'s Engineering Dept. U.S. Census Bureau 1970 (1880), roll T9_791, p.69.2000, image 0266, (Freehold, Monmouth, N.J.); TAE to Richardson, 10 Aug. 1883, Lbk. 13:22 (*TAED* LB013022); Richardson to TAE, 14 Aug. 1883; Ruf to Insull, 30 Aug. 1883; both DF (*TAED* D8342ZAK, D8313ZBC); see Doc. 2488.

3. Edison routinely drew red lines on maps to show where he wanted the street conductors to go. The maps were sent to Menlo Park for modeling and "determining" the specifications of the lines. At an earlier stage, he sometimes marked in red the likely boundaries of central station service in which a canvasser was to work. See, e.g., Doc. 2441; Insull to Arthur Johnson, 12 July 1883; TAE to Electric Tube Co., 25 Oct. 1883; LM 14:143B, 15:516A (*TAED* LBCD1143B, LBCD2516A); TAE to Johnson, 10 Aug. 1883; TAE to George Wilber, 11 Sept. 1883; TAE to Frank Sprague, 18 Sept. 1883; all DF (*TAED* D8352P, D8316AUT, D8316AVW).

4. Edison had queried the day before whether he should send Frank Sprague to help set up the Louisville exposition plant. Insull replied quickly that it was "no use" to do so because "switches regulators & indicators not ordered   I do not understand what they want & isolated company cannot tell   answer." TAE to Insull, 9 July 1883; Insull to TAE, 9 July 1883; both DF (*TAED* D8366F, D8340ZBC).

5. An allusion to Nye 1882, to which Insull referred in Doc. 2466.

6. The Edison Electric Light Co., Ltd.'s 3 July request to "Cable shortest time delivery London two thirteen by thirteen Engines Criterion Theatres" started a flurry of about a dozen telegraphic and postal messages among Insull (some in Edison's name), Gardiner Sims (and Armington & Sims), and the London company. When Sims initially promised the order by 1 August, the London company urgently asked to advance that date. Sims was noncommittal, but Insull pressed him to "not to allow one moments delay"; it is not clear if he met that deadline. Edison Electric Light Co., Ltd., to TAE, 3, 9, 10 July 1883; TAE to Edison Electric Light Co., Ltd., 10 July 1883; LM 1:318F, 319E; LM 2:5A, 5C (*TAED* LM001318F, LM001319E, LM002005A, LM002005C); Sims to TAE, 3 July 1883; TAE to Armington & Sims, 11 July 1883; both DF (*TAED* D8322ZBB, D8316AFS); see related correspondence in T. A. Edison—Outgoing Correspondence, and Electric Light—Foreign—U.K.—General; both DF (*TAED* D8316, D8338); and Letterbook 17 (1883–1884), Lbk. (*TAED* LB017).

7. Joshua Bailey had cabled on 9 July for the "date and price will ship

two certainly perhaps four Jumbos Genoa   Quote Lowest possible."
Insull retransmitted the message to Edison that evening along with the
suggestion to "sell machines at present in shop" for $9,250 each, refer-
ring to those built for the British Edison company but held in New York
when it ran into financial problems (see Doc. 2456 n. 12). After Edison
wired his approval the next morning, Insull offered (in Edison's name)
the dynamos and Armington & Sims engines to Bailey within forty days
of order (presumably to allow for construction of the engines). The
Italian station eventually received as many as five additional Jumbo dynamos from New York, though the editors have no information about
the dates or details of the arrangements. Bailey to TAE, 25 June and
9 July 1883; TAE to Bailey, 10 July 1883; LM 1:317C, 319D; LM 2:5D
(*TAED* LM001317C, LM001319D, LM002005D); Insull to TAE,
9 July 1883; TAE to Insull, 10 July 1883; both DF (*TAED* D8337ZCX,
D8361ZAU); Clarke 1904, 55.

8. Charles Rocap (c. 1856–1932) was the secretary and bookkeeper of
the Edison Machine Works. Doc. 2293 n. 2.

9. The letter to which Insull referred has not been found, but Rocap had recently complained that the price of $375 apiece for the new
100-light T dynamo was too low even in large manufacturing lots. TAE
to Rocap, 5 June 1883, Lbk. 17:58 (*TAED* LB017058); Rocap to Insull,
6 July 1883, DF (*TAED* D8334W).

10. Two-thirds of the price was to be paid in cash, the remainder
applied to Edison's subscription for stock in the Edison Co. for Isolated Lighting. Insull to Frank Hastings, 14 July 1883; Insull to Charles
Dean, 26 July 1883; both DF (*TAED* D8316AGQ, D8316AJN).

11. Sherburne Eaton went to Babylon, N.Y., on Long Island's south
shore, until at least 15 July. He may have gone with Edward S. Tobey,
the postmaster of Boston from 1876 to 1886, whom Spencer Borden
described as "one of the best friends we have anywhere." Tobey reportedly was trying to install Edison lights at the Post Office; early in 1884,
the Edison Electric Light Co. secured a contract to install a 500-lamp
system there. Eaton to TAE, 12 July 1883; Borden to TAE, 2 July
1883; both DF (*TAED* D8327ZAO, D8325ZBD); Obituary, *Boston
Daily Globe*, 30 Mar. 1891, 1; "Faits Diverse," *La Lumière Électrique* 11
(26 Jan. 1884): 219.

12. William Farley Brewster (1846–1923), a childhood friend of Edison's, went into business after serving in the Union Army and, after the
Civil War, as a government surveyor in the West. He was manager of
the Halls Safe & Lock Co. in New York when he arranged with Edison
to become a Special Agent to the President of the Edison Electric Light
Co. for promoting the central station business. In that role, he was involved in efforts to electrify larger towns and cities, including St. Louis.
Brewster subsequently worked for General Electric and Commonwealth Edison in Chicago. "Brewster, William," Pioneers Bio.; Obituary, *Chicago Daily Tribune*, 10 Nov. 1923, 6; Brewster to TAE, 7 Dec.
and 23 May 1883, both DF (*TAED* D8340ZHU, D8340T1); see also
Doc. 2758.

13. That is, to rush headlong into a confrontation without regard for
consequences. *OED*, s.v. "Bald-headed," b.

14. The engine lubricators ran dry during the 3 July test, melting
the babbit bearings. The engine was hurriedly repaired overnight to be

ready for the station's 4 July start. Sprague 1904 (566) recounts this incident. Gearhart 1936, 177; William Andrews to Harry Keefer, 15 Feb. 1916, EGF (*TAED* E1633AC; *TAEM* 266:644); Andrews to Insull, 11 July 1883, DF (*TAED* D8361ZAY).

15. Insull scolded Hanington for submitting a bill (not found) for a total of eleven days of labor by twelve different men. Hanington argued that Sunbury was "very much mixed from the start . . . and I dont think more than ½ of the people that had a hand in it understood it . . . [and] I was not the only one to go to school on this job." Insull to Hanington, 10 July 1883; Hanington to Insull, 10 July 1883; both DF (*TAED* D8316AFE, D8340ZBD).

–2486–

To George Bliss

[Menlo Park,] July 13th. [188]3

Dear Sir:—

I have your favor of the 6th. inst,[1] and would have replied to it earlier but have been out of town for the last ten days

My Canvasser at the present time is at Davenport, and if you wish to communicate with him, please address C. H. Cooke,[2] Kimball House, Davenport, Iowa.

I thoroughly agree with you as to its being a very bad policy to work up a great deal of excitement in a town, and then have to wait several weeks for an estimate. I think the best way would be for you to order a canvass of a town, whenever you consider there is a reasonable prospects of a company being formed, if the estimates &c. are satisfactory.

I do not think you need fear opposition very much. I have some little experience with central station work, and I have come to the conclusion that it is hard enough for us to conduct the business ourselves, and I have very little fear of opponents who know nothing whatever of what they are dealing with. This is a business in which I know they cannot touch you. They may talk, but that is all they will do. They will never attempt to put in a central station plant, or if they should attempt it, their first experience would be sufficient.

If you will let me know exactly what information you want, I will be happy to supply you with it. I do not think that there has been any attempt to withhold information, as I infer you imagine there has been from a paragraph in your letter.[3] Yours very truly,

TL (carbon copy), NjWOE, DF (*TAED* D8316AGA).

1. Edison may not have composed this reply himself, having instructed "Insull attend" on Bliss's 6 July letter. Bliss had responded to Edison's 3 July letter, which was based on the draft in Doc. 2476. Bliss

directed that the canvasser, after finishing in Davenport, Iowa, should travel to Appleton, Neenah, and Milwaukee, Wis. He went on to argue that it was "bad policy, to work up interest in a town, unless the party can have immediate attention. After the interest has once cooled, it requires three times the effort, to revive again." He also worried that the interval would give competing companies time to gain a foothold. Bliss to TAE, 6 July 1883, DF (*TAED* D8364ZAG).

2. Charles H. Cooke (1847?–1914) was a civil engineer who canvassed for central stations until about August 1884. Cooke had worked on construction of the Union Pacific Railroad and had spent several years with the Long Island Railroad. He applied again to Edison for employment in 1910, when he was already engaged in a long effort to build a hydroelectric dam on the Delaware River. Obituary, *NYT,* 3 Aug. 1914, 11; Cooke to TAE, 25 Oct. 1910, DF (*TAED* D1020ACG).

3. Bliss had written (see note 1) that "With your organization, you ought to be able to do the work cheaper than any one else, but what this and the local Co's require, is such information as will make them competent to appreciate the situation. If information is with held, it will look as if you were making undue profit."

-2487-

From William Hammer

Berlin July 16 1883[a]

My Dear Mr Edison

In a recent letter to Mr Rathenau you requested that you might be kept acquainted with the work which the German Edison Company had in progress and under consideration from time to time[1] and Mr Rathenau has requested that I shall assume this responsibility which I do with pleasure and I will endeavor to keep you posted upon all matters relative to our company's operations which will prove of interest to you. At present our operations are not very extensive as the practical organization of the Company has but just been effected but we look for great things in the near future & believe we have the practical monopoly for the Incandescent Llight in the German Empire. The company has but recently completed several installations in different parts of Germany and has now in hand a number of others both inside & outside of Berlin a further reference to which I will make at another time as there are many important topics which I shall have to refer to.[2] We have a very fine though small Installation of the Edison Light at the Berlin Sanitary Exhibition[3] and it is one of the principal features of this successful & largely patronized Exhibition. We shall send you further information of this also. Next we have here in Berlin two important Club houses the Resource and the Union Club[4] lit up with the light the former employing a "Z" the latter an "L" Dynamo   We have just completed the plans for a central station which will be commenced upon this week[5] which will take in both these clubs which will employ they light hereafter throughout the entire buildings and also take in the Berlin Aquarium[6] and an adjoining private house, besides this small installation we have also under consideration a larger installation situated in the centre of a large and important block of buildings which on account of the character of the business carried on i.e. (Cafe's Restaurants, Hotels Stores & Private houses) will give us a large number of lamp hours   the property we have in view for the central station will allow of our employing three (3) of the "C" or central station Dynamos. There is also a project under consideration which is I believe the largest of its kind as yet brought forward although it is almost too early to make any definite statements concerning it. It consists of a project for establishing a central station for fifty thousand 50,000 Lamps which will take in one of the most important centres of Berlin and include[b] some of its principal Governmental buildings   There are a number of other projects on foot for central station & Isolated Lighting of which I will let you know hereafter, but I am certain in a short time you will see that the Edison people in Germany do not lack enterprise and that the Company is in the hands of men of energy and ability. The Co. has recently taken very fine offices upon one of the principal corners in Berlin[7] has appointed agents throughout the Empire and is now engaged in a systematic course of advertising in the principal Technical, Trades & other Journals in order to bring the Edison light before the people still it is considered of f vital importance that the company should have installations and[c] above all a central station in operation in Berlin as a practical demonstration of the capabilities and commercial value of the Edison Light, this done in Berlin we fear nothing elsewhere. Now leaving the further[b] discussion of the Co's projected work for a future occassion I will speak of a number of important matters which relate to other affairs.

Plans, Blue Prints &c[d]   I   Will you kindly have a set of all drawings, Blue Prints &c relative to Central Station work, Town Plant work &c sent to us at once   You will see the importance of this from my letter and I beg to state that in these matters the company desires to look to New York rather than to Paris & other companies and Mr Rathenau requests me to say that the Co here feels these things are its due but they are very willing to bear any expense consequent upon the preparation & completingon of them

Estimates &c[d] II We should like also copies of any estimates for contruction & maintenance of central stations of various numbers of lamps including any statements you may possess relative to the central station in New York for our guidance in our work here.

Distribution System.[d] III We should like to know if you consider it necessary in the Event of our starting a central station such as I have herein mentioned that we should lay it out similar to your plan of German silver coils made at Menlo Park[8] & since made by Mr Batchelor or do you consider that if plans, distances, approximate distribution of lights &c, were sent either to New York or Paris such an Estimate of proper distribution, main feeding points &c could be made

Three wire system Samples &c[d] IV We have just sent to Paris for a complete set of samples of the new three wire system of distribution and we shall be glad to hear from you the results of any experiments, Estimates &c based upon it.[9] I will also say that Mr Rathenau wishes that at any time as new features and improvements are made that you will send us at once samples of same and inform us (by cable if necessary) of any changes desirable. I have mentioned this in a recent letter to Bergmann & Co[10] & we trust we shall always get at once the lastest & the best and I might also add that we hope our recent order will be filled and forwarded as soon as possible. Forwarding Goods[d] Please see that all B & Co's cases and all from Machine Works are carefully addressed, classified, & numbered with some distinguishing marks theron for our guidance I remember we had at times much trouble from want of this care in London more especially in material received from Goerck St.

Bulletins &c[d] V. Please have a package of all the back Bulletins sent us and as each number is issued please see that we receive a number. I have mentioned this in a letter sent to Major Eaton[11] but mention it again to you as we have none of the numbers at present

Lamp Factory[d] VI. Another important undertaking the Company now has in view is the Establishing of a Lamp Factory and the Directors desire that you will send us at once the best man you have who is capable of organizing & taking charge of our factory and if you have such a man whom you can spare but for a time the company would like him at least to remain until things are in running order and we have a man of our own in charge.[12]

VII The Company desires also an estimate of what you

would furnish complete for an Installation Lamp Factory for Germany capable of supplying five hundred (500) Lamps per diem.[13] This to include the Archimedes Pump,[14] & piping couple, Resistance Boxes for Pumps, Regulating devices, and in fact all the machinery tools & appurtenences for the Glass Blowing, Plating, Mounting, Plaster fitting & all departments save in the carbonizing & Fibre[e] — as we shall probably get our carbons ready for mounting direct from Paris. This estimate it is scarcely necessary for me to add we desire at earliest possible date and in the event of our at once securing a man from you he must come with all necessary plans samples &c in order that our company may not repeat the great delays and enormous outlays of the Co in France & while we trust greatly to profit by both their & your experience we prefer as far as possible to deal directly with you in these matters and feel assured of your assistance.

Lamps.[d] VIII   July 17-83   I have just shown this letter to Mr Rathenau and he has approved of it and requests me to add one or two things further, firstly please send us at once an estimate for supplying in large quantities the outside bulbs[f] The inside parts[15f] with platina wires & clamps already prepared for mounting carbons in so   and also what you will furnish us in large quantities the   carbons[f] all ready for mounting as we desire if possible to make such an arrangement with you if possible for this will facilitate greatly the starting of our lamp factory & getting it in good running shape

Man.[d] IX. Please cable when we can expect man from America[16]

Cash.[d] X   Mr Rathenau says we can at any time deliver you cash in New York through representative bankers there upon statement of delivery of goods

Dynamos.[d] XI. The new "K" machine is on way now to Siemens & Halske[17] will test before taking apart   other machine not heard from as yet. As we have no stock of Dynamos & must have at once please hurry up other models as fast as possible

Steam[d]   XII. Mr Rathenau suggested the idea of using in a central station a separate condenser for all the engines which are is[b] driven by an independent engine   he estimates a saving of 25% at least thereby. What do you think of it? Also do you not consider that there is a valuable income to be derived from employing steam in heating during winter months in connection with lighting thus effecting a great saving in running expenses in the day.

<u>XIII</u>. What do you think of compound wound machines? Nearly all the principal makers are using them & we believe there is much in them.[18] Dr Siemens in a recent paper publishes tests showing the perfect regulation[19] & Crompton[20] of London also has   Have you made any experiments upon this line and do you not think it cheaper more reliable & less complicated than the automatic regulator   I will now close with very sincere regards from all the people here   remaining very truly Yours

Wm. J. Hammer[21]

ALS, NjWOE, DF (*TAED* D8337ZDC). Letterhead of Deutsche Edison Gesellschaft. [a]"188" preprinted. [b]Interlined above. [c]Repeated at end of one page and beginning of next. [d]Heading written in left margin. [e]"Fibre" written in margin. [f]Multiply underlined.

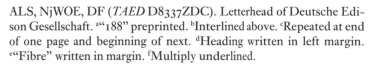

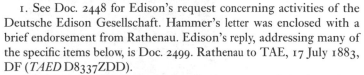

William Hammer's sketch of the first flashing "Edison" sign for the 1882 exhibition at London's Crystal Palace.

Hammer's March 1880 sketch of a "flash interrupter."

1. See Doc. 2448 for Edison's request concerning activities of the Deutsche Edison Gesellschaft. Hammer's letter was enclosed with a brief endorsement from Rathenau. Edison's reply, addressing many of the specific items below, is Doc. 2499. Rathenau to TAE, 17 July 1883, DF (*TAED* D8337ZDD).

2. A resume of installations made by the company is in the Edison Electric Light Co. Bulletin 22:14–15, 9 Apr. 1884, CR (*TAED* CB022). These included a number of Berlin theaters and private clubs, such as the Café Bauer. The Café Bauer was a well-known coffee house, one of the first in the Viennese style in Berlin, and it became a model for others. Petras 1994, 33–70.

3. The Berlin Sanitary Exhibition, devoted to the prevention of disease and the improvement of public and personal health, opened on 12 May. The restaurant on the grounds was lighted by 250 Edison lamps. More dramatic was an illuminated sign devised by Hammer that flashed "EDISON" in light bulbs over the entrance to the Edison pavilion. Hammer adapted its motor-driven commutator from a smaller, hand-driven flashing Edison sign that he had designed for the 1882 electrical exhibition at the Crystal Palace in London. His sketch of the 1882 display differs from an engraving published in a London newspaper (cf. *TAEB* 6:833), the only other representation of it that the editors have found. Hammer apparently conceived the idea at Menlo Park, where he in 1880 sketched "Edison" signs actuated by a clockwork mechanism. "The Berlin Exhibition of Hygiene in 1882–83," *Science* 6 (1885): 36–37; Hughes 1983, 69; Edison Electric Light Co. Bulletin 19:27, 15 Aug. 1883, CR (*TAED* CB019); WJH, Ser. 4, Box 81, DSI-MAII neg. 2003-35552; WJH, Ser. 1, Box 19, Folder 6; Ser. 1, Box 13, Folder 1; Ser. 4, Boxes 99 and 97 (*TAED* X098HAL, X098E28A, X098E01).

4. The Resource Club (established 1797) and the Union Club were exclusive social organizations on the Schadow Strasse in central Berlin, near a number of banks, monuments, and important civil buildings. Newnham-Davis 1908, 170; Baedeker 1890, 23.

Edison pavilion at the Berlin hygiene exposition, with the prominent flashing sign designed by William Hammer.

5. After months of negotiations, the Berlin municipal authorities and DEG reached an agreement in February 1884 allowing the use of city streets for underground conductors in two central station districts. Hammer to TAE, 9 Nov. 1883, DF (*TAED* D8337ZGH); Hughes 1983, 71–73.

6. The Berlin Aquarium, opened in 1869, was located at the intersection of Schadow Strasse and Unter den Linden, roughly a quarter mile north of the social clubs mentioned above. *American Cyclopaedia*, s.v. "Brehm, Alfred Edmund"; Baedeker 1890, 23.

7. The letterhead used by Hammer gave the company's offices at 96 Leipziger Strasse, which was about two-thirds of a mile south of the Aquarium on a major east-west thoroughfare. The editors have found no other particulars of this location.

8. Hammer referred to the physical modeling of distribution systems performed by Hermann Claudius at Menlo Park. See Doc. 2466 esp. n. 4.

9. Edison promised without comment in Doc. 2499 to send samples of the three-wire system, though the editors have not determined specifically what these were.

10. Not found.

11. Not found.

12. Edison had refused requests from Charles Batchelor and Joshua Bailey in March to designate someone (such as Francis Upton) to start a lamp factory in Germany. He also rejected a similar request from the Société Électrique Edison (see Docs. 2406 and 2499 esp. n. 5). Instead, Alfred Chatard, a managing director of the French manufacturing firm, assigned James Hipple, who went to Paris from the Menlo Park lamp factory in 1881, to fulfill the French company's obligations to the German investors. Chatard expressed serious misgivings that in Paris "nobody has been initiated to all the work of the lamps, so that we are in reality in the hands of Hippel." Chatard to TAE, 9 July 1883, Batchelor (*TAED* MB091B); see also Doc. 2554.

13. The capacity to make 500 lamps per day, which Hammer desired, was roughly half the output of Edison's factory in Harrison, N.J. See Doc. 2343 (headnote).

14. Edison adopted an Archimedean screw in late 1880 to recycle mercury to the vacuum pumps in his lamp factory. See Doc. 2010 esp. n. 2.

15. Hammer referred to the glass tube used to support the carbon filament and lead-in wires. It was fabricated separately from the "outside bulb." See Doc. 2097 (headnote).

16. See note 12.

17. Siemens & Halske, founded in 1847, was the Berlin manufacturing arm of the constellation of Siemens family enterprises and a major manufacturer of telegraphic, telephonic, and arc lighting equipment. Doc. 1851 n. 1.

18. A compound winding consists of electrically separate wire coils (not necessarily of the same resistance) in a single magnet winding circuit. One coil typically was energized by the dynamo's own current and the other from an independent source, such as a battery or second machine. This arrangement will, at a particular armature speed, produce a uniform voltage regardless of changes in electrical load. Priority for inventing such a machine, commonly called a "self-regulating dynamo," was claimed by several individuals, but theoretical advances in 1881 by Marcel Deprez and 1882 by Gisbert Kapp made it possible to calculate the critical armature speed with respect to other design criteria. Compound windings were thereafter adopted rapidly and widely by manufacturers. Edison also tried such an arrangement but abandoned it, reportedly because of its great sensitivity to variations in armature speed. Schellen 1884, 481–91; Thompson 1902, 17, 54 56; Kapp 1882–83, 114–15; "Letters to the Editor," *Electrician* 10 (23 Dec. 1882): 135–36, 179; Alglave and Boulard 1884, 345–47; see also Doc. 2499 esp. n. 7.

19. Hammer probably referred to Richter 1883, which gave a general mathematical treatment of dynamo self-regulation and graphical interpretations of experiments with compound windings by Siemens & Halske in the latter half of 1882. It appeared in April 1883 and was quickly republished in English. The translation closed (p. 21) with the statement that "The compound system has been since so developed by Siemens and Halske for smaller and larger machines, even to driving 200 A lamps of Edison, that with a constant speed any number of lamps can be put in or out of the system, without affecting the burning of the others."

20. Rookes E. B. Crompton was a British electrical engineer and manufacturer. He patented (with Gisbert Kapp) a compound-winding dynamo and is credited (with Kapp) for coining the term (Doc. 2342 n. 12; Thompson 1896, 227; Thompson 1902, 57; Bowers 1969, 13–14). Hammer probably referred to Kapp 1882–83 which, from trials of Crompton's machine, laid out a theory of compound machines.

21. William J. Hammer (1858–1934) an Edison associate since 1879, went from London to Berlin as chief engineer of DEG earlier in 1883. Doc. 1872 n. 7 and *TAEB* 5 App. 2.

–2488–

Samuel Insull to Charles Campbell[1]

[New York,] July 17th. [188]3

Dear Sir:—

We find it necessary that somebody should be placed in charge of the Engineering Department, and Mr. Edison has decided to put you in that position.[2]

I shall be glad if you will proceed to fully organize your disorganized establishment, somewhat on the following basis:—

FIRST: The form of estimate, which I prepared with you a short time back, should be got in order as quickly as possible.

SECOND. A book should be prepared giving exactly the cost price of everything we use in connection with our work, until these figures can be kept in the book. I think that a memorandum should be prepared, classified and kept in proper order. This I am sure can be done in a temporary manner.

THIRD. I desire that your Department shall take charge of all maps and blue prints used by us, whether of central stations the maps of town [---][a] showing lots used on the electrical determinations.

For the present, the canvassing and mapping department will remain entirely distinct from your Department, although you and Mr. Richardson[3] will have to agree as to the records being kept together and in proper shape.

The form of register is now in the hands of the printer, which will take care of the recording of this class of data.[4]

Any correspondence that it may be necessary to conduct in connection with your Department, I desire shall be done through my office, or else brought to my notice before it is sent out. The best way would be for you to give me a memorandum of what you want and where you want it from, as otherwise it is impossible for me to keep a run of matters going on in your office. Yours truly,

TL (carbon copy), NjWOE, DF (*TAED* D8316AHL). [a]Canceled.

1. Charles H. Campbell (b. 1854?) was a civil engineer and draftsman who had been working under Gustav Soldan at the Edison Machine Works. He had recently become involved with making drawings and plans for the Construction Dept. Campbell had an undefined connection with A. Campbell's Sons & Co., makers of printing presses in New York City. Doc. 2272 n. 3; Jehl 1937–41, 930; U.S. Census Bureau 1970 (1880), roll T9_855, p. 353.4000, image 0713 (Brooklyn, Kings, N.Y.); letterhead of Campbell to TAE, 9 Aug. 1884, DF (*TAED* D8413ZAX).

2. Insull explained to Charles King, who apparently had also been considered for the position, that Edison chose Campbell "in consequence of the long time he has been with us and the Light Company, and his full acquaintance with the requirements of our work." Campbell remained in charge of the Engineering Dept. until Edison dissolved it in January 1884. Insull to King, 17 July 1883, DF (*TAED* D8316AHP); see Doc. 2595.

3. Charles F. Richardson (b. 1845?) of Freehold, N.J., was the son of two educators and in 1880 was himself a teacher and a member of the town commission. Edison hired him to oversee the growing volume of canvassing and mapping activities, preliminary to the design work that would fall to the Engineering Dept. Richardson started work on 2 July 1883; his major duty seems to have been processing canvass reports, though he was also called upon to make some maps. He remained in the position only until 10 August, when Edison terminated him on the grounds of inability to "make maps & records of our Canvasses with great rapidity & in an accurate manner." U.S. Census Bureau 1970 (1880), roll T9_791, p. 73.1000, image 0273 (Freehold, Monmouth, N.J.); "New-Jersey Charter Elections," *NYT*, 4 May 1880, 5; Richardson to Frank Hastings, 27 June 1883; Richardson to TAE, 14 Aug. 1883; both DF (*TAED* D8327ZAI, D8342ZAK); TAE to Richardson, 10 Aug. 1883, Lbk. 13:22 (*TAED* LB013022).

4. The form was most likely that designed by canvasser George Wilber and sent to Insull on 29 June. Edison tendered a sample to stationers Arthur & Bonnell on 12 July to be printed as a "Proof in Book form." At least a few of the new books were ready by late September, but they were not generally distributed to canvassers before the end of October. Wilber to Insull, 29 June 1883; Paul Dyer to Insull, 27 Sept. 1883; Insull to Dyer, 29 Sept. 1883; TAE to Wilber, 24 Oct. 1883; all DF (*TAED* D8344O, D8342ZBN, D8316AXL, D8316BCR); TAE to Arthur & Bonnell, 12 July 1883, Lbk. 17:161 (*TAED* LB017161).

–2489–

To Richard Southgate[1]

[New York,] July 19th. [188]3

Dear Sir:—

Please reserve suit of three rooms, fronting on the sea, for the first Saturday in August, for Mrs. Edison,[2] her sister[3] and my daughter.[4] They will probably require the rooms for about a week. Kindly let me know per return, whether you can do this.[5] Very truly yours,

TL (carbon copy), NjWOE, DF (*TAED* D8316AII).

1. Richard H. Southgate (1847–1912) took over management of the Long Beach Hotel at Long Beach, Long Island in 1881. A veteran of the hotel business in Saratoga and Montreal, Southgate immediately undertook expansions and improvements. At this time, the establishment had 500 rooms and was considered a fashionable destination. Long Beach is located on a narrow island in the Atlantic Ocean just off Long Island, some two dozen miles from Manhattan. Letterhead of Southgate to TAE, 20 July 1883, DF (*TAED* D8314J); Obituary, *NYT,* 4 Mar. 1912, 11; "Long Beach Hotel Leased," ibid., 17 Feb. 1881, 8; "South Shore Resorts," ibid., 29 May 1883, 2; *WGD,* s.v. "Long Beach 3"; on the development of a vacation infrastructure in the second half of the nineteenth century, and particularly the class and gender implications of resort hotels, see Aron 1999, chaps. 2 and 3.

2. Mary Stilwell Edison, married to Thomas since 1871 (Doc. 218 n. 5). She was in Menlo Park, where she had spent an unknown amount of time since at least early June. She had recently corresponded with Samuel Insull about a violent incident between her brother-in-law Cornelius Van Cleve, an Edison employee, and another worker at Menlo Park. Mary Edison to Samuel Insull, 6, 15, and 25 June 1883; Insull to Mary Edison, 6 July 1883; all DF (*TAED* D8314G1, D8314H1, D8314I1, D8316AFA); evidence of the family in Menlo Park can also be found in the following bills: J. F. Donnel & Co., 4 and 12 June, 1883; J. Brockie & Co., 31 July 1883; J. A. Hopper's Sons, 1 Sept. 1883; all Cat. 1164, Accts., NjWOE.

3. This was Mary's sister Alice Stilwell Holzer (1853–1932) whose husband, William Holzer, came out with Edison for a visit of "1¾ days." Southgate Hotel bill, 13 Aug. 1883, Cat. 1164, Accts., NjWOE.

4. Marion ("Dot") Edison, Thomas and Mary's oldest child, born in February 1873 (d. 1965). *TAEB* 1, chap. 11 introduction.

5. Southgate replied that he would hold a reservation. On 4 August, the first Saturday of the month, Mary Edison telegraphed from Menlo Park for Samuel Insull to "send carriage and express wagon to Courtland St ferry to meet train that leaves here at three twenty." The family stayed in Cottage C for nine days and Edison joined them for "1¾ days." The bill, including a room for the Holzers, extra meals to the rooms, liquor, laundry, and a maid, totaled $264.55. Southgate to TAE, 20 and 27 July 1883; TAE to Southgate, 26 and 28 July 1883; Mary Edison to Insull 4 and 11 Aug. 1883; all DF (*TAED* D8314J, D8314K, D8316AJE, D8316AKG, D8314K1, D8314L1); Southgate Hotel bill, 13 Aug. 1883, Cat. 1164, Accts., NjWOE.

–2490–

From William Andrews

Central Station, Sunbury   July 20th/83.

Dear Sir

Your last budget to hand.[1] There was a bug in the dynamo connections that I sent you,[2] which I did not perceive till yesterday. I therefore herewith enclose a plan[a] of new way of Connecting spare dynamo, to the other two. I have carefully looked this over, and am satisfied that it is all right, and I enclose with sketch a copy of directions which explain themselves, and which will I think meet every necessary requirement.[3] With a spare dynamo there is no need of a break-down switch such as we have at Sunbury[4] Connecting red and blue wires together because it is improbable that two out of three dynamo should become disabled. Still if you think it best, it is easily added without altering any of present arrangements

I do not want to use a <u>bank of resistances</u> such as you write about.[5] It is an unnecessary and costly complication. All I want is simply 3 H resistance boxes—one for each dynamo   What need is there for more?

I have made the directions as simple as possible, and by going over them a few times any man of average intelligence would get them off by heart[6]

Please look them over and see if you can find a bug in them. You will perceive that instead of a two point field switch I now require a 4 point one   This will be only a small additional expense, and is absolutely necessary in order to make and reverse field of spare dynamo

I will write to you further on the Shamokin business tomorrow   It is 11 PM now, so I must close this in order to catch mail— Yours very truly

W. S. Andrews[7]

⟨Dont you see that it is essential that if the spare is on one side that the two boxes must be regulated simultaneously   you cannot raise Volts on that side by raising one as it will run the other as a motor. suppose you had 5 machines on [---][b] A side & 5 on B side dont you see that all the field resistances must be worked simultaneously on all the Dynamos on the B side if volts are to be varied on that side   How in hell could we have seperate boxes— The banks are ordered & will be sent⟩[8]

ALS, NjWOE, DF (*TAED* D8361ZBH). [a]"a plan" interlined above. [b]Canceled.

1. Andrews presumably used "budget" in the figurative sense of a long and newsy letter; the editors have not found such a letter from Edison at this time. *OED*, s.v. "budget," 3.

2. Not found, but see note 7.

3. Andrews may have referred to his sketch of a switch from this date. The extant drawing was torn from a larger page, which has not been found; see also note 7. Andrews to TAE, 20 July 1883, DF (*TAED* D8343M).

4. Edison's experience with the so-called "breakdown switch" at Sunbury reportedly prompted him to begin a competition among his assistants for a better design. The winning switch, invented by Andrews, became the standard at similar stations for many years. Dyer and Martin 1910, 426.

5. On 16 July, Andrews expressed his interpretation of Edison's plan to ground a central station during storms. He sketched his understanding that he should ground each dynamo "through 100 ohms resis. on each of the three wires." Edison replied that he had "no such intention" to do so: "All I do is to ground the three wires of the omnibus main through a hundred ohms of iron spiral resistances." Andrews to TAE, 16 July 1883; TAE to Andrews, 19 July 1883; both DF (*TAED* D8361ZBE, D8316AIE).

6. Andrews referred to the "Instructions relating to Dynamos &c," which Edison had solicited; see Doc. 2500 n. 1.

7. Edison had removed Andrews from the Testing Room at the Edison Machine Works and appointed him chief electrical engineer of the Construction Dept. about 1 June. "Andrews, William Symes," Pioneers Bio.

8. Samuel Insull sent a reply based on Edison's marginalia a week later, explaining that he and Edison were "so busily engaged within the last week, that it has been quite impossible to answer your several letters." In the meantime, Andrews devised "A more simple method of connecting up spare dynamo," a drawing of which he sent to Edison (Insull to Andrews, 27 July 1883; Andrews to TAE, 24 July 1883, enclosing 24 July sketch; all DF [*TAED* D8316AJS, D8361ZBI, D8361ZBJ]). However, these connections remained a problem. Frank Sprague and Andrews took up the matter again during the fall in correspondence with Edison (Sprague to TAE, 5 Sept. [p. 3], 10 Nov. 1883; Sprague drawing, 8 Nov. 1883; Andrews to Insull, 5 Nov. 1883; all DF [*TAED* D8343ZAU, D8343ZCO, D8343ZCM, D8348ZAN]). A number of undated 1883 drawings (some by Sprague) of different connections for the spare dynamo were collected in Undated Notes and Drawings (c. 1882–1886), Lab. (*TAED* NSUN08A).

–2491–

From A. H. Seymour

NEW YORK, 23rd July 1883[a]

Received of Thos. A Edison the sum of one hundred dollars being subscription to Fund being raised by "Telegraphers Brotherhood" for relief of the Telegraphers Operators now on Strike[1]

A H Seymour[2]   Chairman Rlf Com

DS, NjWOE, DF (*TAED* D8303ZEP). Letterhead of Thomas A. Edison; written by Samuel Insull. [a]"NEW YORK," and "188" preprinted.

1. The Brotherhood of Telegraphers called a nationwide strike of operators and linemen against Western Union and several lesser companies on 19 July. Though antagonism between the two sides was of long standing, the immediate issues were demands for shorter shifts, abolition of compulsory Sunday work, and equal pay for men and women. The Brotherhood had put away substantial strike funds and could count on aid from the Knights of Labor, to which it belonged. The strikers also enjoyed widespread public sympathy, and donations and benefits fattened strike funds in many cities. Despite its auspicious

start, the strike began to falter in early August and collapsed a few weeks later. In August, Josiah Reiff, Edison's one-time associate, proposed using him as an intermediary in a clandestine effort to broker a settlement with the Baltimore & Ohio Telegraph. Edison specifically endorsed this idea, but nothing came of it. Gabler 1988, 5–16, 19–25, 205–6; Reiff to Robert Garrett, 15 Aug. 1883, Garrett Papers.

According to one news account in the walkout's first days, Edison "telegraphed the brotherhood in New York to draw on him for any amount needed to sustain the strike." He reportedly sent another check a week or two later, prompting the *Newark Advertiser,* in a notice reprinted by a New Brunswick (N.J.) paper, to observe that "Edison's fortune was largely derived from the Western Union Company, and the officials of the latter are not delighted at this use of the funds which were once the company's." Edison received at least one personal request for help, from a telegraphic acquaintance seeking employment. He declined it, instructing Samuel Insull to write and "Say I am doing all I can here for the boys   Cant very well do more." "The Situation in this City," *Washington Post,* 21 July 1883, 1; *Daily Times* (New Brunswick, N.J.), 6 Aug. 1883, 2; TAE marginalia on Charles Higdon to TAE, 30 July 1883, DF (*TAED* D8303ZES).

2. Alfred H. Seymour (1836?–1901), a thirty-year telegraphic veteran, was chief operator for Western Union. He represented the Brotherhood before Congress during the strike. He later managed the Bankers and Merchants Telegraph Co., then worked for the United Press and the *Brooklyn Times.* Obituary, *NYT,* 18 Jan. 1901, 7; Gabler 1988, 92–93, 205–206.

–2492–

William Lloyd Garrison, Jr., to Samuel Insull

Boston, July 23, 1883.

Dear Sir,

As the work at Brockton is progressing satisfactorily, does it not seem desirable to assure ourselves of some customers? Else how can we make any sort of an exhibition when the work is complete? In accordance with Mr. Edison's views we have made no effort to get customers. The cost of wiring being so much in excess of gas piping not a single individual will do it at first at his own risk.

We are not ambitious to get a large constituency at once but do feel it necessary to make a beginning in order to show the light.— Shall we not be obliged to offer to wire some important building at our own expense with the understanding that eventually the wiring is to be paid for if the light proves satisfactory?

Please advise me.— Very truly,

W. L. Garrison, Treas.

⟨We can wire 2 or 3 prominent places a week before we start with the understanding that if the Light is OK [&?][a]= we shall have to stand cost this wiring but ~~with understanding that~~ will make out a bill & simply receipt it, so there will be no further free wiring & this cant be used against us— ~~from~~ Experience in Sunbury Pa we are getting out an Entirely new line of devices to cheapen wiring way down so there will be no complaints)[1]

ALS, NjWOE, DF (*TAED* D8346T). [a]Canceled.

1. The editors have not found a formal reply to this letter. Edison probably did not have any specific "2 or 3 prominent places" in mind. Among the Brockton plant's first customers were the City Theater, reportedly the first to be lighted from a central station, and the fire station, which was fully electrified, including power-operated doors. The Whipple house, at 42 Green St., was the first private residence in Brockton lit with electricity. Carroll 1989, 53; Edison Electric Illuminating Co. of Brockton list of customers, 17 Nov. 1883, DF (*TAED* D8346ZBN1).

-2493-

To Theodore Waterhouse

New York, July 24th 1883.

My Dear Sir:—

I have read with attention your favor of the 7th. inst[1] and have received full explanations from Mr. Johnson of the terms of the paper submitted to me respecting the proposed amalgamation with the Swan Company.[2]

I have to thank you for the clearness and fullness of your statement of reasons supposed to be in favor of adopting that plan.

It would be now idle and out of place to debate any longer whether such an amalgamation ought at any time to be made.

I yield to the claim which is made with great apparent reason, that the better knowledge of your directors of the local and temporary circumstances gives them the right to have their opinion prevail.

The increase of capital and of interest thereon, accompanied by a decrease of my deferred interest on "B" shares must, as you will allow, make it a serious question for me to consider whether in any case I shall personally gain by this change. I learn from persons accustomed to English rates of interest that seven and a half per cent is regarded as a very high rate and have also heard the opinion expressed that a cumulative dividend of seven and a half per cent. may practically debar the "B" shareholders from realizing,[3] for a long time at least, any dividend which will compensate for my labors or for the actual outlay (exceeding what any shareholder in your company has paid for shares) incurred in producing the inventions which you now own.

I cannot judge whether this is so; I am however strongly desirous to aid the holders of "A" shares in realizing the profits which they have anticipated. I have therefore, after consultation with Mr. Fabbri, Mr. Lowrey[4] and Mr. Johnson,[5] decided to withdraw all opposition to a practical amalgamation if it can be done with the following modifications of your plan:—[6]

First:— The preferred annual dividends upon "A" shares shall be six per cent. cumulative; the division after that payment to be as proposed.

Second:— The Company shall be called the Edison Electric Light Company; or at least shall be distinguished by my name in its title solely. The company may at all times and in any manner put forward the fact that it is the owner of the Edison, Swan and Hopkinson or other patents. To this I have no objection.

The last condition is made solely on business grounds of great importance to me, and not at all upon any such feeling as might be naturally imputed to me of wishing to gain in this way a concession as to the disputed claims of Mr. Swan and myself concerning lamp patents. I leave out of view the fact that we are disputing these claims; but the property which this company is to own and operate is an Electric Lighting system, to which I have contributed other inventions as indispensable to commercial success, my title to which is not as I understand disputed anywhere. Moreover I am under contract with your company in respect to all my future inventions relating to Electric Light. I expect continually to make additions, some of which may be of great value.[7] I am also largely interested in every other country in these inventions and in their good fame. If because Mr. Swan and I are disputing in respect to one element of an Electric Lighting system, I should consent to have all the other great variety of inventions made or to be made by me merged and their identity lost in a common name, the pecuniary loss to me might be greater than any gain which I can expect from even the most successful operations on your part in England.

It would be better for me to lose at once and for ever the interest which I now have in your company, and be free to make other connections in respect to future inventions, than to lose the advantage, in other countries, of having my system clearly recognized. I remain in this country, and wherever else I can, as large an owner as possible in my inventions. I have never parted with any of my holdings except when compelled in order to carry on my various works. Other holders are free to divest themselves of their holdings when they can do so with a profit; but I am bound by pride of reputation and by pride and interest in my work, to remain interested in the business. I expect to be a large owner in all companies employing my[a] inventions after most persons now interested shall have sold out and retired with their profits.

You will hardly expect me to continue working to build up by new inventions and improvements a business in which the identity of my contributions has been lost; especially when the preservation of that identity is so necessary to the value of my similar interests in all other countries.

If Mr. Swan remains interested as I do; and if his contributions heretofore have been as extensive and various as mine; and if he intends continuing to devote an active life to making constant additions to the property of the amalgamated company; and if his pecuniary interests in other countries are subject to be affected as mine are by the use or non-use of his name in England; then his friends may with equal force say what I have said. In that case there would be an insurmountable obstacle to an amalgamation, since neither of us ought, as I think, to give away whatever advantage is liable to accrue from proper credit being given in the course of the development of our respective labors.

I believe the force of these reasons for declining to favor or participate in any amalgamation which does not conform to the above conditions will be admitted by yourself and by the directors of your company; and I hope also by the men of business who compose the Swan Company. In all other respects I accept what is proposed.

Should these conditions be conceded I shall then wish to have the specific form of the amalgamation and of all the necessary papers submitted to me for approval before being adopted.[8] Yours very truly

TL (carbon copy), NjWOE, DF (*TAED* D8338ZAZ). A typed draft is in DF (*TAED* D8338ZBA). [a]Interlined above in unknown hand.

1. Doc. 2483.

2. Edward Johnson returned to New York from London late on 15 July; see also Doc. 2467. The "paper" to which Edison referred is presumably the statement (not found) mentioned in Doc. 2483. Samuel Insull to Charles Batchelor, 16 July 1883, DF (*TAED* D8316AGT).

3. John Lubbock later inferred that Edward Johnson had drawn on conversations held with him in England to advise Edison on this point.

Writing as chairman of the Edison Electric Light Co., Lubbock explained to Edison that "well established undertakings . . . may obtain money by offering a six per cent and even a five per cent preference dividend. But this is not so with commercial undertakings of a novel and speculative character, and unless a rate of interest is given which will attract capital to the undertaking both A and B shareholders will suffer." Although the company had considered an eight percent preferential dividend for the A shares on the paid-in capital (before allocation of remaining profits among A and B shares), it now agreed on seven percent, in deference to Edison's opinion. Lubbock to TAE, 16 Aug. 1883, DF (*TAED* D8338ZBI); see also Doc. 2483 esp. nn. 3–4.

4. Grosvenor Porter Lowrey (1831–1893), a principal in the New York law firm Lowrey, Stone & Auerbach (successor to Porter, Lowrey, Soren & Stone), had been a patent attorney for Edison in the 1870s and became a pivotal figure in the formation of the Edison Electric Light Co. He provided legal counsel to that firm and was also involved in Drexel, Morgan & Co.'s disposal of Edison's foreign patents ("Grosvenor P. Lowrey Dead," *NYT,* 22 Apr. 1893, 11; see Docs. 695 n. 2, 1459, 1471, 1504, 1520, 1558, 1612). Lowrey and Edison also seem to have enjoyed a warm social relationship in the past (see, e.g., Docs. 1711 and 2173 esp. n. 23).

5. The extent of Egisto Fabbri's counsel during July is uncertain. He was at the time "a good deal under the weather, owing to his brother's death," in the words of his partner, Pierpont Morgan (Morgan to Jacob Rogers, 21 July 1883, letterbook vol. 2:397, NNPM). It is also doubtful that Fabbri knew of Edison's decision at this time. As of 7 August, he still had "a copy of the paper embodying the proposition still on my desk as left by Mr. Johnson," and he asked whether Edison had given "any definite answer" to the London company about the proposed amalgamation. Edison replied by sending a copy of this document, explaining that "the letter is of Mr. Lowrey's dictation" and should have been shared with Fabbri immediately but had been delayed by a clerical error. For his part, Edward Johnson claimed that he had been "only able to get him [Edison] to accede to the terms conveyed in his letter through the assistance of Messrs. Fabbri and Lowrey" (Fabbri to TAE, 7 Aug. 1883; TAE to Fabbri, 10 Aug. 1883; Johnson to Arnold White, 24 Aug. 1883; all DF [*TAED* D8302A, D8316ANL, D8338ZBJ]; cf. the draft version of this document in DF [*TAED* D8338ZBA]).

6. See Doc. 2514.

7. Edison's original 1882 agreement with the British company (article nine) set no limit on the period in which the firm was entitled, without additional payment, to his patents for further inventions or improvements. In the telephone business, by contrast, the Edison Telephone Co. of London had rights to Edison's improvements only during the life of the original patents. When it merged with the rival Bell company in June 1880, Edison expressed no objection to the elimination of his name from what became the United Telephone Co., Ltd. TAE agreement with Drexel, Morgan & Co., Grosvenor Lowrey, Fabbri, and Bouverie, 18 Feb. 1882, CR (*TAED* CF001AAE1); cf. TAE agreement with Edward Bouverie and Edison Telephone Co. of London, 14 July 1879 (article five), DF (*TAED* D7941ZCV).

8. According to Arnold White, Edison's letter was "considered

by both boards in detail," and "Some sort of decision" was expected at their respective meetings on 14 and 15 August. White to Johnson, 9 Aug. 1883, DF (*TAED* D8338ZBF); see Doc. 2514.

–2494–

To Roscoe Conkling

[New York,] July 26th. [188]3

My Dear Mr. Conkling:—[1]

You may doubtless have noticed the recent decision, given by the Primary Examiner in the Patent Office, on the various applications for patents in connection with the telephone.[2] I propose to appeal to my cases to the Board of Examiners, and I should very much like to have you argue the appeal.

I understand from my Patent Attorney that there is about sixty days allowed us to give notices of the appeal, so that the matter would not require your attention just at present.

Will you please let me know if you would be willing to act for me in this matter.[3] Very truly yours,

TL (carbon copy), NjWOE, DF (*TAED* D8316AJD).

1. Roscoe Conkling (1829–1888) was a former U.S. Senator from New York and a major figure in the Republican Party. Having resigned his Senate seat in 1881 over a patronage appointment and failing to be elected again, he devoted himself to a private law practice which included patent cases. He had previously acted on behalf of Western Union in the Quadruplex Case and had been retained since early 1882 by the Edison Electric Light Co. to defend an Edison carbon filament patent application in a Patent Office interference proceeding. *ANB*, s.v. "Conkling, Roscoe"; Conkling 1889, 671–74; see Docs. 577, 1096, and 2508 n. 2.

2. The Patent Office had declared two related groups of interference cases in March 1878 and August 1879 to sort out the competing priority claims of several telephone inventors (see Docs. 1270 and 1792). Some cases were decided in June 1881, but the chief examiner did not issue rulings on eleven remaining questions until 21 July 1883 (only two of them favorable to Edison). His decision (reportedly over 300 pages) was then appealed to the Examiners-in-Chief, who considered the matter until October 1884. They upheld his decisions in all cases except regarding priority for the telephone receiver, which they awarded to Bell. Edison subsequently appealed that verdict to the Commissioner of Patents. Lemuel Serrell to TAE, 28 June 1881, DF (*TAED* D8142ZBF); "Decision Regarding the Telephone Inventions," *NYT*, 22 July 1883, 5; "The Telephone Interference Case Decided," *Sci. Am.* 49 (4 Aug. 1883): 64; Decisions of the Examiners-in-Chief, Miscellaneous Interferences, TI 5 (*TAED* TI5 [images 39–63]); Brief for Edison, Telephone Interferences, Case G, TI5 (*TAED* TI5 [images 64–66]).

3. Lemuel Serrell notified Edison on 24 July of the examiner's decision and suggested that Conkling be retained for the appeal. Conkling's reply was delayed by his absence and by his efforts to avoid a conflict of interest with another party to the interferences, but he ultimately

agreed to assist Serrell in preparing the brief for the appeal. Due to the fact that Western Union, for whom Edison had conducted his telephone experiments, had sold its telephone patents to the Bell Telephone Co., there was confusion until 1885 over who would pay Conkling's fees. Serrell to TAE, 24 July 1883; Conkling to TAE, 25 Aug. 1883, 2 Mar. and 14 Apr. 1885; all DF (*TAED* D8370ZBI, D8370ZBW, D8544G, D8544K); TAE to Conkling, 14 Apr. 1885, Lbk. 20:241 (*TAED* LB020241A); Brief in Behalf of Edison, Telephone Interferences, TI 5 (*TAED* TI5 [images 9–39]).

–2495–

To Sherburne Eaton

New York,[a] [July 26, 1883][1]

Major

I want to write a letter[b] to a lawyer in Buenos Ayres giving him power of attorney in my name to apply for permission from the Municipality to lay tubes in the streets (all) of Buenos for the distribution of E[lectricity] for Light H[eat]. & .P[power].[2] I dont know how to do it   Can you help me out[3]

Edison

ALS, NjWOE, DF (*TAED* D8336Y). Letterhead of Thomas A. Edison, Central Station, Construction Dep't. [a]"New York" preprinted. [b]Obscured overwritten text.

1. Date supplied from a docket notation, probably made by Alfred Tate.

2. The earliest steps toward introducing Edison's electric light in Argentina occurred in 1881, when Fabbri & Chauncey ordered machines for South America (see Docs. 2048 and 2144). An Edison electric lighting plant started in Buenos Aires in late August 1882 and supplied light four hours every night for no more than two months, when the dynamo operator became too ill to work (Edison Electric Light Co. Bulletin 14:15, 14 Oct. 1882, CR [*TAED* CB014]; Eaton to TAE, 24 Oct. 1882, DF [*TAED* D8237ZAH]).

The Argentine Edison Electric Light Co. was incorporated in New York on 17 May 1883. It proposed that "Buenos Ayres affords the most lucrative field for electric lighting of any city in the world" because of the late closing time of stores and the high price of illuminating gas ($6.50 per thousand cubic feet). The business was to be capitalized at $500,000, of which the Edison Electric Light Co. would receive $300,000 of stock in payment for central station equipment and license fees. Sherburne Eaton reported to the parent company's executive committee that "Mr. Edison is in favor of developing the business in the Argentine Republic" on this basis, "but Messrs. Fabbri & Chauncey do not approve of the terms proposed." Edison planned to present his views at the committee's 2 July meeting, when the directors ratified this arrangement. The impetus for the formation of the Argentine company appears to have been the arrival from Buenos Aires of George W. Sherman, who is mentioned along with the Consul General of Argentina, Carlos Carranza, in the company's prospectus (see Doc.

2688 regarding Sherman's negotiations with Edison). Argentine Edison Electric Light Co. certificate of incorporation, 17 May 1883, Miller (*TAED* HM830181); Argentine Edison Electric Light Co. prospectus, 17 May 1883; Eaton memoranda to Executive Committee, 15 June and 2 July 1883; Eaton to Sherman, 8 June 1883; all DF (*TAED* D8336O, D8327ZAB, D8327ZAM, D8336U).

3. Eaton suggested that he should prepare a power of attorney that Edison could sign and have authorized by the Consul of the Argentine Republic. Edison evidently gave a power (not found) for this purpose to one Julian Balbin, who presented it to the mayor of Buenos Aires with a petition for authority to lay underground conductors for private lighting of residential and business houses. In his petition to the mayor, Balbin specifically disclaimed an intent to establish street lighting, which Edison considered an unprofitable enterprise. On 1 March Balbin received authorization and on 5 March George Sherman cabled from Buenos Aires "have secured grant for the whole city." Eaton to TAE, undated [c. 27 July] 1883; Balbin to Torquato de Alvear, 16 Feb. 1884; Torquato de Alvear to Balbin, 1 Mar. 1884; George Sherman to TAE, 2 Oct. 1883 and 5 Mar. 1884; all DF (*TAED* D8303ZIL2, D8434K, D8336ZAE, D8337ZAH).

–2496–

To Frank McCormick[1]

[New York,] July 26th. [188]3

Dear Sir:—

I am in receipt of your favor with relation to the action taken by your Company, in connection with the inside wiring at Sunbury.[2] I think that your Directors were not well advised in pursuing the course they have in relation to this matter, and would point out to you that it would have been better for them to have consulted me first as to it, inasmuch as I had taken the trouble to go to Sunbury, and spent upwards of a week there, and had consequently made myself thoroughly familiar with the requirements of the case.[3] I was perfectly aware, before I left Sunbury, that our wiring was too expensive for small towns like Sunbury. This experience, however, could only be obtained by actual work.

Immediately on my return, I started my assistant, Mr. Johnson, experimenting with a view to getting up a cheaper method of wiring and cheaper fixtures, to comply more closely with the resources of our consumers. This work is now going forward, and Mr. Johnson has already obtained excellent results.[4] So far as fixtures are concerned, the wiring will take a little longer [----][a]

Knowing from [my experience?][b] at Sunbury, that our system of wiring is too expensive for small towns, immediately I got back from Sunbury, I started Mr. Johnson experimenting, with a view to cheapening the inside wire work, and also the fixtures. Since then, he and I have been incessantly engaged upon this work, and we have already gone a long way towards cheapening up the work referred to. I have to-day seen a two light chandelier, wired and with sockets complete, which Mr. Bergmann informs me he will be able to sell for $2.[5] This fixture is very much better than the average gas fixture which I saw at Sunbury. This chandelier can be enlarged for any number of lights, that may be desired, from one fixture, at the rate of $1. per light. Mr. Johnson is now occupied the whole of his time in getting up a cheaper system of inside wiring, in order to remedy the defects, the seriousness of which I fully recognize, and did so in fact within a day or so of my arrival in Sunbury. Sending you a man to do this work will not cheapen it, inasmuch as it is impossible to cheapen it as long as the labor and material now required are necessary.[6] It was my intention that Bergmann & Co's Wiring Department should instruct some local man in each town, how to do the inside wire work, so that when they leave, there will be some one left in the town to do the work properly.[7] I have no man that I can supply to the Shamokin Company, nor do I think that any advantage would be gained in supplying such a man, if I had one.

You asked me what I would suggest to counter-act reports of the unsatisfactory character of the work at Sunbury. In reply, I would say, please communicate with me before endorsing those reports, by proposing some new course of action.[8]

As to the lightning, that trouble has been fixed.[9] Mr. Andrews has got at Sunbury now, everything that is necessary to prevent any further trouble in the future. I am very sorry to hear that you have been sick, and hope that you will soon be all right again. Very truly yours,

TL (carbon copy), NjWOE, (*TAED* D8316AJB). [a]Canceled. [b]Illegible.

1. Frank McCormick (1857–1935), a Williamsport attorney in practice with two of his brothers, was the founding president of the Edison Electric Illuminating Co. of Sunbury. "McCormick, Frank," Pioneers Bio.; Lloyd 1919, 715.

2. McCormick wrote on 25 July that the men wiring buildings in Sunbury "have caused a great deal of complaint because of the manner in which the work is done and the conduct of the men doing the work." They reportedly had placed unconcealed wires "over the walls and ceilings with no regard whatever for the appearance of things" and cut private telephone wires lying in their way. They also charged "exhorbitant prices for putting in lamps, in some cases as high as $3.75 per lamp." In reaction, the Sunbury directors decided on 24 July to halt the Wiring Department's men and to undertake the work themselves. The

local newspaper quickly published the directors' resolutions on these two points, prompting Edison and Insull to complain separately about this publicity. Even one of Edison's own men blamed poor workmanship on the wiring for his slow progress in connecting meters. McCormick to TAE, 25 July 1883; TAE to Shaw, 2 Aug. 1883; Thomas Conant to Insull, 31 July 1883; TAE to McCormick, 2 Aug. 1883; all DF (*TAED* D8361ZBK, D8316ALD, D8361ZBQ, D8316ALA); Insull to Shaw, 30 July 1883, LM 14:199 (*TAED* LBCD1199).

3. After inspecting the Sunbury system, Phillips Shaw reported to Edison that the plant ran well "but the wiring makes me Sick I certainly Shall be ashamed to Show this work to people of other towns." Like McCormick, he complained about the cost. The following day, he sent a letter (not found) to the same effect on behalf of the Shamokin company and soon received an answer similar to Edison's reply to McCormick. Edison also sent a brief acknowledgment the same day stating his "opinion that the method pursued at Sunbury should be generally adopted, as to inside wiring, as otherwise half the expense will be 'saddled' on our local companies." Seemingly wounded by Edison's attribution to him of concerted action in both Sunbury and Shamokin, Shaw subsequently sent a more nuanced explanation of the problems and his role in calling them to Edison's attention. Edison asked him to reassure the Shamokin company that the problems would be resolved ahead of wiring buildings there. Shaw to TAE, 23 and 31 July 1883; TAE to Shaw, both 26 July and 2 Aug. 1883; all DF (*TAED* D8361ZBH1, D8340ZBX, D8316AIZ, D8316AJA, D8316ALD).

4. Edward Johnson returned to New York from London on 15 July, and five days later he was working in Shamokin, Pa., with a "Gang," presumably on wiring. Samuel Insull explained to William Andrews soon after that Johnson's plan for "cheapening up the wiring itself" involved "putting two wires together, covered in almost the same manner as the present Ansonia wire is covered, and making a cable of the two. If you have any ideas on this subject, Mr. Edison would very much like you to mention them to him." Johnson was already a co-patentee in Britain on a form of telegraph cable, though his involvement seems to have been financial rather than technical in nature. By the end of July, Johnson was in Louisville, helping with the Edison lighting installation. Insull to Charles Batchelor, 16 July 1883; William Rich to Insull, 20 July 1883; Insull to Andrews, 27 July 1883; Insull to Archibald Stuart, 30 July 1883; all DF (*TAED* D8316AGT, D8343N, D8316AJS, D8316AKM); Doc. 2187 n. 4.

5. This price would be substantially lower than that of any comparable fixture in the current catalog of Bergmann & Co. in April or May 1883 (see Doc. 2410 esp. n. 4). The editors have found no other information about this fixture.

6. In accord with the Sunbury company's resolution, McCormick asked in his letter to Edison (see note 2) whether Charles Story, a member of the Sunbury company's board and "well acquainted" there (and later active in the Ohio Edison Electric Installation Co.), could be detailed to supervise the wiring by local men. Failing that, he requested Edison to send a "competent man" for the job. See also Bell 1981, 496.

7. Edison made such an arrangement in August on behalf of the Sunbury company. He also passed along the opinion of Edwin Greenfield,

head of the Wiring Dept., that carpenters were best suited to learning this type of work. Bergmann & Co. closed down the Wiring Dept. in September, it "not having come up to the expectations which we had formed for it." Insull to McCormick, 7 Aug. 1883, LM 14:288 (*TAED* LBCD1288); Insull to Andrews, 14 Aug. 1883; Bergmann & Co. to TAE, 5 Sept. 1883; both DF (*TAED* D8316ANZ1, D8324ZAE).

8. In a reply the next day, McCormick disavowed any intent to embarrass Edison by acting hastily, noting that the Sunbury company "did not know that the inside wiring was being done by your direction, as our contract with you was only for the installation of the plant and the outside wiring." Edison forwarded to him a letter from Charles Hanington defending the actions of Bergmann & Co.'s crew and the charges for wiring. Edison promised to fire any employees who had behaved rudely. McCormick to TAE, 27 July 1883; Hanington to TAE, 30 July 1883; TAE to McCormick, 30 July 1883; all DF (*TAED* D8361ZBL, D8361ZBN, D8316AKL).

9. Sunbury was hit by an intense thunderstorm shortly after the central station began operating. William Andrews reported that lightning had "been snapping most viciously around our light fixtures" in the City Hotel there, producing a few cracks "as loud as the firing of a gun cap" and leaving "Some of the folks here . . . quite scared." Several test wires at the central station were burned off. (The village plant at Roselle reportedly also suffered lightning damage in a separate storm about the same time.) Edison instructed Andrews to ground the Sunbury system through a high resistance in daylight hours and during thunderstorms, and also to "put omnibus to dead ground when not running storm or no storm." He soon afterward filed two patent applications for connecting overhead distribution systems to ground in order to protect them against lightning. One of these included a switch (activated by an electrical surge) for changing the ground connection from high to low resistance. According to a later reminiscence attributed to Andrews, this incident prompted Luther Stieringer to devise "insulating joints" for the contacts between wires and gas lines, where the static sparking had occurred; the National Board of Underwriters reportedly required these joints well into the next century. Andrews to TAE, 11 and 16 July 1883 (with TAE marginalia); TAE to Andrews, 19 July 1883; all DF (*TAED* D8361ZAV, D8361ZBE, D8316AIE); U.S. Pats. 304,084; 476,529; Keefer and Keefer 1927; see also 2446.

–2497–

To Archibald Stuart[1]

[New York,] July 30th. [1883]

Dear Sir:—

From our conversations with relation to the erection of central stations, I beg to give you below details of our method of operations in connection with this department of our work.[2]

When you indicate to us that you want a central station plant erected in a certain town, we have to send a canvasser to take an exact record of the number of lights burning per half hour in said town. We do this in order to get at the average amount of gas used and on this information we base our estimates.

On the receipt of the canvass at our office, we make a correct map of the town and enter on said map the results of the canvass made by our man. After this work is completed a determination is made of what copper conductors will be necessary to supply the town with incandescent light.

The above data enables us to put in a plant in the most economical manner which fully justifies the delay in starting work which this somewhat lengthy operation would seem to indicate. In the estimates that we shall furnish you we put everything at the lowest possible price compatible with covering us for the risks we run and for the expense incurred in connection with this branch of our work. Mr. Shaw informed you that we should be willing to cut out any part of our estimate and allow you[3] to do that portion of the work provided you could do it cheaper and would not interfere with our portion of the estimate. From experience at Shamokin, we find that this course is very inconvenient and causes us considerable [trouble?][a] loss. We therefore find it necessary to withdraw from this position and to undertake work only in accordance with our estimate. If, however, you can show us that we can put in part of our contract at a price less than that named in our estimate it will always afford us very great pleasure to inquire into the matter and make a revision of our estimate so as to give you the advantage of any such reduction in price. Of course if your company desires to put up part of the work themselves and leave us to put up the remainder we shall be glad to furnish you with an estimate for such part of the work as you may desire us to do. But, as we have above stated, it is impossible for us to allow our estimate for the whole work to be accepted only in piece meal.[4]

In case of an estimate being accepted, we shall require 25 per cent of the amount of the contract on the signing of the agreement, 50 per cent when the installation is ready to supply light, and the balance (25 per centum) after the installation is ready to supply light 30 days. At present it is impossible for us to take a time contract for the performance of the work but the terms of payment that we propose are a guarantee that we shall get through our work as quickly as possible in order to save us the necessity of carrying a large amount of material for any considerable time as our contacts for material are made on the same basis as our contracts with our companies for doing the work.

Our estimates provide that the necessary expert expenses for starting a station and teaching the permanent employees whom I desire should be chosen by myself or my representative, in order to insure that the station shall be run in accordance with the experience that we have gained in connection with this work. We are preparing examination papers which we shall expect the permanent employees to give answers to before we pass them as being competent to take care of the work they have to perform.

In conclusion we should add that it will always be our first endeavor to make your installations a perfect success as a failure would be as disastrous to us as to your company. Very truly yours,

TL (carbon copy), NjWOE, DF (*TAED* D8316AKO). ªCanceled.

1. Archibald Stuart (1846–1938) was secretary of the Ohio Edison Installation Co. in Cincinnati. His family had moved from London to Kentucky in 1849. At a young age, Stuart entered into a succession of business endeavors, including printing, banking, and tobacco processing, in Kentucky and Ohio. In 1881, he became associated with his father-in-law's prominent tobacco warehouse business in Cincinnati. Stuart later helped establish the Central Thomson-Houston Electric Co. in Cincinnati; he remained connected with electric lighting until 1893, when he went into the newspaper business. Doc. 2350 n. 4; Goss 1912, 3:572–73; Ohio Dept. of Health Death Index, p. 01666, certificate 6917, accessed on Ancestry.com, 16 Apr. 2008.

2. Edison and Samuel Insull had apparently met with Stuart in New York before 19 July and promised him a letter on this subject. Stuart to TAE, 19 July 1883; TAE to Stuart, 26 July 1883; both DF (*TAED* D8353D, D8316AIY).

3. That is, the Ohio Edison Installation Co.

4. In Circleville, Ohio, Stuart's company did undertake the construction work itself. In at least one other Ohio locale (Piqua) the company planned to operate the plant after its completion by Edison's workers. In general, the Edison Electric Light Co. was to pay the Ohio Edison Installation Co. a commission of 20 percent on each plant instead of the 15 percent usually granted its agents. Ohio Edison Installation Co. agreements with TAE, 16 May and 21 Feb. 1884, Miller (*TAED* HM840216, HM840210); Sherburne Eaton memorandum, 2 July 1883, DF (*TAED* D8327ZAM).

–2498–

To Frank McCormick

[New York,] July 31st [188]3

Dear Sir:

I beg to enclose you herewith statement of account for the installation of a Central Station Plant at Sunbury amounting to $11 968.00.[1]

This installation has been made in accordance with the agreement between your Company and myself, dated the Eighth day of May,[2] and I shall be glad if you will please send me by return mail cheque for $5984.00 being one half of the total cost[a] of the plant and which payment was due when the installation was in condition to supply light.[3]

Referring to the items on the a/c for Engines and Dynamos I beg to state that the Dynamos installed at Sunbury are what is known as our old "L" machines as explained to your Mr Shaw at the time the contract was made with me. These Machines were installed subject to my right to substitute for them new and improved machines which are now being designed

These new Machines will have[a] the same capacity as the "L" but were it my intention to leave the "L" Dynamos permanently at Sunbury I should be compelled to charge from four to five hundred Dollars more for the two machines; I have however put them in at the same price as the new form of machine which will replace them as soon as we have built some.[4]

In addition to the items specified in the enclosed account it will be necessary for your Company to purchase three 3[b] pressure Indicators[5] and Five 5[b] house Changing Switches[6a] the cost of which will not exceed 100$; these are now being made and will be sent to Sunbury as soon as ready and billed to your Co'y in addition to the a/c now enclosed.

As this plant has been installed at prices which allow of no profit whatever to myself I shall be glad if you will comply with my request and send me cheque for the amount due by return.

I will let you know in the course of a day or so when I shall be prepared to hand the plant over to your representatives and when the remaining balance on the enclosed a/c will be due. yours truly

Thomas A. Edison.

ALS (letterpress copy), NjWOE, LM 14:215 (*TAED* LBCD1215). [a]Repeated at end of one page and beginning of next. [b]Circled.

1. Enclosure not found.
2. Agreement not found.
3. McCormick answered that he was "a good deal surprised at the amount," as he had expected it to be about $9,000 and not in excess of $10,000. McCormick feared that the company, even with a recent increase of capital stock, would be unable to pay Edison, its indebtedness to the Edison Electric Light Co., and its operating expenses. Edison

proposed that Insull meet with the company's board to settle the matter quickly, "inasmuch as I have lost on the work and consequently cannot afford to be out of my money." Insull went to Williamsport, where most board members lived, on 4 August, and the company made two payments soon thereafter. In September, with the company showing a ten percent return on its capital, Edison agreed to accept stock shares in payment of the remaining balance of about $6,400. However, the debt remained on Edison's books a year later. McCormick to TAE, 1 Aug. 1883; TAE to McCormick, 2 and 4 Aug., both 10 Sept. 1883; Edison Electric Illuminating Co. of Sunbury memorandum of payments, 7 Aug. 1883; Insull to McCormick, 19 Nov. 1883; all DF (*TAED* D8361ZBR, D8316ALB, D8316ALV, D8316AUG, D8316AUI, D8361ZBX1, D8316BHJ); Insull to Frank Sprague, 10 Sept. 1883, LM 15:41 (*TAED* LBCD2041); see Doc. 2709.

4. The old dynamos remained in service at Sunbury for about twenty years. It is not clear which new model Edison had in mind, as none of the machines being designed at this time had the same 150-lamp rating as the type L (*Edisonia* 1904, 141–42; see App. 3). In at least one previous instance, Edison's wish to upgrade central station equipment (in Milan, Italy) caused months of wrangling over who should pay for the new machinery (see Doc. 2399).

5. The voltage indicator was a form of galvanometer, such as that used at the Pearl St. station in New York. The indicator was modified by a "great deal of experimenting" in July, and at least some produced at this time included a level to help insure uniform operation. Problems persisted with the sensitive instruments, however, and even William Andrews, a seasoned electrician, had trouble getting consistent readings. Doc. 2242 (headnote); Insull to Andrews, 27 July 1883; Andrews to Insull, 16 Aug. 1883; both DF (*TAED* D8316AJS, D8361ZCT); see also Docs. 2584 and 2592.

6. These devices presumably functioned to balance the electrical load across the three-wire distribution system by switching houses (or groups of them) to one side or other; see Doc. 2484 n. 7.

–2499–

To William Hammer

New York, August 3rd. 1883.

Dear Sir:—

I duly received your letter of July 16th.[1] and am very much obliged to you for the information it contains with reference to the development of the Edison system of electric lighting in Germany. What you say is by far the most encouraging information that it has yet been my pleasure to receive from Europe in relation to the development of our enterprise. I am sure I wish your company every success in their present undertaking and also your good self in the capacity of their chief engineer. Referring to the numerous questions you ask, I will answer them seriatim

PLANS, BLUE PRINTS & c. I am having prepared a full set

of all the drawings, blue prints &c. relating to our central station work and will have them sent forward to you in the course of a week with full explanations as to the same. With relation to the various plans I sent you I may remark that in most instances the plant are being put in by myself as contractor. I take a contract to do the whole work, including the building of the central station, erection of the plant, street conductors & c. and turn the whole thing over to the local people after running it 30 days in such condition that they can go ahead and do practical business without experimentation of any kind. Mr. Batchelor is fully familiar with the method in which this work is done, but my impression is that he is not acquainted with the system we now have of canvassing a town in order to see the amount of light used and enable us to prepare data on which to make an electrical determination of the machines and feeders necessary for the work. I will, therefore, give you briefly the way in which the work is done.[2] Suppose we take Newark. Our canvasser goes to Newark and the first night he is there walks over the best part of the town taking in the business portion and the better residential portion; and makes this examination until 12 or 1 O'clock in the morning. The next day he gets a map and picks out the district which in his judgement is the best for us to light with our station and numbers each block on the map for easy reference afterwards. He then goes round and marks in a book the number of the house and nature of the businesses, the quality of the residence on say 20 or 30 blocks facings. At night just at dusk, when the lighting commences, he starts at one end of block and looks into each store and counts the number of lights burning, makes a note of this in his memoranda book and the time at which he counted such lights. He then goes to the second store and performs the work there and continues to put the time down every five minutes keeping on in the same direction until he has been noting the burning for about a half an hour when he returns to the store that he first examined and does the same work over again continuing this every half hour until everybody is closed or until there are only such[a] lights burning as will burn all night. As to this latter he is obliged to enter the store and inquire of the consumer as to how long he will continue to burn his lights. Thus he is enabled to make a record of the actual number of light burning in each store during every 30 minutes of the evening. He is very careful only to canvass in one evening just enough territory to enable him to get back to his starting point every half an hour. On the next night he takes another lot of block facings and performs the same task as he did on the previous night and so on night after night until the whole district which it is proposed to light is canvassed. Besides taking in the good lighting he takes in a small area round this so as to enable us to judge in the office where the good lighting in a town commences and where the poorer lighting commences. In the day time he goes over the residential portion and marks the number of each house and its position on the block and notes them as being either "palatial" "first" "second" "third" or "fourth class" residences, according to the nature of the town, and this information enables us to judge here in the office of the average number of the lights that a certain house of certain character would consume. Utica, N.Y. was canvassed in about 8 days. Renova, Pa. in about 2 days. Lowell, Mass. in about 5 days. We have 4 men out on this kind of work and they send us in from 6 to 10 canvasses a week.[3] No notice whatever is taken of factories unless they are run very late at night or through the greater portion of the night. Churches, halls & c. may not be lighted when the canvasser goes his rounds and he is therefore compelled to make the necessary inquiries during the day time about the number of lights actually used when they are lighted and the number of times they are lighted a week The whole of the canvas notes (each canvasser has a separate method of his own so far as details are concerned, only being confined to points above indicated) together with map having numbered block facing thereon is sent to my office here where the canvass is plotted on a map, of which copies will be sent to you, and this enables me to take a birds eye view of the lighting of each town where it is proposed we should install an electric light plant. I forgot to mention that most of our canvassers have on their canvass form reports three small columns for cellar, first floor and second floor in order to provide for cases where there is more than one consumer in the building. In making our map here in New York we figure all the lighting back to December the 1st. This we are enable to do so by tables [showing?][b] we have showing what light there would before December 1st. providing there was a certain number of lighting on June the 1st. Of course this varies according to the locality as the factor of how soon darkness comes on is to be considered. Now suppose we take 8 o'clock in the evening in June   No. 51st. street burns no lights at 8½[c] o'clock; 10 lights at 9; 10 at 9-30; 10 at 10; and none 10-30. We find that this man closes at 10 o'clock. Now if it were in December instead of June he would commence lighting at 5. Hence we mark on our map that he has 10 lights maximum with a 5 hours average. On the map a straight line in drawn over the block facing the number of lamps placed at the end of the line; each one an eighth of an inch in length means one hour burning therefore on the map there would be a line five eighths of an inch long with 10 marked at the end of the line. In determining our street conductors we do not allow for stores & c in which the maximum number of lights does not in 312 days of a year give an average of 2 hours, but for all lights over this we allow street conductors. At the central station the dynamo, engine, boiler & c. investment must be enough to carry every light at the heaviest part of the year. Therefore you will see that if a store having 50 lights maximum and closing at [--]^b 6 o'clock was to be connected we should have to carry those 50 lights and consequently make an investment in dynamos, conductors & c. to do so in the Summer we would have to earn no money on such a customer inasmuch as he would close before it was time to light and in the Winter he would use very little indeed as his average would be so very low. Suppose from November to February 650 lights burned one hour each day say 110 nights. This would be 5500 lamps hours. If the investment were $25 per lamp in burners, machines & c. and we received one cent per lamp per hour the investment would be $1250 and the receipts $55. Now suppose we had a man who closed at 10 and had only 5 lights, he would give us 3 hours average the year round. That is for 312 days or 9560 lamp hours. The investment for such a customer would be $250 and the receipts $9360. A man who closes at 10 and burns only 4 lights is more profitable to us than one who closes at 6 o'clock and burns 50 lights. In starting central pioneer stations, we want the dream of the lighting. When we have shown them we can earn big dividends on this we can take the market afterwards and at such a time [--]^b people will be content if we can earn then 7 per cent on their money whereas in the first case, until confidence is established they want us to show a dividend earning capacity of a very much larger amount.

If at the same time you were to send the exact measurement of different plots of ground that could be got for [the?]^b a central station and give us the price of the same with details of any building they may have on them, we could from our great experience here give you the best arrangement of boilers, Engines & c. both for economy of investment in the station itself and economy of wages. If I can assist your company in any such work as this I would be very glad to do so and would simply charge them with the bare cost to me of my engineer's time and thus give them the benefit of our experience here without putting them to any extra expense.

ESTIMATES. At the same time that the blue prints are sent forward to you we will send you copies of our various estimates as requested.

DISTRIBUTION SYSTEM. My remarks above will answer your question as to this. If however you people want to do this work yourselves it will be necessary for you to start some such establishment as we have at Menlo Park.

THREE WIRE SYSTEM SAMPLES. We have ordered samples for you and they will go forward to you with the other material.

FORWARDING GOODS. I notice your remarks as to forwarding goods. Previous shipments have all been made by Bergmann & Co. I will, however, see that in future you have no trouble in this connection.

BULLETINS.[4] I will speak to Major Eaton with reference to copies of Bulletins.

LAMP FACTORY. I have written to the Societe Electrique of Paris stating that it is impossible for me to spare you a man to start a lamp factory.[5] I am short handed myself and our companies here require all our good men, and inasmuch as Mr. Batchelor has been here and knows everything concerning our factory he can start you going in this connection. Personally I think it would be a great mistake for your people to start a lamp factory if you can import lamps from here with an advantage. We have been making great strides in lamp manufacture within the last few months and the more lamps we make here the cheaper we can turn them out and the more money we can afford for experimenting work in improving lamps. I am quite sure that our company could lay down in Berlin 50,000 lamps of a better quality, more even volts, longer life and certainly as cheap if not cheaper that they could be manufactured in Berlin, provided of course there is no prohibitive tariff. You will have quite enough to attend to without the company burdening itself with a risky and uncertain business such as lamp manufacture is as we well know from the enormous amount of money our company has sunk in experimenting.

MACHINES. You will have received from us before the arrival of this letter both the new H machine and the new 100 light machine. We are now designing machines of 25 lights and 50 lights and 250 lights.[6] Are we to send these models as we get them out? I patented and used a long time since compound wound field magnets. All I did was to wind a couple of layers of heavy wire over the regular field wire and place it in the main circuit.[7] I calculated the matter out and in practice the volts were the same if all the lamps were on or off. The only reason we did not use it in the isolated business is that if the engine is unsteady the variation in the candle power is more apparent than if the machines were would in the manner we now adopt.[8] An increase of speed would double up and run the lamps very high. Of course most of our isolated installations have a constant load and even regulation of the engine, but if they have not we put in an automatic regulator which we are constantly improving from results obtained in every case.

CONDENSER FOR ENGINE. I do not believe in the use of condensers for engines as they are so very [---][b] liable to get out of order. Of course if coal is very expensive some slight saving might be made, but inasmuch as in a central station you are never called on to work your engines to their full capacity for more than two hours a day during the darkest days of the year, I believe it is much better to dispense with such things where there is the slightest possible chance of their causing us trouble. Very truly yours,

TL (carbon copy), NjWOE, DF (*TAED* D8316ALG). [a]"half hour until . . . are only such" typed on last line of first page and top line of next page. [b]Canceled. [c]"1/2" written by hand.

1. Doc. 2487.

2. The procedures outlined here are similar to those detailed in twenty-three numbered "General Instructions to Canvassers" printed by the Edison Electric Light Co., which also included a sample canvass form. Edison Electric Light Co. instructions, [undated 1882–83?], CR (*TAED* CD003).

3. The canvassers to which Edison referred were probably Charles Cooke, George Wilber, Paul Dyer, and R. M. Hedden. See Electric Light—TAE Construction Dept.—Canvassing, DF (*TAED* D8342) and, regarding Hedden, Electric Light—TAE Construction Dept.—Stations—Kentucky—Louisville, DF (*TAED* D8345).

4. The Edison Electric Light Co. periodically published bulletins about its activities and news items pertaining to gas and electric lighting. See Doc. 2268 (headnote) and Electric Light Companies—Domestic—Edison Electric Light Co. Bulletins (1882–1884), CR (*TAED* CB).

5. The Société Électrique Edison, among the companies incorporated in February 1882 for the lighting business in France and much of Europe, installed isolated plants (see Doc. 2182 n. 2). On behalf of the German company, the firm asked Edison to designate a "good Director for the Lamp factory." Edison declined on the grounds that "Our business has increased so enormously in this country, that we are taxed to the very utmost to supply men for our own work"; he suggested that Charles Batchelor might provide someone for the purpose. Foreign manufacture of Edison lamps (at this time using carbons imported from

the Harrison factory) occurred at Ivry, France, under Charles Batchelor's supervision. Similar assembly operations were carried out on a smaller scale under the direction of Charles Stilwell, Edison's brother-in-law, at a workshop in Hamilton, Ont. (relocated from Montreal in early 1883), in order to protect Edison's Canadian patents (Société Électrique Edison to TAE, 12 July 1883; TAE to Société Électrique Edison, 26 July 1883; both DF [*TAED* D8337ZDA, D8316AJF]; see Docs. 2404 and 2554; "Stilwell, Charles F.," Pioneers Bio.; *Legal News* [James Kirby, ed.] 12 [30 Mar. 1889]: 98–99).

6. See Doc. 2419 (headnote) and App. 3.

7. Edison had at least three U.S. patents pertaining to regulating dynamos with compound windings. One arrangement, apparently the one to which he referred here, had a set of high-resistance coils in a shunt circuit plus a second set of low resistance formed by the main conductor. He explained in the specification that "When translating devices are first put in circuit the magnet is sufficiently energized by means of the [shunt] circuit . . . ; but as their number is increased the resistance of the main circuit is lowered, so that more current flows through the [main] conductor . . . and the magnet becomes more and more energized" (U.S. Pat. 264,668). Another patent covered two shunt windings having different resistance. The third concerned coils wound in opposite directions on the same core so that the magnetizing effects of one would balance the other (U.S. Pats. 264,671 and 264,662). Edison's practice in commercial installations was to regulate according to changes in line voltage registered by instruments external to the dynamo (see Docs. 2036 [headnote], 2242 [headnote], and 2538 [headnote]).

8. A compound winding with one coil excited from an outside source will, in general, provide a more rapid response to changing line conditions than a single winding in a shunt from the main line. For that reason, however, the effects of fluctuating engine speed will be magnified when the exciting machine is also subject to those oscillations. According to one later authority, the shunt dynamo (the form used by Edison), "if well constructed, is . . . nearly constant in its voltage; the pressure at the terminals falls off very little at full load. With such a dynamo, but a small increase of exciting power is needed to make up for the lost volts at full load" (Thompson 1902, 226). Shunt machines with a very low resistance in the armature relative to that in the outside circuit most closely approached this ideal (Kapp 1882–83, 114–15). However, constant voltage at the dynamo terminals was not necessarily desirable in circumstances (such as central station operation) in which the voltage drop through the lines varies with the load (Jenks 1885, 9; see Doc. 2505 [headnote]).

–2500–

To William Andrews

[New York,] Aug 4th [188]3

Dear Sir:

I send you copy of Examina[tion][a] papers I propose to use in hiring our Central Station men—[1]

Please number Each question, and answer them very fully on other sheets giving as many answers as you can to each question and return both to me at once—

I am sending these to several of the boys to get a variety of answers— Yours truly

Thos A Edison

LS (letterpress copy), NjWOE, LM 14:260 (*TAED* LBCD1260). Written by Alfred Tate. ªIncompletely copied.

1. This document was evidently part of a concerted attempt to codify knowledge essential for the reliable and economical operation of central stations far from the resident expertise in New York. The effort resulted in one catechism on meters, another entitled the "running of dynamos," and a third on "the running of engine and boiler." These were grouped with Doc. 2530, at some later date, as Edison Construction Dept. instructional materials, CR (*TAED* CD001); see also Doc. 2484.

Neither the copy that Edison sent to Andrews at Sunbury, Pa., nor a direct reply has been found, but see Doc. 2505 n. 1. Andrews did prepare extensive undated "Instructions relating to Dynamos &c," and the exchange likely contributed to the eighty typed questions and answers "relating to running of dynamos" mentioned above. These were apparently printed in September for distribution to central station engineers. Edison also drafted two sets of questions and answers about the operation of boilers and engines (the longer group partly in unidentified handwriting). These memoranda, with lengthy responses by William Rich or another Construction Dept. employee, were incorporated into the catechism of fifty-nine typed "questions and answers related to running of engine and boiler." Employees who answered the questions satisfactorily were apparently entitled to a certificate of competence. Andrews memorandum, undated [1883]; TAE memoranda, both undated [1883]; Rich memorandum, undated [1883]; TAE to Andrews, 8 Sept. 1883; Charles G. Y. King to TAE, 13 Dec. 1884; all DF (*TAED* D8343ZFZ, D8343ZFS, D8343ZFU, D8343ZFU1, D8316ATZ, D8403ZIX).

On the same date as this document, Edison sent a similar letter to Thomas Conant, also at Sunbury, about questions for meter men. The editors have not found that enclosure, either, but it presumably informed the catechism of ninety-two "questions relating to meter department" attributed to Harry Leonard. Hired by Edison in June to take over the Construction Dept.'s meter business, Leonard had been studying at the Testing Room of the Edison Machine Works. Charles Clarke labored to make Conant's answers "standards of information for learners who are yet to come"; George Grower also contributed. In the middle of August, after reviewing Conant's answers, Edison sent him four handwritten pages of critique. TAE to Conant, 4 and 15 Aug. 1883, LM 14:261, 388 (*TAED* LBCD1261, LBCD1388); Leonard to TAE, 18 June 1883; Charles Clarke to TAE, 6 July 1883 and undated [1883]; Leonard to Samuel Insull, 1 Aug. 1883; all DF (*TAED* D8313W, D83290, D8329ZAZ, D8330ZAS); see also Ira Watts memorandum, 17 Sept. 1883, DF (*TAED* D8343ZBB1).

Before preparing these teaching and examination materials, Edison consulted with a divisional superintendent of the Pennsylvania Railroad, Robert Pettit, about that company's testing of engineers. Pettit reported that although there was no system of written examinations, engineers were quizzed verbally by the road foreman before taking their positions (Pettit to TAE, 3 July 1883, DF [*TAED* D8343J]). Already keenly aware of the lack of training programs in electrical science (see Docs. 2428 and 2434; also Rosenberg 1990, esp. chap. 2), Edison could have found few—if any—models for certifying his employees' competence to take up positions of responsibility. These types of judgments traditionally were made by masters in the skilled crafts or by foremen in industrial workshops, neither of which had a place in the village plant system. The Baltimore & Ohio Railroad, considered a leader in industrial education, established a Technical School for employees only in 1885 ("Miscellaneous," *Van Nostrand's Engineering Magazine* 32 [1885]: 352).

–2501–

From Charles Batchelor

Ivry-sur-Seine, le 7th Augt 1883[a]

5 1 1[1]

My Dear Edison

New Dynamo.[b] In the ordinary Z. machines we have 60 ft. lbs of energy (about) on every square inch of cutting surface. The same in the K & the others. This is equivalent, calculating from the Z., to 230 circular mills of section for every ampere current passing.[2] As many of my machines run all night I find that this is as much as I dare put through them unless we use a blower.

Your new H., as I learn from a letter from you to E.H.J.[3] gave, for a short time, 400 amperes[4] and, from measurements given, I find that it had only 148 circular mills per ampere so, in order to be safe this ought to be a 320 ampere machine, eh?

⟨Our H gives 400 amperes— we are winding with larger wire & will get 450 @ 475[5] amperes easy without any other change— Say we are running at Louisville 14 H machines with constant load every evening of 400 amperes⟩[6]

The first machine I made was 120 light; this was a well insulated B armature[7] made 3 centimetres longer, with 8 inch cores 28½ in[c] long wound with 9 layers of No 16 wires. At 1250 turns we have 105 volts with 90 amperes on. I can take 150 amperes out of this for a short time but sell it for 90 to be perfectly safe for long running.[8]

After these measurements I have made 2 L machines for ships, which give their 150 lamps with 475 turns, this number being absolutely required by the Austrian navy.[9]

Hopkinson.[b] I must have your opinion on the value of Hopkinson's improvements; he is continually at Bailey about them, and Bailey seems as if he desperately wanted to give Hopkinson a royalty or buy what he has got. We have just a letter from him in London in which he says Dr H. has a K which gives 600 lights easy and they want me to go to Manchester to test it.[10] Pray the alteration of the length of magnet is not patentable and H. cannot claim anything on it & therefore it is useless to make a contract for a thing that is open to all.

To help stop their mouths I am making my K with an armature a little larger in diameter which easily allows me to give a machine of 400 amperes that will run day and night & I believe this will equal what he has got.

⟨Tell him to say to Bailey that that[d] we will discount Hopkinson & go him 10 pct better on anything he can do—taking in costs—⟩

I understand H. has said he should make a bar armature for this machine and take out 1000 lamps. If he takes them and in the same manner as Sprague's test of his dynamo[11] I would not give much for it in practice at that number.

If you see anything in this that wants correcting tell me so quickly;— I dont feel satisfied with the Berlin model because I am afraid that Bailey will make some arrangement with H. if I dont have something always as good.[12]

Lamp tests.[b] Received 923, 926, 938, 941, & 955;[13] these are magnificent. We are now making long carbonization & will send you tests. You must remember that our fibres as yet are only 6" long[14]

Arc lights.[b] I have received your bill for 6 arc lights complete with globes and carbons: value $113.20. What is this? Is it the Drs arc light?[15] because, if so, it was understood between Navarro[16] & myself that he would send these 6 as experimental ones and I did not expect to pay anything for them. Yours faithfully,

Chas Batchelor

⟨Insull= Explain this   Write B[ergmann] & Co about Arc Lights⟩[17]

LS, NjWOE, DF (*TAED* D8337ZDL). Letterpress copy in Batchelor, Cat. 1331:121 (*TAED* MBLB3121); letterhead of Société Industrielle et Commerciale Edison. [a]"Ivry-sur-Seine, le" preprinted. [b]Heading written in left margin. [c]Interlined above. [d]Obscured overwritten text.

1. Batchelor began marking his outgoing letters with sequential numbers in late 1882. See Doc. 2376.

2. The circular mil area of a wire is the square of its diameter as expressed in mils (thousandths of an inch). Batchelor was referring to the aggregate cross section of all the induction wires in a particular armature coil. Each coil (though not necessarily each wire, depending on the design) carries the entire armature current in turn as it cuts through the magnetic field, except for the brief instants when commutator brushes

make contact between adjacent coils. The induction capacity of a wire is determined by the surface ("cutting") area, and its current capacity by the sectional area. The current capacity of armature windings was not standardized for several years. Thompson 1888 (135), the first edition of that standard reference to make a recommendation, advised one square millimeter (1,973 circular mils) for every 3 amperes of current.

3. Not found.

4. The number of lamps this current could operate at 110 volts would depend on the resistance of the lamps. According to a chart of circuit resistance prepared by Batchelor at about this time, he seems to have assumed 140 ohms in each lamp. Although the resistance of actual lamps varied, at that arbitrary value the current would be sufficient for about 510 lamps. Undated table [2 Aug. 1883?], Cat. 1235:17, Batchelor (*TAED* MBN012017a).

5. Edison occasionally used the "@" symbol to express a numerical range; see e.g. Docs. 2314 and 2322.

6. Edison's marginal notes were incorporated into a short reply to Batchelor on 8 September. DF (*TAED* D8316AUA).

7. The B configuration of any Edison armature was intended to operate 55-volt B lamps. It was designed to produce twice the current at half the voltage as the standard A armature for 110-volt lamps. See Doc. 2126 (headnote) and *TAEB* 6 App. 3.

8. Batchelor tested a dynamo like this on 12 June, a few weeks after returning from New York, where he had helped Edison redesign his dynamos (see Doc. 2419 [headnote]). Like Edison (and John Hopkinson), he focused on the design of the magnetic circuit. The major alteration he made from a standard Z dynamo was to increase the mass of wire in each field magnet by about 50 percent. Eight days later, he tried the same machine with a thicker zinc base ("to prevent the carrying of magnetism") and the magnet wires "wound on without paper insulation (except one thickness next to the core) and passed through a hot bath of resin and paraffin." He noted a gain of 4.4 volts over the previous experiment. He again tested a machine on 2 August made "as near as possible to the dimensions and price as the old 60 lamp [Z] machines of Edison" but having the new magnet windings. It ran 120 lamps at 108 volts (equivalent to about 93 amperes) for an unspecified period. Cat. 1235:4–5, 19, Batchelor (*TAED* MBN000412, MBN0005, MBN0020).

On at least one occasion, the redesign of dynamos had unintended consequences in lamp manufacture. This occurred as the filaments were heated by an electric current while the bulbs were exhausted. Batchelor noted on 9 July that "Our new engines and dynamos giving us so much more current and E.M.F our men blackened all the lamps so badly that the majority of the first and second day's run had to be thrown away." Cat. 1235:10, Batchelor (*TAED* MBN012010).

9. The editors have found no information about the Austrian naval specifications. On 30 July, Batchelor made preliminary notes on a "Dynamo for 150 lamps 475 turns for steamboats." It was to have magnet cores about twenty-eight inches long (compared with the standard Edison design of about forty-two inches). In an experiment on 1 August, he compared the voltage produced by a uniform distribution of field windings throughout the cores against that produced by concentrating the wire at the end or through the middle of the cores. In this trial,

shortening the effective length of the magnets produced higher voltage under conditions of load and no-load. Cat. 1237:50; Cat. 1235:15–16, Batchelor (*TAED* MBN006050, MBN012015A).

10. John Hopkinson had a family connection to the Manchester engineering firm of Mather & Platt, which built several of his experimental machines. The letter from Joshua Bailey has not been found. Batchelor wrote him in Milan on 21 July that he had "been to England 4 times during the last month and can find no machines of Hopkinson's that are doing what is claimed in the report of Sprague." Arapostathis 2007, 13; Batchelor to Bailey, 21 July 1883, Cat. 1331:102, Batchelor (*TAED* MBLB3102).

11. Sprague 1883a (298) reported several trials of Hopkinson's machine lasting from thirty minutes to about one hour, with as many as 230 lamps "probably not quite up to candle-power." Drawing Bailey's attention to these circumstances in his 21 July letter (see note 10), Batchelor pointed out that "it is quite a different thing to run that machine for 5 hours (which it will be necessary to do) at such a number of lamps. . . . I cannot therefore say anything about Hopkinson's machines until he can show us them doing actual work."

12. Batchelor presumably referred to the prototype H dynamo sent to Berlin in June (see Doc. 2480 esp. n. 3).

13. These are manufacturing lot numbers of Edison "lamps tested at Newark" whose electrical characteristics Batchelor recorded in a notebook on 18 September. Cat. 1235:24, Batchelor (*TAED* MBN012023).

14. Six inches had been the standard length (uncarbonized) of filaments for Edison's 16-candlepower lamps, but those recently sent from the Harrison, N.J., were a half inch longer (see Doc. 2085 n. 4). In September, the Ivry factory produced some lamps using seven-inch filaments for an electrical exhibition in Vienna. Cat. 1237:107, Batchelor (*TAED* MBN006107).

15. The bill has not been found. "Doctor" was a nickname of Otto Moses (1846–1905), an analytical chemist whom Edison hired at Menlo Park in 1879. He went to Paris in 1881 to assist with the Exposition Internationale de l'Électricité, during which his working relationship with Batchelor deteriorated (see Docs. 1754 n. 6, 2106 n. 1, 2120 [and headnote], 2128, 2133, 2173 n. 18, and 2196 n. 7). He became interested in arc lighting and, with the support of Edison and Sherburne Eaton, tried to develop an arc lamp that would work in the same circuit as incandescent lights. He had a handful of patents issued or pending that he hoped to sell to Edison companies, but Edison found him and José de Navarro, a financial partner, "so impracticable, that we have been unable to do business with them here." Bergmann & Co. had made prototypes by mid-July, some of which evidently went to Batchelor, who was favorably impressed. Edison asked to discuss the lamps with Moses in the middle of October, but the editors have not determined what decisions he made subsequently (Doc. 2286 n. 7, *TAEB* 6 App. 5.C; Eaton to TAE, 10 Jan. 1883; TAE to Batchelor, 17 July and 8 Sept. 1883; TAE to Moses, 19 Oct. 1883; all DF [*TAED* D8370L, D8316AHD, D8316AUA, D8316BBP]; Batchelor to Moses, 10 Dec. 1883, Batchelor [*TAED* MB100]; see also note 17).

16. New York financier José Francisco de Navarro (1823–1909) was

an original stockholder in the Edison Electric Light Co. and holder of Edison's patent rights in Cuba. Navarro helped organize construction of both the Metropolitan Elevated Railroad and, at this time, a path-breaking apartment complex in New York. The editors have not learned particulars of his arrangement with Moses. See Docs. 1920 n. 2, 2232 n. 3, and 2516.

17. In response to inquiries from Edison (not found) and Samuel Insull, Bergmann & Co. declared that Batchelor had placed a verbal order for the lamps. Samuel Insull to Bergmann & Co., 4 Sept. 1883; Bergmann & Co. to Insull, 5 Sept. 1883; Bergmann & Co. to TAE, 7 Sept. 1883; all DF (*TAED* D8316ASY, D8324ZAD, D8324ZAF).

–2502–

To Edison Lamp Co.

[Menlo Park,] August 8th. [1883]

Gentlemen:—

Will you please ship some samples of very small lamps to W. R. Pope,[1] 131 West Baltimore street, Baltimore Md. He wants to experiment with a view to using a battery to light them the lamps to be worn on the head of some men who are to take part in a procession.[2] Very truly yours,

TL (carbon copy), NjWOE, DF (*TAED* D8316AMG).

1. William R. Pope (b. 1856) was an electrician and electrical supply dealer in downtown Baltimore; he was also an agent for the Western Electric Co. Woods 1883, 786, 1164; U.S. Census Bureau 1980? (1900), roll T623_1175, p. 21A (Hastings-On-Hudson, Westchester, N.Y., enumeration district 69).

2. After meeting in May with Edison, Sigmund Bergmann, and John Ott, Pope asked for two small lamps "such as were used in production of Opera of Iolanthe," and also for two larger ones. He inquired how soon Edison could supply 1,000 to 1,500 of either type. Francis Upton provided four samples and promised that the factory could provide 1,500 within three weeks of order. The editors have not determined what prompted Edison to request more samples on Pope's behalf at this time (Pope to TAE, 25 May 1883; Upton to Samuel Insull, 31 May 1883; both DF [*TAED* D8332ZBB, D8332ZBC]). Gilbert and Sullivan's fairyland operetta *Iolanthe* opened simultaneously in London and New York in November 1882. At the Savoy Theatre in London (which had Siemens electric lighting throughout), the production included electric costume lighting produced by small batteries concealed in the performers' garments. New York producers planned to introduce a similar effect in January, but they reportedly could not get the equipment to operate well and dropped it from the staging. Similar equipment did appear in New York in a ballet scene in Balfe's opera *Satanella,* which opened in May ("'Iolanthe' at the Savoy Theatre," *Electrician* 10 [9 Dec. 1882]: 88; "Music and Musicians," *NYT,* 12 Nov. 1882, 8; "Music and Musicians," ibid., 14 Jan. 1883, 7; "Amusements," ibid., 5 May 1883, 4).

Pope's original inquiry was likely on behalf of the Oriole Festival in Baltimore, which was to begin with a ceremonial welcome of Lord

Baltimore to the city. By the time the event approached in September, the plan for using batteries had been abandoned. The procession from the waterfront to City Hall was to include "a military company bearing electric torches" and another "company armed with electric lights fed from a wagon containing a generator." The event was twice postponed by heavy wind and rain, and it is not clear if it came off as planned. "The Oriole Festival," *Washington Post,* 7 Sept. 1883, 8; "Lord Baltimore Lands," ibid., 14 Sept. 1883, 1.

The Edison Electric Light Co., with Edison's approval, arranged a similar spectacle for a New York parade in support of Republican presidential candidate James Blaine in October 1884. The delegation of about 300 insurance men, each wearing a helmet with a 16-candlepower lamp, was described as the highlight of an otherwise lackluster event. Power came from a wagon carrying a dynamo and 40 horsepower engine, supplied with steam through flexible pipes from a fire truck. Employees of the various Edison manufacturing shops reportedly bore the cost of the set-up and it was rumored that workers would be discharged if they did not join the Blaine procession, but the Edison Electric Light Co. denounced this as "a lie, pure, simple, and direct." See chap. 7 introduction; "An Electric Torchlight Procession," *Sci. Am.* 51 (15 Nov. 1884): 310; "Arranging the Parade," *NYT,* 31 Oct. 1883, 5; "Reviewed by Mr. Blaine," ibid., 1 Nov. 1883, 2; Jehl 1937–41, 1000.

Edison reportedly took a strong personal interest in setting up the isolated plant and stage effects used in a New York production of *Excelsior,* an allegorical ballet of Progress, including a scene with the Brooklyn Bridge. The final scene featured ballet dancers carrying wands each with a small lamp at the tip, and "festoons of lamps" were lowered into view (Jehl 1937–41, 999; "The Electric Light in Theaters," *Sci. Am.* 49 [1 Dec. 1883]: 344; Manzotti 1883; "'Excelsior' at Niblo's Garden," *NYT,* 22 Aug. 1883, 5; see also Doc. 2214 n. 4). The electric lighting of individual dancers became increasingly common throughout the decade (Gooday 2008, 105–9; Cordulack 2005, 148 n. 12). Such theatrical effects were the basis for a facetious 1884 article in the *Scientific American* (attributed to the *New York Times*) about "illuminated ballet girls":

> There is . . . a great future awaiting the grand idea of incandescent girls, and there is reason to believe that in a very short time private houses will be lighted by girls instead of stationary electric lights. The formation of the Electric Girl Lighting Company is an event second in importance only to the invention of electric lights. This company proposes to supply girls of fifty candle power each in quantities to suit householders. The girls are to be fed and clothed by the company, and customers will, of course, be permitted to select at the company's warehouse whatever style of girl may please their fancy. ["Electric Girls," *Sci. Am.* 50 (17 May 1884): 312]

From Pitt Edison

Dear Bro[1]

I see by the papers that you are a going to be a buisness Man for a year[2]   I suppose you heard that I sold out The Road or rather gave it away for since the G[rand] T[runk] R[ailway] Trains Comenced running dow the river bank[3] it has almost became a dead letter for it took More than one half of its buisness away and it was lucky for me that I bot the farm   I am agoing to make a stock farm of it but it will take me about two yeas to get it in shape for it must all be seeded down   I seeded 37 acres this spring and will seed down about 45 acres this fall and 40 acres in the spring   I have already bot some young stock   I pick it up here & thare where I can buy it cheap   I keep a good man on the farm so it is not nessesary for me to be thare much of the time for a year at least   now al can't you place me somewhere for that time   thare is lots of work in me yet   I would not care whare I was placd in New York or any whare Else   would rather go west than East so see if you can think of some place that I could attend to for that time or longer   You[a] know that I would take more intrest in you buisness than a stranger would but if you have nothing for me to do I would like to know soon so I can look up something Else[4]

WPE[5]

we are all well

ENCLOSURE[b]

[New York, August 1, 1883][6]
EDISON AND HIS LIGHT.[c]
THE FAVORITE SON OF PORT HURON WILL CONFINE
HIMSELF SOLELY TO BUSINESS.[c]

In a recent interview with a New York journalist Port Huron's favorite son, the great Edison, said:

"I am going to be simply a business man for a year. I am now a regular contractor for electric light plants, and I am going to take a long vacation in the matter of inventions. I won't go near a laboratory.[7] I've sold out my electric locomotive to a railroad company, who, I believe, are going to take the thing up and push it.[8] At the expiration of the year I will renew my experiments at Menlo park. The experiments will all be confined to electricity—no more phonographs or things of that kind." He has for some time been engaged in cheapening the work of the installation of electric light plants, and says: "I have succeeded so well that I can build and equip a station to do the same work as that done in Pearl street for less than one-half the money."

"The chief part of our business now consists in taking contracts to fit out small or large towns with the Edison electric light. Among the contracts signed are those for Sunbury, Pa., where we are putting in 500 lights; Shamokin, Pa., 1,600 lights; Brockton, Mass., 1,600 lights; Lowell, Mass., 1,200 lights; Lawrence Mass., 4,000 lights; Haverhill, Mass., 1,600 lights; Renovo, Pa., 500 lights; Erie, Pa., 1,600 lights; Hamilton, O., 1,600 lights; St. Louis, Mo., 9,600 lights; Minneapolis and St. Paul, Minn., 6,400 lights; Appleton, Wis., 1,600 lights; Davenport, Io., 1,600 lights; Watertown, Pa., 500 lights, and Danville, Pa., 1,600 lights. We are also in negotiation with the authorities of 30 other towns where they want electric lights.[9]

"We contract for a certain sum to fit a town with electric light in every detail. We put up our own buildings, set the engine and dynamos, run the wires through the streets, carry them into the houses, put up the fixtures, set the machinery agoing and furnish light for 30 days, at the end of which time if everything is satisfactory to both sides we turn the system over to the town and pocket our money. We will agree to fit up a town with 2,000 electric lights inside of three months from the time the contract is signed, and sooner than that if necessary. I honestly believe that every one of the gas companies in the towns where electric light is established, will have to go out of the business of furnishing light. I am so convinced of the system's success that, as I said before, I have given up inventing and taken to business pure and simple."

ALS, NjWOE, DF (*TAED* D8314N). [a]Obscured overwritten text. [b]Enclosure is a PD. [c]Followed by dividing mark.

1. Pitt enclosed this letter with his reply of the same date to Samuel Insull, who evidently had asked him to find a copy of a newspaper published long ago by Edison (likely the *Weekly Herald*). Pitt Edison to Insull, 11 June and 12 Aug. 1883, both DF (*TAED* 8314H, D8314M); Doc. 2 (headnote).

2. See enclosure.

3. Pitt was a longtime stockholder of the Port Huron (Mich.) and Gratiot Street Railway Co. and superintendent of its struggling successor, the Port Huron Railway Co. See Docs. 175–76, 1414, and 1766 esp. n. 6.

4. Pitt asked Edison for financial assistance on multiple occasions, including a December 1882 request for help with the mortgage on his farm (see Doc. 2385). Edison replied on this occasion that he did "not think that I can find you anything in connection with our business here, and I think the best thing that you can do is to look out for some-

thing where you are (TAE to Pitt Edison, 14 Aug. 1883, DF [*TAED* D8316AOA]).

5. Edison's brother, William Pitt Edison. *TAEB* 1 chap. 1 introduction, n. 4.

6. This is the date of the original article in the *New York Evening Post* ("The Electric Light," p. 1). Pitt evidently enclosed a clipping from a local Port Huron paper of what presumably was a reprint of the interview; the editors have not determined the date of its republication there. An archivist's spurious notation of 13 March 1883 may have originated from an incomplete advertisement on the back of this otherwise undated clipping (DF [*TAED* D8314N1]). The interview was also republished in almost its entirety by the *Omaha (Neb.) Daily Bee* on 8 August (p. 4); a handful of newspapers printed excerpts from or references to the interview in early August without attribution (see, e.g., "Gotham Items," *Chicago Daily Tribune*, 1 Aug. 1883, 2; "Thos. A. Edison. Great Inventor Will Invent No More for One Year," *Cedar Rapids [Iowa], Evening Gazette*, 3 Aug. 1883, 1; "Thos. A. Edison. Great Inventor Will Invent No More for One Year," *Decatur [Ill.] Review*, 4 Aug. 1883, [n.p.]; and *American Gas Light Journal*, 3 Sept. 1883, Cat. 1016:11F, Scraps. [*TAED* SM016047b]).

7. In June, Edison reportedly told a Chicago paper that he had "closed my laboratory, and I have now got the system worked down to the minutest detail. There is nothing more in electric lighting to be invented or required. It has been shown to be fully capable of replacing gas scientifically and commercially under all conditions, and we have started in to do it." These statements recalled Edison's declaration upon moving to New York from Menlo Park in March 1881 that he had "left the laboratory . . . and am now a man of business." "Edison," *Chicago Daily Tribune*, 19 June 1883, 8; "Edison," ibid., 20 Mar. 1881, 7.

8. The Electric Railway Company of the United States; see Doc. 2431.

9. See App. 2.

–2504–

Samuel Insull to Charles Rocap

New York, August 13th. 1883[a]

Dear Sir:—

Referring to the enclosed bill, dated the 25th. July, for Babbit bearing boxes sent to Shamokin,[1] I beg to draw your attention to the fact that you have charged $30. a set for these boxes, which is the same price that you charge for phosphor bronze boxes. It seems to me that such a price is out of proportion to the value of the boxes. So far as I understand it they are simply cast iron pieces lined with Babbit.

It is my impression that when settling up with the Shamokin Company, if we presented any such charge there would naturally be a "kick," so return you the bill assuming that there must be some error in billing them at such a rate. Very truly yours,

Saml Insull   T[ate]

⟨Insull, There is only a difference in wt. of 1½ lbs. between a set of Babbit & Phsp. bronze boxes. The Bronze costs us 32¢ lb. the Babbit 36¢. a set of boxes costs us $25. Rocap⟩[2b]

⟨If a set of boxes cost the Machine works $25, there must be something damnably rotten at the Machine works=⟩[3]

TL, NjWOE, DF (*TAED* D8334Z). Letterhead of Thomas A. Edison, Central Station, Construction Dept.; first letter signed for Insull by Alfred Tate. [a]"New York," and "188" preprinted. [b]Marginalia written and signed by Rocap.

1. The Machine Works shipped ten babbit bearings to the Shamokin company on 25 July. Eight were to replace phosphor bronze bearings; two were extras requested by Edison. Charles Dean to TAE, 25 July 1883, DF (*TAED* D8334X).

2. Rocap's reply was received on 14 August, when the letter was stamped at the 65 Fifth Ave. office.

3. This was not the first instance of excessive charges by the Machine Works to come to the attention of Edison or Insull; rather suspicious of its books, Insull had already obtained financial oversight of the shop (see Docs. 2321, 2343 [headnote], and 2400). Responding to a 4 August letter from Insull that has not been found, Charles Batchelor admonished: "I had an idea from the time you told me that Dean had made so much money by [inside] contract, that there was something very crooked about it and if I had had time when in America, I should certainly have asked you to open up the books for me. Tell Edison that, if he finds anything of this sort he ought to take very strenuous measures as there is nothing we can lose more money in than a dishonest manager" (Batchelor to Insull, 20 Aug. 1883, DF [*TAED* D8334ZAD]; see Doc. 2512).

DISTRIBUTION SYSTEM REGULATION
Doc. 2505

Edison wrote this document near the end of a sequence of at least eleven letters he exchanged with William Andrews about voltage regulation in the three-wire central station system. Proper regulation was vital for commercial success of the Edison system because it determined the amount of light produced by the voltage-sensitive incandescent lamps. It also directly affected the durability of the relatively expensive lamps.

Andrews had little or no central station experience when Edison placed him temporarily in charge of the Sunbury, Pennsylvania, plant, but his working knowledge of electricity put him ahead of those whom the local company there (or in any central station town) could expect to hire permanently.[1] Edison's cursory instructions, and Andrews's inability to grasp them, foretold problems at Sunbury and subsequent plants. Andrews conceded that excessive voltage probably contributed to the "enormous" number of broken lamps recorded in the first weeks at Sunbury.[2]

The loss of electrical energy in a conductor varies as the square of the current multiplied by resistance (I^2R). Losses in Edison's distribution system were calculated on the basis of the maximum allowable current, and they would decline disproportionately as electricity consumption dropped. The losses were also unequal throughout the system, with the feeder conductors from the generating plant sized for 15 percent attenuation and the mains, to which customer service lines were connected, for only 3 percent. In accordance with Ohm's Law ($V=IR$), maintaining the proper voltage (V) at any point in the system required changing the amount of current produced (I) as resistance (R) varied inversely with the number of lamps in use.

The three-wire system required balancing the voltage in the positive and negative sides of the system with respect to the neutral wire connected between them, a task quite apart from regulating voltage at the dynamo terminals (see Docs. 1950 and 2242 [headnotes]). This was a dynamic network in which altering the electrical load on only one side would cause a current to flow through the neutral line, thereby changing its potential with respect to the other side. Because lamps were connected between the neutral and either the positive or negative wires, they could experience a change in voltage quite independently of variations in the load on their side of the circuit. Edison employed equalizers—essentially resistance boxes—to regulate the feeder lines (the mains, or local distribution wires, remaining out of his control).[3]

Simple in principle, feeder regulation depended on reliable instrumentation to indicate line conditions to an operator back at the station. Properly interpreting signals from instruments in both branches of the network, especially when those devices acted inconsistently, was a difficult art to master.[4] Edison continued to seek better instruments,[5] but later plants in other towns saw heavy lamp breakage. Seasoned by the experience of opening several plants, Andrews cautioned in early 1884 that problems were "the almost inevitable consequence of starting up new Stations, and running the same by guesswork."[6]

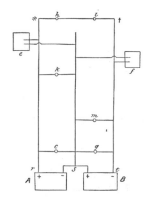

Schematic drawing of three-wire system showing dynamos A and B connected in series to positive (r) and negative (t) conductors. Individual lamps (c, g, h, i, k, m) or entire buildings (e and f) are connected between one feeder and the central neutral wire.

1. See, e.g., Doc. 2563.
2. Andrews to TAE, 12 Aug. 1883, DF (*TAED* D8361ZCJ).

3. "The Edison Central Station System," *Electrical Review* 4 (23 Aug. 1884): 1–3; Latimer 1890, 36–46; Howell 1886.

4. As the correspondence between Edison and Andrews shows, the difficulties lay in interpreting signals from voltage indicators on the feeders to understand conditions on customers' lines, and then managing the relationship between voltage there and at the dynamo terminals. The three-wire design introduced the complication of balancing the load and voltage on both sides of the system, although Edison noted disparagingly that "It requires but a small amount of judgment to make a comparison between the brightness of the lamps on the two sides." TAE to Frank Marr, 13 Nov. 1883, DF (*TAED* D8316BFQ); see Doc. 2484 n. 7 regarding switches for balancing the load between branches of the three-wire system.

5. See Docs. 2526, 2538 (headnote), 2584, 2592, and 2628.

6. Andrews to Edison Construction Dept., 16 Feb. 1884, DF (*TAED* D8442ZBH).

–2505–

To William Andrews

[New York] 15th Aug [188]3[a]

Andrews—

Your answer to questions received havent looked over them yet=[1] about right volts to carry as[b] the load gets lighter you can easily work that out as well as any one=[2] Knowing the number of consumers & their lights & the time you can always tell in a small town like sunbury what load you have on within 10 per cent. [You?][b] The drop in the house and on the service dont count one way or the other when there is a heavy or light load because if everybody but one place was closed there would still be the drop for that place on the service & inside wiring   Sunbury is wired for 15 & 3 drop hence you can make a table=[3] You never should make a sudden change in volts ie[c] run along for a week on one[d]

the village Lamps we are making now are far better in life—[4] is Kirby[5] was packed up in Sunbury   learned in a few days its simply absurd to pay him 25 and Expenses. You can take hold of a likely young fellow at Shamokin for [10$?][e] per day & no expenses— We must run this biz on an economical basis or it wont stand the racket [--- ---- ---][b]

E

ALS (letterpress copy), NjWOE, DF (*TAED* LBCD1378). [a]Date written by Samuel Insull. [b]Canceled. [c]Circled. [d]Last line(s) of page not copied. [e]Illegible.

1. Andrews completed fifty handwritten pages of answers to the examination questions discussed in Doc. 2500 on 12 August. He sent them separately, but they have not been found. Andrews to TAE, 12 Aug. 1883, DF (*TAED* D8361ZCJ).

2. See headnote above. The chain of correspondence was complicated by letters crossing in the mail, and Edison and Andrews appear to have been talking past each other at times. The sequence began with an inquiry by Andrews regarding one of the examination questions (see Doc. 2500) about what change, if any, the station engineer should make in the voltage at the end of the feeder lines when the system dropped from full load to half load. Andrews observed that in his experience

a lower pressure when there are only a few lamps on will keep these few lamps up to a normal candle power, but then if the pressure at end of feeders has to be varied with the load, how is the engineer to decide about his pressure unless he has an ampère meter on red & blue lines as well as on compensating lines [that is, the third wire]? Please explain this to me— He can tell roughly of course by position of neutral point on commutator but this does not seem to me to be sufficiently absolute. [Andrews to TAE, 9 Aug. 1883, DF (*TAED* D8343W)]

Edison provided rather abstract instructions to calculate electrical losses in the distribution lines based on "the maximum lamps in use at any one time and not for the lamps installed, as it is very seldom that the consumer uses all his lamps at once." In response to a subsequent clarifying question, he explained on 14 August that "one volts higher at the terminals would make the feeders all right. You can easily figure this out for yourself." Baffled by this statement, Andrews pleaded the next day for practical advice. Edison made a dismissive marginal note to the effect that "you could figure it yourself," but he nevertheless composed this document as a more considered answer (TAE to Andrews, 10 and 14 Aug. 1883; Andrews to TAE, with TAE marginalia, 15 Aug. 1883; all DF [*TAED* D8316ANI, D8316ANZ5, D8361ZCQ]). In reply, Andrews explained his intuitive method of regulating voltage by keeping "the pressure a little higher for about the first hour after starting and then gradually drop down—then at half past nine when the store lamps are put out, we drop a little lower, and again a little lower at about 11.30 when many of the hotel lamps are put out." On 17 August, Edison reiterated that "it is a matter which you can easily calculate as to what the drop is on the mains. If you get the pressure at the terminal of your feeders, the heaviest drop on the mains would of course occur when the whole load is on, namely, 500 lamps, and this drop would be three per cent. What you have to calculate is for a drop of say 120 lamps, which would be the most you could possibly have on with the present number you have connected, and certainly with this the drop could not be more than one per cent." Although this letter, too, did not offer specific instructions for making the necessary calculations, neither did it— finally—elicit further questions (Andrews to TAE, 16 Aug. 1883; TAE to Andrews, 17 Aug. 1883; both DF [*TAED* D8361ZCU, D8316AOK]).

3. See headnote above. Andrews had interpreted the Sunbury blueprints to indicate a 10 percent voltage drop on the feeders and 2 percent on the mains. Andrews to TAE, 11 Aug. 1883, DF (*TAED* D8361ZCH).

4. After admittedly having run the Sunbury lamps too brightly at first, Andrews lowered the voltage. "[P]eople don't like it," he reported, "but it must be done or the business won't pay." Of the seventy-five

filaments already burned out, the great majority broke at the same point near the bottom. He observed that "If you hold a lamp up to the light edgeways, you will notice that the carbon vibrates very rapidly so that the loop appears fan-shaped . . . and the vibration appears to start from just the point at which the break generally occurs." Matters had scarcely improved two weeks later, despite Andrews keeping lamps at or below their rated intensity "according to my judgment." Puzzled, Edison inquired if some Sunbury customers were connected directly to the feeder conductors rather than the distribution mains. He noted that "the most excellent accounts of the lamps" were received from "everywhere else." Breakage continued to plague the Sunbury station throughout the fall. Andrews to TAE, 12 Aug. 1883; Andrews to Samuel Insull, 24 Aug. 1883; TAE to Andrews, 30 Aug. 1883; Thomas Conant to TAE, 4 Nov. 1883; all DF (*TAED* D8361ZCJ, D8343ZAG, D8316AQS, D8360ZCC); see Doc. 2546 esp. n. 3.

5. J. F. Kirby helped the Wiring Dept. of Bergmann & Co. in Sunbury but was not considered experienced enough to remain there as an expert for the local lighting company. He did wiring for isolated plants in the latter half of 1883, then worked for the Construction Dept. in several central station towns. Frank McCormick to Samuel Insull, 10 Aug. 1883; Kirby to TAE, 26 Dec. 1883; Insull letter of recommendation, 27 May 1884; all DF (*TAED* D8361ZCF, D8313ZCK, D8416BRH).

–2506–

To Richard Dyer

New York[a] [August 15, 1883][1]

Dyer

In making English Complete on 3 wire system put in[2]

biz & draw claims so Hopksns pat wont interfere with them[3]

E[dison]

⟨Aug 15 83   Too late to put this in 2d Eng case on compensating system. There have been two English cases on this Subject, and proper care has been taken to avoid claiming matters covered by Hopkinson's patent— Dyer—⟩[b]

ALS, NjWOE, DF (*TAED* D8370ZBQ). Letterhead of Thomas A. Edison. [a]"New York" and "188" preprinted. [b]Marginalia written and signed by Dyer.

1. Date taken from Richard Dyer's marginal note; form altered.

2. Edison referred to the provisional specification of his British Patent 2,857 (1883), filed on 7 June. It explicitly extended and elaborated on his first British patent on the three-wire distribution system (Brit. Pat. 6,199 [1882]). The June provisional covered methods of regulating, both automatically and by command from the central station, the variable load across both sides of the three-wire system in order "that as little current as possible should ever traverse a compensating conductor

so that such conductors may be made as small as possible." The specification is similar to two U.S. patents (U.S. Pats. 283,893 and 283,894) that Edison had prepared in the spring (see Doc. 2484 n. 7).

The rough sketch in this document represents a method of placing conductors in a distribution system so as to obtain a uniform voltage drop throughout the network under all operating conditions; such a system would tend to balance itself and require a minimum amount of copper. This design was the subject of a U.S. patent application executed by Edison on 15 November. One figure in the patent that issued from this application in August 1884 (U.S. Pat. 304,085) shows this plan in detail. Edison arranged all the conductors and lamps or motors so that "all the translating devices will be equidistant from the source of electrical energy, and hence will be affected alike without the necessity of running feeding-circuits to various parts of the system, in order to equalize the electro-motive force." Edison divided the entire distribution system into a number of smaller circuits connected in series with each other. Between each of these branches, he would run a compensating conductor back to the dynamos. His specification pointed out that this plan required only a single set of mains and compensating conductors, "while the translating devices are all equidistant from the source of electrical energy. This relation is always preserved and is not affected by differences in the number of translating devices in the several divisions of the circuit, or by the total stoppage or neutralization of the current in one or more divisions."

The final specification of Edison's British patent was filed on 7 December 1883. It included a variation on the equidistant distribution system discussed above. Edison would obtain the effects of equidistance by placing compensating conductors in series between groups of lamps or motors, as well as in parallel with the main conductors. The third (and final) claim of this patent was for "The mode of applying the compensating principle to systems of general distribution employing feeding circuits and intersecting and connected positive and negative main conductors" (Brit. Pat. 2,857 [1883]).

Drawing from Edison's U.S. Patent 304,085 showing compensating conductors to keep a distribution system electrically balanced by maintaining all the lamps or motors at a uniform distance from the generators.

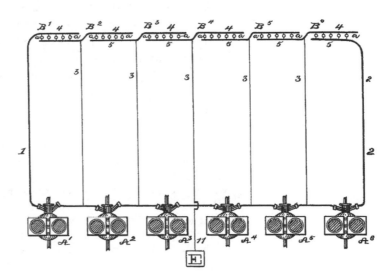

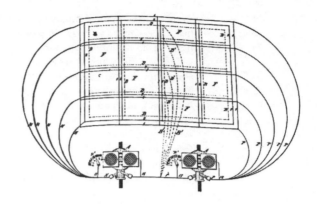

Drawing from Edison's British Patent 2,857 of 1883 showing a feeder and main system modified with compensating conductors to provide equal distance and uniform voltage drop between lamps or motors and the generators, regardless of the electrical load.

3. Edison referred to John Hopkinson's British Patent 3,576 of 1882 for the three-wire distribution system. News of that specification had been received as something of a shock by Dyer and Charles Batchelor, if not by Edison himself. See Docs. 2407 esp. n. 4 and 2414.

–2507–

To Spencer Borden

[New York,] August 18th. [188]3

Dear Sir:—

I am much obliged for your letter of the 16th. inst, and in reply would state that I shall be glad if you will give the gentlemen, who are going out to China to set up the machinery in Shanghai Cotton Mill,[1] as much information as is possible considering the limited time they have to give to the work.

If I get an order to light up the Mill referred to, I certainly feel inclined to borrow a man, either from your Department or the home office of the Isolated Company, to do the work.

I have the plans of the Mill, and would very much like to get an estimate from your Department as to what the plant would cost put down in say a New England mill.[2] Would you tell me what you would charge for preparing such an estimate, and whether a blue print of the general arrangement of the Mill would be sufficient to enable you to estimate correctly. Very truly yours,

TL (carbon copy), NjWOE, DF (*TAED* D8316AON).

1. Edison initially wrote to Spencer Borden on 10 August that he was "arranging to light up some cotton mills in Shanghai" and had promised to give two employees of the plant some experience in a New England mill before they left for China on 15 September. Borden replied that one man had not yet shown up, while the other (E. L. Holbrook) did not want "to take charge of wiring the mill nor take any responsibility beyond getting a general knowledge of wiring." Borden found that plans had already been sent to China and the general information

that Holbrook possessed was "entirely worthless" for planning the wiring. He proposed instead "to select one of our most experienced foremen, post him thoroughly in all the details of figuring, and prepare him for such difficulties as he would be likely to encounter." TAE to Borden, 10 Aug. 1883; Borden to TAE, 16 Aug. 1883; both DF (*TAED* D8316ANE, D8336ZAB).

The Shanghai Cotton Cloth mill was among the first joint-stock manufacturing enterprises organized in China under official patronage. When it was established in 1882, China had no mechanical means of producing thread or cloth to counter rapidly growing imports from the United States and Britain. Feuerwerker 1958, 208–11; Ji 2003, 60.

2. Jefferson Borden, Jr., working with his brother, Spencer Borden, in the New England Dept. (of the Edison Electric Light Co. and Edison Co. for Isolated Lighting), replied to Edison that his firm could prepare an estimate from a "scale plan showing the arrangement of the machinery." Having already received mechanical plans from a steam contractor, Edison sent blueprints to Boston on 21 August for an estimate to be made as if the plant were to be in Massachusetts. A. W. Danforth, the mill's manager, to whom Edison had written on 8 August (not found), wrote from Shanghai in September that he could not "get the mill ready for your lights till next autumn." The company made contracts for machinery and began excavating in 1883, but Danforth apparently did not pursue a lighting plant until 1889; the mill began operating the following year. Jefferson Borden, Jr., to TAE, 21 Aug. 1883; Providence Steam & Gas Pipe Co. to TAE, 16 Aug. 1883; TAE to Spencer Borden, 23 Aug. 1883; Danforth to TAE, 26 Sept. 1883 and 16 Apr. 1889; all DF (*TAED* D8325ZCA, D8325ZBW, D8316APP, D8336ZAD, D8941AAM); Feuerwerker 1958, 212–16.

-2508-

From Roscoe Conkling

[New York,] Aug 20 '83

My dear Mr Edison:

[Think?]ing[a] you may not have [the?][b] opinion of the Ex'rs in Chief in Sawyer & Mann interference, I send it for you to run over.[1]

Please return it in the course of say two or three days, as I need it in preparing additional Brief.[2]

Should any criticism or[b] suggestion occur to you likely to instruct me, please send it when you return the opinion.[3] Cordially yours

Roscoe Conkling

ADS, NjWOE, DF (*TAED* D8370ZBU). [a]Obscured by ink blot. [b]Repeated at end of one page and beginning of next.

1. The copy of the decision by the Examiners-in-Chief that Conkling sent to Edison has not been found, but its text was appended to his Brief and Argument (see note 2). "Brief and Argument for Edison," with Appendix, *Sawyer and Man v. Edison*, RG 241, MdCpNA (*TAED* W100DDB000, W100DDB023).

2. The Edison Electric Light Co., in consultation with Edison, retained Conkling in 1881 to work on this patent interference case (TAE to Uriah Painter, 30 Sept. 1881, UHP [*TAED* X154A3BF]). The Patent Office had declared the interference in September 1880 between Edison's application for a patent on the paper carbon horseshoe-filament lamp (Case 187, filed in December 1879; see Pat. App. Casebook E-2536:44, PS [*TAED* PT020044]) and one filed by William Sawyer and Albon Man in January 1880. At issue was "the incandescent conductor for an electric lamp formed of carbonized paper." After a hearing, the Examiner of Interferences awarded priority of invention to Sawyer and Man in January 1882. Conkling appealed for a new hearing on the grounds that Edison had discovered fresh evidence proving his priority. The Examiner at first refused, but was compelled by the Commissioner of Patents to hear the case again. After doing so, however, the Examiner returned the same decision in June 1883. The Board of Examiners-in-Chief overturned that ruling on appeal on 28 July 1883, awarding priority to Edison. Edison claimed at the time that the outcome was immaterial because he no longer used filaments of carbonized paper. Sawyer and Man "do me no harm," he reportedly said, "and I don't care to prosecute them," but he had contested the interference "on principle." The decision was appealed to the Commissioner of Patents and reversed on 8 October 1883 (see Doc. 2555). Conkling prepared Edison's appeal to the Secretary of the Interior, who determined that a recent Supreme Court decision deprived him of jurisdiction in the matter. Edison then applied to the Commissioner for a rehearing, but the patent was granted to Sawyer and Man in May 1885 as U.S. Patent 317,676. Its broad first claim for "carbonized fibrous or textile material and of an arch or horseshoe shape" failed to withstand a long court battle over Edison's basic carbon-filament lamp patent (U.S. Pat. 223,898), however, and the Sawyer and Man specification was restricted to the use of carbonized paper (U.S. Patent Office Commissioner's decision, 8 Oct. 1883, in *Edison Electric Light Co. v. U.S. Electric Lighting Co.* [1885–1892], Defendant's Depositions and Exhibits [Vol. IV], 2390–2414, Lit. [*TAED* QD012E2390]). Regarding Conkling's involvement in the case see also Samuel Insull to Painter, 2 Dec. 1881, Lbk. 9:395 (*TAED* LB009395) and Zenas Wilber to U.S. Commissioner of Patents, 18 Feb. 1882, *Edison Electric Light Co. v. U.S. Electric Lighting Co.*, Lit. [*TAED* QD012E2271]). For the patent interference and related litigation see "Carbon in Electric Lamps," *NYT*, 29 July 1883, 5; "Notes from Washington," ibid., 9 Oct. 1883, 2; "Decision Adverse to Edison," *Chicago Daily Tribune*, 11 Oct. 1883, 6; Copy of Appeal, *Sawyer and Man v. Edison*; TAE to Commissioner of Patents, 17 Oct. 1883, *Sawyer and Man v. Edison*, p. 2; Defendant's Depositions and Exhibits, 4:2256–2319; all in *Edison Electric Light Co. v. U.S. Electric Lighting Co.*; Lit. (*TAED* QD006198B, QD012E [images 91–121]); *Sawyer and Man v. Edison*, RG 241, MdCpNA (*TAED* W100DDA, W100DDB, W100DDC); Bright 1972 [1949], 67, 87–89; and Wrege and Greenwood 1984, 44–47.

In a separate action in June 1883, the Consolidated Electric Light Co., which owned the patents of Sawyer (now deceased), filed lawsuits in United States court alleging patent infringement by the Edison Electric Light Co. "Events in the Metropolis," *NYT*, 14 June 1883, 8.

3. The following day Edison replied, "I have read the [opinions?] of the Board [of Examiners]. It is pleasing to get a little justice at times. I have nothing to add." Lbk. 17:197 (*TAED* LB017197).

-2509-

To Gardiner Sims

[New York,] 21st Aug [188]3

Dear Sir,

Your engines at the Central Station are still giving a great deal of trouble and a great dissatisfaction is caused thereby[1]

In about six weeks our capacity will be taxed to its utmost[2] and I greatly fear trouble which will prejudice our business and give your engine a black eye. You <u>must</u> take this matter up <u>yourself</u> and deal with it vigorously and effectively   Yours truly,

Thos A Edison

PS   If this trouble is not got rid of immediately I shall have to stop ordering [engines for my?][a] construction dept as I would[3a]

LS (letterpress copy), NjWOE, LM 14:438 (*TAED* LBCD1438). Written by Samuel Insull. [a]Postscript written vertically in left margin and incompletely copied.

1. Regulating the speed of the $14\frac{1}{2} \times 13$ engines at the Pearl St. station had been a problem since at least July, when an Armington & Sims employee had found the regulator of one engine "in bad condition from want of attention and care." Gardiner Sims believed the difficulty was resolved at that time, but in reply to this letter from Edison, he conceded that "The regulation of our Central Station engines is not satisfactory at present." Workers at the firm's shop were able to replicate the problems by restricting the supply of steam to an engine, then applying a heavy load. Sims therefore promised that if Edison made several specific changes to the station's steam pipes and valves, "we can make them regulate." Sims to Samuel Insull, 8 July 1883; Sims to TAE, 23 Aug. 1883, both DF (*TAED* D8322ZBE, D8322ZBZ).

2. The six Jumbo dynamos at Pearl St. were rated collectively for 7,200 lamps (or about 8,400 for a short period). According to the Edison Electric Light Co., the plant was near capacity in mid-August, with nearly 8,000 lamps wired to the network and building wiring for 2,300 additional lamps in place. The number of lights actually in service increased to 10,164 by late October. Edison Electric Light Co. Bulletins 19:4 and 20:3, 15 Aug. and 31 Oct. 1883, CR (*TAED* CB019, CB020).

3. In his defense, Sims questioned why Edison's dissatisfaction at Pearl St. should "influence your future orders, as you do not complain of our engines that belt to your dynamos." Just at this time, though, Miller Moore protested that the regulation of Armington & Sims machines at the Southern Exposition in Louisville was "very bad," especially at low boiler pressure. Edison told Moore that he was "having a 'hell' of a time" in New York and "Armington & Sims do not seem to be able to

July–September 1883   229

fix the trouble. You had better warn Sims to give the matter his personal attention, or his engines will get a bad 'black eye.' I am wanting to use a great number of $14^{1}/_{2} \times 13$ engines, but feel rather afraid of them." Sims to TAE, 23 Aug. 1883; Moore to TAE, 21 Aug. 1883; TAE to Moore, 25 Aug. 1883; all DF (*TAED* D8322ZBZ; D8366ZAB, D8316AQC).

–2510–

Draft Patent Application: Incandescent Lamp

[New York, August 21, 1883[1]]

The object of this invention is to prepare blanks from which filiments may be cut or punched for producing flexible filiments of Carbon for incandescing Electric Lamps[2]

The invention consists in employing thin pure & evenly made tissue paper rubbing[a] soaking several pieces of the same in a thick mass of gum tragacanth, ~~Resin &~~or other suitable Carbohydrate, or viscous substance carbonizable without entire volatilization   then place 2 or more sheets of paper together & drying the same under strain & pressure to produce a perfectly even blank—[3] ~~The grain~~ paper having a grain one sheet is laid on the other so the grain shall be at right angles. a modification of this invention consists in treating each sheet simultaneously with Hydrofluoric acid & then immediately putting them together as stated, drying under strain & pressure.

Claim   A blank for cutting filiments for Carb[oni]z[atio]n from consisting of 2 or more thin sheets of paper secured together by a carbonizable cementing Compound for the purpose

2nd— Drying under strain and pressure.

3   across the grain—

producing the cementing by acting on the sheets themselves

~~Use of~~

[Witness:] H. W. Seely

ADf, NjWOE, Lab., Cat. 1149 (*TAED* NM018AAT). [a]Interlined above.

1. Henry Seely signed and dated the document on this date when he received it from Edison.

2. On 14 September, Edison executed a patent application based closely on this draft. The object of the invention was to produce filaments "for electric lamps which shall be of even density and resistance." The procedures described in the specification would yield "a homogeneous sheet. From this sheet the filaments are cut or punched. . . . By thus placing several sheets together the defects in any one sheet are counteracted or compensated for by the other sheets, each defective portion extending through only a part of the entire sheet, from which

the filaments are cut." Patent Office examiners twice rejected the claims on the basis that these means of treating materials were not new, and any products resulting from them could not be patented. After Edison substantially rewrote the claims, the specification issued in 1892. Pat. App. 470,922; see also Doc. 2511.

3. John Ott began related experiments on this day, when he "Bought tissue paper to make carbon filaments of." He recorded a few experiments on "Parchementized tissue paper" on 27 August, in early September, and at least two occasions in October. On 22 November, he made a composite of eleven sheets, from which he cut filaments that were sent to the lamp factory. They were carbonized and tried in lamps the next day, when Ott recorded their electrical characteristics and number of minutes in use. He sporadically made similar experiments, using various solutions of alcohol, sugar, and resins, into December (N-82-12-21:113; N-82-12-04:103, 109–19; Lab. [*TAED* N150113, N145103A, N145103B, N145109A–N145119]; see also William Holzer to TAE, 20 Aug. 1883, DF [*TAED* D8333L]). Edison had tried to obtain uniform lamp filaments by punching or cutting them from sheets of plumbago in 1881. He had even embarked on a commercial manufacturing process before finding them too unreliable (see Doc. 2085).

–2511–

*Draft Patent
Application:
Incandescent Lamp*

[New York, August 21, 1883[1]]

Patent

The invention consists of forming a cylindrical filiment for carbonization by Cutting out a blank from paper or other organic tissue with broadened Ends & twisting the same with a cementing material to form a cylindrical filiment with thickened Ends[2]

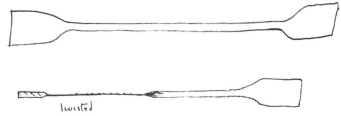

twisted

twisted

The flat blank of tissue paper is covered with a carbonizable cement such as gum tragacanth,[3] & twisted together by machinery in a even manner   An alternative is to twist without the Cementing material & afterwards gelatinize tha portion of the Cellulose of the paper by immersion in a solution of Hydrofluoric acid to cement the whole into a homeogenous mass.

Claim: The filiment of for carbonization made sub[stantially] as herein specified—

Will[a] that form of claim cover[a] better than broad <u>specific</u> claims??—

[Witness:] H. W. Seely   E[dison]

⟨twisted tight into homogs paper tissue⟩[b]

ADfS, NjWOE, Lab., Cat. 1149 (*TAED* NM018AAS1). [a]Obscured overwritten text. [b]Marginalia written by Henry Seely.

1. Henry Seely signed and dated the document on this date when he received it from Edison.

2. This memorandum resulted in a patent application that Edison executed on 14 September. The specification explained that a filament formed by this process

> is a homogeneous one, not made up of a number of separate strands, like the conductors of thread which have been sometimes used, while it possesses the advantages of such thread conductors, being cylindrical and of small area and radiating surface. Being all in one homogeneous piece, all its parts contract and expand evenly, and all are of the same texture and resistance. The strips of paper can be cut with great exactness, so that all will be of precisely the same size, and, being cut from the same quality of paper, very even blanks are produced. [U.S. Pat. 297,585]

The specification included seven claims, as well as one drawing based on the second figure in this document. The editors have not found records of experiments directly related to this application, but see Doc. 2510.

3. Edison had previously tried coating filaments, including cotton thread, with gum tragacanth or its derivatives before carbonizing them. See Docs. 1879, 2411, 2415.

–2512–

Samuel Insull to Joseph Ferrell

[New York,] August 23rd. [188]3

My Dear Mr. Ferrell:—[1]

I duly received your favor of the 20th.[2]

Last Saturday Mr. Edison was compelled to make sudden changes at his Machine Works, which resulted in the old staff severing their connection with us.[3] I am engaged this week in finding out exactly how their affairs stand,[4] and immediately I get through, your claim against them shall be the first one adjusted. I think I shall be able to write to you definitely about it about it on Monday or Tuesday next at the latest.[5]

Will you please see my letter to Mr. Richards of this date[6] Very truly yours,

TL (carbon copy), NjWOE, DF (*TAED* D8316APK).

1. Joseph L. Ferrell (d. 1904) became general manager of the Southwark Foundry & Machine Co. (which manufactured the Porter-Allen steam engine in Philadelphia) in late 1882 or the first half of 1883, about

the time Charles Porter was forced from the company. It was likely Ferrell whom Porter later described as "an oily-tongued man who had never seen a high-speed engine." Ferrell had been senior partner of the Enterprise Hydraulic Works in Philadelphia where he, according to one historian (Dawson 2004, 228–29), presided indifferently over the firm's decline. One apprentice pattern-maker during Ferrell's tenure was the young Frederick Winslow Taylor, who later attempted to find a scientific basis for shop management (Letterhead of Charles Richards to TAE, 1 Nov. 1882; letterhead of Ferrell to Insull, 20 Aug. 1883; both DF [*TAED* D8233ZEI, D8334ZAE]; Porter 1908, 323–24; Kanigel 1997, 109; Miller 1991, 49–50; "Scenery Fireproof Now," *NYT*, 22 Jan. 1904, 16).

2. Ferrell asked Insull to settle the account of the Edison Machine Works with the Southwark Foundry. Ferrell to Insull, 20 Aug. 1883, DF (*TAED* D8334ZAE).

3. On Saturday, 18 August, William Anderson sent Insull a letter containing specific allegations of financial malfeasance at the Edison Machine Works; this was not the first time that questions had been raised about practices there (see Doc. 2504). The editors have not otherwise identified Anderson, but he apparently had intimate knowledge of the shop. He accused superintendent Charles Dean ("aided and abetted by partys in office in Goerck St") of demanding kickbacks from suppliers and skimming money from the payroll or inside contracting system. He also claimed that "The woman D—n keeps has several thousand dollars in the bank which she received from D—n." Anderson promised that several named individuals, including Ferrell, could substantiate his charges. Anderson to Insull, 18 Aug. 1883, DF (*TAED* D8334ZAC).

Insull reported to Charles Batchelor three days later that

> For some time past, in fact for about a year I had very great suspicions as to the honesty of those conducting the Machine Works business and again and again reported to Edison on the subject on mere heresay information. . . .
>
> About two months ago Edison instructed me to go ahead and see whether I could get any definite information. I found that Dean had been taking a commission from practically every body he bought from and had borrowed money right and left and when a tradesman would not lend it Deane would threaten to take the trade away from him and in fact would do so if such a man persisted in refusing to meet Deane's demands.
>
> Edison at first would not listen to these charges against Deane but so absolute and unanswerable were the points that I brought before him that at last he decided to discharge Deane as soon as I could bring him certain affidavits as to the corruption. The last six months balance sheets from the Machine Works seemed to have a wonderful effect upon Edison. It showed quite a heavy loss on the six months work and the fact that Edison could get nothing turned out rapidly brought him to the conclusion to act immediately without waiting for the affidavits he had called on me for, so on Saturday afternoon Edison went to Goerck Street and simply told Deane and Rocap that he proposed running the business himself. . . . He then ordered the Works to be shut down for a week and we are now en-

gaged in trying to find out the exact state of affairs. [Insull to Batchelor, 21 Aug. 1883, LM 3:173 (*TAED* LM003173)]

Francis Jehl later embellished this story. He claimed that Edison, cognizant of Dean's notoriously violent temper, delegated the firing to Insull, who took two burly police officers with him to Goerck St. (Jehl 1937–41, 676–78).

Edison installed Gustav Soldan as superintendent, having (according to Insull's account to Batchelor) "borrowed him from the Light Company." Soldan's authority was more circumscribed than Dean's had been. Somewhat later, Insull reminded Joseph Hutchinson that "the business of the Edison Mach. Works is conducted by me in this office [65 Fifth Ave.]. . . . Mr. Soldan's duty is to build machines for the Edison Machine Works." Insull to Hutchinson, undated 1884, DF (*TAED* D8416BUT).

4. Edison immediately arranged for an independent audit of the Machine Works. The accountants found some procedural deficiencies but concluded that "the book-keeping has, on the whole been well done, and we do not believe that there has been anything wrong done through the books." Insull later conceded that he would never have solid proof but remained as "confident as ever that Mr. Dean had a large 'divy' with his men on contract" and that Rocap, though not guilty of criminal wrong, had "shut his eyes to certain irregularities" and impeded Insull's investigation. Insull to Harding & Burnap, 22 Aug. 1883; William Harding to Insull, 23 Aug. 1883; Insull to Batchelor, 5 Nov. 1883; all DF (*TAED* D8316APF, D8334ZAG, D8316BEG); Harding and W. H. Palmer to TAE, 4 Sept. 1883, Miller (*TAED* HM830191).

Edison also quickly retained attorney John Tomlinson in the matter. Charles Rocap made a claim of $1,350 against Edison, apparently on the basis of an 1882 contract (Doc. 2293). Insull acknowledged the debt, but objecting to the threat of a lawsuit, he advised "getting even" by settling on terms no more favorable than Rocap could win in court (Insull to Tomlinson, 27 Aug. 1883, Lbk. 13:23 [*TAED* LB013023]; Insull to Tomlinson, 3 Oct. 1883, LM 15:310 [*TAED* LBCD2310]; Tomlinson to TAE, 23 Aug. 1883; Insull to Tomlinson, 19 Oct. 1883; Insull to Batchelor, 5 Nov. 1883; all DF [*TAED* D8303ZFL, D8316BBS, D8316BEG]). Rocap settled his claim for $900, and Edison and Insull each wrote recommendations for him in 1884. Edison described him as "in all respects competent and reliable" and stated that his "connection with us was severed owing to changes in the management of our business" (TAE agreement with Rocap, 8 Dec. 1883, Miller [HM830204]; Insull to E. Myers & Co., 18 Mar. 1884; TAE to Charles Warner, 12 May 1884; both DF [*TAED* D8416AXZ, D8416BON]). Dean also sought money due him under his contract with Edison (Doc. 2293). Insull claimed to have obtained a release, but Dean ended up filing suit. The dispute was settled by the end of 1885, but the editors have not determined on what terms (Insull to Charles Batchelor, 4 Sept. 1883; Insull to A. B. Pearce, 5 Apr. 1884; Batchelor to Tomlinson, 5 Dec. 1885; all DF [*TAED* D8316ASU, D8416BCS1, D8531ZAF]); Insull to MacPherson Willard & Co., 3 June 1885, Lbk. 20:316 [*TAED* LB020316]).

5. The editors have not found the letter promised by Insull. South-

wark officials continued to correspond about the outstanding bill until at least early October, when Ferrell complained that he had made Insull's pledges "the basis of absolute promises to my people, and I am sorely harassed about the matter." Southwark Foundry to Insull, 29 Aug., 20 Sept., and 6 Oct. 1883; Insull to Southwark Foundry, 5 Oct. 1883; all DF (*TAED* D8334ZAK, D8334ZAY, D8334ZBA, D8316AYP); Insull to Southwark Foundry, 18 Sept. 1883, LM 15:96B (*TAED* LBCD2096B).

6. The letter from Charles Brinckerhoff Richards (1833–1919), superintendent of the Southwark Foundry in Philadelphia, has not been found. Richards had devised a high-speed steam engine indicator in conjunction with Charles Porter. He was assistant superintendent of Colt's Armory from 1861 to 1880, the year he became a charter member of the American Society of Mechanical Engineers. He left Philadelphia in 1884 to become head of the mechanical engineering department of the Sheffield Scientific School at Yale, a position he held until 1909. *DAB*, s.v. "Richards, Charles Brinckerhoff"; Doc. 1936 n. 2.

—2513—

To Gardiner Sims

[New York,] August 29th. [188]3

Dear Sir:—

I understand that you informed Major Eaton that you would[a] be prepared to build dynamos at a much lower price than that charged by the Edison Machine Works.

Will you please let me know exactly what you stated on this subject, as it is a matter in which I am ~~much~~ naturally[b] interested?[1] Very truly yours,

TL (carbon copy), NjWOE, DF (*TAED* D8316AQO). [a]Typed twice. [b]Interlined above by hand.

1. Edison wrote again two days later after receiving a letter from Sims on another matter to inquire whether or not Sims had "stated that you could build Dynamos cheaper than the Goerck St. Shops do & that you would be wiling to do so." Sims finally replied in a 5 September letter to Insull:

Please say to Mr. Edison in answer to his letter that I did not say that I would be willing to build his dynamos. I did however say that we could build them cheaper here than they had been built. I did not say this as I wrote him to interfere with, or [depreciate?] his business, as I often have the same said of our engines. I was pained to think it should be said that I did not have his interest at heart or that any person should so infer. Sam, the time has come when all personal interest in the Edison Electric Light Corporations should be lost sight of, and all work together to make a grand success of the best opportunity of a life time. Your business cannot go along and succeed if jealousy is to have this preference. If I can assist you at any time at Goerck St. I should be only to glad to do so.

Edison wrote on this letter to Insull "You see we got the young man by the balls   now explain that there was no jealousy but we wanted to put him on his guard against talking too much of our inside biz." No letter relaying Edison's comment has been found and it is likely that Insull did this in private conversation with Sims, a personal friend. TAE to Sims, 31 Aug. 1883, LM 14:498 (*TAED* LBCD1498); Sims to Insull, 5 Sept. 1883, DF (*TAED* D8322ZCJ).

–2514–

To John Lubbock[1]

[New York,] Sep 3rd [188]3

Dear Sir:

I received in due course of mail yours of the 16th ulto[2] and in compliance with your request cabled to you on the 1st as follows:—[3]

"Accept terms but greatly dissatisfied"

This cable substantially answers your requirements but I think it due to you and myself in acknowledging your letter that I should explain my reasons for the concluding words of the cablegram sufficiently to show whether[a] my dissatisfaction be justified or not and that the expression of it did not originate in mere petulance.[4]

The agreement with your Company was in two respects founded on a principle never reached by me from any other Electric Light Company or individual and which was always offensive to me. I refer to the provision of the contract which seeks to bind me in respect to all of my inventions of the character already assigned to your Company, and yet leaving your Company free to reject what they please and thus throw the burden of patent expenses upon myself.[5] The Company[a] which controls my Electric Light patents in this Country and which has paid all the heavy expenses attendent upon my expensive experiments have a contract for only five years.[6] The European Continental Company only demanded my inventions for five years with a condition entitling them to all after that time at their option on paying a price to be fixed by arbitration in case of disagreement.[7]

I can fearlessly say that I have always been most liberal in overperforming my contracts. I am constantly doing for our Company here services which my obligations to[a] them do not call for and I do this because the spirit of our agreement stimulates me to such Concessions. A contract which binds the inventor in such a way that at some time he is likely to think advantage has been taken of his necessities is never a prudent one to make for reasons inherent in human nature and which there is no necessity for me to allude to further. To me such engagements are most objectionable because they paralize my interest in the subject and chill that feeling of goodwill which with me are most potent factors in stimulating[a] me to further exertion in behalf of those I am connected with. Mr Lowrey explained this to Mr Waterhouse at the time of making the contract with you (if I am not mistaken) but the condition was insisted upon and I yielded only because the personnel of your Company was of such a character that I anticipate nothing but success in which case the success would cover up all such questions.

I have been this lengthy in respect to this original and continuing source of dissatisfaction because now when asked to work under this[a] obligation for the benefit of Mr. Swan and his friends at a reduced percentage of interest coming after an increased preferred dividend I find all the objections to this condition amplified and emphasised.

I did not overlook that the arrangement which makes £20,000 an absolute payment nor the fact that in demanding an increase of 25% in "A" shares over and above the allowance to Mr Swan's Company (on the consideration I suppose of the superior value of my patent and position) that allowance was taken wholly to the "A" Shareholders.[8] Were this[a] a matter open to further discussion I could give reasons satisfactory to my own mind for thinking that the change of the £20,000 from a conditional to an absolute payment is not a sufficient consideration for my having no share in the increased allowance of the Edison Company and for submitting me to a reduction of my interest in profit from 50% after 5% preferred to 25% after 7% preferred.[9]

Having indicated, as I hope, sufficient to satisfy you that my expression of dissatisfaction was at least made in good faith, and not as the result of[a] feeling I will drop that subject and proceed to the substantial conditions expressed in your letter as I understand them in connection with my letter to Mr Waterhouse[10] to which I understand it to be an answer:

First: I take as absolutely conclusive your opinion that less than 7% will not obtain the requisite capital.[11] Of course if we are to go on we must do so upon conditions adequate for success and one of such conditions is that the amalgamated Company should have money sufficient to enable them to develop their business[a] properly.

Second: In respect to the name I assume that you have not overlooked the terms of my letter and consider that I am now bound to interpret your statement that my argument is "double edged and in Mr Swan's mouth are as unanswerable against the omission of his name" as equivalent to saying that the condition of Mr Swans past contributions to this property; his obligations for the future; the probability that the performance of these obligations will lead to still larger additions to that property[a] on his part and the conditions of his retained interest in his inventions in England and elsewhere are such as to Enable him to make such an agreement. I admitted in my former letter that under the same state of facts Mr Swan would be entitled to use the same arguments. I added that in that case there would be an insurmountable obstacle to an amalgamation with my consent.[12] I fully meant what I then said and my present consent does not arise from any change of opinion or feeling[a] as to what is due and should have been conceded to me. I have yielded to the seriousness of your words in recommending me not to take the responsibility of defeating an arrangement which is satisfactory to you and others. These words, and the strong desire which I again express to see all "A" Shareholders reap the profit which they anticipate and have not yet received, have overborne my fixed determination at the time of writing my letter to Mr Waterhouse, never to accede to what I have now consented.[13] With[a] relation to the patents taken out since the signing of the agreement with the London Company I desire that an immediate settlement in connection with the same be made with me. I will have my accounts prepared and forwarded to you in the course of a mail or so and trust you will see that immediate payment for the same is made as the charges are onerous and embarrassing to me without any sufficient reason for my bearing same.[14]

As to future patents Some provision must be made by which all expense in connection with same shall be borne by the United Company, as to me it seems unfair that I should be compelled to make a considerable investment in patents when I reap no personal benefit in connection with such investment

Trusting that the new enterprise will prove a success   I remain   Dear Sir Yours truly,

Thomas A Edison

LS, NjWOE, Lbk. 17:217 (*TAED* LB017217). Written by Alfred Tate. [a]Repeated at end of one line and beginning of next.

1. This document is the final version of a letter drafted by Grosvenor Lowrey in his own hand on 1 September. The draft and final version are

essentially the same except as indicated in the notes below. DF (*TAED* D8338ZBM).

2. Lubbock wrote in reply to Doc. 2493 and to report the action taken by the Edison Electric Light Co., Ltd., board regarding Edison's views. Lubbock also reiterated the need for a merger between the Edison and Swan interests in light of the financial depression resulting from the collapse of the so-called Brush Bubble (see Doc. 2317 n. 8):

> the unnatural inflation of electric lighting companies 18 months ago has had a most disastrous effect upon the financial position of really sound undertakings. The public subscribed largely to undertakings which were not sound. Companies sprang into existence anxious to do business on almost any terms and the demand for electric lighting being at the time extremely limited it was impossible to do business at a profit. The necessary result was a re-action and now when many of the unsound undertakings have broken down and for the first time a wholesome demand is arising the public hold back and the funds necessary to develope an extensive business cannot easily be obtained. [Lubbock to TAE, 16 Aug. 1883, DF (*TAED* D8338ZBI)]

3. Edison sent this cable on 31 August; its text is quoted in full (LM 2:11D [*TAED* LM002011D]). In a handwritten letter advising Egisto Fabbri of his acceptance, Edison said he had done so "solely for the reason that you and your firm have a large amount of money in this English Co and I do not want to interfere in the slightest degree in a way that might cause you a loss or a regret for having gone into anything with me. Of course all can get out at any time, but I am tied up for years" (TAE to Fabbri, 31 Aug. 1883, Lbk. 13:24 [*TAED* LB013024]).

4. Johnson reported to Arnold White, secretary of the English company, around the date that Edison would have received Lubbock's 16 August letter:

> Edison is of the opinion that we have all made a very fatal mistake in amalgamating with Swan. He says this problem of electric lighting is one so vast and so complicated, that every Tom, Dick and Harry who essays it does not succeed, and that our amalgamation with Swan can do us no other than temporary good, and he does not believe that your expectations in the matter of raising money, by virtue of amalgamation, will be realized. . . . When he finally gave his consent, he turned to me and remarked, "There Johnson I now consider English interests practically wiped out." [Edward Johnson to Arnold White, 24 Aug. 1883, DF (*TAED* D8338ZBJ)]

Edison reflected years later, with a mixture of resignation and bitterness, that he "never got a cent" from the English lighting business (*TAEB* 6 App. 1.B.37).

5. See Doc. 2493 n. 7. Edison decidedly opposed this provision but was not insistent. With Edison's opposition in mind, Samuel Insull called it "the one mean clause" in the proposed agreement and warned that it would cause the British company more loss than gain. Insull to Edward Johnson, 8 Jan. 1882, LM 3:11 (*TAED* LM003011).

6. Doc. 1576 (article two).

7. The five-year limit was embodied in Edison's 1882 agreement with the Paris patent holding company and apparently also in the master contract with the Edison Electric Light Co. of Europe. TAE agreement with Edison Electric Light Co. of Europe, Ltd. (article 2), 15 Nov. 1881; TAE draft agreement with Compagnie Continentale Edison, undated 1882; both DF (*TAED* D8127W9F, D8228ZAY).

8. Lubbock stated in his 16 August letter (see note 2): "The directors wish me to point out what you do not refer to in your letter" (Doc. 2493), specifically, the proposal to make the £20,000 advance to Edison an outright payment in exchange for raising the preferential dividend rate. Under terms of undated August and September draft agreements and the executed version, the new Edison and Swan United Electric Light Co. was to allot 45,000 A shares (£5 each) to the Edison shareholders and 61,400 to Swan shareholders. Edison and Swan United Electric Light Co. agreements with Edison Electric Light Co., Ltd., Swan United Electric Light Co., Ltd., and George Black, undated August 1883 (articles 4 and 5), undated September 1883 (articles 4 and 5), and 1 Oct. 1883 (articles 3 and 4), all DF (*TAED* D8338ZBK1, D8338ZBS2, D8338ZBS3).

9. On the diminished dividend rights of B shares, see Docs. 2483 n. 4 and 2493. Theodore Waterhouse pressed, at the last minute, for a reduction in the dividend rate. Though he was unsuccessful, he promised that special care had been taken to protect Edison's interests in the new company. He enclosed a draft of the agreement on which legal counsel to the Edison company had noted: "I have perused this agreement in the interests of the B shareholders & approve the same" (Waterhouse to TAE, 2 Oct. 1883; Edison and Swan United Electric Light Co., Ltd., agreement with Edison Electric Light Co., Ltd., Swan United Electric Light Co., Ltd., and George Black, undated September 1883; both DF [*TAED* D8338ZBS1, D8338ZBS2]). The United Co.'s A shares did not yield dividends until 1888 or 1889 (Macrosty 1907, 316).

10. Doc. 2493.

11. See Doc 2493 n. 3.

12. Following the passage quoted by Edison, Lubbock had proposed that

> to place your name in the first place and to treat Mr. Swan as a subsequent inventor will convey to the mind of the British public in the clearest possible manner that your inventions are recognised as preeminent. I should be the last to suggest the adoption of any arrangement which would appear to undervalue the importance of your inventions as compared with those of others in the same field but Mr. Swan's character and reputation stand high among scientific men in this country, and the directors submit that it would be throwing away a valuable element of goodwill if when applying to the British public for funds to support the undertaking Mr. Swan's name were to be omitted from the title. Their proposition is therefore that the new company shall be called "The Edison and Swan United Electric Light Company Limited." [Lubbock to TAE, 16 Aug. 1883, DF (*TAED* D8338ZBI)]

On 2 October, the respective boards approved terms of the combination forming the Edison and Swan United Electric Light Co., Ltd.

The transfer of assets to the new firm was antedated to 30 June 1883 (Theodore Waterhouse to TAE, 2 Oct. 1883, DF [*TAED* D8338ZBS1]; "Railway and Other Companies," *Times* [London] 3 Oct. 1883, 11). An official report by the new company on rationales for the merger is in "City Notes, Reports, Meetings, &c. Edison and Swan United Electric Light Company, Limited," *Teleg. J. and Elec. Rev.* 15 (15 Nov. 1884): 397; for lists of stockholders see Summary of Share Capital & Shares in the Edison Electric Light Co., Ltd., 13 June 1883, 11 Nov. 1884, and 14 Aug. 1885, CR (*TAED* CF001AAO, CF001AAP, CF001AAQ).

13. In Lowrey's draft (see note 1), this text was followed by a closing paragraph in which Edison promised to "turn my back upon" dissatisfaction and to "loyally enter upon the effort . . . to aid you in making the business of Electric Lighting a success in England." This paragraph was crossed out and succeeded by another, longer one, probably written by Alfred Tate and bearing a date of 3 September. Despite some similarities of phrase, the latter draft was decidedly more assertive than Lowrey's original. It expressed Edison's wish to "have the active and friendly support of yourself and your associates in giving value to any further contributions which my investigations may lead me to make to the patent property of the Company . . . & which I trust will enhance the value of the Interest which my friends & myself retain in the Company."

14. The final two sentences of this paragraph concerning patent fees were adapted from text inserted by Lowrey midway through his draft (see note 1). Lowrey's addition quoted a cable message to London (further record of which has not been found) that Edison would "expect to be relieved of all advances or charges for patent fees, which are onerous & embarrassing to me without sufficient reason. You must arrange to pay all such fees immediately." The paragraph that follows in this document, concerning future patent fees, was not in either Lowrey's 1 September draft or the 3 September additions to it (see note 13).

–2515–

Agreement with Ansonia Brass & Copper Co.

[New York,] September 5, 1883[a]

Agreement made this 5th[b] day of September,[b] A.D., 1883, by and between the Ansonia Brass & Copper Co.,[1] a corporation organized under the laws of the State of Connecticut, and located at Ansonia, Conn., and at New York, party of the first part, and Thomas A. Edison, of New York, party of the second part.

During the continuance of this agreement the party of the second part shall purchase from the party of the first part exclusively, all the copper wire to be used by the party of the second part for the purpose of electric lighting, subject to the following terms and conditions, viz:[2]

When the party of the second part shall contract to furnish a plant for electric lighting, a copy of such contract shall be furnished to the party of the first part, together with full information concerning the responsibility of parties with whom such contract shall be made[3] and an estimate of the quantity of wire required to complete such contract. And the party of the second part shall, upon placing the order for such wire, assign to the party of the first part such proportion of any and all sums that may become due to the party of the second part under such contract as may equal the value of the wire to be furnished, such assignment to be in form as per schedule hereunto annexed, marked "B."[4]

Such assignment is to be held by the party of the first part as security for the payment of any and all sums due from the party of the second part for wire furnished; but when such sums shall be fully paid to the party of the first part, then such assignment shall be cancelled and become void.[5] Should the party of the second part at any time contract with parties who in the opinion of the party of the[c] first part are not responsible, the party of the first part may demand other and satisfactory security, and should such security not be given, may refuse to furnish wire, and the party of the second part may then purchase wire elsewhere—but only such wire as the party of the first part will not supply.

The parties of the first part agree to sell to the parties of the second part, under the conditions hereinbefore named, Pure Copper Wire, guaranteed to be not less than 97 per cent. electric conductivity, upon the following terms:

The price of Bare Wire, not finer than No. 20 B[ritish]. W[ire]. G[auge]. to be nine cents (9c.) per pound above the average price of Lake Ingot Copper for the month in which the order is given. For Wire straightened and cut to No. 10, 2 cents per pound additional; finer than Nos. 10 to 15, 4 cents per pound additional; Nos. 16 to 20, 6 cents per pound additional.

The average price of Copper to be taken from the official reports of the Secretary of the American Brass Association, at Waterbury, Conn.[6] The wire to be furnished in coils and the rods in cases, and delivered in New York City. Interest to be charged at the rate of 6 per cent. per annum after 30 days from the date of invoice.

Payment to be made as follows, namely: Seventy-five per cent. of the amount to be paid when the installation contracted for is ready to supply light and the remaining twenty-five per cent. within thirty days thereafter. The party of the second part agrees to make such payments immediately upon receipt of payments from the Illuminating Companies with which he contracts, and should such payments for any reason be deferred for more than twelve months from the date of the delivery of the wire to the party of the second part, then the party of the second part shall pay for such wire in full without further delay.

If at any time the amount due the party of the first part for wire under this agreement shall exceed the sum of Fifty Thousand Dollars, the party of the first part may refuse to supply wire except for cash on delivery.

When payment is made and the conditions of this contract fully complied with the party of the second part shall receive a rebate of two (2) cents per pound upon all wire purchased under this agreement.

This agreement shall terminate on December 31st, A.D., 1884.

The Ansonia Brass & Copper Co	Thomas A Edison
by A A Cowles[7] Secy	Witness Saml Insull
Witness. Thos. L. Fowler[8]	

TDS, NjWOE, Miller (*TAED* HM830192). [a]Date from document; form altered. [b]Handwritten in space provided. [c]"party of the" interlined above.

1. Anson Phelps founded the Ansonia Brass & Copper Co. in 1854 as a subsidiary of Phelps, Dodge & Co. By 1883, the firm had been supplying and collaborating with Edison for several years and was one of the largest producers in the United States. Lathrop 1909, 56–57, 99, 124; Docs. 1191 n. 3, 2084 n. 3, 2351 n. 8.

2. This agreement regulated relations between Edison and one of his most important suppliers for Construction Dept. work. During the summer, his indebtedness to the Ansonia Brass & Copper Co. had exceeded $10,000; coincidentally or not, Edison had complained about the firm's slowness in filling his orders. Edison promptly sent a copy of this agreement to another major vendor, Babcock & Wilcox, with an unidentified proposal. Ansonia Brass & Copper Co. to Samuel Insull, 3 Aug. 1883; TAE to Ansonia Brass & Copper Co., 24 July 1883; TAE to Babcock & Wilcox, 6 Sept. 1883; Nathaniel Pratt to TAE, 7 Sept. 1883; all DF (*TAED* D8321P1, D8316AIP, D8316ATN, D8323S).

3. Samuel Insull had subscribed since March 1883 to regular reports by the Bradstreet Mercantile Agency, the credit reporting firm, which could have been adapted for this purpose. Edison occasionally asked for special reports from Bradstreet on parties with whom he did business. Insull to Bradstreet Mercantile, 24 Mar. 1883, Lbk. 16:11 (*TAED* LB016011); Insull to Bradstreet Mercantile, 16 July 1883; TAE to Bradstreet, 5 Dec. 1883; both DF (*TAED* D8316AGW, D8316BOB).

4. The sample assignment has not been found.

5. An undated report identifying the proportions assigned to the Ansonia firm from nine of Edison's contracts is the Edison Construction Dept. Record Book (1883–1884), n.d. [1884], CR (*TAED* CD004).

Statements of Edison's indebtedness to Ansonia on account of ten village plants as of 14 March and 23 April 1884 are in Billbook No. 1:136, 176, Accts., NjWOE). In the latter half of 1884, Edison was still having difficulty collecting all the money owed him by various local illuminating companies (see, e.g., Docs. 2708, 2709, 2737, and 2734).

6. The brass producers of Connecticut's Naugatuck Valley, which held a monopoly on production in the United States, formed a cartel under the name of the American Brass Association about 1853. The secretary's position was held from 1865 to 1884 by Augustus Milo Blakesley (b. 1830), a Waterbury banker and president of the American Pin Co. Lathrop 1909, 121–23; U.S. Census Bureau 1965 (1870), roll M593_112, p. 767, image 655 (Waterbury, New Haven, Conn.); Osborn 1907, 3:53–54.

7. Alfred Abernethy Cowles (1845–1916) was secretary of the Ansonia Brass & Copper Co. and a longtime friend of Edison. Doc. 2343 n. 7.

8. Thomas L. Fowler (b. 1856?) was identified in the 1880 census as a clerk at a copper works and a resident of the Queens County town of Jamaica. U.S. Census Bureau 1970 (1880), roll T9_918, p. 547.2000, image 0309 (Jamaica, Queens, N.Y.).

–2516–

Draft to Theodore Adams[1]

New York, September 6th. 1883[2a]

⟨Wants firm Estimate & thought Navarro Flats 59th St & 7th Ave⟩[3]

⟨Consult Tudor & Co[4]—about Condensing our exhaust= They do the heating of Navarro flats=⟩

~~Friend Adams:—~~

~~My engineers have worked out the estimate pretty carefully, and I think it is within three to four per cent of being correct~~.

The total cost is (~~with ten per cent added to pay the Illuminating Co. for the right~~) ~~$96,500~~ 89,850.[b] This includes wiring of all the flats for about 11,000 lights, also wiring for telephones electrical Bills[5c] and district telegraph call to a central station in the flats, & to restore the plastering, where the necessary cutting of the walls & ceilings is to be made[d] ⟨Greenfield[6] restores plastering⟩ the furnishing of switch board and fitting up central station, but without the telephones, ~~which will have to be furnished by the telephone Co.~~ & to furnish & put in all speaking tubes, &c[e]

The electric light plant will supply at one time 4,000 ~~lights~~, ten candle lights[f] ~~each equal in illuminating power to the ordinary five foot burner, supplied with coal gas, the electric burners giving~~ aspherical illumination ~~equal to ten standard candles~~. which is about equal to 5 foot of Coal gas as burned in practice[g]   The plant can be forced on occasions to 4,500 lights. ~~We~~ The Illuminating Co[h] will guarantee ~~this and also that~~ the life of the lamps shall ~~be over~~ average[i] 600 hours, when they are run at their rated power of 10 candles each,[j] and also that ten lamps will be obtained for every indicated horse power. ~~We could commence at once, and the electric light would be ready to light the first apartment~~.

The running expenses would be about as follows:[k]

DAILY[7]

Electrician and meterman	3.50
One engineer, seven months	
Two engineers, five months[l].	4.20
One fireman	2.00
Coal 4 lbs. H.P. per hour at $4, being	
$2\frac{1}{2}$[m] tons year round	9.00
Oil, waste	.50
Water	1.60
Lamps at $1. each.	15.00[n]
Insurance.	1.50
Depreciation (long)	11.50
Repairs.	1.50
Stationery	.30
Meter supplies	.25
Fire grate bars	.20
Taxes	4.50
Carting ashes	.25
Extras.	2.00
	$57.80

Above based on a daily average of three hours use of the 4,000 lamps during the year.

Taking into calculation that parties be absent two months in the year and that only 4,000 out of 11,000 lights is figured on, it is believed this average is correct.

Adding five per cent interest on say 100,000, which for 365 days is $13.60 per day, the daily running expenses will be $70.90, or $25,878.50, yearly.

4,000 burners burning equal to five cubic feet of coal gas would be 20 Ms[8] per hour or 60 Ms per day of three hours average, the running expenses being $70.90 daily, would bring the cost of each M equivalent to $1.18. Buying gas at $2.25, would make a loss of $64.10 a day to the occupants of the flats.[9]

Interest at five per cent being already allowed, this $64.10 would pay twenty three per cent more on the 100,000, or altogether twenty eight per cent on the investment, and the investors would obtain a vastly superior article. Very truly yours,

T A Edison

⟨with[o] 128. apartments this would be $750. for each, as investment & Light bills would be about 16.75 per month⟩

TDfS, NjWOE, DF (*TAED* D8340ZCV). Letterhead of T. A. Edison. Handwritten additions not attributed to Edison made in an unknown hand. [a]"New York," and "188" preprinted. [b]Interlined above by Edison. [c]"electrical Bills" interlined above by hand. [d]"& to restore . . . is to be made" interlined by hand. [e]"& to furnish . . . &c" added by hand. [f]"ten candle lights" interlined above by Edison. [g]"which is . . . in practice" interlined above by Edison. [h]"The Illuminating Co" interlined above by Edison. [i]Interlined above. [j]"of 10 candles each" interlined above by hand. [k]Followed by dividing mark. [l]"One engineer, . . . five months" enclosed by right brace. [m]"½" written by hand. [n]Multiplication of 365 by 15 written in right margin but not transcribed. [o]Marginalia written by Edison; "with" interlined above in unknown hand.

1. Theodore Adams was superintendent of construction for the Navarro Flats apartment complex. He represented the investment of James Jennings McComb on the board of directors of the Central Park Building Co., its construction-finance company. "How the Flats were Built," *NYT*, 12 June 1887, 10; "M'Comb Well Secured," ibid., 16 July 1887, 8.

2. Edison's marginal notations in this draft, and at least some of the emendations, were likely made about 15 October, when he was advised to meet with Adams because "other parties are making great efforts to take away from us the lighting of the Flats." On 18 October, a formal version of the estimate was sent to Adams in two parts, each based closely on this draft. One gave the "probable running expenses in connection with the electric light installation"; the other provided "an estimate for the installation of an electric light plant and other electrical work at the eight Navarro Flats." The installation proposal also included Edison's terms, which were 25 percent of the total at these intervals: upon execution of the contract, when the engines and dynamos were in place, upon completion of the lighting plant, and thirty days thereafter. The editors have not found a contract or other evidence that Edison undertook the work. James Russell to TAE, 15 Oct. 1883, LM 15:386 (*TAED* LBCD2386); TAE to Adams, 18 Oct. 1883, both DF (*TAED* D8316BBN, D8316BBO).

3. The Navarro Flats, variously known as the Central Park Apartments or the Spanish Flats, was a palatial group of eight apartment house buildings under construction in New York City, on the block extending east from Seventh Ave. between Fifty-ninth and Fifty-eighth Sts. In a letter to its principal developer, José Francisco de Navarro, Edison had recommended Bergmann & Co. for the wiring installation contract and invited him to inspect their work in "over 1000 places," including the Mills Building (see Doc. 2318 n. 2) and the Dakota Apartment house, which was eventually wired for 5,000 lamps. Landau and Condit 1996, 135; TAE to Navarro, 31 May 1883, Lbk. 17:4 (*TAED*

LB017004); Edison Co. for Isolated Lighting Bulletin 6:17–18, 27; 25 July 1885; CR (*TAED* CC000 [images 24–25, 29]).

Navarro was an original stockholder in the Edison Electric Light Co., as well as the Edison Electric Light Co. of Cuba and Porto Rico, the Edison Electric Light Co. of Havana, and the Edison Spanish Colonial Electric Light Co. (Docs. 1571 n. 2, 2232 nn. 1–2). He was also a director of the Manhattan Elevated Railway, the New York Loan and Improvement Co. (builders of elevated railroads), and the Equitable Life Assurance Co., as well as a founding incorporator of the Metropolitan Opera-House Co. Tilden Blodgett to TAE, 7 June 1883, DF (*TAED* D8315G); "Jay Gould's New Railroad," *NYT*, 9 July 1881, 8; "Elevated Railroad Cost," ibid., 27 Mar. 1883, 3; "The Proposed New Opera-House," ibid., 11 Apr. 1880, 5.

4. A manufacturer, contractor, and innovator in the steam-heating and ventilation business, F. Tudor and Co. was founded in Boston in 1877 by Frederick Tudor (d. 1902). A civil engineer, Tudor invented a significant modulating valve (also known as a fractional or graduated valve) for steam heating systems (see U.S. Pat. 294,982) in 1883. The firm relocated to New York in 1879 and operated branches in additional cities. Its recent steam-fitting and ventilation contracts included the new Metropolitan Opera. "Failure of Frederick Tudor," *NYT*, 16 Aug. 1887, 8; "Funeral of Frederick Tudor," *Boston Daily Globe*, 1 Nov. 1902, 6; see also Doc. 2445.

5. In the final proposal sent to Adams (see note 2), this phrase was correctly rendered as "electric bells."

6. Edwin Truman Greenfield (1848?–1920), former Union Army colonel and a nascent electrical inventor, was general manager of the Wiring Dept. of Bergmann & Co. When that group disbanded, he did central station wiring work directly for the Edison Construction Dept. After leaving Edison's employ a few years later, Greenfield made significant inventions in electrical cables. Doc. 2125 n. 7; U.S. Census 1992? (1920), roll T625_1227, p. 51B, image 442 (New York City [Manhattan Assembly District 23], New York, N.Y.).

7. This itemization is the same as the final version sent to Adams (see note 2), except for the omission there of "½" from the tonnage of coal consumption.

8. The "M" was the amount of energy that would give the same amount of light as that from 1,000 cubic feet of gas burned in a standard gas lamp. Francis Upton began using this unit in Edison's calculations in 1880 as a convenient way of comparing the cost of electric and gas illumination. See Docs. 1897, 1958, and 2008.

9. Within fourteen months, New York gas declined to about $1.75 per 1,000 cubic feet, at least in some locations, because of new competition, electric lighting, and the general "financial troubles of the year." "The Effect of Cheap Gas," *Electrical World* 4 (15 Nov. 1884): 192.

–2517–

From James Pryor

New York, 6th Sept., 1883.

My dear Mr. Insull,

Inasmuch as Mr. Edison has been a month late with several of the payments of rent upon 25 Gramercy Park, I do not hesitate to ask that the rent for October should be paid now, with that for August and September.[1] I shall be much indebted to you, if you will see that this is done. Very Truly yours,

James W. Pryor[2]

⟨Dont pay only to Sept 1   would get out of the dam-d hole if could—⟩[3]

ALS, NjWOE, DF (*TAED* D8303ZFU).

1. James Pryor had written on at least three occasions, most recently on 24 August, about rent overdue by a week or more for the Edisons' residence in New York. Pryor to Insull, 9 Nov. 1882, 29 June 1883, 24 Aug. 1883, all DF (*TAED* D8204ZIP, D8303ZEE, D8303ZFN); see also Pryor to TAE, 30 Jan. 1884, DF (*TAED* D8403P).

2. Attorney James Williamson Pryor (1858–1924) and his sister Caroline Pryor leased the house at 25 Gramercy Park to Edison for two years from October 1882. Doc. 2341 n. 1.

3. The editors have not found a formal reply to Pryor, but see Doc. 2528. After another plea, Insull evidently promised to send the money but failed to do so. Pryor to Insull, 13 Sept. 1883, DF (*TAED* D8303ZFZ).

–2518–

Francis Upton to Samuel Insull

EAST NEWARK, N.J., Sept. 6 1883[a]

Dear Sir:

Your letter of the 6th inst. has received careful consideration.[1]

Holzer stands on a different basis from the others to whom shares have been given in the business in that his name appears already on our books.[2]

The only really satisfactory way to arrange matters is to make a stock Co. out of the Lamp Co. and to give each his share, and then to hold all the stock in trust to be held by trustees and not to be sold, except two thirds of the remainder of the stock-holders are willing.[3]

There are Cos. made on the close corporation principle,[4] we could model after these.

Until some such step can be taken I think all that Mr. Edison can do is to allow the matter to stand. It is distinctly understood by me that he is willing to give Lawson,[5] Bradley[6] and Howell[7] each 1% of the total capital of this Co. from his share, and I have assured them to this effect.[8]

I cannot devise any letter that I would advise Mr. Edison writing, without further consultation.

The next opportunity I have I will ask Mr. Edison regarding the matter. If you desire I will look up the matter of close coporations. The Fairbank's Scale Co.[9] is one that I know of. Yours Truly

Francis R. Upton.

ALS, NjWOE, Miller (*TAED* HM830192A). Letterhead of Edison Lamp Co. [a]"EAST NEWARK, N.J.," and "188" preprinted.

1. Insull inquired about the financial interests in the Edison Lamp Co. promised to William Holzer, John Howell, John Lawson, and James Bradley. Noting that Edison wished to avoid formal contracts, he asked Upton to "give me your ideas as to what kind of letter Edison should sign on the matter." Insull to Upton, 3 Sept. 1883, Lbk. 17:213A (*TAED* LB017213A).

These arrangements, in addition to giving stakeholders a financial incentive to develop Edison's business, also secured the loyalties of a small number of associates with intimate knowledge of his manufacturing operations. Edison and Upton tried to keep out of the lamp factory, in particular, anyone with the ability to understand what went on there, and Upton once rejected a job applicant because he thought it unwise "to employ men that know the value of knowledge gained." This wariness arose at least in part from prior experience with Ludwig Böhm, a skilled glassblower who left Edison's employ on bad terms. Upton to TAE, 17 July 1883, DF (*TAED* D8332ZBX); Doc. 2309 esp. n. 2, 2000 n. 3, and 2022 n. 2.

2. William Holzer (1844–1910), Edison's brother-in-law and a machinist and glassblower, became the manufacturing superintendent in March 1882. At that time, Edison granted him a 3 percent interest in the business, to be paid from retroactive wages and his share of future profits. U.S. Census Bureau 1970 (1880), roll T9_790; p. 273.3000, image 0387 (East Brunswick, Middlesex, N.J.); ibid., 1982? (1900), roll T623 1108, p. 5B (New York City, New York, N.Y., enumeration district 613); Frank Dyer to George Marks, 7 Mar. 1910, CR (*TAED* CL222AAE); see Docs. 2138 n. 6, 2244, and 2260.

3. The idea of incorporating the Edison Lamp Co. had been discussed since at least the middle of 1882, but it did not occur until January 1884; see Docs. 2312, 2536 n. 2, and 2650. On the basis of a draft incorporation document, the editors incorrectly implied in Doc. 2244 n. 3 that it happened in March 1883.

4. A close corporation is one whose shares are held by a small group of stockholders who usually are intimately involved in the business. *Black's*, s.v. "Close corporation."

5. John W. Lawson (1857–1924), a self-taught chemist, began working for Edison at Menlo Park in 1879. He became heavily involved in electric lamp experiments and, when the lamp factory moved to Harrison, took charge of the carbonizing and electroplating departments. Doc. 2081 n. 4; "Lawson, John W.," Pioneers Bio.

6. James J. Bradley (1845–1925), a machinist and long-time Edi-

son employee, was the lamp factory's master mechanic and was also in charge of the fiber department. Docs. 1080 n. 4, 2095 n. 3, and 2260.

7. John Howell.

8. See Doc. 2650.

9. E. & T. Fairbanks & Co. of St. Johnsbury, Vt., was a major manufacturer of scales of all types. Brothers Erastus and Thaddeus Fairbanks established a general foundry in 1823 and about eight years later began producing scales. Based largely on Thaddeus's patented inventions and improvements, scales quickly became the firm's signature trade. *DAB*, s.v. "Fairbanks, Thaddeus"; *ANB*, s.v. "Fairbanks, Erastus."

–2519–

To Sherburne Eaton

New York, Sept. 7th. 1883[a]

Dear Sir:—

Referring to my interview with you as to the Utica contract and your recommendation of acceptance of the proposition of the Utica Directors to make a contract with me to put up the plant there for $57,000., the balance of my estimate (about $6,000) to be paid by the Light Company,[1] I wish to place on record my objection to the course being pursued in view of the fact that you propose bringing the matter before the Executive Committee for final settlement.

The proposed arrangement would be a good one for the "Construction Department," but a very objectionable one, in my opinion, for the Light Company to accept.[2]

I consider it far better to allow the Utica Company to fall through, rather than have the Light Company accept such a proposal which would most certainly prejudice their future business, whereas the failure to close the Utica contract this Fall would have, under the worst circumstances, but a temporary effect on our business. Yours truly,

Thos A Edison

TLS, NjWOE, DF (*TAED* D8352T). Letterhead of Thomas A. Edison, Central Station, Construction Dept. [a]"New York" and "188" pre-printed.

1. Sherburne Eaton objected that he had merely urged consideration of this proposal, not its adoption. Concerned because this correspondence would become part of the Edison Electric Light Co.'s official record, he asked Edison to "run your pen through 'your recommendation' and put over them these two words, namely, 'the question.'" On the signed letter that Eaton had received, Edison later struck through "recommendation of" and interlined instead the words "asking my opinion as to the." He also instructed Insull to have the letter retyped (Eaton to TAE [with TAE marginalia], 11 Sept. 1883, DF

[*TAED* D8327ZAX]). A carbon copy of the amended letter, also dated 7 September, is in DF (*TAED* D8316ATS).

Edison had prepared an estimate of about $72,000 for a generating and distribution system of 3,200 10-candlepower lamps at Utica, N.Y. About a week later, on 16 August, he dropped the price to $63,329 by reducing the amount of copper in the conductors and eliminating other items (Construction Dept. estimate, undated [10 Aug. 1883]; TAE to Arthur Johnson, 16 Aug. 1883; LM 14:336, 400 [*TAED* LBCD1336, LBCD1400]). The local illuminating company balked, however, and evidently made a counter-offer, through the Edison Electric Light Co., of $57,000. Edison indignantly rejected this proposal, pointing out that "the estimate represents the cost of the plant with 12 per cent ONLY added. . . . The universal custom of contractors is to add considerably more than 12 per cent, to cover their risks, probable errors and profits" (TAE to Eaton, 29 Aug. 1883, DF [*TAED* D8316AQN]).

The editors have not determined what further action, if any, Eaton took about this matter. He was still corresponding with Utica promoters in early 1884, but the proposed plant there was never built. Eaton to TAE, 28 Feb. 1884, DF (*TAED* D8427T).

2. To facilitate the liquidity of startup illuminating companies facing high construction costs, the parent Edison Electric Light Co. later made it a policy to take an equity interest as partial payment of its license fees. It is not clear when it began to do so, but in early 1884, its standard terms required a fee of 25 percent of the local company's capital, one-fifth of which was payable in cash and the remainder in stock. In at least one case, the parent company took an additional interest in an established illuminating company, becoming the largest stockholder of the local firm in Sunbury by early 1884. Eaton to TAE, 28 Feb. 1883; Samuel Insull to Bergmann & Co., 24 Mar. 1884; both DF (*TAED* D8427T, D8416AZA); see also Doc. 2604.

Edison's personal financial interests in the construction business also diverged from those of the parent company in that his standard contract called for the firm's chief engineer to arbitrate disputes between him and the local firm over the quality of workmanship. TAE to Eaton, 29 Aug. 1883, DF (*TAED* D8316AQN).

FUEL CELLS   Doc. 2520

Edison began working on a system to convert coal directly into electricity in early 1882 (see Doc. 2230).[1] Over the next two years, he focused on what was in essence early fuel cell technology. By then, he had already expended a good deal of time and effort obtaining suitable steam engines for his lighting plants ("steam engineering forms 75 per cent. of the electric light," he reportedly said in 1880).[2] While skeptical of chemical batteries, Edison believed, as one newspaper paraphrased him, that "means will sooner or later will be found to obtain electricity from the earth without the use of any machinery whatever."[3] Unlike primary and secondary batteries, in which the electrodes themselves are consumed in chemical reactions, a fuel cell operates by the replenishment of an oxidizable fuel, such as coal, to produce free electrons. This could be done directly by the reaction of coal with other substances, such as nitrates, or by the creation of an ionized gas.

The development of fuel cells is generally traced to William Grove's experiments in the 1840s on what he called a "gas voltaic battery," but A. C. and A. E. Becquerel were the first to design a fuel cell that generated electricity by the electrochemical combustion of coal with a nitrate. Edison briefly experimented with a carbon battery of this type in the mid-1870s.[4] By the time he began to focus his efforts on fuel cell technology in the mid-1880s, he was familiar with Paul Jablochkoff's patented battery that used fused nitrates on the positive pole and carbon on the negative pole.[5] Edison also had probably read Alfred Niaudet's discussion of Jablochkoff's battery in his *Elementary Treatise on Electric Batteries* (pp. 240–41), first published in an English translation in 1880, in which Niaudet predicted, "It is very probable that a direct transformation into electricity of the heat produced by the combustion of coal may be obtained. Mr. Jablochkoff has already invented a battery cell which fulfils the above conditions."[6]

Edison's first patent application for the direct conversion of coal into electricity (U.S. Pat. 460,122), filed in May 1882, came into interference in the Patent Office with Jablochkoff's 1879 patent (U.S. Pat. 219,056). In his design, Edison sought to generate a current by efficiently oxidizing a carbon electrode by means of an active oxidizing agent, such as a fusible metal, in a heated iron vessel.[7] Edison filed two other applications for direct conversion employing fuel cells in 1883. The first, based on Doc. 2520, was filed in September (U.S. Pat. 435,688). In that process, Edison produced an ionized gas from either a metal or carbon in reaction with an oxidizable substance. He employed a vacuum chamber to maintain the gas in a rarefied state and to prevent reaction with atmospheric oxygen. The second application, filed in November (U.S. Pat. 490,953), described an improvement over his 1882 design. In this arrangement he maximized the production of carbonic acid from the oxidation of the carbon and of the fusible iron oxide in order to oxidize the carbon more rapidly.[8]

Edison made additional experiments on direct conversion in 1884 (see Docs. 2620, 2633, 2634, 2701, 2703, and 2706). During his trip to Florida, he proposed a new approach to direct conversion in which he used finely divided metal and a peroxide (usually manganese) in a solution of sulphuric acid to catalyze the oxidation of the carbon. By the end of that summer, he had "obtained a very strong current" using anthracite coal. He planned to show his system at the Philadelphia Electrical Exhibition in the fall, but it proved to be too dangerous after "all the windows were blown out of his laboratory."[9] About this time, Edison was reported to have said

> The great secret of doing away with the intermediary furnaces, boilers, steam engines, and dynamos will be found, probably within ten years. I have been working away at it for some months and have got to the point where an apparently insurmountable obstacle confronts me. Working at the problem now seems to me very much like driving a ship straight for the face of a precipice, and when you come to grief picking yourself up and trying it again to-morrow. There is an opening in the barrier somewhere, and some lucky man will find it. I have got far enough to know that the thing is possible. . . . I give myself five years to work at it, and shall think myself lucky if I succeed in that time.[10]

In an 1885 essay reprinted in *Scientific American*, Edison predicted a "marvelous revolution" to come from the cheap electricity of direct conversion, but he apparently had largely abandoned efforts to use the catalytic oxidation of carbon after the 1884 accident. He may have conducted a few experiments along those lines between 1886 and early 1888.[11] By 1887, he had adopted an alternative approach to direct conversion using the principle that the magnetic capacity of iron diminishes as its temperature increases. He applied this idea in the design of his pyromagnetic generator and motor.[12]

Edison's earlier approaches were later taken up by William Jacques, an engineer with American Bell Telephone Company, whose work has generally been treated as the most important effort to develop fuel cells of this type.[13] It is unclear if Jacques knew about Edison's work, but he was in contact with Edison in 1884 and 1885 regarding Edison's telephone experiments for the company, including those on an improved carbon transmitter, which may have influenced some of his experiments on direct conversion in July 1884. In the late 1880s and early 1890s, as Jacques was working on his own fuel cell designs, he was also associated with the Edison Phonograph Toy Manufacturing Company. After Jacques's United States Patent 555,511 issued in March 1896, the *New York Herald* asked Edison about it. Edison reportedly replied, "I have patents and applications for patents on the use of heated iron chambers with carbon electrodes, immersed in melted caustic soda, and preferably caustic potash, into which air is blown. I have also patented numerous variations of the method of supplying oxygen to the carbon, such as the use of oxide of iron and oxide of copper, instead of air." After noting Jablochkoff's work, he asserted, "My patents cover the identical process which Dr. Jacques is said to have used. . . . Electricity can be produced from coal, but the results of my tests have not been commercially successful."[14]

1. The editors incorrectly identified Doc. 2230 as relating to Edison's pyromagnetic generator, an alternative approach to direct conversion that he developed in 1887 (see Israel 1998, 310–11). The editors have also found an additional draft patent application for a fuel cell, dated 12 Apr. 1882 in Cat. 1148, Lab. (*TAED* NM017ABC).

2. Doc. 1897 n. 3.

3. "Speed but No Smoke," *Marshall (Mich.) Daily Chronicle,* 7 June 1883, attributed to the *New York Journal.*

4. Grove 1843, Grove 1845, Grove 1849. For the history of Grove's work and of those who followed him, including Becquerel, see Ketelaar 1993 and the Fuel Cell History Project, "Fuel Cell Origins: 1840–1890," National Museum of American History, Smithsonian Institution, http://americanhistory.si.edu/fuelcells/origins/origins.htm (accessed April 2009). On Edison's device, see Unbound Notes and Drawings—Chemical, Lab. (*TAED* NSUN02 [image 92]).

5. Jablochkoff's application on this battery, filed in 1877, issued in 1879 as U.S. Patent 219,056; "Mr. Edison to Devote Five Years to

Edison's carbon battery from the mid-1870s. Carbon packed with finely divided pumice stone was immersed in a nitrate solution; air pumped through a tube would flow through the mixture, liberating hydrogen. Edison suggested connecting "several thousand" such cells together.

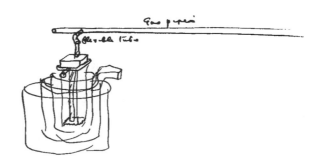

the Elimination of the Steam Engine," *Electrical Review* 5 (11 Oct. 1884): 5.

6. Patent attorney Richard Dyer cited Niaudet's 1880 edition on page 4 of his brief for Edison in the patent interference proceeding. *Weston v. Latimer v. Edison*, Lit. (*TAED* W100DFB).

7. Pat. App. 460,122; see also Doc. 2230 n. 1.

8. Edison's undated drawing for the second application precedes Doc. 2520 in a technical scrapbook. Cat. 1149, Lab. (*TAED* NM018 [image 55]). The issuance of U.S. Pat. 490,953 was delayed until Edison demonstrated that he had successfully tried such a device. Although Edison could not recall "whether in the enormous number of experiments made in the direction of this invention he had ever tested the precise apparatus described . . . he felt entirely certain that his experiments in this direction had demonstrated to his own satisfaction the operativeness of this precise apparatus." In order to meet the demands of the patent examiner, he had Arthur Kennelly make some experiments in 1892 in which he obtained both voltage and current by "providing a chamber having a positive electrode of carbon and a negative electrode of iron with powdered oxide of iron the two electrodes being separated by oxide of iron; the chamber was partially exhausted, and was heated to bright redness." After this patent issued, it was published in the *Scientific American* under the title "Edison's New Art of Generating Electricity" (68 [18 Feb. 1893]: 99).

9. "Producing Electricity Directly from Coal," *Operator* 15 (1 Oct. 1884): 181.

10. "Mr. Edison to Devote Five Years to the Elimination of the Steam Engine," *Electrical Review* 5 (11 Oct. 1884): 5. In another interview shortly after his Florida trip, in April 1884, Edison reportedly said that direct conversion could be applied to steamships and even heavier-than-air navigation. "The Future of Electricity," *Chicago Daily Tribune*, 20 Apr. 1884, 13.

11. Edison 1885, 185; N-86-04-03.1:41, N-87-12-10.2, N-88-01-03.1; all Lab. (*TAED* N315041, NA016 [image 11], NA020 [image 35]); TAE to Richard Dyer, 23 Jan. 1885, Cat. 1151, Lab. (*TAED* NM020AAA); Draft patent application for a pyrochemical generator, 27 Feb. 1888, PS (*TAED* PT032AAV).

12. Israel 1998, 310–11.

13. Edison's earlier efforts have generally been ignored in the history of fuel cells, although Hopkins 1905 (257–58) did treat one of Edison's designs, similar to that shown in his U.S. Patent 460,122, as one of the many fuel cells based "upon the general principle of Jacques."

14. "Direct from Coal," *New York Herald*, 10 May 1896, Cat. 1246:3471, Batchelor (*TAED* MBSB8B3471; *TAEM* 227:844). Important articles about Jacques's work subsequently appeared in the June 1896 *Electrical Review* ("The Jacques Carbon Generator," 38:826–27) and as Jacques 1896.

*Draft Patent
Application: Electric
Light and Power*¹

[A]³

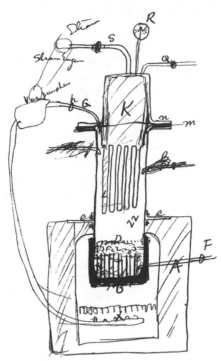

The object of this invention is to produce Electricity from the ~~direct~~ dry[a] oxidation or reduction[b] of a reducible or oxidizable substances such as ~~carbon~~a metal or Carbon & an oxide of a metal.⁴

The invention consists in oxidizing the oxidizable substance in an exhausted chamber formed of two insulated conductors or collectors of the ~~e~~two electricities set free by the decomposition & oxidation.⁵

The invention further consists in assisting oxidation & decomposition by the application of ~~ext~~heat exterior to the chambers

The invention further consists in causing heated electrified gas produced by combustion & decomposition within a closed chamber to become a good conductor of electricity by means of keeping the[c] rarification of the gases of the chamber at a point near a vacuum & thus cause the practical utilization of the electricity produced by dry combustion by causing continuous Current of great Volume to be set up—⁶

The invention further consists in absorbing the waste heat of the chamber by water to form steam to give motion to the steam machinery which keeps of a continuous process of Exhaustion of the gases from the chamber to pump water to the chamber to supply that lost by formation into steam & the utilization of the hydrogen or carbonic oxide given off to heat the chamber[d]

The invention further consists in ejecting water vapor[a] into the vacuous chamber upon the incandescent combustable substances to produce a decomposition of the water by the reducing action of the incandescent ~~Carbon~~Metal & thus set free the two Electricities ~~and the~~ or the equivalent of mixing with the Carbon an oxide of a reducible metal its ox~~ide~~ygen combining with the Carbon to form Carbonic oxide which pa~~p~~ssing to one[a] the insulated chamber giving Electricity to it of one polarity while the reduced metal retaining the electricity of the opposite chamber remains to charge the other chamber

A is a furnace

X grate bars

zz a flanged boiler tube with a thick Cast iron head B. on the inner surface of which are iron rods projecting upwards & between which the Carbon D is packed

K is the other chamber[e] with water tubes L projecting down into the chamber zz   The 2 chambers are insulated from each other at the flanges N by asbestos packing & cement at M. This forms a closed chamber containing the ~~Carbon~~Oxidizable powdered metal   K is filled with water by the pipe Q which is connected to a feed pump   R is a steam guage   G is a metallic pipe passing through K but not insulated therefrom & passes into the chamber zz   This pipe is connected to a vacuum pump worked continuously by steam   f is a pipe for allowing water vapor or steam to pass into the chamber & Come in Contact with the incandescent ~~Carbon~~Metal— The ~~Carbon~~ same[a] being brought to incandescence by the furnace.

~~Claim~~

The ~~V~~

~~Art of~~

Several of these chambers may be connected in series the ~~chamber~~ part[a] zz being connected to the ~~chamber~~ part[a] K of the next chamber, or they may be connected in multiple arc.

When water is not use Oxide[e] of Lead may[e] be mixed with the carbon ~~in this case~~   Anthracite ~~or~~ Coke or Lampblack or charcoal[f] may be used ~~as the~~   all these forms of Carbon including Anthracite Coal become excellent Conductors when incandescent, although the latter is not a conductor when cold. Where an oxidizable metal is used Iron

Claim— Improvement in the art of obtaining electricity sub[stantially] as described

The Method & all the broadest claims that a new & novel method will permit you to make[7]

Witnesses Richd N. Dyer   H. W. Seeley[g]   E[dison]

ADfS, NjWOE, Lab., Cat. 1149 (*TAED* NM018AAV). [a]Interlined above. [b]"or reduction" interlined above. [c]"keeping the" interlined above. [d]"& the utilization . . . the chamber" interlined below; "or carbonic oxide" interlined above. [e]Obscured overwritten text. [f]"or charcoal" interlined above. [g]"Witnesses" and names written by Richard Dyer.

1. See headnote above.

2. This is the date that Richard Dyer and Henry Seely witnessed the document. On 14 September, Edison executed a patent application based on this draft, which was filed at the U.S. Patent Office five days later. The patent did not issue until 1890. U.S. Pat. 435,688.

3. Figure labels are "Steam," "Steam Engine," and "Vac pump here."

4. Edison referred to this process in his patent application as a "dry chemical reaction." The U.S. Patent Office examiner objected to the use of this term due to Edison's use of water in the reaction. In a letter to the Commissioner of Patents, Edison's attorney, Richard Dyer, noted that "'dry reaction' is the ordinary expression in chemistry for such a chemical process," apparently referring to its common use in reference to qualitative analysis using a blowpipe. "In order to avoid any danger from misconstruction of the term," Edison erased it from his claims, which, unlike the description, could be modified. Patent Examiner to TAE, 15 Oct. 1883; Dyer to Commissioner of Patents, 14 Oct. 1887; both Pat. App. 435,688.

5. The patent examiner also initially rejected the patent on the basis that Edison would have to "demonstrate the operativeness of his device" because "each metallic particle itself short-circuits the current developed." Dyer argued that the metallic particles "form one pole of the circuit and are connected with the other pole only through the rarified gases of the chamber. It is known that if two platinum wires are placed, one at the bottom the other at the top of the flame of a burning gas jet a current will be set up between them. This is the same in principle as [Edison's] process" (see note 4 above). This principle was based on a well-known experiment by William Grove (Grove 1854; Watts 1882, 2:431–32); Pat. App. 435,688.

6. Edison was producing gas in a plasma state, which is much more electrically conductive. This fourth state of matter was first clearly observed in 1879 by William Crookes while studying the properties of gases in electrical discharge tubes. Crookes termed it "radiant matter" and it was not until fifty years later that the current term plasma was first used by Irving Langmuir (Pinheiro 2008; Eliezer and Eliezer 2001, chaps. 1–2). Edison produced plasma by using high temperatures to ionize rarefied gases in a vacuum chamber. The patent examiner seems not to have understood the use of rarefied gas and gave as another reason for rejecting the application his understanding that gases would always be present in the vacuum chamber. As Dyer noted in his 14 October 1887 letter (see note 4 above), "In the present case an air pump is continuously employed to maintain the gases in the chamber in a rar-

efied condition. The fact that such rarefied gases exist in the chamber makes it none the less a vacuum chamber, for rarefied gases exist in all vacuum chambers."

7. Edison's initial first claim was for an "improvement in the art of generating electricity, consisting in causing chemical reactions within a suitable inclosing chamber, liberating a gas or gases which charge one pole, the other pole being charged by the non-gaseous product of the reaction." The second claim referred to this as an example of "dry chemical reactions." He subsequently altered these and many other of the original twenty claims. The patent finally issued with seventeen revised claims on 2 September 1890 (Pat. App. 435,688). An article about Edison's design appeared in the 10 September 1890 issue of *Electrical Engineer* ("Edison's Novel Gaseous Generator," 10:281).

–2521–

From Philip Richardson

Woburn Sept. 10" 83

Dear Sir,

Would you kindly give me a little information in regard to the Edison Dynamo—would it be possible to use the current to deposit metal in the electroplating process? I have inquired of a number of Boston electricians and have found no uniformity of opinion So I have taken the liberty to address you in regard to it— Respectfully

Philip Richardson[1]

⟨We have ~~made a~~ changed one of our machines to a plating machine for Mr Keith for metalurgical purposes,[2] it gave 8½ volts pressure & 655 ampiérs of Current. price was $1300.⟩[3]

ALS, NjWOE, DF (*TAED* D8303ZFW).

1. Richardson wrote from Woburn, Mass., using a Post Office box number. The editors have found no reliable personal information about him.

2. Nathaniel Shepard Keith (1838–1925), having trained for medicine, had a long career instead as an inventor and engineer in electricity and metallurgy. After a stint as the electrician of the Fuller Electrical Co., he formed the Keith Electric Co. to manufacture a storage battery of his own design for electric lighting. This proved to be an unprofitable and brief enterprise, and Keith was at this time a consulting electrical engineer in New York. He had recently been working as a contributor and translator for an expanded edition of a major German-language reference work on the generation and transmission of electrical power (Schellen 1884). Doc. 1895 n. 4; *WWW-1*, s.v. "Keith, Nathaniel Shepard"; "Electric Illumination," *NYT*, 15 June 1879, 5; Schallenberg 1982, 56–57; letterhead of Keith to TAE, 20 Mar. and 17 Apr. 1883, DF (*TAED* D8306T, D8303ZCI).

Keith, who had made earlier experiments with a dynamo built to order by Edward Weston, a leading maker of electroplating machines, approached Edison in mid-April about proceeding "with the refining of base-bullion by my electrolytic process. I desire to see you as soon as possible about making a suitable dynamo for the purpose." (Keith's process may be that covered by his 1878 U.S. Patent 209,056 for an "improvement in processes for refining impure lead.") When prompted about the matter on 1 June, Edison noted that drawings for the machine were complete. It was built at the Edison Machine Works and probably completed by 11 September, when Edison asked where Keith wanted it shipped. See also Doc. 2648. Schallenberg 1982, 57; Keith to TAE, 17 Apr.,1 June (with TAE marginalia), and 11 Aug. 1883; TAE to Keith, 11 Sept. 1883; all DF (*TAED* D8303ZCI, D8303ZDQ, D8334Y, D8316AUS).

3. Edison's marginalia was the basis for his answer two days later. Not satisfied with this information, Richardson responded that a friend would allow him to use an Edison dynamo but would not permit him to modify it. He wished to avoid the "considerable expense to ascertain by experiment" if the machine would do what he wanted. Edison drafted a reply, saying "Its hard to tell a person how to do a thing you know nothing about. Try your experiment by passing the current through a lamp & your plating bath= The shock wont kill a fly—its arc light machines that are dangerous." TAE to Richardson, 12 and 15 Sept. 1883; Richardson to TAE, 15 Sept. 1883; all DF (*TAED* D8316AUX, D8305Q, D8305P).

–2522–

To Spencer Borden

[New York,] Sept 11 [188]3

Friend Borden.

Wont you please put more horse power in your collections;[1] am running through large lot[a] different types new machines at Goerck,[2] and Isoltd cant pay. it delays us badly— home dept has collected everything right up[3]   yours

Thomas A Edison

ALS (letterpress copy), NjWOE, Lbk. 17:237 (*TAED* LB017237). [a]Obscured by ink blot.

1. Cf. Doc. 2420.
2. See Docs. 2419 (headnote), 2499, 2529, and App. 3.
3. Presumably, Edison was referring to the main office of the Edison Co. for Isolated Lighting (cf. Doc. 2420 n. 21).

–2523–

To Frank McLaughlin

[New York,] 12 Sept [188]3

Mac.[1]

There is no great difficulty in lighting the tunnel,[2] but it would require that I send one of my best experts there, again it is a great ways off, and hence I prefer to try tunnell work nearer home, and I am now getting bill through Pa legislature making change in mining law about lighting Coal mines so as to enable us to go into that business.[3] I will not undertake anything that I cannot make a success of both electrically and commercially and it is a little harder to make the success in California than in Pa, so I will try Pa first. Will you not have to well light the bend when you draw water off & commence taking out metal. Yours

Edison

ALS (letterpress copy), NjWOE, Lbk. 17:242 (*TAED* LB017242).

1. Business promoter Frank McLaughlin (1845?–1907), customarily addressed as "Major," possibly because of service in the California state militia, had known Edison since about 1875, when both were involved with telegraphy. At various times, McLaughlin acted on Edison's behalf as a mining contractor (prospecting for gold and platinum) and agent (for the electric pen and phonograph); until Francis Upton bought him out in November, he also held a small share in the Edison lamp factory. In 1882, McLaughlin combined with James Logan and George Ladd to obtain an exclusive franchise for Edison's isolated lighting business in California in return for shares of the proposed company, though his subsequent involvement with what was called the San Francisco Edison Electric Light Co. is unknown. Dormanen n.d.; Hodgkins 1979, iii, 16; Conot 1979, 490; American Electrical Directory 1886, 243; Upton agreement with McLaughlin, 1 Nov. 1883, Upton (*TAED* MU063); Docs. 660, 1756, 1844, 1938 n. 7, 2189, 2212 n. 5, 2303 n. 3.

2. The "tunnel" was the Big Bend Tunnel in Butte County, Calif., a project of the Big Bend Tunnel & Mining Co. It was intended to create access to an eleven-mile section of the Feather River by diverting water through the tunnel. McLaughlin seems to have become involved in 1880, but work did not begin until late 1882; by this time it had progressed between 2,000 and 3,000 of a total 12,000 feet. August 1883 saw the smallest advance in excavation, which perhaps prompted the interest in electric lighting. McLaughlin's original inquiry on the subject has not been found, but he prompted Edison about it by telegram from Newark on 11 September. In early October, the construction superintendent, N. A. Harris, sent information to company president Ray Pierce "in reply to Mr. Edison's inquiries about position of buildings, and lamps required" to light dormitories, outbuildings, and the tunnel itself; Harris estimated the company was presently spending $300 per month on candles. Edison put the matter off in November, telling McLaughlin, "the matter is one which I wish to attend to personally" but that he had no time ("The Big Bend Tunnel in Butte County, California," *Sci. Am.* 54 [6 Feb. 1886]: 85; George Cummings to McLaughlin, 1 Apr. 1880; McLaughlin to TAE, 11 Sept. 1883; N. A. Harris to Pierce, 25 Sept. 1883; TAE to McLaughlin, 6 Nov. 1883; all DF [*TAED* D8033Y, D8325ZCK, D8325ZCT1, D8316BEH]). Although the tunnel was successfully constructed, no payable gold was found on the riverbed, and the river was returned to its original course. In 1902, Pierce and McLaughlin converted the holding into the Eureka Power Company with the intent of using the tunnel to supply water for hydroelectric purposes, a plan that was realized by Great Western Power, which acquired the tunnel a few years later (Teisch 2001, 241–43; Edison Co. for

Isolated Lighting, list of isolated plants installed prior to 1 Oct. 1885, PPC [*TAED* CA002E]).

3. Although a major revision of mining regulations for bituminous and anthracite mines had passed into Pennsylvania law on 13 June 1883 (No. 97) and another came on 30 June 1885 (No. 170), neither addressed mine electrification. The editors have no other indication that Edison played a role in the legislative process. Pennsylvania General Assembly 1883, 101–11; Pennsylvania General Assembly 1885, 218–50.

Regulation of mine safety was a matter for the individual states before the U.S. Bureau of Mines came into being in 1910. Pennsylvania passed the first anthracite mine safety laws in 1869, but the effects of this and subsequent efforts were mitigated by the traditional ambiguity of responsibilities among operators and miners, who often considered themselves independent contractors. Although the presence of blasting powder, methane, and coal dust created obvious dangers for open flame illumination, mine electrification was not necessarily recognized as a clear improvement. In fact, electricity became associated with the hazards of increasingly mechanized production. As late as 1899, the chief mine inspector of Pennsylvania maintained that "the use of electricity in any form in coal mines is a menace to life, limb, and property," which suggests the uphill battle Edison faced. Aldrich 1997, 41–58, quoted p. 58.

–2524–

To Eduard Reményi

New York [September][1] 12, 1883[a]

Remenyi[2]

I send call extended.[3] I'm mentally constipated today & overworked. hope to see you in December

Edison

ALS, NjWOE, Miller (*TAED* HM830192C); a letterpress copy is in Lbk. 17:243 (*TAED* LB017243). Letterhead of Thomas A. Edison. [a]"New York" and "188" preprinted.

1. Month supplied from postmark on envelope.

2. Eduard Reményi (1828–1898) was a violin virtuoso and popular concert artist whose late-night visits with Edison included an impromptu performance on at least one occasion (Doc. 2137 nn. 1–2). In April, he had given Edison his highly favorable impressions of Edison's light at the Palmer House hotel in Chicago. Edison had recently received a confidential appeal for money on Reményi's behalf from actress Marie Prescott, whose own funds were tied up in Oscar Wilde's play, *Vera, or the Nihilists*, at New York's Union Square Theatre. Edison declined this request, but he loaned Reményi $100 in April 1884 (Reményi to TAE, 25 Apr. 1883; Prescott to TAE, 10 Aug. 1883; TAE to Prescott, 15 Aug. 1883; all DF [*TAED* D8303ZCU, D8303ZEX, D8316AOB9]; "The Union Square Theatre," *NYT,* 19 Aug. 1883, 10; Reményi note to TAE, 17 Apr. 1884, Miller [*TAED* HM840215A]).

3. In September 1882, Edison had offered a six-month option on ten shares of Edison Electric Light Co. stock for $625, which Reményi failed

to take up before it expired. Reményi wrote on 9 September from Denver enclosing "the call you gave me last year—and I request you to give me another callllll—may be a more fortunate one—I leave it entirely to your judgement — <u>but do it</u>"; he also promised to return to New York in December. Edison amended the original offer with a handwritten and signed notation: "This call is extended until May 1884." Although Reményi had instructed him to reply in care of a Chicago theater, Edison mistakenly addressed this letter and the stock option to Denver; the envelope was subsequently returned undelivered (TAE agreement with Reményi, 20 Sept. 1882, Miller [*TAED* HM820165A]; Reményi to TAE, 25 Apr. and 9 Sept. 1883, both DF [*TAED* D8303ZCU, D8303ZFV]). When Reményi returned to New York in December, he invited Edison to join him one evening at Delmonico's, the famous restaurant. Edison replied in his fine calligraphic hand that it was "Impossible to have the pleasure; must work, work, things are booming. . . . Have closed my belly to all the allurements of Gastronomy" (Reményi to TAE, 5 Dec. 1883, DF [*TAED* D8303ZHT]; TAE to Reményi, 5 Dec. 1883, Edison MS, DSI-MAH [*TAED* X092C]).

–2525–

Edison Co. for Isolated Lighting Memorandum

New York, [September 13?, 1883][1a]

Have 49 Consumers in Roselle as per August Meter account.[2] There bills are classified as follows

Customers whose meters register between

~~one~~ two Dollars[b] and under		12
Between $2 and $3		15
" $3 " $4		9
" $4 " $5		3
" 5 " $6		6
" 6 " $7		1
One large consumer bill $14.		1
Central RR.[3] "	45	1
Street Service[4] "	82.50	1
Total		49

It will be seen that private Consumption is very light— About 50% of income is from the street service and two large consumers Dicks[5] & Cen. R.R.

⟨The winter months give three times the Consumption in present houses⟩[6]

D, NjWOE, DF (*TAED* D8327ZAY). Letterhead of Edison Co. for Isolated Lighting, general manager's office. [a]"*New York*" preprinted. [b]"two Dollars" interlined above.

1. The document was stamped "ANSWERED" on 13 September.

2. Edison had requested similar information about customer bills in Sunbury, Pa., about a week earlier. TAE to Frank Marr, 5 Sept.

1883; Marr to TAE, 8 Sept. 1883; both DF (*TAED* D8316ATG, D8361ZDC).

3. Planned for hauling Pennsylvania anthracite coal, the main line of the Central Railroad of New Jersey (chartered in 1849) ran east-west between Jersey City and Phillipsburg, incorporating an older route through Roselle. The platforms, waiting rooms, and ticket office of the Roselle depot had more than a dozen electric Edison lamps in all. *Ency. NJ*, s.vv. "Central Railroad of New Jersey," "Roselle"; Edison Electric Light Co. Bulletin 17:5, 6 Apr. 1883, CR (*TAED* CB017).

4. The Roselle district included about 150 street lights. Edison Electric Light Co. Bulletin 17:5, 6 Apr. 1883, CR (*TAED* CB017).

5. Not identified.

6. The editors have not found records of wintertime usage.

-2526-

To Frank Sprague

[New York,] Sept. 14th. [188]3

Dear Sir:—

We have the pleasure to acknowledge your favor of the 8th. inst, which would have been replied to earlier but for circumstances which have interfered to prevent this being done.[1]

AUTOMATIC REGULATOR. We do not believe at all in the use of automatic regulators in central stations whilst the regulator is in such a condition. In our opinion the automatic regulator is a necessary evil in isolated plants where there is great variation number of lamps on at any one time and the variations in speed.[2] Both these difficulties are comparatively overcome in our central stations, and I much prefer hand regulation.[3] With relation to the hand regulator, I enclose you herewith a sketch which will show you the method in which Mr. Andrews connected up the regulator at Shamokin, where the same blunder had been made in this connection as the one you now have.[4] You probably can set it right by pursuing the same course as Mr. Andrews did.

With relation to men, if you find it cheaper and more economical to use wiremen, you can of course have them.[5] What we want to do is to have the work done as well and as economical as possible

The connections, such as designed by Mr. Johnson,[6] have been ordered for you and will be shipped immediately they are ready

With relation to the switches, we enclose you herewith a rough sketch of the arrangement of the station at Shamokin[7] All reference in letters written you have been partly based on the method pursued at Shamokin.

Will you please see if you have switches to answer for the

work at Brockton, following generally on the same character of work as that done at Shamokin, and a knowledge of which the enclosed sketches will give you? If you have not we will order such switches as you may intimate as being requisite. It is our impression, however, that you have all the switches, and in fact several over, except possibly several plug switches for cutting the bank of resistance in and out.

The reference in our letter of the 8th. [to?][a] Sunbury was in error, Shamokin should have been mentioned,[8] as the criticism on switches required by you was based on what had been found to suit the purpose at Shamokin. Yours truly,

TAE   T[ate]

TL (carbon copy), NjWOE, DF (*TAED* D8316AVM). [a]Canceled.

1. Sprague had been working on the central station in Brockton, Mass., since at least the beginning of September. The previous day, Edison (or more likely someone in his office) had acknowledged several communications from Sprague and explained that "Mr. Insull, who has these matters in hand, is at the present moment ill and confined to his bed" but would reply immediately upon returning to the office in a day or two. The editors have not found an 8 September letter from Sprague, but this document appears to be in answer to a nineteen-page letter from him on 10 September, which was a direct reply to one from Edison two days earlier. Edison mildly chastised Sprague for exceeding his authority by discussing technical or business matters with the local operators. Sprague defended his actions but also deflected the criticism as likely being "the emanations of your private secretary." TAE to Sprague, 13 and 8 Sept. 1883; Sprague to TAE, 10 Sept. 1883 (p. 1); all DF (*TAED* D8316AVI, D8316ATY, D8343ZAY).

2. Edison referred to the two principal causes of voltage fluctuation in his distribution system: change in electrical load and variation in the speed of the driving engine. Sprague had recounted that William Lloyd Garrison, Jr., treasurer of the Brockton company, had expressed to him a preference for an automatic instead of a manual device, and a willingness to pay accordingly. Sprague suggested a general approach that he thought would work, but Edison scrawled across the paragraph: "What do you mean by automatic Regular= please dont suggest things to Mr Garrison   let us do that otherwise you will raise hell with us" (Sprague to TAE, 5 Sept. 1883, DF [*TAED* D8343ZAU]). Sprague took up the subject again in his 10 September letter, pointing out that in talking with Garrison

I not only did not encourage the idea, but I expressly stated that it was not just yet practicable— Since thinking of it, I see that while we get a variable drop from the brushes to the feeder terminals, we want a constant potential at the feeder terminals, and hence we can make the automatic work just as well as the hand regulator by taking the <u>pressure wires</u> from the terminals of that feeder which most nearly represents the mean condition of affairs, and bring these to the automatic's terminals, making of course the necessary adjustment and allowance for the return drop due to the resistance of the wires— I feel very certain that the automatic will work admirably if so used . . .

I have now to concur very cordially with Mr. Garrison's request, (and I think you will understand that some discussion has been necessary) to write you and urge the replacing by the automatic of the present hand regulator—

Pointing out that he had recently seen the Edison regulator work well— and to wide acclaim—at a Boston exhibition, Sprague added that "Nothing so impresses anyone with the beauty of the Edison system as to see our instrument acting with almost human intelligence to perfectly control the potential existing on the mains—and the amount of life given off all over a district—" (Sprague to TAE, 10 Sept. 1883 [pp. 4–6], DF [*TAED* D8343ZAY]).

3. Edison continued to hold this opinion. Writing on his behalf in May 1884, Samuel Insull explained that "The use of the automatic regulator in a central station system is very dangerous. Some time or other something will go wrong with that regulator, and in an instant the Shamokin company will lose all the lamps. This is the main reason that we have been objecting to the use of these regulators" (Insull to Harry Doubleday, 1 May 1884, DF [*TAED* D8416BMO]). Edison's standard automatic regulator for isolated plants, manufactured by Bergmann & Co., is illustrated in Doc. 2242 (headnote). At a central station, an operator could manually adjust the resistance of the dynamo field circuit in response to the indication of high or low voltage. One such indicator, not necessarily designed for central stations but possibly useful there, is described and illustrated in Doc. 2190, esp. n. 9.

4. The drawing has not been found. Andrews discovered that when the two resistance boxes (for regulating the dynamo fields) were installed back to back for esthetic reasons, turning the control rod in one direction increased the resistance in one box but had the opposite effect in the other. Andrews pointed out the necessity of reversing the electrical connections, but he apparently fixed the situation in Shamokin simply by placing the boxes side by side. Sprague spelled out a method of rewiring the boxes in his 10 September letter. Andrews to Insull, 31 Aug. 1883; Sprague to TAE, 10 Sept. 1883 (pp. 7–8); both DF (*TAED* D8343ZAO, D8343ZAY); Insull to Sprague, 10 Sept. 1883, LM 15:41 (*TAED* LBCD2041).

5. Sprague had taken umbrage at Edison's directive that for wiring the Brockton station, "one man would be able to give you all the assistance you will require. You could 'break in' a local carpenter in a short time, and thus save a great deal of unnecessary expense. When Mr. Andrews was starting the Sunbury central station, he only required one man to assist him, and we presume that you can work with equal economy." With evident irritation, Sprague pointed to his own role at Sunbury and argued that additional responsibilities at Brockton prevented him from properly training what he termed a "wood chopper," or a "carpenter who has not only the lack of knowledge, but has not the manual skill in handling heavy wire than an intelligent wireman would have. The cost of such drilling [instructing] would be more than the wages of one or two reliable men." TAE to Sprague, 8 Sept. 1883; Sprague to TAE, 10 Sept. 1883 (pp. 9–10); both DF (*TAED* D8316ATY, D8343ZAY).

6. This is probably the method of joining overhead conductors without solder for which Edward Johnson filed a patent application in January 1884. It consisted of a rigid joint through which a screw was forced between two conducting wires so as to make contact between them. U.S. Pat. 311,130.

7. The editors have neither found the drawing nor ascertained what type of switch Edison meant. In his previous two letters, Sprague had referred to switches in a variety of places, including dynamos, feeders, and the lightning-protection ground connection. Sprague to TAE, 5 and 10 Sept. 1883; TAE to Sprague, 8 Sept. 1883; all DF (*TAED* D8343ZAU, D8343ZAY, D8316ATY).

8. At the close of a paragraph about orders of various switches for Brockton, Edison's 8 September letter stated: "The bank of test lamps requires 2 large plug switches and 10 small plug switches of different design. These switches were ordered on Mr. Andrews experience at Sunbury." DF (*TAED* D8316ATY).

–2527–

To Frank Sprague

[New York,] Sep. 14th. [188]3

Dear Sir:—

Please impress upon Mr. Jenk's the fact that we must not cut in any customers the total of whose lamps will not give us at least two hours and thirty minutes average for 300 days.[1] If he is going to cut in any one who wants the light, the station will simply be a commercial failure.

The maximum number of lights that a party ever burns at any one time, divided by the total lamp hours in 300 days should give not less than 750 hours per lamp. Of course a party may have 50 lamps connected, but never have more than 16 lighted; if that is the maximum, then these are the lamps to be counted.[2] Do you see the point?   Yours truly,

E[dison]

TLS (carbon copy), NjWOE, DF (*TAED* D8316AVO).

1. Sprague had reported that since the Brockton service area had been canvassed, "within the span of a thousand feet, five important brick buildings have been erected or are in process of construction," that "local people" were inquiring about the modification of wiring for broader service, and that in making such decisions the local management "will not await the action of people in New York." Sprague to TAE, 10 Sept. 1883, DF (*TAED* D8343ZAY).

In his lengthy combined reply to this message and others, dated 16 September, Sprague indicated that he had "tried to impress" the matter of managing service economics on Jenks, and that at startup the station would service about 250 to 300 lights, mostly "in billiard halls clothing stores, &c—," which were lit for many hours a day and therefore profitable to install. Sprague to TAE, 16 Sept. 1883 (p. 11), DF (*TAED* D8346ZAG).

2. The distribution system required a certain level of consumption to offset fixed costs. The editors have not determined how Edison arrived at the rule of thumb given here as the break even point. Edison's policy entailed the installation of lamps beyond the plant's generating capacity in the expectation that even at its peak, demand for electricity would not exceed the supply available, an expectation so far justified by his experience in the Pearl St. district. See Doc. 2420 n. 41.

-2528-

To James Pryor

[New York,] Sep 20/[188]3

Dear Sir:

I very much regret to say that owing to the illness of my wife it has become imperative that she should give up housekeeping in accordance with the Doctors instructions.[1]

My lease of your[a] house does not expire until the 1st day of October 1884,[2] but I write this letter to ask if we cannot make some arrangement which will relieve me of the necessity of carrying the place for twelve months—

The circumstance is most unfortunate but as Mrs. Edison's health depends upon her immediate ~~removal~~ retirement[b] from ~~the city~~ all Household responsibility[c] there is no other course for me to persue but to comply with the necessities of the case and ask your very kind cöoperation and endeavors to release me from the situation I am now in—[3]

Trusting that you will consider the matter favorably, I am Yours very truly

Thos A Edison

L (letterpress copy), NjWOE, Lbk. 17:255 (*TAED* LB017255). Written by Alfred Tate. [a]Obscured overwritten text. [b]Interlined above. [c]"all Household responsibility" interlined above.

1. On 4 October, while Edison was in Brockton, Mass., Samuel Insull reported that he had gone "around to see Mrs. Edison this morning and she seems considerably better. She is going to move this afternoon [see note 3], and the doctor has undertaken to accompany her, in case of any mishap." The editors have not determined the nature or extent of her illness, but cf. Doc. 2517 and see Doc. 2712 (headnote). Insull to TAE, 4 Oct. 1883, DF (*TAED* D8316AYJ).

2. See Doc. 2341 n. 3.

3. Pryor declined to terminate the contract, citing his own obligations and the substantial expenditures made at Edison's request on plumbing and carpeting in the house. Edison and his family, however, set up their household at New York's Clarendon Hotel at the beginning of October. Pryor did agree to waive the clause prohibiting a sublease, and Edison arranged one sublease through February and another until April, a sequence which drew a mild reproach from Pryor. See Doc. 2545; Pryor to TAE, 24 Sept. 1883 and 11 Feb. 1884; Joseph Huppman Valbella to TAE, 7 Apr. 1884; all DF (*TAED* D8303ZGD, D8403W, D8403ZBF).

From Samuel Insull

My Dear Edison:—[1]

Your telegram came to hand asking for chemicals, meters, scales &c., and the various articles enumerated have been shipped.[2] The meters were kept back purposely; the rest of the articles should have been shipped but for a blunder, which I suppose I must take the credit for.

Soldan started to test the 300 light machine[3] yesterday, but found a cross in the armature, and when I left the Works this morning this trouble had not been fixed, but he promises to let me have a report on a partial test this afternoon, but up to this writing it has not yet come to hand. As soon as I get it I will send it to you.[4]

I saw Cowles to-day in reference to the cable, and he says it is impossible to state exactly when they can deliver some of it,[5] and as it is a new thing he is depending entirely upon the factory as to when it will be turned out. If it was an article of ordinary manufacture he could name a time, but says he will push it through as quickly as possible.

I was in the central station to-day and Chinnock was trying a pair of the new brushes.[6] There was quite some heat from them, and one brush had got burned very badly indeed. When I left, however, the brushes were working a little better, and he was carrying 600 or 700 lights on the brushes from the district and not from the test lamps upstairs.

I had a man in here to-day from Fitchburg, Mass.[7] He runs the Thomson-Houston business and wants to get up Edison Companies in about ten towns. He came on here at the suggestion of Sims, and I think that when you get to Brockton next week you can make a "dicker" with him. I am to see him to-morrow again. he is one of those real live fellows, and he would have closed a contract with me on the spot. Of course such difficulties as terms he has got to be talked into.

I hope you are enjoying the bad "grub"[8] and the "twofors."[9] I will go to Delmonico's[10] to-night and eat a double sized dinner, so that you can imagine you have had part of it.

I enclose you the latest emanation of the Engineering Department.[11] It is a most beautiful piece of blue printing, and will no doubt go a long way towards cheering you on in your lonely way in Shamokin.   Yours very truly,

TL (carbon copy), NjWOE, DF (*TAED* D8316AWA).

1. Insull addressed this letter to Edison at Weavers National Hotel in Shamokin, Pa.

2. Edison had recently sent two telegrams. The first, dated 19 Sep-

tember, directed Insull to "Send balance and chemicals and few meaters quick   Send our oil and balance meaters." He then complained the next day, "Engine and boiler questions not <u>arrived</u>   no meters or chemicals   hurry new wire   shall be detained here indefinite if things not rushed forward." Insull wired separate replies to each, promising that the requested items had been sent or would go promptly. TAE to Insull, 19 and 20 Sept. 1883, both DF (*TAED* D8360ZAS, D8360ZAT); Insull to TAE, both 20 Sept. 1883, LM 15:123B, 124 (*TAED* LBCD2123B, LBCD2124).

3. This was the Y dynamo, a model whose two field cores represented a significant design change for Edison (see Doc. 2419 [headnote]). Edison had anticipated making a 300-light dynamo as early as June but did not begin designing it until July. The Edison Machine Works started building a prototype (and at least one other new dynamo model as well) in early September, by which time Edison had ten orders for this machine (TAE to Charles Batchelor, 11 June 1883; TAE to Samuel Flood Page, 19 July and 1 Sept. 1883; Insull to Charles Dean, 26 July 1883; all DF [*TAED* D8316AAC, D8316AHW, D8316ARO, D8316AJN]; Insull to Frank Sprague, 10 Sept. 1883, LM 15:41 [*TAED* LBCD2041]). Charles Batchelor balked at its expected price, but Edison explained that it had to help cover "all the heavy experimental expenses at the Machine Works, as owing to troubles I have had with the Light Company I decided I would not put in any payments to them for this kind of work" (Batchelor to TAE, 4 Oct. 1883; TAE to Batchelor, 19 Oct. 1883; both DF [*TAED* D8337ZEP, D8316BBW]).

4. Later in the day, Insull reported that Gustav Soldan "has tested the 300 light machine. He ran the armature at 1100 revolutions and got 260 amperes out of it at 113 volts. The armature was perfectly cool . . . and the machine had been running $1\frac{1}{2}$ hours. He seems highly elated over the result of the test, as he says he has come out exactly in accordance to his calculations." Notes of tests made the next day indicated that "Magnets warm (50°C)   Very little sparking." Insull to TAE, 20 Sept. 1883; Edison Electric Light Co. Testing Room report, 21 Sept. 1883; both DF (*TAED* D8316AWB, D8330ZAX)

5. Alfred Cowles of the Ansonia Brass & Copper Co. had written the previous day that the cable samples were "now in the works and we expect to be able to furnish them very soon; I cannot say exactly when, but will push them forward with all possible despatch." The cable was for "inside house wiring" and was to have been "wound with two colors of cotton on the outside," though Edison changed this specification to a uniform color. He placed an initial order of about 1,500 pounds with the expectation of needing "a good deal more of this very shortly." Cowles had been working on the specifications of the cable at least since August, when he submitted a sample in which the wire was "wound with Kreusi's tape." He cautioned at that time that it was "only partially fireproof owing to the inflamable nature of Kreusi's tape when large proportion of this is used." Cowles to Insull, 19 Sept. 1883; Cowles to TAE, 13 and 10 Sept., 13 Aug. 1883; TAE to Cowles, 8 Sept. 1883; all DF (*TAED* D8321Z, D8321Y, D8321V, D8321R, D8316ATU).

6. Sparking and overheating of commutator brushes remained chronic problems (see Docs. 2420 n. 14, 2484, and 2551). The editors have found no particulars about the new brushes at this time, or even

the dynamo models for which they were intended. However, a few days after this letter, Charles Chinnock reported that with a load of 3,500 lamps at the Pearl St. station, "the brushes were sparking so bad we were on the point of starting the fifth [dynamo]   We had one set of new brushes that worked well only heated some." Chinnock to TAE, 24 Sept. 1883, DF (*TAED* D8326R).

7. This person was probably Lee Ross of Fitchburg, whom Gardiner Sims planned to meet on this day about electric lighting business. Miller Moore also wrote a letter of introduction to Edison on Ross's behalf, identifying him as general manager of both the Merchant's Electric Light and Power Co. of Boston and of a Fitchburg company. Sims to TAE, 19 Sept. 1883; Moore to TAE, 20 Sept. 1883; both DF (*TAED* D8322ZCO, D8340ZDH).

8. English slang for plain food, or a provision of food simply to feed *DSUE*, s.v. "grub."

9. That is, two of something when only one is expected or purchased. Insull's sense of irony regarding the appeal of Shamokin was undoubtedly whetted by William Andrews's promise to finish his work there "just as soon as possible, if only for the reason that I do not like Shamokin well enough to want to stay here one day longer than is necessary! It is a horrible hole here." *NPDS*, s.v. "twofer"; Andrews to Insull, 31 Aug. 1883, DF (*TAED* D8343ZAO).

10. A leader in culinary innovation and first-class dining since 1827, Delmonico's operated several restaurant and hospitality establishments in New York City. Insull regularly took his morning and evening meals at the café of its flagship restaurant on Fifth Ave. and 26th St., where he was permitted to eat on account when he was short of funds. Edison had introduced Insull to Charles Delmonico soon after his arrival in New York. *Ency. NYC*, s.v., "Delmonico's"; Insull 1992, 45–46; "The Funeral of Mr. Delmonico," *NYT*, 18 Jan. 1884, 8.

11. Not found.

–2530–

Memorandum: Central Station Business

[New York, Summer 1883?][1]

INSTRUCTIONS

By T. A. Edison

Before a consumer is connected, ascertain by passing his place at least every half hour, the exact number of jets actually burning until he closes; from this his gas bill can be almost exactly calculated. Sometimes it will be possible to read his gas meter, keeping the time of each jet actually burning, for, say 20 minutes, and thus ascertain. It is highly important to obtain the total number of gas jets, and their hours of burning, and then when the electric light is connected, the same information as to the hours of burning must be obtained, and if it is found that the consumer is using a greater number of jets, or burning the same number longer, or fails to economise by not turning some jets off when near closing up time, or asks for a higher candle power lamp, or does anything that would tend to cause his bill to be greater than his gas bill, he should be notified at once, and explanations made that his bill will be higher, and to what extent, so when the bill is presented at the end of the month he knows what to expect, otherwise his ignorance will cause him to use more light, and then quarrel over the bill, and perhaps refuse to pay. It will also produce an impression on his mind that the electric light is more costly than gas, whereas, in truth, it might be as cheap, if not cheaper.[2]

To make comparisons, the light must be burnt on the same basis as gas, light for light, hour for hour, and both about the same power. Suppose a customer had been burning 10 lights, each equal to 10 candle power electric, and he had been paying $1\frac{3}{4}$¢ per hour for each gas jet, or $1\frac{4}{10}$ mills per candle, per hour, for 3 hours per night, for 26 days, his bill would be $13.65. Suppose he puts in an equal number of 10 candle power electric lamps at $1\frac{1}{4}$¢ per hour, his bill for corresponding period would be $9.75. Now suppose he wanted 16 candle power lamps, and we put them in without making him understand that it would cost him more than the same number of gas jets, because the 16 c.p. lamps were more powerful, then his bill would be $13.52, the same as gas, he would compare his gas bill, and would then firmly believe that our light was no cheaper than gas, for the reason that in most cases customers cannot make a comparison as to the amount of light, while with his bills he is perfectly competent to compare. Some customers on making a change from gas to electric lamps, put in more electric jets than they previously had gas jets. Before the month is over, and when bill comes in, they generally forget that they are using a greater number of electric lights than they previously used gas jets, and most of them will swear to it, stick to it, and actually believe it, because the fact of adding two or three jets is so small a matter that it does not impress itself on their memory. The consequence is that the bill is more than the corresponding gas bill of last year, or the amount which they should save over gas is lost, and the consequence is that this impression is never eradicated, and the consumer is dissatisfied, while if he were notified, within two or three nights after he had the premises connected, of the fact that he had two or three more electric lamps than he previously had gas, and that his bill would be increased to that extent, he would be warned and led to expect the increase, and this impression would be maintained for all future time.

In private houses it is impossible to walk past every 30 minutes and jot down the lights; in this case, the explanations as to cost, light, &c., should be very full to the owner. Unless these instructions are carried out, 75 per cent. of the consumers will be dissatisfied, many will have it discontinued, or not use it, although it may actually be cheaper, and the business will be an up-hill one.

All the mistakes mentioned have been made at the New York Station, where this experience has been gained in dealing with the public, and the remedies here mentioned have been found to work perfectly in every case, so that all new consumers are perfectly satisfied that they are getting just what we represented they would get.

In no case, unless it is extremely exceptional, should wiring be done at the company's expense, in order to gain a customer after the station is well started. It is always best that the company do no wiring themselves, but turn the order over to regular authorized wiring company or individual who will give estimates, and do the wiring promptly and in a proper manner. If in small towns the company does the wiring itself, it would be necessary to obtain an expert, the bill would be presented by the company and the customer would in most cases use the threat that he would stop using the light rather than pay his wiring bill, in order to coerce the company to either give him the wiring free or to cut down the bill, thus making a loss to the company. On the other hand, the private company, or individual who does the wiring have no difficulty in collecting their bills, as the customer has no excuse, or means of coercion.

In small towns, from 2,000 to 20,000 inhabitants, after the station has been started and three or four prominent places are lighted, stop soliciting orders for light, but run along quietly and wait.[3] The customers will gradually come in and ask to be connected; their judgment and desires having been gained by seeing the light in the prominent places. Customers obtained in this manner give no trouble, and in time 75 per cent. of the whole number counted upon to take the light, will ask of their own accord to be connected. After this result is obtained soliciting will be beneficial, the parties who do not order of their own accord are laboring under a misconception as to cost of light or wiring, or require a little urging.

In nearly every instance, except with the best concerns, parties will prevaricate as to the amount of their gas bills. Many will mentally halve the amount. In many cases customers will produce their gas bills when the electric light bill is much larger, and it is certain that they have burned about the same number of jets and hours. When this occurs, investigation has proved that their gas meter is wrong and has registered too slow. In small towns the gas meters, as well as the management of the gas company, is wretched. There will be many anomalous bills that will cause a great deal of trouble.

Sometimes the gas bill is small by reason of the fact that the parties are using small burners. If the burner is found to be small, and not to give nearly as much light as the regular electric, the fact should be shown to the proposed customer, and the suggestion made that if he wants to keep his bill the same he should put in less electric, or if he wants more light, the same number, explaining that the electric will run his bill up in comparison, by reason of the fact that he is using very small gas burners. In placing a meter always try to put it in a place where it will be convenient to remove or replace the bottles, and inexpensive to place in position, and where boys and others will not be liable to tamper with it. It is also important to select a place where the liquids in the bottles will not be liable to freeze.

When the station is first started and 2 or 3 consumers are connected, it is not actually necessary (unless it can be done conveniently) to put the meters in these 2 or 3 places. They should, however, be put in at once if ready, but thereafter no consumer should be connected, so that he can obtain light, unless his meter is in position and ready to record. For the first month a record should be taken every 15 days and bill should be presented. After the first month, the taking a meter reading every month will be sufficient. The object of taking it twice during the first month is to allow the consumer to see what it costs so that he may be able to answer the enquiries of others, who hesitate until they can ascertain from their neighbors what their bill is and how it compares with gas.

Do not take any customers who close at 6 o'clock, or even 7 o'clock, no matter how many lights they may burn, or how high their gas bill may be. Early closers are not desirable customers, although to a person not versed in electric lighting a store burning three times as many jets as any other concern in town, or a large factory closing at 6 or 7 P.M., would be thought to be very desirable. This is not the case, for the reason that in the winter, for every light burning between 5 and 6 P.M. there must be invested from \$15 to \$25 in apparatus at the Central Station and wires in the street. Hence, if a store which closes at 6 P.M. in December had 100 lights, it would require an investment of say $2,000, and 100 lamps burning for 1 hour only, at 1¼ cents per hour, would amount to $1.25 per day. There are only about 130 nights in the year that any light would be required in a store closing at 6 P.M. and the whole lighting would not amount to over 100 full hours, which, at $1.25 per hour, would amount to $125, whereas if these 100 lights were sold to various parties who closed at from 9 to 11 P.M., the average would be 3 hours of actual burning the year round, Sundays excepted; hence 312 days in the year and 100 lamps, at $1.25 per hour, would give $3.75 per day, or $1,170 per annum for $2,000 invested, as against $125 from the single consumer, who closes at 6 P.M. The latter consumer would not be desirable if he kept open until 10 P.M. only Saturday night.

Factories that work late or all night are very desirable, but those factories that generally close at 6 P.M., but sometimes, when business is good, work nights, are not desirable. Whenever possible the local company should endeavor to sell an isolated plant to the factory, from which the company would derive a profit, both from the original sale and from the continuous sale of lamps to the purchaser.

Private houses are very desirable, owing to the superior effects and beauty and quality of the light, which gives it a greater advantage in this connection than in a store.

When it is once installed in a private house parties will not discard it, even if the price were to be placed far above gas.

Where it is possible get the consumer to purchase permanent electric light fixtures from the manufacturers. They will then have made an investment, and will be more likely to pay their bills promptly.

The mains and poles should be inspected every two or three days. Any moving of buildings, rebuilding, or any changes liable to be made, should be noted and our poles and wires removed or changed sometime before the actual necessity occurs.

In this business, where no stoppage of the light must occur under any consideration, the person in charge must look ahead for danger, or changes going on in the town likely even by chance to produce an interruption.

In the same way, the moment defects in the apparatus, engine, boilers, or anything which is likely to cause a stoppage in the future, is noticed, it must be remedied at once, or means provided, so that when trouble does occur, it will not cause the stoppage of light for an instant.

There should be at least four or five days' supply of coal on hand in case of a snow storm that might render cartage impossible. The water supply also should be looked after in time, so that there shall be no sudden lack of water by failing of the source of supply, stoppage of supply pipe, freezing, &c. Stations running only at night should start engine and put pressure on mains 30 minutes before the necessity occurs for lighting up the darkest part of stores. This may occur sooner than expected in some cases, by reason of a storm.

Lamps are furnished to the consumer free. An account of the lamps supplied to each person should be kept and also of the number of broken ones returned.

Always obtain the broken lamp for every new one furnished. When the consumer is connected inform him that any lamps missing, or broken by carelessness, must be paid for at $1.00 each.

In a place where they burn from 15 to 50 lamps furnish 1 dozen extra; from 8 to 15 lamps, furnish ½ dozen extra; and from 1 to 8, 3 extra lamps, over and above those in the sockets.

In all cases when a stock of lamps is replenished be careful to bring away as many broken ones as new ones supplied, and if any are missing notify the consumer that they will be charged on his next bill.

If they are not properly charged people will get careless and the lamp account will get mixed.

In lighting up a store, &c., where the ceiling is low, and the walls and ceiling white, it is best to put the lamps upright and use no shade. This cheapens the installation, the white walls acting as a reflector.

Where the walls and ceiling are white, but very high, a shade may be used, the lamps being placed upright, but where the ceiling is dark, the best position for the lamps is pendant, shade being also used. A room can be lighted with one-third less jets when placed in the centre than when placed on brackets near the wall. In many stores in country towns, especially the inferior ones, the ceilings are so low that no chandeliers are required, the sockets and shades being connected direct on the ceiling.

TD, NjWOE, CR (*TAED* CD001A).

1. Edison likely prepared this memorandum sometime after the successful opening of the Sunbury, Pa., central station and as he was planning for a large business in village plants. At an unknown later date, it was grouped with other instructional materials prepared during the summer for the operation of central station systems (see Doc. 2500 n. 1). It concerns questions of consumer expectations and billing, which had arisen by summertime from the operation of Edison's central stations in New York City and Sunbury, and from planning for the Shamokin,

Pa., plant. It also relates to billing information that Edison sought from Sunbury and received from Roselle, N.J., by mid-September (see Doc. 2525; Frank Hastings to Charles Chinnock, 11 July 1883; Samuel Insull to William Andrews, 30 July 1883; both DF [*TAED* D8326K, D8316AKI]). Edison's instructions here are consistent with those he issued to his village plant canvassers during this period (see Doc. 2499).

2. Edison's concern with educating customers about the intensity and cost of electric lights was reflected in a statement published in November by the Brockton, Mass., illuminating company to address public expectations that could lead to the "reckless" consumption of electric current. William Jenks statement, 14 Nov. 1883, in Edison Electric Light Co. Bulletin 22:12–14, 9 Apr. 1883, CR (*TAED* CB022527).

3. Cf. Doc. 2492.

–2531–

Samuel Insull to
Gardiner Sims

[New York,] Sept 29th [1883]

Gardner C. Sims

Mr. Edison and self go to Fall River by boat tonight thence to Brockton on nine twenty nine train tomorrow morning.[1] Can you spend Sunday in Brockton with me.[2]

S. Insull   T[ate].

L (telegram, letterpress copy), NjWOE, LM 15:269 (*TAED* LBCD2269). Written by Alfred Tate.

1. Edison had returned early on 26 September from Shamokin, Pa., after having started the central station a few days earlier. He and Insull carried to Fall River an estimate for construction of the village system there. They planned to make arrangements to receive a $15,000 payment from the local company in Brockton, where Edison intended to start the central station on 1 October. Insull to Paul Dyer, 26 Sept. 1883; TAE to A. J. Dull, 26 Sept. 1883; Insull to William Lloyd Garrison, Jr., 27 Sept. 1883; all DF (*TAED* D8316AWI, D8316AWK, D8316AXE); Insull to W. H. Dwelly, Jr., 27 and 29 Sept. 1883, LM 15:226, 268 (*TAED* LBCD2226, LBCD2268).

Edison remained unexpectedly long in Brockton, detained by a dynamo short circuit and by a lack of examination questions and answers (see Doc. 2500) for training the operators there; he left for New York late on 6 October. TAE to Insull, 4 and 6 Oct. 1883, all DF (*TAED* D8340ZEA, D8340ZEB, D8346ZAQ).

2. The editors have not determined Sims's whereabouts on 30 September. He did spend the day with Edison in Brockton on 4 October, when Insull was back in New York. Armington & Sims to Insull, 4 Oct. 1883; TAE to Insull, 4 Oct. 1883; both DF (*TAED* D8322ZCS, D8340ZEA).

–3– October–December 1883

Edison began October in Brockton, Massachusetts, over-seeing the inauguration of the first underground three-wire central station system. He also spent time visiting with steam engine manufacturer Gardiner Sims before returning to New York about the sixth, as his family settled into their new home at the Clarendon Hotel.

Edison kept a characteristically full schedule through the last months of the year, devoting most of his time to the Construction Department. Declining an October invitation to write articles for distribution by the American Press Association, he explained that he could "Hardly get time to sleep." (He did offer to sit for an interview instead.)[1] Perhaps as a consequence, Edison became ill toward the end of that month and had to cancel a planned return visit to Brockton.[2]

The outlook for Edison's central station business certainly appeared favorable. In early December, Edison cabled Charles Batchelor in Paris to "Prepare yourself return— Immense boom" in village plant construction.[3] According to Samuel Insull's tabulations about that time, the Edison Construction Department was overseeing the completion of central stations in eight small cities in Massachusetts, New York, Pennsylvania, and Ohio, totaling approximately $168,000 in contracts. Insull noted elsewhere, however, that the "depressed state of the general stock market and the great shrinkage in values which has been going on now for months" had begun to impinge on the Edison Electric Light Company and the Edison Company for Isolated Lighting.[4] Financial troubles also contributed to the retirement of Edison's long-time supporter Egisto Fabbri from Drexel, Morgan & Company. In order to square its partners' books, in December, the firm asked Edison to repay more than $35,000 in loans.

Edison continued to learn from the early operating experience of his central stations. In particular, Frank Sprague, whom he delegated to Brockton until early December, and William Andrews, who visited a number of different plants, submitted frequent problem reports and their own suggestions for improving the installations. One chronic problem was sparking at the dynamo commutator brushes, which Edison again attempted to address in November experiments and two related patent applications. Even higher on the list of difficulties was the challenge of maintaining a steady voltage throughout the distribution system as the load changed. Insufficient voltage caused the lamps to burn too dimly; an excess caused them to burn out. With Andrews complaining about the unreliability of the standard voltage regulator, Edison deployed John Ott on experiments. In mid-October, however, he took an entirely different approach himself and quickly devised a new voltage indicator. Based on what would soon become known as the "Edison Effect," the device used a modified lamp with an independent electrode in which an electric potential was induced in proportion to the voltage on the filament. The instrument was highly sensitive and apparently reliable, and Edison immediately ordered it put into service. In December, Andrews forwarded an account of one station operator being drunk on the job. The first of several such reports that continued into 1884, this sort of problem was, of course, immune to any sort of technical solution that Edison might devise.

Although the Edison electric lighting company in Great Britain had formally combined in October with the Swan company, Edison retained both financial and strong personal interests in the British business. He dispatched Samuel Insull to London in early December, in part to assess the combined company's prospects. He also wanted Insull to attend to matters on the Continent, including reaching a settlement of Edison's long-running dispute with the Italian Edison company over charges for equipment. Insull also planned to meet with Charles Batchelor about lamp manufacture for the British market and the proposed reorganization of the French Edison companies, and to discourage the idea of a lamp factory in Germany.

Closer to home, Edison hired attorney John Tomlinson in November to prosecute the legal defense of his electric lighting patents against infringers. In December, he entered the politically charged debate on the safety of overhead electric lines, directing Edward Johnson to conduct experiments on the wires' possible hazards to firefighters. He also testified as a plaintiff's witness in a lawsuit against the Illuminating Company concerning its wires. Edison's laboratory records also show that he directed experiments on manufacturing lamp filaments from various gelatinous materials. He also maintained keen interest in ongoing experiments at the Edison Lamp Company in Harrison, New Jersey, but most records of that work in this and subsequent years, including specific instructions he may have given, have generally been lost.

The family's expenses, in addition to weekly charges for rent and meals and the Clarendon Hotel, included tuition for Marion Edison at Mme. Mears's school in New York, where she had enrolled in October as a day student. The family also rented a piano, on which Marion took lessons. In December, perhaps in anticipation of the upcoming holidays, Edison arranged to have his sleigh put in order and shipped back to Menlo Park.[5] At about that time, Edison sought to avoid paying, at least for the moment, the judgment against him by Lucy Seyfert, instructing his representative to "postpone the matter just as long as possible, in fact carry out the same policy that has been carried out right along."[6]

1. Orlando Smith to TAE, 10 Oct. 1883; TAE to Smith, 12 Oct. 1883; both DF (*TAED* D8306ZAR, D8306ZAR1).
2. Samuel Insull to William Jenks, 24 Oct. 1883, DF (*TAED* D8316BCF).
3. Doc. 2565.
4. Insull to Charles Batchelor, 5 Nov. 1883, DF (*TAED* D8316BEG).
5. TAE to Laurence Smith, 8 Dec. 1883, Lbk. 17:326 (*TAED* LB017326).
6. TAE to Josiah Reiff, 3 Dec. 1883, DF (*TAED* D8316BMN).

–2532–

From Abraham Kissell

Knoxville, Tenn., 10/3, 1883.

Dr. Sir,

1[a]   In case I send parent Co. $2,000 & have $16,000 paid into treasury of an Edison E. Illuminatg. Co., of this city within 10 days, can yu. with 30 ds. from such payts. erect all machinery, polls, & wires of a C[entral].S[tation].? Of course Id. secure ground & have buildg. erected within tht. time   2[a]   Can yu. for $16,000 guarantee 800 16 or 1280 10 c.=p. lamps to be run fr. a C.S. where no lamp wld. be over 1600 ft. fr. dynamo, or C.S., & the average distance of all lamps wld. not exceed 800 ft.? The plant wld. hv. to be run fr. sunset to sunrise. If the Co. is organized here, it has already 1000 lamps contracted for & wld. rather yud. proceed w. these than tht. yu. shld. lay out a system for future bus.[1] The C.S. cld. be located centrally to these 1000 lamp consumers & wld. then be central in the city for future bus. 3[a]   I can secure two 1st=class wiremaen & electricians here to aid us. One of them is manager here of W.U. office & supt. of the telephone Co.[2] He has subscribed for stock & is aiding me to get up Co. Hell. superintend our Co. here & get an engineer & a boy to assist him for $100 pr. mo. 4[a]   Hope yull find time to ans. this letr. <u>immediately</u> & not hv. it remain unanswered as my last two letters. 5[a]   Pl. ask Hastings to answer my ¶2 of 9/25 letr.[3] Manly of my replies to lets. are very unsatisfactory. If an agent's courage is to be kept up in this barren country, why replies shld. be prompt & full. Truly,

A. S. Kissell, agt.[4]

⟨The only way we can do this thing is to make a regular Canvas & make a tender= dont want any Electrician   one Engineer & meter man is running Sunbury 500 10 cp ditto 1600 Shamokin ditto 1600 Brockton   2 men can run anything less than 1600 Lights. The meter man can help Engineer from 5 pm until 10 pm when load goes nearly off— Engr runs balance night meter man comes on 10 am does meter work bookeeping collect bills= turns money into one of the Directors Co who acts a Treasurer= Wiring is done by our Expert who Learns Local plumber— Co must do no wiring otherwise Consumers will play the light against bills= If you write Eaton that prospects good for organizing Co he will order a canvass— & whe will do it & make a tender[5]—we will not permit Local people to do anything. we have had Enough of their ignorance & delay which only injures themselves & causes useless Expense to them— E⟩[6]

ALS, NjWOE, DF (*TAED* D8340ZDX). [a]Circled.

1. Kissell had visited Tennessee in August and reported that the Brush Company was trying to sell battery-powered lighting in Chattanooga. In Knoxville and Nashville, he believed, investors could not be induced to "subsc. For C. S. Ltg. Without first seeing an Edison lit. in operation." Edison, noting the proximity of the Southern Exposition station at Louisville, took this as proof that "every Loctn had to stand by itself as far as influence goes." In March 1884, Kissell was still pleading the case that the station needed to be put in and proven commercially

successful before investment capital would flow. Kissell to TAE (with TAE marginalia), 3 Sept. 1883 and 10 Mar. 1884, both DF (*TAED* D8340ZCQ, D8460F).

2. Joseph C. Duncan managed the Western Union Telegraph Co. station in Knoxville; he later became involved in a number of early Tennessee telephone companies. Kissell may also have had in mind as a wireman H. G. Comstock (b. 1850?) of Chattanooga, who worked for Western Union until the 1883 strike. Kissell to TAE, 13 Oct. 1883, DF (*TAED* D8313ZBK1); Rule 1900, 225; U.S. Census Bureau 1970 (1880), roll T9_1259, p. 184.4000 (Chattanooga, Hamilton, Tenn.).

3. Not found.

4. Abraham S. Kissell (b. 1829), a former state superintendent of education in Iowa (1869–1871), was working in 1880 as a fire insurance agent in Chicago, where he somehow made contact with the Edison lighting enterprises by the middle of 1883. At that time, Kissell sought territory as an agent for organizing central station companies. He remained connected with Edison through most of 1884, principally in connection with the prospective Knoxville plant. Kissell addressed this and several other letters to "Friend Edison." Kissell passport application, 31 Jan. 1872, U.S. Passport Applications (1795–1925), online database accessed through Ancestry.com, 23 Jan. 2009; U.S. Census Bureau 1965 (1870), roll M593_415, p. 266, image 532 (Des Moines Ward 6, Polk, Iowa); ibid. 1970 (1880), roll T9_185, p. 518.4000, image 0433 (Chicago, Cook, Ill.); Kiddle and Schem 1883, 474; Kissell to TAE, 26 June 1883, DF (*TAED* D8340ZAS).

5. Knoxville was canvassed in December, and Edison's office completed a construction estimate by March 1884. However, Kissell had difficulty organizing local investors, and the station was not undertaken (Kissell to TAE, 27 Feb., 15 Mar., and 16 May 1884, all DF [*TAED* D8460D, D8460H, D8460P]). The caution of local capitalists was likely compounded by the legal hurdle of the Knoxville Gas Light Co.'s contract for lighting the city for forty years from 1855. In 1886, after successfully negotiating with the gas company, the city issued a limited concession for electric lighting to the Schuyler Electric Light Co. (Rule 1900, 128–29; *American Electrical Directory* 1886, 352; *Whipple's* 1890, 407).

6. Edison's comments formed the basis of the typed reply sent to Kissell on 9 October. DF (*TAED* D8316AZD).

–2533–

From Frank Marr

Sunbury, Pa., Oct 3rd 1883[a]

Dear Sir:

What are the facts in reference to the inclosed— It is a slip taken from the Public Ledger of Phila of Sept. 29th.—[1] It is being used against us—[2] Respt

Frank Marr[3]

⟨There was no fire except in a service pipe 7[b] feet underground where there is no safety Catch= no fire can or has ever

happened in our district & we have reduced the average fires
35 percent in our district in 1st year—[4] Edison⟩

ALS, NjWOE, DF (*TAED* D8361ZDG). Letterhead of Frank Marr,
Attorney and Counselor at Law. [a]"*Sunbury, Pa.*," and "18" preprinted.
[b]Obscured overwritten text.

1. The editors have not otherwise identified the clipping enclosed by
Marr, which stated in its entirety: "A fire in New York the other morning [is thought?] to have been caused by electric [apparatus?] of the Edison System. This is one of the things to make a note of, for, although
it is not difficult to make the system safe, carelessness in insulation or
the use of wires that are too small, is attended with danger from fire in
spite of safety plugs." The report likely originated with an item in the
New York Times of 24 September 1883, which speculated that a fire the
prior day in a beer saloon on South St., on the edge of the First District,
was "caused probably by the electric light wires which enter the building from underneath the sidewalk and run under the flooring of the
saloon." Upon investigation, officials concluded that a broken insulator
had allowed one of the bare wires to make contact with a floor plank.
"Fired by Electric Wires," *NYT,* 24 Sept. 1883, 2; cf. Doc. 2449.

2. Marr did not otherwise identify these opponents or describe how
they were employing the story.

3. Frank S. Marr (1852–1916) was a Sunbury attorney who served
as treasurer and legal agent for the local Edison company. He later operated the Marr Construction Co., which installed Edison central station plants for the Edison Electric Light Co. after Edison disbanded his
Construction Dept. In 1892, Marr cofounded the Helios Electric Co.
(in Philadelphia) with two local businessmen who owned the United
States patent rights to an AC arc lamp developed by Carl Coerper in
Germany. Marr headed the firm through its changeover, in 1906, to
making aftermarket batteries (as the Philadelphia Storage Battery Co.)
and held the position until his death in 1916. That company eventually
became Philco, a major manufacturer of radio equipment. U.S. Census Bureau 1980? (1900), roll T623_1464, p. 5A (Philadelphia Ward 22,
Philadelphia, Pa.); letterhead, Marr to Samuel Insull, 29 Apr. 1885, DF
(*TAED* D8523ZAS); Johnson to Uriah Painter, 8 Apr. 1885, Unbound
Documents, UHP (*TAED* X154A4DI); Wolknowicz 1981, 6–8; Leonard 1922, s.v. "Andrews, William Symes"; Ramirez 2006.

4. Edison's handwritten marginalia became the basis for a typed reply (TAE to Marr, 9 Oct. 1883, DF [*TAED* D8316AZI]). Edison was
alert to the suspicions that other people placed on electric lighting as a
new technology. Following the recommendation of the New York Board
of Fire Underwriters, he had already advised Pearl St. superintendent
Charles Chinnock, "whenever there is a fire in your district to look out
for our wires and to be right on the spot to see that no blame is attached
to us." Chinnock appears to have forthrightly accepted responsibility
for at least one previous fire (TAE to Chinnock, 6 June 1883, Lbk. 17:74
[*TAED* LB017074A]; see Docs. 2449 n. 3 and 2462 n. 2).

[Brockton, Mass.,][1] Oct 4th= 1883.

Insull.

Please get Bordens storage battery from Major Eaton, have it signed. "Electron." and then put Spencer Borden's name at the bottom, not for publication   then write a letter to

Friend Beach,[2] I send herewith an interesting article suitable for your columns on the storage battery written by Mr Borden of Fall River who sent it to me to ask my opinion[3]   I think it a very correct article, Yours TAE   per I[nsull].

Beach is editor Scientific American   Save a Copy![4]

Edison

ALS, NjWOE, DF (*TAED* D8306ZAU).

1. See Doc. 2531 n. 1.

2. Inventor Alfred Ely Beach (1826–1896), coeditor and part owner of the *Scientific American,* was also a patent solicitor for Munn & Co., which had handled several Edison duplex telegraph patents in 1873. *ANB*, s.v. "Beach, Alfred Ely"; Doc. 304 (headnote).

3. Spencer Borden had been experimenting with storage batteries in 1882 (see Doc. 2278), but Edison referred here to a highly critical paper that Borden wrote on the Brush battery. After *Scientific American* failed to publish it, Sherburne Eaton used it in a special issue of the Edison Electric Light Co. Bulletin (21:21–36, 18 Dec. 1883, CR [*TAED* CB021474]; Eaton to Insull, 15 and 22 Oct. 1883; Eaton to TAE, 5 Nov. 1883; all DF [*TAED* D8306ZAS, D8306ZAV, D8306ZAW]). Devoted substantially to storage batteries, the issue included a reprint of Edison's January 1883 disparaging newspaper interview about that technology (see Docs. 2389 n. 1 and 2403 n. 4). It included two other essays on storage batteries. The first, written by Sidney Borden Paine, Borden's nephew, reprinted and analyzed a report by Massachusetts Institute of Technology professor Charles Cross on tests of an experimental Brush battery plant conducted in July and August 1883 (21:3–20, CR [*TAED* CB021456]). The second was George Bliss's paper on the "Brush-Swan Storage Installation at Willimantic" (21:37–43, CR [*TAED* CB021490]). Borden's paper included extensive comment on both the M.I.T. and Willimantic plants and a discussion of the economics of storage batteries for electric lighting. The editors have found no correspondence between Borden and Edison about the paper, but Paine may have contributed to it as well. For a discussion of the British and U.S. contexts for developing storage batteries for use in central station systems, and the effects of Edison's skepticism, see Schallenberg 1981 (729–41).

4. The typed letter to Beach that Insull prepared in Edison's name, dated 5 October, was essentially the same as Edison's draft. DF (*TAED* D8316AYL).

Samuel Insull to A. C. Mears[1]

[New York,] October 5th. [188]3

Madam:—

If you have a vacancy in your school for a little girl ten years old, Mr. Edison would like to send his daughter as a day scholar.[2]

He desires that she should take lunch with the family.

If you have a vacancy, will you please let me know at your early convenience what your terms would be and how soon you would desire Miss Edison to commence her studies? Yours very truly,

S.I.   P[ersonal].S[ecretary].

TL (carbon copy), NjWOE, DF (*TAED* D8316AYO).

1. Mme. A. C. Mears operated a girls' boarding and day school of that name. Mears evidently used her husband's name and the couple moved in high social circles, but the editors have not determined her relationship with the school's founder, Catherine Mears (1800–1877). The Mears school, founded in 1840, was located at 222 Madison Ave., between Thirty-sixth and Thirty-seventh Sts., some eighteen blocks uptown from the Edison family's new residence at the Clarendon Hotel. Its program, though advertised as essentially francophone, also provided "lectures in English, French, and German by eminent Professors." Advertisement, *NYT,* 24 Aug. 1883, 7; "At the Springs," ibid., 5 Aug. 1887, 3; Obituary, ibid., 26 Nov. 1877, 5; "Funeral of Mme. Mears," ibid., 28 Nov. 1877, 2.

2. Insull referred to Marion Edison, who had attended two other schools since the family moved to New York in 1881 (see Doc. 2344 esp. n. 4). The fall term started on 26 September; Edison made a payment of $140.26 to the school on 25 October. Marion Edison was enrolled at the Mears school for several years. Edison journal book (6 Jan. 1881–14 Nov. 1885), p. 308, Accts., NjWOE; see also Doc. 2732.

From Francis Upton

EAST NEWARK, N.J., Oct. 6 1883.[a]

Dear Mr. Edison:

I call attention to the following facts.

A careful statement sent you on July 2, of our financial work for the previous six months, showed that the Lamp Co. had made a profit of $24,628.84 during the six months.[1] [---][b]

Since July 2 to Oct. 2 we have also made money though we have only sold $29,383.79 worth of lamps.

We owed outside parties on

July 1	$33,173.14
Oct 1	29,687.81
	3,485.33

That is we have decreased our liabilities to every one except the money due on my loan and the mortgage $3.485.32[2]

My loan has increased $241.[0]7[c]

	3,485.32
	$241.[0]7
leaving net gain	3,244.25
Bills due to us Oct. 1	$22,708.70
" " " " July 1	14,357.96
Gain.	8,350.74

Total Gain

Decrease liabilities	3,244.25
Increase Assets	8,350.74
Better off Oct. 1 than July	$11,594.99

This of course is due[b] largely money realized on the sale of stock on hand. Yet after making every allowance we have made money about $3000.00—[d] in the last three months.

The orders now on hand will bring us over $11,000 more from sale of lamps in stock.

Conclusion— I think the Lamp Co. should now pay me a salary. It is making money, and my pay will come from the profit of the business. I have been here without salary nearly three years.

I have given my entire time to the interests of the concern. I have done everything in my power to help the Lamp Co. financially, often at a sacrifice.

I value my services at $500 a month. I am 31 years old and have served you for five years, I know faithfully, with the understanding that you would be liberal when my salary could be taken from the profits of the business.[3] Yours Truly

Francis R. Upton

ALS, NjWOE, Upton (*TAED* MU062). Monetary expressions standardized for clarity; letterhead of Edison Lamp Co. [a]"EAST NEWARK, N.J.," and "188" preprinted. [b]Canceled. [c]Followed by dividing mark. [d]"about $3,000—" interlined above.

1. The Edison Lamp Co.'s 2 July statement covered the first six months of 1883. DF (*TAED* D8332ZCB1).

2. Upton had been advancing money to the Edison Lamp Co. since at least early 1882. The partnership's indebtedness to him rose quickly to $12,358.52 by June of that year, to $27,021.62 at the start of 1883, and peaked at $33,404.21 on 1 October 1883. The editors' statement in Doc. 2297 n. 3 about the interest paid to Upton was based on Samuel Insull's ambiguous declaration to him that "Mr Edison says you are entitled

to receive 10% Interest on you[r] money. He does not consider 6% enough"; it is unclear when or at what rate Upton's loans began to accrue interest. Edison Lamp Co. to TAE, 1 Feb. and 1 June 1882, 2 Jan. and 1 Oct. 1883, all DF (*TAED* D8231AAC, D8231AAX, D8332B, D8332ZCU); Insull to Upton, 2 Mar. 1883, Lbk. 15:379 (*TAED* LB015379).

Upton had proposed in April to convert at least part of his credit into an increased ownership stake in the factory. He appears to have done so effective 1 November, when he used about $10,000 in notes Edison signed over to him from the lamp factory toward the purchase of Frank McLaughlin's entire interest (5 percent) for $12,500, a figure based on his own valuation of the business (Upton to TAE, 18 Apr. and 31 Oct. 1883; Insull to Charles Batchelor, 5 Nov. 1883; both DF [*TAED* D8332ZAJ, D8332ZDC, D8316BEG]; McLaughlin agreements with Upton; Frank and Margaret McLaughlin agreement with Upton, all 1 Nov. 1883, Upton [*TAED* MU063, MU064, MU065]). This action coincided with (and likely spurred) efforts to incorporate the company. In preparation for the purchase from McLaughlin, Upton consulted an attorney who subsequently expressed to Edison his surprise at finding that the company's

> affairs were not clearly defined in writing, and that each of the partners depended upon tradition, and the honor of his associates to establish his actual interest in the business; that the personal property was nominally owned by the Company, while the real property was in your name, and in case of your death would descend to your heirs. . . . The Company is in such condition that the death of any member of the firm would bring with it untold and most undesirable results.

The best remedy, the lawyer advised, was to incorporate the business as a stock company (William Curtis to TAE, 25 Oct. 1883, DF [*TAED* D8332ZCZ]).

3. The editors have not determined what terms, if any, Upton arranged at this time. In November 1884, following a discussion with Charles Batchelor, he confirmed that he would "allow the present arrangement as to salary continue i.e. $60—per week for half my time, until such time as the interest on the original investment has been paid back in dividends" (Upton to Batchelor, 18 Nov. 1884, Batchelor [*TAED* MB152]). Upton came from a family of at least modest wealth, and his employment by Edison on the basis of electric light royalties instead of a fixed salary antedated the period of three years indicated in this letter (see Doc. 1762). Upton reported to the R. G. Dun credit agency that he held about $30,000 of stock in various Edison companies, in addition to his stake in the Lamp Co. He had also received from his father, Elijah Wood Upton, a successful Massachusetts manufacturer and bank director who died in 1881, a $25,000 trust, due to mature after seven years (R. G. Dun & Co. report, 18 Dec. 1883, Dun [*TAED* X112CAV]; Hurd 1888, 1057).

From Charles
Batchelor

My dear Edison,

Yours of 26 Sept to hand—[1] 100 light drawings to hand but the most important part of the information is not there that is the size and amount of wire on armature ~~and magnets~~— As you say you change from the 2 coils to 1 I think I shall not change the 120 light—

I see the English Company is asking about prices here for lamps and our company I find out have already sent some there. With the price you have given for 16 and 8[2] I think you will hold that market easy enough—they never asked me to furnish them and I only accidently found out they had been sent.

I have made a lot of experiments and have come to the conclusion that the long magnet is not worth a damn!!![3] excuse me for the strong language!

As I am making an estimate (which is almost finished) for a Station here of 24 000 lights (guaranteed 5 hours per day); I should like to know what is your actual experience with 3 wire system?:[4] Can you in practise work more than 2 amperes per millimetre on the underground conductor:?

If you give 10% loss on conductors with 3 wire system, the short feeders would require to carry ~~many~~ more amperes per section than the long ones therefore would it not be better in your mind to have the station divided in two one half feeding out to the [---- ----][a] with 10% loss and the other half with less loss[5]

You see if you have 1000 lamps at 200 metres on one feeder & you give 10% loss you have a current of about of about 4 amperes per square m.m. but if you have 1000 lamps at 600 meters then in that conductor you have a current of 2.2 amperes per m.m. So it seems to me you have got to put the copper in the short conductors to carry the current & then loose it again in the feeder regulator—[6] But it seems to me better to have one half of such a big station working at 110 volts & ½ at 100 volts   I dont mean outside but inside the station— Another way would be to keep the machines entirely separate! and pump into any feeder with any of two or three machines until your [pressure?][b] tested right— Give me a few points please.

"Batch"

⟨@   send this to Sprague to answer—[7]
write Batch that if he will send a Canvass shewing maximum lights we will agree to send him a determination accurately in 3 days after its arrival— we have a quick offer process⟩[8]

ALS, NjWOE, DF (*TAED* D8337ZER). [a]Document damaged. [b]Illegible.

1. Batchelor referred to Edison's 27 September letter, not that of the 26th about various financial transactions. Edison wrote that he had mailed blueprints of the 100-light T dynamo as presently manufactured, adding that he was "thinking of using only one pair of cores, and should this change be made I will forward you blue prints of same." He had experimented during the summer with one pair of field cores (essentially a single magnet), and he later adopted this construction in the Y but not the T dynamos. TAE to Batchelor, 26 and 27 Sept. 1883, both DF (*TAED* D8316AWV, D8316AXA); see Doc. 2419 (headnote) and App. 3.

2. That is, lamps of sixteen and eight candlepower.

3. Since testing a short core design at the end of July (see Doc. 2501 n. 9), Batchelor had been experimenting largely, if not exclusively, on machines with magnet cores about twenty-eight inches or less in length. On 27 September, after varying the dimensions of the "heads" or faces of the magnet pole pieces, he reached a definite conclusion: "This experiment proves that the length of cores matters very little, but that the section makes all the difference, and I think that the maximum strength will be attained when the sectional area of the core is equal to sectional area of the head. . . . or in other words is equal to the length and height of the 'cutting' field." Batchelor then tried different field winding configurations and, after recording some improvement, he noted: "I always thought that we were very far from saturation point in our machines and it will be very necessary to immediately make characteristic curves of all the sizes." Around this time, Batchelor also articulated in three "deductions" the arithmetic relationships among the number of magnet turns and the weight, height, and sectional area of the core. These showed ways either to increase the magnetism from the same amount of copper and wire, or to maintain the same field strength with less metal. Cat. 1235:25, 30; Cat. 1237:175, Batchelor (*TAED* MBN012025, MBN012030, MBN006175).

4. The editors have not found this estimate. Batchelor had recently tabulated the length of conductors needed for some fifty blocks of a Paris central station district. He had also, on 12 September, sent to Charles Porgès copies of letters from Sherburne Eaton and Charles Clarke documenting the cost of the Pearl St. plant in New York (Cat. 1237: 125–29; Cat. 1331:163; Batchelor [*TAED* MBN006125, MBLB3163]). He did so evidently in response to Porgès's incredulity over figures that Batchelor had previously provided about the cost of reproducing the New York plant in Paris (Eaton to Batchelor, 6 and 21 Aug. 1883; Clarke to Eaton, 18 Aug. 1883; Clarke cost estimate, 18 Aug. 1883; Cat. 1237:83A, 83, 85, 89; Batchelor [*TAED* MBN006083A, MBN006083, MBN006085, MBN006089]; Batchelor to Eaton, 31 July 1883, DF [*TAED* D8337ZDJ]).

5. Batchelor had tested on 2 October how much current a conductor in Edison's underground tubes could safely carry. Finding that 411 amperes heated a conductor of 133 square millimeters enough to melt the insulation, he concluded that "For tubes having conductors with so great a section as this it is sure we must never put 3 amperes per [b] m.m." Cat. 1235:32, Batchelor (*TAED* MBN012032).

6. Batchelor seems to have been talking about an asymmetrical three-wire system in which the load on one side was farther from the dynamos than the load on the other side. To compensate for energy lost in the long conductor and to balance across the neutral wire, such a system would require extra current through the feeder on the short side.

The feeder regulator (or equalizer) mentioned by Batchelor was essentially a resistance box in the feeder circuit at the central station. Although the plans for New York's Pearl St. station called for them, they apparently were not installed there at this time. Feeder regulators were used, however, at the Milan, Italy, plant, and in many (if not all) of those erected in the United States by the Construction Dept., where they helped engineers equalize voltage across the three-wire system. The instruments were made by Bergmann & Co.; Frank Sprague reportedly designed one form, but the editors have learned nothing of its particulars. See Docs. 2269 n. 3, 2505 (headnote), and 2625; TAE to Willis Stewart, n.d. [May 1884], DF (*TAED* D8435ZAL); Samuel Insull to Frank Sprague, 10 Sept. 1883, LM 15:41 (*TAED* LBCD2041); Vail 1916, 16.

7. Edison sent this letter to Frank Sprague on 26 October, asking him "give me full replies to all of Mr. Batchelor's questions." Sprague sent a detailed reply in December, most of it devoted to conductor capacity. He pointed out that the correct answer "depends entirely upon the size of conductor, and the material in which it is laid," but he generally disagreed with Batchelor's suggestions. TAE to Sprague, 26 Oct. 1883; Sprague to TAE, 10 Dec. 1883; both DF (*TAED* D8316BCV, D8343ZEO); Sprague to Batchelor, 10 Dec. 1883, FJS (*TAED* X120CAF).

8. The editors have not found such a letter.

VOLTAGE INDICATORS Doc. 2538

Regulating voltage throughout the distribution system, not just at the dynamo, was a major issue for early Edison installations. It became especially important in the village plants' three-wire system, where interconnected circuits and rapidly changing loads made balancing electrical supply and demand a constant and crucial task.[1] Regulation, whether automatic or manual, required reliable instruments (voltmeters) in the stations and at the ends of feeder conductors to indicate fluctuations that could shorten the life of expensive bulbs (which were the local companies' financial responsibility). The accuracy of indicators and their reliability over time were persistent problems that led to a number of redesigns, including several novel solutions and one, in particular, that saw a brief period of use in the early village plants at the beginning of 1884. Like the major changes in dynamo design between the spring and fall of 1883,[2] these innovations are not clearly reflected in the surviving documentary record; their course has been pieced together from elliptical correspondence, relatively few technical notes and drawings, patents, and published sources.

The indicators used in lighting plants from 1881 to 1883 followed a form designed by Edison and constructed in September 1881 by John Ott.[3] Edison had filed a patent application on this indicator in September 1881; the patent drawing and production version of the instrument are shown in Doc. 2190 nn. 9–10. It used an electromagnet and spring to control the movement of the arm. When the voltage through the electromagnet fell outside the desired range, the arm would move enough to close a contact in a circuit containing an alarm bell that could command the operator's attention to the need for an adjustment. Edison specifically intended this device for isolated stations, which generally had a single circuit, ran relatively few lamps over a short distance, and did not employ trained engineers; operators only needed to know how and when to adjust the field magnets to stay within tolerance. In these circumstances, Edison believed, a good indicator would "verify our statements as to the life of the lamps."[4] This indicator could also be adapted readily to a regulator that would make the necessary adjustments in the dynamo circuit without the operator's intervention. Edison regarded the automatic regulator as a "necessary evil" for isolated plants, but it became a staple of such installations;[5] one form of it is shown in Doc. 2242 (headnote n. 5).

These mechanical devices underwent further improvement and simplification to lower production costs. The indicator initially used at the Pearl Street central station in 1882 gave a visual indication of trouble with red and blue lamps, a useful feature in a large and noisy plant. That model, like most others, had problems with sparking at the terminals, a problem that Charles Clarke and William Andrews sought to resolve in late 1882. Improved indicators, and not merely cheaper ones, would also be of considerable utility to the Edison Company for Isolated Lighting and Edison affiliates abroad.[6]

In January 1883, Charles Bradley, an engineer assigned to the Edison lamp factory in Harrison, patented a wholly new indicator offering finer tolerance. Its principle had evidently been used at the Pearl Street station since late 1882, using a galvanometer and a zero-voltage baseline. Taking advantage of the fact that the resistance of a carbon filament decreases with rising temperature, he placed two lamps (only one in an alternate version with a different indicator) in opposite legs of a Wheatstone bridge. One standard and one adjustable resistance were in the other pair of opposing legs. Bradley filed for a patent on this device in February 1883. His specification made clear that the method of measurement was indirect, "based upon the fact that in certain substances variations in temperature produce appreciable variations in electrical resistance. . . . By noting these variations of resistance the electromotive force of the current may be calculated."[7] The precedent of the older indicator was carried forward in that the needle or the arm would be set to center at the desired voltage, but any increase or decrease would generate corresponding changes in the lamp resistance and alter the voltage crossing the bridge, thus changing needle deflection or electromagnet strength.

This indicator became known as the "Bradley Bridge." Manufacture of these indicators was placed in the hands of Bergmann & Company, which struggled to provide them in time for the opening of the first village plants; some stations likely went into service with pre–Bradley indicators or a mix of the old and new patterns.[8]

The major flaw in the basic Bradley design stemmed from its dependence on the standardization of lamp resistance when hot and the maintenance of that value over the bulbs' operating lives. Each of these was an uncertain proposition, and inconsistencies produced errors in even the most carefully calibrated instruments. Edison's Pearl Street station coped by manually recalibrating the instruments against a standard at

Patent drawing of circuit for the "Bradley Bridge" voltage indicator.

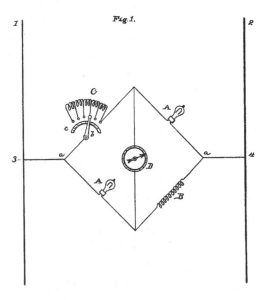

the Edison Machine Works every few days. The New York system, with a high geographic density of lighting and the robust two-wire system, was also less susceptible to voltage fluctuations than the three-wire village systems, which required more deliberate care. The effects of unreliable voltage indication began to appear almost immediately at these smaller plants, with numerous complaints of high costs due to lamp breakage.[9]

Investigations implicated the indicators in short lamp life despite a number of cases where the plant operators were claimed to be at fault. A particularly acrimonious dispute arose in Sunbury. Edison suspected the local manager of deliberately running lights at nearly twice their rated illumination, a suspicion that met with heated denials and counterclaims.[10]

Noting these problems, Construction Department engineers William Andrews and Harry Doubleday began to compare readings from multiple Bradley-type indicators at the Sunbury and Shamokin plants. Doubleday reported to Samuel Insull on 11 August that the indicators "vary considerably an adjustment of ten degrees necessary in one." Four days later, seeing a disparity of five volts between two seemingly identical devices, Andrews advised Edison that "my pressure will be always an unknown quantity until I get my positively standard Instrument." At that time, Andrews and Doubleday believed the issue to be one of poor calibration, not a matter of design or the limits of manufacture. Andrews, in particular, suspected that the instruments had not been level when calibrated, causing gravity to tip the needle. Engineers conducted tests and made modifications of the design intended to correct the problem, but ultimately Andrews identified "two or three bad points" of design that created variable readings. He conveyed to Edison on 4 October his conclusion that the problems were a combination of changing force in the electromagnet cores, variable resistance in the cores created by heating, and the effects of oxidation and heating of the pointer itself.[11]

Within a few days of Andrews's report, John Ott built prototypes of two new indicators that made significant changes to the existing styles. One was electromechanical and appears to have relied upon an electromagnet to raise or lower a lever attached to a vertical scale. The second, designed by Edison, operated on a completely different principle.[12] This was Edison's first attempt to make use of a phenomenon that had been a mere curiosity up to this point, but which was later named

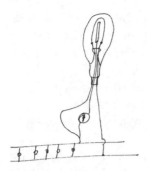

John Ott's 8 October 1883 drawing for a working model of a new indicator based on the Edison Effect phenomenon.

the Edison Effect and became the foundation for generations of electronic regulation and amplification devices.[13]

The Edison Effect generated a consistent difference in electromotive potential between a lamp filament and an independent conductor (usually platinum) in an evacuated bulb at a given voltage. When properly calibrated, such a bulb would impress on the extra conductor a voltage in definite proportion to the voltage on the filament. As Edison described it in the patent specification for the new indicator,

> if a conducting substance is interposed anywhere in the vacuous space within the globe of an incandescent electric lamp, and . . . is connected outside of the lamp with one terminal, preferably the positive one, of the incandescent conductor, a portion of the current will, when the lamp is in operation, pass through the shunt-circuit thus formed, which shunt includes a portion of the vacuous space within the lamp. This current I have found to be proportional to the degree of incandescence of the conductor or candle-power of the lamp. [U.S. Pat. 307,031]

Edison exploited this proportionality by connecting the additional conductor to a galvanometer, whose deflection thereby provided a measure of the voltage on the line. An adjustable resistance allowed the apparatus to be responsive at the desired point, given that the modified bulb only produced current on the third conductor above a certain line voltage.[14] Recognizing that this instrument was "absolutely new & novel," Edison moved quickly to draft a patent application (Doc. 2538; see also Doc. 2556). He also decided almost immediately to adopt the design at his village plant stations, expecting that it would provide an accurate absolute measure of line voltage to the station operators. Insull assured Frank Sprague that the old difficulties would be "entirely overcome" by the new instrument.[15]

Edison's subsequent experience with this indicator deflated the enthusiasm evident in Doc. 2538. William Andrews, who thought it "too fine and complicated an inst. for the use of an ordinary Engineer," was initially skeptical, as was Frank Sprague, who immediately began trying to improve it.[16] Over a dozen instruments were sent to village plants for operational trials between November 1883 and January 1884, while the older Bradley-style indicators were still being shipped to some stations. A number of difficulties soon cropped up, the most fundamental being that the Edison Effect lamps' electrical characteristics changed over time. Edison attributed this to the "deposit of Carbon on surface of platina wire," and he believed that deliberately coating the wires with carbon would solve the problem (see Doc. 2582). It did not, and Edison later blamed this variability for the instruments' unreliability and his consequent decision not to use them in isolated plants.[17] The indicators were also troubled by poor insulation of the center wire and the danger of short-circuits. Breakage in transit of the special bulbs led the Construction Department to have them shipped (or carried by hand) directly from the Harrison factory.[18] They were also comparatively expensive, and Edison soon suggested ways to limit the number of them needed in each station (see Doc. 2584). Reliability remained a fundamental problem, even as a number of the so-called "3-pole style" indicators entered commercial service.

Frank Sprague leveled severe criticism against the new device in early January 1884. He argued that it was impossible to get an accurate "absolute" voltage reading from any sensitive instrument dependent upon a horizontal magnetic field. Because of variations in the earth's own field, identical indicators would produce different readings in different locations.[19] Edison delegated John Howell to help John Ott make systematic investigations in mid-January (see Doc. 2592), and Ott submitted a series of reports. Their experiments showed significant variation among instruments, arising at least in part from differences in temperature and construction. The discrepancies were decisive and apparently insurmountable.[20] By 26 January 1884, Edison abandoned efforts to salvage the indicators and ordered those in use to be returned and altered to a new design.[21]

John Ott experimented on the instrument into the middle of February before he had a satisfactory one, which he sent to Brockton. "We have at last got indicators all right," Edison wrote to Andrews from Florida in the latter part of February. Already in use at the Pearl Street station, this latest design was a modification of the Bradley instrument, with a lamp in the Wheatstone bridge in order to use the "decrease & increase of rcsistance of the lamp to give deflection" to the galvanometer.[22] Giving a comparative rather than an absolute measurement, it was designed to balance with zero voltage across the bridge, so that the galvanometer deflection was based on the polarity and strength of the current instead of a particular calibrated load between two live lamp circuits. The first indicators of this type were placed at the Brockton, Lawrence, and

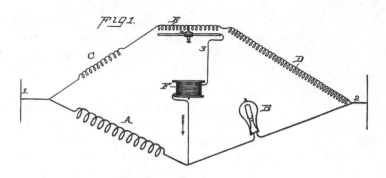

Bridge circuit of John Howell's voltage indicator, taken from his U.S. Patent 339,058. B is a high resistance lamp, F the galvanometer; A, C, E, and D are resistance wires.

Fall River, Massachusetts, stations by the end of February; and more were installed at other village plants in the following weeks. They were not a perfect solution, though. Andrews, for instance, complained about the absence of a scale for the deflector needle that would enable him to run deliberately at high or low voltage, which he sometimes found necessary.[23] However, their shortcomings were evidently more amenable to calibration and cross-checking.

John Howell significantly improved Bradley's instrument a few years later. He applied for a patent in 1886 on an indicator expressly intended to be "cheaper and more compact in construction, and more accurate, efficient, and convenient in operation" than the Bradley design. Placing only a single lamp in the bridge enabled Howell to use less of the costly resistance wires in the other legs, because they did not have to carry the full current for two lamps, and it also reduced the error attributable to heating. Howell also used enough resistance that the high-resistance lamp "is not brought to its full candle-power, and its permanency is thereby assured."[24] Although promoted as more reliable over time than their predecessors, the new Howell indicators were supplied with a separate calibrating indicator. They also came with a spare bulb intended "to correct any *possible* change in a lamp."[25] With greater reliability and redundancy built in for a price similar to the old Bradley style, the Howell indicators became the standard in 1886 and remained so for years.[26]

1. See Doc. 2505 (headnote). Edison's prior work on regulation is discussed in Docs. 2036 and 2242 (headnotes).

2. See Doc. 2419 (headnote).

3. Ott's device, in turn, drew from a simpler design first sketched by Charles Batchelor in June 1880. Ott notebook, pp. 19, 33, Hummel (*TAED* X128B019, X128B033); Unbound Notes and Drawings (1880), Lab. (*TAED* NS80AAK).

4. TAE to Johnson, 16 Jan. 1882, Lbk. 11:85 (*TAED* LB011085).

5. See Docs. 2526 and 2557.

6. TAE to Edison Electric Light Co., Ltd., 1 Feb. 1883, Lbk. 15:22 (*TAED* LB015223); Andrews to TAE, 5 Jan. 1883; Clarke to TAE, 19 Jan. 1883; Spencer Borden to TAE, 7 Feb. 1883; all DF (*TAED* D8330B, D8329C, D8325I). Brooks 1953 (401–2) describes the early Pearl St. instruments.

7. Bradley technical note, 13 Jan. 1883, Kruesi (*TAED* MK002); TAE marginalia on Andrews to TAE, 13 Feb. 1884, DF (*TAED* D8442ZBB); U.S. Pat. 280,563.

8. Alglave and Boulard 1884, 347–48; Dyer and Martin 1910, 1:414; Jehl 1937–41, 1098; Brooks 1953, 403–5.

9. Edison reportedly claimed about this time that the load at Pearl St. varied so predictably that he could regulate the dynamos according to a graph of the field resistance required throughout the day and night (Alglave and Boulard 1884, 347–48). On the Pearl St. indicators, see Doc. 2243 (headnote); on lamp breakage, see Doc. 2505 (headnote) and, for example, Andrews to Samuel Insull, 24 Aug. 1883; Harry Doubleday to TAE, 20 Sept. 1883; Alfred Tate to TAE, 24 Oct. 1883; John Howell to Francis Upton, 29 Dec. 1883; TAE to William Douty, 16 Jan. 1884; E. B. Hubbard to TAE, 13 Feb. 1884; Insull to William Schwenk, 18 Feb. 1884; all DF (*TAED* D8343ZAG, D8325ZCP, D8361ZDL, D8346ZBV, D8416AEL, D8451ZAR, D8416AOM).

10. See Doc. 2546 n. 3. A gradual voltage increase in plants controlled by automatic regulators was also noted in Chicago; see Doc. 2557.

11. Doubleday to Insull, 11 Aug. 1883; Andrews to Insull, 16 Aug. 1883; Andrews to TAE, 15 Aug. and 4 Oct. 1883; all DF (*TAED* D8366X, D8361ZCT, D8361ZCQ, D8343ZBL).

12. N-82-12-04:105–107, Lab. (*TAED* N145105); see also Doc. 2538.

13. See Doc. 1898 (headnote).

14. Doc. 2575 includes Frank Sprague's description of the Edison Effect indicator.

15. Insull to Sprague, 16 Oct. 1883; Insull to Andrews, 15 Oct. 1883; both DF (*TAED* D8316BAZ, D8316BAS).

16. Responding to initial criticism from Andrews, Edison wrote: "I think the more you work with the new indicator the better you will like it Sprague said the same thing." Andrews to TAE, 12 Dec. (with TAE marginalia) and 15 Dec. 1883; TAE to Andrews, 13 Dec. 1883; TAE to Sprague, 8 Jan. 1884; all DF (*TAED* D8343ZES, D8347ZAR, D8316BSL, D8416ACL); see also Docs. 2556 and 2575.

17. Edison testimony, *Edison v. Thomson*, 7, MdCpNA (*TAED* W100DL).

18. TAE marginalia on Harry Doubleday to Edison Construction Dept., 1 Jan. 1884; Bergmann & Co. to TAE, 3 Jan. 1884; Alfred Tate to Bergmann & Co., 4 Jan. 1884; Thomas Conant to TAE Construction Dept., 2 Jan. 1884; all DF (*TAED* D8444A, D8424B2, D8416ABK, D8451D).

19. The influence of terrestrial and other magnetic sources was one of the major obstacles to the design of reliable laboratory meters of current and voltage at this time. Sprague to TAE, 11 Jan. 1884, DF (*TAED* D8442N); Gooday 1995, 243–54.

20. Ott to TAE, 14, 18, 19, 22, and 25 Jan. 1884, all DF (*TAED* D8430F, D8430F1, D8430G, D8430I, D8430L).

21. Bergmann & Co. to Edison Construction Dept., 26 Jan. 1884, DF (*TAED* D8424R). Insull recalled the old equipment from at least seven plants at the end of March for refitting; see, for example, Insull to William Schwenk, 31 Mar. 1884, DF (*TAED* D8416BBD). Most, if not all, of the extra lamps—over 80% of which were broken—and the old instruments were back in New York by mid-April (Andrews to Insull, 2 Mar. 1884; Ott to John Campbell, 4 and 19 Apr. 1884; all DF [*TAED* D8442ZBX, D8467O, D8467Q]).

22. Edison may have based his confidence at least in part on Frank Sprague's satisfaction with the instrument, conveyed by Samuel Insull in Doc. 2608. Ott marginalia on Insull to Ott, 14 Feb. 1884; TAE marginalia on Andrews to TAE, 13 Feb. 1884; both DF (*TAED* D8467F, D8442ZBB).

23. On a letter from Andrews complaining about the new instruments being "useless," Edison wrote: "Too much importance cannot be attached to this matter of pressure indicators as the lamp breakage hinges directly upon it." Sprague wrote a passionate and detailed defense of the new instruments. Andrews to Insull (with TAE marginalia), 2 Mar. 1884; Sprague to Insull, 4 Mar. 1884; both DF (*TAED* D8442ZBX, D8442ZCD).

24. U.S. Pat. 339,058; Edison Lamp Co. instructional pamphlet for indicators, n.d. [1887], 2–8, 17–19; PPC (*TAED* CA041J1). Brooks 1953 explains the advantages of Howell's modifications to the Bradley Bridge. It also describes other design changes made by Howell, though it does not explain when—or if—they entered commercial service.

25. Dyer and Martin 1910, 1:414; Edison Lamp Co. Catalogue of Lamps, n.d. [1887], 24–26; Edison Lamp Co. instructional pamphlet for indicators, n.d. [1887], 2–8, 17–19; both PPC (*TAED* CA041J, CA041J1).

26. The Howell indicator was invoiced to outside operators at $50.00 in 1886; the indicators used in 1882 before the Bradley Bridge were likely about $48.50, although Bergmann & Co. invoiced one modified Bradley (with box) in February 1884 at the internal price to Edison of $35.80. Bergmann & Co. to Insull, 22 Sept. 1882; Edison Lamp Co. memorandum, n.d. [1886]; both DF (*TAED* D8201V, D8626Q); Voucher no. 93 (1884).

*Draft Patent
Application: Electric
Light and Power*

Electrical Pressure Indicator

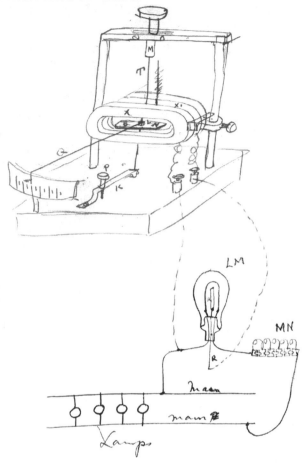

Dyer=

I have just completed a pressure indicator which works perfectly[2]

X X' are the coils[a] copper wire   L is heavy steel needle magnetized   it is secured on a torsion wire T which has[a] its torsion increased or diminished by turning[a] the smooth[a] stud M round   this is held by friction the crosspiece being split & 2 screws 3 & 4[b] used= K is a spring which keeps the wire T stiff[a]   P is a nut for increasing or diminishing the spring— Q is an index straw   The torsion of the wire keeps needle at zero—

LM is a regular lamp

in addition there is a strip of platina R which is secured to a wire & passes through glass. one End is connected to the galvanometer the other End of the galvanometer is connected to the pole of the lamps upon which the blue halo appears.[3]

when the lamp is brought up to incandescence a ~~current~~
constant current of considerable power passes through the
galvanometer wire, & deflects the index— The scale may be
graduated to read Volts or candlepower— A Resistance M.
may be put in with the Lamp & adjusted so as to standardize the lamp to Raise & lower in candle power like those on
the circuit— It is very sensitive ~~up to~~ with regular lamps it
deflects very little until you reach 12 candles ~~after that each~~
~~Volt raise~~ after that it is more nearly proportiond to the candle
power— In fig 2 I show 2 Lamps one runs all the time while
the other is used to be thrown in circuit to see that the regular
one has not varied.

fig 2

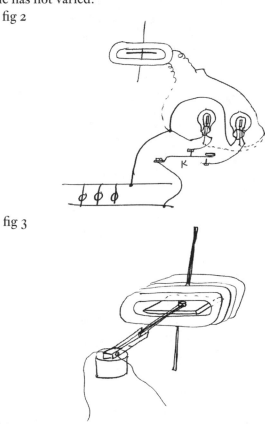

fig 3

fig shows how it can be arranged to close the circuit for
working auto Regulator & perform other things[c]

Want broad claims on using the Current given out by ~~heated~~
Vacuum Lamps for this purpose & other purposes.[4] This is
absolutely new & novel—

[Witness:] H. W. Seely   E[dison]

ADfS, NjWOE, Lab., Cat. 1149 (*TAED* NM018AAW). Document multiply witnessed and dated; all figures drawn on three separate pages. [a]Obscured overwritten text. [b]"3 & 4" interlined above. [c]Paragraph written below drawing on separate page.

1. Henry Seely signed and dated the document on this date when he received it from Edison.

2. See headnote above. John Ott drew this instrument on 8 October, noting that "Mr Edison ordered a working model made to indicate pressure across the line, according to sketch given to me dated Oct 8, 1883." The page of Edison's drawing referenced by Ott, in Menlo Park Notebook No. 204 (p. 67), is dated 8 March and is clearly related to others made that day concerning carbon carrying experiments, including Doc. 2411; N–82–12–04:105; N–82–05–26:65–69; both Lab. (*TAED* N145105, N204065).

3. Edison and his associates first noticed a blue halo or glow around one of the filament clamps in early 1880, during experiments to prevent "carbon carrying," or the deposition of carbon from the filament onto the lamp globe. This was one of several phenomena, including what later became known as the Edison Effect, that Edison observed during this research but could not explain. See Docs. 1898 (headnote), 2029, and 2061 esp. n. 2.

4. Edison executed a patent application, based closely on this draft, on 2 November. It issued as U.S. Patent 307,031 in October 1884. Three of the patent's four drawings closely resemble those in this document, the fourth being a detail of the special lamp itself. While the patent was for an "electrical indicator," the text noted that the arrangement could apply equally to a regulating mechanism. Its eight claims covered "an indicating or regulating apparatus" in circuit with the third conductor of the lamp as well as the special lamp itself, including "the vacuous space within the globe" as part of the circuit.

Drawing of Edison's new indicator, from the patent that resulted from Doc. 2538. A is the special Edison Effect bulb. Its extra conductor, on which a voltage was impressed in the evacuated space inside the bulb, is connected to the galvanometer coils d and d.

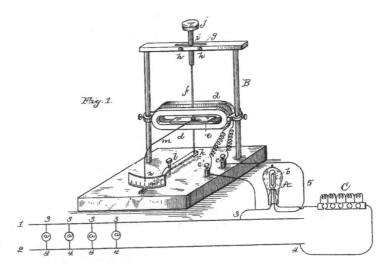

−2539−

To Alfred Tate

NEW YORK, [October 19, 1883?]^{1a}

Tate=

We have received several boxes & one bobbin of Cable wire=² I have not unpacked the boxes but the Coil is <u>Red</u>^b colour. If its all this way it is useless= It would look like hell on a mans white walls= I told Insull to have it <u>white</u>^b= Investigate.³

Edison

ALS, NjWOE, DF (*TAED* D8340ZET). Letterhead of Thomas A. Edison. ᵃ"NEW YORK" preprinted. ᵇMultiply underlined.

1. Tate stamped the reverse of this letter with this date when he received it.

2. This material was likely samples of the new colored cable for interior wiring discussed in Doc. 2529 esp. n. 5. The Ansonia Brass & Copper Co. had advised Samuel Insull to expect samples during this week. Ansonia Brass & Copper Co. to Insull, 15 Oct. 1883, DF (*TAED* D8321ZAB).

3. Insull was in Newburgh, N.Y., having been in Shamokin, Pa., the previous day. Tate wrote on the back of this letter: "Respectfully referred to Saml Insull Esqre." The editors have not found what action, if any, Insull took in this regard. Some red cable was shipped by mid-November to Sunbury, Pa., where customers held out for white. Insull to Tate, 18 and 19 Oct. 1883; Frank Marr to TAE, 15 Nov. 1883; all DF (*TAED* D8340ZEQ, D8340ZES, D8351P, D8361ZDP).

−2540−

From Charles Batchelor

Paris, le Oct 22 1883.ᵃ

My dear Edison,

I telegraphed you a few days ago to give lower prices for lamps to England for 10 & 32.¹ I did not think there was any chance of our people taking an order for the A & B—² The English people have however given them an order for about 7500 lamps A & B which they are making now— These lamps cost in the factory after all expenses including rent, insurance, etc, etc, etc, Ƒ1.50—³ The English people pay them Ƒ2.25 for them—

There are a great many complaints on the American lamps & people begin to ask for Paris lamps.⁴ I don't know whether the lamp factory sends us the best, but I wish you would tell Upton that <u>he must lose on this</u> European business <u>if he slights us</u>— I think you can only get any of this trade by having a <u>better</u> & <u>longer life</u> lamp than us and the lamps do not hold out the magnificent curve of life that I have brought over with me & blowed so much about   Yours truly

<u>"Batch."</u>

October–December 1883   302

⟨Write Batch that he better arrange to have a curve made somehow with our Lamps also to send us 25 of his best Lamps to make a curve   There is some error somewhere as Louisville Expost gave life 2000 hours 5200 Lamps burning tested photometer   E⟩[5]

ALS, NjWOE, DF (*TAED* D8337ZFB); a letterpress copy is in Cat. 1331:188, Batchelor (*TAED* MBLB3188). Letterhead of Société Électrique Edison. [a]"Paris, le" and "188" preprinted.

1. Batchelor cabled: "English going buy ten and thirty two candles here your price   better send them reduced price" (Batchelor to TAE, 17 Nov. 1883, Batchelor [*TAED* MBLB3184]). Edison had recently reduced from sixty to forty cents his price for large lots of 10-candlepower lamps to the Edison Electric Light Co., Ltd. The company placed an order but then evidently balked when Edison would not offer this rate on orders of fewer than 10,000 lamps (TAE to Edison Electric Light Co., Ltd., 12 Sept. and 8 Oct.1883, both DF [*TAED* D8316AUZ, D8316AYX]; TAE to Edison Electric Light Co., Ltd., 3 and 8 Oct. 1883; Edison Electric Light Co., Ltd., to TAE, 4 Oct. 1883; LM 2:20D, 22A, 21A [*TAED* LM002020D, LM002022A, LM002021A]). Edison also came under pressure at about this time not to undercut the Paris factory's prices for European orders (see Doc. 2553).

Edison had developed the 10-candlepower lamp in 1882 at the behest of the English company, which prized their economical operation. Since then, he recommended this type as "quite sufficient" to replace standard coal-gas jets in U.S. towns. Doc. 2339 n. 7; Samuel Insull to Spencer Borden, 11 July 1883, DF (*TAED* D8316AFT).

2. That is, A lamps designed for 110 volts and B lamps intended for 55-volt service. See Doc. 2085 n. 4.

3. Batchelor later made a detailed analysis of the production costs of lamps at the Ivry factory for the month of January, showing a unit cost of F2.09 per lamp. Cat. 1235:64–66, Batchelor (*TAED* MBN012064).

4. Occasional reports of lamp problems (most related to the plaster bases) are scattered throughout Sherburne Eaton's Defect Book Reports, 1883, DF (*TAED* D8328). Edison passed Batchelor's letter to Francis Upton, who assured him that "We have not slighted foreign orders. We have given them the same quality of lamps that are in use in this country. We think that we can stand any competition as to quality." Upton did, however, ask for samples of the French factory's lamps. Upton to TAE, 7 Nov. 1883, DF (*TAED* D8332ZDG).

5. The editors have not found a reply to Batchelor, nor have they determined the source of Edison's information about the Louisville lamps. A scientific jury made a rather rushed set of tests of incandescent lamps in the Exhibition's closing weeks, reporting on lighting efficiency and steadiness but not, apparently, on lamp life. "Electric Light Tests at the Louisville Exposition," *Science* 3 (4 January 1884): 14–15.

From Francis Upton

Dear Mr. Edison:

The price of frosted lamps has been reduced in all cases to a charge of five cents over the price of lamps unfrosted, no matter how large the order.

We are very sorry that the 32 candle power lamps should break so badly.[1] We will send Mr. Jenks five more to replace the broken ones[2]

We take special pains in packing these lamps. Yours Truly,

Francis R. Upton.

⟨Why dont you pack ½ doz Lamps & handle them Roughly in Laboratory ie[b] Experiment on different packing⟩[3]

Edison 32-candlepower lamp.

ALS, NjWOE, DF (*TAED* D8332ZCY). Letterhead of Edison Lamp Co. [a]"EAST NEWARK, N.J.," and "188" preprinted. [b]Circled.

1. Having four parallel filaments, these special lamps were likely especially vulnerable to breakage. Alglave and Boulard 1884, 160.

2. Upton evidently was answering a complaint (not found) about broken lamps received at Brockton, Mass. A few weeks later, Sherburne Eaton advised Edison of "much difficulty caused by the way in which lamps are packed. Many breakages are recently reported." The issue of packaging held considerable importance in view of the Edison Lamp Co.'s growing domestic business and its markets abroad for filaments (shipped to the factory at Ivry) and whole lamps. The fragile lamps survived rough handling in transit on at least one documented occasion (see Doc. 2128 n. 9), and Upton took pride in the factory's track record. Responding to a problem registered earlier by Batchelor, he claimed that in two and a half years of doing business, involving 3,000 packages, "We have never received a complaint of breakage of lamps due to poor packing," although Edison had gotten complaints on at least two prior occasions. Upton also defended the worker responsible for the task as one having "'brains over and above what are necessary for his existence,' as I have found him an exceedingly trustworthy man." Batchelor to Samuel Insull, 14 June 1883; Upton to TAE, 26 June 1883; Sherburne Eaton defect book reports, 3 Feb., 25 Aug., 24 Nov. 1883; all DF (*TAED* D8337ZBY, D8332ZBK, D8328D, D8328D3, D8328O, D8328T).

3. Edison's marginalia was the basis for a typed reply prepared by Alfred Tate a few days later. TAE to Upton, 29 Oct. 1883, DF (*TAED* D8316BDB).

Draft to Harry Leonard[1]

[New York, c. October 29, 1883[2]]

Write him we dont want any Auto Meter Switches, the system is complicated Enough now   Neither do we want the ~~custom~~ circuits thrown together so as to produce the result you speak of=[3] The big Engine & Regular ckt. must be run if even there is a single Customer on the circuit   Dont for a moment attempt[a] anything Else— There is no more difficulty in running the big Eng & 2 dynamos than one Dynamo & the Extra Cost due to friction is <u>nothing</u> compared to the[b] bad[c] results on meters=

The oxidation is[4]

100 mg on 25 light
 50 mg on 12
 25   6

ADf, NjWOE, DF (*TAED* D8346ZBG1). [a]Obscured by ink blot. [b]Obscured overwritten text. [c]Interlined above.

1. Harry Ward Leonard (1861–1915), a native of Cincinnati, graduated from the Massachusetts Institute of Technology in 1883. Carrying letters of reference from his professors and William Lloyd Garrison, Jr., he applied to Edison for work in June. Edison hired him to take charge of the Construction Dept.'s meter business. After a short stint at the Testing Room of the Edison Machine Works, where Leonard apparently helped prepare training materials (see Doc. 2500 n. 1), Edison posted him to the Brockton station at the end of September. In addition to installing and troubleshooting meters, Leonard's duties included providing twice-weekly reports to New York on the number of meters installed, "in whose house they are installed, on which side of the circuit the lights are and the number of lights each consumer is wired for." Leonard remained at Brockton until late November, when he was reassigned to Lawrence, Mass., and eventually to other Edison plants. After a stint as superintendent of the Western Edison Light Co. in Chicago, he held managerial positions at the Edison United Manufacturing Co. and then Edison General Electric Co. Leonard later enjoyed a successful career as an electrical inventor and manufacturer, notably of train lighting equipment and control systems for heavy-duty direct current machinery. *DAB*, s.v. "Leonard, Harry Ward"; Leonard to TAE, 18 June and 22 Nov. 1883; Leonard to Samuel Insull, 1 Aug. 1883; Insull to Leonard, 12 Oct. 1883; all DF (*TAED* D8313W, D8346ZBP, D8330ZAS, D8316BAK); Insull to Frank Sprague, 27 Sept. 1883, LM 15:225 (*TAED* LBCD2225); Leonard bio. file, Edison Assoc.

2. Edison drafted this response on the back of a 28 October letter from Leonard. The editors have not found a formal reply. DF (*TAED* D8346ZBG).

3. Edison had evidently solicited Leonard's opinions for improving the meter portion of the central station business (Leonard to TAE, 12 Oct. 1883, DF [*TAED* D8346ZAU]). In his 28 October letter (see note 2), Leonard ventured somewhat further to offer a suggestion for the operation of the three-wire system. He described how the Brockton station had largely segregated customers desiring light on Sundays and after midnight onto the B side of the circuit, so that current could be supplied to them using a spare dynamo and small engine instead of the regular equipment. However, some consumers on the A side occasionally wanted light during those periods as well. At such times, Leonard

explained, "We then close the switch w'ch converts the circuit into a two wire circuit by joining the + & − wires, making them both negative & making the neutral wire positive." This practice created no difficulties on the B side but would cause meters on the A side to run backward (cf. Doc. 2461 n. 4) so that those customers "are actually making money by burning the light." To overcome this problem, Leonard devised "an arrangement w'ch will always keep the current thro' the meter in one direction, no matter how the current may go. It is simple & I think efficient & cheap," as well as automatic. He asked Edison's opinion before continuing to develop the device (Leonard to TAE, 28 Oct. 1883, DF [*TAED* D8346ZBG]).

4. In his 28 October letter (see note 2), Leonard had asked about the results of experiments at the Machine Works on "<u>oxidation</u> and the curve of resistance of the meter solution."

−2543−

To William Dwelly, Jr.[1]

[New York,] November 1st. [188]3

Dear Sir:—

Your favor of the 31st. ulto. came to hand this morning, and the tone of it surprises me extremely.[2] You refer to Mr. Kruesi[3] subjecting you to delays as if he was doing so intentionally. Cannot you understand that it is impossible for his men to be of any use in Fall River untill he has the ~~tools~~ tubes[a] there ready to lay. We are just as anxious as you are to have the plant installed quickly, but we must protest most decidedly against the frequency and unreasonableness of your complaints.[4]

Allow me to suggest that it is quite premature on your part to ask on the first of November how we propose to get your plant running by the end of November. I would remind [you?][b] that you were warned before the contract[5] was closed that there would be considerable delay unless your Company guaranteed the cost of the tubes, so as to enable us to order same in anticipation of our closing the contract, and I would further remind you that you promised to send us such a guarantee and that you failed to do so, and any complaint that you now have to make as to the delay should be made to yourself for failing to send us the guarantee above referred [to?][b] and not against us.

I fully agree with you [-----][b] that[a] the importance having your tubes completed before frost sets in has been sufficiently enlarged upon and that a repetition is entirely unnecessary, and I would again beg of you to exercise a little patience, inasmuch as only twenty three working[c] days have elapsed since the contact with your Company was closed.

We are unwilling to give any such bond as you ask for. The work will be done in the best manner possible and will be started as soon as possible. The laying would have commenced at least five days ago if the guarantee from you as to the tubes had been sent at the time we asked for it.

To talk of Mr. Kruesi assuming the responsibility of putting off laying the tubes, seems to me an absurdity. It is to my interest and to Mr. Kruesi's interest that the laying should be commenced as early as possible, although my assertions to this effect do not seem to carry any weight with you.

As to none but competent men being employed, I have yet to find that it has been our habit to employ incompetent men, and the troubles that we met with in Lawerence and Brockton are such troubles as we fully expect to meet with at Fall River. If you can tell us how to avoid them I am sure we shall be very glad to receive your advise in the matter. Very truly yours,

TAE   I[nsull]

TL (carbon copy), NjWOE, DF (*TAED* D8316BDR). ªInterlined above by hand. ᵇObscured by smudge. ᶜInterlined above.

1. William H. Dwelly, Jr. (1859–1946) had worked for the Edison lighting interests in New England since 1882. He was an organizer and, at this time, the treasurer of the Edison Electric Illuminating Co. of Fall River, Mass. He held a similar position years later with the American Woolen Co. and was an officer of the Osceola Consolidated Mining Co., a major copper producer in Michigan. *Pioneers News* 1947, p. 50, Pioneers Printed; letterhead, Dwelly to TAE, 29 Sept. 1883, DF (*TAED* D8347I); Gas and Electric Light Commissioners 1917, App. A, 4a; United States Dept. of Labor 1914, 148.

2. Dwelly's letter followed a telegram the same day (not found) complaining that John Kruesi had "ordered his men back to New York for a period of ten days." Dwelly protested what he saw as Kruesi's summary abrogation of an earlier promise to begin laying underground conductors in Fall River as soon as the same work was completed in Brockton. He reminded Edison that municipal authorities would forbid excavation of the streets after frost had set in, and that they had demanded a $25,000 surety bond on the work in any case. He asked Edison to post the bond and to "see to it that none but competent men are employed in connecting and testing the tubes, so that there may not be a repetition of the troubles at Lawrence and Brockton," which involved digging up the same streets repeatedly. In an immediate reply to the telegram, Edison explained that "inasmuch as the tubes will not be ready for ten days, we do not see the necessity of keeping a lot of men laying idle in Fall River." He asked for Dwelly's "kind co-operation which you can extend to us by exercising a little patience and placing a little more confidence in our capability of keeping the promises we have made to you." Dwelly to TAE, 31 Oct. 1883; TAE to Dwelly, 31 Oct. 1883; both DF (*TAED* D8347T, D8316BDM).

3. John Kruesi (1843–1899), a highly skilled machinist, had worked for Edison for many years before becoming manager and treasurer of the Electric Tube Co. upon its formation in 1881. He had made a cost

estimate in July for installing underground conductors in Fall River and was generally responsible for the underground tubes in Edison's village plant systems. Docs. 659 n. 6 and 2058 n. 2; Samuel Insull to Spencer Borden, 11 July 1883, DF (*TAED* D8316AFT).

4. Dwelly's 31 October telegram and letter (see note 2) are the first complaints identified by the editors about progress at Fall River. Edison had promised, when he sent the final construction estimate to Dwelly in early October, that "the work will be pushed on as vigorously as possible," and that crews would begin laying the tubes ("the most troublesome part of the work") by the middle of the month. TAE to Dwelly, 3 Oct. 1883, DF (*TAED* D8316AXZ); cf. Doc. 2547.

5. An unsigned copy of Edison's 4 October contract with the Edison Electric Illuminating Co. of Fall River is in DF (*TAED* D8347J). The executed agreement was returned to Edison on 5 October. Dwelly to TAE, 5 Oct. 1883, DF (*TAED* D8347M).

-2544-

From Drexel, Morgan & Co.

New York Nov 5" 1883[a]

Dear Sir.

On July 1st 1882/ we loaned you $12,240. upon 100 Shares Edison Electric Light Co. Stock

On July 6th and Nov 6th 1882, we loaned you $25,000 on 846 shares Illuminating Co. Stock.

On January 2nd 1883, we further loaned $2,337.80 on 50 Shares Isolated Co. Stock.[1]

As one of our partners retires from the firm on January 1st next,[2] we are desirous of closing up our matters as nearly as possible previous to that date and would therefore, thank you, on or before Decr. 15th to pay us off the above loans.[3] Yours very truly,

Drexel Morgan & Co

L, NjWOE, DF (*TAED* D8302B). Letterhead of Drexel, Morgan & Co. [a]"New York" and "188" preprinted.

1. TAE promissory notes to Drexel, Morgan & Co., 1 and 6 July and 6 Nov. 1882, 2 Jan. 1883, Miller (*TAED* HM820164C, HM820164D, HM820167A, HM830168B); see also *TAEB* 6 chap. 6 introduction.

2. After Charles H. Godfrey retired on 1 January 1884, George S. Bowdoin (of Morton, Bliss and Co.) and Charles H. Coster (formerly of Fabbri & Chauncey) became resident partners of Drexel, Morgan & Co. Carosso 1987, 168.

3. Edison had made some payments on these notes but did not substantially reduce the principal balance until the first week of December, when he sent a check for $7,500. He paid the remaining balance of $35,306.78 (including interest) on 11 December. Drexel, Morgan & Co. to TAE, 4 Dec. 1883, DF (*TAED* D8302C); Drexel, Morgan & Co. to TAE, 10 Dec. 1883, Miller (*TAED* HM830204A); Edison Cash Book (1 Jan. 1881–30 Mar. 1886), p. 221, Accts., NjWOE.

–2545–

Samuel Insull to
Western Union

[New York,] November 5th. [188]3

Dear Sir:—[1]

Will you please have an American District Telegraph call[2] put in Mr. Edison's apartments at the "Clarendon," 18th. Street and 4th. Avenue.[3] If the man you send will ask for Mrs. Edison, she give him instructions as to where the call is to be placed. Yours truly,

SI   P[ersonal]. S[ecretary].

TL (carbon copy), NjWOE, DF (*TAED* D8316BEE).

1. Insull addressed this letter to the manager of the Western Union office at 852 Broadway, near Fourteenth St.

2. American District Telegraph was founded in 1872 to exploit Edward Callahan's district and fire alarm telegraph. Its call boxes allowed home and office subscribers to transmit signals to nearby messengers, private police, and similar respondents. Edison made extensive use of the messenger services of American District and similar companies. Doc. 226 n. 2; Israel 1992, 107–11; American District bills (1883–1884), Vouchers.

3. Edison was living at the Clarendon Hotel by the beginning of October (TAE to R. Tobin, 4 Oct. 1883, Lbk. 17:281 [*TAED* LB017281]). He paid $180.00 weekly for the family's board and rooms, though charges for extra meals, laundry, and external services like cables and cabs often brought the bill above $200 (Clarendon bills [1883–1884], Vouchers). The five-storied Clarendon was opened in the Union Square district of New York in 1851 and soon came under the management of Gerrit Kerner. Kerner's son Charles took over in 1862 and managed the hotel until 1893 ("Clarendon Hotel to Lease," *NYT,* 8 Apr. 1893, 9). According to an 1884 guide, the Clarendon offered

The Clarendon Hotel,
the Edison family's new
residence in October 1883.

accommodation for upwards of one hundred and forty guests, and many of whom are permanent residents here for extended periods. It is provided with a first-class passenger elevator to all floors, inclusive of every possible convenience. Its offices and chief reception rooms are artistically decorated, while its spacious dining hall on the 18th Street side is a splendid apartment, richly frescoed and decorated with mirrors, etc. The whole establishment is in keeping with the requirements of a strictly first-class metropolitan hotel. [Evans 1884, 282]

–2546–

To Frank Marr

[New York,] Nov. 7th. [1883]

Dear Sir:—

I have your favor of the 3rd. of November.[1] With relation to the charge for 32 candle lamps, I would beg to say that the average lighting time for the 365 days in the year is at 6 o'clock in the evening. Hotels want the light until 2 A.M. This makes 8 hours at 1¼ cents per ten candles per hour. The amount would be a 3³/₄ cents per hour or 30 cents for the 8 hours. This would give you about $110 for the 365 days of the year. If you charge $50 per year it would be at the rate of about 1¹/₁₆ per thousand feet of coal gas. You might make a contract at this rate only by the month with the privilege of stopping when you get loaded up and have an opportunity of taking better consumers. I agree with you that the extension of your mains should take place before the cold weather comes. If you will please have your meter man mark upon the charts sent herewith exactly where the present mains run and where the feeders run [--][a] and also mark on the chart where you propose running your new mains, I will have calculations made and let you know exactly what size of copper wire should be run for the extended circuit.[2] I cannot understand why your lamps arc so.[3] Very truly yours,

TAE   T[ate]

TL (carbon copy), NjWOE, DF (*TAED* D8316BEU). [a]Canceled.

1. Marr reviewed the business in Sunbury, particularly Edison's plan to sell 32-candlepower lamps to hotels for $4 apiece. Hotels were "very anxious to have them," but not on Edison's terms. Hotels, the plant's largest customers, already had large bills and some had threatened to turn out their electric lights, which Marr suggested would cripple the local company and damage Edison's prospects beyond Sunbury. He proposed to make a public streetlight demonstration with some of the 32-candlepower lamps and then to offer them to hotels on an annual contract for $50. By that time, Marr noted, "we will have our entire plant in & we can get along without them." Marr to TAE, 3 Nov. 1883, DF (*TAED* D8361ZDN).

2. Marr later supplied a plan for a proposed fifty-light extension. Edison queried whether he should estimate for conductors sufficient to run only the new total of 650 lights or "calculate our copper so as to carry more lights in the future." Marr to TAE, 15 and 23 Nov. 1883, 15 Jan. 1884; TAE to Marr, 27 Nov. 1883; all DF (*TAED* D8361ZDR, D8361ZDS, D8458C, D8316BKR); Alfred Tate to Marr, 28 Dec. 1883, LM 17:276 (*TAED* LBCD4276).

3. Marr reported in his 3 November letter (see note 1) that despite having the system run at 105 volts, in one night eight bulbs had "arced & burn out the plugs— This has not happened to us for some time." Edison forwarded his complaint to Francis Upton (TAE to Upton, 7 Nov. 1883, DF [*TAED* D8316BEN]). Lamp breakage had plagued the Sunbury station throughout the fall, with about 100 lamps failing each month. Edison blamed the local company for deliberately exceeding the capacity of the 10-candlepower lamps. The company reduced voltage to the point that disgruntled customers used gas to supplement their electric lights, but when this did not solve the problem, Edison dispatched John Howell (about 6 November) from the lamp factory to investigate. Edison eventually ascribed the problem, without direct proof, either to

some customer lines having been erroneously connected to the feeder conductors or to the station operator's inattention; Howell reportedly blamed improperly installed voltage indicators (Frank McCormick to TAE, 23 Oct. and 5 Nov. 1883; TAE to McCormick, 26 Oct. and 9 Nov. 1883; Thomas Conant to TAE, 4 Nov. 1883; TAE to Marr, 13 Nov. 1883; Marr to TAE, 3 and 15 Nov. 1883; all DF [*TAED* D8361ZDK, D8361ZDO, D8316BCW, D8316BFH, D8360ZCC, D8316BFQ, D8361ZDN, D8361ZDQ]; see also Doc. 2538 [headnote]).

–2547–

To Archibald Stuart

[New York,] November 7th. [1883]

Dear Sir:—

I am in receipt of your telegram of the 5th. inst. asking why definite work has not been begun as promised, such delays are ruinous.[1] You must really leave to me to decide when I shall send a man to a certain town to start the work locally. Immediately a contract is signed that moment the work commences although it may not be visible in the town where the plant is to be installed. We have quite an amount of machinery, wire &c. to get out before it is possible to do much work locally. This has been proceeding and I do not think you have any cause for complaint as to delay at Tiffin. As matter of fact our Supt., Mr. Rich, left Fall River Monday night for Tiffin, and signs of work will be visible there in a few days.[2] Very truly yours,

TAE   I[nsull]

TL (carbon copy), NjWOE, DF (*TAED* D8316BEO).

1. Stuart had wired from Cincinnati, "Why hasnt Tiffin work begun as promised   such delays are ruinous." Stuart's inquiry was evidently prompted by a telegraphic alert the same day from E. B. Hubbard, the Tiffin company's secretary, that work had not begun. Stuart noted on Hubbard's telegram: "Receive this about once a day on the average." Edison replied quickly to assure Stuart that "Our Superintendent leaves tomorrow for Tiffin to start work there— There is no delay— We are pushing the matter." Edison had executed a standard agreement with the local Tiffin company on 24 October promising to use "due diligence" in constructing the plant. Stuart to TAE, 5 Nov. 1883; Hubbard to Stuart, 5 Nov. 1883; both DF (*TAED* D8356ZAB, D8356ZAA); TAE to Stuart, 5 Nov. 1883, LM 16:74 (*TAED* LBCD3074); TAE agreement with Edison Electric Illuminating Co. of Tiffin, 24 Oct. 1883, Miller (*TAED* HM830197); cf. Doc. 2543.

2. Samuel Insull prepared a letter of introduction for William Rich to J. F. Bunn, president of the Tiffin company, on Monday, 5 November. Soon after, Edison told Bunn the installation would likely be completed in twenty to twenty-five days. Stuart remained anxious enough to suggest starting the station with only about half the total lights wired, but Edison demurred: "This is not a biz where the iron is to be struck while

hot   its a biz that must go on for ever hence must be started carefully."
Insull to Bunn, 5 Nov. 1883, LM 16:78 (*TAED* LBCD3078); TAE to
Bunn, 9 Nov. 1883; Stuart to TAE (with TAE marginalia), 12 Nov.
1883; both DF (*TAED* D8316BFI, D8353ZAO1).

-2548-

From Demas Barnes

New York Nov 7 1883[a]

Dear Sir

You are doubtless aware that I am investigating Electric
rail-way & c. with view to negotiation with yr company. It will
facilitate me if you will briefly give me your best opinion as to
the ~~their~~ methods of applying Electric power to cars.

In the line of mutual interest accept my thanks and believe
me   Yours Truly

Demas Barnes[1]

⟨answer & say will give answer soon. Keep this letter before
me so I can answer soon   E[b]

my opinion on the application of Electricity to cars is that
the motor will in most cases be placed directly under the cars
~~each~~ and that for small roads such as street roads, the Elevated,
& underground roads running from 1 to 10 miles that it is the
coming method, all that is requisite is to put one road in operation so as to gain actual Experience and a chance to surmount
what difficulties & objections there may be found   TAE⟩[2]

ALS, NjWOE, Scraps., Cat. 2174 (*TAED* SB012BCL). Letterhead
of Demas Barnes. [a]"New York" and "188" preprinted. [b]Followed by
"over" to indicate page turn.

1. Demas Barnes (1827–1888), a journalist, politician, and businessman, was a director of the Long Island Railroad Co. His original fortune
was made in the patent medicine business, with products such as Mexican Mustang Liniment and Hagan's Magnolia Balm. He also had been
a Democratic representative from Brooklyn in the 40th Congress, a
founding trustee of the Brooklyn Bridge, and a cofounder of the short-
lived Brooklyn *Argus*. *BDUSC*, "Barnes, Demas"; New York (State)
Board of Railroad Commissioners 1889, 33; Obituary, *Brooklyn Eagle*,
2 May 1888, 4; "Demas Barnes's Death," *NYT*, 2 May 1888, 5.

2. The typed reply sent to Barnes two weeks later was based closely
on Edison's marginalia. TAE to Demas, 22 Nov. 1883, DF (*TAED*
D8316BJF).

[New York,] November 8th, [188]3

To Charles Small[1]

Dear Sir:—

Replying to your letter of the 5th. inst.,[2] I beg to say that it would give me great pleasure to comply with your request and deliver a lecture for the benefit of the Members of your Society[3] were I possessed of the powers of oratory requisite for such an occasion. The fact of the matter is I never delivered a lecture in my life.[4] You of course understand that there is a very great difference between talking and inventing, and though I have made rather a success of the latter the former is an accomplishment which I do not possess, and I must therefore ask you to very kindly accept my regrets together with my best wishes for the welfare[a] of your Society. Very truly yours,

TAE   T[ate]

TL (carbon copy), NjWOE, DF (*TAED* D8316BEW). [a]Obscured overwritten text.

1. Charles S. H. Small was born in England about 1852. He had been assistant superintendent of the Gold & Stock Telegraph Co. for several years, but he wrote to Edison as president of the New York Electrical Society. Reid 1879, 626; U.S. Census Bureau 1970 (1880), roll T9_856, p. 445.4000, image 0074 (Brooklyn, Kings, N.Y.).

2. Small had written Edison, "Many members of the NY Electrical Society have expressed wishes to hear you deliver a lecture." He wished for a "good lecture from a popular man to give us a good send-off this season." DF (*TAED* D8311H).

3. Edison was a charter member of the New York Electrical Society, founded in March 1881. He served as third vice president in 1885 and as a member of the committee on entertainment in 1887, by which time the Society had become the Electrical Section of the American Institute. The Society met at Edison's West Orange Laboratory in 1888 and heard a lecture by Arthur Kennelly, who then had charge of the electrical laboratory there. Edison resigned from the organization in 1894. New York Electrical Society to TAE, 5 May 1881 (enclosing Constitution and Bylaws of 2 Mar. 1881), 13 Mar. 1885, 29 June 1887, 21 Nov. 1888; all DF (*TAED* D8112B, D8112C, D8511B, D8711AAL, D8812ABD); TAE to New York Electrical Society, 29 Jan. 1894, Lbk. 59:61 (*TAED* LB059061).

4. Edison wrote atop Small's letter, "I never did such a thing in my life—in fact I couldnt." Edison's first attempt at a lecture had occurred in 1868 when the principal of a private girl's school in Boston invited him to give a talk on the Morse telegraph. Edison recalled that "I never was so paralyzed in my life; I was speechless, there were over 40 young ladies from 17 to 22 years, from the best families. I managed to say that I would work the apparatus and Mr. [Milton] Adams would make the explanations," although in the end Edison did successfully address the students (*TAEB* 1, App. 1.A.17). Twenty years later, when he was invited to appear on the stage where a professor from the University of the

City of New York was to lecture on the phonograph, Edison declared that he "wouldnt face an audience for 100 dollars" (Doc. 1263). Nonetheless, he did occasionally demonstrate apparatus at public events even if he would not address the attendees (cf. Docs. 690 n. 2, 692 n. 3, 1217 n. 2, 1324 n. 5). The only other prior instance of Edison delivering a lecture occurred at the 1878 annual meeting of the American Association for the Advancement of Science, where he read a paper on his tasimeter experiments during a solar eclipse. He prepared three other papers for this meeting, but delegated their presentation to George Barker. Barker and Francis Upton also delivered Edison's papers at the Association's meeting the following year (see *TAEB* 4, chap. 4 introduction and Docs. 1796 n. 1 and 1810).

–2550–

From Francis Upton

EAST NEWARK, N.J., Nov. 8 1883[a]

Dear Mr. Edison:

I demand that definite instructions be given regarding the responsibility in the financial concerns of the Lamp Co, as regards meeting various obligations as they become due.

I am driven to this by the events of this week.

For two weeks I have been preparing to meet the payments that I knew were to fall due this week by hurrying forward all shipments that would bring money, and by not drawing money with the understanding that efforts would be made to give me funds this week.

Yesterday I found that I was completely done for. When I asked Mr. Hastings for money, he said that he had it but he had instructions from Mr. Insull not to pay any out to any of the Edison concerns until Saturday.[1] On going to Mr. Tait I found that no provision had been made for us on the agency account.[2] This was in the face of the following liabilities not provided for

Wednesday pay roll	1400.00
Thursday note	262.21
" "	2562.50
Friday Interest	900.00
	$5124.71

This has compelled me to draw my money to cover the coming note for $2562.50 as the check from them is not good for three days. I shall also be compelled to draw my check for the Interest due Tomorrow $900.

This is in addition to putting off our pay roll and creating dissatisfaction among our help.

I had to draw on

Due from Hastings	6280.44
Due from T.A.E. Agent	5736.94
	$12,017.38

All I was asking for was $2300—as I was willing to take care of the Corning note[3] and we could draw on our bank balance for the small note.

After such treatment I am compelled ask for decided action. I refuse to have anything to do with providing money for running the Edison Lamp Co., unless the following conditions are agreed upon in writing.

That money due the Lamp Co. from Mr. Hastings is under my control, the same as if [--][b] the Lamp Co. were outside Company.

That the money due from T. A. Edison Agent be paid without request a definite numbers of days after ~~ordering of~~ the getting of bills of lading.

If these conditions are not complied with you will have to make such arrangements as to relieve me of all financial responsibility as regards collecting money or meeting liabilities.

If ~~it~~ these conditions are granted I will become absolutely responsible and personally liable to meet any and all obligations incurred in running the business. In case the business grows too fast or runs behind I will also agree to give sixty days notice of the need of additional money.

In case we have funds and you need them on formal notice to the Lamp Co. to that effect I will agree to do everything in my power to help you.

The amounts of money now at stake here are so large and I am so heavily interested in this company, having more than half of all I have invested here, that I am compelled to ask for a decided action on your part fixing responsibility. Yours Truly
Francis R. Upton

ALS, NjWOE, DF (*TAED* D8332ZDH). Letterhead of Edison Lamp Co. [a]"East Newark, N.J.," and "188" preprinted. [b]Canceled.

1. Saturday was 10 November. Upton referred to Frank Hastings's role as treasurer of the Edison Electric Light Co. Because Insull was in Providence, R.I., Tate telegraphed him on 7 November stating that Upton wanted $1,400 for payroll. Noting that he had received Tate's telegram too late to respond in kind, Insull wrote back the same day telling him to have Hastings give Upton $1,400 "if Upton has bills in" (Tate to Insull, 7 Nov. 1883; Insull to Tate, 7 Nov. 1883; both DF [*TAED* D8332ZDF, D8322ZDD]). As Tate later reported to Insull, Upton needed $1,800 for his payroll and a note, but

Hastings said he could only let him have about $1,000.00 but as this would not meet his <u>pay roll</u> and the other matters as well I considered it best to only give him enough to meet note &c and allow the pay roll to wait till Saturday as he could'nt pay <u>part</u> of it without paying all— Besides he'd have simply appropriated the extra $500.00 and have come around here on Saturday howling for pay roll money thereby getting $500 ahead of us which, considering we wont be able to strike the Lt Co. for much money on Saturday on a/c of notes they have to meet, I consider would be a bad piece of financing! [Tate to Insull, 8 Nov. 1883, LM 16:119 (*TAED* LBCD3119)]

2. Upton referred to Edison's capacity as a purchasing agent for orders, such as those for village or foreign plants, placed through Edison or his office rather than through the lamp works.

3. The Corning Glass Works of Corning, N.Y., supplied bulbs and tubes to the Edison Lamp Co. and was the company's principal creditor. According to monthly statements and related correspondence for the years 1882–1884, the Lamp Co.'s monthly balance to Corning typically fluctuated between two and six thousand dollars but stood at nearly twenty thousand at the end of 1883. Doc. 1971 n. 1; R. G. Dun & Co. report, 18 Dec. 1883, Dun (*TAED* X112CAV); Edison Lamp Co. report, 2 Jan. 1884; Electric Light—Edison Lamp Co.—Accounts; Electric Light—Edison Lamp Co.—General; all DF (*TAED* D8429D, D8231, D8332, D8429).

–2551–

Technical Note:
Electric Lighting

[New York, November 9, 1883?[1]]

[A][2]

[B]

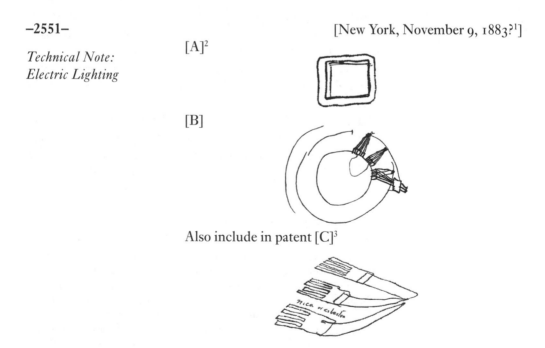

Also include in patent [C][3]

[D][4]

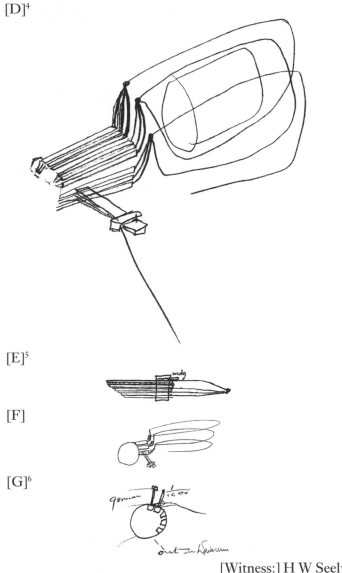

[E][5]

[F]

[G][6]

[Witness:] H W Seely

X, NjWOE, Cat. 1149, Lab. (*TAED* NM018AAY).

1. Henry Seely signed and dated the page when he received it from Edison on this date.

2. This document pertains to two closely related patent applications that Edison executed on 15 November (both issued on 20 May 1884). Each one concerned means to suppress sparking at the commutator, which had continued to cause problems through the summer and into the fall (see Doc. 2529 esp. nn. 4, 6). Edison had evidently tried these ideas before the end of October, when William Andrews (in Massachusetts) relayed to Samuel Insull a report that "Mr Edison has struck a

good thing to get rid of sparking at commr." Andrews to Insull, 29 Oct. 1883, DF (*TAED* D8348ZAI).

Edison focused on the local circuit formed briefly through a collector brush between two adjacent commutator bars. One approach, counter to his previous efforts to increase conductivity through the brushes, was to increase the resistance in the local circuit. Edison proposed doing this by using a commutator brush composed of thin metallic leaves, each insulated from the others by mica or asbestos, and connected together only through a resistance such as german silver. He explained the effect this way: "The local current . . . encounters the resistance of these separate conductors, having to pass through them to and from their point of connection together; and, therefore . . . [it] will be weak and the breaking at the surface of the cylinder of the local circuit will produce little or no spark." A more complete version of Figure A in the patent represented a cross-section of the compound brush, which is shown in Figure C. Figure E is a side view; the separate conductors are joined by solder only at the right edge. U.S. Pat. 298,954.

Edison had executed an application in October for a means to reduce sparking by applying a counter electromotive force through the local circuit at the instant the commutator brush moved from one bar to the next. U.S. Pat. 293,432.

3. Figure label is "mica or asbestos."

4. Figures D and F represent an alternative method of increasing resistance in the local collector circuit, embodied in the second patent application of 15 November. Instead of dividing the collector brushes, Edison divided each commutator bar into three sections, each insulated from the others but connected to the same armature coil. This configuration would provide the highest resistance "when the brush leaves a bar" because "only one connection [to the armature], and consequently the highest resistance, is interposed in the local circuit at the moment when the current-collector leaves the bar." U.S. Pat. 298,955.

5. Figure label is "wedg[e]."

6. Text is "German" [silver], "$^1/_{1500}$," and "out in Drawing."

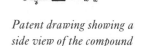

Patent drawing showing a side view of the compound commutator brush with its clamping sleeve **D**. *Individual conducting strips* **b** *are connected to strips of higher resistance,* **d**, *which are united only at the edge* **d′**.

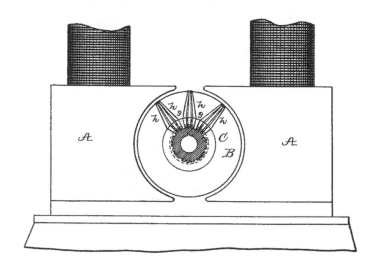

One drawing in Edison's U.S. Patent 298,955 incorporated his sketched Figure B from this document, representing the division of each commutator bar into three separate connectors between the armature coil and commutator

To John Tomlinson[1]

Dear Sir:—

I hereby agree to pay you a monthly salary of Four Hundred and Seventeen dollars ($417) from this date in consideration of your giving your whole time and attention to the making up and preparing the case of my eletric Lamp against Infringers—[2] The salary above stated to be discontinued at my option upon my giving you three (3) months notice.

I further agree (if I am able to prevail upon the Edison Eletric Light Company) to allow you to conduct the case in the Courts even if other Counsel are brought into the case to make arguments in Court & to pay you the sum of Five Thousand dollars or in lieu thereof Fifty shares (50) of the Capital stock of the Edison Eletric Light Company at your option providing you win the suit so that my patent on the filiament Lamp will be construed good against Infringers for a period at least twelve years from the date of issue of patent no 223.898[3] Yours Truly

L, NjWOE, Miller (*TAED* HM830198). Letterhead of Thomas A. Edison. [a]"NEW YORK," and "188" preprinted.

1. John Canfield Tomlinson (1856–1927), a prize-winning undergraduate orator, received bachelor's (1875), bachelor of law (1877), and master's (1878) degrees from the University of the City of New York (later New York University). A principal in the New York law partnership of Ecclesine & Tomlinson, he represented Edison for several years in a variety of business and personal matters, including the Seyfert lawsuit (see Doc. 2662). Tomlinson seems to have entered Edison's orbit about the middle of 1883, when his firm investigated the United States Electric Lighting Co.'s Brooklyn Bridge plant and several other installations. Starting in August, he also represented Edison's interests in legal actions related to the firing of Charles Dean and Charles Rocap from the Edison Machine Works. "J. C. Tomlinson Dies from Heart Attack," *NYT*, 29 Oct. 1927, 17; Ecclesine & Tomlinson to James Russell, 27 June 1883, DF (*TAED* D8327ZAJ); *General Alumni Catalogue* 1916, 48.

2. Tomlinson accepted Edison's offer the same day. Over the next several years, he contributed to the defense of numerous and varied Edison patents in the United States, Europe, and South America, sometimes on behalf of Edison companies. Tomlinson to TAE, 13 Nov. 1883, Miller (*TAED* HM830199); see *TAED* s.v., "Tomlinson, John Canfield."

3. Edison received U.S. Patent 223,898, his first and most fundamental patent on the carbon filament incandescent lamp, in January 1880. The specification was involved in at least one Patent Office interference at this time, a proceeding initiated in December 1881 on behalf of an application filed in July 1881 by Brooklyn inventor Walter K. Freeman. The interference was suspended and Freeman's application rejected in 1885. Commissioner of Patents to TAE, 8 Dec. 1881, Pat. App. 223,898;

Commissioner of Patents to Freeman, 10 Mar. 1885, also related correspondence and Freeman's application, all printed in *Edison Electric Light Co. v. U.S. Electric Lighting,* pp. 2530–41, Lit. (*TAED* QD012E2531).

The period of twelve years specified by Edison had to do with the unusual administrative and legal history of the patent after it was issued. Between the time he filed the application and the date the patent was awarded, Edison submitted similar specifications in Great Britain, Canada, Belgium, Italy, and France. He did not acknowledge the foreign filings to the Patent Office while his application was pending there, having apparently been advised that he need not do so. A subsequent decision in an unrelated case in federal court appeared to contravene this advice, however, and raised the question whether a patent issued under such circumstances might be voided. Edison accordingly petitioned the Patent Office on 15 November 1883 to abbreviate the life of his patent to have it expire with "the foreign patent, having the shortest term, dated prior to the date of the United States Patent." The Patent Office acceded, issuing a Certificate of Correction to this effect on 18 December. (This action, and the relation of Edison's U.S. patents to those in other countries generally, became an issue in 1886 in the Edison Electric Light Co.'s suit against the United States Electric Lighting Co.) TAE to Patent Office Commissioner, 15 Nov. 1883, printed in *Edison Electric Light Co. v. United States Electric Lighting Co.,* defendant's depositions and exhibitions (Vol. IV), pp. 2219–20; United States Electric Light Co. amended plea, 4 Aug. 1886, complainant's proofs (Vol. 1), pp. 16–21; both Lit. (*TAED* QD012E2219, QD012B0016); U.S. Patent Office Certificate of Correction, 18 Dec. 1883, Pat. App. 223,898.

The alteration stood until March 1893, when the Edison Electric Light Co., with Edison's concurrence, again petitioned the Patent Office. Citing more recent court decisions (including *Edison Electric Light Co. v. United States Electric Lighting Co.,* alleging infringement of this patent), the firm argued that the original petition had been in error and that in cases in which the patentee had already accepted the specification, the Commissioner had no authority to change the lifetime except by surrender and reissue of the patent. The Patent Office again agreed; it canceled the Certification of Correction and restored Edison's patent to its original term. Edison Electric Light Co. petition to Patent Office Commissioner, 7 Mar. 1893; U.S. Patent Office recision of correction, 13 Mar. 1893; both Pat. App. 223,898.

–2553–

To Charles Batchelor

[New York,] November 14th. [188]3

My Dear Batchelor:—

Your letter of the 30th. inst. came to hand this morning.[1]

I do not see why we should keep up our prices on lamps if it is going to ruin our lamp sales in Europe,[2] and I shall therefore be very careful what I say in reply to Porges letter, which you say is to come here.[3] Anyway I should by no means be disposed to accede to his request, considering that he is bidding for the English lamp trade as against us.

He With[a] relation to the 100 light dynamo we sent to Berlin, we have not up to this writing made any single core machines.[4] We propose to do so at some future time, but how it will interfere with the manufacture of the German dynamos I cannot see. Our policy is, when we design and build a machine, to be very careful to sell all machines of a given design before we bring out anything better. By this means we save ourselves being "stuck" with a lot of obsolete machines. We have sold in this country all the old forms of dynamos and our new types are now coming out, and we are none the worse off financially for having brought out such new designs. It would be ridiculous, as you well know, to suppose that we could build a machine and say that there will never be any alteration. Your Ivry friends do not seem to understand this matter. Do they think that simply because we are going to turn the 100 light machine into a single magnet machine, that we propose throwing out away[a] all those that now on stock[b] we have built with the double magnet? At this writing we have about fifteen double magnet machines of the 100 light type at Goerck Street, and they will all be sold before we make any single magnet machines. We follow this policy out right through our business. The first "H" machines that were made would run only 350 lights and were of the long magnet type. We built about twenty of them. Then we get out the improved "K", on which we can run 450 lights with ease. The whole of the former were sold with the exception of two, and these two we have just managed to dispose of without any sacrifice whatever, and at the same price at which we sold the improved "H", to a very obliging customer who wants 700 lights and nothing more, although he can get 800 for the same figure.

I am delighted to hear that you propose[c] returning, and the sooner you do so the better.[5] The prospects are that we shall have plenty of work [now?][d] and very few of us to do it. Yours very truly,

TAE   I[nsull]

TL, NjWOE, DF (*TAED* D8316BFZ). [a]Interlined above. [b]"now on stock" interlined above. [c]Obscured overwritten text. [d]Interlined above by hand and copied off top of page.

1. Batchelor wrote after a "stormy" meeting of the Société Industrielle et Commerciale Edison, saying Charles Porgès was "simply crazy" and "very insulting" to him over the issue of Edison's sale of lamps to Germany at prices lower than those offered by the Société's own factory. (Another source of friction was the dynamo models recently sent to Germany, which Porgès feared were now obsolete.) Batchelor advised

Edison that Porgès would "write a letter to you or Eaton to have you stop sending lamps except at the same price as we do in France, which of course would stop all American, as there are extra expenses incurred on yours." In a different context in 1882, Porgès had complained that the price of Edison lamps from the United States was too high. Batchelor to TAE, 30 Oct. 1883 and 22 Nov. 1882, DF (*TAED* D8337ZFG, D8238ZEJ).

2. Seeing the French firms as a gateway to the lamp market for the whole of Europe, Edison had asked Francis Upton at the end of August to give his full "views as to offering lamps to the French Company at a permanently reduced price, in order to control the whole of their trade." Upton came out "strongly in favor" of doing so, even if it meant selling lamps at cost. By analogy with the textile industry, he argued that high volume at the Harrison factory was the key to profitability, standardization, and suppressed competition. "The demand from Europe," he added, "comes exactly at the right time. It is ahead of the demand here and comes in our dull season." The Edison Lamp Co. had been supplying lamp carbons to the French factory at Ivry for some time, though Upton's deviations from a fixed price schedule nettled both Edison and Batchelor. A few weeks later, Edison directed the shipment of 600 carbonizing forms to Ivry, presumably for the manufacture of filaments there. TAE to Upton, 30 Aug. 1883; Upton to TAE, 31 Aug. 1883; Batchelor to Samuel Insull (with TAE marginalia), 29 Aug. 1883; all (*TAED* D8316AQP, D8332ZCM, D8337ZDQ2); TAE to Edison Lamp Co., 27 Aug. and 8 Dec. 1883, Lbk. 19:204, 224 (*TAED* LB019204, LB019224).

3. The editors have not found the letter from financier Charles Porgès (1836–1906?), president of Compagnie Continentale Edison and a founding director of Société Électrique Edison (Docs. 2114 n. 4 and 2133 n. 2). A Parisienne of Czech-Viennese origins, in 1881 Porgès had assembled the working capital for Edison's three French electric companies. Strengthened by the backing of his Banque Centrale du Commerce & de l'Industrie, the original syndicate included his cousin, diamond merchant and art collector Jules Porgès, and the banking house of Ephrussi & Porgès, with which another cousin (Theodore Porgès) was associated. Many among these original subscribers were, like Porgès, representatives of prominent Jewish banking houses, including Ernest Cassel, Seligman Frères, and Jacob Landau. According to Batchelor, Porgès contended at the end of 1883 that he and the Banque Centrale du Commerce & de l'Industrie together "held as much interest in the [French] companies" as Drexel, Harjes & Co., another of the original subscribers. Sherburne Eaton "Memoranda on the Proposed International Edison Company," 5 Oct. 1883; Batchelor to TAE, 25 Dec. 1883; Porgès report to Banque Centrale du Commerce & de l'Industrie, n. d. [1881]; Société Électrique Edison incorporation papers (p. 7), 2 Feb. 1882; all DF (*TAED* D8331ZAD1, D8337ZFZ, D8132ZCJ, D8238Q); Fox 1996, 185, 193; Listes éelectorales de Paris et ses environs, 1891, accessed through Ancestry.com, 11 Mar. 2009.

Batchelor had an enduring dislike of Porgès, whom he recently had described as "the most disagreeable man I ever met" (see Docs. 2196 and 2392; Batchelor to TAE, 20 Aug. 1883, Batchelor [*TAED* MBLB3133A]; also his correspondence in Electric Light— Foreign—

Europe, DF [D8436]). Particularly telling of his regard for Porgès is his unflattering characterization of Jewish businessmen in Batchelor to TAE, 9 Dec. 1883, DF (*TAED* D8337ZFT).

4. Edison meant by "single core machines" those with a single magnet, or pair of cores.

5. In the final sentence his letter of 30 October to Edison (see note 1), Batchelor wrote: "If they dont arrange quick to make a long talked of Central Station here I shall be home soon." Edison commented: "The sooner you return the better." Batchelor had also written to Samuel Insull on 30 October that he expected "to be free here shortly & come back." Batchelor to Insull, 30 Oct. 1883, DF (*TAED* D8337ZFF).

–2554–

To Joshua Bailey

[New York,] November 15th. [188]3

Dear Sir:—

Referring to cables elsewhere confirmed, I beg to inform you that I have cabled to the Societe Electricque Edison, Paris, stating that I have not got a man whom I can send to start the Edison Lamp Factory in Germany.[1] So many of our men have gone to Europe that, in justice to our Home Company here and our manufacturing interests, it is imperative that we should keep what good men we now have to do our work in this country. We have a great deal of it and very few people to do it. It certainly seems to me that the Ivry Factory ought to be able to supply the man required.[2] Very truly yours,

TAE   I[nsull]

TL (carbon copy), NjWOE, DF (*TAED* D8316BGL).

1. Bailey had cabled Edison the previous day to "Answer about german Lamp factory" (see Docs. 2487 esp. n. 12 and 2499 esp. n. 5); the antecedent cables to Edison have not been found. Bailey to TAE, 14 Nov. 1883, DF (*TAED* D8337ZFJ1); TAE to Société Électrique Edison, 7 and 15 Nov. 1883, LM 2:26C, 30C (*TAED* LM002026C, LM002030C).

2. Concerned by the plan to dispatch James Hipple from Paris to establish a German lamp factory, Bailey had recently pressed on Edison the "desirableness of having a competent man who knows all the latest points come from your own factory." He worried about capturing the German market in the face of a nascent lamp production enterprise by the Siemens interests. Bailey conceded that he was

> not an expert in lamps as you know, but I can certify from the reports that I pick up all about the continent, that the lamps turned out from the French factory do not on the whole give satisfaction; some of them are excellent but the quality is very unequal. Hipple has now been absent from the States so long that he is entirely out of the current of improvements and excepting the stray information that gets over to Paris, is working an old stock of ideas. In the

next place he speaks no German, and it is an absurdity to send a man here to instruct Germans who cannot say a word to them except by an interpreter. [Bailey to TAE, 21 Oct. 1883, DF (*TAED* D8337ZFA)]

–2555–

To Deutsche Edison Gesellschaft [1]

[New York,] November 15th. [188]3

Dear Sir:—

Referring to our cables confirmed elsewhere, the statement that the Sawyer-Mann people have beaten me in a patent suit is entirely untrue.[2] Your Patent Attorney will explain to you that the procedure in the American Patent Office is that if two applicants cover the same point in their application for a patent, the matter is thrown into "interference", and the merits of the case decided by the Patent Office, and if they are appealed against the decision is revised by a higher court. The Board of Examiners who are the scientific part of the Patent Office decided the Sawyer-Edison interference in my favor. The Sawyer people then appealed to the Commissioner of Patents,[3] who reversed the decision of the Board of Examiners. We have now appealed to the Secretary of the Interior,[4] but have no idea who will win. Anyway whoever wins, the case will be carried to the Supreme Court. There is really no great question involved in the application, and we fight it out on the general policy that wherever we have dispute in the Patent Office we carry the matter to the highest possible Tribunal to obtain a settlement. The Sawyer interference refers to matters in relation to paper carbon, the use of which is entirely obsolete, and if I gain the case the result will simply be a little gain of prestige on my part, inasmuch as I should never think of using paper for my loop in the present advanced state of the art.[5] Very truly yours,

TAE   I[nsull]

TL (carbon copy), NjWOE, DF (*TAED* D8316BGI).

1. Emil Rathenau organized the Deutsche Edison Gesellschaft für angewandte Elektricität (DEG, or German Edison Co. for Applied Electricity) in March 1883 with a capitalization of 5,000,000 marks. According to the terms of interlocking agreements reached at that time, the DEG gained rights to manufacture, sell, and install equipment covered by Edison's German electric lighting patents, but it ceded practical authority to make lamps to Siemens & Halske. Siemens & Halske, which promised not to contest Edison's patents, also contracted to supply DEG with other items of its own manufacture on favorable terms. German Edison Company for Applied Electricity articles of

association, 13 Mar. 1883 (see esp. article 35); TAE agreement with Edison Electric Light Co. of Europe, Ltd., and the Compagnie Continentale Edison, 13 Mar. 1883; Compagnie Continentale Edison Co. agreement with Société Électrique Edison, Sulzbach Brothers, Jacob Landau, and National Bank of Germany, 13 Mar. 1883; all DF (*TAED* D8337ZAO, D8337ZAK, D8337ZAL); Kocka 1999, 59–64; Hughes 1983, 67–68; Feldenkirchen 1995, 15–16, 489 nn. 8 and 9.

2. Edison referred to the October ruling by the Commissioner of Patents in the Sawyer-Man interference case, which had been reported in the press (see Doc. 2508 n. 2). On 8 November Deutsche Edison Gesellschaft had cabled Edison, "Great excitement because newspapers state that incandescent lamp lawsuit is decided against you in favor of Sawyer— Cable whether true." Edison replied on the 14th, "Not true— Case still before Secretary Interior— Only covers minor details process manufacture." Not entirely satisfied by this answer, Rathenau complained by post on 16 November that DEG had not been properly informed of the interference case. "If on your side matters of such vast importance to our success and yours are not considered worth mentioning even to your representatives on this side," he wrote, "we cannot help but seeing in such proceedings on your part a moment of great weakness of your cause which we have made our own!" Rathenau apparently did not write again on this subject. DEG to TAE, 8 Nov. 1883; TAE to DEG, 14 Nov. 1883; LM 2:27B, 30A (*TAED* LM002027B, LM002030A); Rathenau to TAE, 16 Nov. 1883, DF (*TAED* D8337ZFK).

3. Edgar Marble resigned as Commissioner of Patents effective 31 October 1883. He was succeeded on an interim basis by Benjamin Butterworth, who was confirmed to the position in December. Doc. 2402 n. 1; Marble 1920–1921; "Mr. Marble's Resignation," *Washington Post*, 12 Oct. 1883, 2; Hopkins and Bond 1915, 358.

4. Henry Moore Teller, a former (and future) senator from Colorado, served as Secretary of the Interior from 1882 to 1885. Doc. 2402 n. 3.

5. Edison did not appeal this case to the courts, choosing instead to focus on defending his more important basic carbon lamp patent, U.S. Patent 223,898.

–2556–

From Frank Sprague

Brockton, Mass— Nov. 22. 1883.[a]

Dear Sir:

In testing for polarity on the new indicator, I find it as per following diagram

This would seem to indicate that the platinum was positive to the hot carbon, or that the passage of the current was due to vaporization of <u>platinum</u> instead of carbon   As yet there is no blackening—[1] This test knocks some of my theory about this business in the head, but I'll try to look into it— I think there has been a slight drop on the running lamp, but will not yet be positive— Have not broken the lamp yet— When will new indicator be along— Better send eight finally so that they will all be alike— Need one soon very much—

Everything running smoothly.

Ground not lower than 1000 ohms— Yours very truly,

F. J. Sprague.

P.S. As telegraphed, am certain of the success of the new indicator—[2] Will send house rules to-morrow— F.J.S.

⟨We have put some of the Lamps with wire in center up on test at 64 candles & find they have just as good life as Regulars & platinum dont blacken so its OK   if Bergmann is getting[b] along Rapidly with the 12   about next Thursday we shall have some[3]   E⟩

ALS, NjWOE, DF (*TAED* D8346ZBQ). Letterhead of Thomas A. Edison, Central Station, Construction Dept. [a]"188" preprinted. [b]Obscured overwritten text.

1. The blackening was a result of carbon being deposited on the inside of the bulb. Edison's 1880 experiments to prevent this deleterious process led to the observation of the electrical phenomenon later called the Edison Effect (see Doc. 1898 [headnote]). Edison intended to carry an experimental model to Brockton in mid-October, but he fell ill and apparently did not make the trip. Sprague probably obtained the test indicator when he visited New York at the end of that month. Insull to Sprague, 16 Oct. 1883; Insull to William Jenks, 24 Oct. 1883; Sprague to TAE, 27 Oct. 1883; all DF (*TAED* D8316BAZ, D8316BCF, D8346ZBF).

2. Sprague's telegram has not been found.

3. Sprague wired on 3 December that no additional indicators had arrived and asked whether he should wait for them before returning to New York. He returned to New York in early December and took up the matter with Bergmann & Co., who acknowledged on 15 December that "Mr Sprague informs us that they are very much in need of the New Style Pressure Indicators at Brockton and we do not find that we have any order for them." The order was resubmitted on 17 December, but indicators apparently were shipped in the new year, just as Edison was deciding to abandon the "new" instrument and revert to a modified form of its predecessor (see Doc. 2538 [headnote]). Sprague appears to have overseen their installation in Massachusetts village plants in February 1884. Sprague to TAE, 3 Dec. 1883; Bergmann & Co. to Edison Construction Dept., 15 Dec. 1883; Alfred Tate to Bergmann & Co., 26 Dec. 1883; Tate to William Dwelly, 14 Feb. 1884; Sprague to In-

sull, 16 Feb. 1884; all DF (*TAED* D8343ZDR, D8324ZBF, D8324ZBN, D8416ANT, D8442ZBG).

–2557–

From George Bliss

CHICAGO, November 23d, 188[3][a]

Dear Sir:

Replying to your favor of the 21st.[1] will say that the automatic regulator is so valuable in controlling the candle power of lights, and in decreasing breakage of lamps that we would not think of operating any large plant without it.

There are however some difficulties which we would like to overcome.

I understand there has been complaint elsewhere that the spring in the relay magnet gets weak.

The trouble about which we wrote you is that the candle power gradually runs up when the regulator is in use which would indicate either that the spring grew stronger or that the magnet weakened.

We do not see how the suggestion you make that the magnet coils gradually increase in resistance, owing to temperature holds good.

Our greatest trouble is at the Daily News[2] where they have a 150 light dynamo, which runs constantly except about six hours each week.

The room where the regulator is placed is hot, but the temperature is uniform.

We have moved the magnets up to armature once and have gradually let out the spring until it can go no farther without changing construction.

We have had the same trouble in a less degree elsewhere.

We will give you our future experience, but meanwhile shall be glad to have a remedy for this special difficulty.

The Bell Indicator[3] which we will have in Haverly's Theatre[4] has operated perfectly so far as the relay magnet is concerned, for a year without change of adjustment.

The same principle is involved in this magnet as in the automatic regulator.

We use only the lamps on the Bell Indicator, as the bell portion seems to be of no use. Sincerely yours,

Geo. H. Bliss   Gen'l Supt

⟨Let us know Every defect & Explain it minutely   We are working to perfect it & Every well taken critiscism aids us very greatly   E⟩

TLS (carbon copy), NjWOE, DF (*TAED* D8364ZBZ). Letterhead of Western Edison Light Co., George Bliss general superintendent; "AUTOMATIC REGULATOR" typed in preprinted memo line. ᵃ"CHICAGO," and "188" preprinted.

1. Bliss had complained on 14 November that "The adjustment on the automatic regulators installed by us seems to change gradually so as to cause the lights to run higher." Edison referred the matter to Edward Johnson and, in his 21 November letter, quoted Johnson's response: "I presume this is due to the increased resistance of the magnet coil when heated. The remedy that occurs to me is to occasionally re–adjust the spring. I have not noticed any such action in our own." Early in the new year, Bliss sent "voluminous criticism" of the regulator, but this has not been found. Bliss to TAE, 14 Nov. 1883; TAE to Bliss, 21 Nov. 1883; Samuel Insull to Western Edison Light Co., 27 Feb. 1884; all DF (*TAED* D8364ZBW, D8316BIV, D8416AQD).

The instrument in question likely was the standard electro-mechanical regulator manufactured by Bergmann & Co. for use in Edison isolated plants, illustrated in the Bergmann catalog published in April or May 1883 (see Doc. 2410 n. 4) and in Doc. 2242 (headnote). This device had an electromagnet, in a shunt circuit with the main lines, that acted on

*Patent drawing of the automatic voltage regulator for isolated plants. The electromagnet **B** and adjustable armature c control circuits for electromagnets **C** and **C′**, which move lever **E** to vary resistance in the dynamo field circuit.*

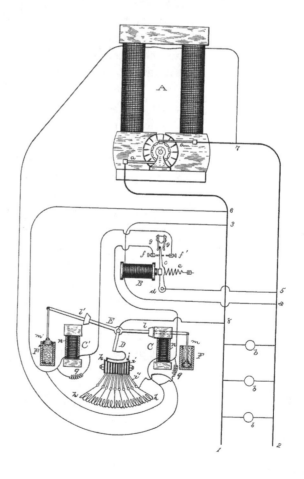

an armature pivoted against spring tension. The armature's movement closed one of two separate circuits to one or another different electro-magnets that, when energized, would move a lever to adjust a variable resistance in the dynamo field circuit. Alglave and Boulard 1884, 347–50; U.S. Pat. 287,524.

2. The *Chicago Daily News*, Chicago's first penny paper, published morning and afternoon editions and was one of the city's most widely read papers at this time. Founded in 1875 by Melville Stone, the paper remained under his editorship until 1888. *Ency. Chgo.*, s.v. "Chicago Daily News Inc."; "Stone Retires as Editor," *NYT*, 17 May 1888, 4.

3. Bliss presumably referred to the standard voltage indicator patented by Edison and manufactured by Bergmann & Co. for isolated installations. Its bell and signal lamps provided both auditory and visual indications when the voltage dropped or rose beyond set limits. See Doc. 2190 n. 9 and illustration following there.

4. The Haverly Theatre, on Monroe St. in downtown Chicago, opened in September 1881. Constructed with gas lighting throughout, the theater was retrofitted by the Western Edison Electric Co. with complete electric service in place of gas; three dynamos began providing power in April 1883. John H. ("Jack") Haverly, formerly Christopher Heverly (1837–1901), a well-known minstrel showman and operator of theaters in Chicago, Brooklyn, New York, and San Francisco, built and initially operated the house, but ceded ownership to a partner when he encountered financial difficulties in mid-1883. Young 1973, 1:281–83; "Wonderful Work," *Chicago Daily Tribune*, 11 Sept. 1881, 7; "Electric Light: Its Introduction at Haverly's a Decided Success," ibid., 10 Apr. 1883, 8; *ANB*, s.v. "Haverly, Jack H."; *DAB*, "Haverly, Christopher."

–2558–

Samuel Insull to Germania Bank[1]

[New York,] November 28th. [188]3

Dear Sir:—[2]

I enclose you herewith the information requested[3] as to the Edison Electric Light Co., The Edison Co. for Isolated Lighting and the Edison Electric Illuminating Co. of New York.

The Light Co. has a controlling interest in the Isolated Co., owning $^{51}/_{100}$ of the Isolated Co's stock.[4] The Light Co. has also a 25 per cent interest in the Illuminating Co.

The paper that we have submitted to you for discount,[5] and which you have kindly obliged us by so doing, was that of the Edison Co. for Isolated Lighting. This Company's business is to install electric light plants in mills and factories, and it obtains its license to do so from the Edison Electric Light Co., which controls it.

The Edison Electric Illuminating Co. of New York supplies light on a meter in the district bordered by Wall, Nassau, Spruce Streets and the East River.

Mr. Edison relations to these companies is that of manufacturer. He supplies them with dynamo electric machines, lamps, and electrical sundries. Our main customer is the Edison Co. for Isolated Lighting, who take a large amount of material from us right along. The Edison Electric Illuminating Co. has its plant erected, and consequently buys very little from us at the present time. The Edison Electric Light Co. is simply a patent owning organization, who do nothing beyond selling licenses to use their patents, and they also defray the cost of Mr. Edison's experiments in connection with any improvements or inventions he may get up on his present system of electric lighting.

Besides the above Mr. Edison acts as a general contractor for the installation of electric light central station plants. Our modus operandi is as follows:—

A company formed in some town gets its license from the Edison Electric Light Co., the local company requires a plant erected, in order to enter into the business of selling light, or in other words, ~~in competing~~ to compete[a] with the local gas company for business, and Mr. Edison himself undertakes the contract to erect a plant for such a local company at a given contract price. He does this work by permission of the Edison Electric Light Co., who are the sole owners of the patents.

Any further information that I can give you I shall be glad to or if you care to make an appointment, I will come down to see you and explain matters further to you, as it is our wish that you should be fully posted in relation to the matter. Very truly yours,

Thomas A. Edison Constn. Dept.   By SI

Enclosure.[6]

TLS (carbon copy), NjWOE, DF (*TAED* D8316BKY). [a]"to compete" interlined above by hand.

1. The Germania Bank of the City of New York was incorporated in 1869 and headquartered at 215 Bowery. Edison opened an account in June 1883 (giving Insull power of attorney to conduct his business), and the bank provided him with discount services and routine checking. Samuel Insull wished to design a personalized Germania check having in the center "where the Revenue stamp ordinarily goes, an Edison Lamp represented as burning and giving off rays of light." The checks were printed, but the editors have not found an example of them. *Illustrated New York* 1888, 61; TAE to Germania Bank, 5 June 1883; Insull to Arthur & Bonnell, 6 June 1883; Lbk. 17:71, 90 (*TAED* LB017071, LB017090); Insull to Germania Bank, 15 June 1883; Arthur & Bonnell to Insull, 14 June 1883; both DF (*TAED* D8316AAV, D8367U2).

2. Insull addressed this letter simply to "Cashier."

3. The editors have not found such a request.

4. Sherburne Eaton summarized the Isolated Co.'s affairs and its current financial relations with other Edison companies in a year-end statement. Eaton to Edison Co. for Isolated Lighting directors, 27 Dec. 1883, Batchelor (*TAED* MB109).

5. That is, the commonplace process of a person or business selling another's promissory note to a third party, before maturity, for less than face value. *OED*, s.v. "discount" v.1.3; Kinley 1913, 359.

6. Enclosure not found.

–2559–

Memorandum:
Llewellyn Park Village
Plant[1]

New York, [November 29, 1883][2a]

If station stops at 1 oclock— Then

Engineer.	3.350
Boy	1.00
Coal	3.00
Oil[b]	.25
Depreciation[b]	2.840
Repairs	.25
Insurance.[b]	~~.40~~30
Lamps[b]	2.40
Taxes	.2~~5~~0
Station	.05
Brushes	.05
	~~13.95~~
	~~1.55~~
	$12.40

1000 feet of Coal gas burned in 5 foot burn[er]s gives a light in practice[c] of 10 candles for 1 hour for each 5 foot, hence 2000 Candles of light can be obtained from 1000 ft gas 1 gas burner burning 200 hours 5 ft per hour consuming 1000 ft gas & gives 10 candles of light in each hour ~~The~~ 1 Electric Lamp burning the same time & giving the same light is the equivalent of 1000 ft gas; if the gas is 2.00 per 1000 then each Electric burner would cost the consumer 1 cent per hour

now if we conclude to sell light at $2 per Equivalent of 1000 cubic feet gas, the receipts at the Park[3] if 800 lights are sold[d] ~~will~~ & average 2½ hours each which they will do providing the majority of the houses are not closed more than 3 months in the year at 1 cent each, 365 days amounts to $7,300 if a 3 hour average the receipts will be $8,760. The street lights are considered to give no <u>revenue</u>

Cost of running on 3 hour average ~~& keeping~~ & allowing for depreciation

       $4,526

Recpts from Houses only on 2½ average   7,300

     net $2,774

Recpts Houses 3 hours average[e]   8,760

Expenses   4,750

                                     4,010

The cost if the street lights were paid for at the rate of $2 per thousand it would amount to $6 per day, or 2190. per year This added onto the lowest House receipts would be

                                   7,300

                                   2,190

                                   9,490

Expenses   4,526

                                   4,964 net profit

$4964. If added to the higher average it would be:

Recpts Houses   8,760

Street Lamps   2,190

                                   10,950

Expenses   4,750

                                   $6,200

Capital $30,000.

    Results $2. per thousand   Capital $30,000.[e]

2½ Hours average House recpts only.   7,300

Expenses[b]   4,526

net   2,774

Dividend—9⅙ pct

3 Hours[f] average House recpts only   8,760

Expenses   4,750

                                   4,010

Dividend 13⅓

2½ House recpts—   7,300

Street Lighting—   2,190

                                 9,490[g]

                                 9,490

Expenses   4,526

                                 4,964

Dividend 16½

3 hours= House Recpts	8,760
Street Lamp—	2,190
	10,950[h]
	10,950
Expenses	4,750
	6,200

Dividend $20^2/$_3$

Lamps are supplied to the consumer free & is charged to running expenses

The ~~average~~ cost of wiring varies according to taste & wants of partis from 1.75 per light to 4.75— This is only for wire safety devices,= Chandliers Sockets shades etc are not included   The Cost of running from mains to houses varies with the length & no of lights   100 feet for 50 lights costs about $75.[e]

ADDENDUM[4i]

[New York, 29 November 1883]

Most of the lights are within 100 feet Dynamo

"	"	"	150	"
"	"	"	200	"
"	"	"	250	"
"	"	"	300	"
"	"	"	400	"
"	"	"	500	"
"	"	"	600	"

AD, MiDbEI, EP&RI (*TAED* X001G2BE). Letterhead of Thomas A. Edison, Central Station, Construction Dept. Monetary expressions standardized for clarity; a miscellaneous canceled calculation has not been reproduced. [a]"New York," preprinted. [b]Followed by line to right column. [c]"in practice" interlined above. [d]"are sold" interlined above. [e]Followed by dividing mark. [f]Obscured overwritten text. [g]Calculation enclosed by brace at right. [h]Calculation enclosed by brace at left. [i]Addendum is an AD.

1. The editors have assembled this document from looseleaf pages found together. A docket notation describes them as "Mr Edison's own figure on Llewellyn Park Plant" (see note 3). An estimate for constructing a station there had been prepared in July. Samuel Insull to Frank Hastings, 9 Apr. 1884, DF (*TAED* D8416BEH).

2. This document is dated from a docket notation of "Thanksgiving Day 1883."

3. Llewellyn Park was a planned residence community in Orange (later West Orange), N.J., about twelve miles from New York City. Searching for a more serene residence than New York City, drug importer Llewellyn Haskell purchased forty acres of land in Orange in 1853. By 1870, the Park had grown to 750 acres and was the site of a

number of homes, most of them Gothic-inspired. Each home site was required to be at least one acre, but the average was six acres. There was also a common park, known as the Ramble, designed by landscape architect Andrew Jackson Downing in a romantic style. In 1886, Edison purchased Glenmont, one of the largest homes in Llewellyn Park, as a family residence. *Ency. NJ*, s.v. "Llewellyn Park"; Israel 1998, 248–49.

4. The page presented here as the addendum was evidently composed at or about the same time as the others, as its reverse side absorbed ink from the front of another sheet. However, the list does not obviously follow from the contents of the other pages, and it is followed by a half page of rough notes and a sketch on armature design, all apparently unrelated to the rest of the document.

—2560—

To Albert Chandler

New York, Nov [December][1] 1 1883[a]

Friend Chandler[2]

Do you know the address of a <u>1st Class</u>[b] foreman of construction who is used to city work.[3] Perhaps WU have some; want a pusher who has brains, who wears an 8⅜ hat & reads Plato in the original Greek. Yours

Thos A Edison.

ALS, NNC-RB, Edison Coll. (*TAED* X090J). A letterpress copy is in LM 16:351 (*TAED* LBCD3351). [a]"New York," and "188" preprinted. [b]Multiply underlined.

1. Edison apparently wrote the wrong month inadvertently. The letter was copied into a Construction Department letterbook (see document sourcenote above) among other correspondence of 30 November. Edison also dated a similar inquiry to a Western Union construction superintendent on that day. Chandler replied from his office at nearby Union Square on 1 December (see note 2). TAE to D. Doren, 30 Nov. 1883, LM 16:350 (*TAED* LBCD3350).

2. Albert B. Chandler (b. 1840) was at this time president of the Fuller Electrical Co., the arc lighting firm in New York, but he spent most of his career in the telegraph industry. He had worked for Western Union for many years before becoming assistant general manager (and later president) of the Atlantic & Pacific Telegraph Co. in 1875. At that time, the company acquired rights to Edison's automatic telegraph system, and Edison worked briefly as company electrician. Chandler later became president of the Postal Telegraph Co. See Docs. 510 n. 5 and 585; Taltavall 1893, 161–62.

3. Chandler replied on 1 December recommending A. J. Brown, the Western Union foreman of construction in New York City, and W. C. Smith, foreman for the Rapid Telegraph Co., as capable telegraph men who might suit Edison's needs, though "Neither of them wears so large a hat as you mention, and I doubt whether either of them ever heard of Plato." He also promised to send anyone else he found suitable for Edison's "inspection." Edison offered positions to both men. Brown initially accepted the job but later changed his mind; the editors have

not determined whether Smith accepted. Edison also had an inquiry at the end of December from Samuel Bogart, the superintendent of cables and underground lines for Western Union, who had heard about the offer to Brown. Edison replied that because Brown had taken the position he was unable to offer it to Bogart. It is unclear who, if anyone, Edison hired, but see his undated memorandum about Jerome Hennesey, a lineman for the Central American Cable Co., who "will work in Small Towns here 65 mo & 1.50 expenses." Chandler to TAE, 1 Dec. 1883; TAE to Chandler, 4 Dec. 1883; Samuel Insull to Brown, 4 Dec. 1883; Insull to Smith, 4 Dec. 1883; TAE to Smith, 13 Dec. 1883; Bogart to TAE, 26 Dec. 1883; TAE to Bogart, 3 Jan. 1884; Brown to TAE, 16 Jan. 1884; TAE memorandum, n.d. [1884?]; all DF (*TAED* D8336ZAL, D8316BNM, D8316BNH, D8316BMX, D8316BRX, D8313ZCL, D8416AAL, D8403E, D8413ZBJ).

–2561–

To Cotton & Gould[1]

[New York,] December 4th. [188]3

Dear Sirs:—

We have frequent inquiries from Boston as to the duplicating ink[2] of mine which was sold extensively some few years back, but of which none has been made for sometime past.

Messrs. J. P. Rogers & Co.[3] have suggested to me that I [-----][a] arrange[b] with you to carry a stock. If you are willing to do so, I will have some made up in my laboratory, and will sell it to you at a reasonable price. I am anxious to pursue some such course as this in order to oblige the people applying to me for it.[4] Very truly yours,

TAE   I[nsull]

TL (carbon copy), NjWOE, DF (*TAED* D8316BNL). [a]Canceled. [b]"arrange" interlined above by hand.

1. Cotton & Gould were prominent Boston stationers; the firm was established in 1883. This letter was mistakenly addressed to Cotton & Guild. "Funeral of H. N. Cotton," *Boston Daily Globe*, 1 Jan. 1916, 11.

2. In 1876, Edison devised a duplicating ink for use with the electric pen and other copying devices. In November of that year, he and Edward Johnson formed the American Novelty Co. to market it and other inventions. Edison transferred his entire interest in the product to George Bliss and Charles Holland in 1877. See Docs. 825, 829, 831 esp. n. 4, 836, and 912 n. 4.

3. John P. Rogers & Co. were exporters and commission merchants in Boston. Letterhead, Rogers to TAE, 11 Sept. 1883, DF (*TAED* D8303ZFY).

4. Rogers, whose firm had used the duplicating ink for years but could not find more, had asked Edison for a supply in September. Unable to recall the formula, Edison wrote to James MacKenzie, who had been in charge of making it at Menlo Park, and then had John Ott prepare a gallon in October. Rogers to TAE, 11 Sept. 1883; TAE to Rogers,

26 Sept. 1883; TAE to MacKenzie, 26 Sept. and 5 Oct. 1883; Samuel Insull to Ott, 17 Oct. 1883; all DF (*TAED* D8303ZFY, D8316AWQ, D8316AWP, D8316AYN, D8316BBK).

After Edison received another inquiry about the ink from H. O. Stratton, a Boston merchant, in late November, he asked Rogers to recommend a stationer who could stock it. Rogers suggested Cotton & Gould on 1 December. Edison then had Ott make up two more gallons, which were ready by the end of the year at a cost of about a dollar each. Stratton sent Edison money but had not received the ink by March, when he inquired again. The editors have not determined what, if any, merchandising arrangements Edison made for this product. Stratton to TAE, 27 Nov. 1883 and 3 Mar. 1884; TAE to Rogers, 30 Nov. 1883; Rogers to TAE, 1 Dec. 1883; TAE to Ott (with Ott marginalia), 19 Dec. 1883; all DF (*TAED* D8305Z1, D8403ZAF, D8316BLY, D8303ZHO, D8316BTY7).

–2562–

Samuel Insull to Paul Dyer

[New York,] December 4th. [188]3

Dear Sir:—

I have your favor of the 26th. inst.[1]

If you are not near home at Christmas time we will allow you the amount of your expenses to get to Washington and back to the point where you are at work. I presume you have no objection to this.[2] Of course we sympathize with a man who goes into a town like New Orleans and buys his experience at the rate of $30 a lesson.

Our business is getting along swimmingly, at the present time we are erecting the following central stations, Fall River, $40,000., Newburgh, N.Y., $40,000., Bellefonte, $10,000., Mt. Carmel, $14,000, and Hazleton $20,000. The last three places are in Pennsylvania. Piqua, Ohio, $13,000., Tiffin, Ohio, $18,000, and Middletown Ohio, $13,000. We shall start in Williamsport and the course of a week, contract $58,000. Brockton, Lawrence, Shamokin and Sunbury are running elegantly, getting all the customers we want and making money. This is the state of business in brief.[3]

I have been going for our boys in the office for not sending you Bulletins, but they do not contain any information [----][a] about our central stations. The New York Central Station is just "booming," collections $10,000. per month. Have more customers then we can possibly carry, and if we get through December without a breakdown it will simply be that the "angels" are our side, as we have too many lights connected already.

I sail for Europe on the 11th. December and shall be away until the 1st. February and if you do not get any word from this office that you can go home to spend Christmas just take "French leave,"[4] and charge your expenses in your weekly statement and if they "kick" refer them to this letter. Very truly yours,

SI

TLS (carbon copy), NjWOE, DF (*TAED* D8316BNN). [a]Canceled.

1. Dyer wrote from New Orleans, where a week of rain had slowed his canvass activities. He asked if there were any work that would bring him close to his home in Washington, D.C., for Christmas. He also requested Insull to send him the Edison Electric Light Co. Bulletins or other printed material that would keep him "posted on what we are doing." Dyer to Insull, 26 Nov. 1883, DF (*TAED* D8342ZDW).

2. A few days later, Insull addressed a letter to Dyer in Galveston, Tex., offering that if he would "work like a race-horse, and get through the towns you now have to canvass South before Christmas, we will give you a 'picnic' after Christmas in the shape of a district in New York to canvass. It will be kind of a vacation to you." Insull to Dyer, 7 Dec. 1883, DF (*TAED* D8316BPN); see Doc. 2585.

3. See App. 2.

4. To take French leave is to depart or take other action without authorization. *OED*, s.v. "French leave."

–2563–

William Andrews to Samuel Insull

Lawrence Mass Dec 5th 1883[a]

Dear Sir—

I saw Kerby[1] this morning and he showed me a letter from his wife wherein it was stated that the Engineer in charge of our Station ~~there~~ in Sunbury Pa[b] William Bateman,[2] got drunk the other night and left Station in care of a boy— Moreover lights are not regulated well there and City Hotel[3] people threaten to go back to gas, unless there is a change for the better— I cannot of course vouch ~~a~~in any way for the truth of these reports, but think it to be my duty to inform you in order that Enquery may be made—[4] Very truly yours

W S Andrews

⟨Write to Marr to investigate this & let me know   Can send another man⟩

ALS, NjWOE, DF (*TAED* D8361ZDY). Letterhead of Thomas A. Edison, Central Station, Construction Dep't. [a]"188" preprinted. [b]"in Sunbury Pa" interlined above.

1. J. F. Kirby.

2. William Bateman held this position at Sunbury until at least February 1884, when Edison sought to have him discharged. See Doc. 2603.

3. The City Hotel was one of the first two establishments to have

electric service in Sunbury. "The Edison Light," *Williamsport (Pa.) Daily Gazette and Bulletin*, 6 July 1883, 1.

4. Samuel Insull asked Frank Marr, treasurer of the Sunbury company, to respond to these allegations. Afer denouncing William Rich at some length, Marr pointed out that the City Hotel had long had problems with its lights, which he blamed on defective wiring. Regarding the other charge, he explained that the engineer, who was routinely on duty from 3 p.m. until 8 a.m. the next morning, was permitted a few hours' sleep in the evening while a young assistant tended the machinery. On the evening in question, he claimed, the boy had simply failed to wake Bateman before leaving for the night. Insull to Marr, 6 Dec. 1883; Marr to TAE, 8 Dec. 1883; both DF (*TAED* D8316BOX, D8361ZEB).

–2564–

Charles Clarke to Samuel Insull

New York, 7. Dec. 1883[a]

Dear Sir

The pressure indicators in question were never recommended by me or through me by the Engineering Dept. The indicator was designed by me, but Mr. Edison in ordering them did so on his own responsibility. The Engineering Department simply tried to embody in the instruments the numerous conditions which Mr. Edison insisted it should have.

He accepted the instrument on his own responsibility. A few were ordered, I think one dozen, and immediately put in use at Sunbury and Shamokin. If they proved entirely unsatisfactory why were numbers ordered afterwards for other places?[1]

Mr. Edison was all along aware of the difficulties which developed but continued to order and use them. According to the resolutions adopted when the Standardizing Committee[2] was organized no innovations on old practices or new apparatus or appliances can be introduced unless authorized by that Committee, or by the Special Committee of that body after having been referred to said committee by the general committee. Latitude is allowed however to individuals to make departures from the established rules when sudden developments require it, but the responsibility of failure on account of such changes rests on the individual taking the responsibility of making the change and he must stand the consequences. This pressure indicator was never passed upon by the committee in any way, and I disclaim any responsibility for myself or the Engineering Department as represented by me.

The Engineering Dept., as far as innovations are concerned is subject to the action of the Standardizing Committee, and this Committee alone can make the Light Co responsible.

If Mr. Edison as Engineer pro tem. of the Light Co recommended the indicator to himself as head of the Construction Co, and can persuade the Light Co that he had a right to do so, then the Construction Co may possibly be entitled to recompense, but even then it is an open question, and dangerous to the interests of the Light Co if decided against it.

The Illuminating Co as well as all foreign companies that have met with any loss due to defects which have developed in plant furnished them or constructed with the knowledge of the Light Co, can demand remuneration amounting to thousands of dollars. I do not see that your claim is any better than theirs. Very truly Yours

Chas. L. Clarke   B[3]

LS, NjWOE, DF (*TAED* D8329ZAU). Letterhead of Edison Electric Light Co. [a]"*New York*," and "*188*" preprinted.

1. The pressure indicators used in the earliest village plants had proved to be, Insull wrote to Clarke on 5 December, "absolutely useless" and would have to be replaced (see Doc. 2538 [headnote]). According to Insull, Edison believed that the Engineering Dept. had approved the design, and he told Clarke that if the instruments had "emanated from your Department we would have the right to claim from the Edison Electric Light Company the amount that we have lost." Insull to Clarke, 5 Dec. 1883, DF (*TAED* D8329ZAR).

2. Edison had created a Standardizing Committee at an unknown earlier date, consisting of the heads of all departments in the Edison lighting businesses. The committee was responsible for "testing and criticising all existing and proposed devices" and considering suggestions and complaints offered by workmen. It was intended both to improve the quality of the Edison system and to educate the department heads about technical details. Dyer and Martin 1910, 379–80.

3. Unidentified.

–2565–

To Charles Batchelor

[New York,] Dec. 8th/83

Eknoside, Paris[1]

Prepare yourself return—[2] Immense boom— Building thirteen central stations—[3] Result returns New York Station which netted over five thousand November.

L (telegram, copy), NjWOE, LM 2:33A (*TAED* LM002033A).

1. "Eknoside Paris" was a registered cable code for Charles Batchelor used since at least 1882. *TAEB* 6 App. 4.

2. Cf. Doc. 2667.

3. See App. 2A.

–2566–

To Egisto Fabbri

[New York,] Dec 8 [1883]

Friend Fabbri

This may interest you.[1] Please show to Mr Wright,[2] and perhaps to Mr Morgan, as he is a "Connoisseur de la balance sheet."[3] Yours

Edison

75 percent of recpts are from customers who burn in day time hence recpts will be high in summer. station is good for $50 000 net next year. Supt tells me has 120 applications for light which he is unable to supply   E

ALS (letterpress copy), NjWOE, Lbk. 17:326A (*TAED* LB017326A).

1. The enclosure has not been found; possibly it concerned the receipts from the Pearl St. central station discussed in his postscript (cf. Docs. 2562 and 2565).

2. A founding partner of Drexel, Morgan & Co., James Hood Wright (1836–1894) held a variety of interests in Edison electric light and power enterprises. His roles included: director of the Edison Electric Illuminating Co. of New York, original stockholder of the Edison Electric Light Co., and investor in the Edison Electric Light Co., Ltd. (Doc. 2187 n. 7; Carosso 1987, 138; Edison Electric Light Co. Bulletin 17:39, 20 Dec. 1882; Summary of Share Capital & Shares in the Edison Electric Light Co., Ltd., 13 June 1883 [1926]; both CR [*TAED* CB015237, CF001AAO]; Edison Electric Light Co. list of stockholders, 29 Sept. 1883, Miller [*TAED* HM830194]). Wright's home in upper Manhattan was among the first residences to be outfitted with an Edison isolated plant for electric lighting (Sherburne Eaton to TAE, 21 Aug. 1882; DF [*TAED* D8226ZBI]; Edison Electric Light Co. Bulletin 18:36, 31 May 1883, CR [*TAED* CB018]).

3. J. Pierpont Morgan carefully monitored the books of Drexel, Morgan & Co. He had a reputation for quick but thorough understanding of financial records and, according to a statement attributed to his son Jack, a "most uncanny ability to find mistakes in the books." Carosso 1987, 145.

–2567–

To William Hammer

[New York,] December 10th. [188]3

Dear Sir:—

Referring to your letter with[out] date,[1] I think if you will look up my patents you will find that one of them covers a patent for taking care of irregularities in a gas machine.[2]

DYNAMOS. The "H" machine is tested for 400 amperes, 500 lights means nothing, suppose they were 50 candle lights. Please talk amperes.[3] We gave Batchelor the carrying power in amperes. Since you received the model "H", we have reduced the resistance of the armature by using larger wire, thus easing it up so that is runs cool with 400 amperes on it.[4] You people were in such a hurry to receive your model, that we were compelled to ship you the first machine we had, and because there was considerable delay in making the shipment you seem to get the the impression that we were not paying attention to you. If you had left us to ship you what we considered our best machine, we should not have shipped the 400 ampere machine, but would have waited at least a month after the date that we made the shipment.[5]

We can send models of a half horse power and one smaller size motor for sewing machines.[6]

With relation to the lamp factory, I note what you state as to the desirability of having your lamp factory in Germany itself.[7] I will send you the apparatus that you may order, but it will be quite impossible for us to supply your Company with any men. Our staff has already been reduced too much by sending out our assistants to Europe, and we must really look to ourselves in relation to this matter. We have now reduced the price of the of the ten candle lamps to 50 cents. As we make more of them we shall probably reduce the price further, but the shrinkage[8] in the manufacture of the lamp is probably greater than in any other. We do not use any "B" lamps at all to speak of in this country.[9]

I will have copies of estimates and plans sent forward to you in the course of a week or so.[10] You will be able to get information in relation to these matters from Mr. Insull, who leaves here on Tuesday.

We do not propose putting the three wire tubes in the down town district, as the whole of that district is now covered with the two wire tubes. We are however, arranging to build another station uptown, which will be run on the three wire system. All our small town plants, whether underground or overhead, are run on the three wire system. Very truly yours,

T.A.E   [--][a]   T[ate]

TL (carbon copy), NjWOE, DF (*TAED* D8316BQB1). [a]Canceled.

1. Edison referred to a letter of 9 November; Hammer did not write the date, but it can be inferred from the contents. Edison's marginal comments formed the basis for portions of this reply. DF (*TAED* D8337ZGH).

2. Hammer had reported (see note 1) the recent completion of two isolated plants in Berlin, "the power being supplied by Otto gas machines, the latter are of the double-cylinder type, 8 horse power, and by means of a governor on the counter-shaft, which was designed by Mr Rathenau, we secure perfect steadiness and regularity in the light. . . . I

think the regulator spoken of, is a big thing, and will give us very much business in Berlin, which we could not handle, except from central station." Their fluctuating speeds made gas engines problematic for driving dynamos. Edison had tried to address this problem in late 1882 with a new regulator on which he received a patent in January 1883. See Doc. 2359.

3. For the relationship between amperes and number of lamps, see App. 3, esp. n. 3.

4. Edison's paragraph up to this point is based closely on his marginal notes on Hammer's 9 November letter (see note 1). Hammer had complained that the H dynamo ran very hot when tested with 500 lamps in Berlin. A later examination revealed charring of the insulation on some armature wires, which he suspected might have occurred during prior testing in New York. In a separate note on that letter, Edison had instructed his office assistant to "Explain to Hammer how we do our biz. so as not to get stuck on machines— The H machine was tested with heavy load— also understand that the machines are tested with 400 amperes of course if you put on 500 Lamps of .92 amperes it would strain the machine   it should have been tested with .75 ampere Lamps   our later H machs are wound with larger wires—ie few large wires in each section instead number of small wires this gives increased capacity."

5. At the request of Emil Rathenau, Hammer had complained in his letter (see note 1) "that we have so been harassed by delay in receipt of models & because of constant changes being made in the Types that our competitors Siemens & Halske & Shuckert are doing the business we should be doing . . . we are now forced to look to Dr Hopkinson who is making slow speed and compound machines &c &c."

6. Bergmann & Co. had been making small numbers of fractional horsepower motors in September, some of which went into commercial use, while others remained experimental. Bergmann & Co. to TAE, 20 Sept. 1883; Charles Clarke to Samuel Insull, 20 Sept. 1883; both DF (*TAED* D8324ZAL, D8329U); see also Docs. 2459 and 2579.

7. Hammer had protested in his 9 November letter that

you people in America do not understand the lamp question here in Germany. Were it not for several reasons we should buy every lamp from you or the French Co=, but the objections are these. First, the German Patent Law demands that, at least the greater portion of a "Patented" article, should be manufactured in the country. Secondly, the sentiment in Germany is perhaps stronger than in any other country on the subject of employing native labour in such manufactures. Thirdly, Siemans and Halske have at present a small Lamp-factory; and perhaps in a few years, when our agreement with them terminates, they will be in the possession of a large and well organized factory, while our Company would be dependant on other countries.

These reasons had persuaded the German company (DEG) "to decide upon going to the trouble and expense of putting up a factory." Following Charles Batchelor's advice, Hammer wished to obtain the factory equipment, such as mercury pumps, directly from Edison. In referring to the legal necessity of domestic manufacture, Hammer probably had in mind the provision in German law under which a patentee who did not adequately "work" the invention within three years risked losing patent protection. Abbott 1886, 206–7.

8. That is, a bookkeeping allowance for waste or theft. *OED*, s.v. "shrinkage" 3b.

9. Hammer had inquired in his 9 November letter about cheapening the manufacture of 10-candlepower lamps, which were "much less troublesome than the B lamp, whose greatest virtue lies in its economy, although the short carbons look very pretty when grouped in large chandeliers."

10. Hammer had urged Edison to send complete drawings of the dynamos, pointing out that mechanics at Siemens & Halske, to prepare their own, had been compelled to disassemble the models "down to the last bolt." Edison had similarly promised full drawings to the Société d'Appareillage Electrique in Geneva, but the editors have not found records of their transmission either to Berlin or Geneva. TAE to Société d'Appareillage Electrique, 21 Nov. 1883, DF (*TAED* D8316BIN).

–2568–

From Fred Catlin

New York Decem 10th '83

Dear Sir

I read with much interest Johnsons letter to The Times published today.[1] He covers the subject completely and floors Prof Morton at every step, and the beauty of it is he confines himself to facts. Professor Mortons opinions will, hereafter, in matters pertaining to electrical science be received very cautiously by telegraph people, and in fact by all persons dealing in artificial lighting, for, to the person of very limited electrical knowledge it cannot but be apparent that he is ignorant on the subject in question or that he is trying to hoodwink the public. In either event according to Mr Johnson he has seriously committed himself.

It would be a nice experiment, and could be accomplished without much difficulty, to measure the resistance of the stream of water and then resistance of the drenched fireman holding the hose, and make a comparison.[2]

It is a grave matter to deceive the public on this all important subject. The person of ordinary electrical knowledge knows the laws of electrical distribution, and will say, put the dangerous dynamic fluid conductors under ground. In the meantime he will pass under them cautiously and breathe free-er[a] when he gets out of their reach.

Examples of electrical distribution present themselves to me every day.

For instance, 4 wires between New York and Boston become crossed or make contact with each other and at the same time make a partial earth contact 20 or 30 miles out. A battery connected at Boston will divide at the earth contact and sufficient quantity reach New York through the different wires to supply each instrument with a strong current, [---][b]

We use as you understand machine currents on all our wires, but it is toned down with artificial internal resistance so as to render it applicable to our work, and this process at the same time makes it harmless as far as danger to humans beings is concerned. 350 volts, being equal to about that number of cells gravity,[3] is the highest we use.—

I am looking every day for an illustration of the law of electrical distribution that will be a convincing one to the non-believers

It is sure to come but I hope it will strike among cats and dogs and that human souls will not be the price of the evidence. Yours Truly

Fred Catlin[4]

⟨My Dear Catlin=

Your recd   Johnson is fixing up hose and a Dynamo giving 2000 volts to give an Exhibition, he has measured the water Stream & finds it about 850 ohms $2^{1}/_{2}$ inch Section per foot. ~~you will hear some~~ [f---][b] ~~pretty soon.~~ E⟩[5]

ALS, NjWOE, DF (*TAED* D8373Y). [a]Obscured overwritten text. [b]Canceled.

1. Edward Johnson's letter to the *New York Times* was his epistolary contribution to a controversy, manifested in a recent spate of articles and editorials, about the safety of overhead electric lines. A 4 November article in the *Times,* "Sudden Death in the Air" (p. 6), left little doubt about that paper's support for a mandate to put wires underground, a position amplified in its pages in subsequent weeks. The underlying problem, however, was not new. A General Committee on Underground Communication had already been appointed to study the proliferation of telegraph and other wires, and it finished its work at about this time. Johnson letter to the editor, "The Electric Wire Problem," *NYT,* 10 Dec. 1883, 2; "Underground Telegraph Wires," ibid., 31 Oct. 1883, 8.

Johnson's letter followed a 10 December fire on Fulton St. that was attributed to poorly insulated wires of the United States Illuminating Co. ("Notes," *Teleg. J. and Elec. Rev.* 14 [12 Jan. 1884]: 31). The spate of articles also coincided with several court cases in which plaintiffs sought injunctions to prevent arc lighting companies from erecting pole lines on or near their property. Edison appeared as a plaintiff's witness in a suit against the Brush Illuminating Co., testifying under cross-examination that passersby might be injured by a stream of water in contact with an exposed wire, although he could not recall such an event. Johnson was a witness in at least two cases, including *Smith v. U.S. Illuminating Co.,* in which he submitted an affidavit endorsing Edison's underground system. Henry Morton of the Stevens Institute, a consultant to the New York Board of Fire Underwriters, attested to the safety of the defendant's system of overhead wiring. Morton reportedly claimed that the underground system was "not successful" and that the Edison company had begun resorting to overhead wires. Morton also pointed to several fires involving Edison wires, which, he feared, indicated the probability of "the ignition and explosion of confined gases" by buried conductors (cf. Docs. 2462 n. 2 and 2533). In addition, the United States Illuminating Co.'s president, Eugene Lynch, swore an affidavit accusing the Edison firm of having "inspired" suits against arc lighting companies ("Pleading for the Wires," *NYT*, 2 Dec. 1883, 3; "Miscellaneous City News. Dangers from Electric Wires," ibid., 15 Nov. 1883, 2; Grosvenor Lowrey to TAE, 2 Jan. 1884, DF [*TAED* D8427A]). Sherburne Eaton, in his own letter to the editor, defended the Edison underground system and denied that his company had anything to do with either the litigation or the newspaper attacks on the arc wires. Johnson followed up with his letter, published five days later, maintaining that the commercial supply of electricity could be safely delivered only by low-pressure underground cables. Morton replied in print by claiming that the danger from high-pressure overhead wiring was exaggerated and that Edison's underground system had, in fact, caused more damage than any other (Eaton letter to the editor, "The Electric Wires Controversy," *NYT*, 5 Dec. 1883, 6; Morton letter to the editor, "Electric Light Wires," ibid., 16 Dec. 1883, 4).

At the end of December, the New York Board of Aldermen approved legislation requiring electric companies to place their wires in conduits below ground. This marked only a stage in the controversy rather than its end, however; arguments continued for years in the pages of newspapers, the Board of Aldermen, and the New York legislature. Regulations for the construction of new underground lines were not in place before 1887, when the Edison Second District was being built. *Thirty Years of New York*, 109–11; "The Electric Wire Nuisance," *NYT*, 20 Dec. 1883, 8; "Wires to Be Buried," ibid., 20 Dec. 1883, 4; "To Bury Electric Wires," ibid., 17 Jul. 1886, 1; see also Sullivan 1995 (chap. 4), which attributes the push for underground wires to esthetic rather than safety concerns but also provides a narrative history of the corrupt Board of Commissioners of Electrical Subways and related political efforts through the 1880s.

2. One potential danger of overhead arc lights cited in newspaper articles was that of injury to a fireman if the water stream from his fire hose made contact with live conductors. Morton, in his affidavit in *Smith v. U.S. Illuminating Co.*, claimed that this possibility was "too absurd to merit serious consideration." Johnson, in his testimony and a letter to the editor, denied that such an accident was out of the question. He theorized that its occurrence would depend on the resistance of the water upstream in the hose compared to that of the fireman himself. Although Edison stated in this document that Johnson tried to produce experimental evidence on this question, the editors have found no other record of this effort. According to the *Electrical Review*, such an experiment was made at the U.S. Electric Lighting Co.'s factory in Newark on 18 December: when water was directed through a metal nozzle

onto bare wires, those holding the hose "were not injured, nor did they feel the least shock." "Sudden Death in the Air," *NYT,* 4 Nov. 1883, 6; "Dangerous Electric Wires," ibid., 13 Nov. 1883, 8; "Pleading for the Wires," ibid., 2 Dec. 1883, 3; Johnson letter to the editor, "The Electric Wire Problem," ibid., 10 Dec. 1883, 2; "The Boomerang Returns," *Electrical Review* 3 (13 Dec. 1883): 8–9; "Are the Arc Wires Dangerous to Property?" ibid. 3 (20 Dec. 1883): 8; "Experiments with Electric Light Wires," ibid. 3 (27 Dec. 1883): 3.

3. The gravity cell—so called because the different specific gravities of copper sulfate and zinc sulfate electrolytic solutions kept them apart—produced a steady voltage and was widely used on telegraphic circuits. Doc. 368 n. 2.

4. Fred Catlin (1848–1911?), a telegrapher, began working for Western Union in New York in the late 1860s. He became acquainted with Edison in 1868 by sending press copy to him in Boston. Catlin eventually became a chief operator and manager at Western Union's New York City headquarters, from which he wrote this document to Edison. He organized annual telegraphers' speed tournaments beginning in 1884. Catlin to TAE, 30 Mar. 1898, DF (*TAED* D9802AAH1); Taltavall 1893, 341; U.S. Census Bureau, 1982 (1910), roll T624_883, p. 2B, image 297 (East Orange Ward 4, Essex, N.J.); Obituary, *NYT,* 16 Apr. 1911, 11.

5. Edison's marginalia was the basis for a brief typed reply. Alfred Tate initialed the carbon copy but presumably signed the original for Edison. Tate to Catlin, 14 Dec. 1883, DF (*TAED* D8316BSN).

–2569–

To Sherburne Eaton

[New York,] December 13th [188]3

Dear Sir:—

I have the pleasure to hand you herewith account amounting to $7,766.74, being amount due my Construction Department, also account amounting to $1,119.00 for sundry amounts paid by me on the Light Company's account.[1]

In addition to the above there is the amount expended by me on electric railroad experiments, for which you already have my account, amounting to $38,541.80, and also about $25,000 expended by me at the Edison Machine Works in improving your dynamo machines, with the result of decreasing the cost on some of the types of dynamos about twenty five per cent, and increasing the number of lights obtained from such machines about forty per cent.

With relation to my railroad experiments, inasmuch as the inventions in connection therewith have been transferred to the Electric Railroad Company, I presume my accounts in relation to this matter must stand over, although by stating this I wish it to be clearly understood that I consider that I have a valid claim against the Edison Electric Light Company for the amount I have expended in this connection.

As to the cost of the dynamo experiments, in consideration of the kindness shown by the Light Company in carrying my stock for me, it is not my intention to ask them to reimburse me to the amount referred to, more especially as the rendering of account of this character is usually the cause of great irritation. In this connection I may mention that I am now experimenting with the object of increasing the capacity of the central station dynamo[2] but inasmuch as my means are limited I cannot conduct the experiments on any such scale as the importance of the matter would warrant, and I would urge upon your Directors the necessity of placing at my disposal $30,000. for this purpose. Of course it is very difficult to estimate the exact cost of such experimentation, and I mention this amount as a limit and not as giving an indication that the experiment would cost as much.

I shall be glad if you will arrange to pay the small account due ($1,119.60), and my Construction Department account ($7,766.74) at an early date, inasmuch as the latter must of necessity receive additions or deductions from month to month.

I would suggest that the payment of $5,000. on account be made immediately. Very truly yours,

T.A.E.   T[ate].

Enclosures.

TL (carbon copy), NjWOE, DF (*TAED* D8316BSG).

1. The enclosures have not been found, but the expenses referred to were probably for central station canvasses, estimates, electrical determinations, and other engineering work, the reimbursement for which Edison and Eaton seem to have had an understanding. The editors also have not determined if or when the Edison Electric Light Co. paid the sums discussed in this document, but see Doc. 2704. Edison likely sent this statement in response to a 5 November request from Eaton, made on behalf of the Electric Light Co.'s Committee on Expenses, for a comprehensive statement of "every claim which you have against the Light Co., down to November 1st. 1883, such as experimenting, testing, making estimates, canvassing, sundry expenses of all kinds &c. &c." Eaton had repeated his request twice, most recently on 8 December. Eaton to TAE, 5 and 10 Nov., 8 Dec. 1883, all DF (*TAED* D8327ZBK, D8327ZBO, D8327ZBQ).

2. William Kennedy Laurie Dickson "made a test as you [Edison] desired on heating of short core 'C' Dyn" on 12 December (Dickson to TAE, 12 Dec. 1883, DF [*TAED* D8330ZBB]). The editors have found no earlier reference to a "C" dynamo modified in the general way described in Doc. 2419 (headnote), but see Doc. 2580.

[New York,] December 15th [188]3

Dear Sir:—

I have your letter of the 10th, inst., and regret exceedingly that a former letter which I wrote you has apparently miscarried.[2]

The painting came to hand safely, and I can assure you that I am delighted with it. It is hanging in my office at the present time, and I consider it one of the finest pieces of work that I have seen in a long time.[3]

Did you not make some request in your former letter in relation to a Phonograph? If you will repeat this I shall be only too happy to attend to it.[4]

Again regretting that the transmission of my thanks has been so long delayed, believe me to be, Very truly yours,

AOT[ate][5]

TL (carbon copy), NjWOE, DF (*TAED* D8316BTS).

1. Emile François Buchel (1862–1902) cofounded the Buchel Machine Works in New Orleans with his brother, Jules (an inventor who worked on an improved Gramme dynamo around this time). "New Orleans, Louisiana Birth Records Index, 1790–1899," Vol. 32:21, online database accessed through Ancestry.com, 24 Oct. 2008; "New Orleans, Louisiana Death Records Index, 1804–1949," Vol. 127:165, online database accessed through Ancestry.com, 24 Oct. 2008; Strelinger & Co. 1979 [1897], 611; Buchel to TAE, 27 Dec. 1883, DF (*TAED* D8303ZIK).

2. Buchel inquired about the painting sent to Edison in October (see note 3), for which Edison sent a letter of thanks on 15 November. Edison's marginal note on Buchel's query formed the basis for this reply. Buchel to TAE, 10 Dec. 1883; TAE to Buchel, 15 Nov. 1883; both DF (*TAED* D8303ZHZ, D8316BGP).

3. Buchel had gifted Edison with a landscape scene of Louisiana cypress swamps, painted in oil by New Orleans artist George David Coulon, a cofounder of the Southern Art Union. Having asked Edison "to show the painting to your many friends and visitors, and put in a good word for the artist," he also suggested that it should be placed "in a good strong light" and advised that it could be varnished in six or seven months. Buchel to TAE, undated [c. 23] Oct. 1883, DF (*TAED* D8303ZGY); Bonner 1982, 41–61.

4. Buchel's initial inquiry asked whether Edison had "made any improvements on the Phonograph lately" as a way of ascertaining the originality of his own improvements. Edison instructed his staff to draft a reply that eventually read, "With relation to the Phonograph, I have been so busy in developing my system of electric lighting for the past few years that I have been compelled to drop all other matters." Buchel to TAE (with TAE marginalia), undated [c. 23] Oct. 1883; TAE to Buchel, 15 Nov. 1883; both DF (*TAED* D8303ZGY, D8316BGP).

5. The original was presumably signed by Alfred Tate on behalf of Edison.

*Draft Patent
Application: Electric
Light and Power*

Case 611[1]

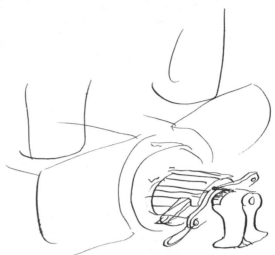

Patent

Scale in amperes. The sparking point of the brushes depends directly on no. of amperes. By seeing where brushes are adjusted you know the[b] load on machine

Espec'ly when 2 machines of diff. capacity and R[esistance]. & E[lectro]M[otive]F[orce] being same they would give same C[urrent]. and consequently one burn out.[2]

Df, NjWOE, Cat. 1149 (*TAED* NM018AAZ). Letterhead of Thomas A. Edison, Central Station, Construction Dept.; date and text except "Patent" written by Henry Seely. [a]"NEW YORK," and "188" preprinted. [b]Obscured overwritten text.

1. Edison's patent attorneys assigned this case number to the patent application prepared from this draft. The application was filed at the Patent Office on 24 January 1884, but it was rejected in March and subsequently abandoned. Four claims and a single drawing survive in Edison's patent records. Casebook E-2538:306; Patent Application Drawings (Case Nos. 179–699); PS (*TAED* PT022306, PT023A).

2. Edison executed a related application on 5 January and filed it, too, on the twenty-fourth. It covered a method for regulating dynamos of the same electromotive force (voltage) but unequal generating capacity in simultaneous use at a central station. Edison noted that such operation "has not been found practicable . . . for the reason that, as the current is proportional to the electro-motive force and resistance, the entire current generated tends to divide itself equally between the two machines, and the smaller machine having armature-coils of less radiating capacity is unequal to the work put upon it, and has its coils injured or burned out." In this circumstance, Edison proposed to use ammeters, such as that described in his Case 611, to indicate the amount of current generated by each machine and to regulate field windings accordingly. U.S. Pat. 298,956.

In a patent application executed in November, Edison noted that he had devoted "considerable experiment" to this general problem. Finding that voltage indicators had to be wired in parallel across the circuits, thereby always giving the same reading "no matter how the current changed or in which direction," he "found it necessary to use ampère-meters which would show the variations in current strength." U.S. Pat. 509,517.

–2572–

From Samuel Insull

London 21st Dec 1883

My Dear Edison[1]

I have your cable as follows:—

"Better see German Co explain our methods   things here satisfactory"[2]

I cabled you yesterday

Is Rotterdam machine shipped. Tell Perry cable sales Batchelors Isolated[3] & I got your reply as follows this am:—

Rotterdam shipped Schiedam ~~17~~seventeenth[4]

I am now just starting for Paris to see Batch as I received a telegram from him this morning[5] requesting me to go over there at once as he has to got to go to Nice Christmas morning. I will ~~then~~ when in Paris consult with Batchelor about going to Berlin & will in all probability go there about the 29th inst.[6]

I saw Lord Anson Mr White & Maj Flood Page[7] (now Secy of the United Co White having resigned—I presume by request). From them I gather that there is little or no Lighting business in England except Shipping & of this character of business they seem to get a great deal. Holborn Viaduct Central Stn. is being run at a <u>loss</u> of £10 ($<u>50.00</u>) per <u>day</u>.[8] what few Isolated Plants they have—they state—are running <u>well</u> but more of this later when I have seen more. All "outsiders" write in stating that Electric Lighting enterprise is dead in this country & the amalgamation of the Swan & the Edison Companies is looked upon as the union of two dying Ducks who by this means hope to get a new lease of life. I asked Flood Page if there was any objection to my going through the Swan Lamp Factory[9] & he said he thought not   If <u>you</u> object please cable me not to go

⟨~~only go through Swan books ascertain Cost, dont go through factory   E~~

Can you ascertain exact cost last year Swan Lamps   Dont go through factory⟩[10]

They are very anxious to know what you are doing in the way of high resistance Lamp. I could not give them any definite information as I do not know enough to tell them about the matter   If you have any information you want me to give them please cable it.

I made the proposition about the Patent Account & I <u>hope</u> that I <u>may</u> be able to get the money out of them early in January. Of course in proposing that, in future, we will only ask them to ~~pa~~ take the applications from us and then deal with them themselves they[a] paying all patent fees & we waving our right to <u>double</u> the cost of experimenting—I make it a first condition that they should pay us the amount of our present account

I will write you from Paris fully   Yours Sincerely

Saml Insull

ALS, NjWOE, DF (*TAED* D8367ZAR). [a]Obscured overwritten text.

1. Insull reached his family's London home on 20 December. He sailed again for New York on 19 January. Insull to Alfred Tate, 21 Dec. 1883; TAE to Insull, 17 Jan. 1884; both DF (*TAED* D8367ZAQ, D8403F).

2. This is the complete text of Edison's cable to Insull on 17 December. LM 2:34 (*TAED* LM002034C).

3. William Sumner Perry (1848?–1933) was a stockholder of the Edison Electric Light Co. of Europe, Ltd., and, in 1881, its secretary *pro tempore* (Obituary, *NYT,* 14 Nov. 1933, 19; List of stockholders of Edison Electric Light Co. of Europe, Ltd., 15 July 1881; Perry to TAE, 24 May 1881; both DF [*TAED* D8127L3, D8127J]). His association with Edison continued into the 1890s, when he was involved with ore milling projects (Israel 1998, 350–55).

Perry apparently was not directly employed at this time, serving instead as an independent business agent. In 1883, he managed some stock transfers on behalf of Edison, Insull, Alice Stilwell, and, as indicated in this document, Charles Batchelor. By early 1884, Perry had not sold any of Batchelor's stock in the Edison Co. for Isolated Lighting because of a "dull" market. TAE to Perry, 16 July 1883; Perry to TAE, 4 Dec. 1883; Insull to TAE, 4 Jan. 1884; all DF (*TAED* D8316AHC1, D8367ZAO1, D8465A); Insull to Perry, 2 Nov. 1883, LM 3:182 (*TAED* LM003182).

4. This is the full text of Edison's cable to Insull, dated in New York on 20 December (LM 2:34E [*TAED* LM002034E]). The dynamo in question was apparently one of the "C" (Jumbo) machines that the Edison Electric Light Co., Ltd., had ordered in 1882 but declined to receive after encountering financial difficulties (Doc. 2374 nn. 4–5). It had recently been used in tests of a new short-core Jumbo at the Edison Machine Works, where it was referred to as the "Holland 'C' Dyn" (William K. L. Dickson to TAE, 12 Dec. 1883, DF [*TAED* D8330ZBB]). The Machine Works billed it to the London firm, which presumably consigned it to the Nederlandsche Electriciteitsmaatschappij (NEM), a company created in June 1882 to promote the Dutch interests of the Compagnie Continentale Edison. It went aboard the SS *Schiedam* (Voucher no. 275 [1883] for Edison Machine Works; Van der

Feijst 1975, 17–18; TAE to Edison Machine Works?, 10 Dec. 1883, Lbk. 19:225 [*TAED* LB019225]).

The editors have not determined what NEM intended for the machine. NEM oversaw a number of isolated installations in Amsterdam and Rotterdam by the end of 1883. One of these plants used the Jumbo dynamo from the 1881 Paris Exposition; after having been employed in Charles Batchelor's factory at Ivry, the dynamo was sold to a factory in Rotterdam despite ongoing questions of ownership rights (Compagnie Continentale Edison Bulletins 2:22, 10 Oct. 1882; 3:15, 15 Feb. 1883; 5:16–17, 15 June 1883; PPC [*TAED* CA007B, CA007C, CA007D]; Doc. 2122 [headnote]; Sherburne Eaton statement regarding Edison Exhibit at the Paris Electrical Exposition, 14 Sept. 1883, DF [*TAED* D8331ZAC]). NEM had much larger ambitions and had planned to erect two small central stations of about one thousand "A" lamps each in Amsterdam and Rotterdam as proofs of the principles of the Edison system. Although it abandoned the Amsterdam plan because of competition from gas, difficulty with permits, and limited underground access, it continued with the Rotterdam plant. The Rotterdam station, at 34 Baan, began operating on 18 December. It had three 20-horsepower engines, four dynamos (probably the new "T" model), and was connected to roughly one thousand lamps. It closed a few months later after experiencing damage to the main underground cable as well as safety and economic problems. Another attempt at central lighting in Rotterdam was not made until 1889, and then not on the Edison system (Van den Noort 1993, 51–53, 55, 60–62).

5. Not found.

6. The editors have not determined whether Insull visited Berlin, but cf. Doc. 2581.

7. Samuel Flood Page (1833–1915) succeeded Arnold White as secretary of the Edison & Swan United Electric Light Co., Ltd., on 4 December. In that capacity, according to Theodore Waterhouse, he would "have the entire responsibility of the management of the Company subject of course to the control of the Board." A retired infantry officer, Flood Page had, as manager of London's Crystal Palace, organized the 1882 Electrical Exhibition. In June of that year, he became general manager of Edison's Indian & Colonial Electric Co., Ltd. He toured the Pearl St. central station in New York in July 1883. Obituary, *Times* (London), 10 Apr. 1915, 10; "Crystal Palace," ibid., 21 June 1882, 13; Waterhouse to TAE, 7 Dec. 1883; Flood Page to TAE, 5 July 1883; both DF (*TAED* D8338ZCA, D8339ZAG).

8. The Edison station for public and commercial lighting along London's Holborn Viaduct began operating in April 1882. Originally planned as a short-term demonstration, the station had its charter extended several times. The Edison and Swan United Co. suspended the plant's operation in mid-1884 but then resumed it until 1886 (Docs. 2180 esp. n. 6, 2261; Haywood 1883, 5–7; Commissioners of Sewers of the City of London, report, Oct. 1883, 8). Concerning Holborn's losses in 1884 and its eventual closure, see Hausman, Hertner, and Wilkins 2008, 78 and "City Notes, Reports, Meetings, &c. Edison and Swan United Electric Light Company, Limited," *Teleg. J. and Elec. Rev.* 15 (15 Nov. 1884): 397 (cited in Hughes 1993, 62 n. 59).

9. Insull presumably meant the Swan company's factory at South Benwell, Newcastle, established there in 1881 in a converted tannery. By this time, Swan's assistant James Swinburne had also opened a factory in Paris. *DBB*, s.v. "Swan, Sir Joseph Wilson (1828–1914)"; Swan and Swan 1929, 73, 82–83; Bowers 1982, 121.

10. Edison cabled to this effect on 2 January (TAE to Insull, LM 2:38A [*TAED* LM002038A]). Acknowledging these instructions, Insull promised to find detailed information but conveyed his understanding that Swan "sold 180,000 last year & made about £8000 ($40,000) but I presume this was at 5/($1.25) per lamp" (Insull to TAE, 4 Jan. 1884, DF [*TAED* D8465A]).

–2573–

Notebook Entry:
Incandescent Lamp

[New York,] Dec [c. 24] 83.[1]

Went to slaughter house to get samples of horn blader[a] and Intestines Grissel, softened them in Linseed Oil, and pressed them in sheets then cut them as samples to send to Lamp Factory to be Carbonized

J. F. Ott

X, NjWOE, Lab., N-82-12-04:123 (*TAED* N145123). Written by John Ott. [a]Obscured overwritten text.

1. This notebook entry, which Ott dated "Dec. 83," falls between his entries of 21 and 27 December. It represents a shift in the materials used in lamp filament experiments during the third week of December. Since August, Ott had been experimenting with making paper into filaments by soaking it in various solutions, usually containing resins (see Docs. 2510 esp. n. 3, and 2511). Now he began attempting to make filaments with gelatinous materials such as those listed here; he also experimented with flour mixed with a variety of materials (N-82-12-04:123–29, Lab. [*TAED* N145125, N145127]). At the lamp factory, William Holzer had begun by 18 December to carbonize materials such as tortoise shell, horn, cocoa fibre, vermicelli, goose quills, bast fibre, and tissue paper. Holzer continued to play a role in these and other filament experiments during 1884 (Holzer to TAE, 18, 19, and 28 Dec. 1883; all DF [D8333Y, D8333Z, D8333ZAA]; Holzer to TAE, passim, Electric Light—Edison Lamp Co.—Lamp Test Reports, DF [*TAED* D8430]). On 4 January, Martin Force and Hugh De Coursey Hamilton also began a series of filament experiments using "starch as a base" (N-82-12-04:131–41, Lab. [*TAED* N145131, N145133, N145137, N145139A, N145139B, N145141A]). Many of Edison's entries in the pocket notebook he took to Florida discuss filament experiments with these and similar materials. On 1 April, after Edison's return to New York, Martin Force began a series of filament experiments using gelatin (see Doc. 2609 [headnote] and N-82-12-04:175, Lab. [*TAED* N145175A]).

From Samuel Insull

London 27th Dec 1883

My Dear Edison

I went to Paris on Friday night last in compliance with a telegram I got from Batchelor.[1] When I got there I found a copy of Baileys letter to Eaton of 8th Dec setting forth the proposed plan for the amalgamation of the French Companies.[2] There is no necessity for me to go into the details of this proposition—Batchelor does that in the enclosed & sets forth fully the true inwardness of the scheme which seems to me a most disastrous one to the stockholders of the European Co[3]

I must say that I feel very much dissatisfied with the reception I received from Bailey. He evidently looked upon my visit as one o[f][a] hostility to himself & in explaining to me the proposed contract he gave me no more information than would an entire stranger & it was quite evident that his intention was to tell me as little as possible   Batchelor noticed this & in fact was the first to draw my attention to Baileys attitude. Bailey is in my opinion afraid of our knowing too much about him. Batchelors letter tells you exactly how he stands & the severe point to which he is pushed. It stands to reason that if Bailey is heavily in debt to these Paris Jews & is obliged to borrow money from them all the time he is a very poor agent for both[b] you and the European Company.[5] It is folly to suppose that they will keep loaning him money without getting an equivalent & that equivalent is in my opinion further concessions from you and the European Co. I do not mean to say that Bailey is dishonest but I do say that that from his unenviable position he is unable to do you and the European Co justice. He is away from Paris all the time; he has no idea whatever of what you have invented or the value thereof; he has but to hear of an alleged new lamp, a so called new Dynamo to come to the conclusion that unless our people purchase them their chances of success have absolutely gone. This is the man you have to depend on to protect your property; en[c] put enthusiasm into your people here & to see that the Edison System is preserved as a System. I must say that it surprises me to have to write this but it is the result of opinions formed after talking to Batch & Porges, & Batch's letter shows that he agrees with me.

Porges treated me elegantly— that is for Porges so Batch says.[5] Porges took me to the Theatre to see Sarah Bernhardt[6] & during the performance gave me his ideas of the business. If you accept all he says as his <u>real</u> opinion you would imagine that our business is dead in Europe but in listening to him you must read between the lines for his real opinion and I feel sure that he thinks he has a good thing & what he wants to do is to get it as cheap as possible—true to his Jewish instinct. The way he and his people will get it cheap is to "work" Bailey in the manner above suggested. That a reorganization is desirable and necessary I feel sure but that such a proposition as Batch's letter sets forth is by no means certain. If we want to have a Central Station started in Paris some modification of the plan suggested in Bailey's letter of 8th inst will have to be accepted but I would advise that you absolutely refuse to accept the last edition of the plan as Batchs letter depicts & Bailey should be given to understand that all modifications must be in the direction of giving more to the European Co & less to the French Companies   Bailey seems to think the opposite is the true policy.

You must understand that the amalgamation in England has materially altered the situation in Europe & I feel sure that in any final Contract you will have to yield to the French people the right to take in Swan although on no such terms of equality as in the case of the English Fusion

If some new Contract is made it is almost certain that a Central Station will be started in Paris which means that the reputation of yourself and & your System in its entirety will be established in Paris. From the English Co you can expect but little but more of this when I have seen Forbes the Chairman of the United Coy.[7]

I confirm following cable sent by Batch & myself from Paris on 23rd[b] inst.

Batchelor urgently desires return. Porges will abandon Central Station project if he leaves abruptly. Batchelor proposes sail January with full central Station data, consult you, decide plans, take Expert Paris start building finally return New York July. Cable Batchelor early Monday your views Baileys letter Eaton eighth for our private ~~guild~~ guidance   Batchelor Insull[8]

London is en fête that is so far as it is possible for a people subject to fogs (which require a hatchet to cut your way through) to be gay, and it is quite impossible to get people to do business this week. For the same reason you must excuse me for not writing more as I can assure you it is difficult to keep ones mind on business with a host of good looking cousins and sisters around. Yours Sincerely

Saml Insull

I have not seen the Sun since I set foot in England   I am getting hungry for a sight of it

ALS, NjWOE, DF (*TAED* D8367ZAS). [a]Paper torn. [b]Obscured over-written text. [c]Canceled.

1. Not found, but presumably this was the message that Insull received on Friday, 21 December, and mentioned in Doc. 2572.

2. Joshua Bailey's 8 December letter has not been found; it concerned the much-discussed proposition to recapitalize and reorganize the Compagnie Continentale Edison (holding company), the Société Industrielle et Commerciale Edison (manufacturing), and the Société Électrique Edison (isolated lighting), which were formed in February 1882 and collectively referred to as the French Companies. The projected reorganization was itself alternatively described as a fusion or as the International Co. Discussion of the proposals' merits and terms had taken place since the summer and at this time involved correspondence among Edison, Bailey, Insull, Charles Batchelor, and Sherburne Eaton (Doc. 2182 n. 2; Bailey to TAE, 10 Aug. 1883; Eaton "Memorandum on the Proposed International Edison Co.," 5 Oct. 1883; both DF [*TAED* D8337ZDN, D8331ZAD1]). According to a fourteen-page typed analysis prepared by Bailey for Sherburne Eaton on 29 December, the reorganization would require two legal steps. One was a set of changes to the bylaws of the Compagnie Continentale Edison, including a provision allowing it to use patents other than Edison's. The other was a contract between Edison and the Banque Centrale du Commerce & de l'Industrie, making the bank the agent for the other parties in carrying out the plans (Bailey to Eaton, 29 Dec. 1883, Batchelor [*TAED* MB103]). Over the short run, these negotiations received the endorsement of the Edison Electric Light Co. of Europe, Ltd., in March 1884, but the talks continued, and the Compagnie Continentale Edison did not receive authority to absorb the other two Paris firms until 1886 (Edison Electric Light Co. of Europe, report of directors, 7 Mar. 1884, CR [*TAED* CE001003]; printed "Notes on the Formation of an International Edison Co." and appended Articles of Association, n.d. [1883?], PPC [*TAED* CA006A]; Bailey and Batchelor to Eaton, 19 Feb. 1884; William Meadowcroft to TAE, 31 Dec. 1884; see also related correspondence in Electric Light—Foreign—Europe; all DF [*TAED* D8436ZAL5, D8428ZAO, D8436]; Hausman, Hertner, and Wilkins 2008, 78).

3. Insull enclosed a 25 December letter that Batchelor had expressly asked him to send to Edison. In ten handwritten pages, Batchelor evaluated the proposal put forward by Bailey for a "new company" to take over the existing French firms and for its unnamed capitalists to buy out a portion of the original investors' interests. After rehearsing details of proposed royalty and profit-sharing arrangements, Batchelor gave his opinion that "these people especially Porges who I have talked with a great deal would give anything to have the control that is now exercised by the [Edison Electric] Light Co [of Europe, Ltd.], in Paris where it can be consulted at a minutes notice. For this they were willing to give the Light Co 25 c[entimes]/ royalty on lamp forgo the payment of the F200,000 of capital and give 40% profits in a new company of 6,750,000 paid up capital." Insull's own negative view was doubtless informed by Batchelor's conclusion that the projected new company would be one "in which the Light Co gets $8\frac{1}{2}$ per cent of the total prof-

its . . . [and one] which in two years if they did as good a business as I have shown would make the Light Co. stock worth about $25" instead of its par value of $100. Batchelor to TAE, 25 Dec. 1882, DF (*TAED* D8337ZFZ).

4. Batchelor criticized Bailey's personal indebtedness to participants in the negotiations, particularly Elie Léon and Georges Lebey, in his 25 December letter (see note 3). Perhaps in response to reports of those affairs, Sherburne Eaton told Edison about this time: "when you took Bailey two years ago, you know Fabbri said you w'd regret it. I guess F. had the best judgment abt that— Personally, I dont think Bailey reliable." Eaton to TAE, 31 Dec. 1883, DF (*TAED* D8337ZGD); see also Doc. 2593.

5. Batchelor had introduced Insull to Porgès at Edison's behest. TAE to Batchelor, 15 Dec. 1883, LM 2:34B (*TAED* LM002034B); cf. Doc. 2553 n. 3.

6. The renowned French-born actress Sarah Bernhardt (1844–1923) was performing in Jean Richepin's *Nana Sahib* at the Théâtre du Porte Saint-Martin. Insull and Porgès attended with Batchelor on 22 December. Bernhardt had visited Edison's Menlo Park laboratory in 1880 to see the new electric light. Edison's daughter, Marion, later recalled that Bernhardt gave Mary Edison an autographed portrait on that occasion and later sent Edison two of her paintings in remembrance of their meeting. "French Plays and Gossip," *NYT,* 7 Jan. 1884; Huret 1899, 102; Batchelor diary, Cat. 1343:179, Batchelor (*TAED* MBJ002179); Docs. 2019 n. 3, 2036 (headnote) n. 1; *TAEB* 5 App. 1.B.25; Oser 1956, 2; Bernhardt's recollection of that 1880 meeting is in Bernhardt 1907 (376–79).

7. James Staats Forbes (1823–1904), a former draftsman for the famed engineer Isambard Kingdom Brunel, was chairman of the Edison & Swan United Electric Light Co., Ltd., having held that position in the antecedent Swan United Electric Light Co., Ltd., since its inception in 1882. He became a leading figure in attempts to amend the Electric Lighting Act of 1882. Forbes was also chairman of several railway companies (*Oxford DNB*, s.v. "Forbes, James Staats"; *Daily Telegraph* [London], 26 May 1882, in Cat. 1327:2197, Batchelor [*TAED* MBSB52197]; Theodore Waterhouse to TAE, 2 Oct. 1883, DF [*TAED* D8338ZBS1]; "Lord Thurlow's Committee on the Electric Lighting Act," *Times* [London], 12 Dec. 1884, 7; Hughes 1983, 62–63). Bailey had been negotiating with Forbes for the participation of the British Swan interests in the new arrangements; by his own account, "something different is talked [about] every day" (Bailey to Batchelor, 14 Dec. 1883, Batchelor [*TAED* MB102]).

8. This is the full text of the cable (LM 2:36A [*TAED* LM002036A]). Edison's reply on Monday, 31 December, is incorporated into Doc. 2593. In a postscript to his 25 December letter (see note 3), Batchelor elaborated on his motivation: "After working nearly 3 years and not receiving enough to board me you will not wonder that I would sooner come over there when the prospect of working two or three years more to make my stock worth $25 stares me in the face."

Frank Sprague to
William Hammer

New York, Dec. 27 1883[a]

My dear old Chum=

Let me wish you a Merry Xmas and a Happy New Year, before the close of which I hope I may have the good pleasure of seeing you again.

I know I owe you all sorts of apologies, and I know too you have long ere this vented in no mild way your vexation at my seeming indifference— But be assured old man, that it has not been because you have been in any way forgotten, or that I have not wanted to hear from you— I have led such a mixed up sort of existence that I have grown into the sad habit of procrastination, and then, too, I have been looking forward to something specific to tell you— Yours of long ago[1] was received with pleasure—and I'll now let you know briefly what has been occupying me so long.

In July, Andrew, Edison, and myself started the first 3-wire overhead system at Sunbury, Penna— In a week or so, I was recalled to New York, and soon after Edison started Clark, Campbell, someone else, and myself in an attempt to mathematically determine the feeders and mains of the town plants without going to the trouble of practical experiment at Menlo Park—[2] After 3—or 4 days blind work, I began to work backwards, and soon had it by the ___.

Then I became expert in the work, and soon could lay out the main and feeder to the right sizes in 2 to 4 hours where it had taken 6 to 7 days— In doing this I proved the Menlo Park work all wrong, and, generally, I made a pretty good bit— This made me pretty solid with Edison, and I kept the thing in my own hands for several weeks. Then I went to Brockton and started the first 3-wire underground system   I spent altogether about 2 months there, learning[b] a great deal by practical experience in the development of this new system— I was often in Boston, and also went home—[3] While there I made for a while all the street determinations   Now they are done here in the office by one of Clark's former men and another—[4] Have you done anything with the 3-wire system over there yet?

We have Sunbury, Shamokin, overhead, and Lawrence, Brockton, & Fall River, underground, in very successful operation, and about a half a dozen others in process of erection— Next year there will be a big number started.

The down-town station is running up to its capacity—700 more lights waiting to be cut in.[5] Receipts about double the expense   Chinnock in charge—

Edison is really running the business, and puts up these stations on contract. Things are pretty dead downstairs because matters to a great extent have been taken out of their hands— While I was in Brockton, the down-town station failed,—cause overloaded feeder, in the stock exchange quarter, and a successive failure of safety catches.[6] I telegraphed reason from Brockton, and E. telegraphed me to come down here, to attend to some work here. The first two days was engaged on calculation down town. Feeders have[b] been changed, and everything is much better now—

Then I went on to a new pressure indicator—a daisy— founded on an old discovery of E's, thus,[7]

carbon of lamp

main +   − main

E galvanometer coil

With above connection of this special lamp, no appreciable current will flow through the galv. coil across from the platinum to the carbon until a fair incandescence is reached, and then the current increases very rapidly— Hence the indicator shows no deflection until you wish it to, and then it begins to move rapidly— It has fine adjustments and is so sensitive that if $\frac{1}{100}$th of the lamps be thrown off, it will show the change in pressure— Its sensitiveness can be made more or less. I came back from Brockton early this month, and am now looking after a new feeder equalizer and a boss ampère meter of my own design—[8] The equalizer is for changing the pressure by introduction of resistance in case of an abnormal distribution of the load in a town— Generally my present duties are practically those of an assistant to E— I doubt whether I go out to get any more stations ready— T.A.E. and I went to Fall River last Tuesday, and the station was started up on Wednesday— Stayed there two days, went to Brockton and tried to kill myself jumping on a moving train, and came back Saturday— Spent Xmas at home—

I have been quietly working up a new motor— I think I explained it to you in London— The U.S. ~~Patent~~ claims[c] have been allowed, and I will pay the final fee as soon as I hear the French and English cases have been received— These are now being prepared. Last Saturday, I tried the motor and an entirely new method of winding an armature—and I proved my claim to run in either direction, at any speed or power up to the maximum on a given circuit— Johnson is going in on the English and French cases— T.A.E. does not know of my work yet— I am going to soon have a good sized motor made, and then he can see it— I have solid claims here on it—and I am absolutely certain of its action— Will send you copies of the U.S. patent when it is out—[9]

Beves[10] is at Fall River— A regular car is being established— B. was well, inquired particularly after you— [Seems?][d] happy, but said seeing me made him a little homesick. The Johnsons are all well, and they have a lovely house— They have not yet got their gas engine—and so they did not have the light on Xmas—[11] "Sammy," I suppose you know, has gone over the water, and probably by this time has gone to Berlin— He[e] and I have had occasional rows, and when in Brockton I wrote about the hottest letter Edison has received for some time, in which I gave Sammy the devil—[12] He is too conceited, and occasionally forgets himself— But it is not well for him[b] to try to step on my toes—for I kick.

Geo. Sonn[13] I have not seen for some time— I shall try to get to Newark soon. I liked both him and Miss Honness[14] very much

I have not seen your parents yet—but shall soon— As I have said, I have heretofore been away a great deal, and occupied with special work. Upton I see occasionally. I have not gotten over to the lamp factory yet, nor have I been to ~~the~~ Menlo Park.

Dean was fired[b] out of Goerck St. for dishonesty, and Soldan is in his place, a good man— The "H," 400 ampères, the "Y," 300, and the "S," 200, are fine machines. The S and Y have one pair of cores only—

Do you get the bulletin? Come to think, I'll find out down stairs—of Miss [Selden?][15d]— I send two or three pamphlets

Saw Frank Locke[16] yesterday. He has written you, and wants your address— I send it to him to-day— He is looking very well, and I think will stay so— He thinks of coming down here— I hope he will— Wish you were here— Wouldn't we have a time! He inquired after you most cordially—

I enclose money order for $25.00, amount I believe I owe you— I am positively ashamed I have let this stand, but I was waiting to write you— At any rate, I am heartily obliged[b] to you for the loan—

If I have made any mistake, please let me know— Should Pierce,[17] who assisted me in London, turn up in Berlin, you will find him a[b] good fellow to have—

My report to the Navy Department is in proof—[18] When in form will send you a copy— Now write and tell me about your own position and work, how[b] you like it, your "affaires de coeur," &c— Can you speak the lingo[b] like a native yet? With best wishes, Yours sincerely,

F. J. Sprague

ALS, NN, FJS (*TAED* X120CAG). Letterhead of Thomas A. Edison, Central Station, Construction Dept. [a]"New York," and "188" pre-printed. [b]Obscured overwritten text. [c]Interlined above. [d]Illegible. [e]Repeated at end of one page and beginning of next.

1. Not found.

2. The unnamed person may be F. Hoffbauer, a self-described inventor, observatory assistant, and instrument-maker of German birth. Having previously left the employ of Edison or one of the Edison companies, Hoffbauer wrote to Edison in April 1883 about a mathematical method and a "mechanical wire calculator" for determining the necessary size of conductors. Edison urged Charles Clarke to investigate, and he offered to hire Hoffbauer to make determinations on a contract basis. Hoffbauer to TAE, 7 Apr. 1883, DF (*TAED* D8313J1); TAE to Clarke, n.d. [1883], Hodgdon (*TAED* B017ADA); TAE to Hoffbauer, 13 Apr. 1883, Lbk. 16:136 (*TAED* LB016136).

3. Frank Sprague was born in Mitford, Conn., but grew up in North Adams, Mass., in the care of an aunt, after his mother died. *NCAB* 24:15.

4. Sprague later described the construction of the physical models as "tedious and unsatisfactory work." Reportedly at Edward Johnson's instigation, he took up the problem of making the determinations mathematically. His Naval Academy training, he recalled, "stood me in good stead" (Sprague 1904, 567–68). The identities of the men listed here are unclear, but see the list of employees in Doc. 2482. Their work supplanted that of Hermann Claudius, who had continued to work on determinations at Menlo Park during the fall. At the prompting of Sherburne Eaton, and over Charles Clarke's objections, Edison decided in late November to dismiss Claudius, having decided that "We have no use for him at present." Edison and Clarke tried to secure him a position at the Ansonia Brass & Copper Co.; at an unspecified later date, Claudius was promoting himself as an independent engineer for making "Electrical Tests and Measurements" (Samuel Insull to Sprague, 4 Sept. 1883; Claudius to Clarke, 26 Nov. 1883; Eaton to Clarke, 14 Nov. 1883; TAE marginalia on Clarke to Insull, 24 Nov. 1883; Alfred Cowles to Clarke, 27 Dec. 1883; Claudius circular letter, n.d. [1884?]; all DF [*TAED* D8316ASN, D8329ZAN, D8329ZAG, D8329ZAL, D8321ZAY, D8420ZAL]).

5. Sprague referred to the Pearl St. station in New York, the capacity of which was increased in May 1884 by the addition of two dynamos (see Doc. 2669).

6. According to a short account in the *New York Tribune*, 4,500 lamps went out in the New York first district at 5:15 p.m. That number of lamps was the greatest used simultaneously to that date, "[o]wing to the dark afternoon," and drew more current than the safety catches were sized to handle. Workers replaced the catches with larger ones, and the lights came on again shortly before 9 p.m. The newspaper reported that "Mr. Edison, who was at home sick, did not consider the accident of any scientific moment. He regretted its occurrence on account of the inconvenience to the public, and says it shall not occur again." "The Edison Lights Go Out," *New York Tribune*, 24 Oct. 1883, 8.

7. This is the new voltmeter based on the Edison Effect; see Doc. 2538 (and headnote). Figure labels are, clockwise from left, "main +," platinum in lamp," "carbon of lamp," "– main," and "galvanometer coil."

8. Sprague executed a patent application for this instrument on 24 January and filed it at the Patent Office two days later; it resulted in U.S. Patent 314,891 in March 1885. The instrument was an "indicator of variations in electric current which shall operate entirely by such variations, and shall not be influenced in any way, or at least to any practical extent, by permanent magnetism." It acted in response to the magnetic field created by current flowing through a conductor and did not require steel magnets, which were subject to the influences of dynamos and terrestrial magnetism. See also Sprague to TAE, 11 Jan. 1884, DF (*TAED* D8442N).

9. Sprague had filed a patent application in May 1883 that issued as U.S. Patent 295,454 in March 1884. The specification was a broad one, with sixteen claims, and Sprague intended it as "the first of a series" of patents for improvements in electric motor design (see also App. 4C). In the specification, he described a general method of shunting current between the armature motor and field coils so as

first, to vary the speed at will; second, to vary the power developed at any given speed; third, to maintain a fixed speed, or one varying only between narrow limits with varying loads; fourth, to maintain a high efficiency—in short, to vary automatically or at will the general arrangement of circuits in a motor so as to meet the various requirements of load and speed, and at the same time to preserve economy.

10. Arthur S. Beves (1857–1945) met Edward Johnson and other Edison associates in his native London in 1882 and was hired to help prepare for the Crystal Palace Exhibition. He soon immigrated to the United States and by 1883 was working at the Edison Machine Works and, by the end of that year, for the Edison Construction Dept., largely in Ohio. Beves later served as treasurer of the Sprague Electric Railway & Motor Co. and the Edison General Electric Co., and as secretary and treasurer of General Electric. Following a personal bankruptcy in 1901, he held executive positions in companies unconnected to electric lighting. Obituary, *NYT*, 24 July 1945, 23; Obituary, *Pioneers News*, 1946, p. 1, Pioneers Printed; Samuel Insull to Charles Chinnock, 23 Nov. 1883; Alfred Tate to John Randolph, 6 Dec. 1883; letterhead, Beves to Insull, 6 Apr. 1886; all DF (*TAED* D8316BKH, D8340ZHR, D8639C); Sprague Electric Railway & Motor Co. circular, 25 June 1887, Unbound

Documents (1887), UHP (*TAED* X154A6AH); "Bankruptcy Notices," *NYT*, 29 Mar. 1901; Trow Directory Co. 1906, 269; *Directory of Directors* 1915–16, 54.

11. Gasoline engines were sometimes used to power isolated lighting plants because of their compact size and their quiet and clean operation compared to steam. Their speed tended to oscillate with the power stroke, however, and this variability commonly caused the lights to flicker, a problem that Edison had tried to address in 1882 (see Doc. 2359). Edward Johnson, seeking to acquire an engine for his family's home at 139 E. Thirty-sixth St. in New York, had recently telegraphed the Sterne Gas Engine Co., a prominent manufacturer, "Christmas coming is engine." As early as 1882, Johnson apparently decorated his family's Christmas tree with Edison lamps that turned on and off as a small motor revolved the tree. In December 1883, he used that experience to design ornamentation for a revolving tree forty-five feet high with 225 red, white, and blue lights at the Foreign Exhibition in Boston. Johnson reprised the domestic display at his home the following year with 120 lights in clear and colored globes on a six-foot tree. Trow's 1884, 844; Johnson to Sterne Gas Engine Co., 10 Dec. 1883, LM 2:33B (*TAED* LM002033B); "A Giant Electric Tree," *St. Louis Globe-Democrat*, 22 Dec. 1883, 16; "In and About the City. A Brilliant Christmas Tree," *NYT*, 27 Dec. 1884, 5.

12. Sprague may have meant his 8 September letter (not found, but acknowledged in Doc. 2526) or one written soon after from Brockton, Mass., in response to an incident discussed in Doc. 2526 nn. 1–2. The latter letter's tone does not fit the angry description given in this document, but in it Sprague blamed Insull for "an undue amount of criticism of whatever I have done or said concerning another at this station." Sprague to TAE, 10 Sept. 1883 (p. 1), DF (*TAED* D8343ZAY).

13. John George Christopher Sonn (1859–1906) had attended Newark High School with William Hammer. He graduated from Yale in 1879 and soon after applied to Edison to become an agent for the telephone abroad; it is not clear if he was so engaged. At the time Sprague wrote, Sonn was teaching political and natural sciences at New Haven High School. "Professor" Sonn became head of Physical Sciences at Newark High School in 1886, a post he held until his death. Williams 1906, 391–92; Sonn to TAE, 22 Dec. 1879, DF (*TAED* D7941ZKA).

14. Ada Dusenberry Honness (b. 1859), married Sonn on 9 April 1884. Williams 1906, 397; U.S. Census Bureau 1982? (1900), roll T623_964, p. 3A, enumeration district 80 (Newark Ward 8, Essex, N.J.).

15. Not identified.

16. Frank Locke has not been identified. Sprague carried on correspondence with someone in Brooklyn whom he addressed as "Frankie." Dalzell 2009 (44–45 esp. n. 44) speculates that this person may have been Frances Scott, a young woman with whom Sprague developed a warm relationship; Middleton and Middleton 2009 (24–25) identify the woman as Frances Seale. Sprague to Locke, 5 Dec. 1880 and 4 Feb. 1881, FJS (*TAED* X120CAB, X120CAC).

17. Not identified.

18. Sprague's lengthy report on jury tests at the 1882 Crystal Palace electrical exhibition, prepared in March 1883, was published as Sprague 1883b; see Doc. 2342 n. 14.

-2576-

From George Bliss

CHICAGO, 28th December, 188[3][a]

Dear Sir:

Dec 5th we received a letter from you signed by Mr. Insull, containing the following paragraph:

"We must confess that we cannot understand why it is, but the troubles with machines seem to come almost entirely from your territory, as we do not hear of any troubles from any other places that amount to anything."[1]

Your favor of the 24th is now at hand, enclosing Mr. Solden's letter, which we return herewith.[2]

We are not familiar with the whole field, but judging from the defect sheets received here, and information from other sources, this is not the only territory where trouble is experienced with dynamos.

You have repeatedly requested me to report all defects in apparatus fully, no matter where where from.

We have endeavored to do this always in a fair way, hoping the facts would assist you in perfecting the apparatus.

You must not forget that we have a considerable territory, with many plants in operation—some of them having run a long time; Furthermore, these plants are not under our control.

If the dynamos were run by our men or men selected by us there would be less difficulty.

The average engineer, especially if the has other duties to perform, oftentimes neglects a dynamo at a moment when with little care, trouble would be avoided, but once started disastrous results follows.

As a rule electric light plants increase the duties of engineers without additional compensation and they will beat the electric light plant if permitted to do so.

A great many of our dynamo troubles come with plants where a number of lamps have been installed in excess of the capacity of the dynamo.

Notwithstanding explicit specifications in our contracts as to the number of lights to be burned at one time the dynamos are often abused, causing troubles for which the purchasers themselves are alone responsible.

We have now adopted the rule of never installing more lamps than the rated capacity of the dynamo.

We have some plants in use where the service is very hard on the dynamos as a full load must be carried for ten, and in some cases, twenty-four hours.

It is my opinion that the capacity of the dynamos furnished has been over-rated when such a service is required.

I know that a 250 light dynamo will often burn even 300 lamps for a limited time, but my experience is against their capacity to carry 250 lights ten hours continuously.

The greatest trouble we have with dynamos comes from the raising of the commutator bars when hot resulting in their sinking when cool.

We almost invariably find the paper under the bars carbonized which causes mischief.

⟨We use mica under bars— if the parties would prevent sparking & thus not heat Com[mutator] the bars would not raise—⟩[3]

I supposed you were now using mica under the bars, but the 100 light dynamo recently received has paper. ⟨ask Soldan if that is so⟩

The dynamo has been carrying 88 lights ten hours, but has not performed well. Bars have raised in the commutator and the dynamo has always run hot.

Mr. Wirt,[4] our inspector, thinks the commutator is too small to take off such a current.

I am not yet prepared to express this opinion.

The dynamo certainly has not done well.

⟨—We test this mach with 125 lights 5 hours & have run 120 lights 11 hours—⟩

Besides the bars raising and sinking it has been hot enough twice to unsolder wires and burn insulation on end of the wires.

This dynamo has had a good deal of attention and inasmuch as we have run commutators nearly two years without a thing being done to them I believe we know how to take care of them.

The Davenport L repaired armature went direct to the Davenport Electric Light Company[5] who could not make it work, although one we loaned them did well.

In trying to make it work they undoubtedly broke off the knife blade between the bars.

This, however, does not prove it to be have been received in good order.

Inasmuch as two to other repaired armatures came back to us in bad condition we felt warranted in making a complaint.

I shall be heartily glad when our machinery is concentrated in central stations under competent management.

Isolated business will always be a great annoyance.

In closing please allow me to call attention to the long time it takes to repair armatures.

We have several now at the Machine Works for which our service is suffering at the present time.[6] Sincerely yours,

Geo. H. Bliss   Gen Supt.

ALS, NjWOE, DF (*TAED* D8364ZCF). Letterhead of Western Edison Light Co., George Bliss, general superintendent; "DYNAMO TROUBLES." typed in preprinted memo line. ᵃ"CHICAGO," and "188" preprinted.

1. TAE to Bliss, 5 Dec. 1883, DF (*TAED* D8316BOO).

2. Bliss had asked for advice about whether to get an ammeter for a brewery lighting plant having "a considerable number of lamps in excess of the capacity of the dynamo." Edison answered that he should obtain one from Bergmann & Co. (Bliss to TAE, 19 Dec. 1883; TAE to Bliss, 24 Dec. 1883; both DF [D8364ZCE, D8316BVB]). The editors have not found the enclosed letter from Gustav Soldan, apparently sent in reply to Bliss's inquiry to Edison about acquiring new commutators. Edison (or Alfred Tate) chastised Soldan for having failed to send the information to 65 Fifth Ave.: "It is impossible for us to maintain the correspondence of this office unless we are allowed to answer communications which are addressed directly to us" (TAE to Soldan, 15 Dec. 1883, DF [*TAED* D8316BTG]).

3. Bliss had written in August about similar problems resulting from charred paper insulation; Edison promised then that "We are going to use mica hereafter." A brief typed reply to Bliss's December complaint stated that "this machine was made before we adopted mica. All machines you get from us in future will have mica backs for the commutators." Edison had for some time used mica insulation between commutator bars (see, e.g., Docs. 2122 and 2149), but he did not file a patent application for using mica to separate the bars from the underlying metal shaft or sleeve until July 1883. Bliss to TAE (with TAE marginalia), 20 Aug. 1883; TAE to Bliss, 7 Jan. 1884; both DF (*TAED* D8364ZAX, D8416ACB); U.S. Pat. 438,302.

4. Charles Sumner Wirt (1858–1924) started working in the Testing Room of the Edison Machine Works in 1881. He was quickly reassigned to the Edison Co. for Isolated Lighting, where he installed lighting plants and also devised an improved commutator brush. In 1882, he went to Chicago, where he did installation work for the Western Edison Co. and developed a portable photometer and compact potentiometer. He remained in the Edison lighting business, through Bergmann & Co. and Edison's West Orange, N.J., laboratory, for several more years. Wirt later became a successful electrical inventor, entrepreneur, and manufacturer based in Philadelphia. "Wirt, Charles" Pioneers Bio.; Obituary, *NYT,* 15 Apr. 1924, 21.

5. Organized in May 1882, the Davenport Electric Light Co. was evidently not an Edison company; formation of an Edison company there was still being discussed a year later. *Souvenir Booklet* 1926, 4; "Edison's System," *Davenport Daily Gazette,* 22 May 1883, 6.

6. Bliss mailed a list the next day of armatures shipped to New York for repair, some as long ago as October. In response, Edison asked him to

attach identifying tags to equipment sent for repairs; he also requested Gustav Soldan to keep records of all such returns. In addition, he asked Soldan for a "full report" on one armature sent by Bliss, which had "portions of the brush sticking to it proving that they had been melted by careless handling." Bliss to TAE, 29 Dec. 1883; TAE to Bliss, 7 Jan. 1884; TAE to Soldan, both 7 Jan. 1884; all DF (*TAED* D8364ZCG, D8416ABN, D8416ACG, D8416ACF).

–2577–

From William Andrews

Tiffin O. Dec. 31 1883[a]

Dear Sir

I find in the Y Dynamos both at Fall River and at Tiffin that there is <u>considerable</u> magnetism in yoke across top of the magnet, and a consequent pole[1] about eight inches below yoke. Does not this entail great loss of energy, and is it not probably due to imperfect fit between said yoke and the iron cores of magnet? Yours truly,

W. S. Andrews

⟨All the new single magnet Dynamo has consequent poles it is due to bad iron— E⟩[2]

ALS, NjWOE, DF (*TAED* D8343ZFQ). Letterhead of Thomas A. Edison, Central Station, Construction Dept.; W. S. Andrews, CHIEF ELECTRICAL ENGINEER. [a]"188" preprinted.

1. A consequent pole is one formed between poles normally found at the ends of a bar magnet. It may be formed in magnetized iron or by irregular windings in an electromagnet. A consequent pole in the field winding of an Edison dynamo would tend to weaken and distort the magnetic field around the armature, as would the partial magnetization of the yoke. Atkinson 1883, 607–8; Caillard 1891, 181–82.

2. Edison's marginalia was the basis for a short typed reply from Alfred Tate on behalf of the Edison Construction Dept. Edison also inquired of Gustav Soldan at the Machine Works at this time "if the new iron gives higher volts." At the beginning of December, Samuel Insull had complained of poor quality iron in forgings supplied to the Edison Machine Works. Tate to Andrews, 2 Jan. 1884; TAE to Soldan, 31 Dec. 1883; Insull to MacPherson, Willard & Co., 1 Dec. 1883; all DF (*TAED* D8416AAE, D8316BXD, D8316BMG).

–2578–

Draft Advertisement: Edison Construction Department [1]

[New York, December 1883?[2]]

Edison Electric Light System[a]

The only system of Lighting by [~~the?~~][b] incandescence[c]

Thos A Edison, contractor for the errection of his[d] complete system of electric lighting by incandescence for public & private ~~lighting~~ purposes, distributed from a central station through all or any of[e] the Streets of ~~the~~ a[d] city either underground or on poles. Houses connected with the mains and supplied with light measured acurately by an Electric Meter, ~~lights of~~ each light independent turned off or on like gas. Lamps of 8 10 16 20 32 & 50 candle power each ~~used~~ supplied from the same system indiscriminately— Eight cities in the United lighted by this system and 37 others now be contracted for.[3] Central station in NYork been in successful operation for nearly one year supplying light to over 750 stores newspaper offices such as the New York Herald Times, Truth, Commercial Advertiser, The New York Stock Exchange. Edison undertakes to install complete system and ~~leaves~~ furnish Experts to run the same guaranteeing the result. Isolated plants for Mills, Sugar Estates, etc furnished with Experts to set them in practical operation. for [~~fur?~~][b] terms & further particulars address Thos A Edison 65 5th Ave N York

ADf, NjWOE, DF (*TAED* D8340ZIK). Letterhead of Thomas A. Edison, Central Station, Construction Dept. [a]Obscured overwritten text; heading multiply underlined. [b]Canceled. [c]"The only . . . incandescence" apparently interlined. [d]Interlined above. [e]"all or any of" interlined above.

1. From its content and tone, the editors conjecture that this document was the draft of an advertisement or circular.

2. The editors have inferred this date primarily from the internal reference to Edison's New York central station having been "in successful operation for nearly one year." The Pearl St. station started supplying electricity on 4 September 1882, but the Edison Electric Illuminating Co. did not begin charging customers—one reasonable criterion of "successful operation"—until mid-January, 1883; see also note 3. Jones 1940, 209, 212.

3. The eighth Edison central station (including the small plant in Appleton, Wis., discussed in Doc. 2352) opened at Fall River, Mass., on 18 December. Approximately four dozen towns and cities had been canvassed by the end of the year; the editors have not determined the contractual status of most of them, but see App. 2.

January–March 1884

Edison began the new year on familiar ground. Although the Edison Construction Department generated a large amount of business correspondence, Samuel Insull handled the routine matters, leaving Edison time to address technical problems. Among those issues was the perennial question of improving and cheapening the incandescent lamp filament. In January, he directed experiments by several assistants to develop synthetic filaments from various gelatinous compounds that could be shaped, cut, and baked into carbons of greater uniformity than those obtained from bamboo or other raw natural fibers. He also oversaw the research of Edward Acheson on a process of electrodeposition of a thin metallic coating in a vacuum. Edison's hope of producing a gold foil, perhaps for decorative purposes, was not realized, but the technique subsequently proved useful in the reproduction of phonograph records and, decades later, in the manufacture of semiconductor circuits.

Edison also had to face again a technical problem that he believed had been overcome: voltage regulation in small central stations. The village plant systems proved susceptible to uneven voltages throughout the network, as well as to fluctuations caused by the station operators' inattention or inexperience. Unfortunately, the voltage indicator based on the Edison Effect lamp, which he had adopted with such confidence in the fall, proved inconstant and unreliable in service. Experiments in January by John Ott and John Howell provided no way to stabilize the instrument's own internal resistance. By the end of February, Edison ordered the instruments superseded by Charles Bradley's modification of the voltage indicator already being used satisfactorily at the Pearl Street station in New York.

Edison invariably believed in his ability to overcome technical problems. The difficulties with voltage regulation at village plants in Pennsylvania, Ohio, and Massachusetts, however, hinted at larger problems in the electric lighting business that were beyond his expertise and, in some cases, outside his control. Poor regulation caused lamps to burn out prematurely, and managers of some of the local Edison illuminating companies complained bitterly about the cost to their profits and their service reputations of these failures. Edison recognized that "Too much importance cannot be attached to this matter of pressure [voltage] indicators as the lamp breakage hinges directly upon it."[1] Inattention by station operators, or their poor training, contributed to the problem. The operators were employees of the local companies and most had no prior electrical experience. Edison, who was accustomed to working intimately with his own subordinates, belatedly took steps to standardize the operators' training, certify their competence, and hold them accountable. Frustrated also by delays and inaccuracies in the engineering work for the Construction Department, he responded by reorganizing his engineering staff in late January.

Managers of several local stations also complained about shoddy construction and equipment failures and, at least once, refused to make a final payment. New construction continued, however, supervised by Edison's men. Two more stations opened in January, two in February, and one in March. These plants alone represented more than $100,000 in outlays by Edison for equipment and labor. Repayment to him came in installments and, in some cases, in stock shares rather than cash.

Edison was not alone in feeling his financial resources pinched. At the end of January, a shareholder of the Edison Electric Light Company and related firms suggested reorganizing the fragmented Edison lighting interests into a single company. The proposal's stated rationale was to raise capital for expansion, notably for the proposed Second District in New York, but it also reflected concern about deteriorating economic conditions generally. The keystone suggestion that the Edison Electric Light Company negotiate a takeover of Edison's profitable manufacturing shops was an idea that Edison had long resisted and would continue to oppose. Perhaps not coincidentally, two of those shops, the Edison Lamp Company and the Edison Machine Works, were legally incorporated in January. In Paris, investors in Edison's French electric light companies continued negotiations, similar but unrelated to the New York effort, to consolidate the three separate but interdependent French firms. Edison continued to hold out for terms that would give him the most cash from recapitalizing the French business. He meanwhile looked to other foreign markets. Indicating a growing interest in South America, he negotiated with the Edison Electric Light Company for the rights to sell isolated plants in Chile and to set up central station plants there in cities other than Santiago and Valparaiso. At the start of February, he designated an agent, Willis Stewart, to conduct business for him in Chile.

Whatever concerns Edison may have felt about his finances, he and his family continued to enjoy a comfortable material life. Mary Edison purchased fine fabrics and clothing, the couple dined well, and they apparently attended the theater on occasion.[2]

In early February, Edison and his wife embarked on a Florida vacation. This was an increasingly common itinerary for affluent Northerners, and Thomas and Mary had made the trip once before, in 1882. From the rail terminus at Jacksonville, they traveled south along the St. Johns River to Sanford and possibly as far as Titusville before returning, with an excursion to St. Augustine. Leaving the children at home, they traveled in the company of Mary's good friend Josephine Reimer (and possibly Josephine's husband, as well). They were away from New York for about seven weeks. Edison fished and hunted. He also let his mind play in a small pocket notebook, where he made numerous entries, many of them undated. Many of his ideas were related to lamp filament material, but he also returned to past subjects like multiple telegraphy and evidently drew inspiration from both the natural world (for artificial materials) and fellow travelers (concerning the uses of bran).

Edison conducted some business during the trip, including making an estimate for a hotel isolated lighting plant. He also sent instructions to John Ott regarding filament experiments, and he telegraphed occasionally with Samuel Insull concerning the negotiations in Paris, his own desire to manufacture steam engines under license from Armington & Sims, and the nullification of key lamp patents in Sweden. However, during the three weeks that Thomas and Mary vacationed in St. Augustine in mid to late March, communication with his New York office came nearly—and uncharacteristically—to a stop.

When Edison returned to New York about 31 March, two familiar faces were no longer there. Charles Clarke, who had played crucial roles in engineering dynamos and central stations, had left the Edison Electric Light Company in mid-January to develop his own inventions. Laboratory experimenter Edward Acheson left in March for similar reasons, taking another step toward what became a brilliant career. In addition, financier Henry Villard, a friend and one of Edison's most enthusiastic backers, was loosening his American business ties in preparation for returning to Germany. Edison had paid a consoling visit to Villard in January, after the transportation magnate's railroad empire collapsed, both wiping out his personal fortune and hinting at the general liquidity crisis that would follow in the spring.

1. TAE marginalia on William Andrews to Samuel Insull, 2 Mar. 1884, DF (*TAED* D8442ZBX).
2. See Vouchers (Jan.–Mar. 1884).

–2579–

From Bergmann
& Co.

New York, Jany 2 1883[a]
Constr Dept
Dear Sir

In reply to yours inclosing letter from Manager of Shamokin Co,[1] asking information concerning the sizes & prices of motors we would say that we make a Motor $\frac{1}{4}$ H.P. the price of which from us is \$35.00 and also a Sewing Machine Motor (about $\frac{1}{8}$ H.P.) the price of which from us is \$30.00. This price does not contemplate the payment of any royalties on our part however, and we do not know under what rule they are to be sold to any other than the Edison Company. We have Endeavored several times to get information & have addressed both Major Eaton & Mr Hutchinson on the subject. We have however had no reply or instructions. Perhaps you can give us the necessary information. Will you please see what can be learned on the subject. Very truly Yours

Bergmann & Co   Klein[2]

⟨Maj E= please write official letter so B & Co can furnish our licencees at same price as he would furnish us—

Say to Bergmann & Co that the price of the Sewing Mac Motor is so high that it will kill the sale entirely   they should be made & sold for 15 dollars   can it be done— E⟩[3]

L, NjWOE, DF (*TAED* D8424B). Letterhead of Bergmann & Co.; monetary expressions standardized for clarity. a"New York," and "188" preprinted.

1. Neither the letter from Edison nor from station manager William M. Brock has been found. Brock (1856–1930) learned telegraphy as a railroad messenger in Shamokin. After having set up a telephone line to Sunbury, he became manager of a local telephone company and installed exchanges in the area. With Edison's approval, the local Edison company hired him to assist William Andrews, and he became the first superintendent of the Shamokin plant. Brock took over management of the Edison station in Lawrence, Mass., in 1886 and later held a similar position in Paterson, N.J., for many years. Obituary, *NYT,* 5 July 1930, 9; "Brock, William M.," Pioneers Bio.

2. Philip Henry Klein, Jr., (b. 1859?) worked for Bergmann & Co. as a bookkeeper and secretary. He later had an electric lamp import and export business in New York. "Klein, Philip Henry, Jr.," Pioneers Bio.; Klein to TAE, 18 Aug. 1882, DF (*TAED* D8201Q); U.S. Census Bureau 1982 (1910), roll T624_1022, p. 16B, image 428 (Manhattan Ward 12, New York, N.Y.).

3. After Alfred Tate expanded Edison's notes into separate letters to Sherburne Eaton and Bergmann & Co., the Edison Electric Light Co. acceded to his wishes. About the same time, Edison apparently declined a request from the Western Edison Co. with the comment that "we are so Crowded to supply light fixtures that Cant go into motors just yet." Tate to Eaton, 4 Jan. 1884; Tate to Bergmann & Co., 4 Jan. 1884; Frank Hastings to Edison Construction Dept., 10 Jan. 1884; TAE marginalia on Bergmann & Co. to TAE, 8 Jan. 1884; all DF (*TAED* D8416ABC, D8416ABF, D8439F, D8424E).

–2580–

To Giuseppe Colombo

New York, January 3rd. [188]34

My Dear Colombo:—

I suppose you know of the financial success we are making of the central station down town, sales for November $10,018., collections $9,000. or $9,200., I forget which, working expenses for November, $4,850., which does not include taxes, insurance or interest on our mortgage of $35,000.

We have 120 customers on our books whom we have refused, as we are loaded up. I have one of the "C" machines set up at Goerck Street with shortened cores. It gave 203 volts with no load, 350 revolutions, and with 1050 amperes gave, I think, 140 volts.[1] The spark is very much reduced. We can carry such a load that with 125 lbs. of steam it will run a Porter engine[2] down to 180 revolutions, so an Armington engine is necessary. With this alteration the future is all right, because I am going to make some higher resistance lamps requiring about 125 volts. The change on each machine will cost you about $575. You could easily have it done in Milan. Our lamps are about .75 ampere, therefore you can carry about 1350 lamps, and if the engine will pull it I believe you could take off 1500 for a while when the heavy load comes on.

If you desire working drawings, showing how to make the change, we can have them prepared for you. With this change which we are going to make down town, I think we should nett $60,000 in 1884.[3]

How does your station show financially?[4] Very truly yours,

TAE[a]

TL (carbon copy), NjWOE, DF (*TAED* D8416AAJ). [a]Initialed by Alfred Tate.

1. The editors have not found the physical dimensions of the modified Jumbo dynamo, but see Doc. 2419 (headnote) on the short core machines' general principles and proportions. Frank Sprague prepared a detailed report on tests conducted in March 1884 on one of the new "C" machines made for the New York Pearl St. plant. It did not give the amount of power Edison reported in this document, but Sprague promised that it would produce 121,000 watts (1,000 amps at 121 volts) in actual service, provided that excessive sparking at the commutator brushes could be curtailed; he accordingly recommended using a thinner brush for better contact surfaces. Sprague also suggested that Edison reduce the proportion of current diverted to the field windings because an excessively strong magnetic field prevented the steam engine from reaching its full speed. Sprague to Samuel Insull, 25 Mar. 1884 (esp. pp. 4–6, 11–14), DF (*TAED* D8442ZCR).

2. The Porter-Allen was a high-speed steam engine rated for 350 revolutions per minute. This speed and its relatively precise governing ability recommended it for electric lighting service, and it was Edison's initial choice for driving his direct-connected Jumbo dynamos. The Southwark Foundry, in Philadelphia, manufactured it. Docs. 1936 n. 1, 1981 n. 1, and 2238 (headnote).

3. The Edison Electric Illuminating Co. of New York installed two short-core dynamos at Pearl St. a few months later (see Doc. 2644), but the editors have found no indication that the machines already there were altered to the new design.

4. Colombo answered that although only about 3,000 lamps had been wired to the Milan station, the success of the Teatro alla Scala installation "has given us more confidence, and now I go on wiring rapidly for 1000 more at least, which will improve our financial situation, not yet settled upon an industrial ground." Colombo to TAE, 5 Feb. 1884, DF (*TAED* D8436V).

From Samuel Insull

My Dear Edison,

Many good wishes for the New Year! I hope better weather has hailed its advent in New York than has been the case in London.

When last I wrote you[1] I referred at length to Bailey & his relations with the Paris Cos. I also spoke of his reserve in dealing with me at Paris   This a.m. I have received a letter from him in which he says (referring to my proposed trip on the continent) "I wanted to get you to come by here to go over the project de contract and also the matter of the Bills of the Paris Exposition. I wish to go into the whole subject & dispose of it"[2]   I shall therefore go to Paris tonight talk matters over with Bailey, see Rau[3] about Milan & then push on to Nice to see Batchelor & from there go to Milan & try & get some of that $6476.00 which they owe you[4] & after that go to Berlin & try to get back here by about the time you get this letter. I understand from Batchelor that you or Eaton have written to Paris partially agreeing to the proposed new Contract.[5] I trust however you will insist on more concessions for the New York Co.[6] I will write fully from Paris or Nice about my interview with Bailey. Batchelor has written Porges a letter undertaking to ~~say~~ stay six months more but saying that he must immediately go to New York to see you.[7] Batch will return to New York finally in June.

England. I understand now more clearly how things stand here. Flood Page is Secretary with large powers as a manager. He is as enthusiastic as ever—what there is behind his enthusiasm you are as well able to judge as I am. I had a very long interviewed yesterday[a] with Mr Forbes the Chairman of the United Edison & Swan Co. He is the Chairman of the London Chatham & Dover R.R.,[8] is a man with a great deal of influence in the City, & appears to me to be a hard headed smart business man whose one object is to make the companies which he runns pay & in his enterprises outside electric lighting he has been very successful so far. He wished me to tell you that although formerly Chairman of the Swan Co, in his present position he knows no preference—his only object is to make money for his Co—& not to uphold Swan against Edison or vice-versa   With relation to your wish that neither Swans name nor yours should form part of the Coys name he said that independent of all Patents your name bore a commercial value to the Coy & likewise did Swans. Mr. Forbes did not conceal his want of respect for the manner in which[b] the Edison Co did their business   He said that when he got on what we would term the "inside track of Arnold Whites office he found that the word "bounce"[9] best expressed the method on which the Edison business was conducted. He also told me that they had not a single contract with relation to the Edison Installations ~~were~~ made here & in fact that these affairs were conducted in a most unbusinesslike manner. Mr Brand[10] who is Chairman of the United Telephone Co & who is interested in both the Edison & Swan Co was present at the interview. He said "Mr Insull do you know the cause of the failure of the Edison Co"   I said I thought I knew. He then added "The failure was caused by the people you Edison folks associate with. It is Arnold White Bouverie & Co— They did the same in the Telephone & it was inevitable that they should do so with the same with the Light"—and there is no doubt this is so. The Engineering work of the Edison Co was simply disgraceful. I should fill a book were I to start out to tell you the many blunders & the folly of the Edison Co so I will not commence. There is no wonder whatever that[c] they should have failed & the wrecking of the valuable property they held is due to incapacity—that old old tale which is at the bottom of so many failures. When I get back from my Continental trip I am to spend a whole day with Forbes & then I shall talk the Edison System till I am hoarse & I think the time will be well spent as Forbes is a business man—& not one of these "Guinea pigs"[11] or in other words he is not an Ornamental Aristocratic Director who joins the Board of a Coy just to get so many guineas for an attendance fee & who are popularly called "Guinea Pigs"—

I have put in the claim for the Patent accounts. My letter is to be considered at a Board meeting on the 8th Jany I do not know whether I shall get the money or not— Forbes says it entirely depends on the state of the Finances—which I fancy are low.—[12]

I asked Brand the Telephone Coys Chairman whether he would put your name on all Transmitters issued. He said he would be glad to do so unless some Contract prevented it. I am to see him about this matter when I get back.

Business really only commenced here last Monday. Everybody was away the previous two ~~wheel~~ weeks I was here & now I have got my various matters here well under way I shall be off to the Continent tonight & see what I can do there [~~between~~?][d] so for the next eight days I must turn myself over to the tender mercies of the slow going Continental R.R.s. Yours very Sincerely

Saml Insull

I wish you could visit the Holborn Restaurant. It is the most gorgeous place of its kind I have ever seen & if the Engineering of the Edison Co was at all decent it <u>would</u> be a magnificent Edison Installation[13]

ALS, NjWOE, DF (*TAED* D8465C). [a]Interlined above. [b]Interlined below. [c]Obscured overwritten text. [d]Canceled.

1. Doc. 2574.

2. Bailey's letter has not been found. On 5 January, Insull cabled Edison from Paris: "Consider letters concerning Bailey private" (DF [*TAED* D8465C1]). Sherburne Eaton's 14 September 1883 "Statement regarding the unpaid claim of The Edison Electric Light Company of Europe, Limited," summarized claims and contracts made on account of the Paris Exposition, which also involved Elie Léon (DF [*TAED* D8331ZAC]).

3. A naturalized Frenchman, industrialist Louis Rau was born in Munich and played a major role in the commercial introduction of Edison's electric light in France and Germany. He was managing director of both the Compagnie Continentale Edison and Société Électrique Edison. See Docs. 2249 n. 5 and 2593; *ABF*, fiche I875:152–53; Lanthier 1988, 1:38, 69–70.

4. Insull planned to seek a settlement of the Milan company's account, long a source of irritation to Edison, who had sought (unsuccessfully) to get a $3,000 payment in September. The company's insistence on a 50 percent discount on a more recent shipment of sockets, which it claimed were defective, only complicated matters. See Doc. 2399; TAE to Charles Batchelor, 30 Aug. 1883, Lbk. 17:209 (*TAED* LB017209); TAE to Comitato per le Applicazioni dell'Elettricità, 30 Nov. 1883; TAE to Société Électrique Edison, 27 Sept. 1883; Insull to Bergmann & Co., 22 Nov. 1883; all DF (*TAED* D8316BLW, D8316AXD, D8316BJD).

5. The editors have not identified such letters from Edison or Eaton, but see Edison's 31 December cable to Batchelor, which is transcribed in Doc. 2593.

6. The Edison Electric Light Co. of Europe, Ltd., formed in 1880, was based in New York City (see Doc. 1736 csp. n. 6). It was sometimes referred to as the "European Company" or "European Edison." The company assigned its patent rights to the Compagnie Continentale Edison in 1882 but retained the right to veto all contracts (Israel 1998, 215).

7. Letter not found.

8. The London, Chatham & Dover Railway served Greater London and southeastern England; James Staats Forbes became its chairman in 1874. *Oxford CBRH*, s.vv. "Forbes, James Staats," "London Chatham & Dover Railway."

9. The word "bounce" conveyed a range of contemporary meanings, from boastful self-assertion or bragging to outright swindle. *OED*, s.v. "bounce"; Barrère and Leland 1967 [1889], s.v. "bounce."

10. A London merchant and member of the firm of Harvey, Brand, and Co., James Brand (1828?–1893) was chairman of the United Telephone Co., Ltd., from its organization in 1880 through its merger with the National Telephone Co. in 1889. In 1883, Brand's business interests overlapped with those of Forbes in the Regents Canal City and Docks Railway Co. Obituary, *Times* (London), 22 Nov. 1893, 13; "Estate Notice," ibid., 6 Jan. 1894, 10; "The Proposed Amalgamation of Telephone Companies," ibid., 5 June 1889, 5; "Regents Canal City and Docks Railway Co.," ibid., 27 Feb. 1883, 13.

11. In British commercial slang, "guinea pigs" was a contemptuous reference to "gentlemen of more rank than means" who served nominally as directors of companies in order to draw income from compensatory fees. *OED*, s.v. "guinea-pig" 3.

12. The editors have not determined the outcome of this question, described in Doc. 2572.

13. An Edison isolated plant was installed in the Holborn Restaurant at 218 High Holborn St., London, in February 1883 during renovations of this London landmark. Many of its lamps were mounted in the decorative electroliers of Messrs. Verity and Son (Doc. 2317 n. 11; "The Edison Light at the Holborn Restaurant," *Teleg. J. and Elec. Rev.* 10 [24 Feb. 1883]: 349; "Notes," ibid. 12 [3 Mar. 1883]: 187). When plans were announced for increasing its capacity to 1,200 lamps, the promoters expected the isolated plant to be the largest in England ("The Edison and Swan United Electric Light Co., Limited," *British Architect* 21 [Feb. 1884]: 57).

–2582–

From William Andrews

Mount Carmel Jan 7/84

Dear Sir

It seems to me that the new style indicator would be a nearly perfect instrument for the purpose intended if the lamps did not Change their resistance, and the following idea has struck me as being worth experiment for if it would work, it would certainly maintain a permanent standard— Instead of the carbon horse shoe and central "idle pole" let there be three small plates of platinum thus:—

According to Crook's experiments where a nearly perfect vacuum is produced inside bulb and electrical connections made[a] "radiant matter"[1] would be projected from the negative pole towards the positive, and the middle or "idle pole" would be strongly positively charged.[2] The only weak point that I fear would hinder the successful working of this arrangement in our practice is that perhaps our EMF is too low. If you have not already tried this, I wish you would do so for as I said before, there would be no "wearing out" about it— Very Truly Yours

W. S. Andrews

⟨You are off your base the only reason we get a Current is the incandescence of the Carbon & until the incandescence reaches a given point we cant detect an Current Even with a Thomson galvamtr—[3] The reason Lamps change is due to deposit of Carbon on platina— we are going to deposit Carbon on it in making it that will remove inconstancy[4] E⟩

ALS, NjWOE, DF (*TAED* D8442G). Letterhead of Thomas Edison, Central Station Construction Dept.; W. S. Andrews, chief electrical engineer. [a]"and electrical connections made" interlined above.

1. English physicist and chemist William Crookes had experimented extensively in the late 1870s with electrical discharges in high vacua, observing unusual luminescent phenomena in the discharge tubes. Following Michael Faraday's speculation, he attributed these effects to the existence of "radiant matter." He postulated this as a fourth state of matter composed of a stream of negatively charged molecules moving at high velocity away from the surface of a negative electrode (cathode). As a corpuscular phenomenon, radiant matter, in Crookes's view, was capable of mechanical action, such as causing phosphorescence of the glass tubes themselves: "It . . . not only strikes the glass in such a way as to cause it to vibrate and become temporarily luminous while the discharge is going on, but the molecules hammer away with sufficient energy to produce a permanent impression upon the glass." This theory met with some determined opposition, but Crookes's experiments were widely cited by scientists in Britain and Europe and were familiar to Edison. See Docs. 1898 (headnote), 1902 n. 3, and 2173 n. 8; Crookes quoted in DeKosky 1976, 51; see also Müller 2004.

2. With its discontinuous circuit and enlarged electrodes, Andrews's device suggests the design of triode vacuum tubes that came decades later. The center plate also recalled experimental attempts at Edison's laboratory to use solid partitions to block carbon (or electrical) carrying across the filaments of incandescent lamps. See Docs. 1898 and 1944 n. 6.

3. This highly sensitive instrument, devised by William Thomson for detecting the weak currents transmitted through submarine telegraph cables, was often used in laboratories for making electrical measurements. It employed a small mirror on a magnetized needle that was suspended within the magnetic field of an induction coil; the mirror deflected a beam of light across a graduated scale. Doc. 1401 n. 2.

4. Alfred Tate closely followed Edison's marginalia in preparing a

brief typed reply. The exchange continued with Doc. 2589. Tate to Andrews, 9 Jan. 1884, DF (*TAED* D8416ADB).

–2583–

From Grosvenor Lowrey

[New York,] Jan 7th 1884

My dear Edison

I thank you for calling on Villard— He told me of it & that your visit gave him more pleasure than anything which had happened since his misfortunes.[1] Hc is[a] honest, & will again be rich or at least as[b] influential & prosperous as before[c]— He told Carl Schurz[2] in my hearing that the Edison Electric Light had a greater field than O.R.&N.,[3] O&T.[4] & Northern Pac together had ever had,— that it was commercially perfect,—& he afterwards told me privately that he was going to give his time to it after this— He added that Fabbri had been his best friend—& Fabbri has told me that mistakes & all Villard is a good & a great man   Sincerely Yours

Lowrey

ALS, NjWOE, DF (*TAED* D8403A1). Letterhead of Grosvenor P. Lowrey. [a]Multiply underlined. [b]Interlined above. [c]"as before" interlined above.

1. Years later, recalling Henry Villard's financial collapse, Edison noted that "When Villard was all broken down and in a stupor caused by his disasters in connection with the Northern Pacific Mrs. Villard sent for me to come and cheer him up. It was very difficult to rouse him from his despair and apathy, but I talked about the electric light to him, and its development, and told him that it would help him win it all back and put him in his former position" (App. 1.B.73). De Borchgrave and Cullen 2001 (339) state that, at this time, Edison gave Villard a draft for $40,000 to repay advances for electric railway experiments, but Edison evidently had repaid Villard's investment in 1882 (see Doc. 2195 n. 1). Villard had invested heavily, using both his own and borrowed money, to bring the Northern Pacific Railroad into his transportation empire and complete its transcontinental rail line. As the railroad's president, he authorized construction expenditures well beyond the company's ability to pay or raise capital. Immediately after Villard's triumphant celebration of the last spike in September 1883, the stock price of the Northern Pacific and its parent holding company, the Oregon and Transcontinental Co., plummeted. The bursting of the speculative bubble wiped out Villard's personal fortune and forced him from control of the Transcontinental Co. in December and, on 4 January, from the Northern Pacific. It also cost him the palatial new home (with Edison lighting) that his family occupied in December 1883 (de Borchgrave and Cullen 2001, 305–38; Buss 1978, 151–61).

2. Carl Schurz (1829–1906), the distinguished German-born military leader, politician, and journalist, was a longtime friend of Villard's. He was a coeditor of the latter's *New York Evening Post* until forced

to resign in 1883, as a consequence of the telegraphers' strike. *ANB*, s.v. "Schurz, Carl"; Kobrak 2008, 34–35; de Borchgrave and Cullen 2001, 124.

3. Villard organized the Oregon Railway & Navigation Co. in 1879 to consolidate his transportation companies in the Pacific Northwest. Among those firms was the Oregon Steam Navigation Co., which operated the steamship *Columbia*; Edison had installed his first commercial isolated electric lighting plant on that ship in 1880. Buss 1978, 87–93; de Borchgrave and Cullen 2001, 307–8; see Doc. 1892.

4. Villard organized the Oregon and Transcontinental Co. in June 1881 as a holding company for both his new Oregon Railway & Navigation Co. and the Northern Pacific Railroad. He also intended it as a construction company for completing the Northern Pacific transcontinental line. Buss 1978, 129, 131.

–2584–

To Frank Sprague

[New York, c. January 8, 1884[1]]

Sprague—

How would it do to have one of the large indicators to regulate by on one of the feeders and use the cheaper detector with lamp as in Volt box for the other feeders[2]   this would save about 24 dollars on Each feeder & help us out greatly= I was thinking of another dodge & that is to use one of the regular feeders Indicators & then on the other feeders connect the positive of one feeder with the positive of another feeder through a $5 detector   if both were of same potential the detector wouldnt deflect & if one higher than the other it would deflect one way or other[3]   is there anything in this idea— I want to cheapen down the stations—

E[dison]

⟨Not worth a d———   Sprague, Jan 9, 1884.⟩[a]

ALS, NN, FJS (*TAED* X120CAI). [a]Marginalia written by Sprague.

1. The editors have conjectured this date based on Edison's 6 January drawing (see note 3) and the marginal notation made by Sprague, who was in New York City at this time.

2. The "volt box" was evidently a portable instrument used to calibrate or check pressure indicators, or for troubleshooting (see Docs. 2605 [addendum] and 2625). Edison apparently sometimes conflated the names "volt box" and "pressure indicator"; the variation "volt box indicator" also appears in central station correspondence around this time. Bergmann & Co. listed two unspecified types of pressure indicators in a September 1884 catalog of their products (Harry Leonard to TAE, 25 Apr. and 9 June 1884; Samuel Insull to William Dwelly, 31 Mar. 1884; all DF [*TAED* D8449ZAK, D8449ZAZ, D8416BAS]; TAE agreement with Bergmann & Co. [Exhibit D], 1 Sept. 1884, Miller [*TAED* HM840230]). These devices may be among those described in Brooks 1953, esp. pp. 406–15.

Edison's 6 January sketch of a distribution system that he hoped would "cheapen down" small central stations by permitting the use of inexpensive voltage indicators on feeder lines.

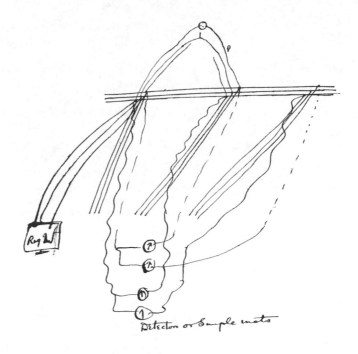

3. On 6 January, Edison sketched an arrangement corresponding to the one described here. Each pair of feeder lines is connected through a "Detector or Simple insts." Unbound Notes and Drawings (1884), Lab. (*TAED* NS84ACE); see also Doc. 2505 (headnote).

–2585–

Alfred Tate to Samuel Insull

[New York,] January 9th. 1884.

My Dear Insull:—

I was very glad to receive your letter of the 21st. December. We have received all your cables and I trust have replied to them as promptly as you desired.[1]

In relation to the Williamsport business, which I mentioned to you in my last letter, as we did not hear from these people in reference to the first payment on account of their contract Mr. Edison deemed it desirable to write and cancel the same, which we did.[2] We had Tomlinson draw up a letter for us, and the Williamsport business is "nill."[a] We do not lose a cent by the transaction, as we had not given an order, neither had Rich done any work in connection with the erection of the building, and we congratulate ourselves upon having gotten out of the matter so easily. I think they would have stuck us for about $15,000.

The Tiffin station was started up successfully and has been running ever since. I cabled you to this effect some days ago.[3] Mt. Carmel starts up on Saturday next and Hazleton and Bellefonte will follow quickly.[4] We have had considerable trouble with Bergmann lately in regard to his shipments. He has been very tardy in getting our orders off, but we have at last prevailed upon him to work at night and thereby we are insured of more prompt delivery. I wired the Tiffin Company to send us the second payment on account of their contract, but received a reply from them stating that all their supplies had not been received, and they do not consider their installation complete.[5] This is a delusion which we should endeavor to dispel in connection with the installation of our plants. As soon as a station is ready to supply light we should receive our second payment, whether all of our supplies are there or not. It is not equitable for a company to hold $15,000. of our money simply because $100. of worth of material has not been received at a station.

We have had Tomlinson working on a contract for some time past and he has gotten out a splendid document to be used in closing [of?][b] our contracts. We have had numerous discussions about this matter, and I suggested that in future we demand fifty per cent upon the signing of the agreement, thirty five per cent when the installation is ready to supply light and the remaining fifteen per cent after it has been ready to supply light for thirty days. This applies to small contracts, that is contracts not over $55,000. Anything above that I think your suggestion of having a payment due when the building is roofed is an excellent one. Now in regard to this fifty per cent upon the signing of the agreement, if we can collect that much money at that juncture it will at once establish the responsibility of the parties with whom we are dealing. We made some calculations based upon the assumption that after a plant was installed the company would fail to make the necessary payment and we should have to remove the installation, being unable to carry it. In these limited contracts we need fifty per cent on the signing of the agreement to save ourselves from loss, and this Williamsport transaction has in reality been a valuable experience to us. We got a list of the stockholders and we found that P. B. Shaw, young Detwiler and Dr. Detwiler,[6] his father, held one third of the stock. Now it was their intention to scrape together the first payment on account of this installation and after it was ready to supply light they would let us wait for our money until they unloaded their stock. None of these parties have sufficient money to warrant their undertaking to carry $20,000. between them, and they simply meant to gamble with us. If we should close a contract before you return I will do all in my power to have the payments made as I suggest. Of course I cannot tell at this writing whether you approve of them or not, but if they are not in accordance with your ideas you of course will substitute other terms when you come back. I, myself, should not feel justified in advocating any other course than that which I have explained.

In regard to finances, everything is moving all right, but this Tiffin business has knocked me out a little and we are not flush. I can, however, keep things moving all right. I paid the Ansonia seventy five per cent of the amount of the copper they supplied on account of the Fall River installation.

MACHINE WORKS. The Machine Works had run out of orders so we gave them instructions to make 10 "H" dynamos, 10 "Y" dynamos, 15 "S" dynamos and 10 "R" dynamos. We had to order 15 "S" machines, as we had requisitions from the Isolated Company for 6 more than Soldan had instructions to make. This leaves us 9 for our village plant business, and at this season it is necessary to anticipate a little. We will not have any trouble financing, as I think we will get orders for most of them before they are finished. I gave Dougherty[7] the usual four months note, and will give Macpherson, Willard & Co.[8] a note from the Isolated Company for the amount of their December bill to-morrow. Anything else I can meet with cash.

CANVASSERS. We have six more towns in Pennsylvania to canvass and have Cooke and Lyon[9] on this work. Wilber[10] is canvassing Lexington, Ky., from which point he goes to Frankfort, Ky., thence to Danville, Va., after that we will put him in Connecticut. Dyer is making a canvass of New York City.[11]

In regard to the Piqua installation, Stuart of Cincinnati has been here the last two days in connection with changing this from a 500 light plant to a 1,000 plant.[12] We have made a new estimate for 1,000 lights, which we will substitute for the old one, and we have also made a bill for the loss we will sustain on account of having to carry the[c] material ordered for the 500 light plant but which is useless in connection with the larger unit. The boiler, pump and some smaller material cannot be used and we are charging him twenty per cent of the cost. The wire which is thrown out of service we are are taking back at sixteen cents per pound. It is only a matter of about $430. The full amount which we are charging as loss on account of this determination of theirs to alter the size of their plant is $928. This will cover everything. The Newburgh business is going along very slowly, and they have not yet got the roof on the building.[13] I think that this is the "last straw that breaks the camel's back," and we should never again allow a local company to erect their own edifice. The payments on account of the material we ordered in connection with this plant came[a] due and we are at the mercy of these local parties and have to finance in such a manner that, if we had more than one installation of the kind, I think we would be seriously embarrassed. You will remember that you used to have some pretty warm arguments with Mr. Edison, who rather favored the idea of allowing local people to erect their own building. He is now thoroughly converted and says he will never again make an agreement of this nature.

I suppose you have heard that the Board of Aldermen of New York passed an Ordinance commanding the electric light companies to put their wires underground within two years.[14]

Clarke, the Light Company's engineer, has resigned his position, and I believe undertakes the commercial development of a recent invention. This invention is an indicator which can be used on shipboard, and when placed in the captain's cabin, with wires running to certain points of the ship, will,[c] presuming he has a cargo of cotton, indicate the temperature of this material and show whether there is any danger of combustion. Of course it can be used for many other purposes, and I have heard it very highly spoken of. I believe he has organized a company to push it.[15]

I hope my letters reach you with sufficient regularity. I am looking forward with much pleasure to your return, as the fact is I have not had a chance to take a respectable meal since you left here. I am also looking forward with a great deal of pleasure to my Canadian tour.[16] It would do you a good deal of good were you to see the cynical smile which appears over the beaming countenance of our joint benefactor Mr. Gilmore,[17] when he hears me talk about going away.

Mr. Gilmore joins me in wishing you a safe and pleasant voyage on your return trip, and hoping to see you on the 28th. (note the 28th.) of this month, I am, Very truly yours,

Tate

TLS (carbon copy), NjWOE, DF (*TAED* D8465D). [a]Obscured overwritten text. [b]Canceled. [c]Interlined above.

1. In his 21 December letter, Insull notified Tate of his arrival in London and his plan to go to Paris (see also Doc. 2572), and he requested that his letters be "answered promptly by cable if necessary" (DF [*TAED* D8367ZAQ]). Subsequent cable correspondence includes Insull to TAE, 28 Dec. 1883, 3 and 5 Jan. 1884; and TAE to Insull, 3

Jan. 1884; LM 20:36A, 20:38C, 20:39B, 20:38D (*TAED* LM002036A, LM002038C, LM002039B, LM002038C).

2. Tate apprised Insull on 28 December that Edison had requested information from the Bradstreet credit reporting agency on the Williamsport company's principal stockholders and found that "None of them are quoted," which Tate took to mean that none of them "have one dollar to rub against another." As Tate noted elsewhere in this document, the Williamsport company was obliged to pay Edison one quarter of the contract price in advance but was not to make second payment until the plant was ready to operate, when half of the total would come due. Edison rescinded his contract with the Williamsport company on 4 January. Tate to Insull, 28 Dec. 1883; TAE to Edison Electric Illuminating Co. of Williamsport, 4 Jan. 1884; both DF (*TAED* D8316BWR, D8416AAP).

3. Not found.

4. See App. 2A.

5. Edison's contract with the local Tiffin illuminating company specified only that 50 percent of the price be paid "when the installation is ready to supply light." The company's secretary replied to Edison's telegram (not found) that "while it might be construed in one sense that the station was ready to supply light yet for lack of service wires, sockets, meters etc. etc. there was no one to supply light to—and all this without any fault of ours." TAE agreement with Edison Electric Illuminating Co. of Tiffin, 24 Oct. 1883; E. B. Hubbard to TAE, 7 Jan. 1884; both DF (*TAED* D8356X, D8451M); see also correspondence related to wiring and materials in Electric Light—Stations—Ohio—Tiffin (1883 and 1884), both DF (*TAED* D8356, D8451).

6. Thomas C. Detwiler (b. 1860?), an 1881 graduate of the University of Pennsylvania, practiced medicine in Lancaster and was a director of the Edison Electric Illuminating Co. of Sunbury. His father, Benjamin H. Detwiler (b. 1830?), was a prominent Williamsport physician. He helped to organize the Lycoming County Medical Society in 1864 after having served in the Civil War as assistant surgeon of a Pennsylvania regiment (U.S. Census Bureau 1970 [1880], roll T9_1153, p. 502.3000, image 219 [Williamsport, Lycoming, Pa.]; Bell 1891, 496; Meginness 1892, 310–11, 461, 792; Detwiler 1881). According to a list of the Williamsport company's stockholders, the younger Detwiler was the among the largest shareholders but his father held only a small stake (Thomas Detwiler to TAE, 18 Dec. 1883, DF [*TAED* D8362ZAG]).

7. Hugh B. and William Dougherty, father and son, were iron founders doing business as H. B. & W. H. Dougherty at 143 and 147 Bank St. in New York. They made an agreement with Edison in mid-1883 to make castings for the Edison Machine Works and to allow thirty days for payment of their bills, for which they promised to accept Edison's four-month notes. According to a Bradstreet Co. report, the partnership was "doing considerable work for Edison" by the end of 1883. Insull to Hugh Dougherty and William Dougherty, 8 Aug. 1883; Bradstreet Co. report, 15 Jan. 1884; both DF (*TAED* D8316AMJ, D8465E1).

8. Macpherson, Willard & Co. did business in Bordentown, N.J., as the Union Steam Forge. They provided castings, evidently of dynamo field cores, for the Edison Machine Works, and they were among the firms that Insull suspected of having paid improper commissions

to Charles Dean as superintendent of the Machine Works. *Directory of Iron and Steel Works in United States and Canada* 1890, 225; letterhead, Macpherson, Willard & Co. to Insull, 20 June 1884; TAE to Macpherson, Willard & Co., 3 Sept. 1883 and 5 Apr. 1884; all DF (*TAED* D8465ZAB1, D8316ASF, D8416BDE); Insull to Macpherson, Willard & Co., 3 June 1885, Lbk. 2:316 (*TAED* LB020316).

9. Charles P. Lyon had done canvassing for the New England Dept. of the Edison Co. for Isolated Lighting. On the recommendations of Spencer Borden and George Wilber, Insull hired him for similar work for the Edison Construction Dept. at the end of 1883. Borden to Insull, 19 June 1883; Wilber to Insull, 25 Nov. 1883; Insull to Lyon, 28 Nov. 1883; Lyon to Insull, 30 Nov. 1883; all DF (*TAED* D8340ZAK, D8342ZDT, D8316BLC, D8342ZDZ).

10. George W. Wilber had worked with Charles Lyon as a canvasser for the New England Dept. of the Edison Co. for Isolated Lighting and was recommended to Insull by Spencer Borden. Edison hired him in June 1883. Borden to Insull, 19 June 1883; TAE to Borden, 23 June 1883; both DF (*TAED* D8340ZAK, D8316ACL).

11. Tate meant Paul Dyer, who likely was working on the projected Second District in New York.

12. Despite the fact that Edison's contract with the Ohio Edison Installation Co. had not yet been signed, William Rich had apparently been ready to start work in Piqua since about the first of the year. Alfred Tate formally tendered a new contract for the larger plant to Archibald Stuart on 17 January. Tate to Rich, 7 Jan. 1884; Tate to Stuart, 17 Jan. 1884; both DF (*TAED* D8416ABW, D8416AEY).

13. Jenkins 1984 describes the organization of the local illuminating company among the "old fashioned and cautious" people of Newburgh (in the opinion of Edison's assistant Charles Hughes) and the generating station's opening and subsequent history. Hughes to Insull, 27 June 1883, DF (*TAED* D8351B); see also App. 2A.

14. See Doc. 2568 n. 1.

15. Charles Clarke's formal resignation was 1 February. He had at least two recent patents related to the electrical transmission of thermometer signals. These were for a "Circuit and Apparatus for Electric Temperature and Pressure Indicators" and a "Telethermometer" (see App. 4C), both assigned to Robert Hewitt, Jr. Hewitt, a noted New York businessman and collector of Lincolniana, was an organizer of the Telemeter Co., in New York, of which Clarke became manager in February 1884 (though Clarke seems to have retained at least a consulting role with the Edison company through the spring; cf. Doc. 2677). Clarke remained with the Telemeter Co. until 1887, when he joined the Gibson Electric Co. as an electrical engineer. He subsequently had a successful career as an independent consulting engineer and patent expert. "Clarke, Charles L.," Pioneers Bio.; Little 1909, 3:1243; Obituary, *NYT,* 7 Oct. 1913, 13.

16. Tate, whom Insull described as "used up by overwork," left New York for his family home in Canada on 16 or 17 February. When he returned to work on 6 March, Insull was in Sunbury and Shamokin, Pa. Tate to Insull, 17 Feb. 1884; Insull to Archibald Stuart, 28 Feb. 1884; both DF (*TAED* D8465F2, D8416AQN); Tate to Insull, 6 Mar. 1884, LM 18:317 (*TAED* LBCD5317).

17. William Edgar Gilmore (1863–1928), whose widowed Scottish-born mother was a druggist, had been a New York grocery clerk before he began working, in 1881, as an assistant to Frank McGowan at the Edison Electric Light Co.'s office. He soon became a stenographer for Samuel Insull. Edison put him in charge of the records and estimates of the Engineering Dept. at the end of January and, when the Construction Dept. was phased out in May 1884, appointed him "stenographer in chief." Gilmore became one of Edison's most important business associates, occupying key positions in several firms, including the Edison Machine Works, the Edison Manufacturing Co., the National Phonograph Co., and General Electric. "Gilmore, William E.," Pioneers Bio.; U.S. Census Bureau 1970 (1880), roll T9_874, p. 561.4000, image 0285 (New York City, New York, N.Y.); ibid. 1982? (1900), roll T623_968, p. 3B (Orange Ward 2, Essex, N.J.); TAE to Gilmore, 24 Jan. 1884; Insull to Gilmore, 1 May 1884; both DF (*TAED* D8416AGZ, D8416BMI).

–2586–

From William Holzer[1]

EAST NEWARK, N.J., Jany 10th 1884[a]

Dear Sir.

Lamps exhausted without applying Current, whilst Exhausting; with side tube containing Cocoanut Charcoal, set up at 64 C.P. are doing well no arc's so far up to time of writing.[2]

Lamps up in Test, Exhausted & heating Bulb with Alcohol Flame using no Current set up at 48 C.P. have an average of 1150 minutes   One Lamp still burning & no arc's

The Experiment of applying Oxide of Magnesia on Fibre & then Carbonize came out very poorly, the Carbons twisted over on side of Bulb. Will try another lot.

I have now under way your Experiment of Burnishing the Fibres   those that I have burnished show a very fine surface. will carbonize them on Monday next.[3]

Also under way an experiment of securing Carbon to Inside part by dispensing with the Shank on Carbon and applying a paste made by dissolving the finest grade of Indian Ink & securing the Carbon direct[b] to the Platinum wire without using any Copper. This is done very neatly and quick & becomes sufficiently hard in a few moments to hold Carbon without binding   I have a few up in test & seem to show well.[4] On Saturday next the 12 inst will try an experiment on Cementing Carbon to Wires by using Sponge Platinum

The Experiment in taking a Lamp that has given a long Life and patching it will receive prompt attention. will have it under way to-morrow.[5]

The Bamboo Paper sent out to be Carbonized was handed to John Ott to deliver to you.[6]

Made a few Lamps for your Indicators with the Platinum
wire in Ring Forms so as to prevent the breaking of the Carbon in transporting. Yours Respectfully

Wm Holzer

ALS, NjWOE, DF (*TAED* D8430D). Letterhead of Edison Lamp Co.
a"EAST NEWARK, N.J.," and "188" preprinted. bObscured overwritten
text.

1. This document is one of nearly a score of letters written by Holzer
to Edison between November 1883 and February 1884 about experiments made at Edison's request at the lamp factory, mostly on carbonization and manufacturing processes. It generally represents this wintertime research, some specifics of which are noted below. Although
Holzer referred frequently to Edison's directives, the editors have
found only one letter from Edison to him in this period. See Holzer
correspondence in Electric Light—Edison Lamp Co.—Lamp Test
Reports (1883 and 1884), both DF (*TAED* D8333, D8430); TAE to
Holzer, 17 Jan. 1884, LM 18:15B (*TAED* LBCD5015B).

2. Holzer had begun working by early January on alternatives to Edison's standard manufacturing process, in which filaments were electrically heated to drive off gases as the bulbs were evacuated (Holzer to
TAE, 5 Jan. 1884, both DF [*TAED* D8430A, D8430B]). Several score
of the lamps experimentally made without electrical heating were tested
in the first half of January, and their durability, lamp "curves," and electrical characteristics recorded on the standard forms used for tests of
production lamps (Electric Light—Edison Lamp Co.—Lamp Test Reports [D-84-30], DF, NjWOE). Charcoal or similar substances were
ordinarily used in the lamp factory's vacuum pumps to absorb moisture
and residual gases; Edison had previously patented a process for substituting chemical action entirely for the mechanical action of the mercury
pump. Edison understood an electric arc between the filament clamps as
an indication of relatively high gas pressure in the bulb (see, e.g., Docs.
1816, 2029 n. 6, 2098 [headnote], and 2219).

3. Several batches of these lamps were tested between 17 and
21 January (Electric Light—Edison Lamp Co.—Lamp Test Reports
[D-84-30], DF, NjWOE). Holzer reported that burnished filaments
showed good durability. Although he hoped "the Burnishing process
will be in order for all types of Lamps," and Edison continued to think
about it during his Florida vacation, the editors have no evidence of
its commercial use (Holzer to TAE, 21, 23, and 25 Jan. 1884, all DF
[*TAED* D8430H, D8430K, D8430M]; see Doc. 2623).

4. Holzer's later report on a 100 candlepower lamp with "Clamp
made of the Indian Ink" presumably referred to this process. When it
was tested at 1,000 candlepower, "particles of the Indian Ink Clamp deposited itself to the side of the Globe and continue so until the Clamp
narrowed down to a small contact" before the filament broke. Holzer to
TAE, 25 Jan. 1884, DF (*TAED* D8430M).

5. Holzer tried Edison's "experiment in taking a Carbon, that has
burned one Thousand (1,000) Minutes . . . and then joining the Carbon together at the parts broken." Using an unspecified "Indian Ink
mixture" to repair several carbons broken at different places, he pro-

nounced the tests "a success." Holzer to TAE, 28 Jan. 1884, DF (*TAED* D8430N).

6. This sentence may refer to bamboo paper that John Ott retrieved from the lamp factory, where it had been carbonized, on 21 December. That batch was for experiments on treated paper described in Doc. 2573 n. 1. N-82-12-04:119, Lab. (*TAED* N145119B).

–2587–

Notebook Entry:
Electrochemistry[1]

New York Jany 10 1883̶4

For producing Gold Foil by Electro-vacuo deposit—[2]

(Copper rods with Gold points)

M N Force

X, NjWOE, Lab. (*TAED* N145143). Document text written by Edward Acheson; dateline and signature written by Martin Force.

1. This notebook entry was written by Edward Acheson (1856–1931). Hired as a laboratory assistant in 1880, Acheson had been in Europe since 1881, working first on Edison's exhibition at Paris and then setting up lighting plants for the Compagnie Continentale Edison. Having resigned that position at the end of 1882 in a salary dispute, Acheson had been conducting—mostly unsuccessfully—his own experiments. A "miser and poor man" in December 1883, he wrote Edison from London seeking employment, pointing out that he had been offered a job there at a lamp factory, which "would require me to make use of the information I received while in your services." Edison made a positive but noncommittal response. Prompted by Samuel Insull, whom Acheson had seen in London, Edison cabled on 3 January that he would pay twenty dollars a week for unspecified work; Insull immediately advanced money for the passage to New York. Acheson resigned in March 1884 for a position in the lamp factory of the Consolidated Electric

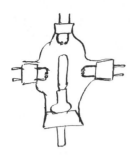

One of Edison's
10 January sketches
for electro-vacuous
deposition of unspecified
material on lamp filaments.

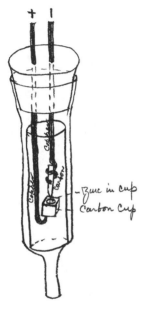

Edison's 11 January
drawing of an arrangement
for the vaporization and
vacuum deposition of zinc.

Light Co. but soon pursued his own experiments again. He later built a distinguished career as an industrial chemist. Docs. 2069 n. 2, 2085 n. 3, and 2235 esp. n. 6; Acheson to TAE, 13 Dec. 1883 and 18 Mar. 1884; TAE to Acheson, 28 Dec. 1883; Insull to TAE, both 4 Jan. 1884; all DF (*TAED* D8313ZCI, D8403ZAO, D8316BWK, D8465A, D8465B); Acheson 1965, 29–30.

2. The editors have not found preceding drawings or laboratory records concerning this process, but see Edison's marginalia in Doc. 2582 about depositing carbon on lamp clamps. Edison made several other small sketches on 10 January for "vac plating" of what appear to be lamp filaments. Edison continued making sketches on this subject until 8 February. These variations encompassed vaporizing zinc from a carbon cup suspended in the electric arc, depositing "alternate layers of Gold and Zinc after which Zinc is eaten out by acid" and, in early February, coating lamp filaments with vaporized copper. Acheson and Force tried these ideas but reported "very poor results" shortly before Edison left for Florida. Having tried many of the experiments repeatedly, they could "get a deposit but it is not what you want, and it seems almost impossible to get it." By mid-March, Force had evidently turned his attention from a vacuum process to more traditional electroplating methods. Unbound Notes and Drawings (1884); N-82-12-04:145–73 (quoted p. 155); both Lab. (*TAED* NS84ACF; N145145–145173); Force to TAE, 9 Feb. 1884; John Ott to TAE, 13 Mar. 1884; both DF (*TAED* D8467E, D8467H).

Edison had in the meantime executed, on 22 January, a patent application for "coating one material upon another . . . by throwing the material to be deposited into the form of a vapor in a vacuum." The transfer was effected by electrically produced heat, whether created from an arc, passage of current through the material itself, or by an external resistance. Edison specified that this general method applied "to the plating of any material whether a conductor or non-conductor of electricity, while in the ordinary process of electro-deposition only conductors can be treated." Electroplating was a well-established industry, however, and Yale professor Arthur Wright had published some preliminary investigations of vacuum deposition phenomena; consequently, the Patent Office took a more stringent view of the novelty claimed by Edison. The application underwent multiple cycles of rejection and amendment before the Patent Office allowed it to issue a decade later. In the interim, Edison added text to restrict his use of the term "electro vacuous deposition" to the process of "depositing in an exhausted chamber by electrical vaporization by means of a continuous current or a continuous arc, as distinguished from an intermittent current and from a series of sparks." He also dropped one of the seven drawings and entirely rewrote the broad original claims, in part to restrict them to a process using continuous current. Figure 2 of the published specification resembles that shown in this document. U.S. Pat. 526,147; Pat. App. 526,147; Boxman 2000, 1–2.

Edison asserted in his specification that "The uses of the invention are almost infinite, for coatings of any material and of any desired thickness may be formed. Metal sheets so fine as to be transparent and yet even and homogeneous can readily be produced." He pointed particularly to coating mirror glass by this method. Also on 22 January, Edison completed a patent application (not filed until April) for making lamp filaments by vacuum electrodeposition: "I either deposit carbon, silicon, boron, osmium, or other refractory high-resistance material in sheets, from which filaments for the incandescing conductors are cut, stamped, or otherwise formed; or I deposit the material directly in the filamentary form" (U.S. Pat. 395,963). Edison additionally filed a patent application in April for applying a "coating reflective of light" of "Silicon or equivalent material" to lamp filaments by "electrically volitalizing" the material in a vacuum (Case 616, Patent Application Casebook 2538:314, PS [*TAED* PT022314]). Edison subsequently used a similar process for reproducing phonographic records, a sample of which he provided at the Patent Office's request while his application was pending there. (This was the subject of Edison's U.S. Patent 713,863, filed in 1900.) The process became important in a variety of later industries, including the manufacture of semiconductor circuits (Richard Dyer to Commissioner of Patents, 8 Mar. 1890, Pat. App. 526,147; Waits 2001, 1668–71).

–2588–

Draft Patent Application: Electric Light and Power

New York, [c. January 10, 1884][1a]

Dyer—

Take out following patent[2]

[A][3]

[B]

[C][4]

[D][5]

[E][6]

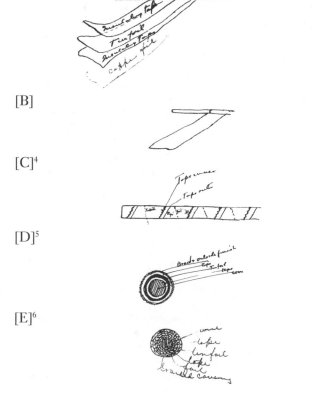

I first wind insulating tape strip on the wire overlapping it, then this is smoothed down and tin ~~fo~~ or lead or ~~any metallic~~ other metallic foil [--][b] is wound as a strip overlapping, the Center of the strip being over the lap of the tape= over this again is lapped another tape the center of which is over the lap of the foil. the whole is then preferably braided with Cotton but before the braid is put on another layer of foil may be put on & the braid placed over this which has a tendency to make it fire proof especially if the foil is copper foil.

Claim   Insulating wire by first coating the wire with an insulating flexible[c] tape overlapping then a layer of overlapping metallic foil then tape [~~foil?~~][b] & braid

also with double metallic foil—& braid also, with inner tin or lead foil & outer of Copper or other high melting point foil.

Describe that the tape is strips of paper or woven fabric saturated & Coated with an insulating compound

Kruzi says the insulation[d] of this wire is astonishing

E[dison]

ADfS, NjWOE, Lab., Cat. 1150 (*TAED* NM019AAA). Letterhead of Edison Machine Works. [a]"New York," preprinted. [b]Canceled. [c]Interlined above. [d]Obscured overwritten text.

1. Richard Dyer marked the document "Jany 84" and filed a patent application for Edison based on this draft on 24 January. The amount of time taken by Edison's patent attorneys to complete an application for his signature and then transmit it to Washington, D.C., varied widely.

2. The application based on this draft (Case 614) was filed on 24 January and rejected by the Patent Office on 11 February. Only the application's five claims for variations on insulating coverings and its two drawings, corresponding to the figures labeled A and E, remain. Patent Application Casebook 2538:312; Patent Application Drawings (cases 179–699); both PS (*TAED* PT022312, PT023 [image 118]).

3. Figure labels are "Insulating tape," "Tin foil," "Insulating Tape," "Copper foil," and "wire."

4. Figure labels are "Tape inner," "Tape outer," "Tin foil," "Tape," and "foil."

5. Figure labels are "Braid & outside finish," "Tape," "Tin foil," "tape," and "wire."

6. Figure labels, possibly written at a different time, are "wire," "tape," "tin foil," "tape," "foil," and "braided covering."

Figure 1 from Edison's patent drawing clearly shows four layers around the wire. Though unlabeled, the coverings presumably were tape, tin foil, tape, and copper foil, as in the document's Figure A.

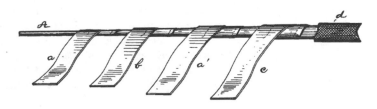

–2589–

To William Andrews

[New York,] January 14th. [188]4

Dear Sir:—

I have your letter of the 11th. inst. and in reply beg to say that I think radiant matter a humbug.[1] Edlund's[2] theory of the conductivity of a perfect [✗-----][a] vacuum,[b] the apparent resistance being due to polarization at the surface of the electrodes, perfectly explains the action because the current varies with every different metal being lowest for zinc and highest for carbon.[3] The resistance also changes with the current showing polarization.

The pamphlet you speak of has not yet come to hand. Yours truly,

TAE   T[ate][c]

TL (carbon copy), NjWOE, DF (*TAED* D8416ADY). [a]Canceled. [b]Interlined above by Alfred Tate. [c]Initialed by Alfred Tate.

1. After receiving Alfred Tate's 9 January reply to Doc. 2582, Andrews wrote again about his proposed new indicator lamp. He also promised to send a pamphlet that he hoped would illustrate its principles, but this item apparently did not reach Edison. Andrews pointed out that Edison's "present [indicator] lamp will give no current from 'idle pole' until incandescence is arrived at, because there is a <u>closed circuit</u> inside lamp through carbon loop, and in this instance it seems to be heat that produces the 'radiant matter'— . . . There is something very wonderful in all this, and to me it is an intensely interesting subject— but your indicator is evidently the first <u>practical</u> application of 'radiant matter.'" Edison made notes on the letter that Alfred Tate used almost verbatim in preparing the first paragraph of the present document. Andrews to TAE, 11 Jan. 1884, DF (*TAED* D8442M).

2. Erik Edlund (1819–1888), a physicist of the Swedish Academy of Sciences in Stockholm, studied electricity and associated phenomena throughout his career. Edlund gained further renown posthumously for having supervised and promoted the doctoral research of chemist Svante Arrhenius, who won the Nobel Prize in 1903. *Gde. Ency.*, s.v. "Edlund (Erik)"; *DSB*, s.v. "Arrhenius, Svante August."

3. In an 1881 paper (republished in English in 1882), Edlund argued that the conductivity of a gas increased as its pressure diminished and that a vacuum (or highly rarified gas) was actually a good electrical conductor. To explain the well-known impossibility of forcing a spark between electrodes in a nearly perfect vacuum, he distinguished between the resistance of the residual gas itself and the resistance between the gas and the surface of the electrodes. In a successive paper (translated in 1883), he noted that one could induce a current in a vacuum through which a spark would not jump. Edlund argued that the passage of current between the metal and the gas was increasingly resisted as the gas pressure diminished. He attributed this result to a counter electromotive force on the surface of the metal whose strength varied inversely with gas pressure. This process is defined as polarization, a familiar phenomenon in batteries (the accumulation of ions on electrodes op-

poses the main current), although Edlund referred to it instead as a "disjunction-current." Edlund presented his conclusions in support of a much broader unitary theory of electricity. "Electricity in Vacuo," *English Mechanic and World of Science* 34 (30 Dec. 1881): 395–96; Edlund 1882 [1881], esp. 14–20; Edlund 1883 [1882].

–2590–

From Francis Upton

East Newark, N.J., Jan. 15, 1883[4][a]

Dear Sir:

I submit to you the following report for the year[b] 1883.

Table A shows in detail the assets of the Lamp Co. on Jan. 2, 1883 and on Jan. 2, 1884.[1]

Jan. 2, 1884	$170,891.83
Jan 2 1883	145,472.81
Gain in assets	$ 25,419.01
Less gain in bills due to Lamp Co.	8,852.44
Leaves gain in material	$ 16,566.57[b]

On Table B is given in detail the profit of the business as well as the amount for the year 1883. I consider that the amount shown as profit to be a fair estimate.[2]

If however you cut the assets still more sharply, there is a large margin over the claim that I make: which is, that the Edison Lamp Co. during the year 1883 earned ten per cent on a capital of $250,000—and if we had this amount of money as working capital we could have easily paid a dividend to this amount, besides putting by [------][c] ~~amount~~ as surplus five per cent more under a reasonable estimate of assets.[3]

The business is now in such shape, that we can hold our own and earn interest on our capital, on the sales of lamps in this country alone, at present prices and in the same amounts as last year.

During the coming year I am anticipating an increased demand for lamps and hope by Jan. 2, 1885 to bring our indebtedness down to running bills.[4] Yours Truly

Francis R. Upton.

ALS, NjWOE, DF (*TAED* D8429C). Letterhead of Edison Lamp Co. [a]"East Newark, N.J.," and "188" preprinted. [b]Obscured overwritten text. [c]Canceled.

1. Upton enclosed two tables, each occupying one closely written page. Table A detailed lamp materials and supplies as of January 1883 and 1884, including the real estate, buildings, and tools that constituted the factory. Edison Lamp Co. statement, 2 Jan. 1884, DF (*TAED* D8429D).

2. Table B contained year-to-year lists of significant liabilities, value of lamps sold, and aggregate cash paid in by the partners. What Upton termed the year's "profit" (nearly $38,000) was the decline in the factory's running loss since its inception to $45,574.30. An increase in lamp sales (from $91,229.52 in 1882 to $139,758.24 in 1883) largely accounted for the change. Edison Lamp Co. statement, 2 Jan. 1884, DF (*TAED* D8429D).

3. See Doc. 2733 concerning the Lamp Co.'s first dividend and its finances for 1884.

4. A credit report in December 1883 had mentioned the difficulty of getting "from unbiased sources" an accurate account of the factory's profitability because "of the experiment being so new, but many have faith in it. Their sales are constantly increasing & it is thot. they will ultimately get on better footing. At present their cash is not at all plentiful & Cr[edit]. should be limited." R. G. Dun & Co. report, 18 Dec. 1883, Dun (*TAED* X112CAV); see also Doc. 2733.

–2591–

Sherburne Eaton
Report to Edison Ore
Milling Co.

[New York, January 15th, 1884[1]]

To the Stockholders of the Edison Ore Milling Company, Limited:[2]

Your Board have little to report as to the operations of the Company during the past year. We expected to have a number of the magnetic iron ore separators in operation in Canada last Spring, but owing to the fact that successful means of smelting the ore had not been devised, the owners of the deposits were not willing to go to the expense of separating ore if they could not find a market for it.[3]

This was our own trouble when we were operating a separator at Quonocontaug Beach, as stated in the last annual Report.[4] In that Report it was stated that we had ceased work and closed our business on that beach, because we could not find a market for our product. The Cost to the Company of this experiment was $4,832.44.

The financial resources of the Company are low. There is owing to Mr. Edison about $1,650 for machinery, experiments, &c. The cash advanced to the Company by the General Manager, Mr. S. B. Eaton, amounting to $1,577.89, which was owing to him a year ago, is still unpaid; and this amount, together with some other small advances for petty cash expense, is still owing to him. We also owe $548.17 for taxes and $250 to the Farmers Loan & Trust Co.,[5] for acting as Registrar of the Stock. There is no cash in the Treasury with which to pay theses debts, nor have we any cash to pay our current expenditures.

Since the closing of the operations at Rhode Island, Mr

Conley,[6] who was employed as Superintendent there, has devoted his time and attention to devising means for the successful reduction of the ore to a marketable product.[7] After spending over a year in numerous experiments, Mr. Conley believes he has invented a process which will accomplish this result, and has raised some capital for the purpose of erecting a furnace in the spring, when he expects to commence the manufacture of iron and steel from the magnetic iron ore. If Mr Conley's process proves successful, it will open up a large demand for magnetic iron ore, and, consequently, for the ore separator.

Mr. Edison's experiments as to gold and silver ores have been limited, owing principally to want of funds, there having been no money in the treasury for the last year and a half. He has conducted a great deal of experimenting at his own expense, paying for it himself, but it is not probable that any further considerable experiments will be made until the treasury is supplied with adequate funds.[8]

D, NjWOE, CR (*TAED* CG001AAI4); typed transcript in DF (*TAED* D8466Z). Written by William Meadowcroft.

1. Eaton presented this report at the company's annual stockholders' meeting on 15 January.

2. The Edison Ore Milling Co. was organized in December 1879 to exploit Edison's inventions for concentrating ferrous and precious metal ores by electromechanical processes. Sherburne Eaton was general manager. Edison Ore Milling Co. association papers, 9 Dec. 1879, CR (*TAED* CG001AAD); Doc. 1844 n. 5.

3. The owner of the Canadian deposits to whom Eaton referred was evidently W. J. Menzies of the Pennsylvania Salt Manufacturing Co., who had corresponded with Edison in 1882 about using the separator to process iron sands in Moise, Quebec. Menzies hoped to experiment on "reducing the magnetic iron sand by a current of hydrogen gas, or hydrocarbon gas, instead of mixing it with charcoal." By the fall of 1882, he had received his first delivery of iron sands from the Ore Milling Co., presumably from the Quonocontaug Beach, R.I., deposits. Like another recipient of that sand, described in Doc. 2393, Menzies was dissatisfied with his initial twenty tons. Questioning whether the machine could not do better, he stated that "I must say I am not surprised now that you have difficulty in selling the sand you separate. With 4 per cent of impurity it is unfit for any direct process for making steel, and only comes under the category of ordinary magnetic iron ore." Menzies to TAE, 19 May and 6 Nov. 1882; Menzies to Edison Ore Milling Co., Ltd., 6 Nov. 1882; all DF (*TAED* D8245M, D8246Q, D8246S).

In May 1883, Edison suggested to Menzies that he likely could arrange for the Edison Ore Milling Co. "to contract to separate at Moisie, Canada, within three months after operations can be commenced, not less than ten thousand (10,000) tons of magnetic iron ore from the Moisie deposit belonging to you; the ore to be cleaned within one per cent of impurity." Menzies replied that although his recent visit to Goerck St. had convinced him of the efficiency of Edison's improved separator, he would not for several weeks have authority to use a Siemens process needed to work the purified ore. The matter seems to have rested there. TAE to Menzies, 28 May 1883, Lbk. 17:41 (*TAED* LB017041); Menzies to TAE, 28 May 1883, DF (*TAED* D8368L).

4. Doc. 2393.

5. The company was founded in 1822 as the Farmers Fire Insurance & Loan Company. After the New York City fire of 1835, it stopped underwriting fire insurance and in 1836 changed its name to the Farmers' Loan & Trust Co. In 1883, the firm had $21 million in assets and over $18 million in liabilities, making it one of the largest trust and mortgage companies in New York State. Its offices were located at 20–22 William St. in Manhattan ("Trust Company Profits," *NYT,* 14 Jan. 1884, 3; letterhead, Farmers' Loan & Trust Co. to Samuel Insull, 24 Dec. 1884, DF [*TAED* D8466X]). Farmers' Loan and Trust merged in April 1929 with National City Bank, the largest bank in the United States. National City underwent a series of mergers and name changes in successive years and became Citibank, N.A. in 1976 (Lanier 1922, 187, 275, 278; Cleveland and Huertas 1985, "Note on the Bank's Name" facing p. 1, 22, 133).

6. Michael R. Conley (b. 1843?) became Superintendent of the Ore Milling Co.'s operations at Quonocontaug Beach in November 1881. He later obtained several patents on furnaces and processes for iron ore. He identified his occupation as "Electrical Engineer" and his workplace as a laboratory in the 1910 federal census (Docs. 2175 n. 3, and 2246 n. 4; U.S. Census Bureau 1982 [1910], roll T624_979, p. 4B, image 179 [Brooklyn Ward 26, Kings, N.Y.]). Among Conley's patents was one granted in 1889 (U.S. Pat. 404,344), which he assigned to William Bell, president of the Ocean Magnetic Iron Co. That firm had signed an 1885 agreement to use Edison's separator, but the agreement expired before the company was able to get its operation running and it was not extended (Edison Ore Milling Co., Ltd., agreement with Ocean Magnetic Iron Co., 11 Dec. 1885; Ocean Magnetic Iron Co. to TAE, 19 May 1887; M. R. Conley to TAE, 10 June 1887, all DF [*TAED* D8543G1, D8748AAF, D8748AAJ]; TAE to Ocean Magnetic Iron Co., 21 June 1887, LM 5:336 [*TAED* LM005336]).

7. Conley had apprised Frank McLaughlin in September 1883 of "estimates for building experimental furnace, and making steel from magnetic sand." The equipment was expected to produce a ton of steel per day, but the editors have found no details of Conley's experiments. Conley to McLaughlin, 14 Sept. 1883; Conley cost estimate, 14 Sept. 1883; both DF (*TAED* D8368X, D8368Y).

Eaton had pointed in Doc. 2393 to the need for a new process of reducing ore to a marketable product because the "difficulty is not with the ore . . . but there is no furnace in existence which will successfully and continuously smelt this ore, both on account of its fineness and of its being exceedingly hard and tough." This statement overlooked not only the nature of the difficulties experienced by the Ore Milling Co.'s only customer at the time (American Swedes Iron Co.), but also the long histories of furnaces capable of smelting ferrous sands and of magnetic separators used in New York, New Jersey, and Pennsylvania (Hunt 1870, passim; Gordon 1996, 95–96). Both Menzies and the American

Swedes Iron Co. (see Doc. 2393 n. 8) seemingly were attempting so-phisticated techniques—bringing iron to an advanced state of refine-ment or even alloyed into steel without intermediate steps such as pig iron—which may have had a direct effect on the usability of the Edison company's ore for their purposes (Menzies to TAE, 28 May 1883, DF [*TAED* D8368L]).

Menzies's Canadian deposits had already been proven amenable to similar processes. The magnetic iron sands at Moisie had been suc-cessfully separated and processed into a high quality iron in the 1860s and 1870s by William Markland Molson's Moisie Iron Works for the United States and Canadian markets. That firm used a process devised and patented in 1869 by François A. H. La Rue, a professor of chemis-try at Laval University in Quebec, in which the ore was extracted from the sand and then, in a single step, manufactured into cast iron or steel. Some of the resulting products were exported to the United States until the company was liquidated in the mid-1870s (Can. Pat. 17; Hunt 1870, 286–87, 292; Bartlett 1885, 233–34; Gordon 1996, 99; cf. the descrip-tion of Edison's separation process in Eaton to R. B. Bristed, 3 Feb. 1882, LM 5:108 [*TAED* LM005108]).

8. The editors have found no mention of Edison experimenting with gold separation at this time, but see Doc. 2673. See also see Doc. 2112 regarding the rationale for shifting from gold to iron as the focus for Ore Milling Co. operations.

–2592–

To Francis Upton

[New York,] Jan 16th 1884

Dear Sir:—

Will you be kind enough to have Mr. Howell give every as-sistance to Ott, and also try to eradicate the sources of error in our new pressure indicators,[1] it is very important for our busi-ness that we have good indicators, and I have no one who can get it on a solid basis quicker or better than Mr. Howell. Please use all possible dispatch. Yours Truly.

~~Thos. A. Edison Constn Dept. By~~. TAE   T[ate][a]

TL (carbon copy), NjWOE, DF (*TAED* D8416AEM). [a]Initialed for Edison by Alfred Tate.

1. The new voltage indicators were the Edison Effect instruments introduced at the end of 1883. John Ott and John Howell experimented with the three-pole lamps during the next few weeks, trying variations in the composition of the center conductor and its distance from the fila-ment, and noting changes in the output current with respect to time and temperature. Ott's results showed, evidently, more variation than Edi-son found acceptable. Edison abandoned this form of instrument and returned to a modified version of the earlier Bradley indicator. Howell significantly improved the Bradley indicator in 1886 without changing its core operating principle. See Doc. 2538 (headnote) esp. nn. 21 and 26; Ott test reports, 14, 18–19, 22, and 25 Jan. 1884; all DF (*TAED* D8430F, D8430F1, D8430G, D8430I, D8430L); Brooks 1953, 406–15.

Paris, le 16 Jan 1884[a]

*From Charles
Batchelor*

My dear Edison,

I acknowledge reception of cable as follows:—

"Get largest lamp royalty for founders no matter what lamp. Also portion cash from licensees and two hundred thousand francs immediate cash for on these terms will surrender veto power and allow other lamps dynamos   leave details you Bailey— Edison Eknoside"[1]

Since my return from Nice as Insul will tell you we have been hammering at the two points "cash from licensees" and Fr200 000 advance. As regards the cash I find Porges will not listen to it because he does not see where it is to come from. Rau and Leon are entirely willing to find some means of providing it but they agree that it cannot be put in the contract and that they must find a way of doing it as a loan on a small portion of the shares of founder belonging to the Light Co— I wanted it not as a loan but as an advanced royalty which in any case should never be paid back, but these people here object to that as they will own pretty nearly half the founders shares and they say that it depreciates their value for them— Insul says that you would probably have difficulty in raising the Fr200,000 over there on the 'founders shares' I therefore do not press the point as to whether they shall be paid interest or not, as it seems to me the "Light Co" must have the money to pay its obligations—

Now as these people will undoubtedly raise this themselves it is very probable they will ask Eaton to let Wallerstein[2] look over the whole matter and see the condition of the Light Company's books. Wallerstein is coming over here shortly and his report here will influence Rau and all his friends, who I need not say are a large part of our Co. Rau also is a large holder of European Stock also Leon— Puskas[3] stock having all been put on the market. At prices just what they would fetch—

I am afraid the cash payment 200,000 has been touched on very lightly instead of it being a 'sine qua non,' as in talking with Porges at Nice, when I remarked that it was possible that the Light Co would want some money, as it had obligations and it was not able to wait 2 years to receive a decent sum; he replied that he did not see where they could get it, and certainly the company would never pay it

The other item of cash to the Light Co from licensees I believe cannot now be got from these people. The fact is the negotiations are finished as far as Bailey is concerned, and in interviews with Rau and Leon at different times that I had with him he argues with them to show me the utter absurdity of the Light Co asking such a thing   I feel sure of this that if we press that it will break up this present arrangement—

In my former letter[4] I gave you a few people to whom Bailey owed much money and asked you not[b] to use it however when Insul & I had an interview with Porges at Nice he gave us a full corroboration of all I had said and much more. He furthermore told us that he had taken all power of negotiations out of his hands— He said he owed money wherever he had made negotiations and he (Porges) would not[b] have a man in that position where the company's interests were concerned—[5] I shall join today if necessary in sending a cable to say that cash from licensees is impossible   Yours

'Batch'

⟨Maj Eaton   Batch in Cable today has changed mind about 200,000 fr⟩[6]

ALS, NjWOE, DF (*TAED* D8436A). Letterhead of Société Électrique Edison. [a]"Paris, le" preprinted. [b]Obscured overwritten text.

1. Edison's 31 December cable, addressed to "Ecknoside   Paris," instructed Batchelor to obtain "two hundred thousand francs immediate cash for us"; its text is otherwise as transcribed here. "Eknoside" was a cable code for Batchelor. TAE to Batchelor, 31 Dec. 1883, Lbk. 17:345A (*TAED* LB017345A); see *TAEB* 6 App. 4.

2. Henry ("Harry") Wallerstein (1853?–1915) was Louis Rau's brother-in-law and a stockholder in the Edison Electric Light Co. of Europe, Ltd., as well as the financial agent and business advisor in New York for the Compagnie Continentale Edison and the Société Électrique Edison. ("Dies While Visiting Friend," *NYT*, 6 Feb. 1915, 18; Obituary, *New York Tribune*, 6 Feb. 1915, 1; Lists of Stockholders of Edison Electric Light Co. of Europe, Ltd., 15 July 1881, 22 Apr. 1882, 23 Feb. 1883, and 25 Feb. 1884; Société Électrique Edison to TAE, 18 Jan 1884; all DF [*TAED* D8127L3, D8228R2, D8331N, D8428E, D8436D]; Bailey and Puskas to TAE, 31 Oct. 1881, LM 1:82B [*TAED* LM001082B]; Norcross 1901, 133–34). A former banker and member of the Stock Exchange, Wallerstein had succeeded his father as head of D. Wallerstein, a leather wholesaler and importing firm in lower Manhattan; the firm had close ties to the Lyonnaise tannery of inventor and industrialist Simon Ullmo, an original subscriber in Edison's French electric lighting companies (Banque Centrale du Commerce to TAE, 9 Jan. 1882, DF [*TAED* D8238J]; Krause 1880, 358; Cahen 1870, 422). During a meeting in Paris with Insull and Batchelor around this time, Rau asked for Wallerstein to examine the European Co.'s books and make a statement of its indebtedness; this task was not yet accomplished by late February. In June 1883, Wallerstein had been mentioned as a potential additional member of the Edison Electric Light Co. of Europe's board of directors; in August 1883, he requested the installation of a few electric lights in the sub-basement of his business at 174 William St. (Samuel Insull to Sherburne Eaton, c. 29 Feb. 1884; TAE to Batchelor, 11 June 1883; both

DF [*TAED* D8416ARC, D8316AAC]; Insull to Charles Chinnock, 16 Aug. 1883, LM 14:404 [*TAED* LBCD1404]).

3. Hungarian promoter and inventor Theodore Puskas (1844–1893) had been involved with commercializing Edison's inventions in Europe since 1877 (Docs. 1153 nn. 4–5, 1164 n. 8). He had helped to organize the Edison Electric Light Co. of Europe, Ltd., in 1879, and he was involved in negotiations leading to formation of the Compagnie Continentale Edison and its subsidiaries in 1882. Puskas sat on the original management committee of the French manufacturing company, the Société Industrielle et Commerciale Edison (see Docs. 1736, 2182 nn. 1–2, and 2249).

4. Batchelor had reported in December: "In my opinion he owes a great deal of money to [Elie] Leon— He told me the other day that he was in a fix and Leon & [Georges] Lebey had guaranteed to arrange for him to borrow another F25,000 to help him over, he did not say what return he gave for this. I know he has tried to borrow money from Porges." Batchelor to TAE, 25 Dec. 1883, DF (*TAED* D8337ZFZ); see also Doc. 2574.

5. In a conversation on 28 January, Porgès reportedly again refused to consider Bailey for membership on the board of directors of the fused French companies. Batchelor to TAE, 29 Jan. 1884, DF (*TAED* D8436L).

6. Edison may have been referring to a 28 January cable from Batchelor and Bailey, which in part reads: "porges refuses absolutely before formation company engagement advance or loan   says considers loan easy afterward   answer quick whether accept" (DF [*TAED* D8436J]). Attempts to secure an advance on royalties continued unsuccessfully within negotiations about the fusion of the French companies, even as the prospective financial backing switched in February from Porgès to French metals manufacturer Pierre-Eugène Secrétan of the Société des Metaux (see, for example, TAE to Batchelor, 15 Feb. 1884; LM 2:46C [*TAED* LM002046C]; Bailey and Batchelor to Eaton, 7 and 16 Feb. 1884; Batchelor to TAE, 8 Feb. 1884; all DF [*TAED* D8436Z, D8436ZAL2, D8436ZAE]; Ratzel 1998, 37, 41). The matter remained unresolved at year's end, but see Doc. 2667.

–2594–

To William Dwelly, Jr.

[New York,] January 18th. [188]4

Dear Sir:—

I have the honor the introduce to you Mr. William Siemens,[1] son of Dr. Siemens of Berlin, whose name no doubt is well known to you.

Mr. Siemens would like to inspect the Fall River station and to receive from you as much information in detail as you can very kindly give him. Any attention you show him will be very greatly appreciated by Mr. Siemens and also by myself.[2]

I beg to remain, Very truly Yours,

TAE   I[nsull]

TL (carbon copy), NjWOE, DF (*TAED* D8416AFJ). Initialed for Edison by Samuel Insull.

1. Wilhelm von Siemens (1855–1919), second son of Werner von Siemens, joined Siemens & Halske in 1879 and took an ownership stake in April 1884. He became a general partner when the firm converted to a limited partnership in 1890, exerting a dominant influence after his father's death in 1892. Feldenkirchen 1999, 41–43; Feldenkirchen 2000, 77; Siemens 2005 [1892], 282.

2. Edison wrote a similar letter to the manager of the Brockton, Mass., central station. TAE to William Jenks, 18 Jan. 1884, DF (*TAED* D8416AFR).

–2595–

Edison Construction Dept. to Charles Campbell[1]

[New York,] January 24th. [188]4

Dear Sir:—

In view of the unsatisfactory manner in which our work in the Engineering Department has been handled since you have had charge, and the seemingly unnecessary delays attendant upon turning out estimates, also in view of the very incomplete filing and recording of data, we have deemed it advisable to dissolve the present organization and issue instructions to our engineers directly from this office.[2]

Each engineer will be detailed to perform certain work in connection with the preparation of estimates and drawings, and will be held responsible for the correctness and despatch exercised in completing them.

On and after the 30th. of this month Mr. Gilmore will be given charge of the office records of the Engineering Department, and will receive all special reports, and enter the dates of the arrival of canvasses, which canvasses are handed to the Mapping Department upon their reception at this office, and any papers therein pertaining to the Engineering Department will be handed to Mr. Gilmore by Mr. Guimaraes.[3]

In addition to this Mr. Gilmore will prepare all estimates obtaining from the engineers that portion of the estimatcs which must necessarily be filed by them.

All calculations made in computing the details of the cost of central station buildings must be handed to Mr. Gilmore, together with all calculations and details of the cost of street installation, which should show each item separately and in such a manner as to enable us to check any and all parts comprising the whole of any estimate prepared in the Engineering Department, and those particulars will be filed with the papers pertaining directly to them. In this manner we hope to very materially increase our capacity for turning out estimates, and we desire each of our engineers to co-operate with this end in view. Yours truly,

Thomas A. Edison Constn. Dept.   By

TL (carbon copy), NjWOE, DF (*TAED* D8416AGY).

1. This letter was probably drafted or dictated by Samuel Insull.
2. Edison had recently planned to suspend Campbell for two weeks on account of tardiness and absences. Campbell blamed ill health, and the suspension seems not to have been enforced. After Edison made the change announced in this document, Campbell continued to make calculations and drawings for central stations until at least early July. A month later he was out of work. His wife desperately sought his reinstatement in September, but Edison considered him too "unstable." Years later, when Campbell used Edison as a reference for an engineering position in Syracuse, N.Y., Edison described him as "smart & probably competent but Erratic." TAE to Campbell, 2 Jan. 1884, LM 17:312 (*TAED* LBCD4312); Campbell to TAE, 3 Jan. and 9 Aug. 1884; TAE marginalia on William Andrews to Edison Construction Dept., 7 Jan. 1884; Insull to Campbell, 9 July 1884; Mrs. Charles Campbell to TAE (with TAE marginalia), 17 Sept. 1884; Edward Powell to TAE, 6 Apr. 1893; all DF (*TAED* D8313A, D8413ZAX, D8456C, D8444ZAA, D8413ZBC, D9307AAO).
3. Henry Guimaraes took charge of the Canvassing and Mapping Dept. in early August 1883, replacing Charles Richardson. Guimaraes was still working for Edison in late 1884, but a year later he asked for an employment reference. He sought to work for Edison again in 1888 as a matter of "the utmost necessity." Guimaraes to TAE, 24 Oct. 1883; Guimaraes to Samuel Insull, 1 Oct. 1885; Guimaraes to Alfred Tate, 4 Nov. 1888; all DF (*TAED* D8342ZCO, D8513T, D8814AEW).

–2596–

To Moses Belknap[1]

[New York,] Jany 25/ [188]4

Dear Sir:

I am today in receipt of a letter from my Superintendent Mr Rich[2] in which he informs me that work in connection with the Newburgh building has been advanced to such a stage as to allow us to go ahead with our portion of the installation which we will do with the utmost dispatch

I regret to say that on account of the great length of time which it has been necessary to consume on your portion of the plant, I am in this letter compelled to make a request of you which, had the work been all done by me, I should have avoided—

You no doubt understand that my position in relation to the installation of these plants is simply that of a contractor, which position I have assumed on account of being more thoroughly versed in the details and construction than any one else can be, owing to the superior advantages ~~my~~[a] I possess as the inventor of the system, and from the fact that I should not feel safe at this Early date to trust the success of my system to any person possessing less Knowledge in regard to it than myself.

You are also no doubt aware that my Construction Department is not a corporation existing with a capital, but that I install these plants and meet the bills for material furnished on account of them and labor performed in connection therewith, by the payments on account of my contracts which I arranged in such a manner as to mature at proper intervals to take care of the above mentioned bills.

Thus you will see that owing to the number of times I turn my money a payment in my business delayed <u>one</u> month is as detrimental as if it were delayed six months in almost any other business.

With the exception of a few hundred dollars worth, all the material for Newburgh was ordered by me over two months ago and consequently the greater portion of the bills have matured— As I stated before had I done all the work in connection with your plant I would have—and from the foregoing you will see that <u>from necessity I must have</u> proceeded in such a manner as to secure to myself the second payment on account of the Contract—due when the station is ready to supply light—long ere this—

I now have to ask if you will very kindly send me a check for $7500.00—or as near that amount as possible—to be accepted by me as part payment of the second installment on account of the contract.[3]

I trust you will understand my explanation of this matter and decide favorably in regard to my request and I also trust that on account of the said request being an unusual one in cases of this nature you will accept this back as my apology for taking up your valuable time in the perusal of this lengthy letter— I am   Yours Very truly

Thomas A Edison.

LS (letterpress copy), NjWOE, LM 18:102 (*TAED* LBCD5102). [a]Canceled.

1. Moses Cook Belknap (b. 1832?) was secretary and treasurer for the Edison Electric Illuminating Co. of Newburgh, N.Y., and was associated with the Highland National Bank in that city. In 1860, when Belknap was a bank teller in Newburgh, he claimed real estate valued at $20,000 and a personal estate valued at $6,000. In 1880, Belknap, by then

a bank cashier, was married with five children and a small household staff. Belknap to Edison Construction Dept., 31 Oct. 1884; Samuel Insull to Belknap, 8 Apr. 1884; both DF (*TAED* D8446ZBC, D8416BDR); U.S. Census Bureau 1963? (1850), roll M432_573, p. 121, image 244 (Newburgh, Orange, N.Y.); ibid. 1967? (1860), roll M653_834, p. 206, image 258 (Newburgh, Orange, N.Y.); ibid. 1970 (1880), roll T9_911, p. 133.1000, image 427 (Newburgh, Orange, N.Y.).

2. Letter not found.

3. Belknap sent a check for this amount to Edison on 31 January. In acknowledging its receipt, Edison offered to run the station for an additional thirty days if the company was not satisfied after the initial thirty days of operation. TAE to Belknap, 1 Feb. 1884, DF (*TAED* D8416AIT).

–2597–

To James Harris[1]

[New York,] Jany 25 [188]4

Personal

Dear Sir.

The whole of your troubles comes from being "too quick" while it may seem curious to you our experience has shown that it is better for many reasons not to wire a single place before the station is ready.[2] in every instance where local people wanted to push ahead, so many mistakes were made and so much extra Expense occured that the little gain in recepts at first was nothing as[a] compared to the loss; this business is to run for many years; and no mistake should be made to save a month or so. It is things feverish haste that disarranges our methods of procedure and makes one think that we have a poor system   we are not prepared for it because we do not believe in the method. Yours Truly

Thomas A Edison

ALS (letterpress copy), NjWOE, LM 18:98 (*TAED* LBCD5098). [a]Repeated at end of one page and beginning of the next.

1. James Harris (b. 1832), a hardware merchant in Bellefonte, Pa., since 1864, was the founding secretary of the Edison illuminating company in that town. Edison Electric Illuminating Co. of Bellefonte resolution, 3 Aug. 1883, DF (*TAED* D8357K); Bureau of the Census 1970 (1880), roll T9_1112, p. 221.2000, image 0444 (Bellefonte, Centre, Pa.); ibid. 1982? (1900), roll T623_1391, p. 11B (Bellefonte, Centre, Pa.).

2. Harris had repeatedly telegraphed and written, with growing impatience, for Edison to send supplies for wiring buildings in Bellefonte and an expert to instruct a local contractor how to do the work. As he put it in an 11 January letter, the station itself was nearly complete but the company could not give specific information about wiring to prospective customers, who "will not be content with looking at the Engine & Dynamos— We want to make light and get the returns" (Harris

to TAE, 11–12, 14, 16–19, 21–23 Jan. 1884; all DF [*TAED* D8453C, D8453D, D8453E, D8453F, D8453G, D8453L, D8453N, D8453P, D8453S, D8453T, D8453V, D8453W, D8453Y]). Harris wrote again on 24 January about having heard that Edison would not supply the necessary wire. Alfred Tate answered a few days later that Edison would, in fact, send the material but that it fell outside the Bellefonte company's construction contract and would be billed separately. Tate also took the opportunity to point out that as the plant was then "ready to supply light," 50 percent of the contract price was due immediately (Harris to TAE, 24 Jan. 1884; Tate to Harris, 28 Jan. 1884; both DF [*TAED* D8453Z, D8416AHO]).

–2598–

From Charles
Batchelor

Ivry, le 27 Jan 1884[a]

My dear Edison

I confirm my cable of this morning as follows:—

"German patent office decided all incandescent lamps with carbon filaments are subject to Edisons patents."[1]

The Kölnische[b] Zeitung of 25 Jan had a telegraphic item from Berlin as follows:—

"Today the German patent office decided[c] in a case of 'demand of nullity' made by Swan against Edison's principle g̶German patent. The attack of Swan was caused by the declaration of the German Edison Co that all incandescent lamps having carbon conductors are subject to Edisons patents   Swan demanded that the fundamental patent should be made void or at least restricted in such manner that other people could also make[d] carbon conductors independently from Edison's patent. The Patent office rejected the "demand for nullity" as also the "demand for restriction" ~~declaring~~ and condemned Swan to costs— It furthermore declared that all incandescent lamps with filaments of carbons are exclusively the property of Edison's patents[2]   Dr Katz[3] and director Rathenau in Breslau represented Edison in that law suit—

How is that for high?[4] I believe this will be the death of the Austrian lawsuit also[5]

Batchelor

ALS, NjWOE, DF (*TAED* D8436I). Letterhead of Société Industrielle et Commerciale Edison. [a]"Ivry, le" and "188" preprinted. [b]Obscured overwritten text. [c]Written in left margin. [d]Interlined above.

1. This is the full text of Batchelor's cable to Edison. LM 2:43F (*TAED* LM002043F).

2. The Swan United Electric Light Co., Ltd., had brought a motion in the German Imperial Patent Office to nullify or at least severely restrict Edison's first and most fundamental incandescent carbon lamp

patent in Germany (notwithstanding the merger with the Edison forces in Great Britain and proposals, such as for a Continental Co., to do so elsewhere). The defendant was the Edison Electric Light Co. of Europe, Ltd., which owned the specification (12,174 of November 1879). The Swan company claimed that crucial portions of Edison's invention were anticipated by earlier patents or publications, including those of St. George Lane-Fox and E. A. King (but not specifically those of Joseph Swan). The Patent Office rejected these contentions as well as the plaintiff's general legal argument that the insufficiency of description alleged in Edison's specification was itself grounds for annulment. German Imperial Patent Office ruling in *Swan United Electric Light Co., Ltd., v. Edison Electric Light Co. of Europe, Ltd.*, 24 Jan. 1884, reprinted (in English translation) in *Edison Electric Light Co. v. United States Electric Lighting Co.*, Complainant's Proofs, Vol. 1, pp. 248–60, Lit. (*TAED* QD012B0248); "Notes. Swan *versus* Edison," *Teleg. J. and Elec. Rev.* 14 (12 Apr. 1884): 314.

Edison reportedly welcomed the ruling as being "in the highest degree important, not only there but here and in every other country, for it practically affirms that every other incandescent lamp is an infringement upon the Edison patent." At least one historian points to the outcome of the German case as contributing to the Edison Electric Light Co.'s decision to bring a number of infringement suits in the United States in 1885. "Edison Wins a Victory," *NYT*, 25 Jan. 1884, 3; Passer 1953, 151.

Though unsuccessful at this time, the challenge by the Swan interests continued in German courts until 1890. It succeeded then, when Edison's German patent was essentially limited only to a manufacturing process. Heerding 1986, 2:3.

3. This was probably attorney and legal scholar Paul Alexander-Katz (b. 1854). Originally from Silesia, he earned the degree of Doctor of Law from the Königliche Universität, Kiel, in 1883; he apparently also represented the Edison interests in another proceeding before the German Imperial Patent Office at about this time. Singer 1910, s.v. "Alexander-Katz, Paul"; German Imperial Patent Office ruling in *Edison v. Naglo Bros.*, 9 Mar. 1885, reprinted (in English translation) in *Edison Electric Light Co. v. United States Electric Lighting Co.*, Complainant's Proofs, Vol. 1, pp. 266–79, Lit. (*TAED* QD012B0266).

4. "How's that for high?" is an American slang phrase meaning, "What do you think of it?" Farmer and Henley 1970, s.v. "High."

5. The Swan company had initiated an Austrian lawsuit on similar grounds in October or November, soon after the Compagnie Continentale Edison considered undertaking its own actions for infringements there. When Louis Rau informed him of the pending Austrian case, Edison drafted a reply stating that "these steps are part of a system of harrassing, & coercion to scare you into a consolidation which should never be made." Edison and Sherburne Eaton took some role in preparing the defense. According to a British journal, the Swan interests also planned to sue in France. Rau to TAE (with TAE marginalia), 15 Oct. 1883; Joshua Bailey to TAE, 2 Nov. 1883; Porges & Moeller to Compagnie Continentale Edison, 3 Sept. 1883; TAE to Compagnie Continentale Edison, 21 Nov. 1883; all DF (*TAED* D8337ZEW, D8337ZFH,

D8337ZDY, D8316BIG); "Notes. Swan *versus* Edison," *Teleg. J. and Elec. Rev.* 14 (12 Apr. 1884): 314.

–2599–

From Anthony Thomas

New York, January 29th, 1884.

Thomas A. Edison Esq[a]   DIRECTOR.

Dear Sir:[1]

The Edison system of lighting has been for a year past an accomplished, practical, working fact, but the financial results, as shown by the balance sheets of the three Companies,[2] are disappointing, if not disheartening.

In my opinion we are drifting toward destruction.[3]

More money will be required, and in the present state of public feeling Stockholders will not respond to our calls on a mere promise of large dividends at some indefinite period in the future.[4]

Now that we have, so to speak, passed the experimental period and are on an established working basis, it is our duty as Directors and Officers to so manage the business that the best financial results shall be secured for the Stockholders.

Whilst I am not willing to assent to any niggardly policy, I must insist on proper economy, and a wiping out of all interests that, in the nature of things, are antagonistic to the best welfare of the several companies.

With these explanations I submit the following suggestions for the purpose of bringing out a general discussion and expression of opinion as to what is best for us to do to develop the Edison System, and place it where it ought to be, a blessing to mankind, and a profit to those who, in their confidence in Mr. Edison, have so freely and liberally advanced their moneys to enable to elaborate and complete his wonderful inventions.

I beg you to understand that, feeling the necessity that something should be done, I give these suggestions merely as an initial or starting point, or in other words to open up the subject.

The Junior Companies have practically lived up to their agreement with the Parent Company, whilst it, so far as I can learn, has never taken any steps towards protecting them in the rights for which they paid so heavily. On asking the reason, I am told it will cost $300,000 and take years of time. Further, that opposition Light Companies make no secret of using our patents and boast they will continue to do so whenever they can make them available.[5]

To merge these differences I suggest to consolidate into a new Company at about present market value, say   COMPANY (give *it a short name), Paid-up Capital $3,120,000, with privilege of increasing to $ .000.000, divided as follows:*

> *Light Company $1,080,000, at 1½% new shares*
>   *for each old share, ... $1,620,000*
> *Isolated,   1,000,000 at share for share, ...   1,000,000*
> *Illuminating,   1,000,000 at 1 new share*
>   *for 2 old ones, ...   500,000*
>   *Total, ... $3,120,000*

Calling up unpaid instalments on old shares will give us money for our wants for some time. As the Parent Company now has in its treasury a large amount of Isolated and Illuminating Shares, the exchange of these into new stock will leave in the new treasury some $800,000 of stock which could be used in building one or more new stations up town, and in buying out the present manufacturing companies, besides leavings a surplus in the treasury.

Make a new contract with Mr. Edison equitable and just to all parties. I am confident that gentleman will treat in a most generous manner the claims of those (the Stockholders) who, through their implicit confidence in him, have advanced nearly two millions of dollars[6] to enable him to make very costly experiments, and thereby complete his inventions through which he has earned and gained an imperishable fame.

All outside manufacturing companies controlled by any of our people should, on a fair basis, be brought into the new Company.[7] In the nature of things their interests are antagonistic to ours. It is only human for them to strive for the largest profit practicable, leaving the smallest for us. At all events they will be charged with doing so even if they work without profit.

As a matter of fact public opinion is becoming aroused to this very thing in other corporations, and I do not believe our Stockholders will hold us harmless unless we make some change in respect to this. So far, under force of circumstances beyond our control, we are fully justified in all that has been done in this manufacturing business. There has not been any wrong in it. It was an absolute necessity for the proper development of the system, but the time is at hand for us to take measures for a general reorganization of all these matters.

The cost of introducing the light into buildings must be materially reduced; there is not any valid reason why the plainest and most simple wiring should cost three or four times as much as gas piping for similar service. To do this the business should be thrown open to competition, giving any one the right to wire and supply fixtures, provided he complies with the rules of the Board of Fire

Underwriters. Any royalty we may make out the present system is more than lost in the detriment to our business through enhanced cost to customers.

My faith in the Edison Light is stronger than ever, but to make it a financial success it must be supplied at a price, every thing included, reasonable enough to make it available for the masses.

I believe it can be done.

As we are going on now we shall be worse off financially at the end of the year than we are to-day, but if we can arrange to re-organize our business, take advantage of the hard times to build and install two stations uptown[8] at a reasonable cost to the Company and to the consumer, we will have done a great deal toward bringing the market value of our Stockholders' property, much nearer than it is to-day, to what I believe to be its intrinsic value.

In closing I beg to repeat the these suggestions are merely given as an initial or starting point for something better, and are conveyed to you in this manner, rather than verbally at a meeting, for the purpose of giving ample time for the investigation and thought that is necessary for such an important matter, as any radical change should only be made after a mature, careful consideration and interchange of views.[9] Respectfully,

Anthony J Thomas[10]

PDS, NjWOE, DF (*TAED* D8427I). [a]Edison's name written, probably by Thomas, on line provided.

1. Francis Upton received his own copy of this document (Thomas to Upton, 29 Jan. 1884, Upton [*TAED* MU067]). In an undated memorandum (see note 3) challenging many of its key contentions, he referred to it as "Mr. Thomas' letter to the directors of the Edison Cos."

2. As indicated in the text below, Thomas meant the Edison Electric Light Co. (the "Parent Company"), the Edison Co. for Isolated Lighting, and the Edison Electric Illuminating Co. of New York.

3. In an undated memorandum concerning Thomas's letter, Francis Upton summarized information about technical improvements and the central station and isolated businesses to support his own more sanguine views. "Drifting toward destruction" was one of several phrases to which he took particular exception. He pointed out that

> During the past year the Isolated Co. have made a small dividend in spite of the great dullness among textile manufactures. The Illuminating Co. have turned the corner, and are now receiving from the regular sales of ~~lamps~~ light about twice the cost of making the light. The Light Co. have taken in stocks worth par and cash over $100,000 from various sources. . . . The various sub-Cos. are feeling that that they have done well in taking hold of Edison Light." [Upton memorandum, n.d. [30 Jan. 1884?], Upton (*TAED* MU068)]

4. Thomas alluded to the general economic contraction that began in 1882 and worsened in 1883. The alarming rate of business failures

in the latter year contributed both to the sense that the U.S. boom of 1879–81 was over and to the analysis that "the great bulk of the disasters were due to blind and amateurish ways of trading, speculation, and the overstraining of credit." Toward the end of 1883, Insull attributed the sluggish share-price performance of Edison's electrical company stocks to "the troubles in Wall Street, where everything has been going to the 'dogs' for sometime past." "The Year's Trade Disasters," *Chicago Daily Tribune,* 1 Jan. 1884, 4; Insull to Charles Batchelor, 5 Nov. 1883, DF (*TAED* D8316BEG).

Commercial institutions linked to the Edison lighting enterprises registered losses as well. At Drexel, Morgan & Co., annual profits dropped by 59 percent in 1883; losses the next year amounted to $41,000 (Strouse 1999, 243–50). In London, J. S. Morgan and Co. found its capital reduced by £41,078 at the end of 1884 (Burk 1989, App. 2). Also notable in 1883, though not necessarily as a result of the general economic conditions, stock of the North Pacific Railroad dramatically collapsed (and Henry Villard with it), while Fabbri & Chauncey lost much of their business and disbanded in the next year (James 1993, 187).

5. The Edison Electric Light Co. did not make a policy of initiating patent infringement lawsuits until 1885, when (in May) it simultaneously filed a number of cases in federal court against the makers and users of lighting devices allegedly covered by Edison patents. Defendants included the Maxim, Weston, and Swan interests. "Edison on the Warpath," *Teleg. J. and Elec. Rev.* 16 (23 May 1885): 470.

6. The editors have not determined Thomas's basis for this figure.

7. Thomas referred to the Edison Lamp Works and the Edison Machine Works, which Edison controlled, and the Electric Tube Co. and Bergmann & Co., in which Edison held a large stake (see Doc. 2343 [headnote]). The Edison Electric Light Co. had considered as early as October 1882 whether to take control of the manufacturing operations, but Edison and some of his principal partners firmly rejected the idea (see Doc. 2395 esp. n. 3). The company again deferred the issue at its annual meeting in October 1883 and explained to stockholders that it was gaining "valuable experience . . . so that when the moment arrives for finally deciding this question, it will be a comparatively easy matter" (Edison Electric Light Co. annual report, 23 Oct. 1883, in Edison Electric Light Co. Bulletin 20:48, CR [*TAED* CB020442]).

Francis Upton valued the capital and assets of the Edison Lamp Co. at about $240,000 in late 1883, a figure that Samuel Insull regarded as too low by a quarter. Also according to Insull, Bergmann & Co. had made a profit of $55,000 on the year through November, all of which had been returned to the business (Insull to Charles Batchelor, 5 Nov. 1883, DF [*TAED* D8316BEG]; see also Docs. 2536 and 2590). An internal statement placed the total investment in the Edison Machine Works near $206,000 at the end of 1883 (Edison Machine Works statement, n.d. [31 Dec. 1883?], DF [*TAED* D8334ZBU]). In response to Sherburne Eaton's inquiry in February 1884 (see Doc. 2638), the Electric Tube Co. stated its total investment at $50,000 (Electric Tube Co. statement, n.d. [Feb. 1884], DF [*TAED* D8433J]). Doc. 2368 contains comparable information for the manufacturing shops as of November 1882.

8. By December 1883, the Edison Electric Light Co. had begun to

consider dividing the long-anticipated uptown second district centered on Madison Square (for which preliminary planning work began in 1882; see Doc. 2223 esp. n. 3). One portion would remain around Madison Square; the other would lie further north, somewhere between Thirty-fourth St. and Central Park, to encompass theaters, restaurants, and "the best residence district." Edison Electric Illuminating Co. of New York report to stockholders, 11 Dec. 1883, in Edison Electric Light Co. Bulletin 22:22, 9 Apr. 1884, CR (*TAED* CB022531); see Docs. 2445 and 2669.

9. See Doc. 2638.

10. Railroad financier and executive Anthony J. Thomas (1826–1903) was a shareholder in the Edison Electric Light Co., and he held stock or an office in the Edison Electric Illuminating Co. of New York, the Edison Co. for Isolated Lighting, and the Electric Tube Co. Thomas had worked for Drexel, Morgan & Co., first as managing clerk, from around 1870 until 1881, when he became a vice-president of the Northern Pacific Railroad (Obituary, *NYT,* 25 Apr. 1903, 9; U.S. Census Bureau 1982? [1900], roll T623_1114, p. 13B [Manhattan, New York, N.Y.]; Edison Electric Light Co. list of stockholders, 1 July 1882; Sherburne Eaton to TAE, 21 Nov. 1883; both DF [*TAED* D8224ZAP1, D8325ZDA]; Edison Electric Light Co., list of stockholders, 29 Sept. 1883, Miller [*TAED* HM830194 (image 10)]; Electric Tube Co. certificate of incorporation, 1 Mar. 1881, NNNCC-AR [*TAED* X119TA]). He often represented the Morgan interests in the management of railroad companies. With Egisto Fabbri, Thomas had recently helped to conduct an internal investigation of the Oregon and Transcontinental Co. after the collapse of Henry Villard's portfolio ("Villard's Great Schemes," *NYT,* 1 Jan. 1884, 1).

–2600–

From William Preece

[London,] 30, 1, 84

My dear Edison

The British Assocn meets in Montreal on Aug 25th.[1] I sincerely hope to meet you there and to hear something from you. All original papers—papers purely Scientific—are read in Section A under the Presidency of Sir W. Thomson[2]—all practical papers—papers relating to the applications of science are read in Section[a] "G" under the Presidency of Sir Frederick Bramwell.[3] I am a "G" man first and an "A" man second. Now whatever you do for "A" I hope you will do something for "G"— We always have an electrical day and anything from you will make it the red letter day of the meeting. I shall probably do something and so will Thomson but we are going to be elective and selective— We count on seeing and hearing you and also at least hearing your prophet.[4a] With Kindest regards, I am   Yours sincerely

W. H Preece[5]

⟨[D̶r̶a̶u̶g̶h̶t̶?̶]ᵇ   Draft Letter [tho?]ᶜ   Say that I will try & get
some of my new traps together if he P will read it as I should
be so frightened that I wouldnt go within 3 miles of the meeting if I had to say anything before an audience⁶   E⟩

ALS, NjWOE, DF (*TAED* D8403O). Letterhead of the Whitehall
Club. ªObscured overwritten text. ᵇCanceled. ᶜIllegible.

1. The British Association for the Advancement of Science was
founded in September 1831 by the Reverend William Venables Vernon
Harcourt and others who, disillusioned with the elitist and conservative attitude of the Royal Society, proposed an organization that would
"promote the intercourse of the cultivators of science with one another,
and with foreign philosophers" (Morrell and Thackray 1981, 223–24;
Report of the British Association for the Advancement of Science 1831, 22).
The Association invited and supported inventors and scientists who
shared their populist leanings, including many (such as Edison) who
were not British, and cultivated relationships with like-minded foreign
scientific societies as well. The 1884 annual meeting, held from 28 August to 2 September, was the first held outside of the United Kingdom,
and many British attendees also took part in the American Association
meeting in Philadelphia as part of a "North American Tour" (Pancaldi
1981, 156–63; Worboys 1981, 173–74).

2. William Thomson (1824–1907), created Baron Kelvin in 1892,
was Professor of Natural Philosophy at Glasgow and a prominent
chemist, physicist, and science publisher. He did important work in the
mathematical analysis of electricity and thermodynamics, including the
early recognition of a need for an absolute thermometric scale, which
was eventually named after him. Thomson, who wrote extensively on
electrical theory, was also the driving force behind the establishment of
the British Association's Electrical Standards Committee in 1861. Doc.
2173 n. 8; *Oxford DNB*, s.v. "Thomson, William, Baron Kelvin (1824–
1907)"; Brock 1981, 106.

3. Sir Frederick Bramwell (1818–1903) was a mechanical and railroad engineer widely recognized as an expert on legal and technical aspects of municipal engineering. He was one of the authorities (William
Thomson was another) to whom Edison's financial backers in Great
Britain had turned in 1882 for an independent assessment of the economic feasibility of the Edison system and the legal soundness of its
underlying patents (see Docs. 2203 [headnote], 2211 n. 19, 2258 n. 16).
Bramwell presided over Section G of the Association, which focused
specifically on "Mechanical Science" ("Engineering" after 1901) as opposed to the more theoretical focus of Section A, "Mathematical and
Physical Science," under Thomson (Brock 1981, 93–94).

4. Preece referred to an appellation given by the London *Daily News*
in 1882 to Edward Johnson for his role in the Crystal Palace exhibition: "There is but one Edison, and Johnson is his prophet." "The Electrical Exhibition at the Crystal Palace," *Daily News* (London), 8 Apr.
1882, 6.

5. Welsh engineer William Henry Preece (1834–1912) was electrician
and assistant chief engineer for the British postal telegraph network. He
became a Fellow of the Royal Society in 1881 and, at this time, was a vice

president of the British Association's Section G (Doc. 803 n. 7; *Oxford DNB*, s.v. "Preece, Sir William Henry"; *Report of the British Association for the Advancement of Science* 1884, lviii). He and Edison suffered a bitter and public rift in 1878 in what had been a collegial relationship (see Docs. 1348, 1370, 1385, 1482, and 1602). Their relationship improved in 1881 and early 1882 after Preece publicly praised the Edison lighting system, and Preece became an important proponent of the system in British circles (see Docs. 2119 n. 1, 2203 esp. n. 34, 2217, and 2225). After the Montreal meeting, Preece toured the United States and arranged to meet Edison at his Bergmann & Co. laboratory on 10 or 11 October. In a lecture about his travels, he observed that despite their previous differences, it would be "impossible for any man to receive another with greater kindness and attention than Mr. Edison received me" (Preece to TAE, 8 Oct. 1884, DF [*TAED* D8403ZHN]; TAE to Preece, 9 Oct. 1884, Lbk. 19:297 [*TAED* LB019297]; Preece 1884, 590).

6. Edison's marginalia formed the basis of a typed reply sent after his return from Florida in early April (TAE to Preece, 5 Apr. 1884, DF [*TAED* D8416BCS]). Edison had personally given a paper to the 1878 annual meeting of the American Association for the Advancement of Science, but even on that occasion he delegated the reading of three others to George Barker. The next year, he had Francis Upton make a presentation on his behalf to the same body (see Docs. 1401 and 1796). Edison evidently did not draft a paper for the British Association in 1884. At that body's meetings in 1884 and 1885, Preece gave papers to Section G on "Domestic Electric Lighting" related to "the lighting installation of his house in Wimbledon" (*Report of the British Association for the Advancement of Science* 1884, 893; ibid. 1885, 1197–98).

–2601–

To Benjamin Vicuña Mackenna

New York January 31 1884

Senor Don B Vicuña Mackenna.[1]

Permit me to thank you for the courtecy[a] and honor of your photograph and the sentiments of esteem with which it is accompanied.[2] My library contains no work more valued than those which your genius has produced, the possession of which I owe to your courtecy.[3a]

Allow me to thank you for the interest you have always shown in the efforts of your colaborer in science and the considerate opinion you have always expressed of my inventions. As a slight return for your esteemed favor I take the liberty of emulating your courtecy[a] by enclosing a photograph of myself. Thanking you again for your sentiments of friendship, I am with the renewed assurance of my most profound consideration   Very Respectfully yours

Thomas A Edison

ALS, NjWOE, DF (*TAED* D8403Q); letterpress copy in Lbk. 18:10.
[a]Obscured overwritten text.

1. Benjamin Vicuña Mackenna (1831–1886), a lawyer, journalist, and historian, participated in and was imprisoned during Chile's Civil War of 1851. He escaped to California and did not return to Chile until 1856, when he became editor of *La Asamblea Constituente*. An active critic of Conservative president Manuel Montt Torres, Vicuña Mackenna was again banished in 1859. Returning five years later, he was elected deputy to Congress. He later served as a special envoy to the United States and to Europe, founded the Partido Democrático Liberal, and was appointed Mayor of Santiago in 1872, where he promptly began major public works projects. Briefly a candidate for president in 1876, Vicuña Mackenna withdrew for fear of provoking renewed civil conflict. At this time, he was a senator from Coquimbo. Many of Vicuña Mackenna's numerous books and articles, including a five-volume national history completed in 1882, concern the Chilean struggle for independence. *HDC*, s.v. "Vicuña Mackenna, Benjamín"; *DPC*, s.v. "Vicuña Mackenna, Benjamin"; Marley 2005, 2:733–34.

2. The editors have not found prior correspondence with Mackenna. A few months later, Willis Stewart, an Edison agent in Chile, reported that Vicuña MacKenna had recently returned and sent his thanks for both this letter and the Edison photograph it enclosed. Vicuña MacKenna asked Stewart to "say in answer that 'He would rather receive this letter from you than to receive the Order of the Garter from the Queen of England.'" Stewart remarked that "You will now have to go him one better on the taffy [flattery] question." Stewart subsequently mailed Edison's letter back to him with the explanation that Vicuña Mackenna had "perpetrated another of his eccentricities, & had the letter which you sent him lithographed"; the phrase "AUTOGRAFO DE ALMA EDISON" appears at the top of the page (Stewart to TAE, 9 May and 8 Sept. 1884, both DF [*TAED* D8435ZAD, D8435ZBG]). Edison's evident regard for Vicuña Mackenna's work recalls his personal affection for another liberal and former revolutionary, Eduard Reményi (see Doc. 2137).

3. Edison acquired Vicuña Mackenna 1881 and Vicuña Mackenna 1882 for his library at an unknown date. The editors have not identified a photograph of Vicuña Mackenna among Edison's collections.

–2602–

To Willis Stewart[1]

[New York,] Feby. 1st. [18]84.

Dear Sir:—

Having made a contract with the Edison Electric Light Company giving me control of all Isolated work in Chili and Central Station work outside of Santiago and Valparaiso, you are hereby appointed my agent for that territory.[2]

Although this contract would undoubtedly prove very remunerative to me, yet, I greatly desire that the Santiago Station[3] should be kept in operation if possible, without prejudice to any interests involved, and that it should eventually acquire the entire right to use the Edison system of lighting in Chili, as was originally contemplated when the Company was formed;[4]

and you will first do your best to secure such a result, consulting me by mail or cable as may be necessary. You will also give such aid as may be necessary to Messers Ed. Kendall & Co.[5] in any effort that may be made looking toward the organization of a new Company to control either Santiago alone, or all Chili.[6]

Recognizing the fact that the present Central Station plant at Santiago is both difficult and expensive to operate,[7] you will give the present Local Company the benefit of any advice or suggestions which owing to your recent experience here may tend to make such Station a dividend paying one, if so desired by said Company or its representatives; it being understood that all efforts shall be directed toward replacing of the present machinery by a plant fully up to the latest improvements.

In your efforts to re-organize the present Santiago Co. you may assure my friends in Chili[8] that I will use my best endeavours to aid them with the Edison Electric Light Company, and that having full confidence in the future of the business in Chili I strongly urge them to unite with you in a permanent arrangement.

In case you are able to re-organize the Company in Santiago, you will make such studies of the actual situation of the lighting business in that city as shall enable me to furnish a plant suitable for existing needs at the lowest possible price   Yours truly,[a]

TAE

TL (carbon copy), NjWOE, DF (*TAED* D8416AIW). Initialed for Edison by Alfred Tate. [a]"Yours truly," written by Tate.

1. Willis N. Stewart (b. 1854), trained in dynamo installation and repair at the Edison Machine Works, went to Chile in 1881 on behalf of Edison lighting interests and worked there until 1889 (Doc. 2279 n. 5; U.S. Dept. of State [n.d.] Passport Applications, issued 24 Sept. 1881; Stewart to TAE, 13 July 1889, DF [*TAED* D8905AEI]; Dyer and Martin 1910, 1:444). Stewart regularly corresponded with Edison and the Edison Electric Light Co. about the electrical business in Chile and South America. In 1881, he had expressed the desire of Santiago investors to form a national company quickly, before gas companies could secure franchises in other Chilean cities. At the end of 1882, he reported the likelihood "that a large Company will be formed here to control the use of electricity in this country in every form, by the purchase of all valuable patents" for incandescent and arc lighting, telephones, and electric power (Stewart to TAE, 11 Nov. 1881, DF [*TAED* D8131Q]; Stewart to TAE, 1 Dec. 1882, Scraps. [*TAED* SB012BBO]). Stewart was active on a number of fronts in the early 1880s, including efforts to expand the Santiago central station (originally under contract with Kendall & Co.), negotiating for village lighting installations in several Chilean towns, and supervising an extensive lighting installation for Dona

Isidora Goyenechea de Cousiño. He visited the Shamokin, Pa., plant with representatives of the Chilean company in October 1883 (Stewart to Sherburne Eaton, 21 May 1883; Eaton to TAE, 5 Nov. 1883; Thomas Conant to Samuel Insull, 21 Oct. 1883; all DF [*TAED* D8336Q, D8327ZBM, D8360ZBP]; *TAEB* 6 App. 1.F.8; Cousina "Notes," *Electrician* 12 [8 Dec. 1883]: 74; Edison Electric Light Co. Bulletin 20:21, 31 Oct. 1883, CR [*TAED* CB020]; Compañia Electrica de Edison pamphlet, 1885, CR [*TAED* CA011B]). Stewart's activities in 1884, including canvasses for village plants, prospective central stations, electric railroads, and plants for private residences, are represented in scores of communications among himself, Edison, and Edison's New York associates in Electric Light—Foreign—Chile, DF (*TAED* D8435).

2. This offer likely was drafted by John Tomlinson in consultation with Sherburne Eaton (TAE to Tomlinson, n.d. [1884]; TAE to Eaton, n.d. [1884]; Stewart memorandum of instructions [with TAE marginalia to Tomlinson], n.d. [1884]; all DF [*TAED* D8435ZBK, D8435ZBJ, D8435ZBM]). Under terms of a contract dated 9 February, Edison appointed Stewart as his agent in Chile for two years, with power to sell isolated plants and machinery, set up central stations, and organize local companies. In May 1884, Edison invited Stewart to play a similar role in Boliva, where he was "not particularly impressed by Fabbri & Chauncey's agents"; Prevost & Co. were entitled to the concession until October 1884 but had yet to exhibit Edison's electric light system (TAE to Stewart, both 23 May 1884; Frank Hastings to TAE, 8 July 1884; all DF [*TAED* D8416BQW, D8416BQY, D8434ZAC]). This relationship flowed from a 26 January 1884 agreement in principle between Edison and the Edison Electric Light Co. that effectively licensed Edison to make these arrangements on a royalty basis. The company ratified the agreement in February; the editors have not found the contract, but its terms are summarized in Doc. 2677 (TAE agreement with Stewart, 9 Feb. 1884; TAE to Eaton, 5 Feb. 1884; both DF [*TAED* D8435B, D8416AJR]).

These arrangements were part of Edison's more active involvement in South and Central American lighting markets. His increased involvement began with Argentina in mid-1883, by this time he had also arranged for an agent in Bolivia and hoped to do the same in Mexico. His continuing dissatisfaction with the business in those regions was evident in September 1884, when he referred a thoughtful proposal from Sao Paulo, Brazil, to Edward Johnson and "the Busted-Disgusted South American Dept." Insull to Stewart, 23 May 1883; R. S. B. Davids to TAE [with TAE marginalia], 18 Sept. 1884; both DF (*TAED* D8416BQT, D8434ZAK); TAE to David Cone, 5 June 1884; Lbk. 18:75 (*TAED* LB018075); Edison Electric Light Co. Bulletin 20:49, 31 Oct. 1883, CR (*TAED* CB020); see also Doc. 2677.

3. The Santiago central station opened on 4 November 1882 in a two-story brick building near the "Grand Plaza" (probably the Plaza de Armas) and business center. Like similar installations, its first customers included a score of shops, stores and commercial houses, and a cafe and billiard room. The first district encompassed an area within the boundaries of calle de Mapocho on the north, Alameda de las Delicias on the south, calle del Estado on the east, and calle de los Teatinos on the west. Originally outfitted with three Z dynamos, the station's capac-

ity was increased in early 1883 by the addition of six K machines. By late that year, it served about 100 customers with 2,500 lights, and the Electric Light Co. in New York had authorized estimates for the station's further enlargement. Doc. 2279 n. 5, Edison Electric Light Co. Bulletins 15:32, 20 Dec. 1882; 17:9–10, 6 Apr. 1883; 20:37; 31 Oct. 1883; all CR (*TAED* CB015, CB017, CB020); "Chile. Noticias de Valparaíso," *La Voz del Nuevo Mundo* (San Francisco), 13 May 1882, 1; Eaton to TAE, 5 Nov. 1883 (*TAED* D8327ZBM).

4. Although formation of a national company had been discussed at least since 1882, the only firm for Edison electric lighting in Chile seems to have been the local one in Santiago, La Compañía de Luz Eléctrica Edison-Santiago. Formed in 1882, its capital stock was valued at \$112,000 in 1884 (Edison Electric Light Co. report to stockholders, 24 Oct. 1882 [pp. 5–6], DF [*TAED* D8224ZBJ]; "Chile. Noticias de Valparaíso," *La Voz del Nuevo Mundo* [San Francisco], 13 May 1882, 1; handstamp on Stewart to TAE, 27 Oct. 1882, Scraps. [*TAED* SB012BBI]; Edison Electric Light Co. of Santiago memorandum, n.d. [Apr. 1884], DF [*TAED* D8435Q]). Although its affairs were, according to Stewart, "a most fearful mess," there was pressure to reorganize or suspend business because all contracts would expire on 7 April 1884. Additional confusion revolved around a recent contract between the New York company and a Chilean gentleman known as the Ovalle (sometimes spelled Ovalli); the editors have not been able to identify him among the large Ovalle family cluster in Chile (Stewart to TAE, 28 Mar. 1884 [p. 1]; Stewart to Eaton, 28 and 13 Mar. [pp. 3, 9] 1884; DF [*TAED* D8435H, D8435I, D8435D]).

5. Edward Kendall & Co. (Eduardo Kendall y Ca.) was the business correspondent of Fabbri & Chauncey in Valparaíso. The company held Edison's power of attorney for electric lighting in Chile as early as January 1880; at the end of that year, it secured Edison's electric lighting patent rights in Chile for an eight-year period. Like other agents appointed by Fabbri & Chauncey, the exact nature of Kendall's authority remains unclear, but the firm dominated the operational and financial affairs of the Santiago company to such an extent that it could threaten to close the station. Kendall's financial management vexed Stewart, who considered it both inept and, because of reported collusion with the gas company, dishonest (Fabbri & Chauncey to TAE, 16 Feb. 1881; Eaton to TAE, 20 Nov. 1882; Stewart to TAE, 21 Mar. and 23 Apr. 1884; Stewart to Eaton, 13 and 28 Mar. and 10 July 1884; all DF [*TAED* D8131D, D8237ZAJ, D8435F, D8435Y, D8435D, D8435I, D8435ZAT]). Kendall & Co. likely suffered when Fabbri & Co.'s shipping business on the west coast of South America was absorbed in 1883 by the Merchants' Line (a W. R. Grace concern). By the beginning of 1885, the firm was insolvent (James 1993, 187; Stewart to TAE, 10 and 31 Jan. 1885, both DF [*TAED* D8534A, D8534D]; "Money-Market and City Intelligence," *Times* [London] 5 Jan. 1885, 11; "Money-Market and City Intelligence," ibid., 6 Jan. 1885, 11).

Edward Kendall was a founding director of the Valparaíso Sporting Club (1882) in the seaside resort of Viña del Mar; when an Edison isolated plant was installed in the Gran Hotel there, more than 1,300 feet of conductors were run to supply power to a few lights in Kendall's home. Mayo 1987, 216; Vicuña Mackenna 1877, 59; Silva Vargas 2008,

143; Edison Electric Light Co. Bulletin 10:11, 5 June 1882, CR (*TAED* CB010); Compañia Electrica de Edison pamphlet, 1885, CR (*TAED* CA011B).

6. Stewart reported in March about the possible closing of the Santiago station and liquidation of the parent company on account of Kendall & Co.'s "gross and criminal (I mean it) mismanagement" (Stewart to TAE, 14 Mar. 1884, DF [*TAED* D8435E]). With support from like-minded stockholders, he negotiated a reorganization plan in which he would take personal responsibility for the station and the formation of a new company. A version of this plan was offered to the Edison Electric Light Co. in April, but Sherburne Eaton made an unfavorable point-by-point review of it (Stewart to Eaton, 13 Mar. and 9 Apr. 1884, p. 4; Stewart to TAE, 9 Apr. 1884; Eaton memorandum, 21 May 1884; all DF (*TAED* D8435D, D8435R, D8435P, D8435ZAH]). Stewart also prepared, at an uncertain date, a variant "Plan for the Reorganization of The Edison Electric Light Co. of Santiago," with fuller details for converting the Santiago company into a holding company to operate throughout Chile; it is not clear that he presented it to the Electric Light Co. (Stewart memorandum, n.d. [Apr. 1884?], DF [*TAED* D8435Q]; see also Docs. 2669 and 2711).

7. Stewart blamed a doubling of the station's coal consumption on the excessive voltage drop caused by "the injudicious extension of mains beyond the points to which they were calculated." He suggested increasing economy by substituting 10-candlepower lamps already on hand for the 16-candle lamps used in the system and promised Eaton that the plant could return $1,000 a month over its expenditures (Stewart to Eaton, 28 Mar. 1884, DF [*TAED* D8435I]). In May 1884, Stewart ordered a 300-light dynamo and an engine to expand service to an important city block in Santiago and to two daily newspapers (Stewart to TAE, 30 and 31 May, 5 Aug. 1884, all DF [*TAED* D8435ZAP, D8435ZAQ, D8435ZAU]).

8. These friends may have included Enrique Lanz and Luis Ladislao Zegers y Recasens, who soon extended $10,000 in new credit for Stewart to order dynamos and other equipment. An engineer and professor of physics, Recasens had exhibited an E dynamo and fifteen Edison lamps at the University of Chile in 1882. Zegers also promised in March to help raise more capital for reorganizing the Santiago company and, at Stewart's request, Edison sent him a phonograph in May. By November, however, he was threatening legal action over Stewart's loose accounting of the earlier monies (Stewart to TAE, 14 Mar. 1884; Stewart to Eaton, 28 Mar. 1884; Zegers to TAE, 29 Nov. 1884; TAE to Bergmann & Co., 28 May 1884; all DF [*TAED* D8435E, D8435I, D8435ZBI, D8416BRN]; Edison Electric Light Co. Bulletin 13:11, 28 Aug. 1882, CR [*TAED* CB013]; Gutiérrez and Gutiérrez 2006). Lanz was a founder (1883) of the Sociedad de Fomento Fabril, an industrial development organization, and a financial backer of Edison's electric railroad in Chile. He had installed an E dynamo and thirty B lamps at his flour mill near Valparaíso in 1882 (Ceppi and Correa 1983; Stewart to TAE, 14 Dec. 1883; Lanz to TAE, 8 Aug. 1885, both DF [*TAED* D8372M, D8534P]; Edison Electric Light Co. Bulletin 13:15–16, 28 Aug. 1883, CR [*TAED* CB013]).

–2603–

To Sherburne Eaton

New York, [c. February 3,]188[4][1a]

Maj

Read this over—[2] Then I want to arrange it to bounce the men at Sunbury & put better & cheaper men in   I believe with your stock & <u>ours 41 shares</u> we can do what we please[3]

Edison

ALS, NjWOE, DF (*TAED* D8458E). Letterhead of Thomas A. Edison, Central Station, Construction Dept. [a]"New York," and "188" preprinted.

1. Edison wrote this document after receiving the 2 February letter discussed in note 2. Sent from the central Pennsylvania town of Bellefonte, some two hundred miles from New York, that letter likely reached Edison on 3 February.

2. Edison enclosed William Rich's four-page response to Alfred Tate's request to fix the Sunbury, Pa., central station's leaky roof and make "such other repairs as you may deem necessary." Tate had also asked for Rich's opinion of the engineer, noting a report of burned-out engine bearing boxes. Rich replied that he had inspected the Sunbury plant prior to receiving Tate's request, and his observations provided a broad indictment of its operation. Windows were broken or painted over, a sheet of metal covered a hole in the roof directly over the voltage indicator, there was extensive exterior corrosion, and the interior was generally in a "filthy condition." Rich pointed out that the original driving belts betrayed little wear, having been replaced because they produced noise that "disturbed the slumbers of the engineer (but still he slept on)." He also related an incident like the one mentioned in Doc. 2563 in which engineer William Bateman was absent overnight and left the station in the hands of a young assistant. Tate to Rich, 28 Jan. 1884; Rich to TAE, 2 Feb. 1884; both DF (*TAED* D8416AHN, D8458F).

More damning information about Bateman was soon provided by William Andrews, who corroborated Rich's account of the station's condition and operation. Andrews reported that Bateman was in debt "all over Sunbury" and had "made the station a regular rendevouz for women— I found a couple of doz. empty beer bottles behind boiler." The situation, he concluded, "shows the evil of leaving a station entirely in charge of an engineer, with no one else in the town that knows anything about Station matters, or has authority to act." Andrews to TAE, 9 Feb. 1884, DF (*TAED* D8442ZAV).

Accounts of intoxicated engineers also came in at about this time from the Hazelton and Bellefonte, Pa., plants. Andrews suggested replacing the Hazelton man with another whom he described as "rather a rough hand at work but very faithful and sober— We Cannot do with drunkards in this business that is certain." Sherburne Eaton to Samuel Insull, 18 Feb. 1884; Samuel Insull to James Harris, 19 Feb. 1884; Andrews to Edison Construction Dept., 14 Feb. 1884; all DF (*TAED* D8439U, D8416APC, D8455ZAG).

3. Edison and the Edison Electric Light Co. were the two largest shareholders in the local Sunbury illuminating company, a consequence of their separate decisions to accept stock in partial payment of con-

struction costs and licensing fees, respectively. Samuel Insull to Bergmann & Co., 24 Mar. 1884, DF (*TAED* D8416AZA); see also Docs. 2424 (headnote) and 2519 n. 2.

Edison apparently acted quickly in this case. Bateman telegraphed on 6 February that he and the meterman were resigning on account of a wage reduction. Edison instructed him to "run station til relieved afterward may get you position." A new engineer was on duty by 9 February. Bateman to TAE, 6 Feb. 1884; TAE to Bateman, 6 Feb. 1884; William Andrews to TAE, 9 Feb. 1884; all DF (*TAED* D8458H, D8458I, D8458L).

–2604–

To William Chamberlaine[1]

[New York,] Feby. 4th. 1884.

Dear Sir:—

I am in receipt of your favor of 30th ultimo, and in reply beg to say that I will refer same to the Edison Electric Light Co. for their decision.[2]

In order to aid you in the formation of a company in Norfolk, I beg to say that the Edison Electric Light Co. do not [~~wish~~?][a] to work under any other name. They have refused to do so several times. In Harrisburg Pa. they objected to do so with a local Fuller Company.[3] In most cases parties have organized a separate company in the interests of the Edison Light, and have then arranged with the Edison Company's contractors to put both systems in one station in order to effect an economical management.

The Edison Electric Light Company charge 25 per cent of the capital of the Edison sub-companies for the license, no matter whether the capital is actual or watered. One fifth of this amount is payable in cash, and the balance taken in stock. They derive no profit from the machinery or other supplies necessary for the installation of the plant, but have arrangements with other parties who take the contract of installing the plant complete and operate same for 30 days, the contractor being required to furnish and erect all materials at actual cost, plus 12 per cent, which latter amount is to cover his risk and profit.

If there is a probability of your doing anything, the Edison Electric Light Co. after being requested by you to do so, will have your city canvassed in order to obtain data for the ⟨Wish they had added "at your expense."— SBE   Feb 6. [-]⟩[4b] contractors, and those latter parties will make a firm tender for the installation of a plant of from 500 to 10,000 lights capacity as the case may necessitate, working the system either overhead or underground.

TL (carbon copy), NjWOE, DF (*TAED* D8416AJO). ªCanceled. ᵇMarginalia written by Sherburne Eaton; fragmentary word illegible.

1. William Wilson Chamberlaine (1836–1923) was a Norfolk, Va., banker and former Confederate officer who worked in the railroad industry and had been involved in running the municipal waterworks for many years. He was instrumental in building the first light and power plant in Norfolk in 1884 and was the president of the Norfolk Electric Lighting Co. (the local Brush-Swan Electric Light organization) for several years subsequently. *NCAB* 42:372.

2. Chamberlaine's letter has not been found, but Samuel Insull enclosed it with a letter to Frank Hastings on 4 February. DF (*TAED* D8416AJI).

3. Chamberlaine probably referred to the Harrisburg Electric Light Co., one of whose backers, A. J. Dull, had communicated with Edison agent Henry Clark. Clark reported in June 1883 that the company had "a few Fuller arc lights burning. I do not think they are giving the greatest satisfaction. The Co. have decided to adopt the Edison system but this is an office secret." (The Fuller Electrical Co., headquartered in New York City, manufactured and installed arc lighting equipment under the patents of James Billings Fuller, an inventor and electroplater, and James Wood. The company had won a contract in 1879 to light Madison Square Garden and was a member of the Gramme Electrical Co. patent consortium.) Edison promised to meet Dull in November 1883 about Harrisburg; the Edison Construction Dept. completed a canvass of the city later in February, apparently without Dull's involvement. "Testing Electric Illumination. Madison-Square Garden to Be Lighted To-night by Electricity," *NYT*, 16 June 1879, 8; Passer 1972 [1953], 374 n. 58; Clark to TAE, 12 June 1883; Clarke to Samuel Insull, 5 Nov. 1883; Alfred Tate to TAE, 7 Nov. 1883; Insull to Charles Lyon, 11 Feb. 1884; all DF (*TAED* D8340ZAG, D8340ZFL, D8340ZFN, D8416AMG).

4. The editors have found no evidence of a canvass.

–2605–

Draft Agreement for Central Station Engineers

[New York, c. February 5, 1884[1]]

Engineer at station.[2]

Bond of ___ 2 suritis for faithful perfmc duties, ~~to~~

Retention of wages to pay for loss due to Carlessness [-]ª at time accident balance Collectable from suritis.

~~That he is skilled Engr & fireman & believe of ___ years Experience and believes himself capable of attending to performing the duties of fireman E~~

Keep allᵇ machinery &ᵇ is good condition & the [~~stato?~~]ª

Must clean boiler tubes at least Every 3 months.

It is agreed and understood that ~~he is to~~ his duties and hours shall be as follows blankᶜ

Supplementaryᵇ Clause at end of agreementᵈ which is to be signed after he has run 30 days in which he states that he considers himself competent to & does undertake to run the ~~station B~~ Electrical & steam apparatus Economically[b] in a proper manner ~~the~~[c]

Supt.   to have charge of the station &[b] —Engr is[b] under him   Duties to ~~keep~~ attend to the meters make out & collect bills,—keep ~~the~~all the necessary books reports & records of the station, account of stock, &[b] order necessary material for running the station, Solicit~~or the~~ consumers of the light, to[d] be held responsible for the running of the station kept neat good order, run economically.

to ~~assist in~~ be on duty and attend to the ~~Electrical~~ Dynamos & electrical regulation & running of the system ~~between the~~ from ___ until ___ each day in the year.

shall be responsible for all monies etc.

shall turn in all monies to the treasr of the Co within 3 days after its Collection.

In case any [~~failure?~~][a] accident or depreciation of the apparatus or monies or property of Co due his his negligence he shall have same deducted fm such[d] monies as may be due him at the time bal collectable from his suritis—

that he will cause a monthly statement to be made out not later than the 5 of following month, of the amount bills presented Collected.

also the monthly Expenses sheet, and report as to condition of the property of the Co— That he will see that the contract made with the Engr is carried out.

~~Salary   He A~~ salary shall be payable [~~two?~~][a] blank[c] ~~day~~ monthly &[f] days after the same shall be due—

ADDENDUM[g]

New York, [February, 1884?][h]

1[i]   Stack shall be inspected at least once a month

2[i]   All steam piping ~~and~~ flanges shall be thoroughly inspected ~~every~~ [-][a] at like intervals—

3[i]   All electrical connections in station shall [---][a] be carefully inspected [---][a] Each [----][a] days before starting up—

4[i]   ~~Each station should be supplied with a volt box and pressure[b] indicators tested every two weeks~~—

ADf, NjWOE (*TAED* D8439ZAA1). [a]Canceled. [b]Interlined above. [c]Written in underlined gap. [d]Obscured overwritten text. [e]Paragraph written below "Every 3 months" and enclosed in box; line drawn to indicate intended insertion point. [f]"monthly &" interlined above. [g]Addendum is an ADf written on letterhead of Thomas A. Edison. [h]"New York," preprinted. [i]Circled.

1. Edison likely drafted this contract in early February in conjunction with Doc. 2603 and the circumstances around it. Having evidently turned the matter over to attorney John Tomlinson before leaving for vacation on 10 February, Edison sent additional instructions for Tomlinson from Florida in Doc. 2625.

2. A more complete and polished but undated draft, written in an unknown hand, is in DF (*TAED* D8439ZAA2 [images 1–3]). It contained a large blank space for an enumeration of the engineer's duties. Three other closely related documents were apparently drafted about the same time: a bond agreement for the engineer (images 4–6) and the station superintendent's contract (images 7–11) and bond agreement (images 12–14).

–2606–

Memorandum to Richard Dyer

New York, Feby 6 1884[a]

Dyer=

Please take out patent on armature in series & multiple arc in the field of force, preferably multiple arc. I find that with the present machine, if[b] a large bar of iron be placed across the field when armature is running that It does not bring down the Volts 10 percent. The short circuiting of the lines of force through the bar is certainly as great as would be caused by the addition of another armature in multiple arc   I don't mean multiple arc Electrically only in relation to the field ie[c] magnetically thus I shall be able with very little addition of iron & Copper to run 2 armatures from one field at[b] full volts=

Regarding series in Experiments in boring out field I found by using long field magnet that boring out so as to give twice the space between the iron of the armature & the field piece that it only lowered Volts 15 p cent hence by little[d] more iron I could work 2 armatures. There is a Curious gain here that I dont fully Comprehend— Make broad claims to the increased utilization of the Lines of force.[1]

fig 1   Multiple arc

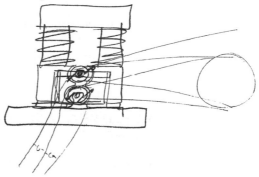

Fig 2[e]

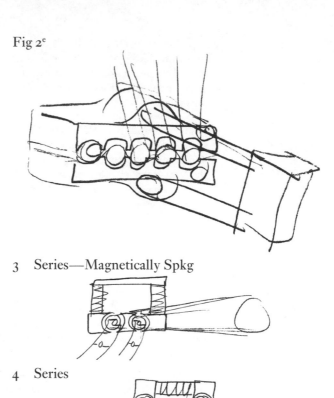

3   Series—Magnetically Spkg

4   Series

AD, NjWOE, Lab., Cat. 1150 (*TAED* NM019AAD). Letterhead of Edison Electric Light Co. [a]"*New York*," preprinted. [b]Obscured overwritten text. [c]Circled. [d]Interlined above. [e]Figure drawn on separate sheet.

1. These ideas were embodied in a patent application (Case 619) filed for Edison on 5 April. It had four drawings, corresponding closely to those in this document, and five claims. The application was unsuccessful, for reasons unknown to the editors, and Edison abandoned it (Patent Application Casebook E-2538:320; Patent Application Drawings Case Nos. 179–699; both PS [*TAED* PT022320, PT023 image 120]). Jordan 1990 elucidates how changes in contemporary conceptions of the magnetic circuit were shaped by the magnetic theories of Faraday, which had deeply influenced Edison.

FLORIDA VACATION, FEBRUARY–MARCH 1884
Doc. 2607

On 9 February 1884, Edison drew a $1,500 check on the Construction Department account for "Expenses South," and on the next day he and Mary Edison left New York for an extended vacation in Florida.[1] It was their second excursion to the state but their first without their children, who had accompanied them in March 1882.[2] Mary's good friend Josephine "Josie" Reimer accompanied the couple on their trip, and her husband, Henry C. F. Reimer, joined them for at least the latter part of the trip.[3]

The Edisons' itinerary followed a route already well-trod by New Yorkers of their financial means. After gaining statehood in 1845, Florida quickly became a destination for invalids and others seeking its temperate winter climate. Early travelers had faced a daunting several days' journey on steamships and trains; but, by the early 1880s, rail service delivered passengers from New York to Jacksonville in thirty-six hours with various overnight comforts, such as Pullman Palace sleeping berths and on-board dining.[4] By the mid-1880s, well over 100,000 tourists yearly visited Florida during the winter months, and one contemporary observer noted that it seemed almost a Northern community. President Chester A. Arthur followed much the same route as most tourists during his Florida vacation in the winter of 1883.[5]

Upon their arrival in Jacksonville, most late-nineteenth-century visitors began their tour of Florida's interior on steamboats plying the main north-south thoroughfare, the St. Johns River. Passengers took in views of plantations and estates along riverbanks lined with southern magnolia, cypress, and palm trees. During the day, men and women could amuse themselves by shooting from the boat decks at a variety of wildlife, including Florida's famed alligators. At night, passengers had a very different view of the river, lit by bonfires in large metal barrels on the boats' foredeck.[6]

In this tradition, the Edisons took a steamboat twenty-eight miles south to the Magnolia Hotel and resort on 12 February. After sending a telegram (Doc. 2607) to announce their arrival, Thomas and Mary spent the next twelve or fourteen days at the Magnolia Hotel.[7] A riverfront resort surrounded by 400 acres of orange groves and woodlands, the Magnolia Hotel was built in 1882 by Isaac S. Cruft. Like the Maplewood, Cruft's summer resort in the White Mountains of New Hampshire, the Magnolia was an elegant four-story wood-framed structure with mansard roofs, Queen Anne–influenced towers, and spacious verandas. In addition to such modern amenities as elevator service and bathrooms, it was outfitted with gas (presumably for lighting) and steam heat in the public rooms. Guests could also enjoy hunting, fishing, and boating.[8] The Magnolia Hotel and resort, and nearby Green Cove Springs, boasted sulphur springs and baths, which contemporary travel guides promoted for their healthful effects for those suffering from neuralgia, nervous prostration, rheumatism, and liver and kidney problems.[9] In the past, Edison had suffered from neuralgia and Mary had experienced unspecified nervous disorders and "uterine troubles."[10] They may have been attracted by the purported healing properties of the waters, and, indeed, Thomas noted spending twenty dollars on "baths" in a cumulative expense list compiled later in the trip (Doc. 2618).

Sometime between 24 and 26 February, the Edisons left the Magnolia and traveled further south on the St. Johns River to areas known for hunting and fishing. Their first stop was a brief stay at the Putnam House hotel in Palatka, forty-seven miles from Magnolia. Built in 1875, the three-story Putnam House featured three towers with mansard roofs, Italianate roof brackets, and a Stick-style veranda across the front of the first floor. The Edisons planned to leave Palatka on 28 February for the 120-mile steamboat trip south to Sanford, a resort town on Lake Monroe with a reputation for good hunting and fishing.[11] The town was founded by diplomat Henry S. Sanford, who also built the Sanford House, where the Edisons took accommodations. A three-story standard frame construction adorned with shutters, Italianate roof brackets, and a wrap-around two-story porch, the Sanford House had opened in 1880 and hosted President Chester A. Arthur during his Florida vacation in the winter of 1883.[12]

The Edisons' itinerary during the last week of February and first week of March remains vague. There was little pressure from Thomas's business interests to return quickly. In a letter dated 22 February, Samuel Insull encouraged Edison to prolong his stay in Florida: "Everything otherwise is going along all right," he noted, "& you need not hurry home. Stay away as long as you feel like it."[13] On 27 February, Mary wrote to Insull that she and her husband intended to leave Palatka for Sanford the following day, instructing him that he could "send all mail to Magnolia and it will be forwarded." From Sanford they intended to travel eastward to the Indian River, where there would be "plenty of Game."[14] The Edisons' time on the Indian River would have been relatively short, if, indeed, they made the trip, and the editors have found no confirmation that they did so. The Edisons left Sanford for St.

Augustine on 4 or 5 March, giving them only five to six days in the Sanford or Indian River areas.[15]

The trip north to St. Augustine probably included a 149-mile steamboat voyage on the St. Johns River to Tocoi, from where the Edisons would have taken the St. Johns Railway fourteen miles east to St. Augustine.[16] They arrived there by 6 March and took up residence at the Magnolia House hotel.[17] The Magnolia House, constructed circa 1848, had been remodeled in the early 1880s in the fashionable Queen Anne style. At this time, St. Augustine had about 2,500 inhabitants, but during the winter tourist season, the population swelled to three or four times that number. Tourist activities ranged from the more active—boating, yachting, and fishing—to the more relaxed, such as health-promoting promenades along the sea wall, visits to Fort Marion (originally Castillo de San Marcos), touring the old Spanish and new American sections of the city, and taking advantage of the city's free library and club house.[18]

The editors have found very little documentation of the Edisons' activities during the approximately three weeks they spent in St. Augustine. Insull continued to encourage Edison to prolong his vacation until the end of March. "There is no reason from a business point of view why you should come home," Insull wrote on 11 March, "and there is every reason from the point of view of your health, why you should stay away, the main one being that we are having cold weather here just now, and I am afraid if you come back, you will catch a severe cold, as you did two years ago."[19] Edison apparently had no objection to staying in Florida; his correspondence with his New York office nearly came to a complete stop. In addition to enjoying St. Augustine's diversions, Thomas also took advantage of his extended time in the city to let his mind play with experimental ideas. Freed from the worries of his Construction Department, Edison continued to fill a pocket notebook with his thoughts and research plans on a range of subjects, from multiple telegraphy to artificial materials to lamp filaments.[20] He also seems to have struck up an acquaintance with Theodore H. Whitney, a machinist-turned-electrician who witnessed his signing of a power of attorney on 6 March. Edison engaged the young man to search for fiber materials further up the Indian River. Whitney and his brother, with Edison's rifle in hand, paddled and portaged as far south as St. Lucie and sent several boxes of fibers back to New York by late April. Edison and Whitney stayed in touch for several years for both personal and business reasons.[21]

The only detailed account of the Edisons' time in St. Augustine is a newspaper report, dated 28 March, relating the fantastic story of "Edison's Electric Shark Hunt."[22] According to the article, Edison heard the legend of the "great St. Augustine 'demon shark'" that had for years plagued local fishermen and even, it was rumored, consumed several residents. Edison reportedly determined to catch or kill the shark and, after days of preparations, including repeated visits to the home of a local scientist, he hired a yacht for a "fishing excursion." Accompanied by Mary, Mr. and Mrs. Reimer, a former mayor of St. Augustine, and the captain of the yacht, Edison brought with him a mysterious basket. On the water, Edison pulled from the basket "a long gutta-percha line— a regular insulated telegraph wire. There was no hook visible, but it was evidently imbedded in a huge chunk of something red that looked like meat, yet was not. A powerful electric battery [presumably a small dynamo] remained in the basket." When the line gave a jerk, the captain grabbed it, and Edison "worked like an organ-grinder at the electric crank." The captain began hauling in the writhing shark, and within fifteen minutes the animal was dead. The fifteen foot, 700-pound shark was towed to the dock and later taken to the local museum of one Dr. John Vedder, where it was put on display with the tag: "The Demon Shark, Caught by T. A. Edison, With Electric Bait."[23]

Region of the Edisons' 1884 Florida vacation.

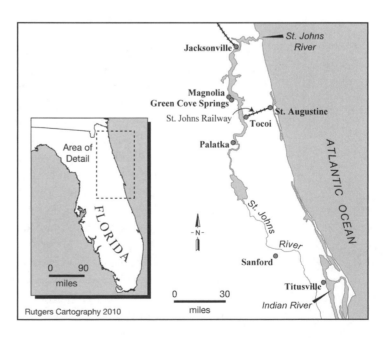

According to the above article, the Edisons left St. Augustine on the evening of 28 March. They probably took the St. Johns Railway to Tocoi and a steamboat up the St. Johns River to the Magnolia resort, where they would have collected their correspondence. From Magnolia they would have returned by steamboat to Jacksonville and then via railroad to New York. They evidently did not tarry on the return trip, and Edison was back at his office on 31 March.[24]

1. Cash Book (1 Jan. 1881–30 Mar. 1886): 233, Accts., NjWOE.

2. See Docs. 2234 n. 2 and 2241 esp. n. 2.

3. See Doc. 2608 n. 11.

4. The Edisons traveled in a Pullman car on their way from New York; see Doc. 2618. Regarding rail travel times and accommodations, see Ramson 1884, 322 and Braden 2002, 78.

5. The observer was Joseph Medill, editor of the *Chicago Tribune*; his comment, attributed to the 12 April 1883 *New York Tribune*, appears in Richardson 1964, 41–42; see "Florida: The State of Orange-Groves," *Blackwood's Edinburgh Magazine* 138 (Sept. 1885): 319; on Arthur's 1883 trip, see also "The President's Vacation," *NYT*, 11 Apr. 1883, 1; and "On the Coast of Florida," ibid., 17 Apr. 1883, 1.

6. Braden 2002, 78. The St. Johns River flowed south to north, from a flat marsh near Vero Beach to Jacksonville. By the late 1880s, the rapid expansion of local and regional railroads and improved harbors for large coastal steamers both cut into the steamboat trade; at the same time, cooler than normal winter temperatures caused freezes that impeded river traffic. Bass 2008, 54–56.

7. The Edisons originally planned to stay three weeks. TAE to Samuel Insull, 15 Feb.1884, DF (*TAED* D8403X1).

8. Classified advertisement, *Frank Leslie's Illustrated Newspaper*, 10 Feb. 1883, 426; "The Magnolia Hotel," brochure, 20 Mar. 1882, private collection (copy on file).

9. Tyler 1881, 11–12; Lee 1885, 125; see also Doc. 2618 n. 15.

10. On the Edisons' health generally see Doc. 2712 (headnote). Regarding Edison's experiences with neuralgia, see Docs. 1523, 1531, and 1539. Concerning Mary's health issues regarding nervous disorders, see Docs. 1394, 1402, and 2213; for "uterine troubles," see Doc. 2213.

11. Lee 1885, 141–42; Braden 2002, 83–84.

12. *ANB*, s.v. "Sanford, Henry Shelton"; *Appleton's Illustrated Handbook* 1877, 18; Braden 2002, 83.

13. Doc. 2608.

14. Doc. 2611.

15. A trip from Sanford to the northern end of the Indian River would have taken the Edisons on a combination of steamboat and railway to reach its main northern access point, Titusville (the distribution and shipping entrepot for southeast Florida). The Indian River, technically, was not a river, but a shallow saltwater lagoon with two inlets from the sea. It stretched some 150 miles along the eastern coast of Florida, from above Titusville in the north to St. Lucie port in the south, separated from the Atlantic Ocean by a narrow strip of land. Given the character

of the Indian River, the variety of boats that provided transportation across it for passengers and freight were of shallow draught and relied on sails because of the lack of a current. Though it was considered very malarial by northerners, it was a popular area for tourists seeking a rustic area to camp and to hunt and fish the wide variety of fauna. Henshall 1884, 12, 14–16; Lee 1885, 209–16.

16. The St. Johns Railway made the thirty-five minute trip twice per day. Lee 1885, 141–42, 145.

17. TAE to Insull, 6 March 1884, DF (*TAED* D8403ZAI).

18. Braden 2002, 143; Lee 1885, 97, 104.

19. Insull to TAE, 11 Mar. 1884, DF (*TAED* D8416AVR); see Docs. 2263 and 2264.

20. See Doc. 2609 (headnote).

21. Whitney (b. 1861?), a New Jersey native, had worked for at least the latter part of 1878 at Edison's Menlo Park laboratory, where he did various jobs such as pressing carbon buttons for telephone transmitters. Whitney appeared in the 1880 federal census of New York City as a machinist; he identified himself as an electrician in the Florida enumeration five years later. U.S. Census Bureau 1970 (1880), roll T9_867, p. 191.3000, image 0094 (New York, New York, N.Y.); Florida State Census (1971), roll M845_11, p. 24 (Precinct 5, St. Johns County); Florida State Census 1971 (1885); TAE power of attorney to Samuel Insull and Sherburne Eaton, 6 Mar. 1884; Laboratory Time Sheets [1878]; Whitney to TAE, 31 Mar. and 25 Apr. 1884; all DF (*TAED* D7817AA, D8428F, D8420K, D8420N).

22. "Edison's Electric Shark Hunt," *Lincoln (Neb.) Daily State Journal*, 4 Apr. 1884; "Edison's Electric Shark Hunt," *Omaha (Neb.) Daily Bee*, 8 Apr. 1884, p. 9.

23. John J. Vedder (b. 1819) operated a "Museum and Menagerie" devoted to living and dead specimens of Florida's fauna, located at the corner of Treasury and Bay Sts. in St. Augustine. "Dr. John Vedder," *National Police Gazette*, 27 Apr. 1895, 7; "A Florida Museum," *Forest and Stream* 25 (14 Jan. 1886): 484; "Remarkable Old Man: Dr. John Vedder's Life Reads Like a Romance," *Middletown (N.Y.) Daily Argus*, 18 Sept. 1896, 2.

24. Two entries of 31 March in Edison's cash book read: "Thos. A. Edison   Paid his note as per Const check No 842   160.00" and "Thos. A. Edison   Paid him as per check on Germania drawn by him while in Florida   300.00." Cash Book (1 Jan. 1881–30 Mar. 1886): 243, Accts., NjWOE.

–2607–

To Samuel Insull[1]

Magnolia Fla Feb 13 1883 [1884][a]

Arrived last night   All well

Edison

L (telegram), NjWOE, DF (*TAED* D8403W1). Message form of Western Union Telegraph Co. [a]"1883" preprinted.

1. See headnote above.

[Providence,][1] Birthington's wash Day 22nd Feby [188]4
My Dear Edison,

Thinking that you might imagine that I had disappeared from this mundane sphere it is perhaps well to let you know that I am still here below. I came on from New York last night to celebrate today with Sims.[2] I go back to New York tonight.

We have closed the 1000 light Contract for[a] Piqua— Stewart promises me another Ohio Contract within the next ten days—it will probably be Circlesville.[3] We have got out North Adams, Hillsdale, Hudson, Syracuse New Orleans, Cedar Rapids, & about three others estimates   Work at the Central Station is going [------ --- ------ -------][b] of next week. The engines will be at Goerck St by then.[4] Sprague reports that the new Indicator works splendidly.[5]

I am having a good deal of trouble at Bellefonte   I cannot at present get my money there. I think it was a <u>big mistake</u> to put 9½ × 12 in there. ~~You~~We should have put <u>11 × 14</u> instead. I have been trying to fix matters up by putting smaller pulleys on Dynamo & running 9½ × 12 at 300 as 350 has proved too high for these engines considering the "talent" employed to run them.[6]

Everything otherwise[c] is going along all right & <u>you need not</u> hurry home. Stay away as long as you feel like it—at least give me till <u>1st April</u> before you show your face in New York   I am conceited enough to want to try & get some work single handed for Const Dept

I got out of Hutchinson an order for $23,000 of Dynamos yesterday. I worked on his feelings told him I would have to shut down at Goerck St[d] if he did not give it to me & so managed to get it out of him.[7] I promised I would not ask for a <u>cent</u> of cash on the order but would take <u>all paper</u>. This the Germania Bank[c] will discount   They are carrying $16 000 [paper?][e] now for us. How is [that?][f] for credit after five months business [---][f] thcm—

The Engineering Dept is doing well now. I have actually caught up to the mapping Dept.

Eaton has been favouring me with some [------][g] memos lately but no matter my skin is thick now

<u>Important</u>. Do you want to build[c] ~~the~~ the Armington & Sims ~~enng~~ engine for Edison Light business at Goerck St or are you more impressed with Straight line Engine.[8] I feel sure I can make a deal through Sims & Moore[9] for a license. Please wire me your ideas. Howard is going to Europe & I think the matter should be fixed before he goes & before Armington gets

back. Sims told me this morning that if you wanted to build Engines at Goerck St for our Companies it could be arranged & that he would help us in every way poss[ible]ᵉ (after the License was obtain[ed)]ᵉ in turning out the work at Goerck St. If you want me to start negotiations wire me[10]

Regards to your wife & Mrs. Reimer[11]   Yours very Sincerely

Saml Insull

Will dictate long letter from New York tomorrow or Sunday

If you want money wire me

ALS, NjWOE, DF (*TAED* D8439W). ªObscured overwritten text. ᵇTop line of page incompletely copied. ᶜInterlined above. ᵈ"at Goerck St" interlined above. ᵉIncompletely copied. ᶠFaint copy. ᵍIllegible.

1. Insull wrote to a New York associate on 22 February (George Washington's birthday) from Providence, where he met Gardiner Sims, that "I cannot possibly get away from here till tomorrow morning." Insull to John Campbell, 22 Feb. 1884, DF (*TAED* D8465H2); see also note 2.

2. Gardiner Sims asked Insull to come to Providence for a night and a full day to discuss unspecified business (but see note 6). Insull planned to arrive on the evening of 21 February. Many cities and towns marked George Washington's birthday with public festivitites. Sims to Insull, 19 Feb. 1884; Insull to Sims, 20 Feb. 1884; both DF (*TAED* D8465G, D8416API); "Duly Celebrated Elsewhere," *NYT*, 23 Feb. 1884, 3.

3. On Edison's behalf, Insull signed a contract with the Ohio Edison Installation Co. to build and equip a central station in Piqua, Ohio. Edison's agreement with the same company regarding Circleville, Ohio (executed in May), specified that he would furnish materials and supervise the Installation Co.'s construction of the plant. TAE agreements with Ohio Edison Electric Installation Co., 21 Feb. and 16 May 1884, Miller (*TAED* HM840210, HM840216).

4. These engines were probably for the two additional dynamos installed at the Pearl St. station during the spring (see Doc. 2669). The first one was to have been shipped to New York on 21 February. John Campbell to Armington & Sims, 19 Feb. 1884, LM 18:244A (*TAED* LBCD5244A); Gardiner Sims to Insull, 19 Feb. 1884, DF (*TAED* D8422S).

5. John Ott had finished the first of the new voltage indicators, designed to replace the Edison Effect instrument (see Doc. 2538 [headnote]), and sent it to Brockton, Mass., by about 14 February. It may have been installed there or at Fall River, from where Sprague wired Insull on 16 February: "indicators working well here." Insull to John Ott (with Ott marginalia), 14 Feb. 1884; Alfred Tate to William Dwelly, Jr., 14 Feb. 1884; Sprague to Insull, 16 Feb. 1884; all DF (*TAED* D8467F, D8416ANT, D8442ZBG).

6. John Perry, who was in charge of starting the Bellefonte, Pa., station and training the engineer there, had reported "considerable trouble" governing the Armington & Sims engine at what he took to be

its rated speed. Forced to operate it more slowly, Perry then had difficulty wringing sufficient voltage from the dynamos (see also Doc. 2615 n. 1). On learning that the engine could only run at 275 or 300 revolutions, instead of the 350 he had expected, Perry asked Insull to provide a different pulley to run the armature faster, which Insull agreed to do. After numerous letters and telegrams among Perry, Insull, and Gardiner Sims—who dispatched one of his company's men to Bellefonte—Sims concluded that Perry and his trainee had "undoubtedly wanted more speed so they immediately screwed down the springs causing them to set and become worthless making it impossible for the engine to regulate. This is a fatal mistake that all of your men make." Sims later told Insull he planned to institute design changes that would make it "impossible for your men to screw down the regulator springs and ruin them for further use. . . . To make changes in the speed of our engines, hereafter, it will be necessary for your men to take our regulator to pieces and taking certain parts of it to a machine shop and make some changes." Perry to TAE, 8 Feb. 1884; Insull to Armington & Sims, 11 Feb. 1884; Perry to TAE, 14 Feb. 1884; Sims to TAE, 13 and 18 Feb. 1884; all DF (*TAED* D8453ZAN, D8416ALW, D8453ZAV, D8422O, D8422Q); Insull to William Andrews, 14 Feb. 1884; Insull to Armington & Sims, 15 Feb. 1884; LM 18:224B, 225B (*TAED* LBCD5224B, LBCD5225B).

7. Two months later, when Gustav Soldan asked Insull to place an order so as to keep certain men employed at the Edison Machine Works, Insull told him that "Our policy is to shut down the shop, and dispense with every class of labor possible as soon as we get through with our work. To give the order referred to, simply means to continue a heavy general expense." Insull to Soldan, 22 Apr. 1884, DF (*TAED* D8416BJM).

8. John Edson Sweet, a pioneer in high speed steam engines and formerly a master mechanic and professor of practical mechanics at Cornell, organized the Straight Line Engine Co. in Syracuse, N.Y., in 1880, serving as its president and manager. The firm's engines, based on Sweet's original 1872 design, featured a novel three-point frame, a precise valve train, and a shaft governor whose close regulation was suited for dynamo work, as Sweet had demonstrated with a Gramme dynamo at the Centennial Exhibition of 1876 (*NCAB* 13:54; *DAB*, s.v. "Sweet, John Edson"; "The Straight Line Engine at the Philadelphia Exhibition," *Electrical World* 4 (20 Dec. 1884): 249; Thurston 1939, 513–14; Smith 1925, 47–59). Edison approached the firm in November 1883 to ask "what royalty will you permit me to build your engine to be used & sold only by the Edison Light Co's in connection with their dynamos   They now use the Armington & Sims engine exclusively and the substitution of your engine would not interfere with your present market." Edison quickly accepted the company's proposed terms and commissioned a custom 10 × 12 engine with two pulleys, designed to run between 240 and 270 revolutions per minute, for tests and as the basis for drawings and patterns (TAE to Straight Line Engine Co., 3 Nov. 1883, Hodgdon [*TAED* B017AH]; Straight Line Engine Co. to TAE, 10, 15, and 26 Nov., 6 Dec. 1883; TAE to Straight Line Engine Co., 5 Dec. 1883; all DF [*TAED* D8334ZBJ, D8334ZBK, D8334ZBM, D8334ZBP, D8316BOR]).

In any case, the Straight Line Co. did not finish its engine until mid-

April. Gustav Soldan, who would have supervised engine manufacture at the Edison Machine Works, put off going to Syracuse, and it remained there until 15 May, when the builders shipped it to New York. Charles Batchelor worked with the engine at the Edison Machine Works during the summer. According to the Straight Line firm, the fact that Edison did not ultimately adopt their model led to unfavorable publicity about it, which Edison disavowed. Edison delayed paying for the drawings and patterns, which the company had made at its own expense, until at least September 1885. Straight Line Engine Co. to TAE, 21 Apr. and 15 May 1884, 23 Apr., 28 Aug., and 26 Sept. 1885; TAE to Straight Line Engine Co., 22 and 28 Apr. 1884; Insull to Straight Line Engine Co., 9 May 1884; Charles Batchelor to Harry Livor, 18 July 1884; Straight Line Engine Co. to Charles Batchelor, 11 May 1885; all DF (D8431I, D8431P, D8531D, D8512G, D8512H, D8416BJK, D8416BLP, D8416BOE, D8431ZAJ, D8531G); TAE to Straight Line Engine Co., 5 May 1885, Lbk. 20:257D (*TAED* LB020257D).

9. Miller Moore, an agent for Armington & Sims since late 1883; see Doc. 2420 n. 22.

10. See Doc. 2616.

11. Josephine "Josie" Stucky Reimer (1852–1915) and her husband accompanied the Edisons on their Florida trip. Like Mary Edison, Reimer was a native of Newark, N.J, where she lived at this time. She had attended Wells College in Aurora, N.Y. in 1868–1869 and married Henry C. F. Reimer in 1873 in Essex County (Newark), N.J. Henry Reimer was a bookkeeper or accountant who worked for the Edison United Manufacturing Co. a few years after this time. Decades later, Marion Edison remembered Josephine Reimer as a close social companion of Mary in Newark and New York. See Docs. 2611 and 2618; "Edison's Electric Shark Hunt," *Lincoln (Neb.) Daily State Journal,* 4 Apr. 1884, 2; U.S. Census Bureau 1963? (1850), roll M432_448, p. 357, image 105 (Newark West Ward, Essex, N.J.); ibid. 1967 (1860), roll M653_688, p. 178, image 181 (Newark Ward 2, Essex, N.J.); ibid. 1965 (1870), roll M593_879, p. 181, image 363 (Newark Ward 2, Essex, N.J.); ibid. 1970 (1880), roll T9_776, p. 111.2000, image 0586 (Newark, Essex, N.J.); ibid. 1982? (1900), roll T623_968, p. 12A (East Orange Ward 2, Essex, N.J.); Marion Edison to TAE and Mina Edison, 7 July 1915, CEFC (*TAED* X018A5AY); Wells College 1894, 55; marriage record of Henry C. F. Reimer and Josephine Stuckey, Reference Number Bk. BM: p. 265, Index to Marriage Records, 1848–1878, N.J. Department of State, (https://wwwnet1.state.nj.us/DOS/Admin/ArchivesDBPortal/Marriage1867.aspx, accessed on Ancestry.com October 2009); Samuel Insull to Henry Reimer, 28 Dec. 1887; Alfred Tate to Reimer, 6 Apr. 1888; both DF (*TAED* D8757AFB, D8818AIB); Oser 1956, 4–5.

EDISON'S FLORIDA NOTEBOOK   Docs. 2609, 2612–2613, 2618–2623, and 2631–2635

Edison took this small, unpaginated pocket notebook on his trip to Florida in February and March 1884, when he let his mind run freely. Although the book provides a comparative wealth of evidence of his inventive ideas at this time, its character is quite different from the laboratory notebooks where he and his assistants carefully recorded their experimental work. Edison dated only a few of his entries, in some cases retrospectively, probably when he was concerned about establishing claims to specific ideas or devices. He seems to have made most of the entries, however, after arriving in St. Augustine on 5 or 6 March; the exceptions are three written from 25 to 28 February while he was traveling by boat between Magnolia, Palatka, and Sanford. Edison made dated entries out of chronological order, and he apparently used the book in both directions. Some pages are blank and others are loose. Edison also intermittently used extraordinarily small handwriting that renders some words all but unintelligible.

The earliest dated entries are circuit diagrams for multiple telegraphy that take up approaches he last explored in 1877. Other undated drawings are related to Edison's 1878 design of a way duplex that enabled signals to be sent simultaneously between intermediate offices as well as over the main line. Most of the entries in the book relate to Edison's ongoing lamp research, especially methods for producing artificial filament fibers from organic materials. He also proposed ways of coating filaments. Some of Edison's notes on filaments became the basis for experimental instructions he sent to John Ott during the trip and for experiments conducted at the Lamp Factory and the New York laboratory after his return.[1] Edison's interest in artificial materials led him to write a relatively lengthy entry regarding the use of bran as a "Base Material for all uses" (Doc. 2622). His notes on this subject seem to have been spurred by meeting with or learning about the work of John Stevens of Neenah, Wisconsin, whose non-cutting roller mill produced a significant improvement in the wheat milling process. The notebook also contains three entries dealing with Edison's ongoing efforts to produce electricity directly from coal.[2] The longest of these entries discusses a new approach to the problem that he would pursue in July. Near the end of the book, Edison recorded his traveling expenses through about 5 March.[3]

The editors have selected nearly the entire notebook but have not transcribed the pages described in this paragraph. The book's first page contains a brief note, presumably written before Edison left New York, about a letter from George Yingling of Tiffin, Ohio, regarding the need for a branding iron that would not hurt cattle. Another short note written between the pages of Doc. 2613 states: "Write Holzer to make some filaments of Bamboo with the outer silica surfaces left on." The editors have neither found such a letter nor determined if this experiment was done. Immediately after the last significant entry in the book is a memorandum about writing to one William Denton, located near Fort Yuma in California, about a rush described as a "Marvellous fruit fibre." Edison left a space for the address but he never filled it in and apparently did not write to Denton. At the end of the book are some miscellaneous rough notes and calculations that have also not been transcribed. One of these appears to be about costs for a central station and another may concern the electric current required for separating iron ore. There are also a few rough sketches that may be related to Edison's notes on telegraphy.

1. Edison's notes to Ott are Docs. 2627 and 2630. For experiments at the lamp factory see William Holzer to TAE, 3 June, 17 July, 15 Sept., and 19 Nov. 1884, all DF (*TAED* D8430R, D8430S, D8430Z, D8430ZAB). In a 1 April 1884 laboratory note, Martin Force stated, "Experiments on combinations of different substances for the purpose of making carbon filiments," but no other notes related to these experiments have been found (N-82-12-04:175 [*TAED* N145175A]). For some of the lamp experiments being conducted prior to Edison's trip, see Docs. 2573 and 2586.
2. Docs. 2620, 2633, and 2634.
3. Doc. 2618.

–2609–

Notebook Entry:
Incandescent Lamp[1]

[Palatka, Fla.][2] Feby 25 1884

India ink
Silkened Cotton.

gelatin impervious H_2O draw hot spin wheel.[3a]

TAE

X, NjWOE, Lab., PN-84-02-25 (*TAED* NP020C). [a]Followed by dividing mark.

1. See headnote above.
2. See Doc. 2607 (headnote).

3. Edison drew a line from the word "gelatin" to the shaded area on the left side of the drawing. He was apparently proposing to draw the hot gelatin into a fiber by a spinning wheel.

-2610-

To Samuel Insull

PALATKA, FLA., [c. February 25, 1884][1a]

Insull

Dont forget to have Tomlinson draw up that contract with the Engineers and Supt (meter man) for stations.[2] Let me know how the new pressure indicator[3] works also if John Ott got the standard Volt box and if any have been made.[4] I think Hutch[5] will want some— How you progresg Newburgh—& Down town.[6] Tell Hutch I saw Shippen presdt of the Hoboken Ferries,[7] he will light all his boats 9 in number if it can be done cheap enough and operated cheap. He says he has the Supt who used to be supt for the Penna boats[8] & he states they have to have an Extra man, and the cost of operating was altogether too much— I told him he better investigate the NJ Central boats,[9] that I was under the impression that The fireman run the plant & that there was no Extra Expense for Labor   Am I right   also have Hutch see about this & get[b] a letter from the NJ Central people giving the Cost of operating. tell Hutch Shippen will be in N York in a week & that he better call on him personally & make a very low figure. 45 Lights Each is all they want= The steam runs down as low as 10 lbs on the long trips & against head wind—perhaps [a safety p?][c] an upright Engine will answer & thus cheapen it.[10] Tell Hutch not to fail in attending to this personally   they have had gas & had trouble now use Kerosene.

Tell Tomlinson to work up my Exhibits & case on the Disk Dynamo[11] also on plating the filament to the clamp Case[12] which old man Dyer[13] has & which I have never taken any testimony= How about Kruzis Brooklyn racket did it go through.[14] I would keep Hastings & Eaton hot on Establishing agencies to get us more towns= did you send Stewarts documents[15] also receive the cypher=[16] How about that Demerara sugar Estate Estimate[17] also the Pernambuco or Para Estimate[18]   Has mather & platte Dynamos gone.[19] How about the workings of the feeder regulators. How is Chinnock getting on plenty dark Days[b] & receipts I presume= I think I have struck a way of utilization of all the surplus power of our stations when not needed for light[20]   If John Ott hasnt that volt box perfect when I get back Im going to raise hell= When does

Batch return= Has Gouraud fixed the Colonial Co yet— I met Anothony J Drexel[21] at Magnolia & now know him well he is a splendid man ~~tell~~ ask[d] John Ott[b] if he has done as I told him in working up the new method of pressure indicator for the automatic regulator for Isolated Co— If he has done all tell him to keep working on a good stock for making filaments from   yours

Edison

ALS, NjWOE, DF (*TAED* D8439X1). Letterhead of Putnam House. [a]"PALATKA, FLA.," preprinted. [b]Obscured overwritten text. [c]Canceled. [d]Interlined above.

1. The editors have conjectured this date based on references to the new pressure indicator and the Edisons' departure from Palatka on 28 February (see Docs. 2607 [headnote] and 2611). At some later time, Edison wrote on this letter "Dont know the date."

2. See Doc. 2605.

3. See Doc. 2538 (headnote).

4. As discussed in Doc. 2584 n. 2, the editors have not conclusively identified the volt box. John Ott had been experimenting with "volt boxes" in January but was not yet satisfied with the design. When Samuel Insull inquired about the "volt indicator boxes" on 14 February, Ott replied that one had been finished and sent to Brockton, Mass., and that Sigmund Bergmann would complete five more in the next week. Ott to Alfred Tate, 11 Jan. 1884; Insull to Ott (with Ott marginalia), 14 Feb. 1884; both DF (*TAED* D8467C, D8467F).

5. Edison referred to Joseph Hutchinson in his capacity as secretary of the Edison Co. for Isolated Lighting.

6. See Doc. 2644 n. 8.

7. William W. Shippen (1827–1885) was president of the Hoboken Land and Improvement Co., the owner and operator of the ferry franchises that connected Hoboken with terminals at Barclay and Christopher Sts. in Manhattan. The company existed as early as 1811 and was incorporated in New Jersey, following the 1838 death of distinguished steam engineer and inventor John Stevens, as a private holding company for the Stevens family estate. In 1884, its capital was worth $1.47 million and its property included the Hoboken waterfront on the Hudson River. In mid-1884, the company also received a franchise for ferry operations to Fourteenth St. in Manhattan. Shippen was a civil engineer, a graduate of the U.S. Naval Academy, and a co-incorporator of the Stevens Institute of Technology. Cudahy 1990, 361, 380; "Will of John C. Stevens," *New York Tribune*, 24 June 1857, 5; Dilworth 2005, 121; "More Tax Returns," *Trenton Times*, 18 June 1884, 4; "Taking a Fresh Start. Astonishing Improvements Along the Hoboken Water Front," *New York Herald*, 14 Aug. 1882, 6; Obituary, *NYT*, 3 Sept. 1885, 5; "The New Hoboken Ferry Franchise," ibid., 4 July 1884, 8; "An Act to Incorporate the Stevens Institute of Technology," *Acts of the Ninety-fourth Legislature* 1870, 169–70.

8. To Edison's consternation, at least some of the Pennsylvania Railroad's ferries operating between the Jersey City rail terminal and Man-

hattan had been lighted by the United States Electric Lighting Co. since 1881. See Doc. 2136.

9. The New Jersey Central Railroad ordered an Edison isolated plant (consisting of 60 A lamps and one Z dynamo) for the *Fanwood,* a Hudson River ferryboat, about August 1883. Edison Electric Light Co. Bulletin 19:5, 15 Aug. 1883, CR (*TAED* CB019).

10. Edison referred to an engine with a vertical (or upright) cylinder, an advantage in marine applications where horizontal space came with a high price.

11. In February 1881, Edison applied for a patent on a dynamo armature with a core constructed of thin discs insulated from each other with tissue paper. The insulated discs reduced the production of eddy currents and wasteful heat in the armature. Just before issuing the specification, the Patent Office declared an interference with an application filed by Edward Weston on 31 May 1881. The interference case went forward in January 1882; in the meantime, Edison amended his application, which ultimately issued in 1890 (see Docs. 1875 and 1899 n. 1; Pat. App. 431,018; U.S. Pat. 256,778 [Maxim]; *Weston v. Edison,* Lit. [*TAED* QD010]). Fredrick Betts, a patent attorney for Edison, conferred in February 1884 with John Tomlinson about exhibits and witnesses for the case. In early May, Edison resisted Eaton's urging that he testify immediately (Eaton to Tomlinson, 23 Jan. 1884; Betts to Edison Electric Light Co., 16 Feb 1884; Eaton to TAE, 7 and 8 May 1884 (with TAE marginalia); all DF [*TAED* D8468A, D8468M, D8468ZAV, D8427ZAX]).

12. Edison was party to several patent interference cases at this time, and the editors have not positively determined the one to which he referred here. Of the known possibilities, the one closest in subject matter involved Edison's January 1881 application (Case 283; see Patent Application Casebook E-2536:260, PS [*TAED* PT020260]) for electroplating the junction between the carbon filament and metallic lead-in wires, as was standard practice at the Edison Lamp Works (see Doc. 2050 n. 1). In April 1881, while George Dyer was still working as a patent attorney for Edison, the Patent Office declared an interference with Lewis Latimer, and it subsequently added Edward Weston. After Weston presented testimony (Latimer having practically withdrawn), Edison moved for a dismissal on grounds that Weston's patent had been anticipated in France. In November 1883, the examiner agreed, dissolving the interference. Weston's appeal of that ruling was pending at this time; he won a reversal in May 1884, and the case resumed. Edison apparently did not testify in the matter until June 1885. "Brief for Edison," 2 Feb. 1885; Edison's testimony, 9 June 1885; both *Weston v. Latimer v. Edison* (*TAED* W100DFB, W100DFA001).

13. George W. Dyer (b. 1824?), Edison's patent attorney from 1880 to 1882, was the father of Edison associates Philip, Richard, and Paul Dyer. U.S. Census Bureau 1970 (1880), roll T9_122, p. 73.3000, image 0149 (Washington, D.C.); Docs. 2120 n. 5 and 2203 n. 25.

14. See Doc. 2624.

15. See Doc. 2602 n. 2 regarding agreements with Willis Stewart.

16. Not identified.

17. John Tilley, an engineer for a sugar firm in Demerara, British Guina, had requested an estimate for a 100-light plant. The Edison &

Swan United Co. complied but then, noting that the colony fell within the jurisdiction of the New York–based Edison Electric Light Co., referred the request to Edison (Tilley to Edison Electric Light Co., Ltd., 16 Oct. 1883; Edison & Swan United Electric Light Co., Ltd., to Tilley, 14 Nov. 1883; Edison & Swan United Electric Light Co., Ltd., to TAE, 1 Jan. 1884; all DF [*TAED* D8434B, D8434C, D8434A]). Edison made an estimate and then lowered it, based on the anticipated volume of his business in South America under the new agreement with the Edison Electric Light Co. Tilley declined in favor of Brush arc lights, citing the prevailing low price of sugar (TAE to Tilley, 30 May 1884, LM 19:307 [*TAED* LBCD6307]; Tilley to TAE, 25 Apr. and 13 July 1884, both DF [*TAED* D8434R, D8434ZAE]).

18. This estimate (not found) may be related to a tobacco company in Pernambuco, Brazil, about which Insull corresponded with Charles Coster. Insull to Coster, 25 June 1884, DF (*TAED* D8434ZAB).

19. Mather & Platt, a major engineering and manufacturing firm in Manchester, had exclusive rights to manufacture dynamos under Edison's patents in Great Britain. Edward Hopkinson, John Hopkinson's younger brother, was the firm's chief engineer (Greig 1970, 19; Arapostathis 2007, 10–11). In response to Mather & Platt's request for "types of the newest machines," the Edison Machine Works sent an unknown number of dynamos in March. The Edison and Swan United Electric Light Co. evidently sought the manufacture of Edison's dynamos in Britain. Having already made arrangements to produce the machines designed by John Hopkinson, Samuel Flood Page, the United Co.'s secretary, admitted to Edison in March that "this machine is at present far from perfect and it is leading us into trouble. The machine as it now is cannot be relied on, and is not nearly so useful practically, as your own." In November 1884, Mather & Platt requested Edison's permission to use his dynamos for locomotive purposes (Insull to Mather & Platt, 13 Mar. 1884; Flood Page to TAE, 27 Mar. 1884; Mather & Platt to TAE, 26 Nov. 1884; all DF [*TAED* D8416AWR, D8437M, D8437ZAK0]).

20. Edison may have had in mind the electrochemical and electrometallurgical processes that he later planned to discuss with Nathaniel Keith (see Doc. 2648).

21. An heir to the Philadelphia banking house Drexel & Co. established by his father, Anthony Joseph Drexel (1826–1893) joined with J. P. Morgan in 1871 to form Drexel, Morgan & Co. Carosso 1987, 133–39; *ANB*, s.v. "Drexel, Anthony Joseph."

–2611–

Mary Edison to Samuel Insull[1]

Palatka Fla Feby 27th 1884

My Dear Insull

I received your letter last sunday.[2] I am delighted to hear that Dottie is getting along so well. she seems to be perfectly happy and contented.[3] We are having a very pleasant time. the weather is quite cool. tomorrow we leave here for Sanford which is over two hundred miles from here. and then we are going over to the Indian river which is three hundred[a] miles further south consequently it will be much warmer.[4] We have quite a good many Strawberries here. the hotel is very good. I sent you a box of Orange Blossoms this morning let me know what condition they were in when received   I am anxious to know as I sent them by a new method. I received a letter from Mme Duval she is very much in need of money.[5] You can send all mail to Magnolia and it will be forwarded. Mr Edison is well and will be happy when he gets on the Indian River as there is plenty of Game etc there.[6] Mrs Reimer is well and sends love to you. I am glad that you have time to take Dottie out occasionally. I am sure I can never thank you enough for your kindness to her. hoping to hear from you soon I am   Yours Sincerely

M. Edison

P.S. please give my kind regards to Major Eaton Major McLaughlin and Mr Johnson   M.E.

Mary Edison's stylized initials on her New York stationery.

ALS, NjWOE, DF (*TAED* D8414B). Letterhead of Mary Edison. [a]Interlined above.

1. Mary Edison wrote on personal stationery with her stylized initials and a heading of the Clarendon Hotel, where she and Edison were then living in New York.

2. Not found.

3. The Edison's oldest child, Marion (nicknamed "Dot"), had just turned eleven. Marion's whereabouts are unclear, but she often spent time with her aunt, Jennie Stilwell, who was at Menlo Park with Mary's mother. Doc. 338 n. 23; Eugenie Stilwell to Insull, 12 Mar. 1884, DF (*TAED* D8465N1).

4. See Doc. 2607 (headnote).

5. Mary probably meant Mabel Virginia Duval (b. 1860?), who was connected with the dressmaking firm of Mme. A[nna]. Duval on 21st St. in New York. The firm submitted an invoice of $391.90 on 29 January for the dress discussed in Doc. 2683 n. 10, and Edison's office paid this amount to M. V. Duval on 4 March. U.S. Census Bureau 1970 (1880), roll T9_870, p. 76.2000, image 0157 (New York City, New York, N.Y.).

6. Insull told William Pitt Edison that Edison was "in Florida shooting alligators" at this time. Edison had fished and hunted while traveling in the West in 1878, and likely shot at least several birds during his 1882 Florida vacation. Insull to Pitt Edison, 26 Feb. 1884, DF (*TAED* D8416AQB1); *TAEB* 4 chap. 4 introduction, Doc. 2241 n. 1.

–2612–

*Notebook Entry:
Telegraphy*[1]

[Palatka, Fla.,] Feby 27 1884

Dep[o]s[i]t[io]n between plates Cop[per]   get Au & Ar in mind— ascertain black mat conductivity—[2a]

Sextuplex.[3]

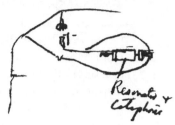

Definite note. bk ckt for not[e][4]

TAE

X, NjWOE, Lab., PN-84-02-25 (*TAED* NP020D). [a]Followed by dividing mark.

1. See Doc. 2609 (headnote).

2. This is apparently related to the experiments discussed in Doc. 2587 n. 2. See also Doc. 2627.

3. Figure label is "Resonator & telephone." In Edison's sextuplex designs from 1877, he combined his quadruplex with devices from acoustic telegraphy (see Doc. 877 [headnote]). In this design, Edison appears to be using a telephone and resonator in a Wheatstone bridge arrangement, probably as a receiver sensitive to a "definite note" in place of the acoustic reed or tuning fork he had used in 1877.

4. Edison probably meant a circuit operated by the back contact point of a relay.

–2613–

*Notebook Entry:
Telegraphy*[1]

[on the St. Johns River, Fla.?] Feby 28 1884[2]

The thing is to throw over both[a] the main & artificial   I didnt do this before hence back kick[3]

good. the thing I left out before

Octduplex[4]

Send constant vibration to keep reed going both Ends   have sp[rin]gs on each side cutting in relays—also controlling putting on extra battery.[5]

TAE

X, NjWOE, Lab., PN-84-02-25 (*TAED* NP020D1). Document multiply signed and dated. [a]Obscured overwritten text.

1. See Doc. 2609 (headnote).

2. Edison and Mary took a boat approximately seventy miles upriver from Palatka to Sanford this day (see Doc. 2607 [headnote]); the editors conjecture he may have made this note in transit.

3. The artificial line was intended to match the electrostatic capacity of the main line in order to counteract the effect of "static return charge" or "back kick" in multiple telegraph circuits so that each receiving relay would respond to only incoming signals. It is unclear what Edison meant here by "throw over," but presumably the main and artificial lines would jointly respond to each transmitting device. For Edison's earlier work on multiple telegraph circuits, see *TAEB* 1–3, passim.

4. In 1877, Edison added a second set of acoustic instruments to his sextuplex in order to increase the number of simultaneous messages from six to eight. He referred to this as an "octuplex" or "octoplex." See Docs. 881 and 919.

5. Figure labels are "extra points" and "Control the battery put." The extra points are presumably the springs that brought in the extra battery.

–2614–   *Magnolia, Fla.,*[a] [February 1884?][2]

Memorandum:
Isolated Lighting
Plant[1]

Wiring 543 Lits, at 3^3	1629
Fares	200
.Pockets,	200.
2. 3200. Cp dynamos 400 16[b]	
or 700 10 c[andle] p[ower][4]	4000
2 $8\frac{1}{2} \times$ 10 Engines.	1700
Foundations for ditto.	275
Steam piping	350
Exhaust pipe.	60
Boiler	1600
Setting boiler	500
Lamps & Extra Supply 700[c]	700
Sockets. 545 $1—	545.
Extra Engine parts	100.
Extra armature	425
Oil Cans small tools.	25
Belts.	40
Extra brushes	10
Shades & Holes.	318[d]
Labor Erecting Dynamos etc. &	
running 60 days.	450*
	12 127 $12 127[e]

other wing

14 Sleepers, Servants	14
40 Sleepers, Regular guests	40.
Corroders	30.[d]
Kitchen—	10
Pantry	10
Engine room	3
Dining Hall.	60.
30 Servants rooms under dining Hall	30.

Laundry   25 Lights
Rooms.   225 inclusive of Servants Rooms.
40. 10 in parlor of 4 Lights Each

	33
	90——on 5 stories
	70
	193.

33 Sleepers	33
Corrodor	10
Girls Bath.	1
Water Closet	1
Landing	1
3 Stories.	46——[f]

Sleepers	29.
Hall	1
Corrodors	4
Billiard	20
Pitcher Room	2
Water Closet	1
Veranda.	4
6 Street Lights.[f]	

1st Story Veranda lights.	20
2nd 1st Story[g]	

Parlor	40
Small Parlor.	4
Ladies Writing.	2
Ofc	17
Newel	2[d]
Elevator	1[d]
Recption	2
Reading Room	6

Proprs Priv Ofc	2
Bath room	1
3 Sleepers	3
Corrodor^d 4	4
Passage.	1
Managers.	2
3 Sleepers.	3
Stairway.	1
3 Sleepers	3
Corrodor	4

$$
\begin{array}{r}
90 \\
107 \\
138 \\
67 \\
20 \\
\underline{98} \\
520 \\
\underline{25} \text{ Laundry} \\
545 \\
\underline{3} \\
1635
\end{array}
$$

AD, NjWOE, DF (*TAED* D8425B). Letterhead of Magnolia Hotel; miscellaneous calculations not reproduced. ^a"*Magnolia, Fla.*," pre-printed. ^bInterlined above. ^cCircled. ^dObscured overwritten text. ^e"$12 127" written in right margin. ^fFollowed by dividing mark. ^g"2nd 1st Story" written in large letters in the right margin, referring to the following list.

1. This memorandum may be an estimate for installing an isolated plant at a hotel. The Magnolia Hotel in Magnolia, Fla., where Edison evidently wrote this document, was the only one at which the Edisons are known to have stayed that would have been large enough to reflect the 225 guest and servant rooms in these calculations. After a fire in 1881, Isaac Cruft and builder James McGuire reconstructed the Magnolia Hotel. An 1885 guidebook noted that the hotel, located twenty-eight miles south of Jacksonville on the St. Johns River, boasted "a commanding site and superb river view" and was "unusually striking and attractive." The hotel was lit by electric lights (apparently not Edison's), making a "dazzling view from vessels passing up the river." Advertisements in New York newspapers in 1886 for the Magnolia Hotel note that it "accommodates 400." Braden 2002, 84–85; *Waugh's Blue Book* 1907, 155–56, 172; Lee 1885, 141, 143, 145; advertisement, *New York Daily Tribune*, 24 Jan. 1886, 7; advertisement, *NYT*, 10 Mar. 1886, 7.

2. The editors have inferred this date based on the Edisons' stay in Magnolia, Fla., in the later half of February; see note 1 and Doc. 2607 (headnote).

3. That is, three dollars per light.

4. Edison probably meant two 3,200-candlepower dynamos each powering 200 16-candlepower lamps or 350 10-candlepower lamps.

–2615–

William Andrews to Edison Construction Dept.

Dear Sirs—

Referring to a former letter in which I stated that the capacity of all our plants has been over-estimated, I beg to enclose you a few figures on another sheet which I should be glad if you would hand to one of your mathematical expirts to examine—[1]

You will see that whilst 55 H.P. will run 800 10 CP. Lamps on a circuit that has no resistance in conductors, the addition of 15% resistance in the latter necessitates a much higher EMF at machines which require a correspondingly higher power to drive them—(The Volt-ampere being a unit of H.P.) and the required power is raised from 55 to 65 H.P. If the figures I have advanced are correct, it is easy to see where all our H.P. and coal is going to, which we have hitherto been unable to satisfactorily account for—

It becomes also a question whether it is economy to put feeders in with a loss of 15% when it thereby becomes necessary to considerably enlarge boilers and engines, speed up dynamos, and burn more coal—

It may be, and indeed I should think it most probable, that you have already gone over, and well considered all the facts I have herein dwelt upon, or it is possible that there is some big bug in my figures, but I must say that the facts that have forced themselves under my observation, in the practical working of plants, prove that there is a large excess of steam power being consumed in some way, that has not yet been carefully considered—

I shall be very glad to hear further from you on this subject— Very truly Yours

W. S. Andrews

P.S. You will note that I have only considered the loss of 15% on feeders whereas the total loss on all conductors is over 20% loss, which would of course raise the H.P. required, in proportion

⟨You probably dont catch on to the fact that the 1st investment is the trouble in pushing our biz— The Local people must suffer a continuous loss in operating for their lack of

Confidence   we are fully aware that down East where Coal is ~~Cheap~~ dear 5 percent is the most that should be lost yet we could not start a Company if we put this amount of Copper in— we long ago made a curve on the critical point between the Cost of Coal & interest on the Copper investments   we[re] confidence in E Lighting universal this rule would be carried out   Edison⟩[2]

ALS, NjWOE, DF (*TAED* D8442ZBY). Letterhead of Thomas A. Edison, Central Station, Construction Dept. ª"188" preprinted.

1. Andrews enclosed two undated pages of calculations and "Notes on the extra H.P. reqd at Stations to provide for 15% drop on Feeders." In at least two recent letters, he had raised the problem of insufficient power for maintaining the required voltage (see Doc. 2608 n. 6). Complaining of routinely having to run the Y dynamos at Bellefonte, Pa., well over their rated speed, Andrews calculated that a 15 percent voltage drop on feeder lines, 3 percent on mains, 1 percent on service lines, and 2 percent on interior wiring "amounts to a drop of about 20%, or in other words if the lamps used are 108 Volts, the pressure at Dynamos in Station should be about 128 Volts— This practical drop of EMF is never considered when testing machines and therefore their proper running speed is always rated too low." Writing again on 26 February, Andrews gave several examples to illustrate his contention that there had "been hardly one station that has not given more or less serious trouble and dissatisfaction . . . due in a large measure, to over-rating power capacity of engines and boilers." Andrews to Insull, 14 and 26 Feb. 1884, both DF (*TAED* D8442ZBC, D8442ZBP).

2. Edison drafted his response on the back of Andrews's enclosure. Andrews calculated that 15 percent loss on feeder lines translated into a need for an additional 17 percent of steam power to run 10-candlepower lamps at 109 volts. Referring to Edison's decision to use relatively small feeder lines, effectively trading high operating efficiency for lower initial cost, Andrews concluded: "Now it simply becomes a question of figures whether this cutting down of feeders is real economy when it involves heavier machinery, and a constant Extra Expense for coal." Andrews enclosure to TAE, 2 Mar. 1884?; TAE to Andrews, 2 Mar. 1884?; both DF (*TAED* D8442ZBY, D8442ZBZ).

Andrews returned to this subject the following day, pointing out that the load of an 800-lamp plant, including total electrical losses of more than 20 percent, would require 76 horsepower from a $9\frac{1}{2} \times 12$ engine rated for only 48 horsepower. The extra power could "be dragged out of the Engine for a limited time . . . but how long will it last, and what kind of satisfaction will this policy eventually give to local people?" Edison noted on the reverse that Armington & Sims actually rated this engine for 90 horsepower and guaranteed its reliable operation well above that level. He added that "the strain on the A&S Engine parts according to size is 6 times less than on a Baldwin passgr Loco indicating 750 hp— their usual strain." Andrews to TAE, 3 Mar. 1883; TAE to Andrews; both DF (*TAED* D8442ZCA, D8442ZCB); see also Docs. 2608 and 2625.

Sanford Fla 3 Mch 1884.[a]

To Samuel Insull

Telegrams just recd[1]   If you think best close with Howard[2]   ask Holser[3] if he thinks it[b] better arrange to get a regular
supply lotus[4]   telegrams will reach me here tomorrow

Edison

L (telegram), NjWOE, DF (*TAED* D8465J). Message form of Western
Union Telegraph Co. [a]"1884." preprinted. [b]Obscured overwritten text.

1. Insull had prompted Edison about rights for manufacturing steam
engines on 23 February, when he telegraphed for instructions about
negotiating with Armington & Sims. Though acknowledging that
Armington & Sims would insist on a relatively high royalty, he urged
Edison not to "lose this opportunity. Straight Line would be experiment   Armington engine we know." Insull wired Edison on the third,
inquiring if he would be in "Sanford long enough for letters to reach
you or shall we mail to Magnolia." The only other known telegram from
Insull around this time was on 27 February (see note 2). Insull to TAE,
23 Feb. and 3 Mar. 1884, LM 18:257B, 307B (*TAED* LBCD5257B, LB-
CD5307B).

2. Edison had begun to explore a license for manufacturing engines
from the Straight Line Engine Co. in November 1883 and with Armington & Sims in February (see Doc. 2608 esp. n. 8). In response to Insull's query on the matter, Edison telegraphed from Florida on 23 February that he would not pay Armington & Sims more than a six percent
royalty for manufacturing rights and would not "agree to any exclusive
business." Insull, who was managing the negotiations from New York,
replied on 27 February that "Howard wont budge a cent   think you
better let me close preliminary contract on a basis equal to ten percent"
of the sale price of the engines. Edison reconsidered and agreed. Company president Henry Howard was too ill to complete the negotiations
immediately, but Gardiner Sims proposed on his behalf a five-year contract that would provide Edison the patterns and special tools to make
the engines, as well as the license to do so. Sims urged Insull to "think
this matter over carefully . . . I will in turn be frank and fair with you,
knowing that we can hitch up in a very short time." Insull pointed out
that "Licenses as a rule are made for the life of the patent, and not for
a given period of years." TAE to Insull, 23 Feb 1884; Sims to Insull,
4 Mar. 1884; Insull to Sims, 10 Mar. 1884; all DF (*TAED* D8403Y1,
D8403Y2, D8422ZAB, D8416AVO); Insull to TAE, 27 Feb.1884, LM
18:281A (*TAED* LBCD5281A); Insull to Armington & Sims, 4 Mar.
1884, LM 18:311A (*TAED* LBCD5311A).

The matter rested there until 14 June, when Sims sent a proposal (not
found) setting the royalty on the basis of $1.30 per horsepower instead
of a percentage of price. Insull promised that if "Howard will consent
to the plan outlined by you it would prove throughly acceptable to Mr
Edison." He also offered to have Edison propose these terms directly
to Howard. Sims thought this a good idea and, evidently expecting a
contract to be signed soon, offered his personal assistance to help start
the manufacturing process. Although the editors have found neither
additional correspondence on this subject nor evidence that a contract
was signed, the Edison Electric Light Co. continued to use Arming-

ton & Sims engines for central station service. Insull to Sims, 16 June 1884, Lbk. 18:086 (*TAED* LB018086); Sims to Insull, 17 June 1884, DF (*TAED* D8422ZBQ); "The Armington-Sims Engine," *Electrician* 19 (16 Sept. 1887): 389–90.

3. Presumably William Holzer.

4. The editors have found no other correspondence about acquiring lotus and do not know Edison's intent. Of the numerous varieties of lotus plants, many (including natives of North America) were valued as a source of edible (and fermentable) starch whose fibrous nature may have been the object of Edison's interest (see Doc. 2609 [headnote]). They have also been prized historically for their blossoms' beauty and fragrance and, at least in legend, for mood-altering and sedative properties, such as those that induced Odysseus's crew to lose interest in their homeward journey. *OED*, s.v. "lotus"; Spencer 1884, 1:804–5.

–2617–

Sherburne Eaton to Samuel Insull

[New York,] March 4th. 1884.

Mr. Insull,

These letters are very interesting. Thanks for the chance of reading them. Mr. Stuart's lengthy letter of the 26th. about Tiffin, is especially interesting.[1] I guess he must have used a fine tooth comb to get at all these subjects of criticism

Mr. G. R. Markle[2] was here last night, and told me there plant had not run three consecutive days in any time without breaking down since it started. He was good natured about it, but a good deal disappointed. He said they thought of writing to Mr. Edison complaining of certain of his representatives that have been at Hazelton in connection with the work, but that they disliked to complain and ~~did~~ had[a] not do so. I urged him to do it, and assured him the complaints would be faithfully attended to. I am very sorry this plant has not run well. You know that Markle means J. Hood Wright,[3] and for that reason, as well as for others, I urged you to have this plant given especial attention to. But it seems to have been peculiarly unfortunate in running. Mr. Markle mentioned at least 5 different accidents that have happened to is since the start.[4]

S. B. Eaton   per Mc.G[owan].

TL, NjWOE, DF (*TAED* D8439ZAC). [a]Interlined above by hand.

1. Eaton had asked Insull generally to inform him of central station problems because officers of the local companies "think I ought to know what is going on. I think so too" (Eaton to Insull, 18 Feb. 1884, DF [*TAED* D8439U]). Accordingly, Insull had forwarded a handful of letters about foreign and domestic matters in recent days. In one of these, Archibald Stuart described his visit to Tiffin, Ohio, where he noted the contrast between public enthusiasm for the Edison light and the local company's dissatisfaction with its new plant. In seven pages, Stu-

art itemized difficulties with its construction, operation, and supplies; problems included an incomplete roof, inadequate tools, a worn engine, excessive lamp breakage, and unpaid bills (Stuart to TAE, 26 Feb. 1884, DF [*TAED* D8451ZAX]). Insull passed the construction-related issues to William Rich and consulted Francis Upton about the lamps. Upton blamed "the arcing of lamps, a fault we cannot wholly avoid" despite extensive experimentation. Insull relayed the responses to Stuart (Insull to Rich, 28 Feb. 1884; Upton to Insull, 29 Feb. 1884; Insull to Stuart, 1 and 7 Mar. 1884; all DF [*TAED* D8416AQT, D8429N, D8416ASA, D8416AUK]).

2. George Bushar Markle, Jr. (b. 1858), was a son of George Bushar Markle, retired by this time from business but formerly a prominent Hazelton, Pa., coal operator, Philadelphia banker, railroad officer, and a veteran of the campaign to suppress the Molly Maguires. U.S. Census Bureau 1965 (1870), roll M593_1365, p. 660, image 592 (Hazelton, East Ward, Luzerne, Pa.); *NCAB* 24:138; Novak 1978, 42.

3. In addition to their ties through the railroad industry, the families of Markle and Drexel, Morgan & Co. partner James Hood Wright also became linked by marriage. John Markle, George Jr.'s brother and by this time the manager of their father's enterprises, married Wright's stepdaughter, Mary Robinson, in April 1884. *NCAB* 18:153, C:525.

4. On 18 February, only a week after the the Hazelton station started operating, Eaton conveyed Markle's report that the engineer "has been intoxicated for several days" and that the plant had already broken down three times. He soon dispatched the following brief letter to Insull: "Merely as a word of general precaution, let me urge upon you the importance of having the Hazelton plant work all right. The Hazelton people propose to invest a good deal of money in our business, as permanent investment, and are watching their own plant at Hazelton in order to satisfy themselves of the safety of the business, from a financial standpoint. Let us make a special effort to have the plant run all right." Eaton to Insull, 18 and 20 Feb. 1884, both DF (*TAED* D8439U, D8455ZAL).

Insull apologized to Eaton for the trouble at Hazelton, noting that he had given "special instructions" that the plant "should be carefully nursed." He left immediately to visit several of the Pennsylvania plants, and he soon forwarded a 5 March report from a Construction Dept. employee that the Hazelton plant was operating smoothly then. Insull to Eaton, 4 and 7 Mar. 1884; Thomas Spencer to Edison Construction Dept., 5 Mar. 1884; all DF (*TAED* D8416AST, D8455ZAV, D8455ZAW).

–2618–

Memorandum: Travel Expenses

[St. Augustine, Fla.?, c. March 5, 1884][1]

96	Fare[2]
36—	Pullman[3]
16	Eating
246	Magnolia[4]
4	fare Mag[nolia][5]

4	Augustine[6]
6	Palatka[7]
67	Paltaka Hot[el][8]
67	Sanford[9]
1̶0̶3.50	Turnips[10]
13	Shells
16	Waiters
16	going boat[11]
20	Coming boat[12]
20	Josie[13]
40	Mame[14]
[+?][a]	
[20?][a]	Baths[15]
8[b]	Extras
688	

688
600
40
————
1328

1500[16]
1328
————
172
12
————
160
25
————
135

AD, NjWOE, Lab., PN-84-02-25 (*TAED* NP020H). [a]Canceled. [b]Obscured overwritten text.

1. The Edisons left Sanford for St. Augustine on 4 or 5 March. The editors have conjectured the place and date from evidence in the document that indicates it was written soon after the Edisons arrived in St. Augustine. Inferences about the monetary amounts and notations listed in this document are based on contemporary transportation routes and the limited documentary evidence of this trip. See Doc. 2607 (headnote).

2. Fare for railway travel from New York to Jacksonville, Fla.

3. Fare for Pullman Palace sleeping car (and possibly Buffet car) on the trip to Jacksonville. Pullman Palace sleeping cars ran on all fast trains between New York and Jacksonville. Lee 1885, 43.

4. Accommodations at the Magnolia Hotel in Magnolia, Fla.

5. Fare from Jacksonville to Magnolia via steamboat on the St. Johns River, a 28-mile trip. Lee 1885, 141.

6. Fare from Tocoi on the St. Johns Railway fourteen miles east to St. Augustine; the train made the 35-minute trip each way twice daily. Lee 1885, 145.

7. Fare for the 47-mile steamboat trip on the St. Johns River from Magnolia to Palatka. Lee 1885, 141.

8. The Edisons likely stayed at the Putnam House in Palatka, on whose stationery Edison wrote Doc. 2610.

9. Cost of accommodations at the Sanford House hotel in Sanford, Fla.

10. See Docs. 2630 and 2632.

11. Steamboat fare from Palatka to Sanford, 120 miles on the St. Johns River.

12. Fare for the 149-mile steamboat trip from Sanford to Tocoi.

13. Josephine Reimer.

14. Mary Stilwell Edison. Edison and his wife used this common nickname for "Mary."

15. The Edisons likely visited the baths at Magnolia, promoted by the Magnolia Hotel as the "largest sulphur baths in Florida." Green Cove Springs, two miles south of Magnolia on the St. Johns River, and the Westmoreland Hotel in Palatka, also boasted hot and cold sulphur baths. Advertisement, *New York Daily Tribune*, 24 Jan. 1886, 7; advertisement, *NYT*, 10 Mar. 1886, 7; Lee 1885, 123–25, 133, 141.

16. On 9 February 1884, Edison drew a check for $1,500.00 on his Construction Dept. account for "Expenses South." Cash Book (1 Jan. 1881–30 Mar. 1886): 233, Accts., NjWOE.

–2619–

Notebook Entry:
Incandescent Lamp[1]

[St. Augustine, Fla.,] Mch 5 1884

Try Gelatine mixed with Phosphate Lime,[2] also phosphate ammonia & other alkaline phosphates in different proportions with view prevent melting & Swelling in Carbonization   also Chloride Ammonia which will volatize before red, also with Iodine & other solid Volatile things   try Anilines Sol in H_2O mixed with gelatine, this aniline going off below red—

Also try various agents   try sulphur crystals with glue

E

X, NjWOE, Lab., PN-84-02-25 (*TAED* NP020D2).

1. See Doc. 2609 (headnote).

2. On the following page, in different ink and likely at a different time, Edison wrote: "Mix gelatine with Lime also phosphate Lime also Lime water   think twill carbonize." PN-84-02-25, Lab. (*TAED* NP020D1A).

–2620–

Notebook Entry:
Direct Conversion[1]

[St. Augustine, Fla., c. March 5–20, 1884][2]

It strikes me after much analysis that the future scheme for direct conversion is the placing of finely divided metal on one plate & a peroxide on the other by rapid mechanical means   the placing of these plates in Vats & the reduction of

the perox & the oxidation of the metal giving the Electricity—
afterwards raising the lower oxide to a per[oxide] by chemical
means & the reduction of the monoxide to metal by gaseous
reduction with heat & lot of machinery for Easy handling &
preperation of plates & rapid reduction & ~~ox~~peroxidation so
one man can handle apparatus for a 1000 Li[gh]t station.[3]

Experiment. press freshly prepared peroxide Lead on
roughened lead plate with holes through it. use heavy pressure & face it with hair cloth— other plate press gently on
roughened & holy plate. Litherage[4a] & reduce by gas & then
press gently. also try Lead plate in porous Cup packed with
Lead reduced by Zinc. also make finely divided lead by electrolysis in Keith sol[5] from another lead plate= also by blowing
air & also by stirring.[b]

Also try holy plate with red oxide Lead[6] in holes for reducing the H.[b]

In this theory both[a] the active materials should preferably
be insoluable.

give this placing on sheets the material to give the E a good
long & exhaustive trial before giving it up   Ascertain the theoretical amount of Lead to Mono & perox to mon in weight
per hpower per hour

Try all other sols than Sul Acid. Try the fine Lead in Mercury as amalgam and the red Mercury[7a] on amalgamated
plate.[b]

Try iron salts as the cheapest.[b]

Carbon plate & Lead plate with finely divided Lead— the
Lead raised to Mono afterwards reduced by gas— The H lost
& dont get full current but by heat & gas agitation prevented
from ~~de~~ polarization   perhaps the Carbon may have perox
manganese in it & air absorb after

Exhaustion. platinized surface to Carbon. ~~See if~~ perhaps
red mercury on Carbon

X, NjWOE, Lab., PN-84-02-25 (*TAED* NP020E). [a]Obscured overstruck text. [b]Followed by dividing mark.

1. See Docs. 2520 and 2609 (headnotes).

2. This entry appears after Doc. 2619 and before Doc. 2631.

3. Edison's design of the electrodes and of the reduction and oxidation processes in this cell draw on his battery research in the fall of
1882, work that he intended to patent (see Docs. 2377 and 2388). Edison
sought to catalyze the oxidation of the carbon by using a finely divided
metal and a peroxide in a solution of sulphuric acid. He carried out experiments on his new scheme for direct conversion in July (see Docs.
2701, 2703, and 2706).

4. Known more commonly today as litharge, this is a protoxide of

lead (PbO) formed by oxidizing molten lead in air. It has a variety of uses, including in storage batteries. Hawley 1987, s.v. "litharge."

5. This was a solution developed by Nathaniel Keith consisting of five pounds of water, one pound of acetate of soda, and as much sulphate of lead as would dissolve in the solution at 50 to 60 degrees Fahrenheit. The solution was to be neutral to litmus paper and was to be used at an operating temperature of 100 to 110 degrees Fahrenheit. Keith to TAE, 18 May 1882, DF (*TAED* D8204ZBL); see also U.S. Pats. 209,056 and 273,855.

6. Red lead oxide (Pb_3O_4) is an oxidizing agent derived from the heating of litharge in a furnace with a current of air. It is used for several purposes, including storage batteries. Hawley 1987, s.v. "lead oxide, red."

7. Mercuric oxide (HgO) presents itself in two colors—red and yellow. It is converted into red oxide when heated for a long time near its boiling point in air. Red mercuric oxide is also produced by heating mercurous nitrate, the process used more commonly for commercial purposes. When heated to a sufficiently high temperature both red and yellow mercuric give up all their oxygen; it was Lavoisier's experiments with red mercuric oxide that led to his discovery of oxygen. Remsen 1903, 571; Hawley 1987, s.v. "mercuric oxide, red"; Wilkinson 2004, 253.

–2621–

Notebook Entry: Incandescent Lamp[1]

[St. Augustine, Fla., c. March 5–20, 1884][2]

try common print cloth dipped in Solution of Linseed oil in which is mixed iodide Cadmium crystals fine, then dry by centrifugal to give a Silk surface

Soak ~~m~~flax & manilla in water day or so blot & then freeze to bust it all into filaments.[a]

Look into Experimental research book to see what essential oil it was that dissolved paper.[3]

X, NjWOE, Lab., PN-84-02-25 (*TAED* NP020E1). [a]Followed by dividing mark.

1. See Doc. 2609 (headnote).

2. This notebook entry appears on the last page of Doc. 2620 and was likely made at the same time.

3. Edison was probably referring to experiments in February 1877 in which oil of white thyme gave the best results of the oils that he tested for dissolving paper (see Doc. 858 n. 1).

–2622–

Notebook Entry: Miscellaneous[1]

[St. Augustine, Fla., c. March 5–20, 1884?][2]

~~J Step~~   J Stevens   Neneh Wis—[3]

Change Color brand[4]

to pure white for grinding into flour—

$\frac{1}{2}$ cent allowable. in addition to $\frac{1}{2}$ ct there would be 2 c profit

Bran worth ½ cent pound. germ goes into bran, but can be seperated.

gathering Butter. Little Falls man[5a]

Artificial Tortoise.

Utilization of Bran as base material. Cheap ½c lb—[a]

Grind the oxide Lime or Magnesia with the gelatin, in paint mill for filiments.[a]

Making Sheets bone by Rubbing layer gelatin then flow Bichromate K   [to?][b] stay 2 mins— sun it = dry then over it again gelatin & so on—

preferable way would be to bleach Epedermis of the wheat on grain

Utilization of Bran combined with Cementing Material & Hydraulic pressure to make Base Material for all uses.[6]

Act on bran with SO_4—ClZn & also HFl—also other Conc[entrated] liquids. find its solvent, also analysis to see if Eperdermis is abnormally Constituted.

Mix with thin glue water then put in Centrifugal & throw most of it off then press & dry— Try Rubber also tragacanth also Resins such as Dammar = Cannot paper be made of it by cutting in exceedingly narrow strips, then felting.

X, NjWOE, Lab., PN–84–02–25 (*TAED* NP020D3). [a]Followed by dividing mark. [b]Canceled.

1. See Doc. 2609 (headnote).

2. This is the first entry in a group of pages that Edison appears to have used in reverse order. In that order they precede the dated entry transcribed as Doc. 2631.

3. John Stevens (1840–1920) of Neenah, Wis., developed the roller mill process of flour production in the 1870s. This process significantly increased the output and quality of high-grade flour by enabling the use of harder-grained spring wheat from the western plains. This improvement made it easier to strip the adhering starch and gluten from the hard outer layer of the wheat grain known as bran, thus producing more high-grade white flour, which sold at much higher prices. Stevens became quite wealthy from his patented process. Lawson 1908; "John Stevens Is Dead," *Oskosh (Wis.) Daily Northwestern*, 6 Aug. 1920, 4.

4. Edison probably meant "bran."

5. David Hamlin Burrell first wrote to Edison about developing an improved method of separating cream from milk in November 1882. They corresponded about this subject for several years, but Edison apparently did not take it up in earnest. See Doc. 2361.

6. Bran was abundantly available: a contemporary estimate put U.S. production at one million tons annually. In 1883, Edison had suggested using hydraulic pressure to compress bran in response to the commercial need for more economical ways of transporting and storing it (see Doc. 2411 and "Compressing Bran," *Manufacturer and Builder* 15 [June 1883]: 129). In this notebook entry, Edison proposes using bran as a

base for creating artificial materials such as tortoiseshell. The development of artificial materials was a subject of periodic interest to Edison (see Docs. 579 and 1701 and Israel 1998, 268).

–2623–

Notebook Entry: Miscellaneous[1]

[St. Augustine, Fla., c. March 5–20, 1884?][2]

Try experiment of using Linseed oil mixed with finely divided Electrotypers plumbago[3] between Carbon plates agitating the Liquid so as not to allow actual contact & keep adding plumbago as long as it comes down in resistance. Another way is perhaps better is to use finely divided metal as an acid will take that out of the Linseed easy—[a]

Try the Effect on Linseed & Cotton seed oil of Various oxidizing agents with & without heat,[4] thus Nitrate Potash, Hot Sulphuric acid.—permanganate of Potash Vanadate Ammonia.

Chromic acid—peroxide of metals other than Lead.

Ferrid[5] cyanide Potash. pass chlorine gas through it with acetate[b] Lead in Linseed—

Boil some Linseed in a Vacuum tube to expel the volitile constituents.

Mix Linseed oil with every kind of solvent to see if some volitile constituents will not come out.

also agitate it with various powders to see if the pores will[b] not take up the volitile constituent

also put Nitrate Potash in it powder   stir rapidly & then put Sul acid[a]

[~~Rubb~~?][c] Rub glass with it & expose to fumes of nitric acid rub again & so on see if cant get layer[a]

Make mixtures of Rubber with various things such as linseed, also rubber with Resins to get a flexible Sheet also Celluloid.[6a]

See Coles[7] about furnishing Ar-Cu stabs[8] to try Experiment with.

Study up running Cuperferous ores into rough matte.[9a]

French polish filiments with various things to get shining surface of carbon.[10]

See if pumice stone rubbed on filiment will make smoother surface on filiment

Mix with the [f--?][c] size or stuff to polish with 50 or more per cent of magnesia oxide which is not reducible by Carbon. grinding the latter up fine with the size— apply as in french polish—

See if Bladder intestine etc horn can be tanned or effected by Tannic acid—

ascertain by experiment everything that will precipitate glue from its water solution.[a]

also if glue is soluable in any other liquid than[b] water.

As Tragacanth is a compound of Lime see if more lime cannot be mixed with it—

Try[b] the cellulose experiment with ceric solutions[11] of all kinds of compounds organic & inorganic

See if Silk can be disolved & if it will combine with glue, rubber, resins etc.[a]

X, NjWOE, Lab., PN-84-02-25 (*TAED* NP020D4). [a]Followed by dividing mark. [b]Obscured overwritten text. [c]Canceled.

1. See Doc. 2609 (headnote).

2. This entry follows Doc. 2622 and precedes Doc. 2631.

3. Wax molds used in electrotyping were coated with plumbago (graphite) to make them conductive. The plumbago had to be free of grit and was "carefully sifted through muslin or a fine wire sieve." Urquhart 1881.

4. In June 1882, Edison had executed a patent application for forming flexible filaments by "carbonizing an oxidized drying-oil," such as linseed or cottonseed oil. The drying oil was "formed into a tough flexible sheet or membrane by drying or baking," then individual filaments were cut or punched from the sheet and "carbonized by heat under strain and pressure." In the original application, Edison stated that the oil could be combined with another carbonizable material, but the patent examiner restricted him to the oil alone because of the prior art in relation to arc-light carbons (Pat. App. 365,509). Edison himself had described making arc-light carbons in this manner in a caveat he drafted in June 1882 (Doc. 2307). Edison's application finally issued in June 1887 as U.S. Patent 365,509.

5. "Ferrid" is an older form of "ferric"; it refers to iron compounds in which iron manifests more than its minimal combining power or valence. See Doc. 370 n. 9.

6. Edison first proposed making filaments from sheets of nitrocellulose, including celluloid, in June 1882. Doc. 2291; U.S. Pat. 543,985.

7. Edison often used this spelling of Alfred Cowles, secretary of the Ansonia Brass & Copper Co., a major supplier to his manufacturing shops. Doc. 2343 n. 7; see e.g., TAE marginalia on Cowles to TAE, 27 Oct. 1883, DF (*TAED* D8321ZAH).

8. Stakes or posts. *OED*, s.v. "stab, n. 3."

9. Account records indicate that Edison conducted an experiment along these lines immediately upon his return to New York (Ledger #5:567, Accts. [*TAED* AB003]). Most copper ores are sulphide ores that also contain iron sulphide; the roasting and smelting process produced a copper matte containing cuprous sulphide (Cu_2S) and ferrous monosulphide (FeS) in varying proportions. Subsequent oxidation and reduction processes produced metallic copper free of iron and sulphur. At this time, copper matte, particularly that produced at mines in the region around Butte, Colo., was often melted into pig copper and shipped to eastern refineries (Dewey 1885; Hoffman 1914, 60–65; Levy 1912, 8–12, 35–51).

10. French polish is a term for a variety of shellac compounds, as well as for a method of applying the shellac. Widely used at the time, French polish was composed of a simple solution of shellac in either methylated spirit or wood naphtha, to which was sometimes added gums or resins to render the polish tougher. Other materials were added to change the color of the polish. The process of French polishing involved using a "rubber" made of rag and cotton wool instead of a brush. The polish was poured into the cotton wool and squeezed out through the rag, which acted as a strainer to prevent contaminants from scratching the surface. A small amount of linseed oil was sometimes used to lubricate the "rubber." The final part of the process, called "spiriting-off," involved using methylated spirits in place of the shellac to soften and smooth the polish in order to burnish it. Bitmead 1910; Cooley 1880, 762; Hasluck 1908, 48–65.

11. Edison presumably meant solutions of the ceric salts.

–2624–

To Samuel Insull

St augustine Fla Mch 6 1884.[a]

Samuel Insull

Use your judgment about Australian Colonial[1]   tell Kruzi no objection to Cassedys two years lease[2]

Edison

L (telegram), NjWOE, DF (*TAED* D8403ZAI). Message form of Western Union Telegraph Co. [a]"1884" preprinted.

1. Edison had received Drexel, Morgan & Co.'s solicitation of his opinion about a plan to reduce the capital of (and his stake in) Edison's Indian & Colonial Electric Co., Ltd. The firm was a London concern organized in 1882 to promote the Edison system in British possessions of the Indian Ocean basin (see Doc. 2280). It had done little business by this time, and George Gouraud's plan for partially liquidating the firm was apparently an effort to settle with or thwart "malcontented share holders." After having discussed the matter with Gouraud in London, Insull expressed his (and presumably Edison's) willingness to "take hold of the Australian business and in fact the whole business of the Indian Colonial Co. in some way or other." Drexel, Morgan & Co. to TAE, 26 Feb. and 7 Mar. 1884; Gouraud to Drexel, Morgan & Co., both 9 Feb. 1884; Insull to Gouraud, 4 Feb. 1884; all DF (*TAED* D8402C, D8402E, D8402D, D8402D1, D8416AJE).

On the strength of Edison's authorization, Insull promised to confer personally with Drexel, Morgan & Co. The bankers assented, and shareholders approved the capital reduction in June (Insull to Drexel, Morgan & Co., 7 Mar. 1884; Drexel, Morgan & Co. to TAE, 7 Mar. 1884; both DF [*TAED* D8416AUC, D8402E]; Edison's Indian & Colonial Electric Co. circular letter, 4 June 1884, Miller [*TAED* HM840220]). Insull subsequently suggested that the company's territory could be folded into an agreement with the Edison Electric Light Co. for Edison "to actively push the lighting business in Mexico, Central America, and South America." Late in 1884, Drexel, Morgan & Co. inquired about wholly liquidating the Colonial firm, to which Edison enthusiastically agreed rather than have its remaining capital "fritted away" by the directors. The firm lingered, however, until it was absorbed by the Australasian Electric Light Power and Storage Co., Ltd., in 1886 (Insull to Gouraud, 6 June 1884; Drexel, Morgan & Co. to TAE, 22 Nov. 1884; Edison's Indian & Colonial Electric Co. circular letter, 2 June 1886; all DF [*TAED* D8416BTB, D8402M, D8630ZBA]; TAE to Drexel, Morgan & Co., 24 Nov. 1884, Lbk. 19:404 [*TAED* LB019404]).

2. After having considered at least one other property in Brooklyn, John Kruesi contemplated relocating the Electric Tube Co. to 18–24 Bridge St. in Brooklyn, near the East River waterfront (Kruesi and Insull to TAE, 18 Feb. 1884, LM 18:234A [*TAED* LBCD5234A]). This address occupied a corner lot at John St., a property owned since 1881 by Edward J. Cassidy, the adult son of Patrick Cassidy. The elder Cassidy had sold machinery and scrap iron at 4–10 Bridge St. during the 1860s. He was also among a delegation of local interests who had sought in 1881 to restore ferry service from the Bridge St. docks to Manhattan, and his family owned several lots in the neighborhood during the 1880s (U.S. Census Bureau 1970 [1880], roll T9_895, p. 582.1000 [New York (Manhattan), N.Y.]; "Advertisement," *Brooklyn Daily Eagle*, 29 Oct. 1862, 1 and 17 Dec. 1862, 2; Block 20 records of 21 Feb. 1881, 21 Apr. 1884, 15 Apr. 1889, 21 Mar. 1890, Land Conveyance Collection, NyBlHS; "The Sinking Fund Commission," *NYT*, 19 Feb. 1881, 3). Insull answered Edison by letter on 8 March, asking whether Kruesi should "take Cassidy's place or wait till you return then arrange for part Goerck street shop." It is not clear what role Edison played in the decision, but the Electric Tube Co. moved to Bridge St. in April (Insull to TAE, 8 Mar. 1884; LM 18:325B [*TAED* LBCD5325B]; Electric Tube Co. to TAE, 22 Apr. 1884, DF [*TAED* D8433O]). The Tube Co. did not immediately need all of its new space. At the end of the year, William Hammer hoped to induce Phillips Bros., a major wire manufacturer in London, to set up a branch operation in the Brooklyn shop. Edison endorsed the idea; such an operation presumably would

The new Brooklyn home of the Electric Tube Co.

THE ELECTRIC TUBE COMPANY'S FACTORY, BROOKLYN, L. I.

have competed with a major supplier, the Ansonia Brass & Copper Co. Nothing seems to have come of the plan, however, and Phillips Bros. eventually manufactured in conjunction with Bergmann & Co. (Hammer to Phillips, 15 Dec. 1884; Hammer reminiscence, [n.d.]; both WJH [*TAED* X098A023b, X0980J image 37]).

–2625–

To Samuel Insull

St. Augustine, Fla., Mch 8 *1884*[1a]

Insull,

In sending any one around to stations to report, be sure they Examine the boiler tubes. they probably complain that the boiler isnt big enough as at Sunbury   I bet there is at least ½ inch of scale on the tubes ~~as the~~ Local Engineers won't clean their boilers and Tomlinson must put in contract with Engineers that the boiler must be freed from scale once a month=[2] Regarding $9\frac{1}{2} \times 12$ at Bellfonte I don't see why it isnt sufficient[3]   you Know that we should carry 120 pds boiler pressure   now they start the station at 80 lbs & bye & bye they find Engine Laboring. Andrews & Rich should be told that they must start & maintain 120 lbs pressure= I understood that 285 or 300 Revolutions was enough at Bellfonte to give proper volts if not [~~then?~~][b] do not increase speed of Engine although 300 is slow for that Engine but Send ½ inch larger diameter field magnets & replace old ones   this will bring Volts up at slight Expense while the ones taken off are good for Isolateds   another way is to increase size of pulley—

~~I~~ [---][b] this costs much more— I am sure that if the Engineers sign the contract we will not have any trouble with a place like Bellefonte because in Every Case trouble may be traced to carlessness   The use of the blower & small Engines in future installations will be a great help. We must not run up the cost of these stations but putting in too great a capacity, a great deal of trouble must be looked for not on account of machinery not being of sufficient capacity but of ignorance prejudice & lack of confidence in our own men. It seems so ridiculous to say the a $9\frac{1}{2} \times$ [c] 12 at 300 is too fast for reliability when thousands of Locomotives developing[c] 5 to 700 horsepower run at 300 @ 350[4] & Even our 200 hp Engine at station placed on stilts under the very worst condition imaginable runs at 350[5] of course it requires attention greater[c] than can be Expected of our small Stations but on the other hand the Engines at our stations are smaller & do not run as high= If we are Ever to cheapen these stations down it must be done by high speeds high boiler pressure high resistance lamps, high volts— because we are having a little trouble we must not go back on this but run the trouble to Earth ascertain about it fully & remedy it. none of our mechanism is perfect and nothwithstanding it has not done bad we must get it improved= I am glad the pressure indicators are wkg so well hope you will fix up Sunbury & Shamokin with the new Volt boxes   Andrews men could go around during heaviest load at night and make a record of pressure at several different parts of the town— Then with the map & possition of Each customer & Lamps burning Gamaris[6] could figure out where to put the feeder ie[d] change them— One thing I want to call your attention to & you might call the Local peoples attention also is that the pressure will necessarily be uneven while ~~the station is~~ customers are being connected as one part of the town may take to the light quickly and most of the Lamps be at that point   this will necessarily thrown it out of balance but when the station has its capacity the pressure will be very even all over. it is during the transition period between starting & getting full capacity that the Lack of Even pressure will show themselves   of course feeder regulators will help this immensely— I don't understand why Bergmann has been so long getting[c] out those feeder regulators   I spose the 1st ones will be sent to fall River= We could prove to the Bellefonte people the capacity of our plant by taking on several hundred Lamps in a bank & running them up to Candle power= Will do as you say regarding the inquiries about the shops=[7]

What is Eaton doing about putting out agents all over the country to get these towns started= Has shaw turned up with any more towns. How about, Auburn, Louisville Haverstraw Hudson, & other towns any prospect.[8] Tell Soldan there seem to be some complaints about the oil running through the commutator   I think the space should be filled up, as the little air that gets through don't do much good= I also think that [--][b] on the 200 & 300 Light the cores should be ½ inch greater diameter so we can get more volts—

Edison

ALS, NjWOE, DF (*TAED* D8439ZAI). Letterhead of Magnolia Hotel. [a]"*St. Augustine, Fla.,*" and "*188*" preprinted. [b]Canceled. [c]Obscured overwritten text. [d]Circled.

1. Edison wrote the date using a different ink than in the text, perhaps postdating the letter as he apparently postdated notebook entries made during the Florida trip (see Doc. 2609 [headnote]).

2. Edison's draft of the contract for central station engineers is Doc. 2605; see esp. n. 2.

3. See Docs. 2608 and 2615 concerning problems at the Bellefonte, Pa., plant.

4. Edison occasionally used the "@" symbol to express a numerical range.

5. The editors infer from the context that Edison meant the 125-horsepower Armington & Sims (11×16) engines that rested on an iron framework over the boilers at the Pearl St. station, where they drove the "Jumbo" dynamos through a direct connection at 350 r.p.m. Doc. 2243 (headnote); Lieb 1904, 65.

6. Henry Guimaraes.

7. See Doc. 2638.

8. See App. 2.

-2626-

From Charles Batchelor

PARIS, le Mch 9 1884[a]

Cable rec'd. "Do everything possible Stockholm save patents"—[1] Cant get information from London[2] on these but from friend in Stockholm[3] I find

1 patent 8 Feb 1879
1   "   24 Sept 1879
2   "   5 Mch 1880
1   "   25 June 1880
1   "   29 July 1880

all cancelled for ~~p~~want of proper working;[4] Trying to find out now whether dead for ever or not![5]

"Batch"

Probably shall have to send up there and make a few lamps to save the other patents— "Batch"

⟨ask Serrell what patents were worked in these countries & by whom— E⟩[6]

ALS, NjWOE, DF (*TAED* D8436ZBN). Memorandum form of Société Électrique Edison. [a]"*PARIS, le*" preprinted.

1. This is the full text of Edison's message of 8 March (DF [*TAED* D8468ZAC]). Batchelor had cabled to New York on 26 February to alert Edison, who was in Florida, about a telegram received from one Mr. Olan, a Stockholm party unknown to him, advising that the Swedish patent office had voided Edison's electric lighting patents (Olan to Batchelor, 23 Feb. 1884; Batchelor to TAE, 26 and 28 Feb. 1884; all DF [*TAED* D8468O, D8436ZAW, D8468U]).

2. Batchelor promptly wrote to Brewer & Jensen, who, as Edison's former patent agents in Britain, had filed his specifications on the Continent, and to Thomas Handford, his present British agent. Brewer & Jensen disclaimed any knowledge of the Swedish specifications subsequent to 1882; no reply from Handford has been found. Batchelor sent Brewer & Jensen's negative reply to Edison, who noted on it: "show

Maj Eaton." Batchelor to Brewer & Jensen, 26 Feb. 1884; Batchelor to Handford, 28 Feb. 1884; Brewer & Jensen to Batchelor, 27 Feb. 1884; Batchelor to TAE, 28 Feb. 1884; all DF (*TAED* D8468R, D8468T, D8468S, D8468U).

3. Batchelor received the following list from Charles Dreydel, a Stockholm agent with whom Edison had worked since 1882. (Dreydel's firm published a three-page booklet in 1882 entitled *Edison's Foreign Electric Light and Motive Power Co.*) Dreydel blamed Joshua Bailey for neglecting prior warnings about Olan, a possible competitor whom Dreydel now thought to be "at bottom of all this trouble" for having tipped off the Swedish patent office. Batchelor promised Dreydel to "do all in my power to save those patents that are cancelled (as you say) or those that remain to us. I can send up there material for manufacturing the lamps in small quantities and could even come myself if it can do any good." Swedish law required patents to be worked within two years of issue. Dreydel to Edward Winfield, 8 Sept. 1882; Dreydel to Batchelor, 3 Mar. 1884; Batchelor to Dreydel, 28 Feb. and 9 Mar. 1884; L. A. Groth & Co. to Charles Dreydel, 13 Mar. 1884; all DF (*TAED* D8238ZCX, D8468Y, D8436ZAY, D8436ZBM, D8468ZAO).

Edison's basic lamp patent in Canada was later invalidated under similar circumstances. The Edison Lamp Co. had set up a small factory in Montreal in 1881 whose operations were somewhat enlarged and moved to Hamilton, Ont., in early 1883. Superintended by Charles Stilwell, Edison's brother-in-law, the plant assembled a relatively small number of lamps from carbons and glass globes imported from the United States. An administrative judge ruled in 1889 that such assembly work did not meet patent law requirements for domestic manufacture in Canada. "Annulment of the Edison Lamp Patent in Canada," *Electrical Engineer* [New York] 8 (Apr. 1889): 199–202; "Stilwell, Charles F.," Pioneers Bio.; Stilwell to TAE, 11 July 1883; Sherburne Eaton to TAE (with TAE marginalia), 1 Oct. 1883; both DF (*TAED* D8313ZAB, D8332ZCV).

4. A Stockholm patent agent subsequently provided a list of eight Edison specifications that had lapsed by the middle of 1883: one each for telephone and phonograph, three for electric lamps or their manufacture, one for generating and measuring electric power, and two for electric lighting generally. Swedish law required patents to be worked within two years of issue. L. A. Groth & Co. to Charles Dreydel, 13 Mar. 1884 (and copy), both DF (*TAED* D8468ZAO, D8468ZAF).

Fearing that termination of the Swedish specifications could affect the life of Edison's U.S. patents, Batchelor later instructed Charles Dreydel (against his recommendation) to petition the Swedish king for their reinstatement. He also held open the possibility of filing anew somewhat amended specifications. Batchelor to TAE, 26 Mar. 1884; Dreydel to Batchelor, 19 Mar. 1884; Batchelor to Dreydel, 26 Mar. 1884; all DF (*TAED* D8468ZAM, D8468ZAN, D8436ZBT).

5. On another version of this document, also in Batchelor's hand but without Edison's marginalia, Batchelor wrote: "I am trying to find out whether these are forever dead or whether we can resuscitate them— Who has charge of them?" DF (*TAED* D8468ZAD).

6. The editors have not located subsequent correspondence on this subject with Lemuel Serrell. Samuel Insull apparently discussed the

matter with Richard Dyer, prompting Sherburne Eaton to reproach Insull that he had learned this "highly important information . . . by chance from other sources than your office." Eaton to Insull, 1 March 1884, DF (*TAED* D8427U).

–2627–

To John Ott

[St. Augustine, Fla.,? c. 10 March, 1884][1]

John

Have Acheson if not busy with the plating go ahead and experiment on disolving cellulose in Cupric Ammonium[2]   let him find out a way of making good Cupric Ammonium and in quantity so it will disolve paper or cotton perfectly, it is very difficult to make it just right   you might try Cotton twine, also the soft pulpy cellulose of Carrots Cauleflower assparagas, also cotton[a] treated with potash and wash and dried. I feel convinced that we can make a good stock by disolving Cellulose in some form by Cupric Ammonium

How are ~~you~~ they progressing with the plating in vacuum— Have you succeeded in using the bridge and lamps with the Automatic Regulator for Isolated Dynamos[3]   Have you done any thing with films of Linseed Oil

Edison

L (copy), MiDbEI, EP&RI (*TAED* X001J1AS). Copied onto letterhead of Thomas A. Edison by Alfred Tate. [a]Obscured overwritten text.

1. Alfred Tate did not include a dateline when he copied this letter; the original from Edison has not been found. The editors have inferred an approximate date from John Ott's reply to Edison on 13 March. The Edisons were in St. Augustine at the time, although as Tate noted on his copy, letters and telegrams were to be addressed to them in Magnolia. Insull to TAE, 8 Mar. 1884, LM 18:325B (*TAED* LBCD5325B); Ott to TAE, 13 Mar. 1884, DF (*TAED* D8467H); see Doc. 2607 (headnote).

2. Edison had some past creative experience with cellulose and its solvents (see Docs. 583, 586, 645, 655, 667, 791, and 813). In a June 1882 patent application for forming filaments from nitrocellulose, he described "the use of unstructured cellulose, which may be dissolved in cuprammonic hydrate" (U.S. Pat. 543,985; see also Doc. 2291). His idea at the present time evidently was to form flexible and uniform sheets from which lamp filaments could be cut, a process related to those discussed in Docs. 2346 (esp. n. 14), 2510, and 2511. Edward Acheson, whom he had commissioned in 1881 to produce filaments from sheets of pressed plumbago (see Doc. 2085), was on the cusp of leaving Edison's employ. Ott reported in his 13 March response (see note 1) that Acheson had not been present for twelve days, having "gone down town to hurry through a patent on an invention in telegraphy."

3. The editors have not found records of Ott's efforts to adapt the regulator for this use, but see Doc. 2628.

Samuel Insull to Sherburne Eaton

[New York,] March 11th. 1884.

Major Eaton:—

Referring to your memo of the 10th inst, as to the experimental accounts.[1]

I hand you herewith statement showing you the disbursements of labor account as requested.[2]

The automatic regulator experiments are carried on at Avenue B.

Just before Mr. Edison left he gave instructions for certain experiments to be made, with a view to adapting the principle we use in the new pressure indicators for small town plants to the requirements of the isolated business, the idea being to get a more reliable automatic regulator for isolated purposes.[3]

The carbon experiments referred to have been carried on for a very long time. I imagine that this week the work has been mainly upon some new fibre, which Mr. Edison has sent up from Florida.

The gas bill is for the laboratory at Bergmann's, gas being mainly used for heating purposes in carbon and other experiments.[4]

I hope the above explanation will be satisfactory.

Saml. Insull   By. H.[5]

TL (carbon copy), NjWOE, DF (*TAED* D8416AVS).

1. Not found.
2. Not found.
3. See Doc. 2538 (headnote) regarding the "new pressure indicators" with which Edison had recently decided to replace the "Edison Effect" instruments in small central stations. In a short typed note to Eaton on 10 March, Insull stated that "All experiments are conducted at Avenue 'B.' I believe that part of the pressure indicator experiments have been conducted at the Lamp Factory, because there are facilities there in the way of instruments and current which we have not, and which would require some fixing up to get, at Avenue 'B.'" DF (*TAED* D8416AUV).
4. That is, for a carbonization furnace.
5. Possibly Charles Hanington.

From Western Edison Light Co.

CHICAGO, Mch. 12th 1884[a]

Dear Sir:

Parties interested in mining have made inquiries of us relative to utilizing the Edison current for precipitating copper in treating crushed refractory ore.

They ask if the current used for lighting at night can be applied successfully for their purposes during the day.

Will you please advise what answer we can make them? obliging   Yours truly

Western Edison Light Co.   Hurlbut[1]

⟨It can be used but best way is to mix the Copper ore with Coke & run it into plates of Black matte[2] then the Copper can be deposited pure & gold & silver remain in the mud— can be done day times at central station= I have six vats being made now to try it   E⟩[3]

L, NjWOE, DF (*TAED* D8462W). Letterhead of Western Edison Light Co. ᵃ"CHICAGO" and "188" preprinted.

1. H. R. Hurlbut was evidently a clerk or stenographer for the Western Edison Light Co. The surname appears frequently on correspondence from the company in 1882–84; see, e.g., George Bliss to TAE, 25 Oct. 1882, DF (*TAED* D8241Z).

2. See Doc. 2623 esp. n. 9.

3. Edison's marginalia, written after his return from Florida in April, was the basis for a typed reply prepared by Samuel Insull (TAE to Western Edison Light Co., 3 Apr. 1884, DF [*TAED* D8416BBV]). The editors have found no records of the experiments to which Edison referred, but see Doc. 2673. Electro-deposition processes, like the one to which Edison alluded here, were widely investigated around this time for removing valuable substances from refractory ores (Watts 1887). Edison's generic marginal notes on two similar letters of inquiry likewise provide no clues (C. Cosgrove to TAE, 21 Apr. 1884; John Blattau to TAE, 28 July 1884; both DF [*TAED* D8466M, D8466S]). Edison's efforts to separate precious metals and iron ores dated to 1879 (see, e.g., Docs. 1776, 1844, 1921, and 1938). In 1881, Edison had made some effort to extract copper from an abandoned mine near Menlo Park, but he found the deposits too lean (see Doc. 2130).

–2630–

To John Ott

St. Augustine, Fla., March 18 1884[1a]

John—

Please get some very clean cotton and[b] put a lot in ounze bottles filled with the following solutions, put corks in the bottles= Strong ammonia with ¼ its bulk of alcohol— 2nd ½ caustic Potash ½ ammonia   3rd Strongest solution Caustic soda; 4th. Strongest solution, chloride strontium, 5th strongest solution Chloride Magnesia, 6th[c] strongest solution of Cyanide potassium, 7th. Nitrate Copper ~~8th~~ strongest solution you can make   8th= Iodide Potassium strong= 9th Iodide of Zinc strong— Tell Acheson try make Cupric Ammonium by the Current with Copper plates in Ammonia—[2]

See if you can disolve Rosin or any hard gums in Collodion   if so make various proportions and pour out on glass dry and save part of the ~~filaments, the~~ piece[c] but send a piece to Lamp factory to be carbonized—

Buy a gross of common wide mouth 3 or 4 oz bottles, as I shall want to use them when I return and they will be handy to try the above experiments with. Let[b] the Cotton stay in the above solutions until I return.[d]

Take boiled linseed oil boil it until it is about twice as thick then get sheets of various kinds biblous paper and soak say 6 or 8 hours in the solutions then hang them up in a place free from dust & let them hang until I return.[3d]

Make a strong mixture of gelatine and make a lot of paper in the same way and hang until I return.[4d]

Also make a solution of Rubber in Benzine or other good solvent [---][e] soak paper in it for some hours then hang up and dry until I return=

If Rubber will mix with Rosin or other gums mix a lot with it & wet cut sheets & hang up to dry

Make strong solution of Rosin in turpentine, soak paper in it for a week then take out & scrape smooth & hang up to dry until I return.

get some good potatoes and make a knife or something that you can cut thin & even slices of the potato & dry them—perhaps you will have to dry them with a gentle weight— out of some of the slices you might use that punch & get some filaments when fresh & also after they are dried. If the potato looks like success you might get some turnips, beets, carots etc[d]

Make a strong solution of Licorice   soak paper in for 8 or 10 hours then scrap and dry until I return=[5]

Edison

ALS, MiDbEI, EP&RI (*TAED* X001A1CA). Letterhead of Magnolia Hotel. [a]"*St. Augustine, Fla.,*" preprinted. [b]Obscured overwritten text. [c]Interlined above. [d]Followed by dividing mark. [e]Canceled.

1. Edison wrote the date at the end of this letter in a somewhat different handwriting than the body of the text, suggesting that this document, like Doc. 2625 and other Florida notebook entries, may have been postdated (see Doc. 2609 [headnote]).
2. See Doc. 2631.
3. See Docs. 2621, 2623, and 2627.
4. See Docs. 2619 and 2634.
5. See Doc. 2632.

Notebook Entry:
Incandescent Lamp[1]

[St. Augustine, Fla.] Mch 20 1884.

Try dry salts of Cupric ammonium & cellulose with heat to disolve for substitute for Celluloid[2a]

See if there is not a Volatile Compound of Calcium Magnesium etc whereby a deposit of Oxide[b] can be put on filiment by Electrical Incandescence

X, NjWOE, Lab., PN–84–02–25 (*TAED* NP020F). Document multiply dated. [a]Followed by dividing mark. [b]Obscured overwritten text.

1. See Doc. 2609 (headnote).
2. Edward Acheson was already working on experiments to dissolve cellulose in Cupric Ammonium; see Docs. 2627 and 2630.

Notebook Entry:
Miscellaneous[1]

[St. Augustine, Fla., c. March 20–21, 1884?][2]

Mix glue & Nitro cellulose together also glue & Cellulose disolved in Cupric Ammonium

See if there arnt substitution Compounds of Cellulose that is soluble in 100 bottle Solutions.[3a]

Way Duplex.[4]

Gold Seperating Mac by Copper disk principle[5a]

Mix glue & organic Carbon Compounds larger proportion of later— Carbonize—

Get large psound potatoes slice them thin   dry under gentle pressure to make Carbons from—

Also Turnips, pumpkin squash, Egg plant, (try Epidermus of Egg plant) Apple Turnip, beet.

Carbonize— Collodion films in Linseed—

Mix Shellac or Lac[quer] with Sol of Colodion   also other things filim[ent]ize & Carbonize

Licorice seems to Carbonize without any Swelling large amount gas goes off— Might be used as binder or mixed with other things roled into sheets to Cut filiments from[a]

X, NjWOE, Lab., PN–84–02–25 (*TAED* NP020F1). [a]Followed by dividing mark.

1. See Doc. 2609 (headnote).
2. This entry appears between dated entries Docs. 2631 and 2633.
3. This term refers to chemical solutions in bottles containing 100 c.c. H_2O. See Docs. 810 and 813.
4. Edison's way duplex enabled telegraph messages to be sent simultaneously over the whole line and between way stations by reversing the connections to create an electrical signal without breaking the circuit. In this manner, the signal would be detectable by a telephone receiver placed at the way stations without affecting the relay receivers used on

the main line. This arrangement became the basis for the Phonoplex system Edison began developing in 1885. See Doc. 1415 esp. n. 2.

5. This principle is probably related to a generator design from 1878 in which Edison drew on a famous experiment by François Arago, which demonstrated that the motion of a copper mass in a magnetic field induces electric current in the copper. In his British Provisional Patent (No. 5,306 [1878]), Edison referred to this phenomenon as the "general principle of the copper disc between the two magnet poles." The generator employed copper rings around an iron core to generate a continuously flowing electric current (see Doc. 1580 n. 1). For Edison's efforts to separate diamagnetic metals such as copper see Doc. 2673.

–2633–

Notebook Entry: Miscellaneous[1]

[A][2]

[St. Augustine, Fla.,] Mch 21 1884

paint filiment with Camel hair brush with acetate of mag also Lime, mag 1st= then get vac & gently bring up then b[rea]k vac & paint again etc several times then get life & Phenomenon

See if a agate or other stone grinding Mill Can be obtained   perhaps a Bogardus paint grinding Mill[3] will do to grind the Oxides with the Carbo[ni]z[a]ble Compounds—E

[B][4]

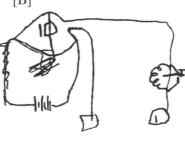

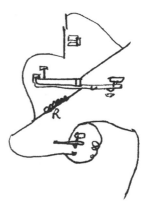

Artificial Mica[a]
    ground safty plug material[5]   Layer like French polish then oil then layer—etc

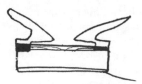

insulating ring on chalk dia[phragm] over it thousands of fine plat 1/1000 wire resting on chalk dia air tight. The H_2O Rushing up & down pores produce Vac & work Dia.[6]

Direct Carbon battery =[7]

Use hot conc sol of Caustic Potash to absorb the CO monoxide to Formic acid— finely divide the C in porous Cup—& use some others Liquid outside—Keep temp high

Bran has a Silica Surface= Hot soda take it off= then by soaking & boiling bleach—possible by displacement can wash— try Electcity for bleaching bet rollers—

Float Linseed oil on Nitric acid. Mixed & unmixed with Sulphuric strong to Keep it Concentrated

Secondary battery with Mercury Electrode & Such Solution as to make Hydrogen Amalgam—[8]

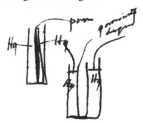

think this will do the biz

X, NjWOE, Lab., PN-84-02-25 (*TAED* NP020G). Document multiply signed and dated. ᵃFollowed by dividing mark.

1. See Doc. 2609 (headnote).

2. Figure label is "Vac."

3. Best known for his cast iron building patent, inventor James Bogardus (1800–1874) patented a universal mill with eccentric, corrugated, chilled-iron plates in 1832 for grinding a large variety of substances, including paint. *ANB*, s.v. "Bogardus, James"; U.S. Pat. 6,903X; "Notice of Bogardus' Universal Eccentric Mill for Hulling, Cutting, and Grinding," *Journal of the Franklin Institute* 11 (3rd ser.; 1846): 337–44.

4. These partial circuit designs are most likely related to Edison's work on multiple telegraphy.

5. The safety plug was an alloy of lead and tin. Swinton 1884.

6. Edison's idea for a telephone receiver was a modification of the electromotograph receiver that he had developed for commercial use in 1878–1879 (see Docs. 1681 and 1784 [headnotes]). In that instrument, a stylus (called a "rubber"), usually made of platinum, pressed against a rotating cylinder (the "chalk") composed of a phosphate compound and moistened throughout with water. The friction between the metal

and the chalk varied with passage of the telephone line current, permitting the stylus to slip against the cylinder and drive a diaphragm that would reproduce the original sound. Edison understood the underlying principle as the passage of a current into damp chalk producing "such an effect on its capillary power as to draw the solution to surface and lubricate it" (see Docs. 1738 n. 5, 1845 n. 1, and 1854). The device in this document, instead of varying friction, would use the water's capillary movement through the chalk "pores" to change air pressure on one side of the sealed diaphragm.

7. This concept is related to Edison's work on direct conversion of coal into electricity (fuel cells); see Doc. 2520 (headnote).

8. Figure labels are "Hg," "Hg," "porous," "Hg," "porous with Liquid," and "Hg."

–2634–

Notebook Entry: Miscellaneous[1]

[St. Augustine, Fla., c. March 21–27, 1884][2]

Conc sol Chl Magnesia & other with glue. Carbonize—[a]

Melt glue or gelatine & mix w mixed with the oxide magnesia or compound thereof in the smallest quantity[b] of water possible & force this out into sheets through a die, blower or otherwise Cooling to set it quick when it comes out.

by this means a minimum quality water can be used—[a]

Try disolving glue[b] in Damp phosphoric anhydride.

Isnt there a Sucunate[3] of Lime or Magnesia.

Tatarate Magnesia will perhaps Carbonize if mixed with little glue

I wonder if starch is ground so as to break all the starch granules if it is moistened & dried in sheets if it wldt work—

Try this grinding with glue—

perhaps Gutta percha mixed with Lime Magnesia etc works—

Try gum Zanzibar & Magnesia melted in Linseed oil & made in sheets a la French polish

I wonder if iodide of nitrogen precip mixed with glue wouldnt carbonize—[a]

Try Asphalt & Magnesia also Compounds of Lime & Magnesia Soluable in Menstrum [----][c] in which Asphalt is soluable in—

Also Tragacanth as a base.

Try Vanadium as an Oxidizer for Linseed oil & Cotton S[eed] oil

Iodide of the high Melting oxides might be used in place of Chlorides as the iodides decomp[b] much Easier.

Use Cerium[b] & the rarer oxides in the Experiments, also, phosphates other than Lime.

perhaps in the Silk Experiment[4] an Elastic precipitate from something might be formed on a platinum point immersed in the liquid & wound up on a platinum Cylinder within the Liquid.[a]

Can a bladder be tanned. if so it would work owing to Oxygenation.

Make a 100 Sticks $\frac{1}{2}$ dia of charcoal at Lamp factory have them turned from white wood & other woods[b] for Experiment in direct Conversion—[5] Make them 8 inches long—

Manganate soda for taking Care H & air revive—

Mix bran with glue having minimum water make into sheets.[a]

Mix Sulphur with glue bath as flower & in compound form also with other bases. think in Carb[oni]z[i]n[g] the S will combine with the H & thus not attack the Carbon[a]

Mix with Collodion an animal matter like glue which will reduce the O of the Nitro-Cellulose in Carbonization. don't think glue is soluable, but some Carbozable Material may be— think Linseed oil is & that wld reduce—

Boil glue in Sulphuric acid see if it don't harden it by taking H_2O away

Cant something be put in Collodion that will be decomped by light such as the iron Solutions hence after Sheet is made Exposure to light decomps the Nitro-C to plain Cellulose[a]

Mix coal tar also Charcoal also crude petroleum with Litherage & also red[b] Lead in Crucible with cover & see if reduce to globule & get proportion & cost[d]

perhaps may have to mix a flux with it such as Salt Borax etc—[a]

X, NjWOE, Lab., PN-84-02-25 (*TAED* NP020G1). [a]Followed by dividing mark. [b]Obscured overwritten text. [c]Canceled. [d]Followed by "over" as page turn.

1. See Doc. 2609 (headnote).

2. This entry follows Doc. 2633, the last dated entry in the book. Edison and his wife began their trip home on about 28 March. See Doc. 2607 (headnote).

3. Edison probably meant succinate, the salt of succinic acid. Succinic acid is distilled readily from amber and certain lignites and also occurs naturally in many plant and animal tissues. A succinate is produced when succinic acid is combined with those alkaline metals or alkaline earth metals, including magnesium and calcium, capable of forming a neutral salt.

4. Possibly the experiment mentioned in Doc. 2623.

5. Edison sought ways to produce a carbonous electrode with high surface area so that it would catalyze efficiently.

[St. Augustine, Fla., c. March 21–27, 1884][2]

Mix magnesia oxide—also Carb with fine telephone Lamp black[3] and gelatine, as much as it will stand of Magnesia[a]— also try Tar magnesia & lamp black— also try Lime, oxide aluminum & other oxides.

Mix gelatine or rather glue with water & the oxides. also with chlorides of the infusible earths.

Try tragacanth and the oxides, the idea being to use as much of the oxide as is possible with the Carbonizable organic material.

If it is possible mix tar with the oxides only

I have a theory that the oxide will not be reduced and that the organic material will be carbonized as a lace work the oxide staying there hence when carrying commences[4] $\frac{1}{2}$ or more will be white oxide hence the carbon will last twice as long and the globe will be no blacker besides I will have a high resistance Carbon.

also try boiled Linseed & the oxides rolled out thin in sheets a la putty.

Sheet gelatine treated with Bichromate of ~~Lime~~Calcium Magnesium Aluminum[5]

X, NjWOE, Lab., PN-84-02-25 (*TAED* NP020G2). [a]Obscured overwritten text.

1. See Doc. 2609 (headnote).

2. This entry follows Doc. 2634 in the notebook.

3. That is, the powdered carbon used in the variable-resistance telephone transmitter invented by Edison. See *TAEB* 3 and 4, passim.

4. That is, "carbon carrying," or the deposition of carbon from the filament onto the lamp globe. See Doc. 1898 (headnote).

5. Based on his handwriting and the difference in ink, it is likely that Edison wrote this final paragraph separately from the rest of the entry.

CIRCLEVILLE, Ohio, Mch 24, 1884[a]

Dear Sir—

I would like your views as to street Lighting. I am unable to see how this Edison Ill Co[1] can attempt street Lighting here   It seems to me an entirely distinct Circuit extending all over the City would be required for Street lamps because they are only lighted in the dark of the Moon, and lights[b] are turned out when the moon rises be that at 9 PM or 2 AM, hence the necessity for a distinct & independent system of lines, this would make the outlay considerable. Have you any Central Station Plant supplying streets with Light on contract with

City, if so where and what advice have you to give us. Our City uses part Coal Oil lamps and 110 Gas lamps, for latter, City pays now 24.00 per Post annually. The City furnishes the lamp posts and keeps them in order and lights and turns out the gas— The contract with Gas Co is out 1st of July and we expect to be invited to bid for street Lighting   If you would not advise our Co to attempt this with Incandescent system, whose Arc system would you recommend and could our Co operate an Arc system without opposition of the NY Edison Co— We would like to get this street Lighting contract and would like to hear from you   Respectfully

S Ward[2]   Secy

⟨You can Easily light the Lamps in the Lines of our mains by Connecting from stores or mains over to the posts & put on & off the same as the gas. wouldnt advise doing this until you have run a year & sold all the light you can to private persons—then you could figure & bid on street Lighting the 2nd year   Arc Lighting in streets is being abandoned as municipalities will not pay for more than 10 or 15 candles only for a short[b] period while the novelty continues[3]   E[dison]⟩

ALS, NjWOE, DF (*TAED* D8448G). Letterhead of Commercial Union Assurance Co., of London, Western Dept., Chicago. [a]"Circleville, Ohio," and "188" preprinted. [b]Interlined above.

1. The Edison Electric Illuminating Co. of Circleville was formed about the beginning of November 1883, though it may not have actually incorporated until 1884. Archibald Stuart, in his dual roles as a director of the company and secretary of the Ohio Edison Electric Installation Co., acted as the primary contact with Edison. By April 1884, according to Stuart, the Ohio Edison firm had decided to "make the Circleville installation ourselves, but we desire you to do the electrical work and furnish the electrical appliances" in order to better respond to local concerns regarding wiring and fixtures. Stuart explained that because the Ohio Installation Co. had "had so many arguments with your folks in relation to the high prices you charge for a building and some other things, we thought we would try our hand at it." This arrangement differed from those governing two other Ohio plants under construction at the time, and it ran contrary to Edison's avowal to undertake all construction work himself after his unhappy experience in Sunbury, Pa. The obligations and rights of this unusual arrangement were fully spelled out in the contract drawn up between the Installation Co. and Edison on 16 May. Paul Dyer to Insull, 7 Nov. 1883; Stuart to TAE, 28 Mar. 1884; Stuart to TAE, 7 Apr. 1884; all DF (*TAED* D8448H, D8342ZCW, D8447W); Doc. 2424 (headnote n. 19); TAE agreement with Ohio Edison Electric Installation Co., 16 May 1884, Miller (*TAED* HM840216).

2. Samuel Ward (1837–1926), an English-born telegrapher, was an insurance agent in Circleville and secretary of the Edison Electric Illu-

minating Co. there at the time of its incorporation. Ward also independently purchased an Edison isolated plant of fifty lights between April 1884 and May 1885. U.S. Census Bureau 1982 (1870), roll M593_1256, p. 268, image 53 (Circleville Ward 2, Pickaway, Ohio); ibid. 1982 (1900), roll T632_1313, p. 10B (Circleville Ward 1, Pickaway, Ohio); Ohio Dept. of Health Death Index, p. 1966, certificate 05279, accessed on Ancestry.com, 22 June 2009.

3. Edison's notes on this letter formed the basis for a formal reply to Ward from the Edison Construction Dept. on 3 April (DF [*TAED* D8416BBW]). In the intervening period, Edison's office apparently forwarded Ward's inquiry to Archibald Stuart with a cover letter (not found) requesting comment or clarification. Stuart replied that the Ohio Installation Co. had told the local Circleville illuminating concern "very positively that we did not want the street lighting, but there seems to be an idea in the mind of every one that the street lighting is a big thing, and it is hard work to get it out of their head." He asked Edison's help to dissuade the local company from selling light to the city at a lower rate than it could charge its private customers. He also took the opportunity to ask about reports that Edison had devised an arc lamp for street lighting to run on the same circuit as incandescent lamps for indoor use, which he thought "would be a great help to us" (Stuart to TAE, 28 Mar. 1884, DF [*TAED* D8448H]).

Although the Circleville company did adopt incandescent lamps for street lighting, they neither waited for the second year of operation to do so nor did they connect them as Edison had suggested. The expansion of the Circleville plant was guided by Harry Ward Leonard, who left the Edison Construction Dept. to become an engineer for the Ohio Edison Installation Co. in April 1884 because his "home and interests" were in Cincinnati. According to Leonard, ninety incandescent street lamps on six circuits were put into use on 7 August 1884, making Circleville "the first town in the world to be lighted solely and entirely by incandescent lamps" (Leonard to TAE, 14 Apr. 1884; TAE to Stuart, 26 May 1884; both DF [*TAED* D8447Y, D8416BRF]; Leonard "Letter to the Editor," *Electrical Review* 4 [16 Aug. 1884]: 9). Leonard saw municipal lighting as a source of significant and reliable income for local Edison companies. He also oversaw the expansion of the plant in Middletown, Ohio, in mid-1884, to supply street lighting in the face of direct competition from the Brush Co., gas interests, and what he believed was wasteful management (Leonard to TAE, 17 July 1884, DF [*TAED* D8447ZBC]).

During the early 1880s, a number of arc-lighting companies competed for city streetlight markets in the United States, often in the same places where Edison erected village plants. The Cleveland-based Brush company was the most aggressive of these, and had by 1884 installed outdoor arc lighting in Cleveland and on a much smaller scale in Cincinnati, as well as other towns and cities outside Ohio (Eisenman 1967, 69–75). Though publicly dismissive of arc lighting, Edison offered to supply arc lamps where local Edison companies insisted on them for street lighting (Tate to Andrew Dull, 28 Jan. 1884; Insull to William Lloyd Garrison, Jr., 14 Apr. 1884; both DF [*TAED* D8416AHS, D8416BGY]).

From John Milliken

Dear Sir:

I herewith mail you copies of two of my last pieces of music.[1] I am about to publish a new polka at Ditsons,[2] which I have named the Dot and Dash Polka, dedicating the same to you with your permission.[3]

Perhaps you will remember me as G.F.M's[4] brother John, when we were all at old 83 State.[5] I am an operator at R. Gardner Chase & Co Bankers & Brokers Equitable Building now.[6]

If I could have afforded the extra expense, and you favored the idea, and I had photos of the children (quite a number of ifs) should liked to have had a nice Lithograph Title Page with vignettes of the children thereon.

If you have occasion to order any of these pieces, please do so through me as I have a number on hand (those already published)—and shall have of the new Polka—and wish to sell them.

Should like to hear from you soon   "73"[7]   Yrs truly

J. H. Milliken[8]

⟨Thank him & send one doz with bill[9]   E⟩

ALS, NjWOE, DF (*TAED* D8403ZAT).

1. The editors have not found the copies sent to Edison. Milliken had several published compositions by this time.

2. Oliver Ditson (1811–1888) began a music publishing firm in 1835 and joined with Samuel Parker in 1836 to form Parker & Ditson. Ditson acquired sole ownership in 1842, when the partnership dissolved; in 1857, he formed Oliver Ditson & Co. Following the Civil War, the firm began a period of great expansion through the acquisition of other firms and by 1884 was among the country's largest music publishers, a rank it held well into the twentieth century. *GMO*, s.v. "Ditson, Oliver" (accessed 12 Jan. 2009).

3. Edison later paid Milliken $3.60 for copies of the music (Voucher no. 212 [1884]). The printed cover of the "Dot and Dash Polka" carried the dedication "To T. A. Edison Menlo Park" (Milliken 1884). Superimposed across the words "Dot and Dash" were alphabetic transliterations of American Morse Code symbols for the number "73," a common telegraphic abbreviation for "Compliments of sender" (Pope 1871, 16). Edison nicknamed his two oldest children, Marion (b. 1873) and Thomas, Jr. (b. 1876), "Dot" and "Dash," respectively, in honor of the Morse alphabet. The nicknames became well-known after appearing in a number of the popular biographical sketches published during and after Edison's *annus mirabilis* of 1878; see, for example, *TAEB* 4, App. 3; Shaw 1878, 490; Bishop 1878, 99; and "Thomas Alva Edison," *Teleg. J. and Elec. Rev.* 9 (5 Nov. 1881): 433.

4. George F. Milliken (1835?–1921), formerly manager of the Boston office of the American Telegraph Co., became manager of the Boston office of Western Union when the two companies merged in 1866. Mil-

*Cover sheet of Milliken's
"Dot and Dash Polka."*

liken invented a much-used repeater and other telegraphic apparatus. He hired Edison to work for Western Union in Boston in 1868. Doc. 44 n. 3; Dyer and Martin 1910, 98; U.S. Census Bureau 1970 (1880), roll T9_559, p. 289.4000, image 0080 (Boston, Suffolk, Mass.); Boston Landmarks [n.d.], 13 n. 4.

5. The main Western Union office in Boston was located at 83 State St. when Edison was an operator there in 1868. *TAEB* 1, chap. 2 introduction n. 4.

6. R. Gardner Chase & Co., a partnership of R. Gardner Chase and Charles E. Legg, was a banking and brokerage firm located in the Equitable Building at 148 Devonshire St. in Boston. The firm officially failed in 1890 due to speculation, though it was discovered that it had effectively been insolvent since 1885. "A Big Crash in Boston: The House of R. Gardner Chase & Co. Goes Under," *NYT,* 18 Sept. 1890, 1; "Small Assets for Heavy Debts: The Money Available to the Creditors of R. Gardner Chase & Co.," ibid., 1 Dec. 1890, 1.

7. See note 3.

8. A native of Maine, John H. Milliken (b. 1841) worked as a telegraph operator in the Boston area until at least 1900. U.S. Census Bureau 1970 (1880), roll T9_546, p. 233.4000, image 468 (Somerville, Middlesex, Mass.); ibid. 1982? (1900), roll T623_662, p. 1A (Malden City, Middlesex, Mass.).

9. Edison asked for a dozen copies of the piece, which Milliken promised to send. TAE to Milliken, 3 Apr. 1884; Milliken to TAE, 10 Apr. 1884; both DF (*TAED* D8416BBX, D8410E).

To Sherburne Eaton

[New York,] March 31st, 84.

Dear Sir:—

Referring to your letters to myself, and the several manufacturing concerns which I control, and which arrived during my absence in Florida,[1] I beg to inform you that I would[a] prefer to postpone giving the information you ask for until I have had an opportunity of discussing the matters in question with Mr. Villard Chairman of the Committee appointed by your directors to consider the subject of ~~our~~ re-organization ~~and~~ of manufacturing.[2] Yours very truly,

TL (carbon copy), NjWOE, DF (*TAED* D8416BAP). [a]Interlined above by hand.

1. Eaton had been seeking financial information on behalf of committees of the directors of the Edison Electric Light Co. On Saturday, 23 February, he wrote to Sigmund Bergmann and the other three manufacturing shops requesting business information in advance of a meeting of the "Light Co's Special Committee on Manufacturing and Reorganization." He asked for each firm's capitalization, an itemization of sales, profits, and losses, and the extent of sales outside of electric lighting. Although Eaton professed no "right" to this information, he welcomed a full reply as "an indication that you are disposed to render the Committee every assistance in getting at the exact status of the four manufacturing enterprises connected with our business." With additional questions on 25 February, he sought information concerning the status of patent ownership, royalties, assignments, and a retrospective statement about investment "mistakes" that might have put "more money in machinery or other plant than would be necessary, in order to develop the productive capacity of your present plant." Eaton urgently requested the answers to be delivered at his office "not later than Thursday morning 28th [February] 1884." The Electric Tube Co. prepared a response, as did the Edison Lamp Co. in April. Eaton to Samuel Insull, 17 and 22 Mar. 1884; Eaton to Bergmann, 23 and 25 Feb. 1884; Electric Tube Co. report, n.d. [Feb. 1884?]; Philip Dyer to Francis Upton, 10 Apr. 1884; all DF (*TAED* D8427ZAD, D8427ZAF, D8427R, D8427S, D8433J, D8429Y).

2. For further action on this matter, see Docs. 2658 and 2690. Henry Villard was reportedly in ill health and withdrawing from business when he took on the duties described in this document. C. A. Spofford to William Norris, 5 Jan. 1884, Villard.

–5– April–June 1884

Edison's return to business at the start of April, following the vacation with Mary in Florida, marked the beginning of a turbulent period in his personal and professional life. One significant change, the death of Mary's father, Nicholas Stilwell, on 9 April after what seems to have been a chronic illness, came unsought and unwelcomed. So, too, did the sicknesses, probably acute colds, that afflicted both him and his wife later in the month, forcing Edison to cancel a trip to witness the start of central station service in Newburgh, New York. Mary's general health, however, was improved enough—or perhaps Edison wished to economize—that the family gave up hotel living and returned to their rented Gramercy Park home at the first of May. Edison had recovered and was attending an electrical exhibition in Worcester, Massachusetts, at the moment, leaving Mary, still experiencing headaches and a sore throat, to manage the move with Samuel Insull.

Outside circumstances contributed to Edison's decision, in mid-April, to close up the Edison Construction Department. Through March, hc had charged the Edison Electric Light Company more than $11,000 for expenses, paid from his own pocket, related to canvassing and estimating for central stations. Much of that sum was still outstanding, and most of those prospective stations were not built.[1] The Edison Electric Company, which Edison had considered a tepid partner in the central station business all along, was itself feeling financial strain, exacerbated by having to take stock in the local illuminating companies to which it sold operating licenses, rather than getting all the cash it originally expected. The Company was evidently also affected by worsening general economic conditions and the recent financial failures of two of its principals, Henry Villard and Egisto Fabbri. For some months, it had been exerting increasing control over Edison's outlays on its behalf through the Construction Department. When Villard, in one of his last official acts before returning to his native Germany, again proposed that the Company take some interest in the money-making manufacturing shops, Edison again balked. Edison additionally declared that because the Company's agents were not successful in contracting for new central station plants, he was "obliged to immediately disband my organization as the expenses in connection with it are too large to allow of my continuing it unless I have work in hand."[2] His decision in mid-May to immediately lay off his engineering staff coincided almost exactly with an acute liquidity crisis that brought the nation's financial system to the "verge of a panic" and led to the failure of several New York banks and brokerages.[3] Sherburne Eaton arranged for the Edison Electric Company to take on the salaries of some men, at least temporarily, but Edison had, in the meantime, been forced to accept the resignation of Frank Sprague and let others go. These events led to a series of negotiations for the reorganization of the Edison lighting business in the United States. No formal agreements were signed until September, but an informal consensus seems to have been reached by mid-June for the Edison Company for Isolated Lighting to take over the central station business and for the Edison Electric Light Company to share in the profits of the manufacturing shops.[4] Among the questions to be settled was how to resolve standing complaints by local illuminating companies against Edison and the Construction Department for defective workmanship.

Even while making these arrangements for adapting to straitened circumstances, Edison was planning for the future. In June, he declined an invitation to attend the opening of a central station in Circleville, Ohio, because, as he noted, "I never had more to attend to than at the present time, and I am compelled to be at my office both day and night."[5] Among his tasks were the engagement of a New Jersey political operative to help secure passage of state legislation favorable to electric lighting companies in small cities and towns. Farther afield, in mid-May, he requested from the Edison Electric Light Company—and later received—authority over the lighting business in South America. Having recently assigned to agents his rights in Chile, Bolivia, and Argentina, he recruited agents for other countries in Central and South America.[6] He also continued to monitor efforts to reorganize and recapitalize the French Edison companies, though little progress was made. He suggested to the Edison & Swan United Company in London a plan to lease, in Britain, its large dynamos lying unused—and unpaid for—in New York. Citing both economies of scale and superior manufacturing skill at his Harrison, New Jersey, lamp factory, he declined the British company's request to set up its own factory. In the face of competition from rival lamp manufacturers in the United States, some of them quite capable, he secured the loyalties of several principals in his own factory by granting them shares in the incorporated business. In June, he installed Charles Batchelor, recently returned from Paris, as general manager of the Edison Machine Works.[7] (Batchelor had already resumed the experiments on short-core dynamo design that he had started in Paris the previous summer.) About the same time, Edison formed the Edison Shafting Company to fabricate shafting, pulleys, hangers, and related transmission equipment under contract with the Edison Machine Works at its Goerck Street shop.

Evidently looking forward to inventing outside his contract with the Edison Electric Light Company, Edison placed attorneys Richard Dyer and Henry Seely on retainer for his personal patent work on 29 May. Immediately after his return from Florida, he had noted that a dried film being tried as lamp filament material was "marvelously sensitive to moisture" and potentially useful in a scientific instrument.[8] Among his other projects were the application of dynamos to telegraph circuits and solicitations for that type of business; applying dynamos to telephone call bells; supervising John Ott's experiments on separating nonferrous metals; a process for simultaneously drawing and annealing wire; and supplying a custom dynamo for Henry Rowland's experimental determination of the ohm at the Johns Hopkins University. He also found time in June to work on a typewriter model.[9]

Edison faced a significant legal challenge in addition to the economic turmoil of these months. Lawyers for Lucy Seyfert, who in 1882 had won a judgment against Edison in a long-running dispute, moved to collect the judgment by levying on his Menlo Park property. Continuing to ignore his own attorneys' advice to settle the matter, Edison instead took steps to disclaim legal ownership of both the real estate and chattel there.

The family's Gramercy Park house figured prominently in a rare newspaper article about Mary Edison, titled "In the

Wizard's Home" (Doc. 2683). Published on 1 June, the article presented Mary's own version of her courtship and wedding, giving her the opportunity to rebut myths whose repetition had evidently rankled her for years.

1. Edison Construction Dept. statement, 25 Mar. 1884, DF (*TAED* D8441K2); cf. App. 2.B.
2. Doc. 2655.
3. "On the Verge of a Panic," *NYT*, 15 May 1884, 1.
4. Concerning proposals for the reorganization of the Edison lighting business in the United States, see Docs. 2658, 2661, 2677, 2685, 2690; Doc. 2725 outlines the terms finally agreed upon.
5. TAE to Archibald Stuart, 10 June 1884, DF (*TAED* D8416BTQ).
6. See Doc. 2688 and TAE to David Cone, 5 June 1884, Lbk. 18:75 (*TAED* LB018075).
7. Batchelor memorandum, June 1884, DF (*TAED* D8431ZAI).
8. Doc. 2645.
9. TAE marginalia on J. M. Robinson to TAE, 27 June 1884, DF (*TAED* D8403ZEG).

–2639–

Telegrams: To/From Mary Edison

April 3, 1884[a]
[New York]

Mrs. Edison
Do you want a trained nurse if so will send one out. Answer[1]

Edison

Menlo park nj

T. A. Edison
Will let you know tomorrow morning

Mame[2]

L (telegrams; first message is letterpress copy written by Alfred Tate), NjWOE, LM 18:460A (*TAED* LBCD5460A) and DF (*TAED* D8414E).
[a]Date from documents, form altered.

1. This message was sent to Menlo Park. The nurse would have been for Mary's father, Nicholas Stilwell (see Doc. 2646 and Israel 1998, 232, 499 n. 7). Also on 3 April, Dr. William O'Gorman, a prominent Newark physician, visited Menlo Park and billed $30 for a "consultation." Voucher no. 388 (1884); "Historical Data," *Journal of the Medical Society of New Jersey* 13 (Aug. 1916): 427.
2. Mary Edison. See Doc. 2618 n. 14.

Magnolia, Fla., 4/3. 1884.[a]

My dear sir:

Enclosed we send you sixteen and 75/100 dollars (16.75) amt charged for the lunch and wine furnished for you to take on your trip.

We are very sorry that you did not receive it, and regret that after so much trouble on your part to have a good lunch provided, that there should have been any mistake about it. We cannot ascertain where, or to whom to attach the blame, but will see that the next time you come here your lunch will go with you

Hoping that you had a pleasant trip home and that you will favor us with a visit to Maplewood[1]   We remain   Very truly Yrs.

Magnolia Hotel[2]   per McGilvray[3]

⟨Acknowledge money & say that [if?][b] I am afraid the recording angel has a special Entry against my account for improper language upon the day of my departure[c]   If I could get the 16.75 to St Peter to balance my account then all would be well. I thought I would leave orders to have the head waiter killed but relented   However next year blood will flow. E[dison]⟩[4]

L, NjWOE, DF (*TAED* D8403ZAY). Letterhead of Magnolia Hotel. [a]"*Magnolia, Fla.,*" and "*188*" preprinted. [b]Canceled. [c]"upon . . . departure" interlined above.

1. The Maplewood Hotel, near the town of Bethlehem in the White Mountains of northeastern New Hampshire, appeared on the Magnolia Hotel's letterhead. Both establishments were owned by Isaac Cruft, a wealthy Bostonian, and managed by O. D. Seavey. Constructed in 1876, the Maplewood was at this time among the region's largest and most prestigious inns, with accommodations for 500 guests in the hotel and adjoining cottages. Johnson 2006, 144; Child 1886, 159–60.

2. Regarding the Magnolia Hotel in Magnolia, Fla., see Doc. 2614.

3. Not identified.

4. The typed reply to the hotel, prepared by Samuel Insull, closely followed Edison's draft. Insull changed the phrase "blood will flow" to "I will be able to avenge myself." TAE to Magnolia Hotel, 11 Apr. 1884, DF (*TAED* D8416BFQ).

[New York,] April 5th. 84.

Dear Sir:—

Referring to your favor of the 26th inst, I beg to say that I am constantly working against the various piratical ~~appeals~~ bills[a] as to patents.[2]

I think it would be a good thing for the recent Inventors Organization to carry the matter into the newspapers.[3] Yours truly,

SI

TL (carbon copy), NjWOE, DF (*TAED* D8416BDD). Initialed for Edison by Samuel Insull. [a]Interlined above by hand.

1. Maine used a return address of Harlan, Allen County, in northeastern Indiana (see note 2); the editors have not further identified him.

2. This typed reply was based closely on the draft Edison wrote on Maine's 26 March letter. Maine urged Edison to act against proposals in Congress to alter the patent system. Such attempts had occurred regularly in recent years, partly at the behest of the Granger movement, though he attributed the present "villainous scheme" to "the Railway Association organized several years since to fight patents," and indeed the Western Railroad Association (a trade group) had been pushing vigorously for major revisions in the patent laws. Maine suggested that Edison, because of his stature and personal interest in patent protection, should reprint a recent *Scientific American* editorial "and mail these to all the newspapers in the United States with a personal request from yourself to publish the same" (Maine to TAE, 26 Mar. 1884, DF [*TAED* D8403ZAS]; Buck 1963 [1913], 118–19; Usselman 2002, 145–63). That editorial, running more than a full column, blamed the railroads for encouraging "the false idea that new inventions and new industries are a bane to the people instead of a blessing." From a number of pending bills, it singled out two in the Senate. One would have reduced the term of patents to five years; the other, like bills considered in 1880 and 1882, offered indemnity from infringement liability to the purchasers of patented articles. Both measures ultimately failed ("The Plot Against Patents," *Sci. Am.* 50 [22 Mar. 1884]: 176; "Record of the Session," *NYT,* 7 July 1884, 2). Edison had opposed similar legislation on at least two previous occasions (see Docs. 1684 and 2287).

3. Edison referred to the "Inventors Protective Association," organized in New York in October 1883 to advance patentees' legal rights. The *Electrical Review* welcomed its formation because "as is well known, powerful influences are now being exerted, especially in the manufacturing districts of the West, against the present patent laws, which furnish at least a slight protection to the inventor." "The Rights of Inventors," *NYT,* 21 Oct. 1883, 3; "Convening the Inventors," *Electrical Review* 3 (1 Nov. 1883): 8–9.

–2642–

To the Société d'Appareillage Électrique[1]

[New York,] April 5th. 84.

Dear Sirs:—

I have your favor of the 2nd ult,[2] with regard to the trouble you have had with the Milan Company,[3] and I very much regret that they should address you in the manner in which they did.

I do not doubt but that it is your intention to fully respect the rights of the Italian Company, and, I have written to Compagnie Continentale to this effect.[4]

I cannot understand why the German Co. should endeavour to do business in your territory, and I have asked the Compagnie Continentale for an explanation.[5]

Referring to your enquiry as to what articles you should furnish outside of Switzerland, and what you should not, I should very much prefer not to enter into this question, as I am afraid by so doing I would only complicate matters.[6]

The Italian Company naturally[a] only desire to make a profit on all articles in connection with the Edison system, and should you supply any of them in competition with them, they will naturally protest against your so doing. I think that more friendly relations would be fostered by your not endeavouring to enter into competition with them, even in connection with such articles as you claim are not patented. At all events, I would much prefer not to be the judge as to what you should, and what you should not export, as should I place myself in the position you ask me to, the only result would be a controversy between myself and the Italian Company. Yours truly,

TL (carbon copy), NjWOE, DF (*TAED* D8416BCX). [a]Interlined above.

1. Organized in Geneva between January and May 1883, the Société d'Appareillage Électrique was incorporated in May 1883 with a capitalization of F500,000. It became the exclusive agent for Edison's lighting system in Switzerland through a transfer of rights from Edison's 1880 agreement with Ernst Biedermann, Antoine Cherbuliez, and Gaspard Zurlinden; Edison received F125,000 for consenting to the transfer. The executives of the new company included Arthur Achard (an engineer) as president; Cherbuliez as secretary; and Zurlinden as legal counsel. Théodore Turrettini served as a consulting engineer but resigned to pursue outside obligations in November 1883. He and Achard were both experts in hydraulic power and members of Geneva's municipal government. See Docs. 1878, 1985 n. 2, 1962, and 2356 nn. 15–16; TAE agreement with Cherbuliez and Ernst Biedermann, 19 Jan. 1883, Miller (*TAED* HM800129); TAE to Société d'Appareillage Électrique, 6 Mar. 1883, Lbk. 15:414 (*TAED* LB015414); TAE to Société d'Appareillage Électrique, 9 Sept. 1883; Achard to TAE, 15 May 1883; Calvin Goddard to TAE, 24 May 1883; Achard to TAE, 1 May 1884 and 16 May 1885; all DF (*TAED* D8316AUC, D8337ZBL, D8337ZBO, D8436ZCL, D8535ZAC); Paquier 1998, 366, 370, 545–48.

2. Achard denied accusations from the Italian company that the Société was selling Edison products in Italy. He enclosed with his 2 February letter copies of correspondence with that company, which had threatened to seize equipment coming into Italy (Achard to TAE, 2 Feb. 1884; Comitato per le Applicazioni dell'Elettricita to Société d'Appareillage Électrique, 25 Jan. 1884; Achard to Comitato per le

Applicazioni dell'Elettricita, 28 Jan. 1884; all DF [*TAED* D8436P, D8436E, D8436I1]). Joshua Bailey commented that "The Swiss Company denies having sold these articles and as far as we are informed they came direct from Mr. Turettini [sic] or some one acting under him. What is unquestionable is that lamps and machines coming from New York to Switzerland have been sold into Italy" (Bailey to TAE, 30 Jan. 1884, DF [*TAED* D8436M]).

3. The Società Generale Italiana di Elettricità was organized at the end of 1883 (and incorporated in January 1884) as successor to the Comitato per le Applicazioni dell'Elettricita Sistema Edison, which built and operated the original Milan central station. Hausman, Hertner, and Wilkins 2008, 80 n. 29; Docs. 2235 n. 5, 2321 n. 5, 2343 n. 14.

4. Edison responded to the Compagnie Continentale's 30 January letter soon after his return from Florida. Pointing out that the Italian and Swiss dispute put him in "an awkward position" and unable to evaluate competing claims, he preferred that the quarrel should be "settled locally." He also blamed the conflict in part on the Compagnie Continentale's impolitic handling of the Société d'Appareillage Électrique's protests against activities in Switzerland by Edison's German agents (see note 5). TAE to Compagnie Continentale Edison, 5 Apr. 1884, DF (*TAED* D8416BCW).

5. Achard had expressed concern that the Siemens & Halske agent in Zurich would get supplies through Deutsche Edison Gesellschaft for resale in Switzerland. Edison had previously assured Achard that the contract between the Société Électrique Edison and Siemens & Halske (and the forthcoming German Edison company) would not grant rights beyond Germany. Edison further promised that he was negotiating with the Paris company to protect the Swiss rights, and that he had not authorized the Paris company to compete there (Achard to TAE, 18 Oct. 1883 and 2 Feb. 1884; TAE to Société d'Appareillage Électrique, 21 Nov. 1883; all DF [*TAED* D8337ZEZ, D8436P, D8316BIP]). Earlier encroachment on the Swiss market by agents of the French and English Edison companies is discussed in Doc 2356 n. 15 (see also Calvin Goddard to TAE, 24 May 1883 [*TAED* D8337ZBO]). The Swiss company, however, had hoped to expand internationally. Its attempt in 1883 to conduct business in Spain was resisted by the Compagnie Continentale, and, with no further success, the Swiss proposed a unification of Edison's European companies during the Vienna Electrical Exhibition of 1883 (Société d'Appareillage Électrique to Goddard, 22 May 1883; TAE to Société d'Appareillage Électrique, 10 Sept. 1883; both DF [*TAED* D8337ZBN, D8316AUF]; Paquier 1998, 548–49).

6. In a separate letter enclosed with Achard's of 2 February (see note 2), the Société inquired about authority to sell non-Edison products beyond Switzerland and asked Edison to clarify which products in the Bergmann & Co. catalogue would be available for them to re-export. Société d'Appareillage Électrique to TAE, 2 Feb. 1884, DF (*TAED* D8436Q).

Brooklyn, Apl 5—1884.

From Howard
Greenman

Dear Sir.

I am a young man who anticipates taking a college scientific course and of making the study of electricity a special one. Before doing so however I think it best to look upon the practical side of the affair and find out if such a course can give a man a knowledge of electricity such that he can take it up as a business and rely upon it, or whether these courses in Electrical Engineering are merely good as giving a man a theoretical accomplishment which he can never apply in practice for his livelihood. Therefore I take the liberty of addressing you to ask for your views upon the general subject of a young man taking up electricity with the view of making it a business and especially of pursuing the study of it in our colleges. By kindly answering the above at your convenience, you will greatly oblige one who comes to you as a practical electrician for advise, which can not be got elsewhere. Yours Truly,

Howard Greenman—[1]

⟨To tell the truth College Electricians are no good   the only way into to learn it where it is commercially carried out. There is always a demand for practical Electricians with business instincts⟩[2]

ALS, NjWOE, DF (*TAED* D8403ZBB).

1. Howard Greenman (1863–1939) was a native and resident of Brooklyn. He evidently did not become an electrical engineer; he instead had a career as an accountant, first in Brooklyn and later in Montclair, N.J. Obituary, *NYT*, 12 Mar. 1925, 19; U.S. Census Bureau 1970 (1880), roll T9_843, p. 442.1000, image 0103 (Brooklyn Ward 23, Kings, N.Y.); ibid. 1982? (1900), roll T623_1061, p. 6A (Brooklyn Ward 23, Kings, N.Y.); ibid. 1982 (1910), roll T624_883, p. 13B, image 835 (Montclair Ward 1, Essex, N.J.).

2. The editors have not found a formal reply to Greenman. Edison had not always held such a dim view of collegiate electrical engineering (cf. Docs. 2427 n. 3 and 2479 esp. n. 3). His ambivalence toward such programs also showed in his responses to other requests for his opinion on electrical engineering schools. In response to an inquiry in late December 1883, for example, he noted that Stevens Institute was "as good as any" to enter to become an electrician. By June 1884, however, he recommended that a 23-year-old considering attending Lehigh University should instead "get into a shop, electrical if possible, & work out your own salvation. Don't go to a College." TAE marginalia on G. W. Euker to TAE, 22 Dec. 1883; TAE marginalia on Horace Engle to TAE, 10 June 1884; both DF (*TAED* D8303ZIJ, D8403ZDW).

Samuel Insull to
John Verity

My dear Verity:—[1]

Your letter of 18th March came to hand a few days ago.[2] I held it, as I wished to speak to Mr. Edison about the matter, and he was away in Florida. I am going to get him to write a letter to Major Flood Page, urging that some arrangement should be made with you.[3] This letter will go off by either this or the next mail.

We do not use steam dynamos[4] for Isolated business. What you should use is a $14\frac{1}{2} \times 13$ engine, running at about 250 or 275 revolutions per minute, and belt direct on to two "H" dynamos without any countershafting at all. This would be better and more reliable. I will send you a plan showing how much space this engine and dynamos take up. If you want to run one machine in the daytime, put in a smaller engine with one machine attached.

We estimate that $7\frac{3}{4}$ lamps are attained per horse power in actual practice. I presume that the 6 lamps you speak of was simply taken as a rough calculation.

All our prices are quoted to you F.O.B. New York, so that all you have to figure on is Insurance and freight. Thirty shillings per ton will cover the freight, but I do not know exactly what the insurance will be. By next mail I will send you a revised price list, quoting prices delivered in London.[5]

Mr. Edison is very anxious indeed that the arrangement you propose should be made by the Edison & Swan Company, and I assure you that everything will done, on this side that is possible, to assist you.

I will get up plans for a 2000 and a 3000 light plant, and send them on to you early next week, and I will also state what I would ship the stuff from New York for.

In installing an Isolated Plant, you should be careful to only put in capacity for the maximum number of lights that will be burning at any one time.[6] That is; suppose the Grand Hotel[7] requires 3000 lights (I mean, wire and outlets for 3000 lights) they could not burn these 3000 lights all at the same time. Perhaps the average would be below 2000, in which case your plant should have a capacity of 1500 or 2000, otherwise you would have 1500 or 1000 lights more than would ever be required. To explain more fully what I mean, I may say that down town in New York, we have 10 000 lights wired for in our district, and the absolute capacity of our Plant is only 6000 lights. The maximum call upon us for lights is about 5500. Now, if every single light in our district was to be turned on at one time, the whole snap would be busted (to use American-ism), that is; part of New York would be turned into darkness, as we have not got a plant sufficient to carry 10 000 lights. But it would require an effort on the part of the Almighty to get the whole lot of lights burning at one time, consequently, we feel perfectly safe so long as we are able to take care of 5 or 6000 lights. I may mention incidentally, that we are now en-larging the capacity of our Plant, so that in about a month we will be able to take on 2400 more lights, which will bring the capacity of the Station up to about 8500 lights. As soon as this work is finished, we will be in a position to supply the demands of about 150 applicants whom we are unable to supply at the present moment. Then, a little later we are going to improve the machines in our Station, and add another 1200 lights to its capacity, simply by a small improvement which will not cost more than 5 or 600 Pounds for the whole outfit.[8]

If you find it impossible to make any arrangements with the London Company, that is, of a permanent character, why not try and make a dicker with them for each particular plant you want put in. If their terms should be so onerous as to pre-vent your making money on your plants, it is just possible that we might be able to make you a little better prices, as we do not want to make money out of our machinery, unless you can make something out of the installation of same. We are very anxious to see our machines operating in London, and put in the shape they would be, if Verity & Sons had charge of the work, with Halloway as their Engineer.[9]

I was up to Johnson's house the other night. Mrs. John-son has been sick. She has either had a swollen sore throat or mumps, or some other complaint to which human flesh falls victim.

We have been having most horrible weather in New York of late. In fact, I thnk I must have bottled up a lot of the London weather. To-day is "just to lovely for anything," as a gushing New Yorker would say.

Hoping that you will be able to do something in this Iso-lated business, and with best regards to all the people at Moor-croft,[10] Believe me, very sincerely yours,

TL (carbon copy), NjWOE, DF (*TAED* D8416BDG).

1. John B. Verity (1863–1905) was educated at University College School and had some experience with the production of incandescent electric lamps in the United States. He became a member of the Institu-tion of Electrical Engineers in 1889 (Obituary, *Journal of the Institution of Electrical Engineers* 35 [1905]: 584; Crisp 1904, 12:66). Verity was ac-

quainted with Insull and Edward Johnson through his family's business, B. Verity & Sons of King St., Covent Garden. This was one of two Verity firms (distinct from Verity Bros.) in the design, manufacture, and fitting of lighting fixtures. A well-established business with a prestigious clientele, B. Verity & Sons made the large chandelier for Edison's Crystal Palace exhibition in 1882 and a similar fixture for the Holborn Restaurant in 1883 (George Verity to Johnson, 1 Apr. 1882; Johnson to TAE, 8 May 1882; both DF [*TAED* D8239ZBG, D8239ZBF]; "The Holborn Restaurant," *The Era* [London], 3 Mar. 1883, 3; Doc. 2342 n. 9).

2. Verity's firm was moving on its own into the isolated lighting business. He had applied to Insull in February for detailed information about Armington & Sims engines, explaining that "We have got sick of the Edison [& Swan United] Co and are going into isolated plants right away. We have as good as got the Hotel Metropôle which will open next Spring and is the largest hotel in England." Insull obliged with information about engines, and also prices and capacities of dynamo models. Verity responded on 18 March with questions about contracting for Edison dynamos through the British company or, preferably, directly from New York. He pressed Edison to help him enter the market, noting that a machine made by rival Schuckert was "all the rage" while those of John Hopkinson built for the Edison & Swan United Co. were "not worth a cent," a belief shared by that firm's secretary, Samuel Flood Page. Verity to Insull, 9 Feb., 18 Mar., and 19 Apr. 1884; Insull to Verity, 1 Mar. 1884; Flood Page to TAE, 27 Mar. 1884; all DF (*TAED* D8437C, D8437L, D8437U, D8416ARW, D8437M).

3. Not found.

4. That is, a dynamo connected directly to the steam engine.

5. The editors have not found such a letter from Insull. In March, Verity had used Insull's earlier list of dynamo prices in the course of his negotiations with the Edison & Swan United Co., prompting Samuel Flood Page to protest the disclosure of this information to an outside party. Verity promised that all subsequent correspondence with Insull on this subject would be "strictly personal" (Flood Page to TAE, 27 Mar. 1884; TAE to Flood Page, 8 Apr. 1884; Verity to Insull, 19 Apr. 1884; all DF [*TAED* D8437M, D8416BDN, D8437U]). By mid-April, Verity reached a license agreement governing royalties to the Edison & Swan United Co. on machines bought directly from Edison in New York. Insull sent him blueprints of dynamo connections and dimensions in May (Insull to Verity, 1 and 2 May 1884, both DF [*TAED* D8416BMJ, D8416BMW]).

6. Edison had written on Verity's previous letter: "Warn him only put in Boiler & Engine capacity for Lamps that will actually burn & not all those installed   this he can get at in various ways Canvass etc." TAE marginalia on Verity to Insull, 18 Mar. 1884, DF (*TAED* D8437L).

7. Frederick Gordon, a former solicitor and restauranteur (connected with the Holborn Restaurant) turned hotel entrepreneur, completed the Grand Hotel on Northumberland Ave., facing Trafalgar Square, in 1881. The seven-storey, 500-room establishment also boasted popular restaurants. *London Ency.*, s.v. "Grand Hotel"; Thorne 1980, 239–40, esp. n. 42; Denby 1998, 143, 241; Verity to Insull, 18 Mar. 1884, DF (*TAED* D8437L).

8. According to the Edison Electric Illuminating Co., 11,272 lamps were wired to the Pearl St. plant at this time, with close to 2,000 more installed and awaiting connection. The company signed an agreement with Edison in February to increase its generating capacity, and it was in the midst of spending about $34,000 to install two short-core C dynamos (modified from the old design), engines, and regulators in the basement of 255 Pearl St. (Insull to Gustav Soldan, 7 Mar. 1884, DF [*TAED* D8416AUJ]; Sherburne Eaton report to Edison Electric Illuminating Co. stockholders [p. 8], 9 Dec. 1884, PPC [*TAED* CA004A]; TAE agreement with Edison Electric Illuminating Co. of New York, 29 Feb. 1884; Edison Electric Illuminating Co. cost estimate, 29 Feb. 1884; both Miller [*TAED* HM840211, HM840211A]; Edison Electric Light Co. Bulletin 22:2, 9 Apr. 1884, CR [*TAED* CB022]; Clarke 1904, 50). Edison insisted that any work that might interrupt service or "run the slightest chance of causing the present dynamo to stop" should be performed on Sundays. The additional dynamos entered service in May. Retrofitting the machines already in service was not part of the contract and evidently did not take place (Edison Construction Dept. to William Rich, 9 Apr. 1884, DF [*TAED* D8416BEK]; see also Docs. 2652 and 2669).

9. James Holloway (1839–1924), an English machinist whom Edison sent to London in late 1881 or early 1882, had been in charge of running the Holborn Viaduct plant. He ended his association with the Edison lighting interests about the beginning of 1884 to take over the Gordon Hotel Co.'s electrical engineering work. He worked in that capacity on the Holborn Restaurant and the Hotel Metropôle. Doc. 2258 n. 6; "Holloway, James," Pioneers Bio.; Verity to Insull, 18 Mar. 1884, DF (*TAED* D8437L).

10. Moorcroft was Verity's country home in Weybridge Heath, Surrey. A photograph and plan of its 50-light installation appear in Verity 1891 (71).

–2645–

Notebook Entry: Hygroscopy

[New York,] April 5 1884

Dried films tragacanth are marvelously sensitive to moisture[1] finger 3 inches away throws it into violent Contortions apply it to Hygoscope—[2]

TAE   M N Force

X, NjWOE, Lab., N-82-12-04:175 (*TAED* N145175B).

1. Edison made this observation four days after Martin Force tried several gelatinous mixtures (but not gum tragacanth) for lamp filaments. N-82-12-04:175, Lab. (*TAED* N145175A).

2. Edison had devised an extremely sensitive form of hygroscope (the odoroscope) on a different principle in 1881 and exhibited it at the Paris Electrical Exposition that year (see Doc. 2111 n. 22). The editors have found no evidence of further experiments for developing such an instrument.

April 7, 1884[a]
Menlo Park, NJ

Telegrams: To/From
Samuel Insull

Samuel Insull

Send trained man nurse who is not afraid of person out of mind[1] Send as soon as possible[2]

Edison

[New York]

Thomas A. Edison

Nurse goes out this morning. Dont fail be here tonight attend Committee meeting

Samuel Insull T[ate]

L (telegrams; second message is letterpress copy written by Alfred Tate), NjWOE, DF (*TAED* D8414F) and LM 18:474 (*TAED* LBCD5474). First message on form of Western Union Telegraph Co. [a]Date from documents, form altered.

1. Edison was referring to Mary Edison's father, Nicholas Stilwell (1822–1884). The son of Nicholas Stillwell and Jemima Aber Stillwell, Nicholas was born in New Jersey and lived most of his life in Newark. Regarding the variant spellings of the family name, see Wright 1939 (229). Both "Stilwell" and "Stillwell" appear for the family members in census records and archival correspondence, although the preferred spelling seems to have shifted to "Stilwell" with Mary's generation and afterward.

Census records indicate that Mary was the second child of Nicholas and his second wife, Margaret Crane (1831–1908). Nicholas and Margaret were also parents of Alice Stilwell Holzer (1853–1932), Charles F. Stilwell (1861–1939), Eugenie L. "Jennie" Stilwell (1868–1942), and Margaret Stilwell (b. 1871). Nicholas and his first wife, Ann Leake (Elizabeth Leek?), were parents of George G. Stilwill (b. 1842), Harriet "Hattie" Stilwell (b. 1844), and Caroline Stilwell (b. 1846). Nicholas earned his living as a sawyer (not as a lawyer, as stated erroneously in Doc. 218 n. 5), but the 1880 federal census noted that he suffered from "General Debility." He had also worked as a paid caretaker at Edison's Menlo Park laboratory for about three weeks in November and December 1883. U.S. Census Bureau 1970 (1880), roll T9_776, p. 68.3000, image 0499 (Newark, Essex, N.J.); Samuel Insull to Nicholas Stilwell, 23 and 28 Nov. 1883; Stilwell to Insull, 26 Nov. 1883; TAE to Stilwell, 28 Nov. 1883; Stilwell to TAE, 29 Nov. 1883; all DF (*TAED* D8316BJV, D8316BLI, D8367ZAN1, D8316BLL, D8303ZHJ); Voucher no. 224 (1883) for Nicholas Stilwell.

Nicholas died at Menlo Park on 9 April 1884. The death certificate attributed the cause to "Cirrhosis of Liver—Pyelitis—ac[companied by?] mania—(Exhaustion from)" (Stilwell death certificate, 9 Apr. 1884, Nj-Ar [*TAED* D8414ZAP1]). Regarding contemporary medical definitions of these conditions, see Quain 1883, s.vv. "Kidney, Inflammation of Pelvis of"; "Liver, Cirrhosis of"; and "Mania." The certificate was signed by Dr. John J. Daly, one of the Edison family's physicians and the Rahway, N.J., City Physician, who served as mayor of that city from 1885 until his death (see Docs. 1383, 1524, and 1525; "Mayor Daly of Rahway Dying," *NYT*, 7 Apr. 1896, 1; Ricord 1897, 140–41).

2. A docket notation of "R. Eagan 314 E 32" on the reverse of Edison's telegram corresponds to the Richard J. Egan identified in a city directory as a clerk living at 314 East 32nd St. in New York. On the day her father died, Mary Edison wrote to Insull on black-bordered stationery asking him to "please pay to Mr Egan thirty-five (35) dollars for Services rendered." The payment was made from a Construction Dept. account (and recorded under the name "Wm. Eagan"). Richard Egan (b. 1863) became a physician in New York City. Trow 1885, 487; Mary Edison to Insull, 9 Apr. 1884, with Voucher no. 194 (1884) for "Eagan"; Edison Cash Book (1 Jan. 1881–30 Mar. 1886), p. 245, 10 Apr. 1884, NjWOE; " U.S. Census Bureau 1970 (1880) roll T9_884, p. 475.2000, image 0668 (New York, New York, N.Y.); ibid. 1982? (1900), roll T623_1119, p. 3B (Manhattan, New York, N.Y.).

-2647-

From Israel Thornal

Metuchen Apr 10th [1884]

Dear Sir

Your Jersey calf which you sent here July 8th 1882 now had a young calf, one week old to day.[1] Please let me know your intentions with regard to it. at your earliest convenience   yours truly

Israel Thornal[2]

⟨Mrs. TAE   What shall be done   TAE⟩
⟨Ask for Bill for Pasture   Will send for them in a few days⟩[a]
⟨OK   R.⟩[3b]

ALS, NjWOE, DF (*TAED* D8403ZBK). [a]Written by Mary Edison. [b]Written by John Randoph.

1. The Edisons, who had owned an unknown number of cows (possibly in conjunction with Mary's half sister, Harriet Stilwell Van Cleve), considered selling them in late 1882 but decided not to do so. Edison's financial records from that time included a "farm account," but the editors have not identified any transactions relating to these animals generally or Israel Thornal in particular. Samuel Insull to John Randolph, 12 Oct. 1882; Harriet Stilwell Van Cleve to TAE, 17 Oct. 1882; both DF (*TAED* D8244U, D8204ZHU); Insull to Randolph, 18 Oct. 1882, Lbk. 14:302 (*TAED* LB014302); Ledger #5:499, 523, Accts. (AB003 [images 246, 258]).

2. Israel Thornal (variously Thornall; 1822–1908) farmed land near Menlo Park that his family had owned since the colonial period. He wrote to Edison from nearby Metuchen. Thornall 1982, 122–26.

3. John Randolph (1863–1908) began working for Edison as an office assistant in 1878. He was now keeping Edison's books for both personal and business matters, including the Edison Construction Dept. Doc. 2121 n. 3; "John F. Randolph, Treasurer of All the Edison Companies,

Puts End to Life by Shooting," *Newark Advertiser,* 17 Feb. 1908, in Unbound Clippings (*TAED* SC2100032a, *TAEM* 221:446).

<table>
<tr><td>

–2648–

To Nathaniel Keith

</td><td>

[New York,] April 12th. 84.

My dear Mr. Keith:—

    I am sorry to say that I shall have to again put off the ap-
pointment for to-morrow that I had with you. My Father-in-
law is dead, and I want to spend the time with my wife and
family.[1]

    I will advise you later, making another appointment. Yours
very truly,

TAE

</td></tr>
</table>

TL (carbon copy), NjWOE, DF (*TAED* D8416BGA). Initialed for
Edison by Alfred Tate.

    1. After a recent conversation with Edison, Keith had sent a draft
proposal for an "Electro-Chemical & Metallurgical Co." for the "Manufacture of chemicals, colors, dye-stuffs &c; the production of works
of art; all by means of Electricity (Furnished by Edison's Electric
Light Plants during off-hours)." He planned to discuss the matter
again on 6 April, but Edison postponed their appointment until the
following Sunday, 13 April. When Keith tried to confirm this engagement on 11 April, Edison wrote on his letter: "Write him say Father
in law dead etc   will make another appointment." Nicholas Stilwell
died on 9 April. Keith to TAE, 4 and 11 Apr. 1884, both DF (*TAED*
D8403ZAZ, D8403ZBL); TAE to Keith, 5 Apr. 1884, LM 18:472
(*TAED* LBCD5472); see Docs. 2639 and 2646.

<table>
<tr><td>

–2649–

To Sherburne Eaton

</td><td>

[New York,] April 14th. 84.

Major Eaton:—

    I am constantly getting enquiries from Europe with rela-
tion to new methods that we have adopted in connection with
the electrical part of our business; requests for drawings or
copies of drawings of new dynamos and other things which
we turn out. In fact, for all kind of information with relation
to the system.

    I am anxious to know what is my exact status with relation
to these enquiries. Am I by agreement compelled to give such
information at my own cost, or have I a right to charge for
same? All I wish to do is to prevent myself from being con-
tinually burdened with these matters, but of course, I am per-
fectly willing to give the information if it is paid for.

</td></tr>
</table>

Will you oblige me by looking into the matter, and advise me as to the position I should take in the matter.[1]

This memo is prompted by an enquiry I have from Berlin to-day, asking for a great deal of information, which it will take considerable time to get out.[2]

Thos. A. Edison.   per McG[owan]

TL (carbon copy), NjWOE, DF (*TAED* D8416BGE).

1. Eaton replied that Edison was required to furnish only drawings and models necessary for taking out patents; those sought for other purposes fell outside his contractual obligations. Edison noted on that letter: "Insull   You can charge for the labor etc. for the infortn requested." Eaton to TAE (with TAE marginalia), 16 Apr. 1884, DF (*TAED* D84280).

2. The Deutsche Edison Gesellschaft (DEG) requested information about Edison's new method of calculating central station conductors without making a physical model. They also inquired about "what kind of measurements for detecting mistakes are used in your central station, and by what instruments they are done." Edison replied that he thought it impractical to explain the necessary calculations in a letter, and that the better course would be to send a man to New York for instructions or pay for someone to go to Berlin. DEG to TAE, 31 Mar. 1884; TAE to DEG, 29 Apr. 1884; both DF (*TAED* D8436ZBW, D8416BLX); see also Doc. 2652.

–2650–

From Francis Upton

EAST NEWARK, N.J., April 14 1884[a]

Dear Sir:

My understanding with Messrs Howell, Bradley, Lawson and Dyer[1] is, as concerns the one share of the Edison Lamp Co.[2] agreed to be sold to each of[b] them:

Mr. Edison sells to the Edison lamp Co for the full fare of the assessments with interest added, four shares of the Lamp Co. He will hold them as trustee for the four men mentioned, until they have paid to the Lamp Co. the money [--][c] paid to Mr. Edison together with the interest. Then the shares will be transferred to the names of the individuals.

With this is enclosed a memo. of the amount due Mr. Edison. We credit him as agent.[3]

You will draw up a trust deed, stating that [-][c] Mr. Edison holds the shares in trust, and that he expects at least[d] five dollars a week to be paid by the individuals to the Lamp Co. In case of failure to pay this amount the shares may be bought back by the Lamp Co. on payment of of the amounts paid in to the date of failure together with interest. Yours Truly

Francis R. Upton   Treas.

ENCLOSURE[c]

East Newark, N.J.,[f] [c. April 1, 1884[4]]

Holzer says that he was promised $50 per week from Aug. 1880, that is from the starting of the Lamp Factory at Menlo Park. He remembers distinctly the conversation and thinks that he can recall it to Mr. Edison.

Messrs. Bradley, Howell and Lawson will each agree to put in $5 per week from their wages toward the assessments.

They understand that the amount of their share is to be held for them in trust. They are to be reimbursed for any money they may put in. as [--][c] at par value, the company to have to opportunity to purchase the stock. at par.

Questions raised:

What will be the value in the future when improvements are made

What term of years will be set before the stock or share belong to the individuals?

Who is to determine whether they are serving the company, and what is to prevent us discharging them at any time?

Who owns the real estate of the here at East Newark?

Mfg.	2777.46[5]	163,011.41[6]
[New?][g] Works	2393.65	.03
		4890.34.23
	2777.46	
Due on Assess.	708.30	
Total due Mr. E.	2069.16	7040
	4890.34	6959.50
	6959.450	80.50

$10,000	Batch	1000
3,600	Upton	1000
6,400	Johnson	500
	MacLaughlin	500
$11,000	Holzer	600
7040		3600
3960	$11,000	1100
		1100
		550
		550
		660
		3960

Partition at the Menlo Park office

ALS, NjWOE, DF (*TAED* D8429ZAB). Letterhead of Edison Lamp Co. Monetary expressions standardized for clarity; column of "11,000" and miscellaneous numbers not transcribed. [a]"East Newark, N.J.," and "188" preprinted. [b]"each of" interlined above. [c]Canceled. [d]"at least" interlined above. [e]Enclosure is an AD on Edison Lamp Co. letterhead. [f]"*East Newark, N.J.*," preprinted. [g]Illegible.

1. Edison had promised to give John Howell, James Bradley, and John Lawson each a 1 percent share in the lamp factory business at least as far back as September 1883 (see Doc. 2518). On 3 April, he finally executed trustee agreements to grant them each one share of Edison Lamp Co. stock on the terms outlined in this document (an agreement with Lawson has not been found). Philip Dyer inquired the next day about his longstanding request for the same privilege. Edison assented a few days later and executed a similar agreement on his behalf, apparently backdating it to 3 April to match the others. Dyer to TAE, 4 Apr. 1884; TAE to Edison Lamp Co., 8 Apr. 1884; TAE to Dyer, 8 Apr. 1884; all DF (*TAED* D8429W, D8416BDJ, D8416BDK); TAE and Edison Lamp Co. agreements with Bradley, Dyer, and Howell, all 3 Apr. 1884, all Miller (*TAED* HM840213, HM840214, HM840215).

2. The Edison Electric Lamp Co. finally incorporated on 5 January 1884 (New Jersey Secretary of State 1914, 200). The editors have not found completed incorporation papers, but a draft document variously dated October, November, and December 1883 identifies Edison as holding sixty-seven shares, Upton fifteen, Charles Batchelor ten, Edward Johnson five, and Holzer three (Edison Lamp Co. draft Incorporation and Association papers, 1883, DF [*TAED* D8332ZDW]).

The Edison Machine Works also incorporated in New York on or about 13 February (with Edison as president). Arrangements for that new legal entity to assume the business of the old partnership were not completed before March. Doc. 2343 (headnote, esp. n. 24); Samuel Insull to John Tomlinson, 4 Mar. 1884; TAE legal statement regarding Alfred Tate, 19 July 1884; Edison Machine Works minutes, n.d. [Feb. 1884]; all DF (*TAED* D8427Z, D8431ZAK, D8431D).

3. At Samuel Insull's request, the Edison Lamp Co. credited the sale of four 1 percent shares in the company to Edison's foreign order account. Each share was valued at $1,879.31 based on the company's capital, with interest. Insull to Edison Lamp Co., 14 Apr. 1884; Edison Lamp Co. to TAE, 17 Apr. 1884; both DF (*TAED* D8416BGG, D8429ZAD).

4. This date is conjectured based on the conditional terms presented below, which are substantially those of the 3 April trust agreements discussed in note 1.

5. Except as noted, the editors have not determined the basis for the values in any of these calculations. The figure of 2,777.46 is fractionally more than 10 percent of Edison's 70 percent ownership portion of the profit the company claimed for 1883. There is, however, no evidence that a dividend had been declared. Philip Dyer to Upton, 10 Apr. 1884, DF (*TAED* D8429Y).

6. This figure was the company's capital, as reported to Sherburne Eaton on 10 April. Three percent of this figure, as calculated here, would represent the total value that Edison planned to transfer to Brad-

ley, Howell, and Lawson. In a calculation below, the figure is added to the amounts due Edison. Philip Dyer to Upton, 10 Apr. 1884, DF (*TAED* D8429Y).

–2651–

To Henry Bentley[1]

[New York,] April 16th. 84.

Friend Bentley:—

Cannot you assist me towards getting the trade of supplying the principal Western Union offices with dynamo current? I could arrange to do the work very cheap, and I am sure my terms would be perfectly satisfactory to the Western Union officials, as I would make the payment depend [-][a] upon[b] the satisfactory performance of the work.[2]

If you can assist me in any way, I shall feel very much obliged to you.[3] With kind regards, I remain, Yours very truly,

TAE

TL (carbon copy), NjWOE, DF (*TAED* D8416BHW). Initialed for Edison, probably by Samuel Insull. [a]Canceled. [b]Interlined above.

1. Henry Bentley (1831–1895) was the founding president of the Philadelphia Local Telegraph Co., established about 1873. A minor inventor himself, Bentley had played an important role in testing Edison's carbon telephone for Western Union in 1878. Obituary, *NYT,* 10 Sept. 1895, 5; Scharf and Westcott 1884, 2132–34; Docs. 548 n. 4, 1194, 1200, 1204, 1223, 1239, 1241, 1247, 1256, and 1257 n. 1.

2. This letter to Bentley was one of several similar ones that Edison wrote to U.S. and Canadian telegraph officials at this time. He pointed out that several months previously, he had installed a small dynamo and gas engine in Bentley's office to run the stock printer, with satisfactory results. Edison had entertained similar ideas in 1882 (TAE to Baltimore & Ohio Telegraph Co., 16 Apr. 1884; TAE to Erastus Wiman, 17 Apr. 1884; TAE to George Ladd, 17 Apr. 1884; all DF [*TAED* D8416BHX, D8416BHZ, D8416BIB]). His correspondence with Bentley was not his first on the subject of applying dynamo current to telegraphic circuits (see Doc. 2326).

3. See Doc. 2676 regarding Edison's ideas for connecting dynamos in telegraph circuits, an idea he had considered in 1882 (see Doc. 2326). Bentley soon reported that several Western Union officials visited his office and were "much pleased" by the "splendid" stock printer arrangement. Edison was invited to submit a bid to Western Union, which he did. In June, Bentley circulated a report showing a savings of $4.43 per day with the dynamo and Otto gas engine. In a notation of thanks on Bentley's cover letter, Edison promised to "paint your name on the bottom of my machine so they will <u>work</u>." Bentley evidently accepted the proposition to power the Philadelphia Local Telegraph Co.'s office and local circuits, and Edison dispatched William Andrews to set up two dynamos there in May. Starting the machines proved troublesome, which Edison ascribed to depolarized field magnets. Bentley to TAE, 23 Apr.

(with TAE marginalia), 6 June (with TAE marginalia), and 9 May 1884; Bentley report, 6 June 1884; Andrews to TAE, 17 May 1884; all DF (*TAED* D8471E, D8471J, D8464E, D8471K, D8464F); TAE to Bentley, 14 and 28 May 1884, LM 19:165, 298 (*TAED* LBCD6165, LBCD6298).

–2652–

To the Società Generale Italiana di Elettricità

[New York,] April 17th. 84.

Dear Sirs:—

Your favors of the 5th of February, and 27th March came to hand while I was absent from New York on an extended trip in the South.[1] I will now take them up, and answer your various questions seriatum.

With relation to the "C" Dynamo,—I send you per express blue prints of the drawings for the modification of the present dynamos. I also enclose you herewith copy of a letter from the Superintendent of our Machine Works, explanatory of the drawings, and I also hand you a copy of the brief of the tests on the two dynamos which we have altered, and which we are now putting in our Central Station.[2] You will notice that we have used on the field magnets exactly the same wire that was originally on the machines. This has been done for the sake of economy in the first investment, and in order to get the volts we normally require, we have altered the field regulating resistance, exact data in relation to which we will send you shortly. If you require any further information in regard to this matter, please let me know.

As advised you elsewhere, the Crank-shaft has been re-shipped to Milan.[3]

I also enclose you a statement showing the general arrangement of our dynamos; the various prices of same, and we hope to send you within a few days detail drawings of all the machines.[4] With relation to your request for information regarding the 3 wire system, I beg to say that although it is possible to connect the 3 wire system with the 2 wire system, yet it is not practical on a small scale. If you are going to install two or three times the amount of conductors that you now have, I would strongly advise you to adopt the 3 wire system; and I think I could then arrange to alter your present two wire system over to the 3 wire system, at a small cost as compared with continuing your present two wire system over a very extended district.[5]

I do not think it would be possible for me to explain by means of correspondence the present methods adopted for calculating out conductors. I would be necessary to send a man to Milan in order to explain the matter. Mr. Batchelor will be here very shortly,[6] and I will consult with him in relation to the matter, as I have a similar request to the one you make from the German Company, and I am rather in a difficulty now as to how it will be possible for me to comply with your request.[7]

The various dynamos you speak of with extra armatures have all been shipped, and you have doubtless received same ere this.[8]

We are making shipments of lamps just as quickly as we can turn them out. As advised you by cable, we shipped 900 ten candle power lamps on the "Guttaedo"[9] on the 12th inst. We shall make another shipment by the first steamer sailing.

We have endeavoured to trace up the error you allege in connection with our shipment of 27th December. We most invariably find that when complaints are made of any extensive short shipment that the barrels have not been completely unpacked. Only a few days ago we received a barrel back at the lamp factory as empty, whereas there was a layer of lamps in the bottom of the barrel. This was in connection with a shipment to some of our friends in this country. Will you please examine the barrel referred to again and have a careful count of your lamps made, and if you still find that there was a short shipment of 200 eight candle power lamps, we will credit you with the amount of same.

Referring to your letter of the 27th March,—we have given you credit for the overcharge on the 750 lamps you speak of.[10]

With relation to your enquiry as to what we will charge for 25,000 lamps delivered to extend over the current year, we beg to quote you prices as follows:—

8 candle power lamps,	40 cents
16	[--][a] 40[b] "
10	50 "
32	60 "
50	$100 "
100	$150 "

If you give us an order for 25,000 we will be willing to forward them in ~~required~~ regular[c] shipments extending over a year at above prices, and all charges such[c] as freight and Insurance paid to Genoa. I trust those prices will be perfectly satisfactory to you.

With relation to your repeated enquiry about the Porter-

Allen shaft,—we have answered this question in the fore-going.

The credit for sockets supplied to Bergmann & Co was an error. We have set this right in our correspondence with the Compagnie Continentale, who will have doubtless advised you of the matter.[11] Yours very truly,

TAE

PS.[d] Statement Re Dynamos sent under separate cover

TL (carbon copy), NjWOE, DF (*TAED* D8416BIL). Initialed for Edison by Alfred Tate. [a]Canceled. [b]Corrected by hand. [c]Interlined above by hand. [d]Postscript handwritten by Alfred Tate.

1. Giuseppe Colombo had written (in his own name) on 5 February, largely in response to Doc. 2580 but referring also to discussions with Samuel Insull in Milan. Having received no reply to his questions, he wrote again on 27 March. Edison noted on the second letter: "Cant all the questions he required in his previous letter be answered   I think we should hereafter be prompt & correct with these people to keep our trade with them which in the event of great success in the future will be a great thing for us." Edison's marginal notes on Colombo's first letter were incorporated into the present reply. Colombo to TAE, 5 Feb. and 27 Mar. (with TAE marginalia) 1884, both DF (*TAED* D8436V, D8436ZBU).

2. Edison referred to the modification of "Jumbo" C dynamos to the short-core design, such as was done on the two machines being installed at New York's Pearl St. station (see Doc. 2644). Samuel Insull asked the Edison Machine Works for "detail information with relation to alteration of Central Station dynamos" to accompany drawings for the Italian company, but none of the enclosures has been found. In the latter half of 1883, the Milan firm acquired two old-style C machines being held at Antwerp by the Compagnie Continentale Edison. Insull to Gustav Soldan, 11 Apr. 1884; Colombo to TAE, 4 Oct. 1883; both DF (*TAED* D8416BFJ, D8337ZEO).

3. The Milan company reported in December 1883 that one of its Porter-Allen steam engines, which had never entered service (and may have been part of a long-running payment dispute with Edison; see Doc. 2399) had a defective crankpin. When the manufacturer, the Southwark Foundry in Philadelphia, confirmed the defect, Insull instructed them to make the repairs and "charge same to us &etc   ship Crank Shaft to Milan." The repairs were completed by 4 April, and Insull so advised Colombo a few days later. Comitato per le Applicazioni dell'Elettricita to TAE, 14 Dec. 1883; Southwark Foundry to TAE, 11 Mar. (with TAE marginalia) and 4 Apr. 1884; Insull to Southwark Foundry, 8 Apr. 1884; Insull to Colombo, 8 Apr. 1884; all DF (*TAED* D8337ZFV, D8436ZBO, D8436ZBY, D8416BDY, D8416BDS).

4. The editors have not found such drawings, but Insull may have sent them in early May when he provided similar information to John Verity in London. Insull to Verity, 2 May 1884, DF (*TAED* D8416BMW).

5. Colombo had discussed with Insull the Milan company's plan to expand its underground conductor network. In his 5 February letter

(see note 1), Colombo stated that Insull "thought it possible to connect a three-wire system of conductors with an existing two-wire network, and promised to send . . . a complete information and sketches of the three-wire system . . . and how I could join the two systems together." Edison commented on this plan: "We found that it wld pay to work the 3 wire in connection with the old." By July 1884, the plant was wired to about 3,500 lamps through roughly three miles of conductors; 2,500 of the lamps were for the house and stage of the Teatro alla Scala. With the addition of dynamos by the middle of 1885, the plant served 5,600 lamps. "The Edison Central Station in Milan," *Electrician* 13 (26 July 1884): 248–49; "Electric Lighting in Milan," ibid. 15 (29 May 1885): 45–46.

6. Batchelor soon communicated to Edison his plan to leave France on 3 May. Batchelor to TAE, 30 Apr. 1884, LM 2:59C (*TAED* LM002059C).

7. See Doc. 2649. Colombo had "hundreds" of requests to light small towns and villages, but he was still awaiting details of the three-wire system in July. Colombo to TAE, 11 July 1884, DF (*TAED* D8436ZDA).

8. The Italian firm had ordered one each of the G and R dynamo; when these were contracted for sale before they reached Milan, Colombo asked for one more apiece. He also ordered spare armatures. Comitato per le Applicazioni dell'Elettricita to TAE, 14 Dec. 1883; Colombo to TAE, 5 Feb. 1884; both DF (*TAED* D8337ZFV, D8436V); Colombo to TAE, 16 Jan. 1884, LM 2:41D (*TAED* LM002041D).

9. This erroneous spelling of *Gottardo* may have arisen from a misreading of Colombo's reference to the ship in his 27 March letter (see note 1). The *Gottardo,* a large iron steamship of the Navigazione Generale Italiana, entered service about the beginning of 1884. Edison cabled about this lamp shipment on 17 April. "A New Italian Steam-Ship," *NYT,* 26 Jan. 1884, 8; TAE to Colombo, LM 2:57C (*TAED* LM002057C).

10. In that letter (see note 1), Colombo discussed the status of a prior order for 3,000 10-candlepower (designated "C") lamps rated for approximately 102 volts. He had received only 1,000, not including 750 8-candlepower ("B") bulbs mistakenly sent in February and charged at the higher price of the more intense lamps. Colombo cabled Edison on 17 April: "Urgent ship ten candle lamps." LM 2:57B (*TAED* LM002057B).

11. Colombo had objected to a shipment of 500 "roughly made" key sockets from Bergmann & Co. He offered to keep them at a 50 percent price reduction; when Edison refused, Colombo returned the entire lot on the authority of Joshua Bailey. Colombo was disappointed to learn that Bergmann & Co. would not refund the full price, which he suspected had been erroneous from the first. The editors have not found contemporaneous correspondence on this subject with the Compagnie Continentale Edison. Comitato per le Applicazioni dell'Elettricita to TAE, 4 Oct. and 14 Dec. 1883; Bergmann & Co. to Insull, 27 Nov. 1883; TAE to Comitato per le Applicazioni dell'Elettricita, 30 Nov. 1883 and 4 Jan. 1884; Bailey to TAE, 25 Mar. 1884; all DF (*TAED* D8337ZEO, D8337ZFV, D8324ZAU1, D8316BLW, D8416ABJ, D8436ZBS).

Baltimore, Md   Apr. 21, 1884.

From Louis Duncan

Dear Sir.

Prof. Rowland[1] wishes me to ask you, if you could have wound for us a dynamo that we could use for charging a storage battery of 50 cells in series. We want it for our experiments on the Ohm[2] and our own machine is worn out. We have an engine of about 5 HP. Yrs &c.

L. Duncan.[3]

⟨Please give Volts required at terminals of machine—also amperes required while charging—[4] E⟩

ALS, NjWOE, DF (*TAED* D8403ZBT).

1. Henry Augustus Rowland (1848–1901), the first professor of physics at the Johns Hopkins University (Docs. 1910 n. 2 and 2391 n. 1), made independent tests on the efficiency of Edison's lamps in 1880. He also had allowed Edison to prepare a dynamo patent application in his name, an effort to circumvent the American interests of Siemens & Halske, but he otherwise declined engagement in Edison's affairs (see Docs. 1880, 1951, 2021, 2033). Rowland had sent a letter on 5 March, while Edison was in Florida, but it has not been found and may not have been answered (Rowland to TAE, 6 Mar. 1884, DF [*TAED* D8403ZAJ]).

2. Rowland, who had made determinations of the ohm in the 1870s, was supervising experiments made by Arthur Kimball with the assistance of Duncan and at least one other person. Their effort to reach an absolute value of the ohm, commissioned by the United States Congress with a grant of $12,500, was done both at Rowland's Johns Hopkins laboratory and at Clifton House, a country residence near Baltimore of the university's late benefactor, where urban vibrations would not disturb their instruments. Rowland referred to a battery of 50 Planté cells, charged by "A small Edison dynamo machine" in the Clifton House basement, to be used in calculating the resistance of standard coils (one of several experimental methods he used). Owing in part to difficulty in constructing the apparatus, Rowland and Kimball did not complete these trials until the end of 1884. *ANB*, s.v. "Rowland, Henry Augustus"; Kimball 1884; Kimball to TAE, 7 May 1884, DF (*TAED* D8403ZDI); Rowland, "Report on the Experiment of the Ohm" [1889?], pp. a, 5–12, Ser. 5, Box 41, HAR.

Standardization of the unit of electrical resistance had had a long and contentious history since the British Association first called for such an effort in 1862. The British sought an absolute determination based on the resistance of standard wire coils. Germans, led by Werner Siemens, preferred a practical value, suitable for engineering purposes, derived from the resistance of a tube of mercury. The French, with an eye to maintaining their position as the center of international metronomy, attempted to arbitrate simultaneous disputes over units of electrical resistance, current, and pressure. Rowland was a delegate to French-sponsored international conferences in 1881 and 1882, neither of which definitively resolved the matter of the ohm. The United States Congress, in order that the nation should not be left out of the debate, allocated $12,500 to the Johns Hopkins University for research under Rowland's direction. Rowland hoped to reconcile the various experimental methods by making new trials, and he planned to present his results to an April 1884 meeting of the French commission on electrical standards (an outgrowth of the 1881 Congress of Electricians in Paris). Unable to complete the work in time, he asked the conference to delay any decision until at least November 1884. The conference, notwithstanding, plunged ahead. Confronted with conflicting results, it adopted an average measurement, which satisfied no one. Rowland believed its value too low, which he subsequently showed to be the case, but he did not report his results to the British Association until 1887. His measurement substantially agreed with earlier findings by Lord Rayleigh and Eleuthère Mascart and, slightly rounded, was adopted in 1893 as the international standard by the International Chamber of Delegates for the Determination of Electrical Units, over which Rowland presided during the World's Columbian Exposition in Chicago. That valuation remained the international standard until after World War II. Kimball 1884; Kershaw 2007, 111–19, 128–29; on the international history of the ohm, see Carhart 1893, Schaffer 1991, and Schaffer 1994; an undated extract of Rowland's 1884 letter (probably spring 1884) to the International Conference for the Determination of Electrical Units is published in Rowland 1902, 217–18.

3. Electrical engineer Louis Duncan (1861–1916), an 1880 graduate of the U.S. Naval Academy, was commissioned an ensign and posted to Johns Hopkins for graduate study in physics and electricity about 1882. He completed the Ph.D. under Rowland in 1885 with a thesis entitled "On the Determination of the Ohm by the Lorentz Method," which the editors have not found. After leaving the Navy in 1886, Duncan taught at Johns Hopkins for fourteen years. During that time he participated as an engineer or consultant in the electrification of rail lines in Baltimore, Washington, D.C., and New York City, and was associated with electric traction pioneer Frank Sprague in the firm Sprague, Duncan, and Hutchinson. He chaired the new electrical engineering department at MIT for two years (from 1902) before returning to private engineering practice in New York, where he organized the firm of Duncan, Young & Co. "Louis Duncan Dead; Noted as Engineer," *NYT*, 14 Feb. 1916, 13; Beach and Rines 1911, s.v. "Louis Duncan"; *Annual Report of the President of Johns Hopkins University* 1886, 68; Duncan application for admission to American Institute of Electrical Engineers, 7 June 1887, Duncan (accessed 15 Oct. 2009 through http://www.ieeeghn.org/wiki/index.php/Main_Page).

4. Edison's marginalia was the basis for a short typed reply to Duncan on 24 April (DF [*TAED* D8416BKJ]). Duncan, writing in Rowland's name, answered that he required "10 volts while running to give 10 ampères" through the battery. Edison promised that "We can wind you a machine that will give 200 Volts and stand a Current of 10 amperes. The volts can be lowered by a resistance box in field   will this answer." Duncan responded affirmatively, again speaking for Rowland, and, as an afterthought in his "great hurry for it," he asked whether Edison had any on hand. Almost immediately, and apparently in reply to a message from Rowland that has not been found, Edison quoted prices and specifications for standard 110-volt dynamos of 45 and 22 ampere capacity.

Rowland requested the smaller machine be sent as "soon as possible," and the Edison Machine Works shipped one with a resistance box on 8 May. Edison subsequently answered separate inquiries from Rowland about the armature speed for each engine, but he did not address the apparent discrepancy, and it is unclear whether Rowland in fact acquired both (Rowland to TAE, 28 Apr. 6, 9 and 10 May 1884; Arthur Kimball to TAE, 7 May 1884; TAE to Rowland, 9 May 1884; Samuel Insull to Kimball, 9 May 1884; all DF [*TAED* D8403ZCE, D8403ZCN, D8403ZCS, D8403ZCV, D8403ZDI, D8403ZCT, D8416BOB]; TAE to Rowland, 5, 10, and 12 May 1884, LM 19:102, 137A, 141B [*TAED* LBCD6102, LBCD6137A, LBCD6141B]).

-2654-

From Archibald Stuart

Cincinnati, April 22nd 1884.[a]

Dear Sir:—

I write this letter asking a little advise, and in order that you may understand the position I will have to make it somewhat lengthy.

When the Middletown plant was put in it was during my sickness in November. Mr. Shaw arranged a kind of temporary organization at Middletown which amounted to nothing, consequently we were compelled to install and run that plant ourselves.[1] We find it very difficult to enlist local influence, for the reason that the Brush Company have had a company there for the last three years. They light the town by means of a tower and three or four scattering lights.[2] As you are aware there is also a Gas Company there,[3] but the main trouble is the Brush Co. who have instead of making money lost three or four thousand dollars. The public seem to be aware of that fact, and placing all electric light in the same category they think we are bound to loose money also, the result is they will not touch electric light stock with a forty foot pole.

We have succeeded in getting in about one hundred and seventy five lights and there the thing seems to hitch. We have some very good friends there, well wishers, etc. but no workers. Some of these friends of ours are in the city council, and one in particular is Chairman of the Committee on light. The Brush Company three years ago commenced to light the town for three thousand dollars per year; last year they had four thousand dollars, this year they want five thousand dollars, consequently the people are getting down on them and they are very unpopular. They are now working very hard at the Storage battery which they have there, trying to get people interested in them, but no one seems to take hold. They succeeded in putting a few of their large lights in some of the large stores, which after a few months trial were thrown out, hence the only revenue they have is from the city lighting having no private consumers at all. Now it so happens that our mains in Middletown are in the main streets, not in the alleys. In conversation with Mr. Andrews—he thinks it would be a very easy matter for us to put up some additional posts, run light mains where we have no main and cover sufficient territory to light the town; and that it could be accomplished by running one more wire where we now have main and utilizing the middle wire of the present system, or as some call it, the compensating wire in connection therewith to make the circuit. This enables us to control all the street lights from the station.

Now the point is; first:— if we can succeed in lighting the town that winds that Brush Company up completely, and they will have to leave. Secondly:— It puts us on top of the Gas Company and makes us at once the leaders in light; establishes public confidence and enables us to get more private consumers, that will at once make the plant pay.

The question is:— Will it work upon the plant above proposed, and what do you think of it? We put a couple of lights up in brackets on the poles at the entrance of the alley going to our station. This seems to have enthused the council committee very much, and they are after us red hot to give them a bid, and offer us a very fair show.

Please answer this as soon as possible [a]s[b] the bid will have to go in very shortly, if we make it. Yours truly.

A Stuart   Secy. A[ltenberg][4]

⟨Idea looks OK   let map be made of where Extra main runs, no of & position of poles & also where Each Lamp is & How new wires connect with our Center wire & we will calculate Cost[c] & what you can afford to do City lighting[5]   E[dison]⟩

TL (carbon copy), NjWOE, DF (*TAED* D8449ZAF). Letterhead of Ohio Edison Electric Installation Co. [a]"Cincinnati," and "188" preprinted. [b]Faint copy. [c]Obscured overwritten text.

1. The Edison Illuminating Co. of Middletown, Ohio, received local authority to install wires on poles and underground in September 1883 and contracted with Edison in November to install the plant (Wilmer 1915, 87; TAE agreement with Edison Illuminating Co. of Middletown, 28 Nov. 1883, Miller [*TAED* HM830202]). By "ourselves," Stuart probably meant the Ohio Edison Electric Installation Co., which later claimed right to ownership of the plant; Edison, however, assigned the title to the Middletown company, with which he had made the contract (George Altenberg to TAE, 19 June 1884, DF [*TAED* D8449ZBF]; Samuel Insull to Stuart, 19 June 1884, LM 19:462 [*TAED* LBCD6462];

TAE agreement with Edison Illuminating Co. of Middletown, 14 July 1884, Miller [*TAED* HM840227]).

2. The outdoor lighting system adopted by the Brush Electric Light Co. (established by inventor Charles Brush) and its local affiliates consisted of clusters of powerful arc lamps mounted on a number of iron or wood towers, each generally more than 100 feet high. In a twelve-month period in 1884–1885, the company erected ninety such structures. One contemporary analyst concluded that "So long as the tower system is retained there can be no competition among electric light companies for public lighting, because of the great expense of building towers, and the power of the Brush company to put their price so low that no other company could think of building towers in opposition." Doc. 2148 n. 7; Moore 1887, 542.

3. The Middletown Gas, Light and Coke Co. was chartered in 1872 to provide gas for public and private lighting. Wilmer 1915, 85.

4. George Altenberg, a clerk or secretary, signed Stuart's name and his own initial on the letter. Stuart to TAE, 12 June 1884, DF (*TAED* D8448ZAL).

5. Samuel Insull prepared a typed reply based on Edison's marginalia. He then sent Stuart's letter to William Andrews, who directed Harry Leonard to determine where the new lines, poles, and lamps should go. Stuart replied that Middletown was "the hardest town we have struck," but that lighting the streets would "forever squelch the Brush Co. there, which will cause our plant to be on a profitable basis in a very short time." Edison noted on a subsequent inquiry from Stuart: "This o[ugh]t to be rushed or He will lose the City Contract for Lighting." The necessary blueprints went to Stuart about 10 May. At the end of 1885, after having installed street lights and expanded the service district, Stuart reported that the Brush company had left Middletown and that he hoped soon to force out the gas company. Insull to Stuart, 24 Apr. and 9 May 1884; Insull to Andrews, 24 Apr. 1884; Andrews to TAE, 27 Apr. 1884; Stuart to Insull (with TAE marginalia), 7 May 1884; Stuart to TAE, 9 Dec. 1885; all DF (*TAED* D8416BKF, D8416BOI, D8416BKA, D8442ZDS, D8449ZAP, D8523ZCA).

–2655–

To Sherburne Eaton

[New York,] 24th Apl [188]4

Dear Sir,

Owing to the fact that I am rapidly finishing up the outstanding work in connection with the Construction Department and that the Agents of the Light Company are not forming any new Companies and that in consequence I am not getting any new contracts I find myself in the position of being obliged to immediately disband my organization as the expenses in connection with it are too large to allow of my continuing it unless I have work in hand.[1]

I very much regret being compelled to take this course and in as much as it has been suggested that the Isolated Company should take in hand the work now being performed by the Construction Dept. I would suggest that the matter be considered and an immediate decision arrived at. Otherwise a staff of the best men in the Service and which has cost considerable time and money to organize will have to be disbanded, and immediately further local Companies are formed all the labor and expense of organization will have to be again incurred.

I shall be happy to supply you with any information you may desire as to the expenses in connection with the running of the Construction Department.

Urging upon you the necessity for prompt action upon this matter,[2] I remain   Very truly yours

Thomas A Edison

LS (letterpress copy), NjWOE, DF (*TAED* D8427ZAL). Written by Samuel Insull.

1. Edison considered the alternative of forming a company to carry on the work of the Construction Dept. on a broader scale, and he went so far as to have papers drawn up to incorporate Thomas A. Edison & Co., Contractors, at Harrison, N.J. The draft papers have blanks for the day and month in 1884, but it seems likely that Edison contemplated this idea in late winter or early spring, after the incorporation of the Edison Lamp Co. (and about the time the Edison Machine Works also incorporated). Edison may also have acted following a legal analysis of his personal obligations under the Edison Construction Dept.'s contracts with local illuminating companies. Though that analysis is undated and unsigned, attorney John Tomlinson would have been an obvious choice for the job. The proposed contracting company would have had an initial capital of $25,000, of which Edison planned to subscribe $24,600, with the remainder divided equally among Francis Upton, William Holzer, Charles Batchelor, and Samuel Insull. The firm would have a broad charter to purchase, own, use, and license patents for the "generation regulation or application of Electricity to light, heat, power telegraphy, [and] telephony" and related manufacturing; and, specifically, "To contract for, and erect install and operate Edison Electric Light Central Station Plants in the various cities towns and villages throughout the United States & other countries." The work of making "estimates determinations and plans" for central stations would be conducted at a New York office. Edison & Co. Contractors draft articles of incorporation, undated 1884; "Memorandum on Contract," n.d. [1884]; both DF (*TAED* D8439ZBO1, D8439ZBR).

2. See Doc. 2661 esp. n. 1.

My dear Mr. Edison:

Since you have recently asked me to take up certain problems relative to the transmission of power, my reluctance to do which you may have noticed, it becomes necessary for me to enter into some explanations and to define my future position.

Before entering your service, in which I hope I have satisfactorily preformed whatever duties you have assigned me, I became interested in this subject, and while I have not had as much opportunity for experimental work as I should desire, I have lost none of my interest in it, and have advanced far enough to wish to keep it entirely apart from whatever duties are owing you, and to make it my especial study, that is to retain intact my individual title & its development. I desire to go into a course of work to settle practically the theories I hold and to compare whatever results I may get with the best work of Duprez,[1] Ayrton,[2] Perry,[3] and Hopkinson—

Such work must be largely mathematical, and a system of distribution for power requires the best work of this character— I feel that I can, and to a great extent in my own mind have solved the question of this transmission— To take up subject in obedience to your request would be simply to make over my own work without due consideration, and a due regard for my future makes it impossible for me to do this.

You, surely, will understand me when I say that I desire to identify myself with the successful solution of this problem, and when I also say that I am actuated by the same spirit with which you attacked the electric light, with the result of making yourself world-famous

You believed that you could solve that problem, and you did it.

But with all the responsibilities which that solution brought you, and with the question of thermoelectricity still remaining, you have stated that you have not the time necessary to give to the transmission of power— Surely it is enough for anyone, and I will have to devote the best of whatever knowledge I possess to make it a success.

As your subordinate, I cannot work with the same freedom as if I take the future into my own hands— Personal reasons, and my relations with others make it necessary that I should look well to the future, and with the confidence I feel, and the example of your own perseverance, I am willing to take upon myself whatever responsibility attaches to my action—

I am aware that the future of the Edison system of lighting must be allied with the transmission of power, and I hope it may be my good fortune to be instrumental in such alliance—

In taking this position, I of course know that you may feel that the salary you are paying me does not warrant you in retaining my service as an electrical expert, but I am ready to relieve you of this burden by tendering you my resignation, if you so desire, to take effect on the 1st proximo.[4]

Lest you may think me influenced by other electric light interests, I will say that I have had absolutely no negotiations with any other company; that I have the same unbounded confidence in the soundness of the principles, and the successful development of the Edison system that I have so long had; and that I shall look forward to that development with the same hope and interest I now feel—

Should it be desirable that I continue any relation with your work, I can only consent to do so in a purely consulting capacity, with a perfect freedom to the time and title of my own inventions—

In regard to the dynamos now being added to the Pearl St. Station, the rewinding of the 50 light machine, and the reducing of the speed of the S, Y, and H machines, I will do what I can, and do it willingly.[5]

In closing let me express my sincerest personal esteem for yourself, and my personal regret that the highest duty to myself and those related to me make it necessary that I should take this step. Very truly yours—

Frank J. Sprague.

ALS, NjWOE, DF (*TAED* D8442ZDM). Letterhead of Thomas A. Edison, Central Station, Construction Dept. [a]"New York," and "188" preprinted.

1. Marcel Deprez (1843–1918) was a French engineer who, after 1875, promoted the use of electric power in industry. In the early 1880s, Deprez experimented extensively on the metering and long-distance transmission of electrical power. He made his first major demonstration at the 1882 exhibition in Munich, where he built an artificial waterfall powered by direct current generated some thirty-five miles away. *DSB*, s.v. "Deprez, Marcel"; Hughes 1983, 131, 335; Gooday 1995, 250–51.

2. William Edward Ayrton (1847–1908), physicist and engineer, had a long history of problem-solving innovation in telegraphy and electric lighting in England, India, and Japan. He was professor of applied physics at Finsbury Technical College in London; in 1885, he became professor of electrical engineering at the City and Guilds Central Technical Institution in South Kensington. Ayrton dealt with various problems of electric regulation and transfer, including the improvement of

Deprez's meters and the design of electric motor systems in concert with his long-time colleague John Perry. *Oxford DNB*, s.v. "Ayrton, William Edward"; Gooday 1995, 250–58.

3. John Perry (1850–1920), a former colleague of Ayrton's in Japan, was appointed professor of mechanical engineering at Finsbury in 1882. Perry worked both separately and together with Ayrton on various electrical problems, helping to devise an ammeter and a voltmeter. *Oxford DNB*, s.v. "Perry, John"; Gooday 1995, 251–56.

4. Dalzell 2010 (55–58) attributes the timing of Sprague's decision, in part, to the young man's recent work in Brockton, where he had "access to a nearby machine shop as well as a supply of electrical materials and components. Restless and ambitious, he surveyed his options and prepared to build." Edison replied immediately in Doc. 2657.

5. In addition to the work mentioned here and his ongoing efforts to improve regulation of the three-wire system, Sprague had recently submitted a report, at Edison's request, on rewinding various dynamo models as motors. Sprague to TAE, 16 Apr. 1884, DF (*TAED* D8442ZDG).

—2657—

To Frank Sprague

[New York,] Apl 24 [1884]

Sprague,[1]

As we are about to close out our construction dept I think the best way is for you to resign on the 1st for the reason that your position would be so curious as to be untenable[2]   Yours
Edison

ALS (letterpress copy), NjWOE, Lbk. 18:46 (*TAED* LB018046); a letterpress version of a copy, probably written by John H. Randolph, appears on the same page.

1. Edison wrote this letter in reply to Doc. 2656.

2. Sprague tendered his formal notice the next day, citing "the reasons given in my letter of yester–date" for resigning as of 1 May. Edison promptly accepted it. Sprague to TAE, 25 Apr. 1884; TAE to Sprague, 25 Apr. 1884; both DF (*TAED* D8442ZDQ, D8416BKL).

—2658—

To Henry Villard

[New York,] 25th Apl [188]4

Friend Villard,

I have concluded not to take hold of the Light Companys business upon the terms indicated by you at our last interview.[1] I have written to the Light Coy. that I would be pleased turn over the skilled men in my Construction Department either to the Light Co or Isolated Co as it is impossible for me to continue to stand the expense of my present organization in as much as the Light Co's Agents are not forming any new Companies and I am consequently not getting any more contracts.[2] If they decide not to take over this organization I shall disband it as quickly as circumstances will permit.

Regarding the question of manufacture my Associates are willing to give the Light Co various percentages in the gross sales and to submit to a limitation of profits in some form to be agreed on. The percentages would vary with each Factory and depend upon the amount of work done in each year. My associates & myself believe however that it would be a great mistake for the Light Co to derive any profit from the Factories. The Company should merely restrict the Shops as to the amount of their profits in as much as the strongest argument the Light Co can make in negotiating with Local Companies as to licenses is that they have no interest whatever in the profits derived from manufacture but that they (Light Co) look, for their profits, to the same source as do the local parties in interest with them viz Dividends on their stock holding in the local Illuminating Companies. We shall however be ready to state the various percentages we are willing to give whenever we are called upon to do so[3]   Very truly yours,

Thomas A. Edison

LS (letterpress copy), NjWOE, DF (*TAED* D8427ZAM). Written by Samuel Insull.

1. The editors have not determined the substance of Villard's proposal, but see Doc. 2638.

2. See Doc. 2655. One gauge of Edison's personal monetary stake in the village plant business is the $8,000 invoice Samuel Insull had recently prepared for a portion of the Edison Electric Light Co.'s liability for canvassing and estimating work done by the Construction Dept. Insull to Frank Hastings, 22 Apr. 1884, DF (*TAED* D8416BJJ); see also Docs. 2704 and 2736.

3. See Doc. 2725. At some unspecified time, probably in mid-1884, Edison itemized the capitalization of his lighting enterprises, including the manufacturing shops. He listed his personal investment as $733,700; combining his stake with that of his manufacturing partners, the total was $1,008,000. TAE memorandum, n.d., DF (*TAED* D8526ZAT).

–2659–

To Archibald Stuart

[New York,] 4/26 [188]4

A. Stuart

Caught severe cold on my return from Florida. Feel too sick to make l so long a journey. Am [---][a] much obliged for the compliment but must ask to be excused under circumstances.[1]

Edison

L (telegram, copy), NjWOE, LM 19:61 (*TAED* LBCD6061). Written by John Randolph. [a]Canceled.

1. Stuart had telegraphed for Edison to attend the opening of the Piqua, Ohio, central station. He began the invitation: "The whole population of Piqua say you must be there next Monday night." Edison also replied in a letter prepared by Samuel Insull (Stuart to TAE, 25 Apr. 1884; TAE to Stuart, 26 Apr. 1884; both DF [*TAED* D8450ZAL, D8416BKX]). Mary Edison also became ill after returning from Florida (see Doc. 2663).

–2660–

Samuel Insull to
William Andrews

[New York,] April 26th. 84.

My dear Andrews:—

Immediately the Piqua Plant starts, we are entitled to a check for $11,500. This check will be payable ~~to~~by[a] Mr. Stuart, and should be sent to us in such shape as to of immediate use to us, by means of a draft on New York. I am in a very bad hole. I suppose you will say this is nothing new, but I want you to again try and help me out, and do your very level best to get Mr. Stuart to send us a draft so that it will reach here not later than Thursday morning next. You have been so successful heretofore in getting money for me that I feel confident you will not fail in this case. Please wire me immediately you have any information as to what the prospects are of my getting money.[1]

With relation to the rumours you have heard about reorganization, and the letter I sent you asking which of your men I should discharge,[2] I beg to say, that we are making arrangements by which all construction work will in all probability be done by the Isolated Company. For the purpose of these negotiations and to produce an effect on those we are negotiating with we find ourselves compelled to relieve ourselves of the greater part of our present staff, although, I presume before their connection with us ceases we shall have made a deal by which I will hand them over to someone else.

Of course the above information is for yourself alone. The outcome will be all right. I will explain more fully to you when I see you in New York.

We were all up to Newburg the other night, and Mr. Edison is simply delighted with the work there.[3] You and your staff deserved the highest possible compliment for the progress you have made since putting in the first station at Sunbury, and I do not mind telling you that we feel quite proud of our Electrical staff.

Now do your best to get money for me, and you will make me happy. Yours very sincerely,

SI

TLS (carbon copy), NjWOE, DF (*TAED* D8416BKY). [a]Handwritten.

1. In a letter prepared by Insull on this date, Edison also appealed directly to Archibald Stuart for prompt payment of the costs of the Piqua, Ohio, station. After starting the plant on 28 April, Andrews went to see Stuart in Cincinnati, some eighty miles south. Stuart did not pay immediately (evidently laying some unspecified blame on Phillips Shaw), but he did promise Andrews to mail about half the balance later that week, before he left for New York. The Ohio Edison Installation Co. sent a check for $5,000 on 1 May; in June, it sent a note for $5,750 with instructions that Edison should apply to the Piqua illuminating company for the remainder. TAE to Stuart, 26 Apr. 1884; Andrews to Insull, 29 and 30 Apr. 1884; Ohio Edison Electric Installation Co. to TAE, 1 May 1884; Stuart to Insull, 13 June 1884; all DF (*TAED* D8416BKX, D8450ZAY, D8450ZAZ, D8447ZAI, D8450ZBD, D8450ZCO).

2. Insull had asked Andrews to identify men whose employment could be terminated because "certain changes are taking place in connection with our business, which render it necessary for us to reduce our staff to a minimum." Insull to Andrews, 23 Apr. 1884, DF (*TAED* D8416BJV); see Doc. 2672.

3. Andrews supervised the start of the Newburgh, N.Y., plant on 31 March. Edison had planned to visit the station on 21 April with Insull, Sherburn Eaton, Francis Upton, and probably Edward Johnson. He was unwell, however, and although the trip was postponed three days, he did not accompany the others. Andrews to Edison Construction Dept., 1 Apr. 1884; Insull to Thomas Conant, 17 Apr. 1884; Eaton to TAE, 21 Apr. 1884; TAE to T. Cornell, 24 Apr. 1884; all DF (*TAED* D8446V, D8416BIA, D8446ZAL, D8416BKI); see Jenkins 1984 on the Newburgh plant's inauguration.

–2661–

*And Edward Johnson,
Samuel Insull, and
Francis Upton Draft
to Henry Villard*

[New York, c. April 28, 1884[1]]

Sir:—[2]

After many discussions upon the subject of the relations of the factories to the Light Co, we the undersigned[3] have reached the following conclusions, and present them with the full expectation that they will be received as honest opinions.

We conceive the prime motive of the Light Co in appointing your Committee to be, <u>first</u>; to provide for the more rapid exploitation of the Electric Light system.

Second; to re-arrange and adjust the relations of the factories to the Light Co that the stock of the latter may participate in the profits of the former where they are the direct outcome of the Company's patents.

Third; To so consolidate the working forces and profits of manufacture as to reduce the cost of installations to such a point as will place the Company in a more favorable position in respect to competition than it now occupies.

In our opinion all these things can be best accomplished by providing for the most economical production and installation of the Company's Plants. We believe that the true principle is to attain the highest possible development and the greatest economy of production, thus rendering the earning capacity of the plants in competition with gas and other forms of artificial light, more sure and more efficient; thereby popularizing the system and creating a demand for licenses.

We believe that any participation of the Light Co in the direct profits of manufacture or installation will defeat their purpose. That their true interests is[a] in the direction of securing in every possible way greater earning capacity from the sale of the product of the plants of which they are by virtue of their contracts part owners,[4] and that through their part ownership they thus derive in a legitimate way an indirect but real participation in the profits of the factories. We are convinced that the personal supervision and self interests of the present owners of the several factories is far more likely to secure economy of production than would result from corporative control and management.

It seems to us that the attention of your committee should be directed to formulating some plan of control and supervision of the cost of manufacture so as to afford your Company a constant guarantee that its goods are produced at the lowest pos possible figure, consistent with a fair manufacturer's profit; and also to the question of the more economical operation of the various installation Departments. This latter is in our opinion of very great importance. The cost of an installation may be may be very greatly enhanced by lack of skill and by bad management. Our suggestion in this respect is that all installations of whatsoever character be made by one Department, thereby concentrating the general charges of installation, and by their diffusion over all the operations of the Company obtain a general considerable reduction in the general charges on each particular job. We suggest for this purpose that the Isolated Company become contractors for installing Central Station Plants, and that they do all the work in relation thereto not now performed by them. It is obvious that the commercial and practical Departments of the Isolated Company in such a case should be presided over by a man of great executive capacity. We therefore suggest that such a man be sought and placed in charge of the affairs of that company.

TDf (carbon copy), NjWOE, DF (*TAED* D8427ZCJ). [a]Interlined above by hand.

1. This undated document was written sometime after Doc. 2658. It is likely that Edison and his associates drafted it in preparation for a meeting of the Edison Electric Light Co.'s Committee on Manufacturing and Reorganization, chaired by Villard, scheduled for 29 April. Sherburne Eaton expected the committee to ask for his opinions on what to do about the Construction Dept. business. He accordingly invited Edison, Samuel Insull, Edward Johnson, and Francis Upton to his office on 28 April to "discuss what should be done with the present business of the Construction Department, on the hypothesis that Mr. Edison may disband his present organization." Expecting the committee to ask his opinions on the matter, Eaton thought it would be "of great value to me to have the benefit of a conference with the gentlemen above named." Villard subsequently prepared his own report on the committee's activities for a meeting of the company's directors on 8 May. He sent a copy to Edison, but it has not been found. Villard largely withdrew from the company's affairs at about that time. Eaton to TAE, 26 Apr. 1884; Villard to TAE, 7 May 1884; both DF (*TAED* D8427ZAN, D8427ZAV); see also Doc. 2725.

2. This letter was addressed to Villard as chairman of the Committee on Manufacturing and Reorganization.

3. See note 1 concerning the likely principal authors; also cf. Doc. 2395. At some later date, an archivist erroneously wrote the names of four putative signers at the end of the document. All were later officials of the Edison Co. for Isolated Lighting, but among them only Edward Johnson participated at this time in Edison's negotiations over the manufacturing shops and central station business.

4. This clause refers to the Edison Electric Light Co.'s practice of taking stock in local illuminating companies in payment of license fees.

–2662–

*John Tomlinson to
Garrett Vroom*[1]

[New York,] April 28th. 84.

Dear Sir:—

Your favor of April 23rd was duly received.[2]

The letter you speak of as written to Mr. Reiff[3] was not received by Mr. Edison until the other day, which accounts for his ignorance that an execution would be issued.

Mr. Edison feels that the judgment is one he should not be compelled to pay as the note on which it was obtained was given to Mr. Seyfert[4] for his accommodation, and was practically liquidated in their negotiations concerning the Automatic Telegraph.

Would the fact Mrs. Seyfert[5] knew that the note was an accommodation note; and that it had been practically settled by the negotiations referred to,—especially, if it was delivered to her after maturity,—in your opinion have any effect upon the disposition of the case in case a new trial could be obtained? We think these facts could be shown, and if there is any opportunity of obtaining a new trial by appeal or otherwise, Mr. Edison would be inclined to make every effort to do so.[6] If a new trial cannot be obtained, or if in your judgment these facts would be of no [-------][a] material[b] assistance to Mr. Edison upon a new trial, I presume nothing is left but to make the best settlement we can.

Mr. Edison has not property in New Jersey sufficient to satisfy the judgment. The property at Menlo Park belongs to the Edison Electric Light Co; and the furniture and personal property in Mr. Edison's house belongs to his wife. The title of the house is in him, but is mortgaged to nearly its full value; so that the plaintiff to satisfy her judgment will probably have to sue upon it here which would entail considerable delay.[7] If therefore, Mr. Edison was disposed to fight this claim as bitterly as possibly, he could delay the plaintiff considerably. Under these circumstances, would she not take less than the face of her judgment.

Mr. Edison has Mr. Seyfert's receipt for $300, which under any circumstances should be deducted from the judgement.

Will you please let me know the exact amount of the judgment, and whether you think any settlement for a less amount could be made; and if so, how much could you get them to throw off. Yours truly,

Mr J C Tomlinson

TL (carbon copy), NjWOE, DF (*TAED* D8416BLF). [a]Canceled. [b]Interlined above by hand.

1. Garrett Dorset Wall Vroom (1843–1914) was a prominent attorney and banker in Trenton. The son of a former New Jersey governor and grandson of a U.S. Senator, Vroom graduated from Rutgers College in 1862. He had served as county prosecutor and, until recently, as mayor of Trenton. At this time, he was also a long-time reporter of the state supreme court and past member of an important judicial commission. Vroom was appointed to the New Jersey Court of Errors and Appeals in 1901. Lee 1907, 785–86; Obituary, *NYT*, 5 Mar. 1914, 9.

2. Vroom warned Tomlinson on 23 April of an imminent effort to enforce a pending judgment against Edison (DF [*TAED* [D8403ZBY]). This tangled legal case, on which Vroom had worked on Edison's behalf since at least early 1881, concerned a $300 note that Edison gave in 1874 to George Harrington, an investor in the Automatic Telegraph Co. Harrington endorsed the note to William Seyfert, who later conveyed it to his wife, Lucy Seyfert. When Mrs. Seyfert tried to collect, Edison disowned the obligation, claiming the note had been given to her hus-

band as part of a personal loan. Lucy Seyfert won a $5,065.84 judgment at trial in December 1882, but Edison resolved to drag out the matter as long as possible. Vroom reminded Tomlinson in his 23 April letter that he had, in November 1883, secured a new trial on technical grounds but this, too, resulted in an unfavorable verdict (see note 6). Having long advised Edison to reach a settlement with Seyfert, he had supposed the matter was closed (see Docs. 516, 2014, and 2382).

3. Railroad financier and telegraph entrepreneur Josiah Custer Reiff (1838–1911) was a longtime Edison associate who had provided most of the funds for Edison's work on automatic telegraphy. It was through his connection with the complex finances of companies formed to exploit the resulting patents that he knew of the contested note held by Lucy Seyfert, and he had been involved in trying to negotiate a settlement of the suit in 1880 (see Docs. 141 n. 7, 452 n. 3, 676, 1713, and 2014 esp. n. 2). Vroom had recently informed Reiff, whom he characterized as having "had charge of this case from the commencement," of immediate plans by Seyfert's attorneys to obtain an execution of the judgment against Edison, the last legal requirement for the seizure and sale of property. In his letter to Tomlinson (see note 2), Vroom also claimed to have given Reiff a statement of the money owed. Reiff reportedly promised to "see that the amount was paid at once" upon Edison's return from Florida (Vroom to Reiff, 31 Mar. and 8 Apr. 1884, both DF [*TAED* D8403ZAU, D8403ZBG]).

4. William M. Seyfert (1822–1902) of Philadelphia headed (since 1848) a major iron and steel firm cofounded by his father and known at this time as the Reading Iron Co. Seyfert had participated earlier in construction of the Croton aqueduct in New York and promotion of the Texas Pacific Railway. Obituary, *American Manufacturer and Iron World* 70 (30 Jan. 1902): 137; U.S. Census Bureau 1982? (1900), roll T623_1470, p. 2A (Philadelphia Ward 29, Philadelphia, Pa.).

5. Lucy Fisher Hunter Seyfert (b. 1829?), wife of William Seyfert, declared her occupation as keeping house in the 1880 federal census but had reported substantial income for tax purposes in the 1860s. U.S. Census Bureau 1970 (1880), roll T9_1180, p. 386.3000, image 216 (Philadelphia, Philadelphia, Pa.); University of Pennsylvania *Alumni Register* 12 (Nov. 1907): 80; U.S. Internal Revenue Service 1977 (1866), roll M787_41, District 8 (Pa.), Monthly and Special Lists (July–Dec. 1866), p. 7.

6. Ruling in November 1883 on Edison's motion for a new trial, the New Jersey Supreme Court brushed aside his argument that the promissory note was uncollectable because William Seyfert had received it as an "accommodation" after its maturity date. Though accepting the plaintiff's view of the facts, the court nonetheless granted Edison a new trial on the grounds that the laws of New York, where the note was made and payable, and of Pennsylvania, where it was transferred to Lucy Seyfert, did not address such an interstate transaction between married parties, hence "the note in suit remain[ed] the property of the husband, the transfer to the wife being a nullity." When Josiah Reiff reported this event to Edison with Vroom's recommendation of seeking a compromise, Edison instructed him to "postpone the matter just as long as possible, in fact carry out the same policy that has been carried out right along, meantime see what kind of a settlement can be made."

The second trial, in February 1884, also resulted in a decision for the plaintiff and a similar judgment (with court costs) of $5,348.64. Opinion of Chief Justice Mercer Beasley, Nov. 1883; Reiff to TAE, 1 Dec. 1883; TAE to Reiff, 3 Dec. 1883; opinion of Justice Edward Scudder, 18 Nov. 1884; Vroom to TAE, 3 Dec. 1884; Strong & Son to Tomlinson, 8 Feb. 1886; all DF (*TAED* D8303ZHN, D8303ZHP, D8316BMN, D8403ZIN, D8303ZHR, D8603ZAA).

At some time during this long process, Edison made three pages of rough notes about the process of appealing to the New Jersey Court of Errors and Appeals, legal requirements for a sheriff's sale, and the idea that "Seyfert probably could be seen in Philadelphia & would settle for less than the amount of judgment— Reiff could [job?] Seyfert." TAE memorandum, n.d. [1884?], DF (*TAED* D8403ZKB).

7. The editors have not established how—or if—Edison transferred ownership of chattel at Menlo Park to the Edison Electric Light Co. John Tomlinson (or another agent for Edison) drafted several affidavits around 13 May stating that various property (carefully itemized) at Menlo Park belonged not to Edison but to his wife and several Edison companies (Mary Edison's affidavit is Doc. 2671). The Edison Electric Light Co. claimed to own the tools, equipment, and supplies in the laboratory and shops. Tomlinson to Vroom, 13 May 1884; Edison Electric Light Co. to Middlesex County Sheriff, n.d. [c. 13 May 1884]; both DF (*TAED* D8403ZCX, D8403ZCZ2); see also Doc. 2671 nn. 3–4.

–2663–

Mary Edison to
Samuel Insull

[New York,] April 30th 1884

My Dear Insull

Will you engage a wagon from Baumans[1] to be here at eleven oclock tomorrow morning. I shall also want a man or men[a] to move furniture from one room to another tomorrow afternoon at two oclock or friday mornings at ten.[2] I am so awfully sick I am afraid I cannot go tonight[3] just before you start for home send one of the Boys over and I will give a decided answer. Just now my head is nearly splitting and my throat is very sore.[4] I will have to have a little money you neglected to send my usual allowance on Saturday and I will need some tomorrow twenty five will do if you are short   Yours in haste

Mrs Edison.

ALS, NjWOE, DF (*TAED* D8414I). Letterhead of Mary Edison from 25 Gramercy Park. [a]"or men" interlined above.

1. A large store for elegant furniture, carpets, and home furnishings, Baumann Brothers was located on New York's Fourteenth St. near Union Square (Advertisement, *New York Herald*, 9 Sept. 1883, 22; Advertisement, *NYT*, 25 June 1884, 8; Spero and Gillon 2002, 34). Although the Edisons had purchased from Baumann's before and would in the future, their engagement this time was for "Moving & setting up of furniture at Gramercy Park." Leaving the Clarendon Hotel, the

family planned to occupy 25 Gramercy Park from 1 May until the end of the lease in September (Voucher no. 252 [1884] for Baumann Bros. for moving; [the vouchers series contains records of various purchases from Baumann's]; James Pryor to TAE, 29 Apr. 1884; Samuel Insull to Pryor, 1 May 1884; both DF [*TAED* D8403ZCF, D8416BMP]; see Docs. 2517 and 2528).

2. Mary indicated Friday, 7 May. Baumann Bros. submitted another bill for moving furniture, hanging curtains, and laying carpet, apparently done on at least two occasions in early or mid-May. Voucher no. 330 (1884).

3. The editors have not identified Mary Edison's plans.

4. Edison was also ill at this time; see Doc. 2659

-2664-

From James Upham

Worcester May 5 1884.[a]

Friend Edison—

Very sorry not to see you again— I called at "Bay State"[1] twice on Friday, but as I feared, some of the folks had carried you off— Could not conveniently get out to Friday evenings exhibition,[2] greatly to my regret, particularly when Mr. Smith[3] told me you got a "breathing spell" from receptions, and had a chance for a quiet chat— Was vexed enough to kick myself, for I wanted to ask you several things, besides indulging in some reminiscences— Smith says you thought you might be here again— If so, let me know please, and I shall present myself, promptly— Was much pleased with what I saw at hall, as part of it had been out of my reach before, except in print—

Where can I get hold of one of those "baby" lamps?— (1 candle power)—[4] Want one for my collection of odd bits—

With best wishes for continued health and prosperity, I am   Yours Faithfully,

James C. Upham[5]

⟨My Dear Upham   I expect if they put through the Worcester Co to come on when I shall have more time

There were but few baby Lamps made & they are all gone   we expect soon to make some more when I will send you one   E⟩[6]

ALS, NjWOE, DF (*TAED* D8403ZCL). Letterhead of New England Telephone and Telegraph Co. [a]"188" preprinted.

1. The Bay State House was a prominent four-floor hotel in downtown Worcester, Mass., constructed in 1856. It hosted conventions and other large gatherings. Morrill, Hultgren, and Salamonsson 2005, 38.

2. Edison participated in public demonstrations of his inventions before some 2,000 people at Mechanics' Hall in Worcester on Thursday,

1 May. Invited by the local natural history society and the mechanics' association, Edison reportedly came "on the sole condition that he should not be asked to speak from the platform." He presented the electric light and motor and related devices, the telephone, microphone, and other inventions as Edward Johnson explained them; Sherburne Eaton also spoke. Afterward, Edison met with some thirty invitees at the Bay State House. One newspaper account of the latter event noted that Edison "is evidently fond of explaining his inventions to appreciative listeners, though no conversationalist on general topics." The lecture and demonstrations were to be repeated the next evening for the city's students. Edison's party in Worcester also included Francis Upton, Spencer Borden, and Luther Stieringer. "Electrical and Scientific," *Electrical Review* 4 (10 May 1884): 4; transcriptions of "Mechanics Hall: A Brilliantly Illustrated Lecture on Electricity," *Worcester Gazette*, c. 28 Apr. 1884 and "Edison's Electrical Exhibit," *Worcester Evening Gazette*, 2 May 1884; both Cat. 116,993, Scraps. (*TAED* SB0210014, SB0210014B; *TAEM* 227:621, 624); transcriptions of other articles from local newspapers about these events are in Cat. 116,993, Scraps. (*TAED* SB021; *TAEM* 227:602).

3. Probably Richmond Smith (b. 1838?), cashier of the Western Union office in Boston when both Edison and Upham worked there in 1868–1869. U.S. Census Bureau 1965 (1870), roll M593_650, p. 28B, image 63 (Chelsea, Suffolk, Mass.); Reid 1886, 732; Smith to TAE, 24 Dec. 1878, DF (*TAED* D7802ZZNR).

4. The editors have not otherwise identified these "baby lamps," which may have been like the small lamps mentioned in Doc. 2502 or used to illuminate the Christmas tree designed by Edward Johnson in 1882. "A Christmas Tree Lighted by Electricity," *Electrical World* 1 (20 Jan. 1883): 38.

5. James C. Upham was a Western Union operator in Boston while Edison worked there in 1868 and 1869. Charles Sanford to TAE, 24 May 1913, EGF (*TAED* E1306; *TAEM* 254:194).

6. Edison sent a formal reply based on his marginalia on 12 May. LM 19:153 (*TAED* LBCD6153).

–2665–

To Horatio Beckman[1]

[New York,] May 6th. 1884.

Dear Sir:—

Mr. Insull informs me that he learned, when at Newburgh yesterday, that your Directors are very much inclined to fix the price of light at one cent per hour per ten candle lamp.[2]

I am strongly of the opinion, from my experience in other cities, that you can just as easily get one and a quarter cents, and I write to urge you to sell your light at the latter figure. It is very easy for you to reduce your price later on if you find that the people in Newburgh will not pay so much, but if you start at one cent you will find it very difficult indeed to raise your price should you come to the conclusion to sell at the higher figure.

At Brockton, Mass., when our station started there, the Directors of the local Company decided to charge at the rate of one cent, they now say that they very much regret that they did not commence charging at one and a quarter cents, as from the manner in which the light has been taken hold of in Brockton, they are convinced that if they had started at a higher figure they would have been able to get just as many lights taken, whereas now they are afraid to increase their price as it might possibly prejudice their business. At Brockton the light was taken hold of much slower than it is being adopted at Newburgh, and the dividend earning capacity of the plant seemed to me at the time to be much less than I think it is at Newburgh.

I write this letter to you not because I think it impossible to make money at one cent per hour, but because I think that your Stockholders should have the extra dividend which the extra quarter of a cent will bring them and which I feel assured your consumers will pay, unless customers at Newburgh are entirely [---- ent][a] different[b] to what they are in other cities.

You should not be over anxious about getting the whole capacity of your plant taken up right away. If you wait a little until your consumers have learned to appreciate what they have got, you will, we believe, find no difficulty in disposing of all the light you can sell practically on your own terms. You are selling a light which is vastly superior to any which is being [-----][a] supplied by your competitors the gas company, and in deciding what price you will charge we trust you will bear this in mind. In thus advising you we not only draw from our experience in New York City but also in every place we have got a station started. Yours truly,

TAE

TL (carbon copy), NjWOE, DF (*TAED* D8416BNE). Initialed for Edison by Alfred Tate. [a]Canceled. [b]Interlined above by hand.

1. Horatio B. Beckman (b. 1830?), a trained machinist and steam engineer, was superintendent of the Newburgh Steam Mills. An organizer of the Edison illuminating company in Newburgh, N.Y., Beckman was its vice president. U.S. Census Bureau 1982? (1900), roll T623_1141, p. 15B (Newburgh Ward 4, Orange, N.Y.); Nutt 1891, 264–65; TAE to Beckman, 12 Nov. 1883, LM 16:149 (*TAED* LBCD3149).

2. Edison sent a similar letter on this subject the same day to the treasurer of the Edison illuminating company in Piqua, Ohio, following a discussion of the plant with Archibald Stuart. See also Doc. 2666. TAE to Henry Flesh, 6 May 1884; Samuel Insull to Stuart, 10 June 1884;

Stuart to Insull (with TAE marginalia) 12 June 1884; all DF (*TAED* D8416BNH, D8416BTZ, D8450ZCM).

–2666–

To Elisha Hubbard[1]

[New York,] May 6th. 1884.

Dear Sir:—

Mr. Stuart, of Cincinnati, has been here for the past few days, and in the course of conversation he submitted to us a proposition which came from you to him, requesting our opinion on the same. It was in reference to the light you are furnishing some hotel in your City, in which, as we understand it, you propose putting about 110 lights for a stipulated sum per year, we think the amount was $375.

Mr. Stuart requested us to write you our views on the subject, and first of all we believe it is a mistake on your part altogether, in connection with the Edison light, to sell it to anybody in lump for so much per year, as it induces reckless burning of the light, it being not at all to their interest to use economy, which you will readily see runs up the expenses on lamps, but the main objection which occurs to us and which we are deeply interested in, is that you are inclined to let anybody have the light, who is willing to put it in, at most any price, and for fear of the gas company. We are of the opinion that it would have been a good plan to let the gas company have it at this price, for the hotel people would have repented of it much sooner than you.

Our experience in other towns warrant us in saying that you should not be too anxious to get your station all taken within the first three, or four, or five months. It is not expected that is to be done, so far from it we would prefer to hear that you were refusing to take lights at reduced prices, and even at your own price, provided it did not have enough burning hours, and by the time the Fall lighting commences you would wish you had selected your customers out of those who have a long average of burning.[2] From what we have heard of Tiffin there is nothing more certain than a fine success with your plant and a good dividend if you will be firm in your terms of price. Very truly yours,

T.A. Edison Constn. Dept.   By TAE

TL (carbon copy), NjWOE, DF (*TAED* D8416BNG). Initialed for Edison by Alfred Tate.

1. Elisha B. Hubbard (1844?–1916) was secretary of the Tiffin Edison Electrical Illuminating Co. A druggist by profession and the owner

of his own pharmacy, Hubbard was also a Democratic representative in the Ohio General Assembly by 1888. U.S. Census Bureau 1970 (1880), roll T9_1065, p. 174, image 0349 (Tiffin, Seneca, Ohio); ibid., 1982? (1900), roll T623_1320; p. 15B (Tiffin Ward 1, Seneca, Ohio); Hafner 1993, 1:755; Howe 1881, 1:172; letterhead, Hubbard to TAE, 16 Oct. 1883; letterhead, Hubbard to Edison Construction Dept., 19 Dec. 1883; both DF (*TAED* D8356K, D8356ZBE).

2. See also Doc. 2665.

–2667–

Edison Electric Light
Co. of Europe Report

[New York, c. May 6, 1884[1]]

TO THE STOCKHOLDERS OF THE EDISON ELECTRIC LIGHT COM-
PANY OF EUROPE, LIMITED:—

A report on the condition of the Company was submitted at a Special meeting of the Stockholders held March 7th. 1884, since which time nothing new has transpired.

At the meeting of March 7th. the stockholders confirmed the action of your Board in giving an option to a Paris syndicate to merge the existing three Paris Companies into one Company and raising their present united Capital, 3,500,000 francs, to 10,000,000 francs.[2] A full statement of the terms and conditions of this proposed merger was made to the stockholders at the above meeting. The option above referred to has not expired and does not expire until June 22nd. We have however had a cable from our Paris representative stating that there is a great probability of the matter being closed very shortly,[3] in which event we presume the capital for the new company will be at once raised and work on the large central station at Paris commenced.

We are informed that the light has made steady progress in Europe during the past year. A central station of 6000 lamps capacity was installed at Milan and has been in operation for nearly a year, lighting la Scala and other theatres, as well as a number of stores and residences. Capital has been raised for a large central station in Berlin, and we are informed that negotiations are pending for the installations of stations in other cities. Mr. Batchelor is now on the way from Paris to New York[4] with the plans and specifications for the Paris Station for the purpose of discussing the same with Mr. Edison, in order that the best possible results may be obtained when the station is installed.

The financial condition of the Company presents a very serious aspect. At the stockholders' meeting of March 7th. a full statement was made as to the Company's bonded indebtedness of $100,000., by which it appeared that the same becomes due November 1st. 1884, and that there was no money in the Treasury either to pay the principal or the half yearly interest due May 1st.[5] The stockholders authorized your Board to pledge a number of our Founders' shares, if necessary, to raise money for the May interest.[6] The Second Vice President thereupon wrote to Mr. J. F. Bailey at Paris regarding the matter, and requested him to obtain a loan[a] for us upon a deposit of a small number of our Founders' shares, to which Mr. Bailey wrote that he would certainly arrange to have the money for us in time to pay our interest. About the end of last month, not having heard from Mr. Bailey in regard to the money, we cabled him urging immediate attention.[7] We received a reply by cable stating that there was the greatest probability that the syndicate would be made up on Saturday May 3rd., and that it would be dangerous to press for the loan now. We cabled in reply saying that we must have the money at once,—but have received a cable from Mr. Bailey stating that the syndicate will be formed and that we shall certainly receive the money this month.[8] We shall therefore be compelled to ask the bondholders to postpone presenting their coupons, for the present until we receive money from Paris.[9]

Inasmuch as we cannot be certain of raising any money in Paris, it would seem that the Company is in a critical position on account of its inability to pay the interest on its bonds. The bondholders, or any of them, can bring suit against us at any time and might possibly apply for the appointment of a Receiver. This, we are assured by Mr. Bailey, would ruin the business of the Companies in Europe, and, consequently, take away from our assets what value they now have.

If the stockholders can furnish any solution of the difficult problem of raising money to meet the interest on the bonds, a task which the present Board of Directors have been unable to successfully perform, they will render a great service to the Company.[10]

In the judgment of your Board there is nothing else remaining to be said except what has been already set forth in the report submitted to the ~~stockholders~~ special[b] meeting, mentioned above.

TD (carbon copy), NjWOE, DF (*TAED* D8428W1). [a]"a loan" interlined above. [b]Interlined above.

1. This document was likely prepared after a 5 May meeting (adjourned from 3 May) of the European company's directors, and in advance of its annual stockholder meeting scheduled for 7 May. It is also dated by internal references to cable correspondence. William Mead-

owcroft to TAE, 3 May 1884; Frank McGowan and Samuel Allin legal statement, 7 May 1884; Meadowcroft to TAE, 24 June 1884; all DF (*TAED* D8428S, D8428ZAB, D8428ZAC).

2. The meeting was called to evaluate the company's situation and to consider acting on "a proposition from Paris for merging the existing three French Companies into one Company." The 7 March report of Sherburne Eaton, second vice president, is in CR (*TAED* CE001003). The option referred to was authorized in anticipation of fresh capital from Pierre-Eugène Secrétan (see Doc. 2593 n. 6). Charles Batchelor to TAE, 20 Feb. 1884; Samuel Insull to Alfred Cowles, 3 Mar. 1884; both DF (*TAED* D8436ZAR, D8416ASP); Batchelor to Francis Upton, 21 Feb. 1884, Unbound Documents (1884), Batchelor (*TAED* MB122).

3. Joshua Bailey cabled to this effect on 1 May. As of 18 June, Secrétan had not officially accepted the option but, according to Bailey, he had promised to do so by 20 June. Bailey to Eaton, 1 May 1884; Bailey to Batchelor, 18 June 1884; both DF (*TAED* D8436ZCM, D8436ZCV).

4. Charles Batchelor planned to leave Paris on 3 May and would have reached New York in eight or nine days. He was at the Edison Machine Works by 16 May. Batchelor to TAE, 30 Apr. 1884, LM 2:59C (*TAED* LM002059C); Cat. 1306, Batchelor (*TAED* MBN011AAO).

5. These were 6 percent debenture bonds with semi-annual interest payable on the first of May and November. Half of the total was held by Frederick Foote, L. G. James, Israel Corse, Dr. Giovanni Ceccarini, and Charles Farley (Francis Upton's brother-in-law). The company's entire capital stock was $2 million, but no profits or dividends had been realized. As of April 1884, the company's liabilities (including the bonds) totaled approximately $137,000 (Meadowcroft to New York Dept. of Taxes and Assessment, 28 Mar. 1884; Meadowcroft affidavit to City and County of New York, 30 Apr. 1884; Edison Electric Light Co. of Europe Ltd., Minutes 16 Dec. 1884; all DF [*TAED* D8428R, D8428Q, D8428ZAJ]). The company's stock, $100 per share at par, traded at $165 during the electric lighting boom but had fallen as low as $40 by April 1883. The asking price in March 1884 was $15, but it reportedly attracted bids of only $3 per share (Meadowcroft to New York Dept. of Taxes and Assessment, 28 Mar. 1884; Frederic Gostenhofer to TAE, 10 Mar. 1884; both DF [*TAED* D8428R, D8428I]; "Electric Light Companies of America," *Teleg. J. and Elec. Rev.* 12 [12 Apr. 1883]: 328).

6. On the instructions of Batchelor and Bailey, 2,000 founders shares of the European company were deposited with Drexel, Morgan & Co. on 3 March. Batchelor and Bailey to Eaton, 1 Mar. 1884, DF (*TAED* D8436ZBD); Eaton to Bailey, 3 Mar. 1884, LM 2:52D (*TAED* LM002052D).

7. The correspondence in March between Sherburne Eaton, the second vice president, and Bailey has not been found. Eaton cabled Bailey on 29 April to warn, as the due date for interest on the bonds drew near, "Company will default and receiver appointed unless you remit." Bailey replied promptly: "Hopeful tomorrow." Eaton report to stockholders, 7 Mar. 1884, CR (*TAED* CE001003 [image 12]); Eaton to Bailey 29 Apr. 1884; Bailey to Eaton, 30 Apr. 1884; both DF (*TAED* D8436ZCJ, D8436ZCK).

8. Bailey predicted that receivership would "explode whole affair." Eaton replied that it was "Impossible raise money here   Receiver threatened by bondholders." Bailey promised on 6 May that formation of a new company within the month was "certain." Bailey to Eaton, 1 and 6 May 1884; Eaton to Bailey, 2 May 1884; all DF (*TAED* D8436ZCM, D8436ZCQ, D8436ZCN).

9. Apprising Edison of these developments and Bailey's promise, Eaton suggested there was nothing to do for the moment "unless somebody sues us. I guess we will have to wait, and meantime put the bondholders off with one excuse or another." Edison appended "Yes" to this note. Eaton to TAE (with TAE marginalia), 6 May 1884, DF (*TAED* D8436ZCP).

10. Edison and other directors of the European company met at his laboratory on 28 November to consider the latest proposal put forward by Joshua Bailey; Batchelor, Samuel Insull, and Francis Upton also attended by invitation. The plan threatened to liquidate the Compagnie Continentale (now unprofitable in part because of expenses associated with its former effort to build a Paris central station) unless the directors approved its combination with the other French companies. Reluctant to surrender the European company's veto powers, the directors voted to reject the proposal if it did not guarantee lamp royalties of $80,000 per year. At Frederick Foote's suggestion, however, they agreed to call a committee of bondholders to study the matter (Edison Electric Light Co. of Europe, Ltd., directors' minutes, 28 Nov. 1884; Bailey draft proposal to Edison Electric Light Co. of Europe, Ltd., 30 Oct. 1884; Bailey to Edison Electric Light Co. of Europe, Ltd., 28 Nov. 1884; all DF [D8428ZAI, D8436ZDH, D8428ZAI1]). The bondholders counseled against the directors' rejection of the fusion. Instead, they unanimously recommended lamp royalties at approximately 2.75 cents per unit and pressed for approval of the French consolidation. Pending such conditions, the bondholders offered to extend their claims on the company until 1 November 1885. Edison reportedly consented to these terms, and the matter was to be taken up by the European company's directors at a special meeting in January 1885, with the expectation that it would be approved (minutes of bondholders committee, 16 Dec. 1884; report of committee of bondholders, 19 Dec. 1884; Eaton to TAE, 23 Dec. 1884; Meadowcroft to TAE, 31 Dec. 1884; all DF [*TAED* D8428ZAJ, D8428ZAM, D8428ZAL, D8428ZAO]; see also Wilkins 1970, 56, 263 n. 41).

–2668–

To Nathaniel Keith

[New York,] May 7th, 1884

Dear Sir:—

I am in receipt of your letter of the 6th inst, and in reply beg to say that I shall be pleased to accept the nomination of your Committee as one of the Vice Presidents of the American Institute of Electrical Engineers.[1] Yours truly,

TL (carbon copy), NjWOE, DF (*TAED* D8416BNZ).

1. Nathaniel Keith, who headed the organizing committee for the nascent American Institute of Electrical Engineers (AIEE), asked Edison on 6 May to stand for election as one of the Society's vice presidents. The stated impetus for organizing electrical practitioners was to receive the many "famous foreign electrical savants, engineers and manufacturers" attending the International Electrical Exhibition at Philadelphia in September, though a sense of rivalry with scientists planning the concurrent Conference of Electricians surely played a role. Edison had signed Keith's petition calling for such an organization of practical electricians, a printed copy of which was circulated to prospective members in April. Edison was elected a vice president (one of six) at an "enthusiastic meeting" on 13 May, and his term was set for two years (which he misunderstood as three years). AIEE petition and prospectus, 31 Mar. 1884; AIEE petition, 11 Apr. 1884; Keith to TAE, 6, 14, and 27 May 1884; TAE to Keith, 31 May 1884; all DF (*TAED* D8411B, D8411C, D8411E, D8411G, D8411J, D8416BSA); on formation of the AIEE, see McMahon 1984, 27–29; also "Historical Preface," *AIEE Transactions* 1 (1884–1887): 1–9; Gibson 1984 (chap. 5) discusses generally the tensions at the Philadelphia conference between those identified as either scientists or practical electricians. Scott 1934 (648–49) includes facsimile reproductions of the signature pages (including Edison's) from Keith's petition.

Edison seems to have missed most, if not all of the Society's initial executive meetings. At one, held on 3 June, he was appointed "Chairman of the Standing Committee of three on 'Incandescent Lamps'" and asked to name its other members. Edison queried on this notice: "what does this mean   why are interested parties put on coms." Keith explained that the committee would deal with routine matters about incandescent lighting, such as proposed papers, coming before the Society. The point of naming an "interested" party to the chair, he continued, "was because it was thought best to put some one who knew something about the subject in the place. All who know anything about dynamos, lamps, etc., are more or less interested. Honorable men can always do justice." Edison marked "Id rather not serve" on the letter and sent a formal reply later. Keith to TAE, 27 May, 5 and 14 June (with TAE marginalia); TAE to Keith, 7 June 1884; both DF (*TAED* D8411J, D8411K, D8411L, D8416BTG); TAE to Keith, 16 June 1884, Lbk. 18:77A (*TAED* LB018077A).

–2669–

To Willis Stewart

[New York,] May 9th. 1884.

Dear Sir:—

We have received your various letters in relation to the Santiago station and the business in general in Chili.[1]

Major Eaton has written you, I believe, that he thinks it would be well for you to go ahead under our contract, at the same time doing all you can for the Santiago people.[2] I do not doubt but what the Santiago matter will settle itself eventually. Your memorandum asking for the shipment of various pamphlets, electrotypes and blue prints will be sent forward to you by fast freight, together with 25 canvass books asked for in your letter of the 21st. March.[3]

The cable key book, together with memorandum regarding Chilian cable key, came duly to hand.[4]

We have got no new contracts of late, in fact our business at the moment is rather dull. This is of course but temporary, and we do not doubt but what as the Summer comes along we shall have plenty of work to do.

The central station down town is running splendidly. The collections keep up above $9,000., and only the day before yesterday we sold $600. worth of light, being the largest amount of light sold in that district, although the maximum number of lights on any one time was only 3,500. The reason for this is that we have been having [---][a] exceptionally bad weather for May, and consequently had a great deal of day lighting. A peculiar circumstance in connection with the running of our plant is that [--][a] as a rule we seldom have now more than 3,000 lights on as a maximum, and that our average lamp hours has gone up from five to six hours. This is owing mainly to the character of the lighting that Mr. Chinnock is now obtaining in the district, as he now has an opportunity of choosing his customers, and in no case does he take a new consumer, unless the average lighting is very high indeed. Within a few days we expect to start on the two new dynamos which have been placed in the basement of the second building at Pearl Street by our Construction Department under contract with the Illuminating Company.[5] This will give an increase capacity to the station of about 2,400 lights, and judging from the results obtained during the first week in May and the month of April, it would seem to indicate that our Summer lighting in the first district is going to be equal in every respect to the Winter lighting.[6]

We are figuring on an uptown plant between 23rd. and 59th streets and 8th. and Madison Avenues, and there is a maximum of 115,000 lights burning at any one time. It is our intention to divide this into two districts, the upper one starting at 42nd. Street and the lower one ending there. In all probability the station for each district will amount to about $500,000., or a total of $1,000,000. At this writing I do not know whether one or both of these plants will be put in this Summer.[7]

Our small town plants are running elegantly, all bugs have

been eliminated. The new pressure indicator is a perfect success. A large party of us went up to Newburgh to view things, and the station there seems absolutely perfect, with first class pressure indicators, feeder regulators we get a result of upwards of 1,000 average[8] with our lamps, and inasmuch as this plant has only been running a month this result is something wonderful, as it is of course expected that the bad lamps will be shaken out during the first month of running.

Your orders shall be attended to promptly and shipments made by us as quickly as possible.[9]

Your proposition to mail us drafts on London will be a satisfactory for the present if you find you are unable to arrange for cable remittances.[10]

We cannot understand why your contracts have not arrived, as they were sent from here on February 9th.[11]

I will see if I can get from the German Company a certified copy of the decision in the suit with the Swan Company.[12] Very truly yours,

TAE   I[nsull]

TL (carbon copy), NjWOE, DF (*TAED* D8416BOC). Initialed for Edison by Samuel Insull. ªCanceled.

1. Stewart covered a number of topics in the dozen or so recent letters that would have reached Edison by this date. Beyond the matters addressed herein and in Doc. 2602, Stewart's recent correspondence concerned the departure of chief engineer Mark Lawrence, a request for personnel with village system experience, lamp prices, and trouble with the Edison dynamo on the steamer *Santa Rosa*. Electric Light—Foreign—Chile, DF (*TAED* D8435).

2. The editors have not located the letter from Sherburne Eaton. Regarding Stewart's plan to reorganize the Santiago company outside the control of Kendall & Co., see Doc. 2602 nn. 5–6.

3. The requested printed materials, together with various catalogs, blueprints, and illustrations, were sent to Stewart on or about 15 May. Stewart to TAE, 21 March 1884 and n.d. [Mar.? 1884], both DF (*TAED* D8435G, D8435ZBN); John Campbell to Stewart, 15 May 1884, LM 19:171 (*TAED* LBCD6171).

4. Stewart mailed these items on 21 March. Stewart to TAE, 21 Mar. 1884, DF (*TAED* D8435G).

5. See Doc. 2644 esp. n. 8. The contract referred to is Edison's agreement with Edison Electric Illuminating Co. of New York, 29 Feb. 1884, Miller (*TAED* HM840211).

6. Edison's optimistic projection for summer lighting was not met. During the spring quarter, the station produced a profit of $7,252. During the summer months, this figure dipped to $3,873 before picking up during the last quarter to $14,000. The Illuminating Co. attempted "to neutralize this falling off" over the summer by furnishing "electric motor fans to be run by current from the central station." Although

they were introduced too late in the season "to secure any considerable results, the fans met with such favor as to warrant the belief that next summer they can be developed into an appreciable and permanent source of revenue." Samuel Allin reports to TAE in Electric Light— Edison Electric Illuminating Co. of New York, DF (*TAED* D8426); Edison Electric Illuminating Co. Annual Report, 9 Dec. 1884 (pp. 6–7), PPC (*TAED* CA004A).

7. In January, Edison told a newspaper reporter that he was planning a 7,000-horsepower station to serve the area between Twenty-fourth and Forty-fourth Sts., and a 9,000-horsepower plant for the area up to Fifty-ninth St. The projected second district had been canvassed and mapped by the end of March. The number of lamps tabulated at that time was about 110,000. The cost of constructing the First (Pearl St.) District had considerably exceeded estimates; Edison hoped to do better with the new stations, presumably in part by building the new district on the three-wire plan. "Open Enmity to Gaslight," *Chicago Daily Tribune*, 9 Jan. 1884, 2; see Doc. 2599 n. 8.

8. Edison referred to the average lifetime of lamps, in hours.

9. Stewart had advised Edison on 1 April that "Business is booming; look out for a dozen orders." He cabled one large requisition and a supplementary one on 22 April; these may have been the orders that Edison remarked totaled about $1,000. On a coded cable message received on 9 May, Edison noted: "Another order from Stewart." Stewart to TAE, 1, 22–23 Apr. and 9 May 1884, all DF (*TAED* D8435L, D8435U, D8435X, D8435ZAC).

10. Finding it impossible to cable money directly to New York, Stewart proposed to give Edison ninety-day notes drawn in London at 6 percent interest. On 23 April, he also mailed a small amount of cash as a deposit on pending orders. Stewart to TAE, 22–23 Apr. 1884, both DF (*TAED* D8435U, D8435X).

11. Stewart had not received the contracts giving him authority as Edison's Chilean agent (see Doc. 2602 n. 2) before 28 March; after their arrival, he required an additional set of copies to be certified by the Chilean Consul in New York. Stewart to Sherburne Eaton, 28 Mar. and 9 Apr. 1884, both DF (*TAED* D8435I, D8435R).

12. See Doc. 2598. Stewart wanted both the German patent decision and explicit authorization from Edison to combat competition from the Brush-Swan interests in Valparaíso at his own expense. Edison promptly requested a certified copy from Berlin and referred the matter to Richard Dyer and Sherburne Eaton, but he advised Stewart that the Edison Electric Light Co.'s directors would have to consider the issue of defending the Chilean patents. Stewart to TAE, 28 Mar. and 6 Apr. (with TAE marginalia) 1884; TAE to Deutsche Edison Gesellschaft, 9 May 1884; TAE to Stewart, 19 May 1884; all DF (*TAED* D8435H, D8435N, D8416BOH, D8416BPQ).

*From William
Meadowcroft*

[New York,] May 10/84

Mr Edison:

The information, as far as I can furnish it, is as follows:[1]

Isolated plants in U.S. and Canada[a] . . . . 59,883 Lamps
N.Y. Centl. Station . . . . 11,500 "
In Europe . . . . 32,000 "
In England & installations by English Co.[a] 12,000 "
In other parts of the world, including So. America,
New Zealand, Australia, Cuba, &c[a]   10,000 "
Total   125,383 lamps.[2b]

To the above should be added the number of lamps in the various Central station plants in the United States, outside of New York City— I cannot supply these— I suppose you have the records of them in your office.

Your mem. did not indicate a wish on your part that I should reply to the letter, so I send you the information.

W H Meadowcroft

⟨1600
1600
1600
500
1600
1600
500
1000
3000
8800
13 800   13 800 Lamps in out town Central Stations[3]⟩

ALS, NjWOE, DF (*TAED* D8420Q). [a]Text enclosed by right brace. [b]Followed by dividing mark.

1. Edison had given Meadowcroft an inquiry from the Philadelphia publisher of the *Acme Cyclopedia and Dictionary* for "the number of your lamps in use, both in this country and in Europe. We have an item in our forthcoming book . . . but have no late data." When it was published in 1884, this "practical compendium of useful information, and book of reference for everybody" reported (under the electric light subheading of its entry for "Electricity") 29,000 Edison lamps in the United States and nearly 20,000 in Europe. These figures were attributed to early 1883. Lantz-Lantz Publishing Co. to TAE (with TAE marginalia), 5 May 1884, DF (*TAED* D8420P); Lantz 1884, 285.

2. Cf. *TAEB* 6, App. 2.

3. The numbers in Edison's list correspond closely, but not exactly, with the capacity of the plants identified in Appendix 2A as operating by this date; the figure 3,000 may represent a subtotal.

-2671-

Mary Edison to Middlesex County Sheriff[1]

[New York, May 13, 1884[2]]

Dear Sir:—

You will please take notice that I am the owner of the personal property mentioned in my affidavit, accompanying this notice, and levied upon by you virtue of a certain judgement rendered against Thomas A. Edison in favor of Lucy Seyfert, and that you will interfere with same at your peril.[3] Yours truly

Mary Edison

ENCLOSURE[4a]

[New York, May 13, 1884]

Mary Edison being duly sworn says— That she is the wife of Thomas A. Edison the defendant in the above titled action:

That she is the owner of the personal property hereinafter mentioned, levied upon by the sheriff of Middlesex County in the State of New Jersey by virtue of a certain judgment against Thomas A. Edison and in favor of Lucy F. Seyfert in the above action   That said property is the sole and exclusive property of deponent and that the said Thomas A. Edison has no right or interest therein, and that the same is and has been in her possession and under her control at Menlo Park in the State of New Jersey. The following is a list of property so belonging to deponent, and levied upon as ~~before described~~ aforesaid.[b]

To wit:— In dining room— one extension table, carpet on floor, seven cane bottoms chairs, six other chairs, one sofa, one side board, one book case and contents, five window shades, one center table, two lamps, table covers, two clocks, one looking glass, one small stand, one sewing machine, one what not,[5] one coal scuttle, one hat rack in hall, carpet, two chairs, four pictures on wall one lamp, one organ

In parlor— Velvet carpet on floor, two rugs, five cushioned chairs one large chair and two rocking chairs[c]—one stand and glass vase, one sofa, one piano, one mirror, six lace curtains and shades and fixtures, two oil pictures on wall & all pictures—

Northwest Bedroom— carpet, Brussels carpet in hall up stairs

Southwest bed room— one crib, one bed-room set seven pieces, carpet and bedding, one wardrobe,

South east bed room— Carpet on floor, one bedstand and bedding, seven chairs, three rocking chairs, wash stand bowl and pitcher, one dresser, pictures on wall, two stand vases, one bureau, one shaving stand.

In hall 3rd   carpet on floor and hat rack,

Center bed room— one bed room suite of eight pieces, bed and bedding and carpet,

South east room— one bed room suite six pieces (marble top) carpet bed and bedding knives and forks, tea and table spoons and dishes lot of tin-ware, pails, wash tubs, ice in ice house, ice box, lot of flower pots in wagon house, two plows, three wagons, poles, three lawn mowers, four pails, lot of forks, two sitting room stoves one table, one screen, one cross cut saw, one ice saw, one stove pipe, one saw, lot of tools, two cutting boxes, one scythe, two saddles, one riding bridle, one set of single harness one double harness, one gray horse, eight blankets, five robes, one carriage, one buggy, one box wagon, two sleighs, wagon jack, one pole and wiffle-tree, one tent, one carriage cover, one rubber blanket, two straps of bells, ~~one neck yoke~~ two neck yokes, fly nets and cushions, two wagon dusters, one looking glass, one cook stove, one bed room set, five set pieces, bed and bedding carpet, lounge, lot of hay one Alderney cow,[6] one brindle cow, on bull, lot of manure, two pigs 17 fowls, coal screen lot of old plank and boards, one iron roller, four ladders, one wheel barrow, lot of rakes and hoes, iron tank, ~~of tar one~~ one dog house two plows

⟨Duck—please sign your name below   TAE⟩

Mary Edison

TDS, NjWOE, DF (*TAED* D8403ZCZ). [a]Enclosure is a TDS; John Tomlinson's notary oath omitted. [b]Interlined above by hand. [c]"one large . . . chairs—" interlined above.

1. Andrew J. Disbrow (b. 1815?), a former judge, served as sheriff. Opinion of Justice Edward Scudder, 18 Nov. 1884, DF (*TAED* D8403ZIN); U.S. Census Bureau 1970 (1880), roll T9_790, p. 393.3000, image 0627 (East Brunswick, Middlesex, N.J.).

2. Date taken from John Tomlinson's notarization of the enclosure to this document, which Tomlinson sent to Garrett Vroom on the same day (see note 3).

3. After Lucy Seyfert's attorney declined to allow Edison more time to pay the judgment against him, John Tomlinson asked Vroom on 13 May to file this document and enclosed affidavit immediately. Tomlinson also drafted similar documents on behalf of the two Edison companies that he claimed owned the remainder of the family's Menlo Park property (Strong & Sons to Tomlinson, 12 May 1884; Tomlinson to Vroom, 13 May 1884; Edison Electric Light Co. to Middlesex County sheriff, n.d. [13 May 1884]; Edison Electric Railway Co. to Middlesex County sheriff and enclosed affidavit, n.d. [13 May 1884]; all DF [*TAED* D8403ZCW, D8403ZCX, D8403ZCZ2]). Told that the documents were not in the form required by New Jersey statutes, Tomlinson drafted them again according to Vroom's instructions (Vroom to Tomlinson, 13 May 1884; Mary Edison to Middlesex County sheriff, with enclosures, 13 May 1884; both DF [*TAED* D8403ZDA, D8403ZDB]).

4. The enclosed affidavit carried a typed heading of "New Jersey Supreme Court   Middlesex County" in reference to *Seyfert v. Edison.* An undated, handwritten draft of this list (likely intended for Seyfert's attorneys) enumerated more items, including "All the Machinery in the Laboratory," as well as railroad cars, which were elsewhere attributed to the Edison Electric Railway Co. (see note 3). Mary Edison draft memorandum, n.d. [May?] 1884, DF (*TAED* D8403ZEI).

5. A wooden stand with open shelves for displaying ornaments, books, or other objects. *OED*, s.v. "what-not" 2.

6. One of the fawn-colored breeds originally from the Channel Island, including Jersey and Guernsey. *OED*, s.v. "Alderney."

–2672–

To Sherburne Eaton

[New York,] May 15. 1884.

Dear Sir:—

I beg to give you below a list of the men I am now carrying in connection with my electrical staff, in anticipation of their services, in all probability, not being required at an early date in connection with central station work. I have no use whatever for these men at the present moment, and my only object in carrying them is, that it would seem to be a very great pity that these men should be lost to our business.[1]

Considering the trouble and expense incurred in enabling them to obtain the necessary experience to efficiently perform their work, I shall be glad to know if the Edison Electric Light Co. are willing to carry these men, pending the decision of the Committee now inquiring into the question of the future conducting of the Manufacturing and Construction business.[2]

NAME.	SALARY.	ALLOWANCE.[a]
H. M. Doubleday[3]	$24.00	$17.50
Thomas P. Conant[4]	25.00	17.50
James Lyman[5]	10.00	17.50
A. S. Beves	15.00	17.50
H. W. Leonard	20.00	17.50
I. Walker (wireman)[6]	16.00	17.50

Very truly yours,

TAE   I[nsull]

TL (carbon copy), NjWOE, DF (*TAED* D8416BOY). Initialed for Edison by Samuel Insull. [a]Heading repeated for continuation of list on following page.

1. See Doc. 2677 esp. n. 3.

2. The Edison Electric Light Co.'s so-called Committee of Three (see Doc. 2690 n. 1) met for the first time on 14 May. Although Eaton, in connection with the Edison Electric Light Co.'s Committee on Expen-

ditures, wanted to "put the knife into our expenses," he agreed to keep skilled workers in the Edison fold and away from competitors. Eaton to TAE, 9, 10 and 14 May 1884, all DF (*TAED* D8427ZAY, D8427ZAY1, D8427ZAZ, D8427ZBE).

3. Harry Mather Doubleday (1855–1934) attended the Brooklyn Polytechnic Institute and the Massachusetts Agricultural College before taking a job at his father's woolen mill in 1872. He remained there until April 1881, when he sought out Edison and was hired to help with a canvass of gas use in New York. Later assigned to the Edison Machine Works, the Edison Co. for Isolated Lighting, and, briefly, the Edison Construction Dept., Doubleday was posted to the Southern Exhibition in Louisville in July 1883. At the end of the year, he returned to the Construction Dept., working at stations in Massachusetts and Pennsylvania. He took on similar duties for the Marr Construction Co., which was formed to install Edison central stations. Doubleday remained connected to the electrical industry until 1906, when he retired to Jamaica, where he served for several years as an American consular officer. Pioneers Bio., "Harry Mather Doubleday."

4. Thomas Peters Conant (1860–1891) came from a distinguished scholarly and literary family that traced its line directly to Roger Conant, a founder of Salem, Mass. Thomas Conant graduated from the Columbia School of Mines in 1882 and soon after entered Edison's employ as a chemical assistant at the Menlo Park laboratory. Since July 1883, he had been helping with central station wiring in Pennsylvania, Ohio, and New York. Conant remained connected with central station work after Edison disbanded the Construction Dept., but he was back in the laboratory in early 1885 making storage battery experiments. He left in March of that year, following the mysterious disappearance of his father, Samuel Stillman Conant, the managing editor of *Harper's Weekly*. After working for various electrical firms, Conant reentered the Edison orbit in 1889 as a construction superintendent for the Edison Electric Light Co. He was the company's eastern district engineer when he died in the care of his mother, the author Helen Conant, after an unspecified illness. Obituary, *NYT*, 25 Feb. 1891, 4; Obituary, *Trans. ASME* 12 (1891): 1056; Johnson and Brown 1904, s.vv. "Conant, Samuel Stillman" and "Conant, Thomas Jefferson"; "List of Graduates 1882," *School of Mines Quarterly* [Columbia College], 6 (Jan. 1885): 186; Samuel Insull to William Andrews, 11 June 1884; Conant to Insull, 15 Mar. and 1 May 1885; all DF (*TAED* D8416BUI, D8513D, D8513K); Unbound Notes and Drawings (1885), Lab. (*TAED* NS85ACS); Watkins 1919, 143–47.

5. James Lyman (1862–1934), an electrical engineer, began working for the Construction Dept. shortly after graduating from the Sheffield School of Yale University in 1883. He helped set up central stations in Pennsylvania, Ohio, and Massachusetts. After Edison broke up the Construction Dept., Lyman went to the Marr Construction Co. Lyman subsequently earned two engineering degrees from Cornell and worked for General Electric, as a district engineer for Chicago General Electric, and for an independent Chicago engineering firm. "Lyman, James," Pioneers Bio.; *NCAB* 28:20; Obituary, *NYT,* 31 Mar. 1934, 11.

6. Isaac C. Walker (1848–1929), a Pennsylvania native, was hired by Edison in 1883 and put in charge of inside wiring for village plant sta-

tions. He remained in Edison's employ until about June 1884. While still working for Edison, he began his own electrical engineering firm in Philadelphia, known after 1888 as Walker & Kepler. Walker's firm initially specialized in the installation of electrical lighting plants but later branched out to include electrical supply and repair. "Walker, Edison Aide in Early Days, Dies," *NYT,* 9 Mar. 1929, 19; U.S. Census Bureau 1982? (1900), roll T623_1406, p. 6A [Folsom, Delaware, Pa.]; TAE to Thomas Conant, 26 Oct. 1883; Walker to Insull, 24 June 1884; both DF (*TAED* D8316BCX, D8442ZEW); *Distinguished Men of Philadelphia and Pennsylvania* 1913, 97.

-2673-

Caveat: Ore Separation

New York, May 15, 1884[a]

To the Commissioner of Patents:

Be it known that I, Thomas A. Edison, a citizen of the United States, residing at Menlo Park, in the County of Middlesex and State of New Jersey, have invented an improvement in Methods and Apparatus for Separating Ores, and desiring further to mature the same, file this my Caveat therefor, and pray protection of my right until I shall have matured my invention.[1]

The invention is designed to act upon gold and other non-magnetic metals in a metallic state, to separate the same by electrically operated apparatus from the gangue or non-metallic substances. This requires that the ore should be thoroughly stamped, so that the division will be fine enough to release all the particles of metal.[2]

The first feature of my invention is an improvement upon the apparatus described in my application No 340 (Ser. No 70,289),[3] the object being to produce a more concentrated magnetic field.

In figures 1 and 2 a winged armature A, is shown located between the poles of magnet B. The armature A is revolved by power to shift the lines of force and to change the trajectory of the metallic portion of the falling powdered ore and thus separate it from the non-metallic portion.

Fig 1.

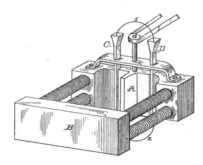

Fig 2

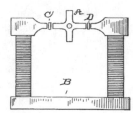

Flat spouts C. D. pass through the two portions of the field and conduct the material. These spouts have a considerable width. The action of the revolving armature is to throw the metallic particles to one side of the spouts, and dividing partitions below the troughs keep separate the two portions of the material. Suitable vibrating sieves are used in the hoppers to break up lumps and to feed the material in proper condition for the action of the apparatus.

The material may be passed through the spouts in a dry state, or with water, acidulated or not as desired: and the ends of the two spouts C. D. may be connected by wires 1 and 2 to complete the circuit. This causes heavy currents to circulate in the material and increases the effect of the action.

The winged armature may have its wings wound with wire connected in a circuit, to polarize it, and increase the field.

Winged electro-magnets E. F. are shown in figures 3, 4, 5 and 6.

Fig 3.

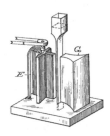

Fig 4

Fig. 5.

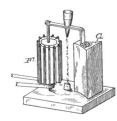

Fig 6.

These I prefer to the construction shown in my application referred to as forming a more uniform and continuous field. The wings may project from a shaft (figures 3 and 4) or from a larger cylinder (figures 5 and 6). The magnets E. F. are revolved by power opposite stationary armatures G. A spout may be used as in figures 3 and 4, or a hopper and partition as in figures 5 and 6. If the spout is used, the ends of the spout may be connected by a conductor outside of the magnetic field to complete the circuit. The revolving electro-magnet may be in the form of a Siemen's C—armature H, as shown in figure 7.
Fig 7.

Instead of changing the trajectory of the metallic portion of the falling ore by passing it through a shifting field of force, the trajectory of the metallic non-magnetic particles may be changed in another way. The material may be caused to pass through a magnetic field, so that an electric current will be thrown into the metallic particles by their cutting the magnetic lines of force.

There will then be a mutual attraction between the current in the particles and the magnet or magnets moving the metallic particles sufficiently to bring them to a different place of deposit than the non-metallic portion of the material.

In figure 8, the magnet I is opposed by a revolving winged armature J and the magnetic field is produced causing the action stated.
Fig 8.

In figure 9 three magnets and revolving armatures are shown as arranged for the same purpose.
Fig. 9.

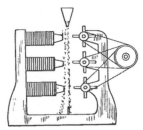

In figure 10 a series of magnets K is arranged close to the gradually receding side of a vertical spout B.

Fig. 10.

No armatures are used, but the spout will be of sufficient length so that the weak attraction will effect the desired object and draw the metallic particles to one side of the dividing partition while the non-metallic portion of the falling body is deposited in the other[b] side of this partition.

Another method is to take advantage of the retardation of the metallic particles in passing through a magnetic field. By making the field long enough a body of metallic and non-metallic particles thrown into the top of the spout will be quite completely separated before it reaches the bottom of the spout. The material as it issues from the spout will be directed into two or more receptacles.

In figure 11, the spout M has magnets N.

Fig 11.

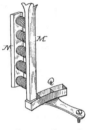

The swinging box O has three chambers. It is arranged to be swung under the end of the spout. The first chambers would receive all non-metallic material, the second chamber mixed metallic and non-metallic material, and the third chamber metallic particles solely.

In figure 12, a sliding scoop P is designed to be brought into the path of the falling body to catch the retarded metallic particles.

Fig 12.

By using water the separation may take place in a trough Q (figure 13) slightly inclined from the horizontal, and a conductor 3 may be used to complete the circuit through the trough.

Fig 13

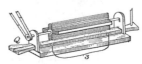

It is evident that with all the methods proposed, a dry or wet separation may be employed, and, if the latter the circuit may be completed by an outside conductor to permit of the circulation of the currents generated through the moving body.

And if desired additional dynamo currents may be passed through the moving body.

It is evident that by repeating the action upon the material the separation can be made practically perfect.

This specification signed and witnessed, this 15th day of May 1884.

Witnesses   Alf. W. Kiddle,[4] E C Rowland[5]   Thos A Edison

DS, MdNACP, RG-241, Edison Caveat 109 (*TAED* W100ABW). Probably written by Alfred Kiddle; all figures drawn on two separate pages. [a]Place taken from oath; date taken from text, form altered. [b]Interlined above.

1. This document was filed on 3 June 1884 as Edison's Caveat 109. The editors have found no evidence of a related patent application.

2. Edison wrote out an undated draft of the basic principles of the caveat. He also made a series of sketches related to it between 3 and 14 May. On or about 8 May, Edison made more complete drawings, which became the basis for figures 1–13 in the caveat. Unbound Notes and Drawings (1884), Lab. (*TAED* NS84ACH, NS84ACI, NS84ACJ); Cat. 1150, Scraps. (*TAED* NM019AAE, NM019AAF).

John Ott conducted a related series of experiments around 14 May using an electromagnet and a series of nonferrous materials. He made a "large Electromagnet and suspended between its poles extensions with silk fiber the following Mettals and compounds," including ivory, aluminum, quartz, glass, zinc, mica, lead, and gold, as well as flasks of muriatic, sulphuric, and acetic acids. Ott included with his notes on the experiments two diagrams, one illustrating his experimental apparatus and the other labeled as a "model for Ore Milling." N-80-04-17: 91–98, Lab. (*TAED* N050091, N050096).

3. Edison executed this application in August 1881 but, due to the malfeasance of patent attorney Zenas Wilber, it was not filed until August 1882 (see Doc. 2323). It issued as U.S. Patent 400,317 in March 1889.

4. Alfred Watts Kiddle (1865–1935) was a frequent attestor or witness to patents or business documents executed by Edison and Richard Dyer, beginning in 1884. Kiddle gained valuable experience in patent law through his association with the firm Dyer & Seeley. He was admitted to the New York bar in February 1887, graduated from the Law School of Columbia College the same year, and worked for a time as a patent attorney for the Edison Electric Light Co. Kiddle developed his

own general law practice, specializing in patent and corporate law, fields in which he became prominent. Obituary, *NYT,* 9 Jan. 1935, 19; Harrison 1900–1902, 2:191–92; "Kiddle, Alfred Watts," Pioneers Bio.

5. Edward C. Rowland (1863–1926) was a native of Metuchen, N.J. He quit private school and joined the Menlo Park laboratory in 1880, where he briefly was involved in the manufacture of incandescent lamp filaments. His skill as a draftsman led to his reassignment to the drafting force, and his name appears on many of Edison's patent applications. He moved to the Fifth Ave. office of the Edison Electric Light Co. in 1881. Rowland later worked for the patent firm of Dyer & Seeley before opening his own drafting office. "Rowland, Edward C., " Pioneers Bio.

–2674–

To Samuel Insull

NEW YORK, [May 20, 1884][1a]

⟨Important⟩
⟨Plan to get more work & sell more E Lamps⟩
Insull—

Write Flood Page and suggest that he might arrange to rent some C machines at £300 per annum which would give him about 10 pct & 5 pct dep[re]c[iatio]n—that if they would merely pay the Engine & extra work of about $2~5~700 per machine they could get them— we want to get rid of them & at the same time help them out= mention that they take up very little room are complete & where party has 800 @ 1000 Lamps they would be cheaply installed & no trouble [---][b] & small space, [--][c] and at the end of a year of whatever period they could be taken out & moved very cheaply where as with any other installation the fixed Expenses & work are very heavy & on removal the[d] losses would be very great, that the rental of 300£ would be very moderate to any one—& that if the party wanted a spare machine which although I consider unnecessary they might Let them have it on payment of £150 Extra per year providing the party would agree to use it only when the other was out of order   thus they would get abt 8 pct & no depcn= [-------][b]   They would make on Lamps & be earning interest on investment instead of letting these machines lie idle here Eating up interest—[2]

E

ALS, NjWOE, DF (*TAED* D8437Y1). Letterhead of Thomas A. Edison, Central Station Construction Dept. [a]"NEW YORK," preprinted. [b]Canceled. [c]Canceled and followed by "over" as page turn. [d]Obscured overwritten text.

1. The typed letter to Samuel Flood Page, based on Edison's instructions in this document, was dated 20 May. DF (*TAED* D8416BPY).

2. Edison may have been prompted to write by the Edison & Swan

United Co.'s recent urgent order of four large isolated dynamos for the International Health Exhibition at South Kensington, London (Edison & Swan United to TAE, 22 Apr. and 1 May 1884, DF [*TAED* D8437W, D8437X]). His suggestions were consistent with those he made in late 1882 urging the antecedent Edison Electric Light Co., Ltd., to develop the isolated lighting business (see Docs. 2374 esp. nn. 4–5, and 2375). The big C dynamos referred to here were probably those remaining in New York from the London company's May 1882 order (see Doc. 2572 n. 4).

Flood Page unequivocally rejected Edison's suggestions, countering that it would be easier to sell than to rent the machines. He also asked Edison to consider repurchasing them from the Edison & Swan United firm. Edison noted on Flood Page's letter, "Insull— Cant we try Colombo on some." Insull apparently promptly wrote to Giuseppe Colombo (letter not found), who was planning for another expansion of the central station in Milan; Colombo then delegated John Lieb to arrange details personally with Edison. The Milan station added two Jumbo dynamos sometime between April 1884 and February 1885 and another two by the start of 1886. Charles Clarke recalled years later that at least one of them came from the Holborn Viaduct plant in London, but it is possible that Edison arranged to transfer one of the British company's machines from New York, much as he had already disposed of one to the Netherlands. Flood Page to TAE, 17 June 1884; Colombo to TAE, 11 July 1884; both DF (*TAED* D8437ZAD, D8436ZDA); "Theatre de la Scalla," *La Lumière Électrique* 12 (5 Apr. 1884): 13; "Station Centrale d'Éclairage Électrique de Milan," ibid. 15 (28 Feb. 1885): 386 (reprinted in English in Edison Co. for Isolated Lighting Bulletin 12:25–26, 18 Jan. 1886, CR [*TAED* CC012025]); "Usine Centrale pour l'Éclairage Électrique à Milan," ibid. 19 (6 Feb. 1886): 241–42; Clarke 1904, 53–54; see Doc. 2572.

–2675–

Memorandum to Richard Dyer

NEW YORK May 21st 1884[a]

Dyer

Please draw up patent for process of drawing wire whereby the annealing of the wire & drawing is a continuous process.[1]

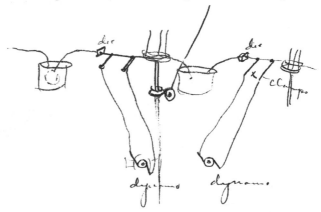

⟨see wire drawing book perhaps in Cyclopedia² E⟩

about 6 inches or a foot of the wires passes between contact rolls clamps or brushes X Aᵇ continuous quantity Current passes through athis section so as to heat it as it passes to the annealing or softening point. it may be heated between every die or every other die. The softening of the wire makes it easier to draw through the die hence there is a saving of power & this saving is nearly enough to run the dynamos— besides by making the process continuous saves a great deal of labor, & increase the output of the plant & also renders unnecessary the complicated annealing process now in use— get a broad claim³

Edison

ADS, NjWOE, Lab., Cat. 1150 (*TAED* NM019AAG). Letterhead of Thomas A. Edison. ᵃ"NEW YORK" preprinted. ᵇObscured overwritten text.

1. Figure labels are "die," "die," "clamps," "dynamo," and "dynamo."

2. Edison may have meant the sixteen-volume *American Cyclopaedia: A Popular Dictionary of General Knowledge,* edited by George Ripley and Charles Dana and published from 1873. Edison acquired the second volume in November 1884, presumably to fill a gap in a set he already owned. The first and last volumes contained entries for annealing and wire, respectively. Each entry referred to the fact that wire, at some point in being drawn, becomes brittle and must be annealed before being drawn further. The editors have not identified the "wire drawing book." Ripley and Dana 1883, s.vv. "Annealing" and "Wire"; Voucher no. 605 (1885) for J. M. Stoddart.

3. Edison sketched this idea for annealing on 23 May and again on 5 June, and a similar process on 24 May (Unbound Notes and Drawings [1884]; Cat. 1150; both Lab. [*TAED* NS84ACK1, NS84ACL, NM019AAJ]). He executed two related patent applications on 2 June 1884. In one, based closely on this memorandum, Edison described annealing "by heating the wire by an electric current of low tension and large quantity at proper intervals between the die-plates" through which the wire was drawn. The specification's single figure, also based on the drawing in this document, clearly indicated pickling reservoirs "ar-

Drawing in Edison's U.S. Patent 436,968 of apparatus for continuously drawing, annealing by electricity, and pickling wire.

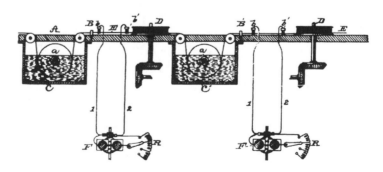

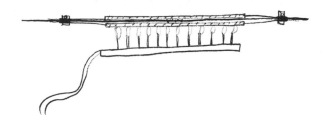

ranged in the line of the wire after each dynamo-connection to remove the oxide" (U.S. Pat. 436,968). The other application covered methods and apparatus for annealing in externally heated sealed vessels, "the wire entering and leaving the chambers through suitable air-excluding stuffing-boxes. The oxygen within the chambers is quickly consumed, so that but little remains therein, leaving the wire surrounded by nitrogen, and hence the heating can be performed with but slight oxidation of the wire" (U.S. Pat. 436,969). Both applications were rejected for conflating separate inventions for a process and machinery. Later changes in Patent Office policy obviated this objection, but the applications were again refused on the grounds that some claims overlapped those of an earlier patentee. After Edison revised and narrowed the claims, a technicality related to the inventor's oath further delayed their issue until September 1890 (Pat. Apps. 436,968; 436,969).

–2676–

Memorandum to John Ott

NEW YORK May 22 1884[a]

John—

please try the following Experiment

I want to work Local Sounders and also work telegraph wires of [400?][b] 2500 to 7000 ohms each from a Dynamo.[1] So arrange it thus & Let me know if the[c] sounders [---][b] Lines dont interfere with Each other[2]   ⟨Worked O.K.⟩[d]

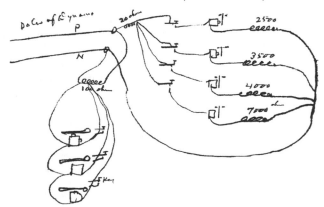

put 100 ohms in & shunt around enough resistance to give good current to the Sounder then see if the sounders interfere with each other[3]

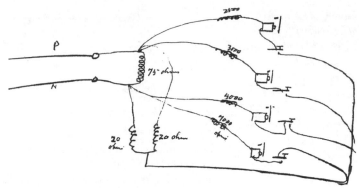

This is arranged so as to send out positive or negative Currents[4] The Lines on top of 75 ohm coil will get Current in one direction while those on bottom of 75 ohm coil will get Current in opposite direction ⟨Worked O.K.⟩[d]

Arrange all these circuits so you can change Relays & Keys etc from one Experiment to the other quickly then Let me know when you are ready to show it to me[5]

Edison   M N Force
J. F. Ott

ADS, NjWOE, Lab., Unbound Notes and Drawings (1884) (*TAED* NS84ACK). Letterhead of Thomas Edison; document multiply signed and dated. [a]"NEW YORK" and "188" preprinted; remainder of date written by Martin Force. [b]Canceled. [c]Obscured overwritten text. [d]Marginalia written by John Ott.

1. Because of their relatively high internal resistance, batteries could operate only a limited number of sounders or lines. As the resistance of a working circuit changed with the addition of lines or sounders, resistance of the complete circuit (including the battery) would not change proportionally. This disproportionate change of resistance resulted in significant variations of current strength, which made it hard to keep instruments adjusted properly. A dynamo whose own resistance was a tiny fraction of that on the outside line, however, would reduce these disparities practically to zero. Dynamos also had the obvious advantages of supplying enough current to operate more devices as well as doing away with tedious battery maintenance. Maver 1892, 27–31, 40–49.

2. Figure labels are (top to bottom from left) "Poles of Dynamo," "100 ohms," "key," "20 ohms," "2500," "3500," "4000," and "7000 ohms."

3. Figure labels are (top to bottom) "75 ohms," "20 ohm," "20 ohm," "2500," "3000," "4000," "7000 ohms."

4. Currents of opposite polarity were required for multiplex telegraph systems, in which a single wire carried more than one signal. Maver 1892, 42.

5. Probably in relation to the events discussed in Doc. 2651, Edison sketched on 23 and 24 May several arrangements of a "Dynamo for W.U.," including one for the quadruplex that "Worked OK" (Unbound Notes and Drawings [1884], Lab. [*TAED* NS84ACK1, NS84ACK2,

NS84ACK3]). In June, Edison and the Edison Electric Light Co. granted Western Union a license under an 1882 patent application, which was still pending in the Patent Office, to manufacture and use dynamos for telegraphic purposes (that use having been specifically excepted from Edison's underlying contract with the Light Co. [Doc. 1576]). The application was probably Edison's Case 404, which had already been rejected, amended, and placed in interference; Edison eventually abandoned it (TAE agreement with Edison Electric Light Co. and Western Union, 13 June 1884, DF [*TAED* D8427ZBI]; Patent Application Casebook E-2537:156; Patent Application Drawings [Case Nos. 179–699]; both PS [*TAED* PT021156, PT023: images 62–63]). On 1 July, Edison filed two patent applications for "Dynamo Telegraphy," one of which pertained specifically to quadruplex telegraphy. He subsequently abandoned these cases as well (Patent Applications: Abstracts of Edison's Abandoned Applications [1876–85], PS [*TAED* PT004:17]).

Edison would have been familiar with the efforts in this area of other inventors, including James Fuller, Orazio Lugo, and Stephen Field. Western Union had adopted Field's dynamo quadruplex circuits in 1883 (see Docs. 634 and 2024 n. 2; "Steam Engines for Batteries," *NYT*, 17 Jan. 1881, 5; Pope 1892, 185–87; *Telegraphy* 1901, 2:143–48; Maver 1888, 67, 79). By 1892, both Western Union and the Postal Telegraph Co. had installed Edison dynamos for local and outside circuits in New York and other stations (Maver 1892, 44, 46, 49).

–2677–

Sherburne Eaton Memorandum

[New York,] May 24, 1884[a]

President's mem. for Executive Committee meeting, Light Company, May 24th. 1884.

CONSTRUCTION DEPARTMENT. TEMPORARY EXPENSES. On May 15th. Mr. Edison notified me that he was unable any longer to carry the electrical staff of his Construction Department. He mentioned 6 men in particular, all, he thought, indispensible to the business.[1] Their salaries and expenses aggregate about $225 a week. I told him that in order to prevent their discharge by him, the Light Company would temporarily pay the expenses, subject to the future action of the Directors. ⟨$250.00⟩[b]

On May 21st. Mr. Edison asked to have the engineering pay roll of his Construction Department also assumed by the Light Co., 7 men, about $200 a week.[2] I told him this also would have to be acted on by the Directors.

Shall the Light Co. assume all or any part of those expenses, aggregating about $425 a week, pending the ultimate decision of what shall be done with the Construction Department?[3]

CONSTRUCTION DEPARTMENT. CONTRACTS. About 2 months ago the Thomas A. Edison Construction Department made a written tender to install a central station plant in Dallas, Texas, for $24,796, provided it were accepted within 60 days. The 60 days have now expired, the company is not yet formed, and Mr. Edison is not in condition to extend the time for accepting his tender. The question arises whether the Light Co. will assume the estimate and authorize our agent to proceed with the formation of the company? There are several other towns situated like Dallas, in this regard.[4]

Shall the Light Co. assume these contracts, pending the final disposition of the construction business?

SOUTH AMERICAN BUSINESS. Mr. Edison asks for authority to make arrangements with such parties as he may select, to exploit our business in the various countries of South America, also in Mexico.

We have already given Mr. Edison the Argentine Republic and Chili. On July 2nd. 1883, Mr. Edison was given authority to at once install a central station plant at Buenos Ayres. Although 11 months have passed, and this has not yet been accomplished, Mr. Edison is confident that before another Winter, he can start a central station there.

On January 26th. 1884, an arrangement was made with Mr. Edison, for Chili, which was ratified by the Directors, Feb. 19th.[5] He is to pay us in cash 25 percentum on the shop price of dynamos, and on village plants 18 per centum on the contract price of the original installation and future increases, all as delivered F.O.B., in New York City. We are also to have a royalty of 10 cents on every lamp, both in isolated and central station lighting. Mr. Edison sent Mr. Stewart to Chili to do business under this contract, and we have thus far received two small isolated orders from him. He expects to do a good business.

Shall we make the same arrangement with Mr. Edison for other countries in South America and Mexico, as we made in Chili?[6]

UNSATISFACTORY INSTALLATIONS OF VILLAGE PLANTS. Several of the village plants installed by the Construction Department are unsatisfactory to the local companies. Mr. Douty, President of the Shamokin company, in a letter dated May 13th. and following up several previous conversations and letters on the matter, formally requests that his plant be made satisfactory and to conform to his contract with Mr. Edison, as Mr. Douty understands it.

Several of those companies have made complaints to us, and insist that either Mr. Edison or our Company should make the installations perfect, to conform, as they claim, to the original contract. There are two sides to this question, Mr. Edison being on one and the local companies on the other. My own opinion is that the installations should be made satisfactory, within reason, but the question arises who shall pay for the necessary changes and additional machinery.

What shall be our policy? Shall we accept Mr. Edison's decision as to what constitutes an adequate plant, or shall we accept the decision of our own engineering department, Mr. Clarke for instance, or any other engineer whom we may select   If Mr. Edison's decision is not accepted by the local companies, and if our engineers do not agree with Mr. Edison, who shall pay the expense of the changes and new machinery? Strictly speaking, it is not our duty to stand this loss. As matter of law, there is probably no reason why we should do so. But there are reasons of policy why the local people ought to be given satisfaction.

S. B. Eaton   President. per W.F.[7c]

TD (carbon copy), NjWOE, DF (*TAED* D8427ZBF). [a]Date from text; form altered. [b]Marginalia written in unknown hand. [c]"S. B. Eaton" and "per W.F." handwritten.

1. See Doc. 2672.

2. The seven individuals identified by Edison were: Charles Campbell and Charles G. Y. King (engineers), Carl Ulmann (draftsman), Joseph Atkins (blueprints), Henry Guimaraes and B. Gmur (mapping), and Paul Dyer (canvasser). TAE to Eaton, 21 May 1884, DF (*TAED* D8416BQG).

3. Eaton (who was concurrently working on the Committee on Expenditures) evidently sent a monetary offer on 26 May concerning both Edison's engineering and electrical personnel, but that document has not been found. Edison rejected Eaton's proposal as "quite inadequate, unless the Executive Committee are willing that the business should be conducted without the Light Co. having an engineering force, or one to which they can refer to when required." Edison promised that Samuel Insull would consult with Eaton, the immediate outcome of which the editors have not determined. TAE to Eaton, 28 May 1884, DF (*TAED* D8416BRP); see also Doc. 2725.

4. The Dallas station was not constructed, but see App. 2B regarding its projected size.

5. See Doc. 2602 n. 2.

6. Edison acknowledged a "perfectly satisfactory" proposal from Eaton about South America and Mexico a few days later. Its terms have not been identified, but presumably they were similar to those for Chile. TAE to Eaton, 28 May 1884, DF (*TAED* D8416BRR).

7. Unidentified.

-2678-

To Everett Frazar[1]

[New York,] May 28th. 1884.

Dear Sir:—

I have your favor of the 27th. inst.,[2] and have given instructions for some instruments to be set up, and as soon as they are ready I will let you know, as I should like to have you and some of your friends see them working. I do not think I could arrange to send any to Corea just at present.[3]

The first time I am down town I will do myself the pleasure of calling on you[4] to see what you have written to General Foote.[5]

I think it would be better to defer the interview with the Japanese Consul until such time as the instruments, above referred to, are in shape, and when I would propose taking him with you over to my laboratory to see them in operation.[6] Very truly yours,

TAE   I[nsull]

TL (carbon copy), NjWOE, DF (*TAED* D8416BRO). Initialed for Edison by Samuel Insull.

1. Everett Frazar (1834–1901), resident New York partner for Frazar & Co. (China) since 1872, received his appointment as Consul General for Korea in New York in April 1884. Frazar's diplomatic career was facilitated by his acquaintance with Lucius Foote, who had recommended him as a host to the Korean Mission to the United States during 1883. Originally from Massachusetts, Frazar had established Frazar & Co. as merchant and commission agents in Shanghai in 1858 and later added branches in Nagasaki, Hong Kong, and Yokohama. Since at least April 1884, Frazar had been guiding Edison's application for an electric light and telephone concession in Korea ("Obituary," *NYT,* 4 Jan. 1901, 7; "In Memoriam," *Asia: The Journal of the American Asiatic Association* 21 [Jan. 1921]: 121–23; Frazar 1884, 4, 49–51; Lucius Foote to Frederick Frelinghuysen, 17 Jan. 1884, no. 53, reel 1, U.S. Dept. of State 1951). The New York house of Frazar & Co. had sought to introduce Edison's electric light in Japan in 1882; the Yokohama branch became the sole agents for his light in Japan and Korea in 1885 (Frazar & Co. [China] to TAE, 16 Aug. 1882; Goddard to TAE, 24 Aug. 1882; Frazar & Co. [Yokohoma] to TAE, 5 Aug. 1885; all DF [*TAED* D8237Z, D8237ZAA, D8534O]; see also Nam 2000).

2. Not found.

3. Frazar was promoting American participation in a proposed industrial arts exhibition in Seoul. He had recently talked with Edison about autographic telegraphy, written samples of which Samuel Insull had promised to send him. At the end of 1884, Frazar & Co. in China requested character printing telegraphic instruments, which it sought to represent in Japan, China, and Korea. "Exhibition in Corea," *NYT,* 16 Dec. 1883, 6; Insull to Frazar, 1 May 1884; Frazar & Co. (China) to TAE, 19 Dec. 1884; both DF (*TAED* D8416BMH, D8471Y).

4. The editors have found no further reference to a meeting. Frazar & Co. (China) had a New York office on the third floor of 73–74 South St.

and shared an additional address with the Korean consul at 123 Front St. "Two Warehouses Burned: The Christmas Eve Conflagration in South Street," *NYT,* 26 Dec. 1881, 1; Frazar & Co. (China) to TAE, 16 Aug. 1882 and 19 Dec. 1884, DF (*TAED* D8237Z, D8471Y); *Guide to the Charities* 1886, 137.

5. Lucius Harwood Foote (1826–1913) was associated with Edison's telephone rights while serving as American Consul in Valparaiso, Chile (Doc. 1823 n. 4). Adjutant General for the state of California from 1872 to 1876, Foote became U.S. Envoy Extraordinary and Minister Plenipotentiary to Korea in February 1883; he then applied on Edison's behalf to obtain exclusive rights for electric light and telephones in Korea (*NCAB* 7:267–68; "Nominations by the President," *Washington Post,* 27 Feb. 1883, 1; Obituary, *NYT,* 5 June 1913, 11; Frazar to TAE, 17 Apr. 1884; Insull to Frazar, 6 May 1884; both DF [*TAED* D8403ZBQ, D8416BNS]). Foote reported in September that Edison was authorized to introduce his lighting system at the palace compound in Seoul; he attributed this accomplishment, in part, to the "observations made by the Corean Envoys while in the United States," referring to the Special Mission of 1883. The same diplomatic mission, headed by Min Yong Mok (president of the Foreign Office), had visited the Foreign Exhibition in Boston and the New York *Herald* building, where they may have seen Edison's lighting. Some members of that delegation returned to Korea aboard the USS *Trenton,* which was outfitted with an Edison isolated plant (one L dynamo and 104 16-candle lights), the first electric lighting installation on a U.S. naval warship (Foote to Frederick Frelinghuysen, 4 Sept. 1884, no. 106, and 17 June 1884, no. 84, both reel 1, U.S. Dept. of State 1951; McCune and Harrison 1951, 34; Thompson 1954.

6. S. K. Takahasi was the Japanese Consul General in New York at this time. "Notes From Washington," *NYT,* 6 May 1883, 14; "A Great Club's Birthday," ibid., 7 Feb. 1883, 2; "Japanese in the Stock Exchange," ibid., 26 July 1884, 8.

–2679–

From Samuel Flood Page

London, 28th May. 1884[a]

Dear Sir,

I beg to inform you that the subject of the use of "Edison" lamps has on several occasions been before the directors. We have a number of Edison installations in the United Kingdom, some of which, such as the House of Commons, the Criterion Theatre, etc., are of great importance; and we have recently obtained a supply of lamps from you in order to replace lamps as required in these and other installations and to enable us to sell lamps in connection with new plants as we make them from time to time. If, however, the company are to continue to supply Edison lamps, and especially if they are materially to increase their use in this country, as we trust may be done, it is evident that we must manufacture them ourselves at our own factory. The superintendent of our lamp factory feels that he could now make Edison lamps and he thinks that in justice to you, to the company and to the lamp itself, he ought to have some one thoroughly competent and familiar with the process of manufacturing your lamps, as a guarantee that the lamps which we[b] may make in this country shall be as good as those that we obtain from you.

I am therefore instructed by the directors to request that you will have the goodness to let us have a foreman in whom you have confidence who would come into our employment either for a short period, or for a considerable length of time, as you, and he, might think best. We are anxious to develope both sides of our business—the Edison, and the Swan—and I am sure you will appreciate the importance of the application that I now make to you. Unless we manufacture our own Edison lamps, I feel sure that the trade in these lamps will not increase.

I shall be obliged if you will let me know whether you can place such a man as I have spoken of at our disposal, what salary he ought to receive and for what length of time he ought to be engaged.

I am glad to be able to say that the lighting of the Criterion Theatre reflects very great credit on those who have carried it out; that it shows the Edison system in an admirable manner, and that it is recognised by all who have seen it, as being quite the best piece of theatrical lighting yet carried out in England.[1] I regret however, to say that we shall lose a considerable sum of money over the installation. I Remain, Dear Sir, Yours Very Truly,

S. Flood Page. Secretary.

⟨Cant send Foreman. The French Co having the only one we had to spare   Suggest they do not go into the business for the reason that if Lamps are made, factory always be better & cheaper, that the cost in our own small French factory is already greater than our price delivered Hamburg & Lamps are not so good— [2] We believe that if all the Lamps are made in one large works that they will always be cheaper than if made in several small factories   We even believe we could give you a bid to deliver Swan Lamps in England for less than the cost to you in your own factory taking every Expense into consideration— E⟩

TLS, NjWOE, DF (*TAED* D8437ZAC). Letterhead of Edison and Swan United Electric Light Co.; typed in upper case. [a]"London," and "188" preprinted. [b]Interlined above.

1. The Criterion Theatre plant began operating regularly on 15 April, running some 600 lights on stage and throughout the house from four L dynamos. There was trouble initially with one of the two engines, which Armington & Sims and Edison jointly blamed on it having been started up improperly. The Criterion installation quickly won favorable reviews, with one observer proclaiming it "perhaps the most perfect example of theatre-lighting in London," not least because the air in the theater remained fresh and below 60 degrees. TAE to Edison & Swan United Electric Light Co., Ltd., 9 May 1884, DF (*TAED* D8416BOG); "The Criterion Theatre," *Teleg. J. and Elec. Rev.* 14 (19 Apr. 1884): 336.

2. Edison's marginalia formed the basis for his 25 June reply, with which he enclosed a copy of Francis Upton's six-page negative response to Page's proposal. Edison pointed out that "we have within the last few months shipped from 70,000 to 100,000 lamps to the Continent of Europe in competition with the French Factory in Ivry-sur-Seine." Upton also referred to the French factory, particularly to problems with quality and high manufacturing costs, both there and at the small Canadian shop, relative to the high-volume, low-cost strategy pursued by Edison's works in Harrison. Upton described William Holzer as the only person qualified to start a new plant but declined to endorse him because of the inadvisability of asking him to compete against the U.S. factory, in which Holzer had a financial stake. TAE to Edison & Swan United Electric Light Co., Ltd., 25 June 1884, Lbk. 18:96 (*TAED* LB018096); Upton to TAE, 19 June 1884, DF (*TAED* D8429ZAO); cf. Docs. 2487 esp. n. 12, 2499 esp. n. 5, 2540, and 2554.

–2680–

From George Bliss

CHICAGO, May 29th, 1884.[a]

Dear Sir:

I trust you will excuse the intrusion of my opinion but having been a consistent Edison[b] man for so many years my anxiety for the more rapid introduction of your light may be permissible.

I know very well how utterly heartless capital is and that it would slaughter your interests remorselessly on the slightest occasion if for its advantage.

It is therefore to be expected that you will look out that your interests and capital are first harmonized.

Having done this the money invested in the Edison companies should be carefully protected and by so doing your own interests will be the best served.

I do not believe that even you realize how generally the idea has permiated the public that the exclusive[c] right to Incandescent lighting belongs to yourself.

This has been an enormous advantage in introducing the Edison system but under the present policy this is likely to be frittered away.

Some six of the Arc companies are now offering to do Incandescent lighting and however inferior their apparatus may be it tends to postpone action, lower prices and divert business.

These competitors are gaining experience and are improving their apparatus constantly.

The public mind is being educated to the idea that Incandescent lighting is an open field.

Now why not put a stop to this by bringing suit on some one of the many points which you control essential to the success of Incandescent lighting.

Capital is proverbially timid and much of it would be frightened from going into the opposition.

The cost of these suits would certainly be much less than the incalculable injury already sustained from the efforts of competitors.

More business at better prices can be done by the Edison companies with suits in progress so that the expenses of litigation can readily be met from increased revenues.

The man who vigorously asserts his rights and contends for them generally succeeds for the majority of mankind do not care to engage in expensive and doubtful contests.

If the Edison policy shall be determined against litigation then it seems to me the only way is to put the whole business on the strictest kind of hard pan basis.

The Edison Co. ought to take all the factories under its control, of course paying you handsomely for your interests, put the price for franchises down very low and sell everything to local companies at such figures as would simply shut out competition.

With the lead which they have in the business this certainly ought to be possible. This is a big country and it is important that there shall be a number of Edison centers established from which the business can be handled promptly and to advantage.

I do not believe this can be well done from New York alone on account of time, distance and expense involved.

When the first installations under your supervision have reached practical success the sooner information is disseminated to these Edison centers the faster future progress will be made.

Your time ought to be altogether too valuable to be expended in ordinary business management when the principles have been throughly established.

I think you can make ten times the money by having others

do the work than by bringing everything directly under your own supervision.

All the evidence and experience coming under my observation goes to show that an enormous business is to be done in incandescent lighting.

With a successful policy your stock interests ought to pay you vastly more profit than can be made out of the vexations of manufacturing and contracting.

The Bell telephone is a most striking example of this and I do not regard the telephone of one tenth the importance of the Edison light even at the present time. Sincerely Yours,

Geo. H. Bliss

⟨Write Bliss in response to this not to get excited, that the arc Lt Cos he speaks of will think a Cyclone has struck them soon—E⟩[1]

TLS, NjWOE, DF (*TAED* D8462ZAM). Letterhead of Western Edison Light Co.; "Business suggestions" typed in preprinted subject line. ᵃ"CHICAGO," preprinted. ᵇInterlined above by hand. ᶜWritten by hand.

1. Edison sent a more anodyne letter of thanks expressing confidence that affairs would soon "be so arranged as to be satisfactory to all," and that he would "be enabled to push ahead vigorously and deal with the business, which I am sure is only waiting to be canvassed for." TAE to Bliss, 5 June 1884, DF (*TAED* D8416BSZ).

–2681–

From Richard Dyer

NEW YORK CITY, May 29th, 1884ᵃ

Dear Sir:

In accordance with the talk you had yesterday with Mr Seely and myself with regard to charges for conducting your soliciting business, I now propose to prosecute your business upon the following terms:[1]

(1)   For the preparation of applications of ordinary length and simplicity including the drawing twenty-five dollars ($25). The case of Mr Livor and yourself on pulleys is an example of this class[2]

(2)   For the preparation of simple applications including drawing twenty dollars ($20). The case on insulating covering for the conductors which you put into my hands yesterday is an example of this class.[3]

(3)   For the preparation of cases more complicated than those of class 1 my charge for preparation will be from thirty to fifty dollars (including drawing) according to the amount of work required cost of drawing &c

(4)   For the prosecution of applications you to pay reason-

able expenses and for time spent provided more than merely formal changes have to be made in the applications.

(5)   For preparation of assignments, two dollars each. This does not include the record fee of Government, which is $1 for 300 words, and $2 for over that up to a number large enough to cover any ordinary assignment.

(6)   The work of Mr Seely to be considered the same as my own work upon all matters we do for you

I consider it for my interest and of course propose to treat your business upon the most liberal basis, giving it my first and best attention and promoting your interests in every way in my power.

While expecting to be paid for work of any considerable length outside of the preparation and prosecution of applications I don't propose to charge for consultations and searches for information which consume but a little time.

You will do me a favor by considering these terms strictly confidential since we hold our services high to the outside world, and propose to maintain in our general business a schedule of fees very much higher than the figures we have given you. The fees of Mr Serrell were referred to in our talks of yesterday. In our general business we hope to merit much higher figures than he charges—in fact as high as the leading firms in the soliciting business.

Hoping that you will appreciate the lower figures we have put upon your work   I remain   Yours very truly

Richd N. Dyer

ALS, NjWOE, DF (*TAED* D8468ZBD). Letterhead of Richard Dyer. a"NEW YORK CITY," and "188" preprinted.

1. This document appears to mark the start of Dyer's partnership with Henry W. Seely, who likely gained entrance to the bar about this time (and certainly had done so by 1885). The firm of Dyer & Seely collaborated closely with Edison personally and with Edison companies for many years. Dyer assured Edison that although he was obliged to attend first to the Edison Electric Light Co., its business was "now so small, and is based almost entirely upon your own inventions," that he did not foresee "any interference with the prompt execution of your personal work." Dyer to TAE, 31 May 1884, DF (*TAED* D8468ZBF); testimony of Cyrus Brackett, 18 June 1885, p. 40, *Weston v. Latimer v. Edison*, MdCpNA (*TAED* W100DFA040).

2. Long-time Edison associate Henry ("Harry") M. Livor (1846–1904) became acquainted with Edison in 1877 as a sales agent for the George Place Machinery Agency in New York. He had also represented other machine manufacturers and arranged credit with one of them on Edison's behalf. Livor came from a well-connected New York family; his father was a noted homeopathic physician, and his three sisters

were all prominent in New York society. One of them, Iphigenia Place (the wife of Harry's original employer), became a leading suffragist in the 1890s. *TAED*, s.v. "Livor, Henry"; Tate 1938, 148; Obituary, *NYT,* 28 May 1904, 9; "Women Ruled the Meeting," ibid., 8 May 1894, 1; U.S. Census Bureau 1970 (1880) roll T9_896, p. 359.4000, image 0722 (New York City, New York [Manhattan], N.Y.); ibid. 1982? (1900), roll T623_1119, p. 5B, (New York City, New York [Manhattan], N.Y.).

The editors have not identified the pulley patent application, but Livor had sent pulleys for Edison's inspection in mid-May. Early the next month, Edison hired him away from the George Place Machinery Agency to become sales agent of the new Edison Shafting Manufacturing Co., then being formed to fabricate shafting, pulleys, hangers, and related transmission equipment under contract with the Edison Machine Works at its Goerck St. shop. Livor became general manager of the new firm almost immediately and received a substantial stake in it; he remained an important Edison manufacturing associate for many years. Livor to Insull, 14 May 1884; Livor agreement with Edison Machine Works, 3 June 1884; Livor agreement with Edison Shafting Manufacturing Co., 14 July 1884; Livor agreement with John Kruesi, 28 Aug. 1884; all DF (*TAED* D8431O, D8431Y, D8432B, D8432G); see also Bazerman 2002, 282.

3. The editors have not found this application; possibly it was a re-drawing of Edison's Case 614, which the Patent Office had rejected in February (see Doc. 2588 n. 2).

–2682–

Technical Note:
Electric Lighting

[A][1]

[New York,] May 29 1884

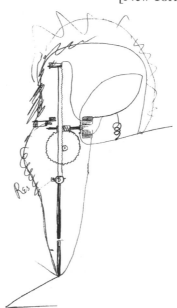

TAE

Witness   Richd N Dyer

X, NjWOE, Lab., Cat. 1150 (*TAED* NM019AAK).

1. Figure label is "Res[istance]." This drawing was the basis for the first of three figures in a patent application that Edison signed on 2 June for an arc lamp regulator. (The specification issued in 1890 with three additional drawings added during the examination process.) The regulator employed a motor with compound windings in the field or armature coils, or both. If the carbons came together, a low-resistance winding, connected in series with the arc, would cause the motor, through a worm gear, to move them apart. This motion would be countered by the effect of a high-resistance winding, in a shunt circuit with the arc, that would drive the motor the opposite way. Edison expected that this design would "maintain an arc of uniform length without flicker of the light" regardless of the lamp's position. Also in early June, Charles Batchelor prepared to adapt the R dynamo for use with arc lights but did not write down any specifics. U.S. Pat. 438,303; Pat. App. 438,303; Cat. 1306, Batchelor (*TAED* MBN011AAV).

MARY EDISON'S STORY OF HER COURTSHIP AND WEDDING   Doc. 2683

The editors' discovery of this article published in the *New York World* on 1 June 1884 has called into question time-honored accounts of the most written-about events of Mary Edison's life: her courtship and marriage in 1871. Following Edison's invention of the phonograph in 1878, newspaper reporters and authors responded quickly to the public's growing curiosity about the inventor's life. One of the most popular narratives to come from the resulting efflorescence of biographical mythology was Edison's courtship of and marriage to Mary Stilwell. The story of his whirlwind romance of a sixteen-year-old employee in one of his shops and his immediate return to the laboratory after the wedding ceremony seems to have first appeared in a front-page article in the *Washington Post* on 26 November 1878. The story, or versions of it, soon appeared in newspapers across the nation and in James McClure's 1879 biography, *Edison and His Inventions.*[1] It was resurrected periodically in the press throughout Edison's life, even after Mary's death and his own remarriage, and it has been treated with a range of acceptance and criticism in a variety of Edison biographies.[2] Aware of the accounts published in her lifetime, Mary evidently was eager to relate her version of meeting Edison and their subsequent courtship.

The *World's* profile of Mary and her home, purportedly in her own words, was apparently unique. The editors have not found other instances in which Mary appeared as the subject of a printed article, nor have they determined how this one came to be written. The article appeared on a women's page over the byline of Olive Harper, the pen name of Helen Burrell Gibson D'Apery, a prolific and well-known newspaper writer and novelist who had endured brief notoriety over the veracity of her early reporting. Harper, who in recent years had mainly written articles about or directed at women, had sought to interview Edison on at least one occasion, in February 1882, when she asked to see him about "the application of your lights as a beneficial change from gas in homes." Edison declined.[3]

When Mary talked with Harper, she vehemently denied having worked for Edison or anyone else prior to her marriage, affirming that she was still a schoolgirl when she met the inventor by chance after taking refuge in his factory during a rainstorm.[4] Far from being a whirlwind romance, she claimed that the courtship lasted several months, beginning in the spring of 1871 and culminating in their wedding on Christmas Day. And contrary to earlier accounts, her portrayal of Edison's courtship was quite conventional, with the suitor making repeated visits to her home and taking care to secure her father's approval of their marriage. A second *New York World* article (Doc. 2718), probably also written by Oliver Harper and published eight days after Mary's death, countered the widely circulated story that the groom had returned to his laboratory immediately after the wedding ceremony to work on his stock ticker.

Edison, for his part, told biographer Francis Arthur Jones decades later that the legend of his going back to work, forgetting that he had just been married, "was nothing but a newspaper story . . . got up by an imaginative newspaper man who knew that I was a bit absent-minded." He did, however, concede that an actual event on his wedding day could have become the basis for the story: "The day I was married a consignment of stock tickers had been returned to the factory as being imperfect, and I had a desire to find out what was wrong and to put the machines right." Far from forgetting his wife, he claimed that he had dutifully consulted her before departing:

> An hour or so after the marriage ceremony had been performed I thought of these tickers, and when my wife and I had returned home I mentioned them to her and explained that I would like to go to the factory to see what was the matter with them. She agreed at once, and I went down, where I found Bachelor, my assistant, hard at work trying

to remedy the defect. We both monkeyed about with them, and finally after an hour or two we put them to rights, and I went home again.

Despite acknowledging that he had gone to work on his wedding day, Edison was adamant "as to forgetting that I was married, that's all nonsense, and both I and my wife laughed at the story, though when I began to come across it almost every other week it began to get tedious. It was one of those made-up stories which stick, and I suppose I shall always be spoken of as the man who forgot his wife an hour after he was married."[5]

1. The story spread quickly in newspapers throughout the country. All of the articles listed below reproduced the original 26 November 1878 (p. 1) *Washington Post* article, in whole or in part; seven gave direct attribution: *Alton (Ill.) Daily Telegraph*, 2 Dec. 1878, 4; *Decatur (Ill.) Daily Republican*, 5 Dec. 1878, 2; *Maquoteka (Iowa) Jackson Sentinel*, 5 Dec. 1878, 6; *Burlington (Iowa) Hawk-Eye*, 7 Dec. 1878, 9; *Decatur (Ill.) Weekly Republican*, 12 Dec. 1878, 6; *Logansport (Ind.) Journal*, 12 Dec. 1878, 6; *Cedar Falls (Iowa) Gazette*, 13 Dec. 1878, 1; *Decatur (Ill.) Daily Review*, 14 Dec. 1878, 3; *Freeborn County (Albert Lea, Minn.) Standard*, 19 Dec. 1878, 4; *Keene (N.H.) Sentinel*, 26 Dec. 1878, 1; *Hagerstown (Md.) Mail*, 10 Jan. 1879; *New Orleans Daily Picayune*, 19 Jan. 1879, 12.
2. Concerning the treatment of the courtship and marriage in Edison biographies, see McClure 1879, 67–68; Dickson and Dickson 1894, 87–88; Jones 1907, 344–45; Simonds 1934, 81–83; Miller 1940, 140–42; Josephson 1959, 97–99; Conot 1979, 46–47; Baldwin 1995, 52–53; Israel 1998, 73–75; Stross 2007, 15–16.
3. Harper to TAE, 24 Feb. 1882, DF (*TAED* D8207N); TAE to Sherburne Eaton, 31 July 1882, Lbk. 7:812 (*TAED* LB007812); on Harper see Doc. 2683 n. 13.
4. See Doc. 2683 n. 5.
5. Jones 1907, 344. According to Simonds 1934 (83), Alice Stilwell Holzer (Mary's sister) also denied the story of Edison returning to work on his wedding night.

–2683–

Olive Harper Article in the New York World[1]

[New York, June 1, 1884]

IN THE WIZARD'S HOME.[a]

HOW THOMAS A. EDISON'S RESIDENCE IS FITTED UP.[a]

One of the Many Elegant Private Homes in Gramercy Park—Mrs. Edison's Deep Love for Her Famous Husband.

An abode of the purest peace, the truest sympathy and the utmost domestic harmony is the home of Mr. Thomas A. Edison, he who is commonly known as the "Wizard of Menlo Park," and whose city residence is at No. 25 Gramercy Park.[2]

The house is rather old-fashioned and without distinct individuality as to exterior, like almost all New York houses, but the inside has in every part a brightness and airy pleasantness entirely its own. The hall is paved with marble and furnished with Turkish rugs and crimson embossed velvet chairs and sofa and has a large hat-rack near the door, whereon hung a coat that bore the impress of the wizard himself. Three handsome parlors are en suite, all thrown wide open and with large windows in front, giving a pleasant outlook on the trees in the park, and with other equally large windows at the back, letting in a flood of sunshine and giving a view of a fine garden in the rear. The front parlor is carpeted with Persian rugs and the rest with Axminster carpets, and the walls are covered with pictures and fine engravings, while every available spot and corner is made to hold a statuette, or vase, or some other article of decoration.[3] The pretty upright piano[4] near the window holds three statuettes of exceeding merit and is covered with an embroidered cloth of peculiar and effective design. Numberless bits of painting and porcelain hang here and there, many done by the busy fingers of Mrs. Edison. The other two parlors are furnished much the same, with pale-blue satin furniture in carved ebony wood and with easy chairs of every design, and nearly all decorated with fancy lace-work, all light, bright, cheerful and pleasant to see. Mr. Edison had just gone out, and so I had a long and delightful chat with Mrs. Edison, who had a particular grievance to complain of, and that is that the newspapers all over the country had published such ridiculous stories of her and Mr. Edison's courtship and marriage, and she averred that they had made her quite unhappy.

"In the first place," she said, "I never worked in any factory, nor for Mr. Edison, nor anybody else in any capacity,[5] and therefore all the stories about his passing along where I was at work Monday evening and proposing to me and setting the wedding for Tuesday morning hasn't a word of truth in it.

"The fact is just this. I was going home from school[6] one afternoon with two of my companions, when it began to rain furiously and none of us had an umbrella, and so we stepped into the hallway of what proved to be Mr. Edison's factory, where he was then making the stock-recorders. A gentleman whom we knew came out and invited us inside, and we went in to be out of the rain. I was about fifteen and a half years old, and was tall for my age.[7] Mr. Edison was at work on one of his machines and this gentleman, whom we knew, showed us around, and I noticed Mr. Edison particularly for two reasons. First,

I thought he had very handsome eyes, and next, because he
was so dirty, all covered with machine oil, &c., and I spoke to
him about the little instrument he was at work on. We talked a
few moments, and as the rain kept up and even grew worse we
concluded it was best to go home. The gentleman brought his
umbrella and took my two schoolmates and Mr. Edison got
his and started to accompany me, first pulling on an overcoat
that hid his dirty clothes. When we got to the house I saw that
he was determined to go in and I had to invite him, and when
my mother came down she asked who that was. I told her and
said that he had brought me home and she went in. I was in
mortal terror least she should ask him to stay, but she did, and
then he got up and took off his overcoat and stayed till nine
o'clock, and then when he went away he asked permission of
my mother and myself to call again. When he got it he availed
himself of it to come almost every evening, and at last after
five months of constant visiting, he made his proposal in this
way, which I tell you, because it is so perfectly his way of doing everything.

"We had been out walking and were coming home when he
said:

"'Have you ever thought you would like to be married,
Mamie?'

"'Why no,' I replied, 'not yet anyhow.'

"'Well, I have and I would like to, and I would like you for
my wife.'

"'Oh, I couldn't,' I answered.

"'Well, and why not? Don't you like me well enough?
Think, now, and try not to make a mistake.' I stammered
out something about the suddenness of it and that I couldn't
marry so young, but he said: 'If you meant no you would say
no, so now I'll see your father to-morrow night, and if he says
yes we'll be married Tuesday. This was Saturday. Sunday he
talked with father and mother and wanted to be married at
once, but father said that he would give him an answer in a
week. Mr. Edison took my hand and said: 'I love your daughter and I'll make her a good husband. I am honest, and I am
good, and I know how to treat a woman. I'll come next Sunday
night.' He did, and father had in the meantime satisfied himself that he need have no fear, and so we were married,[8] and
I have been very happy with him, and I expect to be as long
as I live, for he is good and true and so tender to me and the
children. We have three, Dottie, my daughter, and Tommie,
my big boy, and Willie, my baby. Dottie looks a little like me,

but Tommie is like his father, with the same shaped head and same eyes. Yes, I'm a little in love with my husband's eyes—yes, in fact, a good deal.

"This is his picture. You see how he looks. He is tall and weighs about two hundred now."

And I saw a strong, honest, fearless face, with intentness of purpose and a strength of character most unusual, and with close-set lips and a deep line between the brows; not a handsome face according to the rules of beauty, but one to like and trust.

Mrs. Edison herself is a very handsome woman of about twenty-five, fair complexion, admirably proportioned, tall and probably weighing about one hundred and sixty pounds. She dresses in exquisite taste, although now in mourning,[9] and it is her husband's delight to see her so dressed, and nothing is too good for her he thinks. Mrs. Edison has been called the most extravagant woman in New York as to personal adornment, but it is a mistake, or worse, on the part of those who say it.[10] Mrs. Edison declares she never feels neglected when her husband shuts himself up at Menlo Park for the purpose of making experiments or for invention, and she just waits until he has emerged from his seclusion, only taking pains to see that he has his meals properly, for he forgets self entirely, though he never forgets her.[11] She watches over him and cares for him like a mother, and even has learned to like his smoking, which came hard at first, and she understands his business and his plans and inventions as well as he does, and thus proves the most valuable helpmeet, as in discussing them with her he not infrequently solves a problem that has bothered him and of which the solution always escaped.[12] In this house he has no laboratory other than his busy brain, which is always at work. He is now at work on something which is almost completed and which will be of the utmost importance in telegraphy, but of which I am not at liberty to speak, as it is not covered by patent, and Mrs. Edison complains that the patent laws here are so badly arranged that it is almost impossible to be fully protected, even when you have a patent on anything, and that it only needs to invent something of value to have it immediately pirated.

With all Mr. Edison's success with electricity their home is still lighted with gas, to her disgust and vexation.

Mr. Edison is, as is well known, an entirely self-made man, and he prides himself upon that fact as well he may, for though it is good to be born with a silver spoon in one's mouth, to be born with brains in one's head is very much better. He doesn't care what he wears; how old and worn his clothes are, nor what he eats, nor scarcely whether he eats at all or not, and he never seems to pay the slightest attention to his own needs, but his keen eyes are always upon his wife and children, and his delight is to see them enjoying every comfort and luxury possible, and he is as kind and thoughtful of every one's welfare that happens to be near him. Those who do not know him say, "What a wonderful genius he is!" and those who do know him say, "What a good man he is." It is well to be able to receive both titles given in all sincerity.

OLIVE HARPER.[13]

PD, *New York World*, 1 June 1884, 11 (*TAED* D8414K1). ᵃFollowed by dividing mark.

1. See headnote above. This article was reprinted or extracted in other newspapers, including the *Boston Globe* (11 June 1884, 5); *New York Evangelist* (12 June 1884, 3); and *Atlanta Constitution* (22 June 1884, 2).

2. The Edisons had returned to the Gramercy Park house at the beginning of May (see Doc. 2663). Mary Edison seems to have spent at least part of the summer at their Menlo Park house, as she had during the hot months since the family moved to New York in 1881.

3. Edison received a ten-page inventory of furnishings when he rented the four-story Gramercy house in 1882 (James Pryor inventory, n.d. [1882?], DF [*TAED* D8204ZKY]). The family purchased numerous decorative items, including photographs, small tables, and rugs during their tenure in the house. Nearly one thousand itemized receipts documenting these and other household acquisitions, as well as some of Edison's research and business expenses (all generally from mid-1883 to late 1884) are in the Vouchers series. The editors have generally not attempted to correlate items mentioned in this document with either these receipts or the landlord's inventory.

4. The instrument was rented from Charles H. Ditson & Co. for ten dollars per month. Tuition for frequent half-hour lessons for Marion (forty sessions in three months) was paid separately to Mrs. H. Clarke, at least through May 1884. Years later, Marion remembered Edison sitting "at mother's square piano banging out his self-composed music." Voucher nos. 272 and 640 (1883); Oser 1956, 3.

5. For a brief period, Mary did work for Edison at Edison's short-lived News Reporting Telegraph Co. in the *Newark Daily Advertiser* building at 788 Broad St. (Doc. 205 esp. n. 5). Beginning on 19 October and ending on 25 November, a series of entries appearing in the company's accounts most likely referred to wages paid variously to Mary and her sister Alice: "Cash to Girls" on 19 and 20 October; wages to "Sis Stillwell" on 28 October and to "M. Stillwell" on 2 November are representative entries. These references lead the editors to believe that both Mary and Alice worked in some capacity for Edison as he and his partner, William Unger, set up the fledgling company. From late November until the company ceased operations around the end of the year,

a woman listed as "Miss Hodges" seems to have received seven dollar weekly wages for the work previously done by Mary (Cat. 1213:100–105, Accts. [*TAED* A202]).

Additional evidence that Mary worked for the News Reporting Telegraph Co. came later from P. J. Boorum, one of the company's handful of employees. Writing to Edison for financial assistance in 1878, he sent his greetings to Mary: "How is Mrs Edison? Ask her if she remembers the Surprise Party we were trying to get up when she was learning to transcribe [that is, operate the keyboard of a printing telegraph transmitter] in the old Daily Building" (Boorum to TAE, 29 Aug. 1878, DF [*TAED* D7802ZWR]). Biographer William Adams Simonds asserts that Mary and Alice were together when Mary first encountered Edison and that during the courtship Edison brought Mary to his shop to assist him (Simonds 1934, 81–83).

6. The 1870 federal census enumeration, conducted in June 1870, identified Mary as a student. Mary turned sixteen in September, and her wages from the News Reporting Telegraph Co. in November and December (see note 5) suggest that she did not attend school in the fall. Mary lived at 92 Jefferson St. in Newark with her parents, Nicholas and Margaret Stilwell, and her siblings, Alice, Charles, Eugenie, and Margaret. U.S. Census Bureau 1965 (1870), roll M593_880, p. 423A [77], image 230 (Newark Ward 5, Essex, N.J.); Holbrook 1870, 613; ibid. 1871, 668.

7. Born on 9 September 1855, Mary would have been fifteen and a half in spring 1871. At that time, both the American Telegraph Works, at 103 New Jersey Railroad Ave., and the Newark Telegraph Works, at 15 New Jersey Railroad Ave., were doing work related to Edison's printing telegraph. The latter concern changed its name to Edison and Unger and relocated to 4-6 Ward St. on 1 May 1871, but it did not begin operating until the late spring. See Docs. 92 n. 1, 156 n. 2, and 157; a map identifying these locations is in *TAEB* 1 (148).

8. Thomas and Mary were married on 25 December 1871 at her parents' home at 92 Jefferson St. in Newark (see Doc. 218), "in the presence of her friends and relatives," as Mary's sister Alice recalled years later. Mary Edison Holzer to William Simonds, 2 July 1932, MiDbEI (*TAED* X001D8I).

In 1889, Olive Harper wrote a newspaper article entitled "Edison's Real Courtship," which combined her account in "In the Wizard's Home" with her description of their honeymoon in Doc. 2718. Harper added a few new details, such as identifying "Miss Lucy Hamilton Warner" as Mary's bridesmaid. (The article appeared in the *Atchison [Kan.] Daily Globe* on 23 August and in at least five other papers in the next few days.) Lucy Hamilton Warner (c. 1857–1905) became a well-known children's book author and married Henry D. Tyler in 1893. Lucy's sister, Clara Harris Warner, wrote to Edison in 1917 asking "if you remember the little girl, Clara Warner who used to visit you and my dear friend, the late Mrs Edison at your home in Menlo Park so many years ago?" U.S. Census Bureau 1970 (1880), roll T9_842, p. 288.3000, image 0578 (Brooklyn, Kings, N.Y.); "Notes of the Social World," *NYT,* 7 May 1893, 10; "Weddings Yesterday," ibid., 2 June 1893, 4; "Obituary Notes," *Publishers Weekly,* 22 Apr. 1905, 1156; Clara Warner Fisher to TAE, 3 June 1917, NCB (*TAEM* 278:1007).

9. Mary's father, Nicholas Stilwell, had died on 9 April 1884.

10. Numerous receipts for apparel and household items from Lord & Taylor and other distinctive retailers are in the Vouchers series (1883–1884). Mary's daughter, Marion Edison Oser, recalled years later that

> Mother, being socially inclined, gave many parties. I still have photographs of some of the costumes Lord and Taylor made for her. One unusual one was a red and black brocade decorated with stuffed red birds with black-wings, poor birds! Mother also gave many children's parties for me and when I came home for dinner at noon and found a beautiful satin dress on my bed with slippers and stockings to match, then I knew what the surprise was to be. She must have gone to a lot of trouble to have my dresses unusually pretty. Among those I particularly remember were a nile green satin, decorated with hand painted flowers and a yellow satin decorated with daisies. [Oser 1956, 4]

Mary had a red and black brocade dress with red birds (eight cardinals) made by Mme. A. Duval, a New York dressmaker. The total cost for the dress materials and its construction was $391.90 (Voucher no. 46 [1884]). There is scant direct contemporary evidence of Mary as a hostess, but see Doc. 542 and *TAEB* 6 chap. 3 introduction. Other festive events may be inferred from grocery receipts in the Laboratory Vouchers. In New York, Mary had the material means to gratify her tastes. Nye 1983 (90–97) analyzes the Edisons' responses to their changed material conditions in New York and also discusses the inventor's work habits in relation to the prevailing norms of Victorian domesticity.

11. Marion remembered that her "Mother was not very happy in Menlo Park, as my father neglected her for his work, or so it seemed to her. He never would come to her parties and he would often skip meals and very often not come home until early morning, or not at all" (Oser 1956, 4). Edison employees Alexander Campbell and Edward Ten Eyck, Jr. also remembered that Edison prioritized working in his lab over time with family or socializing at home (Marshall 1931, 56, 108).

12. Regarding Edison's initial negative opinion of Mary's capabilities related to his work, see Docs. 241 and 248.

13. Olive Harper was the pen name of Mrs. Helen Burrell Gibson D'Apery (1842–1915), a prolific newspaper writer and novelist. Born in Pennsylvania's Wyoming Valley, Helen Burrell moved with her family to California in 1851. She married at age fifteen, becoming Helen Gibson, but was widowed with three children a few years later. A prolonged illness left her with anchylosis of the knee joint and dependent on crutches for the rest of her life. It was during her convalescence that she began to write, and she reportedly decided to "own the crutches and not let them own her." Her first publication in *Harper's* was followed by a series of articles in the *Oakland Daily News*. These stories led to a reporting relationship with San Francisco papers that took her all over California by train, stage, horseback, and mule (her children having been sent to boarding school). Her work for the California papers led to an affiliation with the *St. Louis Globe* (soon to become the *Globe-Democrat*). *NCAB* 5:215; "Olive Harper Dies," *NYT*, 4 May 1915, 15; "Olive Harper," *New Brunswick (N.J.) Daily Times*, 24 Aug. 1887,

1; "Miss Olive Harper," *New Orleans Times Picayune*, 7 July 1873, 2; Mighels 1893, 119.

Harper made her name as a newspaper writer, although she eventually authored or translated scores of books, mostly novelizations of mystery-themed or romantic stage plays. She covered the Vienna Exposition in 1873 as the representative of the *St. Louis Globe-Democrat* and other newspapers. While there, she married a French engineer in the Turkish Army, Colonel T. E. D'Apery, with whom she had a son. After the Vienna exhibition, she began a long tour of Europe, including Romania, Russia, Turkey (where she spent a year), and England. *The Writer: A Monthly Magazine for Literary Workers* 23 (Jan.–Dec. 1911): 55; *New Brunswick (N.J.) Daily Times*, 24 Aug. 1887, 1; "Olive Harper

Dies," *NYT,* 4 May 1915, 15; "Offers Fame for Cash: Aged Woman Cripple Writes Melodramas While You Wait," *Washington Post,* 5 Feb. 1911, E4.

Harper gained some notoriety for the carelessness or inventiveness of her reporting in the early 1870s. She caused a minor trans-Atlantic scandal in 1873 when she wrote a gossipy letter for publication in the *New York Daily Graphic* about Americans in European society. In it, she made disparaging comments about the writer and journalist Ambrose Bierce's use of the military title Major in civilian life. Bierce took umbrage, as did the subject of an unrelated Harper article, the British writer Algernon Charles Swinburne. These reports (and perhaps others) led to Harper being parodied in an 1876 satirical essay in the London *Athenaeum* entitled "The Art of Interviewing." About the middle of the decade, Harper's newspaper writing became primarily about and directed at women and their putative domestic concerns. She may also have been the basis for Henrietta Stackpole, a minor character in *The Portrait of a Lady,* which Henry James began drafting in 1880. Grenander 1980, 406–13.

–2684–

From Edwin Houston

PHILADELPHIA, U.S.A.   June 3rd, 1884.[a]

Dear Sir:

As you are doubtless aware, it is the intention of the Franklin Institute to hold an International Electrical Exhibition which shall fully illustrate the state of the science at present.[1] We are receiving great encouragement from all parts of the world and we have every prospect that the exhibit will be an exceedingly successful one.

I have been requested to write to you urging upon you the propriety of making an individual exhibit of your numerous electrical inventions; several of our Institute friends have promised to speak to you concerning this matter. We are frequently asked by the general public as to the names of the probable exhibitors and there is always considerable dissatisfaction expressed on not finding your name among the individual exhibitors. It is true that your company intend making a full exhibit of your electric light apparatus, but this forms so small a part of your electrical inventions that we are exceedingly desirous of having your individual exhibit as complete as possible.

Could you not find it convenient to exhibit the valuable collection you have made for the Exhibitions in other countries?

Trusting you will be able to give us a favorable reply to the above matter, I am   Very Truly Yours,

Edwin J. Houston.[2]

Electrician   International Electrical Exhibition

⟨Write him that it will be very Expensive for me to do it, but I will consider the matter⟩[3]

TLS, NjWOE, DF (*TAED* D8464H). Letterhead of 1884 International Electrical Exhibition, Franklin Institute. [a]"PHILADELPHIA, U.S.A." preprinted.

1. The International Electrical Exhibition in Philadelphia followed electrical exhibitions in Paris (1881), London (1882), Munich (1882), and Vienna (1883) and was the first of its kind in the United States (though one of nine large industrial U.S. exhibitions that year). When Gardiner Hubbard mentioned the idea to Edison in January 1883, Edison offered his assistance but suggested that "next year [1884] or the year after would be quite early enough" after the European and London events. Unlike those exhibitions, the host country predominated in Philadelphia; with the exception of precision instruments, the machines filling the exhibition building (adjacent to the Pennsylvania Railroad depot and near the Schuylkill River in West Philadelphia) were largely American. Opening on 2 September and scheduled to run for about six weeks, the Exhibition overlapped with scholarly or professional gatherings in the city, most notably a meeting of the American Association for the Advancement of Science, and closely followed the British Association's August meeting in Montreal. It also provided the rationale for convening a concurrent federally chartered National Conference of Electricians (organized in part by Edwin Houston), to which Edison was invited (Gibson 1984, 3, 19–20, 39–42, 155–57; Franklin Institute exhibition circular, 2 Sept. 1884; Hubbard to TAE, 27 Jan. 1883; U.S. Electrical Commission to TAE, 20 Aug. 1884; all DF [*TAED* D8303X, D8464Q, D8464O]; TAE to Hubbard, 29 Jan. 1883, Lbk. 15:213 [*TAED* LB015213]). This confluence of events and the expected presence of so many distinguished academic electricians and physicists in the city helped stimulate the formation of the American Institute of Electrical Engineers to ensure the proper representation of self-identified practical electricians (see Doc. 2668).

2. Edwin James Houston (1847–1914) was the founding chair of Natural Philosophy and Physical Geography at Philadelphia's Central High School, which was effectively an undergraduate college. He was a member of the Franklin Institute and the first president of its electrical section; he was also a member of the new Institute of Electrical Engineers (which he served as president, 1893–1894). Houston and his Central High colleague Elihu Thomson made significant inventions in arc lighting and dynamos, and they created the Thomson-Houston Electric Co. in 1879 to commercialize their patents. Houston left the firm in 1882 to focus on educational efforts; it later rivaled the Edison lighting companies and was merged with them into the General Electric Co. in 1892 (*ANB*, s.v. "Houston, Edwin James"). The Philadelphia pair had crossed paths with Edison, their challenge to his announcement of a new "etheric force" in 1876 having touched off a cordial debate over the phenomenon (*TAEB* 2 chap. 11 introduction; see Docs. 718 n. 2, 726, 764 n. 1; Carlson 1991, 57–64). They were angrily accused by Edison in 1878 of seeking credit for a carbon microphone that he claimed to have invented (see Docs. 1374 and 1460; Carlson 1991, 78–80). Late that year,

having conducted an authoritative set of dynamo tests at the Franklin Institute (1877–1878) and with their successful arc lighting experiences in mind, Houston and Thomson argued publicly that it would be impossible to operate economically an incandescent lighting system such as that which Edison was trying to develop (Doc. 1664 nn. 6–7).

3. Edison's marginalia formed the basis for a noncommital reply on 7 June to Houston's "very flattering invitation." Edison evidently made up his mind by 10 June, when he personally requested Babcock & Wilcox to arrange for boiler power at Philadelphia "as we desire to have a large exhibit." When he received an itemized estimate a few days later about the cost of finishing a Jumbo dynamo, he commented: "I think we should make Exhibit." TAE to Houston, 7 June 1884; TAE to Babcock & Wilcox, 16 June 1884; TAE marginalia on Batchelor to TAE, 17 June 1884; all DF (*TAED* D8416BTH, D8464J, D8464K); records of Edison's expenses related to putting up his exhibit are in Vouchers (1884) Box 4.

Edison mounted a comprehensive review of his inventive career, comparable in size to his 1881 display in Paris and similar to the one at London's Crystal Palace in 1882 (see Docs. 2111 and 2226). It contained examples from the full dynamo catalog of the Edison Machine Works (plus the Jumbo), lamps, and ancillary lighting appliances. There were also some three dozen telegraphic and telephonic instruments, the phonograph, odoroscope, tasimeter, and improved motophone. Edison's display was managed by William Hammer, who resigned in June as chief engineer for Deutsche Edison Gesellschaft after a dispute with Emil Rathenau. Though listed separately in the exhibition's official catalog, Edison's exhibit overlapped physically with those mounted by the various Edison lighting companies, and with which it was often conflated in journalistic accounts. The companies (including the Edison Electric Light Co., the Edison Co. for Isolated Lighting, and the three manufacturing shops that Edison controlled) exhibited dynamos, lamps, conductors, meters, and other equipment, as well as a working three-wire system that provided much of the Exhibition's lighting. Among the "large assortment" of lamps provided by the Edison Lamp Co. was "a variety of toy lamps, for stage-jewelry, scarf-pins, etc., and . . . small pocket-batteries for lighting the same." Sherburne Eaton appointed William Hammer to oversee these exhibits as well ("Hammer, William J.," Pioneers Bio.; Jackson 1934, 771; Hammer to TAE, 16 June 1886, DF [*TAED* D8436ZCU]; Franklin Institute 1884, 25–26, 61; "Opening of the International Electrical Exposition, Philadelphia, Pa.," *Sci. Am.* 51 [13 Sept. 1884]: 168; Edward Johnson to John Hoskin, 8 Sept. 1884, Series 1, Box 1, Folder 3, WJH). One of the most popular sites of the entire exhibition reportedly was an Edison lighting spectacle consisting of a tall column studded with 2,100 colored lamps, representing one day's production at the Lamp Works. The various colors were automatically switched in and out of the circuit, while flashing lights successively spelled out the word "EDISON" at the tower's base. Taken as a whole, the International Exposition reportedly contained 50,000 incandescent lamps and 350 arc lights aggregating to more than 1,000,000 candlepower ("International Electrical Exhibition," *Indiana [Pa.] Democrat*, 25 Sept. 1884, n.p.; Stieringer 1901b, 285; "Electrical Wonders," *Chicago Daily Tribune*, 3 Sept. 1884, 1; "Electrical Wonders," ibid., 13 Sept. 1884, 9).

Scientific American's illustrations of Edison exhibits, with lighted tower, at the International Exposition.

From Henry Villard

[New York,] June 4, 1884

Dear Mr. Edison:

In response to another sudden summons from Europe, I sail today.[1]

I have not been able to find time, to see you again, to say goodbye to you.

I expect to return early in August, and hope that in the meantime some progress will be made towards a definite solution of the pending problems. I am not at all discouraged, but believe as strongly as ever in the light.

But our general situation is certainly anything but satisfactory, and we must all act in a spirit of concession.[2] Yours truly,

H. Villard   S[pofford].[3]

L, NjWOE, DF (*TAED* D8403ZDV).

1. The editors have not identified the "summons." Villard's departure came two days before a U.S. Circuit Court permitted a stockholder lawsuit to proceed against Villard and the Oregon Railway and Navigation Co. According to one news account, Drexel, Morgan & Co., to whom Villard had defaulted on personal loans, blocked a plan by the Edison Electric Light Co. to send Villard to Europe as its representative in May. Another report denied, seemingly authoritatively, that there ever had been such a plan. In any case, Villard received warm welcomes in London and Bavaria. He apparently returned to New York on 2 August but departed again on the 20th, and he did not resume full-time residence in the United States until 1886, when he came as a representative of Deutsche Bank. "Western Railways," *Boston Daily Advertiser,* 8 May 1884, 2; "Suing a Villard Company," *NYT,* 7 June 1884, 8; "Mr. Villard Will Not Sail," ibid., 4 May 1884, 2; "Departures for Europe," ibid., 21 Aug. 1884, 3; *Passenger Lists* 1962, microfilm M237_479, line 15, list number 977; *ANB*, s.v. "Villard, Henry"; de Borchgrave and Cullen 2001, 343–50; Kobrak 2007, 35–44.

2. Villard probably referred not only to the difficulties of the Edison Electric Light Co. and the central station construction business, but also to the general economic climate. Generally poor business conditions had turned to near-panic in mid-May with the collapse of several banks and brokerages, starting with the partnership of former president Ulysses Grant. With credit tightening, stock prices fell sharply in the latter part of the month. "On the Verge of a Panic," *NYT,* 15 May 1884, 1; "The National Bank Failures," ibid., 20 May 1884, 1; Editorial, ibid., 25 May 1884, 8; "The Financial World," ibid., 25 May 1884, 9; Strouse 1999, 244.

3. Charles A. Spofford (1853?–1921), an attorney, was Villard's personal secretary and a director of the Northern Pacific Railway and, subsequently, other companies. A native of Ohio, Spofford was the son of the longtime Librarian of Congress, Ainsworth Spofford. Obituary, *NYT,* 7 Mar. 1921, 10; Villard to Edison Electric Light Co. of Europe, Ltd., 29 June 1886, DF (*TAED* D8630ZCA).

[New York,] June 4th. 1884.

Samuel Insull to
Samuel Edison

Dear Mr. Edison:—[1]

T.A.E has requested me to send you four checks, for $50. each, which please find enclosed. You will notice that one is dated to-day, another on the 11th., another on the 17th. and another on the 23rd. I trust these will answer your purpose.[2]

T.A.E. is in very good health, and so is his family. They have left the Clarendon Hotel and are living again at 25 Gramercy Park.

I presume you have heard of the death of old Mr. Stilwell, Mrs. Edison's father.

I was very sorry I could not be here to see you, when you came on last Winter. When are you coming on again?

With my hearty congratulations at your being eighty years of age and all sound,[3] believe me, Very sincerely yours,

SI

TLS (carbon copy), NjWOE, DF (*TAED* D8416BST).

1. Edison's father, Samuel Ogden Edison (1804–1896). Insull addressed this letter to his home in Fort Gratiot, Mich. *TAEB* 1 chap. 1 introduction, esp. n. 2; Obituary, *NYT,* 27 Feb. 1896, 4.

2. Samuel Edison had asked his son in May for $100 or $200 to complete construction of a store building in Fort Gratiot. He expected to rent the new store for $15 monthly and claimed to earn $58.84 in monthly rents from a handful of properties in town, including the office of the *Fort Gratiot Sun.* Samuel pointed out that it had been "some month since I have heard from you or any of yours" (Samuel Edison to TAE, 30 May 1884, DF [*TAED* D8414K]; *History of St. Clair County* 1883, 422). Since 1870 and as recently as December 1883, Edison had occasionally advanced—or offered to send—money to his father (see, e.g., Docs. 99, 129, 178; TAE to Samuel Edison, 21 Oct. 1877, EP&RI [*TAED* X001A1BD]; TAE receipt to Samuel Edison, 22 Jan. 1880; Samuel Edison to TAE, 14 Jan. 1883; both DF [*TAED* D8015C, D8314A]; Insull to Samuel Edison, 21 Mar. 1881, Lbk. 8:94A [*TAED* LB008094A]; Cash Book [1 Jan. 1881–30 Mar. 1886]: 222, Accts., NjWOE).

Samuel Edison had offered to hold all or part of any check from his son. The fact that Edison had the money sent in four separate checks, three of them with different future dates, likely reflected his poor cash situation. Large sums flowed into his two checking accounts, but they flowed out just as quickly. On 1 June, Edison had $18.64 on deposit at Drexel, Morgan & Co. and $3.80 at the Bank of the Metropolis. Cash Book (1 Jan. 1881–30 Mar. 1886): 254, 260, Accts., NjWOE.

3. Samuel had closed the letter to his son (see note 1): "Ever yours   80 years of age and all Sound."

To Sherburne Eaton

[New York,] June 9th. 1884.

Dear Sir:—

As you aware there was a very great necessity of having a bill passed through the New Jersey legislature permitting Municipalities to grant the right of way for electric light companies.[1]

It was quite impossible for us to do any central station business in New Jersey until this bill was passed.

After the failure of the Gramme Company[2] to get a bill through, I had a bill drawn up, and employed Mr. J. F. Fisher[3] to get the bill through. You undertook to be responsible for $1,000. on behalf of the Light Co. in connection with this matter.[4] In addition to this amount there were Mr. Fisher's personal expenses, amounting to $753.00., which I have paid out of my own pocket.[5]

I shall be glad if you will bring this matter before the next meeting of the Executive Committee, or Board, of the Light Co, with a view of their reimbursing me this amount.

I annex a memorandum of amounts paid to Mr. Fisher, duly certified by him.[6] Very truly yours,

TAE

L (carbon copy), NjWOE, DF (*TAED* D8416BTN). Initialed for Edison by Alfred Tate.

1. A bill permitting "cities of the second and third class" to authorize electric lighting companies to put up poles or lay wires under the streets passed the New Jersey legislature in April and was signed by the governor in mid-May. Similar legislation applying only to second-class cities had been introduced in the Senate in January and quickly withdrawn. In its place, the Gramme Electrical Co. found backers in both houses for a more prescriptive proposal requiring municipal consent and direction for any company to place "tubes, wires or conductors for <u>any</u> purpose, electrical or otherwise" (presumably including gas transmission) through public streets. Edison disliked the Gramme bill, reportedly thinking it both unconstitutional and private in nature, or applicable only to specific individual interests. *Acts* 1884, 94; Joseph Fisher to TAE, 16 Apr. and 13 May 1884; Senate Bill No. 84, 28 Jan. 1884; Rowland Hazard to TAE, 8 Feb. 1884; all DF (*TAED* D8403ZBO, D8403ZDG, D8427G, D8427L).

2. The Gramme Electrical Company was a consortium of electric companies organized in 1881 to coordinate patent infringement settlements, legislative efforts, and other shared business interests. It was headed by Rowland Hazard. The Edison Electric Light Co. joined in 1882. Docs. 2103 n. 4 and 2274 n. 9.

3. Joseph F. Fisher (1835?–1885) was a former local official, banker, and postmaster in New Brunswick, N.J. Fisher had been arrested for bank fraud and embezzlement in 1878 and, when similar allegations surfaced in 1881, he was dismissed from the office of Postmaster. The

following year, he was identified as having attempted to bribe a state legislator. Later, in September 1884, Fisher was named as a key member of a long-term conspiracy to defraud the city of New Brunswick; he committed suicide twelve months later. "Death from Escaping Gas," *NYT*, 11 Sept. 1885, 2; "Directors Accused of Fraud," ibid., 24 Nov. 1878, 1; "New Brunswick Postmaster," ibid., 1 Oct. 1881, 1; "Saving the Jersey Shore," ibid., 1 Apr. 1882, 1; "The Bankrupt City," ibid., 9 Sept. 1884, 1.

4. Fisher later complained, in an undated letter to "Friend Edison," that the Edison Electric Light Co. had unfairly deducted $150 from the $1,000 promised him when, in fact, "that amount was for other partys." Edison noted, "OK   I saw Fisher," to whom he gave a check for $150 (charged to the Construction Dept. account) on 3 July. Fisher to TAE, n.d. [1884], DF (*TAED* D8431ZAU); Edison Construction Dept. Cash Book (1 June 1883–28 Feb. 1886): 92, Accts., NjWOE.

As part of his efforts in the state legislature, Fisher helped win a reduction of the state tax levied on the gross income of electric light, telegraph, telephone, and cable companies not owned by railroads. Edison associate Frank McLaughlin had some evident involvement in Fisher's work, having instructed Alfred Tate in January to send Electric Light Co. Bulletins and other printed material to Fisher in Trenton. *Acts* 1884, 232–35; Fisher to TAE, 16 Apr. 1884; McLaughlin to Tate, 29 Jan. 1884; both DF (*TAED* D8403ZBO, D8427H).

5. This amount was reimbursed to Edison, through a Construction Dept. account, on 30 September. Edison Construction Dept. Ledger (1883–1886): 209, Accts. (*TAED* AB033 [image 102]).

6. Not found.

-2688-

Memorandum of Agreement with George Sherman

New York, June 9th 1884[a]

ⱵWe have just heard a recitation of the understanding between Mr. Edison and Col. Geo. W. Sherman[1] in regard to a proposed Electric Light Company to operate in the Argentine Republic which is as follows:—

Mr. Edison and Col. Sherman entered into an agreement to use their endeavors to form a company for the purposes as above stated, Mr Edison having received authority from the Edison Elec. Light Co'y. enabling him to engage in such enterprise.[2]

Mr Edison has stated and Col. Sherman assented, that the mutual understanding between them[b] is that if they are successful in forming the Company they will make a division of the stock, in accordance with the written agreement between them under date May 24th 1884 but if they should be unsuccessful in organizing the Company neither of them will look to the other for compensation but they will simply regard the time as lost which they have spent attempting to form the Company.[3]

They further state that this is the true spirit and essence of the agreement between them above referred to under date May 24th 1884—

A. O. Tate[4]   Saml Insull

The above is in accordance with our understanding[c]

Thos A Edison   G. W. Sherman

DS, NjWOE, Miller (*TAED* HM840221). Letterhead of Thomas A. Edison; written by Samuel Insull. [a]"NEW YORK," and "188" preprinted. [b]Interlined above. [c]Sentence written by George Sherman.

1. George W. Sherman was born in Mexico in 1848 and is identified in the prospectus of the Argentine Edison Electric Light Co. as a civil engineer who had been in Buenos Aires for several years (*Passenger Lists* 1962, microfilm M237_477, line 1, list number 376; Argentine Edison Electric Light Co. prospectus, 17 May 1883, DF [*TAED* D8336O]). Besides his interest in the Edison electric light, Sherman appears to have had or sought business relations with several American companies (Fuller Electrical Co. to TAE and Albert Chandler, 1 Dec. 1883; F. M. Wilder to TAE, 18 July 1884; Sherman power of attorney to Richard Kolb, 1 Aug. 1884; all DF [*TAED* D8336ZAM, D8403ZEV, D8468ZBI]). His first connection with the Edison electric light was in the spring of 1883 in connection with the organization of the Argentine Edison Electric Light Co. (see Doc. 2495). Negotiations between Sherman, Edison, and the Edison Electric Light Co. continued over the course of the next year. During that period Sherman became quite ill twice and ended up in the hospital both times, with Edison paying the bills for his hospitalizations (J. M. Callum to TAE, 14 July 1883, DF [*TAED* D8336X]; New York Hospital bill, 18 July 1883, Cat. 1164, Accts.; New York Hospital bill, 4 Nov. 1883, Voucher no. 150 [1883]). Edison later recalled that Sherman "has a fair Knowledge of Electric Lighting" and was "honest but very erratic & impracticable in business but a fair engineer   He exaggerated also. . . . [and] claimed relationship with prominent persons in Argentine Repub which I knew was untrue" (TAE marginalia on Juan Navarro to TAE, 24 Mar. 1888 and on N. C. Romero to TAE, 19 Apr. 1888, both DF [*TAED* D8805ABT, D8805ACH]).

2. According to the 24 May 1884 agreement between Edison and Sherman, there was a separate agreement between Edison and the Edison Electric Light Co. regarding the assignment of patent rights to the Argentine Edison Light Co. (Miller [*TAED* HM840218]); presumably this is the agreement of 2 July 1883 mentioned in Doc. 2677.

3. The 24 May agreement between Edison and Sherman (see note 2) called for Edison to give Sherman 280 shares in the Argentine Electric Light Co. and to pay Sherman's travel expense from New York to Buenos Aires. In addition, Edison was to use his influence to procure for Sherman the position of consulting engineer to the Argentine company for three years at a salary of $2,000 per year. In exchange, Sherman was "to obtain from time to time all such grants, privileges and concessions as may be necessary to enable" the Argentine company to build central stations in Buenos Aires and elsewhere in Argentina. He was also "to devote his time and energies to the development and suc-

cess" of the Argentine company and "to cause the introduction of the Edison Incandescent Light in the Argentine Republic, so far as he may be able." However, some disagreement appears to have arisen between Edison and Sherman over compensation for the "time and money" Sherman had already expended in regard to the electric light in Argentina, probably a reference to his return to Buenos Aires in early 1884 when he helped negotiate a grant for the whole city (Sherman to TAE, 5 Mar. 1884 and 18 July 1884; George Bidgood to TAE, 5 Aug. 1884; all DF [*TAED* D8337ZAH, D8403ZEU, D8434ZAG]; see also Doc. 2495 n. 3). Nothing further appears to have come of this arrangement (but see note 4 below).

4. Alfred Tate was to have accompanied Sherman as a technical expert and to have shared in his contract with Edison. However, on 3 August Tate expressed surprise in a letter to Insull that he had "received a telegram from Sherman stating that he had arranged things satisfactorily with Mr. Edison and was prepared to sign contracts and adding 'Come immediately I leave on Saturday [August 2].'" Tate wondered "if Mr. Edison and yourself have decided to take him into the business again" and suggested that "you can have contracts drawn such as were at first proposed." Insull telegraphed back that "Sherman left Saturday [August 2]   we have made no arrangements with him." Sherman power of attorney to Richard Kolb, 1 Aug. 1884; Tate to Insull, 3 and 6 Aug. 1884, all DF (*TAED* D8468ZBI, D8465ZAD, D8465ZAE); Insull to Tate, 6 Aug. 1884, LM 21 (*TAED* LBCD8496A); Tate 1938, 82–84.

-2689-

Technical Note:
Electric Light and
Power[1]

No 1[2]

NEW YORK, June 10 1884[a]

show on 3 wire System   Use where water power— Long
distance high e[lectro]m[otive]f[orce]—straggling Village or
Town—
No 2[3]

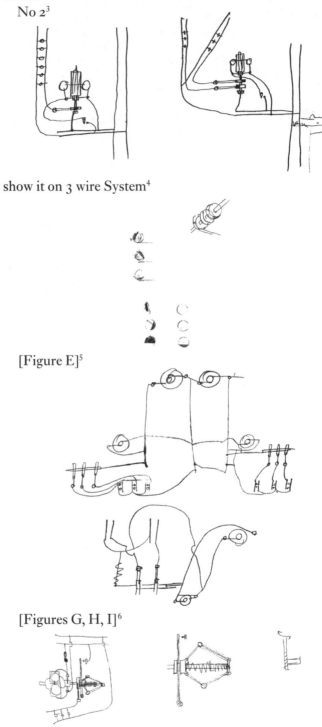

show it on 3 wire System[4]

[Figure E][5]

[Figures G, H, I][6]

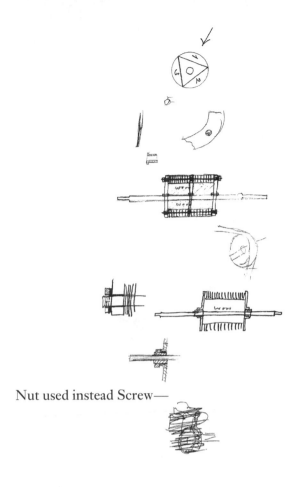

Nut used instead Screw—

TAE

X, NjWOE, Lab., Cat. 1150 (*TAED* NM019AAL). Document multiply signed and dated. First four figures and associated text on letterhead of Thomas A. Edison; final three figures and text on lined paper; remainder on message forms of Western Union Telegraph Co. a"NEW YORK," preprinted.

1. Edison executed a patent application based on this note on 16 July. The resulting patent applied to methods of controlling "the energy consumed by an incandescing electric lamp . . . independent[ly] of the tension of the current and the resistance of the lamp . . . by interrupting the flow of current to the lamp" so rapidly that "the light can be made to appear constant to the eye." Edison intended for current to flow through the lamp circuit "for only a fraction of the time . . . that fraction being inversely the number of times the tension used is a multiple of that required when the lamp is constantly in circuit." He proposed to use "a rapidly-acting circuit controller, which will throw the current first through one lamp-circuit and then through another. . . . This circuit-controller is preferably a revolving shaft carrying circuit-controlling wheels upon which rest suitable springs or brushes," the shaft being revolved by a motor in the high-voltage circuit. U.S. Pat. 391,595.

2. This drawing, corresponding to figure 1 in Edison's U.S. Patent

*In the adaptation of circuit-controlling motors to the three-wire system, motors **B** are in circuit with high-voltage lines 1 and 2 and compensating conductor 3; **D** represents local house circuits.*

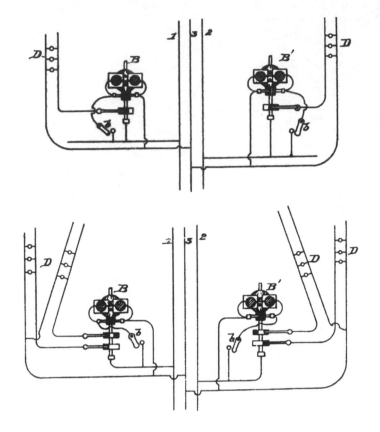

391,595, represents a high-voltage circuit from the dynamo, at bottom, divided among three local feeder circuits through the rotating make-and-break circuit controller so that each circuit is "completed for one-third of the entire time."

3. The two sketches associated with "No 2" represent this system applied to a local house circuit. They correspond generally to figures 3 and 4 in the patent specification. The high-voltage lines (at right) extend directly to the house; the motor would control the lamp circuit (or multiple circuits, as in the second figure) at left, inside the dwelling. In the patent drawing, the motor controlling a single circuit was depicted with a spring governor. U.S. Patent 391,595.

4. Patent figures 5 and 6 depicted three-wire versions of the single and dual house circuits represented in the drawings of "No 2" above (see note 3).

5. This schematic diagram shows the arrangement represented in the first drawing above as it could be applied to the three-wire system. It was the basis for figure 2 in Edison's U.S. Patent 391,595.

6. Edison suggested in his U.S. Patent 391,595 that each motor, whether in a feeder or a local house circuit, should be equipped with a speed governor. The governor would open the circuit when the motor's speed dropped below a certain threshold, in order to prevent high-voltage currents from reaching the lamps for an excessively long interval.

Charles Coster and Grosvenor Lowrey Report to Edison Electric Light Co.

New York, June 18, 1884

Report of the Committee of Three to the Directors of the Edison Electric Light Co.[a]

The Committee of Three appointed at the meeting of the Board of Directors on May 12th. 1884, report:[1]

FIRST: That they have had two full meetings, since which time the illness of Mr. Adams[2] has made it necessary for the other two members of the Committee to conduct negotiations with Mr. Edison and his associates, with whom they have almost daily held either formal or informal meetings.

SECOND. At the outset they determined to submit a proposition which should have the effect of securing to the Company a suitable interest in, and large influence over the manufacturing business; re-organize the business of exploitation in a manner satisfactory to Mr. Edison; and at the same time reduce the expenses of the parent Company to the minimum, but with a maximum of efficiency and benefit, having in view the purpose for which it was originally organized.[3]

THIRD. The proposition was substantially as follows:[4] (a) That the various shops should sell to the parent Company an interest of twenty per cent, and to the Isolated Company a like interest, making an aggregate of forty per cent., and receive in payment therefor, stock of the two Companies at par; the interests in the shop to be appraised. (b) That the Light Company should make a contract with the Isolated Company, conferring upon it power to install central stations and plants, and requiring it to maintain the requisite staff of engineers, officers, &c. and offering to it such inducements as might be proper to lead it to undertake and develop the business.[5] (c) That the Isolated Company should be re-organized by the election of such executive officers as would meet Mr. Edison's desire that the business should be pushed with great energy. (d) That the staff of the Light Company should be reduced to a President, who should also be Treasurer, one engineer, and an Attorney, all to be employed upon salaries; the President to give his attention especially as director in the Isolated Company to its interests there; and in his own proper office as President, to the initiation and prosecution of suits, the perfection of patents, the revision, if desired, of the contract now proposed to local illuminating Companies &c., &c. (e) That Mr. Edison should extend his contract with the Light Company for at least three years,[6] and returning to his laboratory, undertake the perfection of old and the production of new inventions.

FOURTH. This plan was first explained by one of the Committee privately to Mr. Edison, who appeared at the time to make no objection to any part of it, except the extension of his contract, which he exhibited strong disinclination to assent to. Subsequently the paper containing this proposition was considered by Mr. Coster,[7] Mr. Lowrey, Mr. Johnson, Mr. Batchelor and Mr. Edison together; but no definite determination was reached except that it was learned that both Mr. Johnson and Mr. Batchelor were strongly disinclined to part with the interest in their shops upon the ground that those interests furnished to them present means of livelihood which they were indisposed to change for shares of a greater speculative value, but without present income.

FIFTH. Co-incidently with our pressing this plan, the making of contracts with the shops by which their profit should be limited, and they should be led to recognize all the patents of the Light Company, and also to assign to it all patents now belonging to them or which they might hereafter acquire, (except those of Mr. Edison) was discussed.

SIXTH. This method without change of interest was strongly urged by Messrs. Edison, Batchelor and Johnson.

SEVENTH. After many meetings Mr. Edison declared positively his unwillingness at this time to extend his contract, or to enlarge his interest in the parent Company by an exchange for an interest in the shops; and declared at the same time, as the Committee understood it, that his objection was solely based upon his disbelief in the ability of a Company to carry out the projects which he had in view; but that should the reorganization take place with the result of introducing the element into the management, which he thinks is now lacking, he would then feel disposed to consider such a proposition.

EIGHTH. In the meantime a conference with Mr. Edison had shown that it was his desire that Mr. Johnson should take the presidency of the Isolated Company, to carry on its business subject to the control of a small Executive Committee, for the composition of which Major Eaton, Mr. Upton, Mr. Coster, were named. Before Mr. Johnson was named for this place by Mr. Edison, he had announced his willingness to make an exchange of all his interests in all shops for an interest in the Light Company upon terms to be agreed upon under the general principle of appraising, as your Committee understood, all interests in the shop and taking the Light Company's stock at par therefor. Mr. Upton also expressed in a general way his willingness to make a similar exchange. But in respect to the interests of Mr. Upton, no definite conversation has taken place. At the time of writing this report the Committee is still negotiating with the Manufacturers for some partial exchange of interests and will make a verbal report should anything further develop in this connection.[8] Irrespective of such exchange it is the belief of the Committee that if reasonable limit of profit to the Manufacturers be fixed nothing better can be adopted at this time than to make contracts with them in general like the one now submitted, giving to the Parent Company the right to terminate the same at will upon a prescribed notice. The Committee further recommend that the Light and Isolated Companies be re-organized in the manner proposed, and that the Light Company's staff be reduced under the advice of Major Eaton, so that it shall carry the lowest salary and general expense account, consistent with keeping it in full possession of such knowledge of the business as would enable it at a future time to resume all that it may now part with with effect.

A subject requiring early attention is a readjustment of the relations between the Light Company and the Illuminating Company of New York City. A verbal suggestion will be made on this matter.

An outline of a contract between the Parent and the Isolated Companies is herewith submitted; ⟨Copy of this proposed contract is not sent herewith but will be submitted at the meeting.⟩ but in connection with this as with other points all further report can in the opinion of the Committee be best made orally at the Meeting of the Board.

(Signed)   G. P. Lowrey,   C. H. Coster

TD (printed copy), NjWOE, DF (*TAED* D8427ZBJ). Document multiply dated. [a]Heading handwritten and printed lithographically on separate slip.

1. This body apparently succeeded the Special Committee on Manufacturing and Reorganization after Henry Villard, its chairman, reported to the Edison Electric Light Co.'s board about unsuccessful negotiations with Edison (see Doc. 2661 n. 1). As noted in this document, the committee consisted of Charles Coster, Edward Adams, and Grosvenor Lowrey (who had helped Edison negotiate with the bankers forming the Edison Electric Light Co. in 1878). The group planned to convene for the first time on 14 May, and Lowrey apparently sought to meet with Edison a few days later about its business. Eaton to TAE, 14 May 1884; TAE to Lowrey, 19 May 1884; both DF (*TAED* D8427ZBE, D8416BPR).

2. Financier Edward Dean Adams (1846–1931), a partner in the New York banking firm of Winslow, Lanier & Co. (1878–1893), was a trustee of the Edison Electric Light Co. (1882–1886) and a director of the Edison Electric Illuminating Co. (1884–1889). A former student at the Massachusetts Institute of Technology (1865–1866), Adams used his

engineering knowledge in official positions at several railroads around this time and, later, as a principal developer of hydroelectric generation at Niagara Falls. Doc. 2189 n. 3; *ANB*, s.v. "Adams, Edward Dean."

3. The company was organized for the purpose of owning, "making, using and vending and licensing others to make, use and vend . . . all the inventions, discoveries, improvements and devices" of Edison relating to electric lighting, power, and heating. See Doc. 1576.

4. This proposal was probably based on a draft contract that Coster submitted to Edison on 2 June, following a discussion between the two men. Coster acknowledged that he did not "know how the other members of the Committee will view it. I want your ideas first." Coster to TAE, 2 June 1884, DF (*TAED* D8402J).

5. The Edison Co. for Isolated Lighting effectively absorbed the engineering function by July (TAE to Sidney Paine, 22 July 1884, LM 20:51 [*TAED* LBCD7051]). A formal agreement between the Isolated Co. and the Edison Electric Light Co. was executed on 1 September 1884 along with a series of agreements between Edison Electric and the manufacturing firms (Doc. 2725).

6. The contract (Doc. 1576) entitled the company to all of Edison's electric light and power patents for seventeen years from November 1878. The committee likely referred to the initial five-year period during which the company could acquire Edison's inventions or improvements without giving him additional compensation.

7. On the recommendation of Egisto Fabbri, J. P. Morgan selected Charles Henry Coster (1853?–1900) to succeed Fabbri as a partner in Drexel, Morgan & Co. at the start of 1884. Coster was also named to fill Fabbri's place as trustee of the Edison Electric Light Co. and the Edison Co. for Isolated Lighting in September 1883, and he served on the latter firm's executive committee from November. Coster had started his business career as a clerk for a New York shipping company that was acquired by Fabbri & Chauncey in 1872, and he remained with Fabbri & Chauncey for the next eleven years. Coster developed a renown for his ability, like Morgan's own, to analyze and recall highly detailed information. The effort to reorganize the Edison lighting interests was among Coster's first assignments at Drexel, Morgan & Co. From the mid-1880s until his death, Coster engineered the firm's reorganization of numerous railroad lines—a process called "morganization"—and became known as "Morgan's right arm." U.S. Census Bureau 1970 (1880), roll T9_880, p. 89.4000, image 0120 (New York [Manhattan], N.Y.); Obituary, *NYT,* 14 Mar. 1900, 7; "Changes in Banking Firms," ibid., 23 Dec. 1883, 4; "The Business Career of Charles H. Coster," ibid., 18 Mar. 1900, 24; Eaton to TAE, 28 Sept. 1883, DF (*TAED* D8327ZBF); Edison Co. for Isolated Lighting annual report to stockholders, 18 Nov. 1884, PPC (*TAED* CA002D); Carosso 1987, 168; Strouse 1999, 245–46.

8. See Doc. 2725 n. 6.

Charles Batchelor to
John Ott

New York, June 26 1884[a]

My Dear Ott,

Edison wants you to come down to Goerck St early this morning take charge of my experimenting.[1] He says leave the other thing just at present.[2] We are anxious to have a lot of experiments made quick.[3] We have three men there and you can get a good deal of work out of them by directing their work well.

Batchelor

'Come right away'

ALS, NjWOE, DF (*TAED* D8431ZAE). Letterhead of Edison Electric Light Co. [a]"*New York*," and "*188*" preprinted.

1. Batchelor took the title of general manager of the Edison Machine Works soon after his return from Paris, likely at a 21 May meeting called to discuss "the future conduct of the Machine Works." These events followed a complete shutdown of the shop for maintenance. Batchelor also seems to have assumed day-to-day operational responsibilities from Gustav Soldan. By October 1884, the shop's letterhead identified him as general manager and treasurer. Edison Machine Works to Batchelor, 19 May and 10 June 1884; Samuel Insull to Soldan, 6 May 1884; all DF (*TAED* D8416BPL, D8416BTU, D8416BNF); letterhead, Edison Machine Works to TAE, 8 Oct. 1884, Miller (*TAED* HM840231A).

2. The most recent surviving records of Ott's work show he had been experimenting on 5 June with the method of annealing wire for which Edison had already filed patent applications (see Doc. 2675). In late May, Ott had been testing a dynamo on quadruplex telegraph circuits (see Docs. 2651 and 2676). Unbound Notes and Drawings (1884), Lab. (*TAED* NS84ACL, NS84ACK2).

3. The editors have not determined what Batchelor had planned. From the time of his return to New York until early June, he had been making calculations related to dynamos, first to determine the distribution of energy (especially heat) in the machines and then to evaluate possible changes in the winding of field coils, perhaps with slow-speed operation in mind. Records of Batchelor's experimental work (on dynamos and arc lights) during the summer are scant. Cat. 1306, 1235:72–78, Cat. 1234:12; Batchelor (*TAED* MBN011AAO–MBN011AAW, MBN012072, MBN012073, MBN012076, MBN012077, MBN012078, MBN005012).

July–September 1884

Edison spent increasing time in his laboratory in July, leaving to Samuel Insull the day-to-day tasks of closing up the operations of the Edison Construction Dept. This included settling accounts with the local illuminating companies and dealing with complaints regarding the operation of their stations. Edison and Insull also sought to settle accounts with the Edison Electric Light Company, in regard to canvassing and engineering expenses, and with Ansonia Brass & Copper Company, which had extended credit for the cost of conductors for the stations. During this time, Edison also decided to give up his offices on the third floor of 65 Fifth Avenue and to carry out what little business he still conducted there in his library, which he still maintained on the top floor. That business included negotiations between Edison and his manufacturing partners and Edison Electric and the Edison Company for Isolated Lighting regarding the latter company's assumption of the central station business. These discussions culminated in a series of agreements signed on 1 September in which the Isolated Company took over the central station business and Edison Electric received the right of first purchase of stock in the manufacturing shops and of any patents obtained by the shops. These agreements were part of a larger reorganization of the Edison lighting companies that culminated with a proxy fight waged by Edison at the October meeting of Edison Electric.

During the last two weeks of July, Edison was spending most of his time in the laboratory conducting experiments on the direct conversion of coal into electricity. He also experimented with batteries to run small motors and sketched out experimental apparatus related to his growing interest in the conversion of heat and light into electricity and magnetism. In an interview with the *New York Daily Tribune* that appeared on 4 August, Edison noted, "I am keeping pretty busy. I am going into original experimenting again. I'll get out a new crop of inventions during the next year in the electrical line."[1] With his attention again focused on invention, Edison also sought changes in the division of electricity at the Patent Office, which was having trouble adequately dealing with this growing class of patents.

It is unclear whether the Edison family spent much time in Menlo Park during June and July although Edison did place an impressive order for fireworks to be delivered there for the Fourth of July.[2] The family was likely away from Menlo Park on 21 July when a sheriff's sale of the property was held to satisfy Lucy Seyfert's judgment against Edison; Charles Batchelor bought the property and later reconveyed it to Edison. Mary and the children appear to have been at Menlo by the end of July.[3] Why Edison, who was spending most of his time in the laboratory, returned suddenly to Menlo on the evening of 7 August, a Wednesday, is unknown, but it may have been due to his wife's health.[4]

Little is known of the circumstances, but Mary Edison died unexpectedly at their Menlo Park home on the morning of 9 August. The doctor in attendance reported only that she died of congestion of the brain, a general diagnosis based on symptoms that could result from several more specific causes of death. Newspaper reports later claimed that Edison sought to revive his wife using electricity but to no avail. Daughter Marion remembered that she "found my father shaking with grief, weeping and sobbing so he could hardly tell me that mother had died in the night."[5] Marion also recalled driving her father around the countryside at least once a week during the rest of the summer, and she became her father's almost constant companion over the next several months. In September, she accompanied him to Philadelphia; in early October, she also spent time in his laboratory. Marion and her brothers Thomas and William were placed in the care of their maternal grandmother, who helped keep house when the family moved back to New York in September. They did not return to their Gramercy Park home but instead took up residence on the third floor of a house at 39 E. 18th Street.

Edison and Marion traveled to Philadelphia on 4 September to visit the International Electrical Exhibition, which opened two days earlier. Although he had accepted an appointment to the National Conference of Electricians, which sought to develop standards for electrical measurement, Edison left on 6 September and did not attend its meeting from 8 to 13 September. Edison and Marion did return to Philadelphia on 16 September and spent three days touring the exhibition.[6] The *Electrical Review* reported, "He is a very important part of the exhibition, and is welcomed wherever he shows his face. His little daughter, in deep mourning for her mother, is his constant and inseparable companion, and it is a touching and pathetic sight to see them going about hand in hand, the observed of all observers."[7] It is not known if Marion accompanied Edison when he returned to Philadelphia at the end of the month.[8] By then she may have been attending Madame Mears's school on Madison Avenue, between Thirty-sixth and Thirty-seventh Streets. Her brother, Thomas Jr., was enrolled in M. W. Lyon's Collegiate Institute located on Twenty-second Street at Broadway.[9]

The exhibits of Edison and his companies, which were located directly opposite the main entrance, were the largest and most prominent, leading an Associated Press agent to claim they were "the center of attraction."[10] The combined exhibit of the Edison companies featured both the central station and isolated lighting systems, while Edison's own display showed his many electrical inventions, including what became known as the Edison Effect lamp, his telegraph and telephone inventions, and the phonograph. The centerpiece of the Edison exhibits was a tower made up of 21,000 colored lights. During one of his visits to Philadelphia, Edison also sat for a portrait taken by the art photographer William Curtis Taylor with the aid of electric lights, which was then hung in one of Edison's exhibits. A later addition to the Edison exhibit was a bronze bust of the inventor by artist Rupert Schmid, whose own display in Philadelphia included two casts of "Edison's hands in bronze, holding incandescent lamps."[11]

While the acclaim Edison received in Philadelphia was no doubt satisfying, the most gratifying experience may have been reconnecting with his old friend Ezra Gilliland. Gilliland, whom Edison had known since their days together as telegraphers in Cincinnati, was then head of the American Bell Telephone Company's experimental shop in Boston. During their meeting in Philadelphia, Edison "mentioned that his electric light was completed and practically off his hands and he was talking of what would be a good thing to take up next."

This led Gilliland to suggest several ideas to which Edison could turn his attention. Edison was most intrigued by Gilliland's work on long-distance telephone transmitters for the Bell Company, and he decided that this would be "the best thing to turn his attention to at that time and soon after made an arrangement with the American Bell Telephone Company whereby he was to carry on some experiments with the view of improving the transmitter.[12] Telephone technology would become the focus of Edison's efforts over the next few months.

1. "Talking About Electric Motors," *New York Daily Tribune,* 4 Aug. 1884, 8.

2. Edison ordered nearly $90 worth of fireworks to be delivered to Menlo Park. Unexcelled Fireworks Co. bill, 2 July 1884, Voucher no. 371 (1884). Evidence from bills suggest the family was either spending some time or preparing to spend time in Menlo Park during June and early July; see Gilman Collamore & Co., 1 June 1884; B. Y. Ford, 4 June and 1 July 1884; Manning Freeman, 7 July 1884; Alex Ayres, 7 July 1884; Voucher nos. 312, 314, 386, 405, 406 (1884).

3. Edison wrote "my family is away from City" on a letter from Rupert Schmid dated 30 July 1884 (DF [*TAED* D8403ZFE]); see also Doc. 2712 (headnote).

4. On the same day he returned to Menlo Park, Edison had written to Princeton professor Charles Young, "I spend a good deal of my time at my Laboratory . . . so that when you come into Town it would probably be best to call there." Lbk. 18:299 (*TAED* LB018229).

5. Oser 1956, 5.

6. Continental Hotel bill, 19 Sept. 1884, No. 533 (1884), Vouchers; see also Doc. 2732 n. 4.

7. "Notes of the Exhibition," *Electrical Review,* 27 Sept. 1884, 5.

8. On 25 September, Samuel Insull wrote George Forbes that Edison had received his letter "of yesterday's date just as he was leaving for Philadelphia" and that Edison "expects to remain in Philadelphia for the next few days." Lbk. 18:425 (*TAED* LB018425).

9. Doc. 2747 n. 2.

10. "International Electrical Exhibit," *Indiana (Pa.) Democrat,* 25 Sept. 1884, 2; "Our Philadelpia Special," *Frederick (Md.) Weekly News,* 11 Sept. 1884, 1; "The Edison Exhibit at the Philadelphia Electrical Exhibition," *Sci. Am.* 51(1884): 239, 246; *Report of Examiners* 1884, Section 19, pp. 4–5, 7–11, 23; *Report of Examiners* 1886, Section 26, p. 4; *Report of Examiners* 1885, Section 29, p. 53; Banes 1885, 22–23, 25, 27.

11. Clipping enclosed with Chester Pond to TAE, 21 Oct. 1884, DF (*TAED* D8403ZHX); Doc. 2722 n. 3.

12. Gilliland's Testimony on Behalf of Edison, p. 2, *Edison and Gilliland v. Phelps* (*TAED* W100DKB [image 2]).

To Erastus Wiman

My Dear Wiman:—[1]

Some one has sent me a copy of the Chicago *Inter-Ocean*,[2] containing your interview with a reporter of that journal, which I read with great attention.[3]

How much would Western Union lose under the following circumstances? Suppose in sixteen of the largest cities of the United States all the branch offices were closed, and no business was taken in at the main office or branches, the main office being simply used as a repeating office for the general business of the country.

Again, suppose Western Union should refuse to transmit the messages originating in any of the sixteen cities, for delivery within the sixteen cities, what would Western Union lose?

What I believe is, that business originating in the sixteen principal cities destined for delivery within the same is a very small percentage of the general business. Further, that any competing telegraph companies who act on the theory that to get the cream of the business, in a dividend point of view, they should only run between the principal cities, will make a most dismal failure.

Would it not be well for you to ascertain the amount of the telegraph business existing in sixteen of the largest cities *with each other,* and see whether or not your views will find additional confirmation from this suggestion. Truly yours,

Thomas A. Edison.[a]

PL (transcript), Wiman 1884, 33 (*TAED* PA436). [a]Followed by "(over.)" to indicate page turn.

1. Erastus Wiman (1834–1904) was a Canadian journalist associated with R. G. Dun & Co., the mercantile reporting agency. After managing its Canadian business, he moved to New York in 1866 as a partner and later general manager of the firm. In 1881, he became president of the Great Northwestern Telegraph Co. of Canada and also served as a director of the Western Union Telegraph Co. (Edison obtained Wiman's permission to conduct telegraph experiments on the Great Northwestern lines in 1885–86). Wiman was a major figure in the development of Staten Island in the 1880s, establishing the Staten Island Rapid Transit Railroad Co. (on which Edison conducted railway telegraph experiments in 1885–86) and gaining control of ferry service to New York. In 1882, he visited Edison at Menlo Park in relation to the electric railroad and later became a director of the Edison Electric Light Co. Wiman became the first president of the Canadian Club in New York in 1885 and was a strong proponent of commercial union between Canada and the United States. He suffered financial reverses in the 1890s and was convicted of embezzling funds from Dun & Co. Although the convic-

tion was overturned, he lived the rest of his life in straitened circumstances. *DCBO*, s.v. "Wiman, Erastus" (accessed May 2010); Reid 1886, 608–11; Poor 1889, 309; Morris 1898, 235–36, 358, 460, 466; *TAED*, s.v. "Wiman, Erastus"; Israel 1998, 239.

2. Founded as the morning *Chicago Republican* in 1865, this paper was renamed the *Chicago Inter-Ocean* in 1872. After 1895, it became the property of Chicago traction boss Charles T. Yerkes. *Ency. Chgo.*, s.v. "Newspapers" (accessed May 2010).

3. That lengthy interview and Edison's response to it were published in pamphlet form as Wiman 1884. Wiman had enclosed his interview in a 20 June letter to Edison, in which he indicated that "one of your associates yesterday spoke of your opinions about such matters, which seem strongly to confirm the impression I have given." Insull noted on Wiman's letter that "Edison replied w[ith] autograph letter not copied." On 7 July, Wiman asked Edison's permission to reprint his letter along with the interview in pamphlet form. At the same time, Wiman sent a formal reply to Edison's critique (also printed in the pamphlet). Edison agreed but asked to see the proof before publication. Edison's response to the proof has not been found, but Wiman later expressed his "fear that you are repenting the permission you gave me to publish the letters and my reply." Edison evidently relented. The published pamphlet was circulating by 24 July, when Josiah Reiff saw it. Reiff characterized it as a Western Union publication and expressed surprise that Edison would support that company's views. Edison responded that he thought "both the Western Union and the opposition companies are too much of a 'muchness,' one is about as bad as the other." Wiman sent Edison copies of the pamphlet on 27 August. Wiman to TAE, 20 June, 7, 10, and 15 July, 27 Aug. 1884; Reiff to TAE, 24 July 1884; all DF (*TAED* D8471N, D8471Q, D8471Q1, D8471S, D8403ZGB, D8471T); TAE to Reiff, 28 July 1884, Lbk. 19:261 (*TAED* LB019261).

Wiman's interview took place in early June, during the Republican presidential nominating convention in Chicago, in the context of recent Congressional hearings regarding whether the government should build a postal telegraph system. This issue was a long-standing one in post–Civil War American politics, and it would continue to crop up periodically through the 1890s. Wiman criticized the hearings and especially the role of Josiah Reiff, a determined opponent of Western Union, whom he claimed was the person "who had more to do with shaping the Congressional telegraph investigation than any other single individual." The focus of the interview concerned the question of whether it was possible to sustain competition with Western Union or whether the telegraph was a natural monopoly. Wiman discussed the financial standing of competing telegraph companies, especially the Postal Telegraph and Cable Co., which he thought would "make an effort to become identified with the government and thus form part of the Post-Office Department." In fact, Postal Telegraph, after its reorganization in 1886, became the most successful of Western Union's competitors. Wiman 1884, 5, 22–23; "To Urge Arthur's Claims," *NYT*, 29 May 1884, 2; U.S. Senate 1884; Harlow 1936, chaps. 16, 20–21.

The key passage to which Edison was responding was Wiman's assertion that the widespread "belief that the telegraph business between the great cities was the main element of profit" was not only false but the reason for "the long array of failures in competitive telegraphy" (Wiman 1884, 10). Wiman contended that the expenses of this business were disproportionately large due to the necessity of having so many branch offices in each city. In addition, "a still greater mistake is made in supposing that it is the business between the large cities that makes up the business of these places" (Wiman 1884, 11). He argued instead that the local traffic to nearby towns made up about three-fourths of the business. He faulted competing companies for focusing

> business between the great cities, and not the business that ema-
> nates from, and centres in the cities themselves, from the extended
> local areas around them of social, financial, and commercial inter-
> change, which does more to make them great centres than contri-
> butions from other great rivals. This is what gives Western Union
> an enormous advantage, among many others, which any other
> telegraph company without an equally complete system covering all
> small points cannot hope to divide; and even if divided, is too small
> in each locality to yield a profit to the new comer. [Wiman 1884, 12]

Western Union's great advantage was having offices in 13,000 locali-ties while no competitor had succeeded in establishing more than 900. Wiman's 7 July response to Edison, also published in the pamphlet (Wiman 1884, 34–35), took up Edison's suggestion to investigate how much actual business there was between the principal cities. Finding that it made up only about 15 percent of Western Union's total busi-ness confirmed for Wiman his argument about the difficulties faced by Western Union's competitors.

–2693–

From George Bliss

CHICAGO, July 15 1884[a]

Dear Sir:

The United States Co. are blowing about some new form of Incandescent light claiming 3 250 Candle Power lights Per HP. which is to displace arc light & everything else.[1]

Can you tell me what sort of a ———— humbug this thing is.

They talk about delegations going to see it & 500 men being put at work etc

They must be getting in the last ditch. Sin Yr

Geo. H. Bliss

⟨They have been promising this thing a year   last life curve their Lamps set up at 16 c[andlepower] at our factory gave 231 hours average life— E[dison]⟩[2]

ALS, NjWOE, DF (*TAED* D8462ZAO). Letterhead of Western Edi-son Light Co., George Bliss, general superintendent. [a]"CHICAGO," and "188" preprinted.

1. Bliss likely referred to one of the large lamps designed by Edward Weston for the United States Electric Lighting Co. and later displayed

at the Philadelphia International Electrical Exhibition. Rated at up to 130 candlepower, they were favorably compared with arc lights; they reportedly could operate in arc light circuits at 160 volts and were more efficient than ordinary carbon lamps. At an unknown date, William Hammer added two of the 125-candlepower "Mammoth" lamps to his comprehensive historical lamp collection (see *TAEB* 5 App. 3.). "The United States Electric Light Co.'s Exhibit," *Sci. Am.* 51 (18 Oct. 1884): 246; "The Edward Weston Exhibit at the International Electrical Exposition," ibid. 51 (8 Nov. 1884): 287; Woodbury 1949, 133–34; website of Edward J. Covington devoted to Hammer's collection (accessed March 2010), http://home.frognet.net/~ejcov/hammer.html.

2. Edison's marginalia was the basis for a short typed reply in which he noted that "We frequently get hold of some of their lamps" for testing purposes (TAE to Bliss, 18 July 1884, LM 20:31 [*TAED* LBCD7031]). One report of such tests on the United States Electric Lighting Co.'s ordinary carbon lamp is John Marshall to TAE, 11 Apr. 1884, DF (*TAED* D8430P); cf. Doc. 2697.

-2694-

To George Barker

[New York,] 16th July [188]4

My Dear Barker,[1]

I received your telegram dated Chicago with relation to your appointment on the Electrical Commission[2]

Of course I am very anxious that you should be on the Commission but I do not see what influence I can use to bring this about. Even if I had the influence, if I attempted to to use it our friend "the enemy" would bring it up as a proof that I desired you appointed in my interest[3]

Cannot you suggest some other way of fixing the matter[4]

With Kind regards   Yours very truly

Thos A Edison

LS (letterpress copy), NjWOE, Lbk. 18:110 (*TAED* LB018110). Written by Samuel Insull.

1. George Barker (1835–1910), professor of physics at the University of Pennsylvania, had a long and generally friendly association with Edison going back to 1874 (Doc. 500 n. 8). Barker was partly responsible for having renewed Edison's interest in electric lighting in 1878, and he provided some important early assistance and validation for the inventor's research in the field. However, relations between the two men had been strained in recent years (see note 3).

2. Barker had wired Edison on 14 July about a "strong effort being made to defeat my appointment on electrical commission on ground of my being in your interest   if you can bring any influence on the president do so at once   appointment will be made Wednesday." Congress had approved legislation on 7 July authorizing the president to appoint a commission of electrical authorities. The commission was intended to organize a conference of electricians in conjunction with the Exposition

in Philadelphia. Barker was eventually named the commission's corresponding secretary (see Doc. 2720); other members included Henry Rowland (chair), Edwin Houston, Simon Newcomb, John Trowbridge, and Charles Young. Barker to TAE, 14 July and 20 Aug. 1884, both DF (*TAED* D8403ZEQ, D8464O); "Notes and News," *Science* 4 (1 Aug. 1884): 107; "The National Conference of Electricians," ibid. 4 (15 Aug. 1884): 127; *Report of the Electrical Conference* 1886, 3–9, 30.

3. Edison likely referred to a principal rival, the United States Electric Lighting Co., and perhaps particularly to its electrician, Edward Weston, whom he had reason to distrust for both personal and competitive reasons (see Doc. 2479 n. 2). Barker, in conjunction with Henry Rowland, had conducted independent tests on the efficiency of Edison's lamp in 1880 but had since become tangled in the commercial rivalry between the Edison and United States Co. interests. As a consequence, he was not fully trusted by either side, and his friendship with Edison cooled considerably. Stung by unfavorable reports attributed to Barker in the daily press, Edison came to believe by late 1880 that the physicist was "affiliated with the Maxim Co." Barker nonetheless accepted both a retainer from the Edison Electric Light Co. and Edison's gift of electric light company stock shares in 1881. See Docs. 1914 esp. n. 3, 2022, 2033, 2110, 2173 esp. n. 11, and 2188 n. 4.

4. The editors have not found a reply from Barker.

–2695–

From George Dyer

WASHINGTON, D.C., July 18th, 1884[a]

Dear Sir.

I have this morning your letter of the 16th, in regard to the formation of a new class, by division of the class of Electricity,[1] and the placing of Mr George Seeley[2] in charge of it. I found Mr Buckingham,[3] who professed to have authority to speak for the Western Union Telegraph Company, and together we called upon the Commissioner of Patents,[4] who gave us a very frank and quite lengthy interview.

He explained that he had appointed a commission consisting of Examiners Catlin,[5] Stocking,[6] Kintner,[7] and Seeley,[8] to make a thorough inspection of the work of the Examiners, and their several methods of work, and report with recommendations as to methods of work, and also to determine in what way their new classes might best be formed,—at the head of one of which the newly authorized Examiner should be placed, while both of the others should be in charge of First Assistant Examiners.[9] He should wait for the report of his Commission, before he would designate the persons to be put in charge of the new classes. His own impression was from conversation with members of the Commission, that a division would be made of the Class of Electricity, and electrical engines, and some other things put into the new class. He told me that Mr Whittaker[10] did not strike him as the proper person to be put at the head of this new proposed class. He said also that Lyons[11] or George Seeley would naturally appear to best suited for the place, and that he would be glad to talk with me about the matter as soon as the Commission reported.

This Commission is to meet to day, and go to work, and keep at it until completed, and probably will occupy several days. In conversation with them, I find that they are all agreed upon a division of the class of electricity, but are not agreed as to the line of division. I find that Mr Kintner is desirous of having the new class embrace electrical engines, and to have Mr George Seeley at the head of it, and Examiner Seeley agrees with him.[12] I find also, that Mr George Seeley understands that he is not to be appointed an Examiner in any event, and if placed in charge of the new class in Electricity, it will be with his present rank of First Assistant ~~Commiss~~

I think the Commissioner is disposed to be fair, and regards Lyon and George Seeley as equally competent, and really does not care so much about the men, as about the methods of work. He feels proud in having got not[b] only an increase of appropriations, ~~but~~ and[c] an increase in the Examiners force of nearly twenty, but in having crowded out the Indian Bureau from the Patent Office Building, and got the promise of a portion of their rooms.

The Commissioner is very fond of his friends, and thinks very highly of Mr Hiscox of New York.[13] Possibly if this gentleman were coming over here early next week, and received some points from myself, he could do much in inclining the Commissioner to put Mr George Seeley at the head of this new class in Electricity.

Mr Lyon is supposed to be the favorite of the Brush Co and of the Bell Telephone Co. but I see no indication yet of their movements on his behalf. Yours very truly

Geo W Dyer.

ALS, NjWOE, DF (*TAED* D8403ZEW). Letterhead of George W. Dyer. [a]"WASHINGTON, D.C.," and "188" preprinted. [b]Obscured overwritten text. [c]Interlined above.

1. Edison's letter has not been found. In early 1883, Edison had sought to make changes in the Examiner of Interferences at the Patent Office (see Doc. 2402).

2. George Dallas Seely (1838–1908) had earned a degree in chemistry from Yale College in 1859 and, in 1877, joined the Patent Office, where he was at this time a first assistant examiner of electricity. He

was the brother of Franklin Austin Seely (see note 8) and the uncle of Henry W. Seely, the associate of George Dyer's son Richard in their law practice dealing with Edison's patents. "Yale College Commencement," *NYT,* 30 June 1859, 3; Torrey 1885, 17, 22; U.S. Department of the Interior 1883, 93; Yale University 1909, 1129.

3. Charles L. Buckingham (1852–1909) left his position as assistant examiner in the Patent Office in 1880 to become legal counsel for the Western Union Telegraph Co. He held a degree in civil engineering from the University of Michigan and, in 1880, earned an LL.B. from the Columbian University Law School in Washington, D.C. Morris 1896, 89; University of Michigan 1913, 192; MacDonald 1896, 107–9.

4. Benjamin Butterworth (1837–1896), a former congressman from Ohio, took over as acting Commissioner of Patents on 1 November 1883 and was confirmed the following month. Reelected to Congress in November 1884, Butterworth resigned the commissionership to take his seat on 23 March 1885, and he remained in Congress until his retirement in 1891. He resumed his position as Commissioner of Patents in April 1897 at the request of President McKinley, and he served until his death. *DAB,* s.v. "Butterworth, Benjamin"; Hopkins and Bond 1915, 358; United States 1901, 101–2; Dobyns 1994, 195.

5. Benjamin Rush Catlin (b. 1829) was the principal examiner in the Patent Office for gas, metallurgy, brewing, and distillation, having been appointed in 1871. Catlin graduated from Hamilton College in 1851 and attended the Auburn (N.Y.) Theological Seminary. During the Civil War, he was chaplain of the 115th Regiment of United States Colored Troops. Beecher 1883, 104; Berly 1883, 292; U.S. Department of the Interior 1877, 13.

6. Solon Walter Stocking (1834–1905) was the principal examiner for metalworking. He was first appointed as a third assistant examiner in 1873 and continued with the Patent Office until his death. Stocking was valedictorian of the 1855 class at Hamilton College, from which he earned a law degree in 1858. He interrupted his law practice in May 1861 to serve in the Union Army, rising to the rank of captain of artillery. Prior to his appointment to the Patent Office, Stocking had served as a law clerk with the Freedmen's Bureau and as a division chief in the Census Bureau. Stocking 1903, 64, 130–31; Squires 1905, 241–42, 246; Berly 1883, 292; U.S. Department of the Interior 1881, 18.

7. An 1870 graduate of the University of Michigan, Charles J. Kintner (1848–1921) joined the patent office in 1878 as second assistant examiner for electricity and became the principal examiner by 1884. Kintner left the Patent Office in the late 1880s and moved to New York, where he worked as a patent attorney. He also served on the state Board of Electrical Control and was an inventor in his own right, with a number of U.S. electrical patents to his credit. Carter 1921, 22–23; *Polk's* 1922, 285; "Brooklyn's Eligible List," *NYT,* 3 June 1888, 1.

8. Franklin Austin Seely (1834–1895) was the principal Patent Office examiner for trademarks and instruments of precision. An 1855 graduate of Yale College, he served in the Civil War as an officer in the quartermaster's department and afterward was a high-ranking official in the Freedmen's Bureau in North Carolina and Missouri. He joined the Patent Office in 1875 and served until his death. Torrey 1885, 17; Berly 1883, 292; Yale University 1909, 1129.

9. In his portion of the commission's report, Kintner noted that the number of patents originated in the division of electricity had grown from 2,000 in 1880 to 6,000 in 1884, with consequent overcrowding and overtaxing of the workforce. In February 1884, Commissioner Butterworth, in his report to Congress, made special mention of the need to increase the staff in the Patent Office to handle the burgeoning caseload. Kintner 1884, 7–9; "Patent Office Work of 1883," *Sci. Am.* 50 (16 Feb. 1884): 97; Dobyns 1994, 195.

10. Jesse Hadley Whitaker (b. 1843?) was an examiner of household furniture, having joined the Patent Office in 1878 as a first assistant examiner of electricity. In 1883, he formed the Washington, D.C., patent law firm of Whitaker & Prevost with his brother-in-law, George A. Prevost. "Whitaker Funeral Today," *Washington Post,* 28 Apr. 1923, 3; U.S. Census Bureau 1970 (1880), roll T9_121, p. 42, image 0770 (Washington, D.C.); *Congressional Directory* 1884, 140; U.S. Department of the Interior 1881, 19; "G. M. Prevost Funeral Rites Set Tomorrow," *Washington Post,* 8 Nov. 1942, SP8.

11. Joseph Lyons (b. 1842) was a first assistant examiner of electricity. An Austrian immigrant, he was originally appointed in 1878 as a third assistant examiner of firearms, navigation, signals, and woodworking. With Gustav Bissing, he formed the Washington, D.C., patent law firm of Lyons & Bissing. U.S. Census Bureau 1982? (1900), roll T623_159, p. 4B (Washington, D.C.); U.S. Department of the Interior 1883, 93; ibid. 1879, 25; Boyd 1903, 659.

12. The division of electricity was partitioned into two classes by the end of 1884. Class A, headed by Kintner, dealt with telegraph and telephone patents. Class B dealt with electric generation and distribution; George D. Seely was appointed its head in 1886. *Congressional Directory* 1884, 138; U.S. Department of the Interior 1894, 129, 135; Yale University 1909, 1129.

13. Dyer likely referred to Gardner Dexter Hiscox (1822–1908), a hydraulic and pneumatic engineer and for a time a regular contributor to *Scientific American.* From 1886 to 1890, he was an engineer for the Ingersoll Rock Drill Co.; he later published a number of well-received technical books. Fletcher 1909, 372; Obituary, *Mining World* 29 (26 Sept. 1908): 493; "Gardner D. Hiscox," *Compressed Air* 3 (May–June 1898): 439.

–2696–

From Emilie Kellogg

Mount Vernon [N.Y.,] July 168th 1884

Sir.

Our Son entered your Labratory a year ago last March, and we thought it such a good place for him.[1] he was pleased and liked the work. stayed there all Summer without any vacation   we thought little of that as he went late mornings and came home early at night. about the middle of August, he had an epileptic fit, caused as we thought by fright about a fire. he had no more till the first of Oct. then[a] did not know the cause, and let him keep on with his work as he seemed well only his face was much broken out with sores. he had the fits several times during the winter and was nervous and very irritable, always had them before Seven o'clock in the morning at such times going to his work without eating much breakfast. the first of March he had one Sunday afternoon. then we became alarmed and took him from his place, for we knew he was not able to stay. Sent him into the Country, to try the effect of an entire change—but he received no permanent benefit[b] though he had been under a Dr's care all winter.[2] and why? because he had inhaled and[c] absorbed so much mercury and chemicals that his system could not recover itself.[3] we did not know it then. he was and has been[d] handle very strangely. the Dr's could not tell what ailed him besides epilepsy, and its only within a few weeks that the above cause has been found out. it was in his blood. now his nerves and muscles are so affected he cannot control them, are partially numb his mouth is constantly full of water. his mind is weakened so he is like a boy Six or 8 years old. he has a fit nearly every week, but not as hard as formerly. we fear he is ruined for life and wonder who is is <u>responsible</u>? we have been told from the Labratory that he was too young for the place. that boys of his age were apt to be thus troubled why was it not told us at first? it was probably known as well then as now. he worked a year without compensation. that is of small account   we feel as if it was a dreadful thing for him to be so ruined, and would like to have others know of the danger before going into it. I think he wrote you about the first of April.[4] I don't know what, or, if you ever received it. We feels as if you ought to know about him. that you may not take others there to be thus injured. <u>perhaps you did not know it</u>

    Our Son's name is Eddie Kellogg   Respectfully

                      Mrs M. C. Kellogg[5]

    P.S. Since writing the above Eddie has told the Dr. that he had about a pt. of quicksilver poured over him by accident. continued wearing the Same clothes. likely he was careless and did not understand the danger of ~~ease~~ using such things   I feel a Mothers anxiety that other boys should be made to <u>thoroughly</u>[c] <u>understand</u> their danger in handling poisons. <u>perhaps you may know of an antidote in his favor.</u>[6] Mrs M.C.K.

ALS, NjWOE, DF (*TAED* D8413ZAS1). [a]Interlined above. [b]"no permanent benefit" interlined above. [c]Obscured overwritten text. [d]"and has been" interlined above.

    1. Edward Kellogg (1866?–1886) was involved with experiments on chemical paper for recording readings of central station ampere meters.

He tested paper impregnated with mercury, lead, cadmium, barium, and other substances. He also mixed scores of compounds to create insulating materials. Obituary, *NYT,* 4 Jan. 1886, 5; N-82-12-04:39–109, N-82-12-21:115–77, Lab. (*TAED* N145:18–49, N150:58–89).

2. John Ott reported in March 1884 that Kellogg had left work for three months because "His folks want to send him to the country for his helth, as he is subject to Apoplexy." Kellogg died in January 1886 at Red Bank, N.J., near the home of an uncle, George Kellogg. Ott to Samuel Insull, 22 Mar. 1884, DF (*TAED* D8467K); Obituary, *NYT,* 9 Feb. 1889, 5; U.S. Census Bureau 1965 (1870), roll M593_98, p. 656, image 84 (New Canaan, Fairfield, Conn.); ibid. 1970 (1880), roll T9_791, family history film 1254791, p. 314.3000, image 755 (Neptune, Monmouth, N.J.).

3. While mercury was used throughout the nineteenth century as a treatment for various illnesses—it was taken internally, in the form of calomel, or "blue pill"—acute exposure, especially in cases of overmedication, was known to cause excessive salivation, sweating, cramps, paralysis, "mercurial tremors," "gangrene of the mouth," disorientation, and sometimes death. Acute exposure in industrial settings was more controversial. Mercury was used most notoriously in felting hats, but also to bind mirror silver to the glass, to separate gold from other materials, and in electric lamp manufacture. Francis, 1813, 476–519; "Lead and Mercury Poisoning," *Manufacturer and Builder* 9 (Oct. 1877): 232–33; Quain 1883, s.v. "Mercury, Diseases arising from"; "Mercury," *Manufacturer and Builder* 24 (Aug. 1892): 180–81.

4. Not found.

5. Emilie E. Kellogg (1837?–1889) lived in Mount Vernon, N.Y., with her husband, Minot C. Kellogg (1835?–1915). He was the proprietor of a hardware store at 27 Park Row in New York City, which his wife used as her return address. In addition to their son Edward, the couple had two daughters. Obituary, *NYT,* 9 Feb. 1889, 5; Obituary, ibid., 9 Jan. 1915, 11; *Trow's* 1885, 909; U.S. Census Bureau 1970 (1880), roll T9_945, p. 188.3000, image 377 (East Chester, Westchester, N.Y.).

6. Edison's response is Doc. 2702.

–2697–

From John Marshall

East Newark, N.J., July 18 1884[a]

Column 1 shows life of 10 Swan lamps at 160 candles per HP., calculated from a curve of 10 lamps set up 20 candles and giving 226 candles per HP.

Column 2 shows life of 10 Swan lamps at 160 candles per HP calculated from a curve of 10 lamps set up at 60 candles and giving 446 candles per HP.

1	2
Hours	Hours
59	2515
224	2709
279	3547

536	3870
1025*	3902
*These lamps are still burning.	4224
	4482
	7256
	11 061
	<u>18 447</u>
	Avg. 6201

J. T. Marshall[1]

⟨Why is the life so different with the same invoice of Lamps & calculated set up at same candle power.

This seems to show that treated Lamps last longer at high Cp than normal while untreated is just opposite⟩[2]

ALS, NjWOE, DF (*TAED* D8430U). Letterhead of Edison Lamp Co. [a]"East Newark, N.J." and "188" preprinted.

1. John Trumbull Marshall (1860–1909) was identified in the 1880 federal census as a teacher, one of five in his immediate family. Marshall received a scientific degree from Rutgers College in 1881 and that fall took charge of the photometer room at Edison's lamp factory, succeeding his brother William Marshall, who had recently died of typhoid. Marshall became a photometric expert, devising several instruments that were widely used in lamp testing. He followed Edison to the new West Orange laboratory in 1887 and subsequently worked for General Electric until his death. U.S. Census Bureau 1970 (1880), roll T9_790, p. 268.1000, image 0377 (East New Brunswick, Middlesex, N.J.); ibid. 1982? (1900), roll T623_985, p.14A (Metuchen, Middlesex, N.J.); Marshall 1930, 161–62; J. F. Riddle to TAE, 11 Jan. 1910, DF (*TAED* D1016AAF; *TAEM* 195:322); Obituary, *Metuchen (N.J.) Recorder*, 15 Jan. 1910, Unbound Clippings 1910 (*TAED* SC212B; *TAEM* 221:545).

2. The editors have not determined how the Swan lamps were "treated," but Edison may have referred to a new process for making a so-called "squirted" cellulose filament. At the very end of 1883, Joseph Swan filed a British provisional patent specification (5,978 [1883]) for a method of forming filaments "by forcing a mixture or solution of nitro-cellulose in acetic acid through a hole or die in a liquid capable of causing the 'setting' of the filament as it issues from the jet." Although the process met Swan's desire for producing highly uniform filaments, the Edison and Swan United Co. did not put it to commercial use for several years (other manufacturers subsequently adopted squirted cellulose filaments as well). Since about 1880, Swan's commercial lamps had filaments made of cotton thread that was first "parchmentised" by an acid treatment and then carbonized amid powdered carbon. Bowers 1982, 123–24; Swan 1946, 24, 28, 36–39; Bolton 1886, 487; Woodings 2000, 20 n. 3.

Samuel Insull returned Marshall's letter to him with a request for comment on Edison's notes (Insull to Edison Lamp Co., 19 July 1884, DF [*TAED* D8430T]). Marshall explained in reply that

The calculations of life on Swan lamps . . . were made from the law derived from experiments on our own lamps: viz. The life varies inversely as the 3.65 power of the candle-power.

The fact that the life calculated from the 20 C.P. lamps, is so different from that calculated from the 60 C.P. lamps, would seem to show, as you say, that the Swan lamps do not follow the same law, as to candle-power and life, as ours do. [Marshall to TAE, 21 July 1884, DF (*TAED* D8430V)]

–2698–

John Tomlinson to
Samuel Insull

NEW YORK, July 21st 1884[a]

My dear Insull,

The sale of the Menlo Park property takes place tomorrow at 2— It will be necessary for me to have the $600 the first thing in the morning[1]   Yours

John C. Tomlinson

⟨OK   E⟩

ALS, NjWOE, DF (*TAED* D8403ZEY). Letterhead of Thomas A. Edison. [a]"NEW YORK," and "188" preprinted.

1. This sale, arranged to satisfy Lucy Seyfert's judgment against Edison (see Docs. 2662 and 2671), became a protracted and complex affair. Edison took $50 cash and $550 in a check to Tomlinson. This money was evidently an advance to Charles Batchelor, who purchased the property and then returned the money in August (Cash Book [1 Jan. 1881–30 Mar. 1886]: 267, 272, Accts., NjWOE; Conot 1979, 219). Mary Edison gave Seyfert $230 in sale proceeds, identified in a 22 July receipt as 20 percent of the total. A retrospective accounting by Seyfert's attorney listed the proceeds as $2,750, still well short of the judgment due (Lucy Seyfert to TAE, 22 July 1884, Kellow [*TAED* HK069AAC]; Strong & Son to Tomlinson, 8 Feb. 1886, DF [D8603ZAA]). Tomlinson advised Insull on 8 August that Edison should arrange to pay the $2,200 balance. Mary Edison died the next morning, however, evidently without ever having received a deed for real estate that had been conveyed in her name (Tomlinson to Insull, 8 Aug. 1884; Strong & Son to Tomlinson, 15 Aug. 1884; both DF [*TAED* D8465ZAF1, D8403ZFO]). As a result of the uncertainty over the property's legal ownership, the sale proceeds were frozen and an alias writ of execution was issued to authorize a second sheriff's sale on 21 October. Edison again gave Tomlinson $600, but the retrospective reckoning by Seyfert's attorney listed only $11 in proceeds from the second event. The twenty-one Menlo Park lots advertised for that occasion, presumably the same ones auctioned in July, were evidently bought by Charles and Rosanna Batchelor, who held them until they deeded the property back to Edison in 1891 (Opinion of Edward Scudder, 18 Nov. 1884; Strong & Son to Tomlinson, 20 Aug. 1884; Thomas Ecclesine to Tomlinson, 21 Aug. 1884; all DF [*TAED* D8403ZIN, D8403ZFQ, D8403ZFS]; Cash Book [1 Jan. 1881–30 Mar. 1886]: 280–81, Accts., NjWOE; Charles and Rosanna Batchelor agreement with TAE, 30 June 1891, Kellow [*TAED* HK070AAO]).

Edison resisted a resolution of the Seyfert case with bitter determination and at increasing expense. Because of the confusion surrounding Mary Edison's putative ownership of the auctioned property, the plaintiff sought to name a receiver that would allow Mrs. Seyfert access both to the money paid to the sheriff and to "whatever other interest Mr Edison may have" in New Jersey. Edison was accordingly ordered in November 1884 to testify under oath about his property in the state. Refusing this order and a second one in 1885, he was held in contempt by the state Supreme Court (Abraham Schenck to Tomlinson, 8 Jan. 1886; New Jersey Supreme Court decision, 4 Dec. 1885; both DF [*TAED* D8603E, D8503ZEG]). Facing both the contempt citation and his pending remarriage, Edison apparently authorized payment of about $3,200 to settle the Seyfert matter in early 1886 (Strong & Son to Schenck, 20 Jan. 1886; Strong & Son to Tomlinson, 3 and 8 Feb. 1886; Schenck to Tomlinson, 14 Apr. 1886; all DF [*TAED* D8603T, D8603Z, D8603ZAA, D8603ZAS]).

–2699–

To Sherburne Eaton

[New York,] July 22nd. 1884.

Dear Sir:—

I beg to notify you that I have no further use for the third floor offices in this building, as what little business I have now got at 65 Fifth Avenue can as be easily attended to in my library on the top floor.

Inasmuch as the books in my library have been mainly collected for use in connection with the Light Company's affairs, and, furthermore that the library would be of practically no utility to the Company, unless it was kept in this building, I presume that no rent will be charged me for the use of the attic floor. It is not my intention to conduct any considerable business on that floor, as I shall simply leave one clerk, as a rule, with a view to his answering any inquiries in connection with my affairs. Very truly yours,

Thos A Edison

TLS (carbon copy), NjWOE, Lbk.18:141 (*TAED* LB018141).

–2700–

To Sidney Paine[1]

[New York,] July 22nd. 1884.

Dear Sir:—

Referring to your favor of the 19th. inst.,[2] I beg to inform you that I no longer have an engineering force in connection with the Construction Department, as I am closing out that business, and I therefore cannot very well give you the information you desire. I have referred your communication to the Isolated Co. Very truly yours,

Thos. A Edison   I[nsull]

TL (carbon copy), NjWOE, Lbk. 7:051 (*TAED* LBCD7051). Signed for Edison by Samuel Insull.

1. Sidney Borden Paine (1856–1940), a nephew of Spencer Borden, had briefly been in textile manufacturing in Massachusetts and the stove business in Cleveland before becoming an assistant in Borden's Massachusetts agency of the Edison Co. for Isolated Lighting (then called the Eastern Agency) in 1882. In April 1884, Paine succeeded Borden as principal agent of what had been renamed the Isolated Co.'s New England Dept. After the formation of General Electric, Paine was appointed manager of its new Mill Power Dept., where he spent the rest of his career. Doc. 2274 n. 12; "Paine, Sidney B.," Pioneers Bio.; U.S. Census Bureau 1967? (1860), roll M653_491, p. 311, image 312 (Fall River Ward 5, Bristol, Mass.); ibid. 1970 (1880), roll T9_1004, p. 102.1000 (Cleveland, Cuyahoga, Ohio).

2. Not found.

–2701–

Notebook Entry:
Direct Conversion[1]

[New York,] July 22, 1884

[A][2]

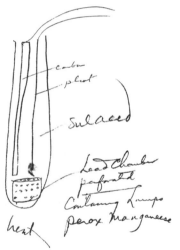

Not strong only 3 deg[ree]s with no resis, shows its not the gas from the Perox Mn but actual contact.

[B]

Carbon & platina Electrodes, filled in between with Lump perox Mn & Sul Acid[3]

Splendid gives 10 deg[rees] through 37 ohms res

We put porous partition powdered Carbon in Contact w[ith] Carbon & Perox mng on plat   OK But porous partition consid[erable] res

We now try pyrusulite[4] against Carbon & nothing on plat, instead of against plat, not much dif—[a]

Sul A[cid] Perox Barium   Carbon & platina gives good deflection but unstable   all Ox comes off

We now try peroxide Lead

TAE   M. N. F[orce].

X, NjWOE, Lab., N-82-05-15:109 (*TAED* N203109). Document multiply signed and dated; miscellaneous doodles and calculations not transcribed. [a]Followed by dividing mark.

1. See Docs. 2520 (headnote) and 2620. Edison had begun experimenting on 21 July with a variety of solutions and oxides for the "direct oxidation" of carbon (N-82-05-15:104–7, Lab. [*TAED* N203104]). A 19 July drawing related to these experiments is in 1884 Unbound Notes and Drawings, Lab. (*TAED* NS84ACM1).

2. Figure labels are "carbon," "plat," "sul acid," "Lead chamber perforated Containing Lumps perox Manganese," and "heat."

3. Edison drew a line from this text to the drawing.

4. Edison probably meant pyrolusite, another name for black oxide of manganese. Essentially manganese dioxide, it is the most common manganese mineral. Hawley 1987, s.v. "pyrolusite."

–2702–

To Emilie Kellogg

[New York,] July 23rd. 1884.

Dear Madam:—[1]

Referring to your favor of the 18th. inst.[2] I [would?][a] state that we have at our lamp factory boys who do nothing but clean mercury.[3] Occasionally they get slightly salivated some and they take a teaspoonful of a weak solution of iodide of potassium every day, and in about three weeks they are all right again.[4] I have had boys much younger than your son in my laboratory for years doing the same work that he was engaged on, and they were never troubled in the way you speak of.[5]

I think that your doctor must be [pushing?][a] the wrong thing altogether, but if he is right, iodide of potassium is the sovereign remedy for mercurial salivation, and in all French factories where mercury is used, the employees are compelled by law to take the above, as any third rate doctor in this country ought to know.

I sincerely sympathize with you in your trouble, and reassure you that I would not for one moment endanger the life of any youth by keeping him on work that would effect his heath.

I may remark that both myself and my principal assistants have for years been engaged in work necessitating the use of all kinds of chemicals, and notwithstanding this the health we all enjoy is something phenomenal.[6]

I really [think that your doctor must be mistaken when?][a] he ascribed your son's sickness to mercury.[7] Very truly yours,

Thos. A Edison   I[nsull]

TL (letterpress copy), NjWOE, Lbk.18:146 (*TAED* LB018146). Signed for Edison by Samuel Insull. [a]Faint copy.

1. The word "Void" was written across both pages of this document, probably by Samuel Insull. There is no indication whether the letter was actually sent, nor is there other extant related correspondence apart from Doc. 2696.

2. Doc. 2696.

3. Mercury was poured through the vacuum pumps to evacuate glass bulbs at the lamp factory. The mercury was cleaned before its initial use and again periodically to remove moisture, sulphur, and other impurities (see Doc. 1950 [headnote, esp. n. 18]). Also, Edison briefly had prescribed the regular application of an amalgam to the commutators of his central station dynamos, a practice he discontinued for health reasons (see Docs. 2149 esp. n. 4 and 2228.)

4. A standard treatment for mercury poisoning was iodide of potassium; this compound was thought to interact with mercury in a way that would allow it to be excreted. Quain 1883, s.v. "Mercury, Diseases arising from"; "Lead and Mercury Poisoning," *Manufacturer and Builder* 9 (Oct. 1877): 232–33.

5. In July 1883, George Stickle, a worker at the lamp factory in Harrison, brought a $5,000 suit against the Edison Electric Light Co. for damage to his health caused by incautious handling of mercury and a company doctor's subsequent prescription of medication. The editors have not determined the outcome of the suit. At least one doctor later identified "the progress of electric illumination" and the manufacturing of glass bulbs for lamps as a cause for a resurgence in industrial mercury poisoning cases. He singled out the Harrison plant as the probable cause of several cases of mercury poisoning. He noted that even with strict safeguards, the process of evacuating glass bulbs made it impossible to avoid breathing mercury vapor, particularly in winter when ventilation was more difficult. He reported that the factory employed its own doctor to provide free treatment to the sick, who may "form not a small percentage of the whole" labor force. He also contended that the condition was sometimes diagnosed there as "malaria, or some other convenient scape-goat of medical terminology to cover the plain fact of quicksilver poisoning." "Health Injured by Mercury," *NYT*, 13 July 1883, 8; Lehlbach 1889, 343–44; cf. App. 1.F.19.

6. Cf. Doc. 2706 n. 6.

7. Edward Kellogg's symptoms, as described in Doc. 2696, were consistent with a contemporary medical authority's description of mercury poisoning. Quain 1883, s.v. "Mercury, Diseases arising from."

–2703–

Notebook Entry:
Primary Battery and
Direct Conversion

[New York,] July 23—[1884]

[A][1]

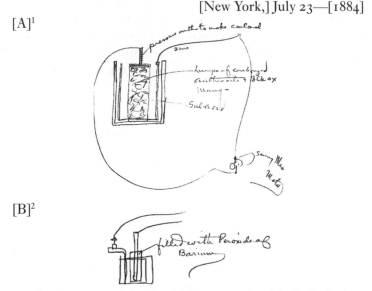

[B][2]

plated carbon with peroxid Mang put it with fresh Carbon in Strong Sul acid gave 10 deg[rees] Cold through 10 ohms— I now heat slight heat   10 deg thru 20 deg

goes down as heat increases   great many bubbles formed again goes up as it gets hotter. nearly boiling. [~~goes bac?~~][a]

then suddenly goes Zero

Reg[ular] Carbon cell[3] but with small porous cup packed with Carb[oni]z[e]d Anthracite & peroxide Lead carbon Electrode, Zinc & Chl ammonium   100 ohms 10° on strap [-][a] with Res box 11½— 1215 pm— 1220 12 deg

Black ox man[ganese] Com[4] packed with powdered Coke— Zinc & Carbon Dilute SO_4—11½ 1225 pm— polarizes goes to 7½ at 1227 stays at 7½ 1230

X, NjWOE, Lab., N-82-05-15:115 (*TAED* N203115). Document multiply dated. [a]Canceled.

1. Figure labels are "pressure on this to make contact," "zinc," "Lumps of carbonized anthracite & Blk ox Mang—," "Sul acid," and "sewing Mac Motor." While this notebook entry is related to Edison's ongoing research on direct conversion, it also marks the beginning of his efforts to develop a primary battery to run the sewing machine motor mentioned in Docs. 2459 and 2579. For his subsequent work on such a battery between 28 and 31 July, see N-85-2-15:147–79 (Lab. [N203147,

N203149, N203159]). These designs appear to be modifications of zinc (anode)-carbon (cathode) batteries, especially the Leclanche, in which manganese was used as a depolarizer, and the Bunsen carbon battery (see note 3). Edison altered the type of carbon and the chemical solutions used in these batteries and also substituted other oxides for magnesium. For these and other batteries see Niaudet 1880.

2. Figure label is "filled with Peroxide of Barium."

3. This was a common name for the battery developed by German chemist Robert Wilhelm Bunsen, who replaced platinum with carbon in the cathode of the Grove battery. Like the Grove, it used a zinc anode in a porous cup filled with a dilute solution of sulfuric acid. Silliman 1871, 579.

4. Edison meant either "common" or "commercial," both of which were frequently used in association with black oxide of manganese.

–2704–

To Frank Hastings

[New York,] July 25th. 1884.

Dear Sir:—

It is of very great importance that my account for canvassing should be brought before the Finance Committee for settlement immediately, as the account has been running now for such a long time. It represents cash paid out by me, and inasmuch as there is no profit whatever on the account, I must press for an early settlement of same.[1] Very truly yours,

Thos. A Edison   I[nsull]

TL (carbon copy), NjWOE, Lbk. 18:161 (*TAED* LB018161). Signed for Edison by Samuel Insull.

1. The editors have not reconstructed details of the canvassing account, which apparently was at least part of the sum that Edison tried to collect from the Edison Electric Light Co. in December 1883 (see Doc. 2569). A March 1884 statement itemized $11,090.63 in expenses for canvassing, mapping, electrical determinations, and general office expenses chargeable to nearly eighty projected central stations. Edison charged the company $1,925.96 for similar expenses in April and $832.99 in May, plus $1,867.86 for work on behalf of the New York Second District (Edison Construction Dept. statement to Edison Electric Light Co., 25 Mar. 1884, DF [*TAED* D8441K2]; Insull to Hastings, 15 May and 13 June 1884, LM 19:191, 425 [*TAED* LBCD6191, LBCD6425]). Samuel Insull asked Hastings to make partial payments of $5,000 on 1 May and $3,000 on 10 May, but the company requested verification of the total. Insull averred that he could not spare the time to produce the documentation but offered to make the books and vouchers available to the Finance Committee chairman (Insull to Hastings, 22 Apr. and 15 May 1884; Hastings to Insull, 6 May 1884; Hastings to Edison Construction Dept., 29 Apr. 1884; all DF [*TAED* D8416BJJ, D8416BPE, D8427ZAT, D8439ZAW]).

Edison's Construction Dept. cashbook noted a number of modest checks received from the Electric Light Co. during the spring and summer, including several loans and payments for the salaries and expenses of "experts." The company also paid checks of $1,500 on 26 April and $1,000 on 21 July. Like most entries in the Construction Dept. book, the purpose of these payments was not indicated. The fact that the book was often used for transactions not directly related to the Construction Dept.'s obligations or credits only compounds the difficulty of understanding these records, but the matter of Edison's full reimbursement remained unresolved in late September. Edison Construction Dept. Cash Book (1 June 1883– 28 Feb. 1886), esp. pp. 73 and 93, Accts., NjWOE; see Doc. 2736.

–2705–

To James Harris

[New York,] July 28th. 1884.

[Dear Sir:?][a]

[I have?][a] your favor of the 24th inst.[1]

[Regarding your?][a] suggestion as to my taking my balance [in stock shares of the?][a] Company in reply I would state that I have a such large capital invested in stocks of the various Edison Companies, and also in the manufactories that produce their material, and furthermore I require such a large floating capital to carry on my business, that it would be quite impossible for me to take the stock in your Bellefonte Company. I should be most happy to do so were it not for the above reasons, as I consider that the stocks in all the local illuminating companies in connection with the Edison System to be an extremely good investment, and should take advantage of same had I any large amount of superfluous capital to invest.[2] Very truly yours,

Thomas A Edison

TLS (carbon copy), NjWOE, Lbk. 18:186 (*TAED* LBo18186). [a]Faint copy.

1. Harris's letter was in reply to one from Edison on 22 July (not found) requesting payment of the unspecified remaining balance on the Bellefonte central station (Harris to TAE, 24 July 1884, DF [*TAED* D8453ZFH]). Harris asked Edison to accept stock in the Bellefonte illuminating company because the firm was short of cash, having not allocated capital to pay for wiring and related expenses. The sum in question was likely the third and final installment on the Bellefonte plant. The local company had made the second payment in February with great reluctance after questioning the station's capacity, but it promptly paid Edison in April for a modest expansion of the distribution network (Harris to TAE, 7, 13, and 23 Feb., 29 Apr. 1884; Samuel Insull to Harris, 16 Feb. 1884; all DF [*TAED* D8453ZAM, D8453ZAT, D8453ZBB, D453ZDN, D8416AOH]).

2. This matter dragged on for months. Edison peppered Harris with

demands and pleas for at least a partial settlement, while the Bellefonte company considered ways to raise the cash and Harris lamented that the business was not as successful as the investors had been led to expect. The company at one point offered $1,000 of its mortgage bonds, a suggestion that Edison firmly refused. The company remitted about $455 in March 1885, but the full balance remained unpaid as of July 1885. Harris to TAE, 11 Dec. 1884, 20 Jan., 28 Mar., and 23 July 1885; all DF (*TAED* D8453ZFP, D8523F, D8523ZAC, D8523ZBK); TAE to Harris, 26 Jan. 1885, Lbk. 20:52B (*TAED* LB020052B); also *TAED* s.v., "Harris, James."

–2706–

Notebook Entry:
Direct Conversion

[New York,] July 28 1884

Direct production E[lectricity] from Carbon Expts.[1]

We sealed Sul Acid, in glass tube with leading in platina wires one pole platina other Wallace Carbon.[2] put Lumps black oxide Manganease crude with 104 ohms in ckt gal[vanometer] Kinney[3a] went to 8 @ 9   9 only after while when tube Exploded violently—it went quickly to 8. There was too much water in acid—going use strongest acid then try=

Tried phosphoric anhydride & peroxide manganese. gal went to 3 without resistance   cell cracked. its very syrupy must try in crucible as glass melts—[b]

Its acts on perox mang to form a violet[c] colored Substance.[d]

We now try caustic soda & also Caustic potash from sticks—

Note= With sul acid I think the action is either 1 of 2 ways: Sul[c] Acid Decomp to SO_2 & O   O combines with Carbon to CO. SO_2 reduces O from peroxide to form SO_3—or water of Sul A decomp & O combines with Carbon to eCO & H reduces O of peroxide Mang to form H_2O.[d]

Caustic Soda & peroxide Mang crude— cracks glass when put in large lumps:— we now powder & put in & it goes up to 12 with lots peroxide, but with 7 ohms only 5. longer heating make it go down nearly to Zero with no Res. There is Evidently great deal Oxy given off from peroxide Mang.

We now try some Nordhausen fuming sulfuric acid[4] with peroxide Mang Crude with plat & Carbon Electrodes heated to boiling in test tube cell. This acid is as bad as 3 square Miles of Hell. Martin [Force] got burnt on the face. ⟨No better than Com[mer]c[ia]l 27@ 30 ohms 10 deg—⟩

Boiler large Scale for direct Conversion Carbon into Electricity[5]

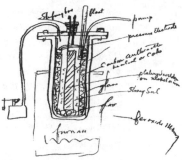

Things to try in direct Conversion Carbon into E.[6]

Peroxide Potassium fuses higher point than Caustic K at white heat decomp into K monoxide & O. Use this.

Also Sodium Dioxide
  "   Calcium Dioxide
      Strontium   "

for withstanding Hot Sul A.

Chromoso–Chromic oxide formed by Electrolysis Solution Chromous containing Chromic Chloride by Current Low intensity   black powder insoluable in any acid Cr_4O_5[7]

TAE

X, NjWOE, Lab., N-82-05-15:127 (*TAED* N203127). Document multiply signed and dated. [a]Interlined above. [b]Followed by dividing mark and "over" to indicate page turn. [c]Obscured overwritten text. [d]Followed by dividing mark.

1. See Docs. 2520 (headnote) and 2620.

2. Little is known of the chemical makeup of carbons used in the arc lights of Wallace & Sons, electrical and wire manufacturers in Ansonia, Conn. However, Edison had tried them in experimental carbon telephone transmitters in 1879. He found them composed of finer particles, with fewer impurities, than those of another manufacturer (see Docs. 1615 n. 1 and 1806). Dredge (1882–85, 1:412) notes that the Wallace firm had for some time had trouble obtaining suitable carbons, as crude retort carbon proved unsuitable.

3. Edison referred to the galvanometer being monitored by Patrick Kenny. Kenny, a former superintendent of the Gold and Stock Telegraph Co.'s manufacturing shops, had begun collaborating with Edison on facsimile telegraphy in the spring of 1878 and started working at the Menlo Park laboratory in December of that year. At this time, Edison and Kenny were working on a chemical stock quotation telegraph for which they had filed a patent application in March 1884; the specification issued in March 1885 as U.S. Patent 314,115. See Docs. 1328, 1388 n. 6, and 1638.

4. Nordhausen fuming sulfuric acid consists of sulphuric acid containing more or less sulfuric anhydride in solution. Its name derives from the small Saxon town where an industry producing the acid first developed. Also known as oleum, it is a thick oily liquid that is distin-

guished from ordinary sulfuric acid by the fact that it fumes strongly in moist air. Lock and Lock 1878, 233; Kolbe and Humpidge 1884, 163; Hawley 1987, s.v. "sulfuric acid, fuming."

5. Figure labels are "Stuf[f]ing box," "float," "pump," "pressure Electrode," "Carbon   anthracite heated or coke," "platinzed [welder?] on nickel or iron," "glass," "Strong Sul," "glass," "Peroxide mang," and "furnace."

6. Edison continued these experiments the following day, concluding with a set of trials that involved heating "in crucibles in forge Sul Acd & Blk ox.   ok but fumes so bad of SO_3 that Martin [Force] spit blood & I was nearly overcome." N-82-05-15:141–45, Lab. (*TAED* N203141).

7. These appear to be notes that Edison took while reading Roscoe and Schorlemmer 1878 (2:162–63). He had acquired the American edition in 1879. D. Van Nostrand to TAE, 28 Jan. 1879, DF (*TAED* D7910A).

–2707–

Notebook Entry: Energy Conversion

[New York,] July 31 1884[a]

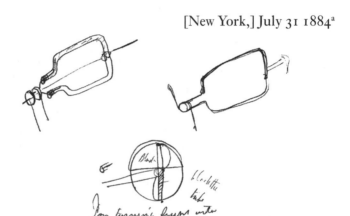

For turning light into electricity[1a]
Ditto

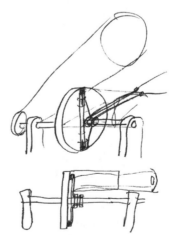

TAE   M[artin]NF[orce]

X, NjWOE, Lab., N-82-05-15:180 (*TAED* N203180). Document multiply signed. [a]Written by Martin Force.

1. Text is "Black" and "black the tube." The design of these instruments is unclear, but this set of drawings is the earliest extant evidence of Edison's interest in the conversion of heat and light into electricity. This effort became part of his search for an unknown force he called the "XYZ." For his work on this force between December 1885 and April 1886, see especially New York Notebook N-85-12-08 and Fort Myers Notebook N-86-03-18, both Lab. (*TAED* N313, N314).

–2708–

*Samuel Insull to
Alfred Cowles*

[New York,] Aug. 1st. 1884.

Dear Sir:—

I enclose you herewith memorandum of amounts due our Construction Dept. by various electric illuminating companies. In addition to this the Edison Electric Light Co. owe us about $17,000., on work in connection with the Construction Dept., and with relation to which we are now arranging settlement.[1]

I will send you a statement of the affairs of the Machine Works in the course of a few days.[2] I would do so immediately, but inasmuch as our books are just being closed for this month, it would be far more easier for me to supply you with this information after the bookkeepers have got out their monthly balance sheet. Very truly yours,

Saml Insull

Enclosure.

ENCLOSURE[a]

[New York, August 1, 1884]

Shamokin, Penna.	$ 1,107.81
Sunbury, Penna	2,411.68
Brockton, Mass.	6,356.75
Fall River, Mass.	9,287.39
Piqua, Ohio	2,337.87
Tiffin, Ohio	1,844.27
Bellefonte, Penna.	2,256.42
Mt. Carmel, Penna.	3,685.88
Middletown, Ohio	627.29
Newburgh, N.Y.	1,276.81
Hazleton, Penna.	1,705.04
Total	$32,897.41[3]

TLS (carbon copy), NjWOE, Lbk. 18:196 (*TAED* LB018196). [a]Enclosure is a TD (carbon copy).

1. Cowles requested the information in this document in July. According to terms of Doc. 2515, the Ansonia Brass & Copper Co., one of Edison's largest suppliers, could claim at least a portion of the payments Edison received from local illuminating companies. An October 1884 statement showed that Edison owed the Ansonia firm about $18,000; the amount decreased to $7,690 by the end of the year. In late December, Edison was trying to negotiate a settlement with the Edison Electric Light Co. and hoped to use those funds to pay Cowles. Insull promised that, if money from the Light Co. were not forthcoming, Edison would "make arrangements to gradually liquidate" his account with Ansonia. Cowles to Insull, 24 July 1884; Ansonia Brass & Copper Co. statement, n.d. [Oct. 1884]; both DF (*TAED* D8421W, D8421ZAB1); Insull to Cowles, 26 Feb. 1884, Lbk. 19:475 (*TAED* LB019475).

2. Not found.

3. As transcribed, these figures sum to $32,897.21. The disparity may be due to an erroneous transcription from the faint and smudged carbon copy.

-2709-

To Frank McCormick

[New York,] 4th Aug [188]4

Dear Sir;

Your favor of 2nd came to hand this morning

If you will read my letter of 1st & substitute Thomas A Edison for the Construction Dept mine of 1st will be correct. I should have used my own name instead of the Constn. Dept in writing you.[1]

I cannot understand what you mean by your being trifled with   I would remind you that I had a Contract with your Co by which you were to pay me in cash for your plant. This you have never been able to do; furthermore a basis of settlement with your Coy was arrived at by your Board of Directors as representing your Coy & Mr. Insull as representing myself came nine months ago & under that settlement you were to pay balance due[a] me in the Stock of your Company. $4100 Stock was delivered to me—the balance was to be paid immediately your Capital could be increased & which was forth with done but I have never received the balance of the Stock & for a very considerable time could get no answer to my letters in relation to same. My accounts as rendered were accepted by your Company & I must confess that if anyone has been trifled with I think that person is myself & I do not feel at all disposed to do anything with relation to the matter [-][b] you complain of until I have some guarantee that the settlement agreed upon by your Board will be carried out. Yours truly

Thos A Edison

LS (letterpress copy), NjWOE, Lbk. 18:221 (*TAED* LB018221). Written by Samuel Insull. ᵃ"balance due" interlined above. ᵇIllegible.

1. Edison responded on 1 August to McCormick's letter of complaint about the Sunbury plant, saying he did not wish to discuss the matter until the Edison Electric Light Co. had made arrangements regarding the business of the Edison Construction Dept. McCormick explained that the Sunbury illuminating company initially had withheld the stock owed to Edison since 1883 (see Doc. 2498 n. 3) because they hoped to sell the shares and pay in cash. More recently, however, local managers concluded that the company had "not received value" for what they considered to be a substandard station, and that ongoing repairs to the plant and equipment would consume all its profits. After enumerating seven of "the most glaring defects" of construction, including a leaky roof and an inadequate steam engine, McCormick charged that Edison's business with the company "looks very much like a swindle." He expressed a desire to settle accounts with Edison and the Edison Electric Light Co., but only on terms that would put the plant in satisfactory condition and reimburse the company its expenses for repairs already made. TAE to McCormick, 1 Aug. 1884, Lbk. 18:203 (*TAED* LB018203); McCormick to TAE, 18 July 1884, DF (*TAED* D8458ZAE).

McCormick wrote again on 2 August after receiving Edison's letter of the previous day. He pointed directly to Edison's ongoing liability for the work of the now-disbanded Construction Dept.:

> I do not understand how any arrangements between the Edison Electric Light Co. and the Construction Department can influence our settlement with you— We have made no contract with the Construction Dept. and are in no way interested in its arrangements with the Edison Electric Light Co.— But we have a contract with you individually for the construction of our plant in a "good and workmanlike manner" and if anything is to be done before the central station falls down it must be done at once— We do not like the way we are being trifled with, and we demand a settlement of our accounts. [McCormick to TAE, 2 Aug. 1884, DF (*TAED* D8458ZAF); see also Doc. 2737]

–2710–

To Eugenie Stilwell

[New York,] August 7th 1884

Miss Jennie Stillwell[1] c/o Mrs T. A Edison
Will out on the five forty train[2]

T. A Edison

L (telegram, copy), NjWOE, LM 21:496 (*TAED* LBCD8496B). Written by Alfred Tate; handstamp of the office of T. A. Edison.

1. Eugenie L. "Jennie" Stilwell (1868–1942) was a sister of Mary Stilwell Edison (though not the youngest, as incorrectly stated in Doc. 2344 n. 3; see Doc. 2646 n. 1). The Edisons had paid for Jennie's schooling since at least 1882 (see Docs. 2344 and 2723). Although her living arrangements are uncertain, she seems to have been at Menlo Park dur-

ing March, when the Edisons were in Florida and her father's health was failing (Eugenie Stilwell to Samuel Insull, 12 and 27 Mar. 1884, both DF [*TAED* D8465N1, D8414D]).

2. The editors have not identified the purpose of this message, specifically whether it was related to Mary's health (see Doc. 2712 [headnote]).

–2711–

From Willis Stewart

Santiago, Chile, Aug. 8th, 1884.

No 22

Dear Sir:—

I have your favor #25 of June 30.[1]

I regret exceedingly that my letters to Maj. Eaton have involved matters regarding the Santiago Station. On the day of my sailing from New York Maj. Eaton particularly and earnestly requested me to write him by every mail regarding this Station & its progress, and of my success in keeping it running. On this account I have regularly sent him details, & have furnished you copies of most of my communications.

Last month Maj. Eaton communicated to me the Co's. decision not to do anything for this Station, which I answered rather sharply, as I felt. I enclose copy of this letter for your perusal.[2] I say that under existing circumstances I shall not form a Santiago Co., although I can do so at any time. My care and labor has completely revolutionized this Station; we are daily adding new & picked consumers; receipts have doubled; confidence is so well restored that we are lighting places previously closed, including the Banks & private residences of the best men in the city; I am now adding a lot of all-night consumers, & the new dynamo is almost ready. The Government is on our side, & I have earnest and influential partners and assistants. This means that so far as Santiago people are concerned everything is ready for large business.[3] But as Kendall & Co. can make $200,000 in an hour by closing the Station when my contract expires, you may be quite sure that they will do it. Ed. Kendall is now here, angry with the Co. for sending me, angry with you for giving me a contract, angry at his cool reception in New York, angry with me because I plainly expose[a] his trickery here & threaten him with the penitentiary, & angry because he cannot play the same game in Valparaiso that has been played here. Do you suppose that a man in this frame of mind will earn a dollar for you or the Co.?[4]

I have Valparaiso nearly arranged,[5] & when you get Santiago I will arrange that also. Tell Maj. Eaton that if he wants to save this Station he must act at once & firmly. If not, there will be a fine scandal here in December or before.

The materials ordered by Mr. Waters[6] should be sent. I will remit for them on receipt of the bill.

My future communications on this subject will be sent only to you.[7] Yours Truly,

W. N. Stewart.

(Enclosure.)

ALS, NjWOE, DF (*TAED* D8435ZBB). Handstamp of W. N. Stewart, Santiago, Representante en Chile de Thomas A. Edison. ªObscured overwritten text.

1. This letter has not been found. Edison sent Stewart letter number thirty-two, also on 30 June, regarding an order of seven ampere indicators, which he considered unnecessary for the Santiago station. Edison's office staff began numbering Edison's correspondence to Stewart on 27 May. TAE to Stewart, 30 June 1884, Lbk. 18:103 (*TAED* LB018103); TAE to Stewart, 27 May 1884, DF (*TAED* D8416BRL).

2. Stewart probably enclosed a copy of his 10 July letter to Eaton acknowledging receipt of a 26 May letter (not found) and unspecified memorandum. The memorandum was likely the three-page critique that Eaton prepared on 21 May in response to Stewart's request for the Edison Electric Light Co. to help finance the operation of the Santiago station, of which Stewart hoped to gain control. Edison recommended Stewart's proposal to Eaton: "I am willing to do personally what Mr Stewart wants your Company to do provided your directors will undertake to give me half of what they receive in stock or money from the Santiago Company. I think it would be a very serious blow to your business if the Santiago Station were allowed to stop and if your Directors are not willing to take the risk necessary to save it." Stewart indignantly replied to Eaton on 10 July that he had "asked you to help me to rescue the business and save its good name by a slight sacrifice, if such should prove necessary, which you decline. I therefore choose to take all the responsibility myself, as I shall take the profits." Stewart to Eaton, 10 July 1884; Eaton memorandum, 21 May 1884; both DF (D8435ZAT, D8435ZAH); TAE to Eaton, 22 June 1884, LM 19:498 (*TAED* LBCD6498).

3. Enrique Lanz, a prospective partner with Stewart, sent Edison on 8 August his views of the station's operations and the intentions of Kendall & Co. Stewart withdrew from direct work for the Compañía de Luz Eléctrica de Edison de Santiago by early 1885. The Santiago station kept operating but, if Stewart's reports were accurate, it lost many large customers in 1885 because Kendall instituted a 60 percent price increase and also because of poor management of its lamp supplies (Lanz to TAE, 8 Aug. 1885; Stewart to Samuel Insull, 26 Dec. 1885; both DF [*TAED* D8534P, D8534X]). While Stewart still hoped to wrest control of the situation in Santiago, he withdrew to Valparaíso and the Compañía Eléctrica de Edison, his general agency for Chile, Peru, and Bolivia. He was assisted there by William J. Clark, a veteran of Edison's telephone business in Chile as well as the Santiago station. Stew-

art suggested that Edison "organize a special bureau for foreign work," but no reply to this suggestion has been found (Compañía Eléctrica de Edison, undated 1885, PPC [*TAED* CA011B]; Clark testimony, 26 Mar. 1881; Stewart to TAE, 14 Mar. 1884 and 31 Jan. 1885; Stewart to Insull, 26 Dec. 1885; all DF [*TAED* D8147J, D8435E, D8534D, D8534X]). During a two-month visit to New York in mid-1885, Stewart secured cooperation from W. R. Grace through its own branch in Valparaíso; Samuel Insull encouraged the alliance, if only to help recover Stewart's considerable indebtedness to the Edison Machine Works. Stewart had fallen out of favor with Charles Coster and the Edison Electric Light Co. by that time (Stewart to TAE, 24 Jan. and 30 Oct. 1885; Stewart to Insull, 26 Dec. 1885 and 28 Apr. 1886; all DF [*TAED* D8534C, D8534T, D8534X, D8630ZAJ]; Insull to TAE, 25 June and 21 Aug. 1885; TAE to Enrique Lanz, 30 Sept. 1885; Lbk. 20:380, 20:444A, 21:35B [*TAED* LB020380, LB020444A, LB021035B]). In March 1886, he was reprimanded by Eugene Crowell, president of the New York company, reportedly for "embarrassing the Santiago Co." and "slandering" the station superintendent (Stewart to TAE, 28 Apr. 1886, DF [*TAED* D8630ZAI]).

4. Stewart continued to complain about Kendall & Co.'s business principles. The firm was later sued by others alleging fraud; it was acquitted in at least one case. In late 1884, Kendall reportedly also initiated negotiations for a takeover of the Santiago station by Brush lighting interests. Riesco 1897, 246–48; Stewart to TAE, 10 and 24 Jan. 1885, both DF (*TAED* D8534A, D8534C).

5. Stewart promised to form a company in Valparaíso after the successful installation of a 100-light dynamo in that city. Stewart to TAE, 8 Aug. 1884; Stewart to Edison Machine Works, 8 Sept. 1884; both DF (*TAED* D8435ZAZ, D8435ZBE).

6. Stewart probably referred to the order for ampere meters mentioned in note 1. George Wellington Waters (b. 1860), a Newark native, had installed an Edison isolated plant at the Prospect House in the Adirondacks, the first hotel to introduce electric lighting throughout its establishment (William Meadowcroft to Edward Babcox, 4 Jan. 1917, Lbk. 115:495 [*TAED* LB115495; *TAEM* 277:668]; Tolles 2003, 81, 180; Edison Electric Light Co. Bulletin, 12:4, 27 July 1882 [*TAED* CB012]; Edison Co. for Isolated Lighting Bulletin 6:4–6, 25 July 1885, CR [*TAED* CC006]). At this time, Waters was the engineer at the Santiago central station. He served as plant superintendent in 1885, when Stewart was highly critical of his work ethic, effectiveness, and loyalty (Stewart to TAE, 23 May 1884 and 10 Jan. 1885; Stewart to Samuel Insull, 26 Dec. 1885; all DF [D8435ZAK, D8534A, D8534X]). He subsequently became associated with Spencer & Waters, a machinery and munitions trading concern in Santiago ("Membership: Candidates for Membership," *Bulletin of the American Institute of Mining and Metallurgical Engineers*, 84 [Dec. 1913]: xxvii).

7. Stewart evidently hoped to avoid working through Sherburne Eaton. Acknowledging a 9 July letter from Edison (not found), he predicted that "The reorganization of your foreign business under care of the Machine Works cannot but be a benefit to all concerned." In September, Edison did consolidate at least the invoicing and bookkeeping aspects of his foreign lighting business at the Goerck St. shop. Stewart

to TAE, 8 Sept. 1884; Samuel Flood Page to TAE, 11 Sept. 1884; Stewart to Edison Machine Works, 8 Sept. 1884; Edison Machine Works power of attorney to Milton Adams, Aug. 1884; Edison Machine Works agreements with Carlos Monteiro e Souza, both Aug. 1884; all DF (*TAED* D8435ZBF, D8437ZAJ, D8435ZBE, D8431ZAN, D8431ZAO, D8431ZAP).

MARY EDISON'S DEATH   Docs. 2712–2713 and 2718

Early in the morning of 9 August 1884, Mary Stilwell Edison died at the age of twenty-nine in her Menlo Park home; the telegrams in Docs. 2712 and 2713 are the first extant acknowledgments of the fact. The latter stated that she succumbed to "congestion of the brain," an explanation repeated in a few newspapers and one that some biographers have since adopted.[1] As it appears nowhere else among the extant documents from 1884, however, congestion of the brain falls short of being a definitive cause of death. In fact, the postmortem medical certificate failed to register any cause, and it also lacked such standard information as the length of illness or the names of attending physicians.[2] Congestion of the brain (also termed cerebral congestion or cerebral hyperemia) was a common, if imprecise, diagnosis in the late nineteenth century, and one contemporary medical dictionary cautioned that "many symptoms have been erroneously ascribed" to it. Another authority pointed out that "Under this name there are to be included several forms of disease very different from each other in the general character of their symptoms."[3] Its bafflingly heterogeneous manifestations have been linked to, among other causes, cerebrovascular apoplexy (apparently its most frequent diagnosis, later attributed to hypertension), and they also found a place in the literature of menstruation, uterine displacement, and other female conditions.[4]

The ambiguity about the immediate cause extends to other circumstances around Mary's death. A number of obituary notices indicated that her demise came suddenly and unexpectedly, after a brief but grave illness, and at least one condolence letter supports this view of events.[5] None of the extant documents suggests a mounting crisis in late July, when Edison often was working in New York,[6] or in early August, when his movements are largely unknown. Unable to reconstruct a context for Doc. 2710, Edison's terse announcement of his planned return to Menlo Park on the afternoon of 7 August, the editors cannot speculate whether that telegram represented anything out of the ordinary. Other evidentiary fragments, though, raise more questions than answers about Mary's final days. One physician, Frank Beardsley Norton of Metuchen, N.J., billed for "medicines" and "visits" to the Edison home on 25 and 26 July and another consultation on 9 August.[7] One news item printed after Mary's funeral pointed out that Charles Stilwell, Mary's younger brother, was en route from Hamilton, Ontario, when she died, but he "arrived a few hours too late to see his sister alive."[8]

After the funeral, a *New York World* article (Doc. 2718) offered the novel and plausible—but unverifiable—explanation that Mary died from an accidental overdose of medicinal morphine. The article claimed that she had been using the opiate in the course of treatment for the chronic pain of "obstinate neuralgia" since 1878.[9] In the newspaper's account, Mary had been coping with a recent bout of gastritis, perhaps caused by the opiate itself, and her effort to escape the pain led to an overdose. The *World's* claim for this particular cause of death, attributed to the sotto voce remark by a "friend of the family," was apparently unique. Still, Mary Edison fit the gender, class, and age profile of the patient for whom doctors most typically prescribed the drug at the time, and her medical history (specifically "uterine troubles") would have indicated its use.[10] Contemporary medical authorities also recognized "congestion of the brain" as a physical symptom of fatal morphine overdose.[11] Many medicinal preparations, particularly cough syrups and also Edison's own analgesic compound, polyform,[12] contained morphine, and Edison's incomplete system of payment vouchers indicates the household's occasional purchase of such compounds. More noteworthy is a purchase record in November 1883 for two one-half ounce bottles of the sulphate of morphia, a form suitable for hypodermic injection.[13] However circumstantial the case for its role in Mary's death, morphine makes a more plausible explanation than typhoid fever, which family members introduced to the lore decades later.[14]

Scattered documents dating back to 1878 touch upon intermittent ailments, both mental and physical, but the underlying state of Mary's health is unknowable. In brief, the surviving record from the summer and fall of 1878 is unusual for its details about Mary's pregnancy with her third child, during which she reportedly experienced considerable anxiety.[15] Edison hastily summoned a doctor for her on the last day of that year, not necessarily a remarkable event in mid-winter for a mother of two young children and an infant.[16] After that incident, the editors have found no further reference to Mary's ill health until 18 January 1882, when Dr. Leslie Ward wrote of her recurrent "uterine troubles" and suggested travel as a way to calm her nervous system.[17] Mary did travel that winter, to Detroit, South Carolina, and Florida; she and Thomas returned to Florida in February 1884.[18] In September 1883, Edison claimed that his wife's illness forced her to give up housekeeping in favor of hotel life.[19] When moving day arrived, Samuel Insull advised Edison that "she seems considerably better," but would be escorted by a doctor "in case of any mishap."[20] Soon after returning from their 1884 Florida trip, both Mary and Thomas became sick, but their reported symptoms point to nothing more than seasonal allergies or ordinary-enough colds.[21] A news report after Mary's death, printed a few days before Doc. 2718, referred to long episodes of illness since 1878.[22]

It is not clear what connection, if any, these incidents have with each other or with those reported in Doc. 2718. From a vantage point more than fifty years later, one of Mary's nieces emphasized her ailments, fancying that her "very earliest recollection" was the sight of family women assisting "Aunt Mame" at Menlo Park after "one of her frequent fainting spells."[23] Yet many contemporary sources also place Mary beyond the sickroom and show her as socially active—giving parties, attending balls, going to the theater—and otherwise engaged in a range of domestic functions. Mary may have been mostly well, or she could have disguised some underlying condition, trying to live as normally as possible. What is clear is that when her health history is examined alongside Edison's—including his own bouts with neuralgia and severe colds—the last years and months of her life do not show a pattern of decline ending in death.[24]

1. Edison biographers citing "congestion of the brain" as the cause of Mary's death include Baldwin 1995 (143), Israel 1998 (233), and Stross 2007 (143). Conot 1979 (219) added to this diagnosis his own interpretation: "apparently a tumor."

2. A sheet of paper that had been attached to the death certificate was torn off at some point in the past, leaving only the words "Menlo" in the upper right corner and "to certify" in the center. Mary Stilwell Edison death certificate, 9 Aug. 1884, Nj-Ar (*TAED* X147B).

3. Quain 1883, s.v. "Brain, Hyperaemia of. Symptoms"; Reynolds and Bastian 1879, 844.

4. Concerning the history and evolution of medical understandings

of "Congestion of the brain" or "hyperæmia of the brain," see Román 1987, passim; and Blustein 1986, passim. For a contemporary understanding of these diagnoses, see Quain 1883, s.v. "Brain, Hyperaemia of"; Minton 1884, 60, 343, 356; Jones 1884, 30, 379.

5. See Doc. 2716 n. 3 regarding published death notices. Dr. Edwin Ruthvin Chadbourne, a physician who had attended the family, wrote in a condolence letter that he was "very anxious to learn what the trouble was that carried her off so suddenly" (Chadbourne to TAE, 13 Aug. 1884, DF [*TAED* D8414ZAJ]). An 1879 graduate of the Columbia College of Physicians and Surgeons, Chadbourne was an attending physician at the New York Foundling Asylum; he also maintained a private practice (Medical Society 1895, 53, 322; "Columbia's New Doctors," *NYT,* 1 Mar. 1879, 2; "Thurlow Weed at Rest," ibid., 23 Nov. 1882, 2). For Chadbourne's invoices to Edison for unspecified medical services, see vouchers for payments of $255 (no. 209 [1883]), $325 (attached to no. 143 [1884]), and $152 (no. 510 [1884]).

6. Edison recorded a number of experiments in his New York laboratory in late July (see, e.g., Docs. 2701, 2703, 2706, and 2707). On 23 July he also attended a meeting of the Edison Shafting Manufacturing Co. trustees at the Edison Machine Works. Edison Shafting Co. minutes, 23 July 1884, DF (*TAED* D8432C).

7. Family members recalled years later that another physician who had attended the Edisons, John Daly of Rahway, N.J., came to the home immediately after Mary's death. Voucher no. 514 (1884); Eugenie Stilwell to John Randolph, 9 Feb. 1895, DF (*TAED* D9506AAG).

8. "Funeral of Mrs. Edison," *Newark Evening News,* 13 Aug. 1884, 1.

9. For a contemporary definition of neuralgia, as well as diagnostic and treatment directives, see Quain 1883, s.v. "Neuralgia."

10. Kane 1880, 173–74; "The Abuse of Morphine in Menstrual Suffering," *American Medical Digest* 4 (May 1885): 68–70; see Courtwright 2001 (9–60) regarding the medical, technological, and culture contexts of opium and morphine in the nineteenth-century United States, and Kandall 1996 (passim) concerning the history of women and opium and morphine addiction in the United States.

11. Quain 1883, s.v. "Opium, Poisoning by."

12. See Doc. 1287.

13. Mary Edison shopping list, 20 Nov. 1883, DF (*TAED* D8314P); voucher no. 44 (1883) for McKesson and Robbins (New York druggists, for purchases of Bull's cough syrup); Kane 1880 (chaps. 2–3) gives contemporary recommendations on the preparation, dosage, and subcutaneous injection of morphine for medicinal purposes.

14. While the attribution of typhoid fever did come from Mary's sister, Alice Stilwell Holzer, and her daughter, Marion Edison Oser, the editors have found no evidence to corroborate this cause (Holzer to William Simonds, 2 July 1932, MiDbEI [*TAED* X001D8I]; Oser 1956, 5). Simonds 1934 (231) and Josephson 1959 (290) both attribute Mary's death to typhoid fever.

15. See Docs. 1394, 1402, 1408, 1523–1525, 1531, 1534.

16. TAE to Leslie Ward, 31 Dec. 1878, DF (*TAED* D7813ZAB).

17. See Doc. 2213.

18. See Docs. 2213 n. 4, 2233, and 2234; see Doc. 2618 (headnote) concerning the Edisons' 1884 Florida trip.

19. See Doc. 2528 and cf. Doc. 2041.

20. Insull to TAE, 4 Oct. 1883, DF (*TAED* D8316AYJ).

21. See Doc. 2663.

22. "Mrs. Edison's Funeral," *New Brunswick (N.J.) Daily Times,* 14 Aug. 1884, 1, reprinted from *Newark Advertiser,* [13?] Aug. 1884.

23. Mary Edison Holzer to Francis Jehl, 27 Apr. 1935, EP&RI (*TAED* X001D8K).

24. Instances of Edison's illness are reported in, e.g., Docs. 1539, 1549, 1644, 1730, and 2263–64.

–2712–

*Telegrams: To / From
Samuel Insull*[1]

August 9, 1884[a]

Menlo Park NJ 10.19 [a.m.]

Sam'l Insull

Minnie[2] and dot[3] went on nine thirty to grammach[4]   will let you know by telegraph if want anything

Edison

New York

Thos A Edison

Have seen Minnie. Have telegraphed Pitt and just going to Newark to see Compton Undertaker[5]   If you have any special instructions telegraph Care operator Market St depot Newark   shall go on to Menlo from Newark

Insull

Menlo Park NJ 12:22

Samuel Insull

Come immediately

T. A. Edison

⟨Wired Mr Edison that you had left for Menlo Park Cooke⟩[6b]

L (telegrams), NjWOE, DF (*TAED* D8414Q, D8414Q1, D8465ZAG). Written on Western Union message forms; second message written by Menlo Park station agent Marcus Hussey. [a]Date from document, form altered. [b]Marginalia written by Charles Cooke.

1. See headnote above.

2. Minnie Dingee was Mary Stilwell Edison's cousin. She was the daughter of Margaret Stillwell Dingee, the sister of Mary's father. Minnie Dingee to TAE, 25 Mar. 1908, DF (*TAED* D0814AAG; *TAEM* 192:290).

3. Edison's daughter Marion.

4. Likely Mary Edison's mother, Margaret Crane Stilwell. She subsequently kept house for the family. Samuel Insull to Mary Wicker, 15 Sept. 1884, Lbk. 18:365 (*TAED* LB018365); Early Recollections of

Marion E. Oser, p. 6, typescript, Box 16, Edison Biographical Collection, NjWOE.

5. Charles W. Compton (1833–1914) was a prominent Newark businessman and second-generation undertaker. The Compton funeral parlor was located at 216 Market St. Leary 1893, 255; U.S. Census Bureau 1982? (1900), roll T623_964, p. 5B (Newark Ward 6, Essex, N.J.); Obituary, *NYT,* 3 Mar. 1914, 9; letterhead of 12 Aug. 1884 Compton invoice (Doc. 2716).

6. Charles Cooke's 9 August telegram to TAE is in DF (*TAED* D8465ZAH).

–2713–

Robert Lozier to
John Tomlinson

NEW YORK, Aug 9th, 1884[a]

Dear Mr. Tomlinson,

Mrs Edison died this morning at 2 o'clock from congestion of the brain, funeral at Menlo Park, N.J. Tuesday, Aug 12th, in the morning.[1] Yours respectfully,

Robert T. Lozier[2]

TLS, NjWOE, DF (*TAED* D841401). Letterhead of Thomas A. Edison. [a]"NEW YORK," preprinted.

1. See Doc. 2712 (headnote).

2. Robert Ten Eyck Lozier (1868–1921) began working for Edison as a clerk at the 65 Fifth Ave. offices in February 1883. After three and a half years, he was assigned to electrical work. Soon after leaving Edison's employ in 1889, Lozier became a consulting engineer and reportedly took out a number of electrical patents. Lozier to Alfred Tate, n.d. [Nov. 1889], DF (*TAED* D8920ABC); "Lozier, Robert Ten Eyck," Pioneers Bio.; Obituary, *NYT,* 23 Aug. 1921, 11.

–2714–

From Grosvenor
Lowrey

[New York,] Aug. 10th 1884

My dear Edison

I learned yesterday of poor Mrs Edison's sudden death, but too late to go out and see you, & yet reach home,[1] where my children are along, their mother being on a visit to Canada— I know the sorrow which you now feel— Nothing is or can be like it; for however the occupations of life may for a time draw a man apart from the woman he loves, & who is the mother of his children, & his most intimate friend, still there is no love like love for her, & no friendship like her friendship for him, & when she dies the greatest loss which a man can suffer has come to him— Having felt it[a] in its bitterness, I offer you my sincerest, warmest sympathy.[2]

When I heard the news, in a moment my heart warmed to

you, as it used to do—& indeed always has done, whenever I have been able to get into relations with your true nature—& I felt that we were brothers in misfortune.

God has brought to me the dearest compensation in the health & happiness of my growing children, & the love of another woman, as nearly like my dear first wife in gentleness of spirit, intelligence, affectionateness and every womanly virtue as is possible— Thus I have been shown how in the order of the creator the sense of loss, is displaced by time, occupation & the ever-springing, new life of the affections.

May God bless you, my dear friend & your little children, and ease the pang which now seems to you without cure   Ever Sincerely Yours

G. P. Lowrey

ALS, NjWOE, DF (*TAED* D8414V). [a]Interlined above.

1. This letter is one of about a score of condolence messages to Edison that were filed in Edison—Family, DF (*TAED* 8414). Correspondents included associates such as Henry Villard and George Bliss, as well as strangers who read of Mary Edison's death in newspapers. Lowrey had telegraphed from New York the previous day: "Dear Edison   I have just heard of your loss and send you my heartfelt Sympathies." His telegram was the first communication known to the editors from him to Edison or Samuel Insull since March 1884. Lowrey to TAE, 9 Aug. 1884; Lowrey to Insull, 17 Mar. 1884; both DF (*TAED* D8414T, D8465P).

2. Lowrey's wife, Laura Tryon Lowrey, died of cancer in 1879. He married again in 1880, to Kate Armour (b. 1855?), a Canadian whose father was Chief Justice of the Court of the Queen's Bench of Ontario. Taylor 1978, 23, 45–46.

–2715–

From Marion Page

Rose Hall [Milan, Ohio] Aug 11th [1884]

Dear Brother

Your telegram[1] arrived too late for me to reach Menlo Park in time for the funeral   I was shocked to hear of Mary's death I supposed she was well and Enjoying herself for you know Alva She and yourself[a] never answered my letters and I only know from outsiders anything about you— I feel hurt sometimes you never write to me or notice me in any way Im your only Sister   Now Alva I want you to come here right away and bring your little children and stay here two or three weeks and make me a little visit   think how long it is Since I have Seen you   I have never seen your youngest Son[2]   do come and bring them all   the old house does Seem so lonesome since

Belle[3] went so far away   kiss the little children for me   now answer this letter wont you   do come   good bye   Ever your loveing Sister

Marion[4]

ALS, NjWOE, DF (*TAED* D8414X). [a]Obscured overwritten text.

1. Not found.

2. William Leslie Edison was born in October 1878; his closest sibling, Thomas, Jr., had arrived in January 1876.

3. Marion Page's daughter Isabella (b. 1852?), known as Belle, married George Washington Ristine. The couple lived in Cleveland in 1880 but did not appear in the city directory the following year. According to a childhood friend of Edison's, Isabella was living in San Francisco in the spring of 1884. By 1888, Isabella and George had apparently relocated to Chicago, where he was general manager of the Erie Dispatch, a freight agency. U.S. Census Bureau 1965 (1870), roll M593_1197, p. 120A, image: 245 (Milan, Erie, Ohio); ibid. 1970 (1880), roll T9_1006, p. 25.2000, image 0054 (Cleveland, Cuyahoga, Ohio); *Cleveland Directory* 1880, 226, 438; ibid. 1881, 453; Alva Richardson to TAE, 13 Nov. 1884; George Ristine to TAE, 10 Nov. 1888; both DF (*TAED* D8414ZBE, D8807ACM).

4. Marion Wallace Edison Page (1829–1900) was Edison's oldest sibling. She married Homer Page in 1849, and the two reportedly bought the home "Rose Hall" and surrounding acreage along the Huron River from the estate of a local farmer-entrepreneur in about 1861. *TAEB* 1 chap. 1 introduction, n. 4; website of the village of Milan, Ohio (http://milan area.com/index.htm, accessed 14 Apr. 2010); Thorndale 1900, 721.

–2716–

Bill from Charles Compton

Newark, N.J. August 12th 1884[a]

For the burial of his wife Mary S Edison

To[1] Crape and Ribbon		3.00
" Iceing and use of preserver and expenses to and from Menlo park[2]		28.00
" publishing death Newark and New York papers[3]		49.25
" best quality drab silk plush covered Casket, draped, tuft lid.		
Satin lined full length and Silk covered extension handles and plates		425.00
" polished cedar Casket Case brass mounted & plate		95.00
" Hearse and team with white drapery		15.00
" 20 Coaches		80.00
" 12 pr 2 button black Kid gloves	1.85	22.20
" notices and serving		2.50

<table>
<tr><td>"</td><td>Flowers. Gate Ajar with Dove, lettered. Mound with Standing Cross & Dove with wreath of buds. lettered. large pillow & Crown, lettered. Smilax & Cut flowers.</td><td>110.00</td></tr>
<tr><td>"</td><td>Receiving tomb Fairmount Cemetery[4]</td><td>20.00</td></tr>
<tr><td>"</td><td>6 Casket Carriers</td><td>24.00</td></tr>
<tr><td>"</td><td>Gloves for Carriers</td><td>3.50</td></tr>
<tr><td>"</td><td>Services of Undertaker & assistance</td><td>30.00</td></tr>
</table>

$877.45[5]

D, NjWOE, Vouchers (1884), no. 492 (*TAED* VC84492). Ruled bill head of Charles Compton, Funeral Furnishing Ware-Rooms. Monetary expressions, in which ditto marks indicated double zeroes, have been standardized for clarity. [a]"Newark, N.J." and "187" preprinted.

1. "To" was used in the accounting sense of denoting a debit entry or specifying an amount paid for goods or services. *OED*, s.v. "to" *prep.* 34.

2. Edison separately paid a bill of $12 to one Alex Ayres for a "woman to lay out body" and for "preserving body on ice for Mrs. T. A. Edison. Dec'[eased]." Voucher no. 509 (1884).

3. Several New York newspapers published the following standard announcement between 10 and 12 August:

> Edison.— Suddenly, at Menlo Park, N.J., on Saturday, August 9, 1884, Mary Stilwell, in the 29th year of her age, wife of Thomas A. Edison.
>
> Funeral services at her late home, Menlo Park, N.J., on Tuesday, Aug. 12, at half past twelve o'clock. Relatives and friends are invited to attend.

New York Tribune, 10 Aug. 1884, 7; ibid., 11 Aug. 1884, 5; ibid., 12 Aug. 1884, 5; *New York Herald,* 10 Aug. 1884, 13; ibid., 11 Aug. 1884, 9; ibid., 12 Aug. 1884, 9; *New York Times,* 10 Aug. 1884, 7; ibid., 11 Aug. 1884, 5; *New York World,* 12 Aug. 1884, 5.

Different death notices also appeared in the *New York Tribune* (11 Aug. 1884, 5); the *New York Sun* (10 Aug. 1884, 5); and the *New York Times* (10 Aug. 1884, 2).

In addition to the formal death announcement, the *New York Herald* also published on 10 August (p. 8) an article entitled, "Death of Mrs. Edison: The Celebrated Inventor's Brief Courtship and Happy Married Life." The article gave one sentence to Mary's death, followed by several paragraphs retelling the traditional story of the Edisons' courtship (see Doc. 2683 [headnote]). The *Herald* article was widely reprinted in the following days. "Inventor Edison's Wife. Romance of Their Marriage Recalled by Her Death," *Sioux County (Iowa) Herald,* 28 Aug. 1884, 7; "Death of Mrs. Edison. The Celebrated Inventor's Brief Courtship and Happy Married Life," *Salt Lake City Daily Tribune,* 24 Aug. 1884, 6; "Current Comments," *St. Paul (Minn.) Daily Globe,* 14 Aug. 1884, 4; "Mrs. Edison," *Titusville (Pa.) Herald,* 30 Aug. 1884, 3; "Gotham Gossip . . . Sudden Death of Mrs. Thomas A. Edison," *New Orleans Daily*

Picayune, 15 Aug. 1884, 3; "Personal Clippings," *Macon (Ga.) Weekly Telegraph,* 14 Aug. 1884, 6; "Mrs. Edison," *Indiana (Pa.) Democrat,* 28 Aug. 1884, 4; "Death of Mrs. Edison—Her Courtship," *Huntington (Pa.) Globe,* 21 Aug. 1884, 4; "Death of Mrs. Edison. The Celebrated Inventor's Brief Courtship and Happy Married Life," *Chicago Daily Tribune,* 12 Aug. 1884, 9; "Personal," *Harper's Bazaar* 17 (13 Sept. 1884), 579.

4. Fairmount Cemetery was incorporated in 1855 in Newark, N.J. Its design reflected the influential landscaped garden cemetery movement that began in the mid-nineteenth century. (Veit and Nonestied 2008, 63–64). Mary's body was later moved to its final resting place at Mount Pleasant Cemetery in Newark (see Doc. 2717 n. 4).

5. A signed receipt for two payments was appended to the bottom of the bill. The first payment was by a $500 check on 15 September. Bergmann & Co. evidently wrote the check to Compton, and Edison credited this amount back to Bergmann under an accounting entry for Compton. The second, on 30 September, was Edison's three-month note for $377.45, payable 2 January 1885. Journal Book (6 Jan. 1881–14 Nov. 1885): 396, 398 Accts., NjWOE.

–2717–

Anonymous Article in the Newark Morning Register

[Newark, August 13, 1884]
The Funeral of Edison's Wife.

Mrs. Mary Stilwell Edison the wife of Thomas A. Edison, the electrician, was buried this morning in this city. Her remains were placed in a casket which rested in the centre of the parlor in the Menlo Park residence preparatory to removal. Around it were ranged many beautiful floral tributes, while cut flowers were strewn on the remains and on the floor underneath.[1] Shortly after eleven o'clock Mr. Edison, who had remained in his room, was conducted to the parlor by Mr. Samuel Insull, his private secretary; the doors were closed, and he was left alone with the dead. At half-past twelve o'clock the Rev. Mr. Mason, of the Presbyterian Church, Metuchen, N.J.,[2] entered, and read the funeral service for the dead. At its conclusion those who were there passed in front of the casket to take a last look at the face of the deceased. The casket was then closed. The pallbearers, ten in number,[3] forming on either side, with relatives to the number of forty, and the throng of friends, the procession moved up to the railroad station, where a train was in waiting to convey the funeral party to Fairmount Cemetery, in this city. There the remains were placed in the receiving vault, pending the decision of Mr. Edison as to their final resting place.[4] There were no services at the cemetery. There were present at the funeral many of the representative business men of New York and Newark, as well as a number of Mr. Edison's employes.[5]

PD, *Newark Morning Register*, 13 Aug. 1884, 1 (*TAED* D8414ZAP2).

1. See Doc. 2716.

2. The Rev. Dr. James Gilbert Mason (1841–1938) was pastor of the Presbyterian Church in Metuchen, N.J., from 1877 to 1930. A graduate of Williams College in 1863 and the Union Theological Seminary (New York City) in 1866, Mason received his Doctor of Divinity from Maryville College, Tenn., in 1884. During his career, he was a Commissioner of the General Assembly of the Presbyterian Church six times and was active in New Jersey politics, focusing on the issue of Prohibition. Williams College 1939, 5; Dudley 1903, 105–8; Spies 2000, 105; Scannell 1917, 344–45.

3. Another news account named the pallbearers as: Sherburne Eaton, Edward Johnson, Frank McLaughlin, Charles Batchelor, Samuel Insull, Henry Reimer, Charles Hughes (an experimental assistant to Edison), William Carman (bookkeeper and office manager), Theodore Carman (a local teamster), and Frederick Price (unidentified). "Funeral of Mrs. Edison," *Newark Evening News*, 13 Aug. 1884, 1.

4. Another newspaper reported that Mary's body would be kept at Fairmount Cemetery "until the vault which is now being built for it is completed" (*New Brunswick, N.J., Daily Times*, 14 Aug. 1884, 2, reprint from *Newark Advertiser*). Undertaker Charles Compton billed Edison $100 on 1 October for the services involved in transferring Mary's body and interring it at Mount Pleasant Cemetery, also in Newark (voucher no. 569 [1884]). Incorporated in 1844, Mount Pleasant Cemetery was the first in New Jersey to reflect the mid-century landscaped garden cemetery movement. Mount Pleasant had become Newark's most fashionable cemetery by the 1870s, and it continued to be the final resting place for prominent citizens into the early twentieth century. Veit and Nonestied 2008, 85–86, 92; *Newark Daily Advertiser* 1872, 65–66.

5. About four hundred people attended the funeral, according to New York press stories. Among them were Edison's brother, William Pitt Edison, and Mary's brother, Charles Stilwell, who reportedly arrived from Hamilton, Ont., "a few hours too late to see his sister alive." Grosvenor Lowrey and José De Navarro were among Edison's business associates present. "Funeral of Mrs. Edison," *Newark Evening News*, 13 Aug. 1884, 1; "Mrs. Edison's Funeral," *NYT*, 13 Aug. 1884, 2; "Funeral of Mrs. Thomas A. Edison," *New York World*, 13 Aug. 1884, 8.

–2718–

Anonymous Article in the New York World *Supplement*[1]

[New York, August 17, 1884]
SORROW AT MENLO PARK.[a]
Last Hours of Mrs. Thomas A. Edison—A Life Full of Joyous Hope Suddenly Goes Out.

The 200 souls who make up the little hamlet of Menlo Park, N.J., are in mourning. They grieve over the untimely death of Mrs. Mary Stillwell Edison, wife of Thomas A. Edison, the famous electrical inventor. She died suddenly on last Friday. Isolated as the village is, without any especial commercial importance other than raising some of the finest fruit and vegetables, the humble farmers felt themselves as important as the citizens of the larger cities on the Pennsylvania Railroad, where Menlo Park is but a side station and trains only stop once or twice a day. The engineers of the accommodation trains treat the hamlet with respect, and never fail to slow up when the youth in charge of the station hangs out the blue and white flag, which means that some farmer, or perhaps Mr. Edison himself, has business important enough to call him to town.

The good people keenly appreciated that in their midst there lives a man towards whom the eyes of the whole civilized world was directed. They are proud to talk about him, never failing to point out to the visitor the little brown-gabled laboratory at the end of the long lane, skirted on both sides with most luxurious vegetation, where Edison drew the vital force from nature and placed a bridle on the electric current. They were proud to discuss electricity, and prefaced their conversation with the story of the mysterious force which had always been a mystery, a curiosity to scientists, until at the hand of their great citizen the current yielded and became the most docile and at the same time the most irresistible power and incalculable aid to man. All this they said was born in the little laboratory at the end of the land. Every tree in the park of the "Wizard's" homestead is held in high esteem, and the sun shines brighter when the wizard himself walks by their cottages and exchanges a friendly smile, and inquires solicitously about their affairs. In short, Edison to them is more than a monarch, and when it was made known that death had robbed him of his wife, the mother of his little children—two bright boys and a flaxen-haired girl—their grief was hardly less than his. Few persons heard of her illness, as only the immediate friends of the family were allowed to approach her bedside during the last moments of her life. Many of them had never seen the gigs of the doctors from miles around driving rapidly through the village on Thursday and stop at the house, but all they could glean was that Mrs. Edison was very sick and not expected to live. The death was as mysterious as it was sudden, and a day after the funeral a life-long friend of the family deepened the mystery by involuntarily remarking:

"She is dead, now, poor thing, but no one will ever know what she died of."[2]

From this story of the past few years of the dead lady's life was revived. To the casual observer Mrs. Edison presented a picture of health. Her face, full and rounded and of a handsome contour, was well set off by bright, sparkling blue eyes. She was not tall, although she weighed in the neighborhood of 200 pounds; but the flesh was so well distributed that if anything it added to her striking beauty.[3] Since the birth of her last child, six years ago, Mrs. Edison seldom enjoyed a day free from pain. She suffered from obstinate neuralgia that refused all manner of treatment. The best physicians were called in, but their remedies proved useless. At last for temporary relief she tried morphine, and soon learned the great palliative powers of the seductive drug—a ready dose of which was always close to her side—and when the premonitory symptoms of an attack came on she knew the value of her white powder. Some said that she became so accustomed to the morphia that she had to be closely watched lest she should take an overdose.

The eminent neurologists whom she consulted advised her to travel, and at the request of Mr. Edison she took a trip to Florida last winter. Instead of obtaining relief she fell victim to gastritis, due to the peculiar atmosphere or perhaps the long acquaintance with morphine.[4] She returned to Menlo Park in a more troubled condition. Her pain intensified, and at times she was almost frantic. Morphia was the only remedy, and naturally she tried to increase the quantity prescribed by the doctors. From the careless word dropped by the friend of the family it was more than intimated that an overdose of morphine swallowed in a moment of frenzy caused by pain greater than she could bear brought on her untimely death. The doctor in attendance said she died of congestion of the brain. When a reporter put the question to him he positively asserted that it was the immediate cause, but about the more remote causes he preferred to remain silent.

The scene during her last moments are truly pathetic. Mr. Edison standing close to the doctors, who checked off the last beats of the pulse as the heart flickered, waited heroically for the fatal moment, when the physician folded the motionless arms across the bosom and tremulously said:

"She is beyond all human aid." Mr. Edison silently drew forth a cabinet and instantly a powerful current of electricity responded to his will. For two hours he kept life from fleeting, but at last he appreciated that his science, like that of the doctors, was powerless.[5] Taking his children by the hand he led them into his study. There they remained for a long time and when he came out his blue eyes glistened and the lids were red and swollen.[6]

The story of Edison's marriage is romantic and truly describes his peculiar greatness.[7] It was told by Mrs. Edison herself not long ago to a friend who told it yesterday. They met by chance, a rainstorm having forced Miss Stillwell and two other young ladies to seek shelter in Mr. Edison's factory at Menlo Park, and that slight acquaintance ripened into friendship and finally, after five months and a half, resulted in their marriage, Mrs. Edison being then sixteen years old. During all this time Miss Stillwell was attending school and only left it a very short time before their marriage. Immediately after the wedding they started on their bridal trip, took the night boat for Albany and spent a couple of weeks away before returning to their home,[8] and the silly pleasantry about Mr. Edison's leaving his bride to wonder at his absence while he returned to his work was simply told as a joke and like many other things of a like nature has been accepted as fact by those who did not know any better. In point of fact, though, Mr. Edison is greatly absorbed in his inventions. He never for an instant forgets that his first duty is to his family and his late wife always attested to his entire devotion and constant watchfulness for their comfort.[9] No woman ever had a more tender or devoted husband, and it seemed as if she could never say enough to prove how keenly she appreciated his goodness to her.

Although Mrs. Edison was a great sufferer she performed the duties of a wife and mother, and went into society, received friends and made no complaint, and few knew her dreadful sufferings. Few, too, knew of her abundant charities and far-reaching kindnesses, but her most intimate friends, and they only knew it as almoners of her bounty. The outside world regarded Mrs. Edison simply as a beautiful woman, with little thought beyond the pleasure of living, of being handsomely dressed and admired, but there was a deep undercurrent of noble character and a mind of unusual power beneath the surface.

She was in all things her husband's true helpmeet, his confidential friend and adviser and his intelligent co-worker, and they were in all things in true sympathy. Mrs. Edison mastered the difficult study of electrical science for the sake of being more to her husband than she could otherwise have been,[10] and she was also ambitious of a literary success. Many a tender and pretty poem that has pleased the people came from her anonymous pen, and she had planned and begun a novel that would have, without doubt given her the prominence she desired, as I saw some of it and recognized evidences of power and beauty, couched in language of simple, womanly grace.[11]

Mrs. Edison's influence in her short life has been gentle and sweet and above all good, but it has been felt further than people generally knew. She died suddenly—taken from the bloom of her youth and beauty—and has left her husband more desolate than any one knows but himself, her children bereft of the tenderest of mothers and her mother deprived of a daughter to be proud of, and her friends, and above all the poor and sorrowing, have lost one of whose gentle ministrations and womanly sympathy never failed them in their need. For every such woman as Mrs. Edison was the world is better and purer, and the tears of all who knew her are eloquent tributes to her worth.

PD, *New York World Supplement,* 17 Aug. 1884, 1 (*TAED* D8414ZAP1). [a]Followed by dividing mark.

1. The editors believe that this unsigned article likely was written by Olive Harper (Helen Burrell Gibson D'Apery), the author of Doc. 2683, who claimed some familiarity with Mary and her family. In mid-September, Harper contacted Edison about a future book that she had intended to dedicate jointly to him and Mary Edison. She sought Edison's approval to dedicate it instead "To Mrs Thomas A. Edison whose untimely death has left a void in the hearts of all who knew her, this book is inscribed with the sincerest affection and as an humble tribute to her many virtues and her pure and noble womanly influence by the author   Olive Harper." She also asked for a loan of $225 toward the $600 needed to publish the book. Samuel Insull declined this request on Edison's behalf, citing the state of "business affairs." Harper to TAE, 14 Sept. 1884, DF (*TAED* D8403ZGM); Insull to Harper, 16 Sept. 1884, Lbk. 18:378 (*TAED* LB018378).

2. See Doc. 2712 (headnote) concerning Mary's health and the circumstances of her death.

3. Both Mary and Thomas Edison had apparently gained weight in recent years; he reportedly added about 40 pounds since 1880 or 1881. Cf. Docs. 1525 illustration and 2683; "Personal," *San Francisco Evening Bulletin,* 6 Mar. 1885, 1.

4. For contemporary understandings of gastritis or inflamation of the stomach, see Quain 1883, s.v. "Stomach, Inflammation of"; cf. Kane 1880, 246, 265.

5. Therapeutic uses of electric current, by physicians and lay persons, had a long history by this time, and a wide variety of devices was available for this purpose (see, e.g., Peña 2003, chap. 3). One contemporary authority specifically recommended applying a "faradic current" in opiate overdose cases (Quain 1883, s.v., "Opium, Poisoning of"). Edison and a partner had manufactured an "Inductorium"—basically a battery and induction coil—in 1874 and promoted it successfully as a cure for rheumatism (see Doc. 435 [headnote]). However, in response

to an unrelated December 1884 inquiry from a self-described invalid, Edison commented: "dont believe in Elec appld to curative Purposes" (TAE marginalia on B. G. Amies to TAE, 29 Dec. 1884, DF [*TAED* D8405V]).

6. Many years later, Mary's daughter, Marion Edison Oser, remembered that "I found my Father, shaking with grief, weeping and sobbing so he could hardly tell me that mother had died in the night." Oser 1956, 5.

7. See Doc. 2683 (headnote) regarding the Edisons' courtship and wedding.

8. Thomas and Mary apparently spent their honeymoon in Boston (Doc. 218 n. 2; Mary Edison Holzer to William Simonds, 2 July 1932, EP&RI [*TAED* X001D8I]; Simonds 1934, 83). Josephson 1959 (99) asserts that the Edisons traveled to Niagara Falls. Olive Harper, in her 1889 article on "Edison's Real Courtship" (see Doc. 2683 n. 8), claimed that bridesmaid Lucy Hamilton Warner accompanied the couple: "They left on the Albany boat, all three, for Niagara Falls that same evening, and remained away a week or so." On the other hand, Simonds 1934 (83) asserts that Mary did not wish to be separated from her sister and insisted that Alice go with them. Concerning possible itineraries that the Edisons could have used, see *Appleton's Hand-book* 1870.

9. Cf. Doc. 2683 n. 11.

10. Cf. Docs. 241 and 248.

11. The editors have uncovered no other indication of such a literary project. If indeed Mary was planning such an undertaking, it conceivably could have been with the assistance of Olive Harper, who was later a prolific ghost-writer in a variety of genres. "Offers Fame for Cash. Aged Woman Cripple Writes Melodramas While You Wait," *Washington Post*, 5 Feb. 1911, E4; *The Writer: A Monthly Magazine for Literary Workers*, 23 (Apr. 1911): 55.

–2719–

To Samuel Flood Page

[New York,] Aug. 22nd. 1884.

Dear Sir:—

You will doubtless remember having some conversation with Mr. Insull, when he was in London, with relation to the patents I have taken out and applied for subsequent to the signing of the agreement between myself and the Edison Electric Light Co.[1]

I beg to enclose you herewith further account, showing total amount expended by me of $11,215.28[2] I am very anxious to make some kind of an arrangement with your Company, by which they will either make some settlement with me in relation to this matter, by which the patents go to your Company, or else I will serve formal notice, in accordance with the terms of my contract with the Edison Electric Light Co., of London, so as to enable me to make other disposition of those patents, if your Company, as the successors of the Edison Electric Light Co., do not require them. I would suggest that the patents should all be turned over to your Company, and that the enclosed account should be passed to your debit with me, as against the amount that the Edison Electric Light Co., of London, advanced on the central station dynamos, which we manufactured on their order, and which we are now holding for their account.[3] Or, in other words, that I should buy from you the central station dynamos up to the amount that I have expended in connection with taking out patents in England since the signing of the agreement between myself and the Edison Electric Light Co., of London.[4] Very truly yours,

Thomas A Edison

Enclosure.

TLS (carbon copy), Lbk. 18:260 (*TAED* LB018260).

1. Under the sixth article of Edison's 18 February 1882 agreement with the Edison Electric Light Co., Ltd., the British company could acquire his patented improvements in electric lighting by reimbursing his experimental expenses and patent expenses (CR [*TAED* CF001AAE1]). If the company failed to do so within three months of written notice from Edison, its rights to these improvements would terminate. These terms would apply to the company's successor, the Edison & Swan United Electric Light Co., Ltd. Samuel Insull later recalled that he had suggested to Page in London that Edison would waive his right to reimbursement of experimental expenses (Insull to Flood Page, 8 Mar. 1886, Lbk. 21:364A [*TAED* LB021364A]).

2. Enclosure not found.

3. See Docs. 2270 n. 4, 2374 nn. 4–5, and 2572 n. 4 regarding the order of six "C" dynamos in 1882.

4. The editors have found no immediate reply from Flood Page but see Doc. 2738. The United Co. seems not to have acted on the matter until prompted by Insull in March 1886. At that time, they offered to compensate Edison for expenses on the fifteen patents they considered valuable by consigning to him one of the "C" dynamos; they were prepared to allow nineteen other patents to lapse. Edison declined this proposal. Insull to Flood Page, 8 Mar. 1886; Lbk. 21:364A (*TAED* LB021364A); Flood Page to TAE, 31 Mar. 1886; TAE to Flood Page, 16 May 1886; both DF (*TAED* D8630Z, D8630ZAP).

–2720–

To George Barker

[New York,] Aug 23rd, 1884

Dear Sir:—

I have your favor of August 20th., and have much pleasure in accepting the appointment as a member of the "National Conference of Electricians."[1] Very truly yours,

TL (carbon copy), NjWOE, Lbk. 18:266 (*TAED* LB018266).

1. Barker had sent Edison a preprinted request to join the Conference, signing it as corresponding secretary of the U.S. Electrical Commission (Barker to TAE, 20 Aug. 1884, DF [*TAED* D8464O]). The Conference met from 8 to 13 September and largely concerned itself with standards of electrical measurement. Edison does not appear to have been in Philadelphia during that period. His name did not appear in the final list of approximately one hundred (mostly American) conferees, a list that included current or former associates Charles Clarke, John Howell, Frank Sprague, and Francis Upton (*Report of the Electrical Conference* 1886, 3–9, 30; TAE to Charles Young, 29 Aug. and 1 Sept. 1884; Samuel Insull to H. Alabaster, 16 Sept. 1884; Lbk. 18:299, 307, 379 [*TAED* LB018299, LB018307, LB018379]; Gibson 1984 [chap. 5] discusses the organization and activities of the Electrical Conference).

–2721–

To Samuel Insull

[New York, c. August 23, 1884[1]]

Insull,

Pryor says that if I will get out on the 1st sept he will take off one months rent. please see Pryor & get this in writing quickly as I will then make arrangements to have Everything out by that time & save the $400.[2] perhaps he would give you till Sept 2nd as [1?][a] reducing payment in proportion

Edison

ALS, NjWOE, DF (*TAED* D8403ZFW1). [a]Canceled.

1. The editors have not found a letter to James Pryor based on Edison's instructions to Insull. A letter was evidently sent to Pryor, however, and it crossed in the mail with one that Pryor posted from Mount Desert, Maine, on 23 August, asking if Edison would vacate the 25 Gramercy Park house on 15 September. Edison had that inquiry in hand by 28 August. Pryor to Insull, 23 Aug. 1884, DF (*TAED* D8403ZFW); TAE to Pryor, 28 Aug. 1884, Lbk. 18:285 (*TAED* LB018285).

2. In response to Pryor's inquiry about leaving the Gramercy Park house on 15 September (see note 1), Insull stated that Edison would do so on that date or earlier, if Pryor would reduce the rent accordingly. Pryor declined to accept an earlier termination date, pointing to an understanding he said Mary Edison had with his sister, Caroline Pryor, that such an arrangement would have to be made before the first of August. Edison ultimately accepted the 15 September date. Pryor to Insull, 29 Aug. 1884, DF (*TAED* D8403ZGD); TAE to Pryor, 28 Aug. 1884; Insull to Pryor, 2 Sept. 1884; Lbk. 18:285, 19:268 (*TAED* LB018285, LB019268).

Meanwhile, on 26 August, while still making plans with Pryor, Edison signed a lease for the 1 September rental of the third floor (east side) of a house at 39 E. 18th St., a few blocks from his present residence. The rent was $1,300 for one year, payable in monthly increments of $108.33. Edison and his children presumably moved to the new apartment on or about 15 September. In October, he contacted Pryor about a velocipede and several other items he believed had been left behind at Gramercy

Park. TAE rental agreement with George Folsom, 26 Aug. 1884; Pryor to TAE, 7 Oct. 1884; both DF (*TAED* D8403ZGA, D8403ZHL).

–2722–

To Rupert Schmid[1]

[New York,] Aug. 23rd. 1884.

Dear Sir:—

I have your favor of August 20th.[2]

I do not desire a bust of Mrs. Edison made at the moment, but will communicate with you later as to the matter.[3] Yours truly,

Thos. A. Edison   G[ilmore][4]

TL (carbon copy), NjWOE, Lbk. 18:267 (*TAED* LB018267). Signed for Edison by William Gilmore.

1. A portrait sculptor from Munich, Rupert Schmid (1864–1932) began his American career in New York in March or April; he later relocated to California. Soon after arriving in New York, Schmid drew on his friendship with August Riedinger, a Bavarian industrialist with ties to the Edison interests in Germany, to request an appointment to show Edison some of his work. He evidently had in mind the upcoming International Exposition at Philadelphia, as he lamented the absence of an Edison bust from the 1882 international exhibition in Munich. Stover 1982, 35; "Fine Arts," *New York Herald,* 1 June 1884, 8; Ludwig von Eisinger to TAE, 5 Apr. 1884, DF (*TAED* D8403ZBC); see also note 3.

2. Schmid asked if Edison would like to order a bronze or marble portrait bust of Mary Stilwell Edison as a family remembrance. During the springtime, Mary had visited Schmid's atelier for a "five <u>minutes</u> portrait" in plaster. The artist had asked her to sit again as recently as 25 July in order to proceed with a clay model. Schmid to TAE, 20 Aug. 1884; Schmid to Mary Edison, 20 May and 25 July 1884; all DF (*TAED* D8414ZAU, D8414J, D8414O).

3. Although Edison apparently did not take up Schmid's offer in October of two versions of Mary's portrait bust, his business with the sculptor was not finished. Edison himself had also had more than one sitting before the end of June, when Schmid reportedly was working on a quarter-length bust of the inventor at "the moment when he made the discovery of the electric light." Schmid ultimately seems to have produced at least two versions of a less idealized likeness, from which Edison was asked in July to select one to be produced in plaster for display at the Philadelphia Exposition. Edison did accept the offer of a bronze for himself, but he asked to postpone its delivery while his family was away in early August and no one would be home to accept it ("Fine Arts," *New York Herald,* 1 June 1884, 8; Schmid to TAE, 20 June, 11, 25, and 30 July 1884, all DF [*TAED* D8403ZEA, D8403ZEP, D8403ZFA, D8403ZFE]; TAE to Schmid, 1 Aug. 1884, Lbk. 18:208 [*TAED* LB018208]; see also Doc. 2741). In addition to the Edison bust, Schmid's display in Philadelphia included two casts of "Edison's hands in bronze, holding incandescent lamps," allegorical figures representing

electricity, and several lighting fixtures (Schmid to TAE, 25 Aug. 1884, DF [*TAED* D8403ZFZ]; Franklin Institute 1884, 57).

4. William Gilmore.

–2723–

From William Bowen

Bordendown, N.J., Aug 25th 1884[a]

Dear Sir

Your letter of the 24th is at hand[1]   We shall be pleased to see Miss Stillwell and sister Sept 1st.

If Miss Stillwell is in the Preparatory Department, her expenses will be as follows. viz

Board &c Tuition	$60.00
Church	1.00
Books about	8.00
	$69.00

These will be used throughout the year. A few Copy books & Blanks besides also needed after first quarter

Should she be in the Collegiate Dept. exp. would be

Board &c.	70.00
Church, 1.00 Books, about 6.00	7.00
	$77.00
	18.
	95.00
Should she take Music add	$18.00
to the other expenses—or Art, add	$10.00

And you will have about the first quarters expense.[2]

Should a language be taken I have not included in cost of books for it in the above items for books.

Should Music or Art be taken there will be some additional expense for Sheet Music & Art Materials.

This will be charged in the next quarter's bill.

I have thus answered as nearly as possible your question. Yours very truly

W. C. Bowen.[3]

ALS, NjWOE, DF (*TAED* D8414ZAV). Letterhead of Bordentown Female College, Rev. Wm. Bowen, president; monetary expressions standardized for clarity. [a]"Bordentown, N.J." and "188" preprinted.

1. Edison's 24 August letter has not been found, but he sent it after receiving a reply to a recent inquiry to the Bordentown Female College (also not found). Edison and his wife had considered sending their daughter Marion and one of Mary Edison's sisters (likely Jennie

Stilwell) to the school in 1882. Marion and Jennie attended school in New York that year (see Doc. 2344), but the Bordentown Female College again raised the question of enrolling Marion, at least, in 1883. In September 1884, Edison paid "Miss Stilwell's tuition and board" at Bordentown (sending the school a $100 check drawn on Bergmann & Co.) but kept Marion in New York City (see Doc. 2732). Bowen to TAE, 19 Aug. 1884, 20 Sept. 1882, and 10 Aug. 1883; Mary Edison to Samuel Insull, 19 Sept. 1882; all DF (*TAED* D8414ZAT, D8214V, D8314L, D8214U); TAE to Bowen, 29 Sept. 1884, Lbk. 18:427 (*TAED* LB018427).

2. The school sent Edison an itemized term bill for $107.86 on 19 November, which was paid on 3 January 1885. Voucher no. 20 (1885).

3. The Rev. William C. Bowen (1832–1891), a Wesleyan University graduate and son of a noted Methodist minister in New York, purchased the Bordentown Female College and became its president in 1875. Established in 1851, the school overlooked the Delaware River from a high bluff near Trenton. Under Bowen's direction, it emphasized music and art instruction but did not offer Greek. Bowen acquired the nearby New Jersey Collegiate Institute in 1881 and began to operate it as a military school for boys. Obituary, *NYT*, 6 June 1891, 5; "Among the Schools," *Christian Advocate* 57 (22 June 1882): 2; Murray 1972, 70.

–2724–

Edison Electric Light Co. of Europe, Ltd., Draft Minutes

New York, August 27, 1884[a]

Meeting of the Directors of the Edison Electric L Co. of Europe Ld, held at the office of R. L. Cutting,[1] Esq. 19 William Street New York City, August 27t, 1884, at 1 P.M., ~~pursuant to the following call~~

~~Insert call A~~

Present   Messr Edison, Eaton and Banker, ~~also~~ Mr. Batchelor was also present by invitation.

The Second Vice President, Mr Eaton, ~~item~~ called attention to a statement ~~of~~ as to[b] the proposed changes in the German Contract, copy of which has been sent to each Director previous to this meeting. This statement is as follows:

(Copy statement)[2]

After a full discussion of the proposed changes in the German Contract and the present situation in Europe, it was on motion

Resolved, that this Company declines to make any further concessions whatever, either in Germany or elsewhere, unless we receive an immediate cash consideration therefor; and be it further

Resolved, that this Company should retain all that is now secured to it under the existing contracts, and take the consequences of doing so, whatever they may be, rather than to make further sacrifices, and to weaken itself by giving up the slight advantages still remaining to the Company.[3]

Mr. Eaton was requested to notify Mr Bailey of the above resolutions, and at the same time to intimate to him that for a proper consideration in cash, the Board might be willing to make some concessions in regard to matters in Germany, although not such considerable concessions as are now asked for.[4] [----][c]

Df, NjWOE, DF (*TAED* D8428ZAG). Written by William Meadowcroft. [a]Place and date taken from document, form altered. [b]"as to" interlined above. [c]Canceled.

1. Robert Livingston Cutting, Jr. (1836–1894) was an incorporator of the Edison Electric Light Co. of Europe, Ltd., the Edison Electric Light Co., and several other Edison firms operating in the United States and abroad (certificates of incorporation for Edison Electric Light Co. of Europe, Ltd., 2 Jan. 1880; Edison Electric Light Co., 16 Oct. 1878; Edison Telephone Co. of Europe, Ltd., 2 May 1879; all DF [*TAED* D8024A, D7820ZAM, D7940R1]; Edison Ore Milling Co., Ltd., 9 Dec. 1879, CR [*TAED* CG001AAD]; Edison Electric Illuminating Co. [New York], 16 Dec. 1880, NNNCC-Ar [*TAED* X119JA]; see also Doc. 2055). Part of a socially prominent family of New Yorkers, Cutting bore the same name as his son (1868–1910) and his father (1812–1887), with whom the editors confused him in Doc. 1728 n. 1 ("Obituaries," New York State Bar Assoc. 1911, 491; Obituary, *New York Tribune,* 26 Feb. 1887). He earned degrees from Columbia College and Harvard Law School prior to joining Robert L. Cutting and Co., the banking and brokerage house in which his father was the senior partner, and he presently led its successor firm, R. L. Cutting, Jr. and Co., at 19 William St. Cutting had been a member of the New York Stock Exchange since 1864 (Harvard Law School 1905, 72; "R. L. Cutting's Sudden Death," *NYT,* 14 Jan. 1894, 7.

2. Sherburne Eaton prepared a seven-page typewritten summary of eight modifications proposed by the Deutsche Edison Gesellschaft (DEG) to its contract with the Edison European Co., as well as seven ancillary points of negotiation among the Compagnie Continentale Edison and Siemens & Halske, co-parties in the original contract (see Docs. 2379 n. 2 and 2392 n. 5). Points under discussion for DEG included the addition of Russia, Denmark, and Austria to its market territory, the removal of restrictions against its sale and licensing of non-Edison dynamos, and permission to manufacture its own dynamos. It also sought to calculate lamp royalties on a fixed rather than a percentage basis. Joshua Bailey, acting as a liaison among the parties, advised that the Compagnie Continentale Edison seemed likely to consent but that Siemens & Halske was more resistant to the proposed changes, particularly with regard to dynamo rights. Siemens & Halske complained that it already suffered heavy losses because DEG procured dynamos and materials from other suppliers. To satisfy Siemens & Halske, Bailey proposed that it participate in the lamp factory, receive a royalty on all dynamos manufactured by DEG, and also get exclusive rights as DEG's supplier

of central station dynamos. Bailey reported that DEG was firmly opposed to granting the Continentale Co.'s desire for a share of its profits. Eaton to directors of Edison Electric Light Co. of Europe, Ltd., 15 Aug. 1884, DF (*TAED* D8428ZAD).

3. These resolutions were consistent with the view expressed by Eaton in his memorandum (see note 2) that

> We are continually called upon to make concessions and are giving away first one thing and then another, so that in the end, judging from what has already occurred, it seems to me that we will have given away everything and will have little or nothing left. Has not the time come to hesitate before we make any further concessions? The status of our business in Europe cannot be much worse, and it may improve, consequently, it seems to me, we might take the risk of refusing to make further concessions so far as they involve a lessening of possible profits.

4. Bailey submitted a modification of the proposed terms in early October, the details of which are not known. The matter was placed on the agenda for a regular quarterly meeting of directors on 1 November and again at special directors' meetings on 26 November 1884 and 5 and 22 January 1885. At the last conference, the board voted to accept the proposed concessions with several alterations. Austria and Hungary would be excluded from the amended DEG contract, and 50,000 German marks would be paid by DEG to the Compagnie Continentale Edison as a nonrefundable premium, with half that amount to be paid immediately to the Edison Electric Light Co. of Europe, Ltd., for the satisfaction of its bondholders. Edison Electric Light Co. of Europe, Ltd., to TAE, 5 and 22 Nov., 31 Dec. 1884; Edison Electric Light Co., Ltd. minutes, 22 Jan. 1885 (pp. 6–8); all DF (*TAED* D8416BUO, D8416BUR, D8428ZAO, D8527H).

–2725–

Edison Electric Light Co. Agreement with Edison Co. for Isolated Lighting[1]

[New York, September 1, 1884][a]

This agreement, made the first day of September, 1884, by and between the Edison Electric Light Company, herein called the Light Co. of the first part; and the Edison Company for Isolated Lighting, herein called the Isolated Co. of the second part, each being a corporation created under the laws of the State of New York, and having its principal office in the City of New York,

WITNESSETH: Whereas the Light Co. has heretofore made a certain agreement with the Isolated Company, dated the 26th of April, 1882;[2] whereby certain rights and privileges are granted to the Isolated Company, relating to the business of exploiting the Edison system of electric light, heat and power, as more fully appears from the said agreement itself, reference to which is hereby made; and

Whereas the Light Co. now proposes to grant to the Isolated Co. still further rights and privileges touching the said business, and, more especially, to assign and turn over to the Isolated Co. during the continuance of this agreement, the Light Co.'s present business of exploiting the said Edison system of electric light, heat and power in all the territory belonging to the Light Co. in both the United States and Canada, as more fully appears in this instrument.

Now, THEREFORE, in consideration of the premises and of the mutual promises herein made, IT IS HEREBY AGREED AND DECLARED by and between the parties hereto as follows, that is to say:

First.— The Light Co. agrees to appoint and hereby does appoint the Isolated Co. its agent, as herein set forth, to exploit its business of electric light, heat and power, both central station and isolated, in all the Light Co.'s territory in the United States and Canada, for the period covered by, and subject to the terms and provisions of this agreement; and the Isolated Co. will and hereby does accept the said agency.

Second.— The Light Co. will and hereby does license the Isolated Co. during the continuance of this agreement and no longer, and for all the territory in the United States and Canada not heretofore, or hereafter, transferred by the Light Co. to other licensees, to sell and install Edison plants for isolated lighting, and for central station lighting, together with all the appurtenances thereto belonging; also to promote the organization of local companies in cities, villages and towns, to introduce into practical use the said Edison system of electric light, heat and power, subject, however, to the provisions of this agreement, and to the approval of the Light Co. as herein provided for.

It is agreed that the license herein given to the Isolated Co., as well as that covered by the said license agreement of April 26th, 1882, is unassignable, and is granted to it as a privilege personal to itself, and is not to be assigned or transferred by it in any way, and that both this license agreement and that of April 26th, 1882, severally, may be forthwith and peremptorily terminated by the Light Co. in the following cases, viz: (1.) Upon the Isolated Co. ever making or attempting to make any assignment of either of said agreements, or of any of the rights or privileges thereby conceded to it, save and except as expressly provided for in the said two agreements, severally; or (2) upon any such assignment resulting by operation of law.

Third.— The prices for all isolated plants shall be fixed by the Isolated Co., in its discretion. But the Light Co. reserves the right, touching the territory covered by this agreement, to determine from time to time, and in its discretion, the prices and terms for supplying central station plants, and the terms, conditions and restrictions for organizing local companies, as aforesaid, including capitalization, territorial area of license, size and type of installation, and all other details therewith connected; and all licenses for such local companies shall always emanate from the Light Co., and be granted by it directly to each local company, without passing through the Isolated Co. While it is the intention of this contract that, in said territory no local company shall be formed, and no central station plant be contracted for or installed, unless with prior written approval of the Light Co., still it is assumed that the Light Co. will exercise its authority in good faith, and will not refuse to grant its approval except for good and substantial reasons. It is further agreed that the Isolated Co. is not compelled to organize local companies and supply central station plants, as aforesaid, if the conditions thus imposed by the Light Co., in its discretion, are not satisfactory to the Isolated Co.

Fourth.— During the continuance of this agreement, the Isolated Co. will, at its own expense, maintain a competent and adequate electrical and engineering staff, will, in good faith, make arrangements for securing sufficient material, and will otherwise in every way make adequate provision for promptly and successfully carrying on the business herein provided for. The Isolated Co. will also at its own expense employ competent executive officers, agents, and salesmen, and enough electrical assistance to keep full and complete records and books of account, and, under the general direction of the Light Co., will attend to all correspondence and otherwise promptly and well dispose of all business that may arise under the provisions and requirements of this contract.

A list of certain existing agency contracts between the Light Co. and its agents, is hereto annexed, marked Exhibit "A,"[3] and the Isolated Co. hereby assumes, during the continuance of this agreement, the various obligations imposed upon the Light Co. by the said contracts, including all compensation to agents therein provided for, and as further consideration therefor, it is agreed that the provisions of the twelfth section herein, giving the Isolated Co. one-quarter of certain percentages accruing to the Light Co., shall apply to all local companies formed by said agents and licensed by the Light Co.

during the continuance of this agreement. The Light Co. will not alter any of the said agency contracts while this agreement lasts, without the Isolated Co.'s consent.

Such canvasses, surveys, determinations and estimates for the central station plants, as the Isolated Co. may voluntarily make from time to time, shall be at its own expense, but the Isolated Co. shall also make them for the Light Co., whenever requested in writing by the Light Co. to do so, and in all such cases a reasonable percentage of profit shall be added to the actual cost of making them, to be paid by the Light Co., as compensation to the Isolated Co.

Fifth.— The Isolated Co. will keep full records and other data of every kind growing out of the transactions provided for in this agreement, and will accord to the Light Co., at all reasonable times, full opportunity to examine the same and to otherwise familiarize itself therewith, including free access to all books of account, records and correspondence, with full and free right to examine the same and make extracts therefrom, relating to the subject-matter of this agreement; and the Isolated Co. will at all times, both during the continuance of this agreement and thereafter, supply to the Light Co. any part or all of such specifications, data, drawings, estimates and electrical determinations, as it may demand, but at the expense of the Light Co.

Seventh.— The Isolated Company shall make written monthly reports to the Light Co., while this agreement continues, setting forth, in such detail, form and manner as the Light Co. may from time to time direct, all transactions and contracts, under this agreement.

The territorial area intended to be covered by this agreement is the whole of the United States of America and the Dominion of Canada, save and except such parts of the United States as are already covered by certain agreements heretofore made by the Light Co., a full list of which is hereto annexed marked Exhibit "B,"[4] copies of which the Light Co. hereby agrees to furnish to the Isolated Co. on demand.

Eighth.— It is agreed that nothing herein contained shall alter, disturb or in any way affect, except as herein expressly provided for, the said certain agreement heretofore made between the Light Co. and the Isolated Co., dated April 26th, 1882, but that, except as specifically provided for in this instrument, the said agreement shall remain in full force and effect, notwithstanding the existence of this contract, just as if this agreement had never been made.

Ninth.— Whereas circumstances not now foreseen may arise, which may make it desirable not only for the Isolated Co. to sell and install Edison plants for central station lighting, but also for the licensee companies of the Light Co., or either of them, to make such installations: it is, accordingly, agreed that the license herein given to the Isolated Co., to sell and install Edison plants for central station lighting, together with all the appurtenances thereto belonging, is not an exclusive one, but that the Light Co. reserves to itself the right to authorize its licensee companies to make them; and it is further agreed that in all such cases, the Isolated Co. will, at the written request of the Light Co., and upon being paid the cost thereof together with a reasonable profit thereon, supply any and all drawings, determinations, plans, materials and other things, that may be required.

Tenth.— No license or agreement to license by the Light Co., herein contained, shall be construed, to refer to, or to embrace by implication or otherwise any license or agreement to license under any letters patent except such as the Light Co. may own during the continuance of this agreement and license.[5] No grants, or licenses, or privileges shall be implied from the licenses and privileges expressly granted in and by this agreement to the Isolated Co., but the rights and privileges of the Isolated Co., hereunder, shall be restricted to those expressly mentioned in this agreement.

Eleventh.— Whereas the Light Co. now owns fifty-one one-hundredths of the capital stock of the Isolated Co., and, under its said contract with the Isolated Co., dated April 26th, 1882, is entitled to receive, without additional compensation, a like proportion of all future increases of said capital, which provision of the said agreement, it is agreed this contract does not in any way alter or disturb; and, whereas, the Light Co. consents, during the continuance of this agreement, and as further and special compensation to the Isolated Co., to make certain concessions touching dividends on its said stock, in favor of the other stockholders, to wit, the holders of the remaining forty-nine one-hundredths (which stock, for convenience of designation, is herein called *cash stock,* as distinguished from the Light Co.'s holdings, herein called *Light Co.'s stock:* it is agreed as follows, that is to say:

1. All net earnings of the Isolated Co., applicable to dividends, shall, during the continuance of this agreement, be applied, first, to paying a dividend of, or dividends aggregating not more than eight per centum per annum on the said cash

stock; second, to paying a like dividend, or like dividends, on the Light Co.'s stock, aggregating not more than eight per centum per annum; and, third, after the said dividends aggregating eight per centum per annum on both classes of stock shall have been paid, any surplus shall be distributed among all the stockholders according to holdings, including both the cash stock and the Light Co.'s stock.

2. As to the said dividends, each year shall stand by itself, and no deficiency in any one year, whether as to dividends on the cash stock or on the Light Co.'s stock, shall be carried over to another year (except as provided for in the twelfth section herein.)

3. Touching the existing profits and losses of the business of the Isolated Co., as they stand at the execution of this agreement, it is agreed that no separate estimate and allowance of them shall be made, but that, whatever they may be, they shall, as regards all questions of dividends herein provided for, be considered as forming a part of the general current business covered by this agreement, just as if they were the outcome of transactions made during the continuance of this agreement. But upon the termination of this agreement, the question what net profits, if any, there then are applicable to the said dividends provided for herein, that is to say the question whether there are any profits, and if so, how much, to be apportioned between the Light Co.'s stock and the cash stock in the manner provided for in paragraph number one of this section, shall be determined by arbitration, as follows, to wit: an arbitrator shall be appointed by the holders of the cash stock, at a meeting of such shareholders, to be specially called for that purposed by the Isolated Co., on not less than six days' written or printed notice stating the object of the meeting, and mailed to every stockholder except the Light Co., whose name and address may then appear on the books of the Isolated Co.; a second arbitrator shall be appointed by the Light Co., and if these two cannot agree, they shall select a third, and the decision of said two arbitrators, or if a third be called in, the decision of a majority, touching the whole subject matter, covered by said question, including not only the amount of dividend to be declared from said net profits, if any, but also when to be declared and paid, shall be final and binding upon both the Light Co. and all the holders of the said cash stock. If the said arbitrators are not able at once and without delay to properly determine the value of assets or other data entering into the subject-matter covered by the said question, referred to

above, and may, for that or other good reason, desire to delay the making of their report, it is agreed that they may take such length of time, within which to make and render their decision as to them, or a majority of them, may seem best. At the said meeting of the cash stockholders to select their arbitrator, as aforesaid, they may vote either in person or by proxy, and a majority of the shares of stock thus voted on shall decide.

4. During the continuance of this agreement, the Light Co. shall not sell, or otherwise transfer its said holding of fifty-one one-hundredth of the stock of the Isolated Co. or any part thereof without the written consent of the Isolated Co.

Twelfth.— In order to stimulate the Isolated Co. to seek and push the business of exploiting central station companies, and more especially to encourage it to do so in localities where ultimate success may seem doubtful, the Light Co. hereby agrees that whenever, during the continuance of this contract, any licensee company shall be formed pursuant to this contract, and a percentage of its original capital, whether in stock or cash, or both, be paid to the Light Co. for a license, one quarter of the percentage of such original capitalization thus paid to the Light Co., shall be immediately paid by the Light Co. to the Isolated Co.; and if in any case the said original capitalization of any such company be increased during the continuance of this contract, the same proportion of the percentage of such increase accruing to the Light Co. shall also be paid to the Isolated Co.; but the Light Co. shall not make any such payment of said one-quarter, where the capital is increased and the Light Co.'s percentage is paid to it after the termination of this contract. The said one-quarter share of stock in any licensee company thus received by the Isolated Co. for its own benefit, shall remain in its treasury till sold, and when sold the proceeds shall be applied as follows, to wit: first, to equalizing all back dividends between the cash stock and the Light Co.'s stock, which are provided for in the eleventh section herein; and second, any balance remaining after all such back dividends shall have been equalized between the said two companies, shall be carried to the general profit and loss account of the Isolated Co.

Thirteenth.— Regarding the Isolated Co.'s privilege to manufacture any and all patented devices of the Light Co., except lamps, required for its business, provided for in said contract between the Light Co. and the Isolated Co., dated April 26th, 1882, and more especially in the fourth section of the said contract, it is agreed, in consideration of the advantages accruing to the Isolated Co. under and pursuant to the provisions of certain contracts heretofore made between the Light Co. and certain manufacturing establishments and the stockholders herein, a complete list of which is hereto annexed marked Exhibit "C"[6] (copies thereof having heretofore been furnished the Isolated Co.) that said contracts, each and all of them, are hereby approved and accepted by the Isolated Co. as binding upon it while the said several contracts continue to exist, touching all questions relating to the Isolated Co.'s privilege of manufacture mentioned above, and that the Isolated Co. will not manufacture any of the devices covered by said contracts, severally, while the several contracts covering such several devices last, but whenever any of said contracts terminate, the Isolated Co shall then be free to manufacture, pursuant to the provisions of its said contract dated April 26th, 1882, and not otherwise, the particular devices covered by said contract. But the Light Co. hereby agrees to use its best endeavors always to enforce the provisions of the said contracts, and more especially to do so at any and all times, whether this agreement shall then be in force or not, upon the written request of the Isolated Co., and if any controversy ever arises between the Light Co. and the Isolated Co. as to whether said provisions are enforced, or ought to be, or can be, it shall be left to arbitration in the same manner as provided for in the eleventh section hereof.

It is further agreed that during the continuance of this agreement the Light Co. will not terminate or alter the said agreements mentioned in Exhibit C, or any of them, without the written consent of the Isolated Co.

Fourteenth.— Whereas radical and important changes in conducting the business of the Light Co. and Isolated Co. are imposed by this contract, wherein all possible contingencies may not now be foreseen, but which, although contrary to the present expectation, it may be of interest to either or both of said companies to terminate at an early day, thereby making it proper that this agreement, at least in the first instance, should be made for only a very limited period, terminable at the option of either party at the expiration of that time; therefore, it is agreed, either party may terminate this agreement after one year from the date thereof, but not before, upon giving at least six months' prior written notice to the other party.

In Witness Whereof the parties hereto have severally caused these presents to be executed by their officers thereto expressly authorized, and their respective corporate seals to be affixed and attested, at the City of New York, the day and year first above written.

PD, NjWOE, Lit., *Edison Electric Light Co. v. United States Electric Lighting Co.*, Defendant's Depositions and Exhibits, 4:2371–84 (*TAED* QD012E2371). ªDate from document, form altered.

1. This agreement had been proposed as part of the reorganization plan for the Edison lighting companies, which was presented in the June report of the Committee of Three to the Directors of the Edison Electric Light Company (Doc. 2690). One of a complex set of contracts executed the same day, it formally transferred to the Isolated Co. many of the functional duties and responsibilities that Edison had exercised through the Thomas A. Edison Construction Department under his informal understanding with the Edison Electric Light Co. (see Doc. 2437 [headnote]). It also contained important provisions related to the manufacturing shops; the related agreements between the shops and the Edison Electric Light Co. are attached as Exhibit C to this agreement (see article 13 and note 6). Grosvenor Lowrey seems to have played a major role in negotiating these agreements, and he later expressed deep reservations about Edison's understanding of the conditions under which they were approved (Lowrey to TAE, 19 Aug. 1884, DF [*TAED* D8427ZBO]; see Doc. 2748). All of the parties to these agreements, including the shareholders of the shops (see note 6), convened at 10 a.m. in Sherburne Eaton's office to execute them (Eaton to Samuel Insull, 29 Aug. 1884, DF [*TAED* D8427ZBQ]).

2. Under this contract, the Edison Co. for Isolated Lighting had rights to install plants in the United States, except within "the municipal limits of any town, city, village, or other territorial municipality wherein illuminating gas was or had, prior to" 1 January 1882, "been supplied for purposes of lighting to more than ten customers." It also entitled the Isolated Co. to install isolated plants within gas limits until notified that the Edison Electric Light Co. had granted a license to another party. Edison Electric Light Co. agreement with Edison Co. for Isolated Lighting, 26 Apr. 1882, Defendant's Depositions and Exhibits, 4:2363–70, *Edison Electric Light Co. v. U.S. Electric Lighting Co.*, Lit. (*TAED* QD012E2363).

3. Exhibit A identified agreements with the following individuals or companies, giving the party's name, date of contract, and territory allocated. The list is summarized here, in the order given, by the contracts' effective date: Western Edison Light Co. (Illinois, Wisconsin, and Iowa); George S. Ladd (California and Nevada); E. A. Mexia (Mexico); Phillips Shaw (Pennsylvania except Philadelphia, Harrisburg, Reading, Erie, and Pittsburgh); Spencer Borden (New England east of the Connecticut River); Charles T. Hughes (nine southeastern New York counties outside Long Island and the New York City environs); Henry A. Clark (Maryland, District of Columbia, Philadelphia, Erie, and Harrisburg, Pa.); Ohio Edison Electric Installation Co. (Ohio); T. P. Wilson (Minnesota); Charles Benton (specified counties encompassing much of eastern New York except the New York City vicinity, Long Island, and Albany); John Hoskin (Philadelphia); John R. Markle (Michigan); William Hix ("general contract"); Thomas Edison ("Republic of Chili");

Abraham Kissell ("general contract" for Indiana, Missouri, Texas, and Tennessee); William Tyler (Oregon and Washington Territory); A. Pizzini (Richmond, Va.); Zachry & Thornton (Atlanta, Ga.); Charles Hughes and Wilson Howell (New Brunswick, N.J.). See Doc. 2437 (headnote) regarding the role of agents in connection with the central station business in 1883–1884.

4. Exhibit B listed the licensee contracts held by the Edison Electric Light Co. and transferred to the Isolated Co. This list included all of the Edison illuminating companies established by July 1884 (see App. 2), including the Edison Electric Illuminating Co. of New York. A parenthetical note indicates that the Edison Illuminating Co. of Mount Carmel had been licensed by the Isolated Co. rather than Edison Electric because it was a non-gas town. Also included in the list were the Appleton Edison Light Co., Ltd., Western Edison Light Co. of Chicago, Ohio Edison Electric Installation Co., George S. Ladd of San Francisco, the Des Moines Edison Light Co., and the Roselle, N.J., plant (which belonged jointly to the Edison Electric Light Co. and the Isolated Co.). The list included the contract dates (except for Roselle) and the capital of each licensee (except Ladd and the Roselle and Mount Carmel contracts). The date of license also appeared for some companies, but in each case this date was identical with the contract date. The list also was attached as Exhibit C to the 1 September 1884 agreements between the manufacturing firms and the Edison Electric Light Co. (see note 6).

5. This clause limited the privileges specified in article four of the 26 April 1882 agreement, which had allowed the Isolated Co. to license patents not owned by the Edison Electric Light Co. The restriction presumably was a concession to the shareholders of the Edison Machine Works, Electric Tube Co., and Bergmann & Co. in connection with the 1 September agreements between the shops and the Edison Electric Light Co. (see note 6) and the manufacturing arrangements detailed in article thirteen below.

6. Exhibit C listed two related sets of agreements between Edison Electric and three of the manufacturing companies: the Edison Machine Works, Electric Tube Co., and Bergmann & Co. (the Edison Lamp Co. was covered by an 1881 agreement [see Doc. 2039 n. 1]). Edison Electric Light Co. agreements with Edison Machine Works, both 1 Sept. 1884; Edison Electric Light Co. agreements with Bergmann & Co., both 1 Sept. 1884; Edison Electric Light Co. agreements with Electric Tube Co., both 1 Sept. 1884 (one revised 19 Feb. 1885 as noted below); all Miller (*TAED* HM840229, HM840229B, HM840230, HM840230B, HM850242, HM850242A1).

The first set of agreements was with the shareholders of each manufacturing firm. The shareholders bound themselves, when disposing of stock, to offer it first to the Edison Electric Light Co. on the same terms given to other parties. (A shareholder list was attached as Exhibit A to each contract.) Shareholders who made inventions related to the Edison lighting system also had to give the Edison Electric Light Co. the first option on the rights. The shareholders and their shares were Edison Machine Works: Edison (1,599), Charles Batchelor (400), John Kruesi (1); Bergmann & Co.: Edison (898), Edward Johnson (998), Sigmund Bergmann (998), Batchelor (100), Philip H. Klein, Jr. (6); Electric Tube Co.: Edison (89), Batchelor (10), Samuel Insull (1), Kruesi (50), J. Pierpont Morgan (50), James Hood Wright (49), Anthony J. Thomas (1). The agreement with shareholders of the Electric Tube Co. was modified on 19 February 1885 by adding a provision (Exhibit E) governing manufacturing royalties on patents assigned by the Electric Tube Co. to the Edison Electric Light Co. Miller (*TAED* HM850242).

The second set of agreements, made with each manufacturing firm for a period of two years, was intended "to secure increased advantages and protection in connection with the manufacture of certain articles applicable to the Edison system of electric light, heat and power." (A copy was nested as Exhibit B in the corresponding shareholder contract.) The Edison Electric Light Co. gave exclusive and nontransferable manufacturing licenses to the three firms, with the relevant product line of each enumerated in Exhibit D and attached to the corporate contracts. The manufacturers agreed not to sell goods related to the Edison system to any parties in North or South America that were not licensed by Edison Electric, and to mark "For Export" all goods destined abroad. The firms also had to assign to the Edison Electric Light Co. any invention or manufacturing process related to the Edison system, but the parent company was prohibited from using such patents for "telegraphy, telephony, or other purposes not connected" with the Edison system. The manufacturers were limited to "an average semi-annual profit of twenty per centum on the cost of the manufactured" articles. An invention made by any of them, whether patented or not, that reduced production costs would entitle the firm to an additional royalty equal to one-fifth of the cost savings. In addition, any "distinctly new class or type of manufactured article" would entitle a shop to an additional ten percent. All experimental costs would be borne by the manufacturers, with some provisions for treating them as a special charge upon profits; the same terms applied to acquiring outside patents and licenses. The shops reserved the right to manufacture for other customers items not related to the Edison light and power system. The corporate agreements included two exhibits of their own: a list of monthly salaries (Exhibit A) and the licensee list described in note 4 above (Exhibit C).

–2726–

To Leo Ehrlich[1]

[New York,] Sept. 4th. 1884.

Dear Sir:—

Referring to your favor of the 2nd. August, I have tried the electric Aurophone, which you so kindly sent me, but it does not seem to answer my particular case.[2] I could not hear any better with it than without it.

Is there any explanation that you can offer, or if you can suggest any experiments for me to try in connection with it, I shall be happy to do so.[3] Very truly yours,

Thos. A Edison   I[nsull]

TL (carbon copy), NjWOE, Lbk. 18:321 (*TAED* LB018321). Signed for Edison by Samuel Insull.

1. Leo Ehrlich (b. 1847?), a native of Hungary, was a former life insurance agent in St. Louis, where he also had been associated with the Humane Society. He later identified himself in the federal census as an inventor, and he received several U.S. patents during his life, among them grants for paper cutters and a safety razor. "Aetna Life Insurance Co. vs. Ehrlich," *Insurance Law Journal* 13 (May 1884): 383–97; U.S. Pats. 338,047; 405,402; and 409,028; U.S. Census Bureau 1970 (1880) roll T9_734, p. 372.2000, image 0274 (St. Louis, Mo.); ibid. 1882? (1900), roll T623_900, p. 4A (St. Louis Ward 25, Mo.).

2. Ehrlich, who claimed to have met Edison several years earlier in New York, explained that his own hearing loss had prompted him to experiment with hearing aid devices. He sent for Edison's trial his "Electric Aurophone," which he claimed had largely restored his own hearing. The instrument consisted of a ¾-inch tube with a "cone shaped vibrating spiral inside." The spiral was connected with a small battery so as to provide an electric current to the outer ear. On 28 August, Edison replied to an inquiry (not found) from Dr. Theodore Parker, of Hart's Island Hospital in New York. Edison promised that he would investigate Ehrlich's device and report to the doctor, but the editors have found no further correspondence with Parker. Ehrlich to TAE, 2 Aug. 1884, DF (*TAED* D8401A); TAE to Parker, 28 Aug. 1884, Lbk. 18:282 (*TAED* LB018282).

Ehrlich's instrument also occasioned an Edison interview in a New York newspaper (reprinted in Chicago). Headlined "Edison Prefers Being Deaf," the report quoted Edison saying that he did not wish to be "cured" of impaired hearing: "I am not very deaf. There are lots of things I don't want to hear. Now I don't have to hear 'em." Among a litany of sounds he preferred to avoid, Edison named those made by carts and vendors: "My poor wife used to be kept awake all night when we moved to New York. The contrast with the quiet of Menlo Park was very great. She never got accustomed to the rumble of wheels. But, as for me, I could sleep soundly through it all." Edison reportedly also expressed gratitude for an excuse not to use the telephone, but he conceded that his deafness was a liability at the theater, where "All plays are spectacles to me. But the advantages are so great that I can stand the disadvantages" ("Edison Prefers Being Deaf," *Chicago Daily Tribune*, 30 Aug. 1884, 10). During 1878, Edison had worked on an acoustic hearing device he named the auriphone; in the rush of publicity around his invention of the phonograph, he received numerous inquiries about the device and announced plans to market it commercially (see Docs. 1228, 1298 esp. nn. 1–2, 1326 esp. nn. 2–3, 1361, and 1464).

3. The editors have found no other correspondence with Ehrlich.

–2727–

To Samuel Insull

Electrical Ex Philada Pa Sept. 5[1] 188[4][a] 12:34

Want photographs machine works lamp factory bergmann Kruzi also all the stations[2]   hurry them forward   tell Batch hurry compound machine & arc lights[3]

Edison.

D (telegram), NjWOE, DF (*TAED* D8464R). Message form of Western Union Telegraph Co. ᵃ"188" preprinted.

1. The numeral is unclear but other evidence places Edison and his daughter Marion in Philadelphia on September 4–5. Continental Hotel bill for "Board 2 days," 6 September. Voucher no. 495 (1884); "At the Electrical Exhibition," *Philadelphia Inquirer*, 5 Sept. 1884, 2.

2. Edison paid the studio of photographer D. N. Carvalho $13.00 for 14" × 17" photos of the Edison Machine Works and Edison Lamp Co. and 11" × 14" prints of Bergmann & Co.'s building, plus the negatives, all of which were made in fulfillment of a 23 September order from John Randolph (voucher no. 546 [1884]). The photographs of these factories, plus the Electric Tube Co.'s new Brooklyn shop, may have been the basis for drawings used to illustrate an 1885 catalog of the Edison Co. for Isolated Lighting (1 Oct. 1885, PPC [*TAED* CA002E]). Some of the photos may have been displayed at the 1904 World's Fair in St. Louis (*Edisonia* 1904, 20–22). Photos of the Edison Machine Works, Bergmann & Co., and the Edison Lamp Works, and a drawing of the Electric Tube Co. were variously reproduced in *Edisonia* 1904 (162) and Jehl 1937–41 (766, 807, 843, 957); see also Doc. 2343 (headnote). Photographs of the village plant central stations from this period are in the Historical Photograph Collection at NjWOE. Images of the Roselle, N.J., Sunbury, Pa., and Brockton, Mass., plants were reproduced in Jehl 1937–41 (1093, 1097, 1117), and of Sunbury in *Edisonia* 1904 (140).

3. The exhibition's official catalog did not identify a compound-wound dynamo among the models in the various Edison displays, nor did it list an Edison arc light. Edison had applied for a patent on a regulator mechanism for arc lights in June 1884. Franklin Institute 1884, 25–26, 61; U.S. Pat. 438,303.

–2728–

To William Dwelly, Jr.

[New York,] Sept. 8th. 1884.

Dear Sir:—

Your favor of August 8th.[1] has remained unanswered owing to the fact that certain negotiations were pending between the Edison Electric Light Co. and myself, with relation to the stations built by me, and which resulted in Mr. Johnson being appointed a Committee to visit same and report on them.

As, however, it seems probable that there may be some considerable delay in any conclusions being reached with relation to Mr. Johnson's report, and as I am very anxious to adjust the differences with your Company as early as possible, I shall be glad to know what you would consider a fair allowance for the error in connection with the erection of your stack?[2]

You requested a proposition of me, and I made it, which does not appear to have been satisfactory to your Directors. Cannot you make some counter proposition in return? What I wish to arrive at is an early settlement of the matter and to obtain from you a check for the balance of account still due me. Very truly yours,

Thos. A Edison   I[nsull]

TL (carbon copy), NjWOE, Lbk. 18:337 (*TAED* LBo18337). Signed for Edison by Samuel Insull.

1. Not found.

2. According to the enclosure to Doc. 2708, the Fall River illuminating company owed Edison about $9,200 at the end of July. The dispute about the smokestack arose from Dwelly's contention that the one Edison had installed at Fall River was inadequate for the station's present needs, as well as from Edison's admitted failure to have provided a stack that would accommodate the plant's future expansion to 400 h.p. Edison sent a blower to increase the stack's effectiveness and offered a rebate of $600 in July to remedy his original error. In November, the local Fall River illuminating company counter-offered with a request for a $1,200 allowance, to which Edison agreed. Insull subsequently presented an adjusted outstanding balance of $1,087, but the company claimed additional charges against Edison and seems to have paid only about $288 by the end of the year. Babcock & Wilcox to Edison Electric Illuminating Co. of Fall River, 7 May 1884; Insull to Dwelly, 3 June 1884; both DF (*TAED* D8423Q, D8416BSK); Insull to Dwelly, 18 July 1884, LM 20:35 (*TAED* LBCD7035); Insull to Dwelly, 19 and 28 Nov. 1884; TAE to Dwelly, 26 Dec. 1884; Lbk. 19:394, 420, 464 (*TAED* LBo19394, LBo19420, LBo19464).

–2729–

To Charles Bailey

[New York,] Sept. 9th. 1884.

Dear Mr. Bailey:—[1]

I have your favor of September 1st., and shall be glad to hear from you at length with relation to the new system of covering wire.[2]

Please send on a sample of the work, and I shall then be prepared to talk "Turkey" with you.

Batchelor has just returned from a stay of several years in Paris.

Griffen[3] has not been with me for the last three or four years, and I have no idea where he is just now. Very truly yours,

Thos A Edison   I[nsull]

TL (carbon copy), NjWOE, Lbk. 18:342 (*TAED* LBo18342). Signed for Edison by Samuel Insull.

1. Charles E. Bailey (b. 1845) served as treasurer of the Edison Speaking Phonograph Co., in New York, from June 1878 until the end of that year. A native of Rhode Island, Bailey moved to Providence about the beginning of 1879 and entered the cotton trade there. He became one of the original incorporators and the founding president of the American Solid Leather Button Co. of Providence in 1881. U.S. Census Bureau

1970 (1880), roll T9_1213, p. 474.2000 (Providence Ward 2, Providence, R.I.); ibid. 1982? (1900), roll T623_1506; p. 17B (Providence, Providence, R.I.); Bailey to Edward Johnson, 14 Jan. 1879, ESP Scraps. (*TAED* X154CAU); Rhode Island General Assembly 1881, 233; *TAED* s.v. "Bailey, Charles E."

2. In a 1 September letter, Bailey apprised Edison of "a new system for covering unlimited lengths of insulated wire with lead or other metal; each wire of cable being kept apart from the others, all absolutely without seams or soldering." He offered that, if Edison were interested in the new process, he would send a cable sample "and be prepared to 'talk turkey.'" Edison wrote a marginal note on that letter instructing Charles Batchelor to reply that "we are ready to talk turkey." After receiving Edison's reply, Bailey subsequently offered to arrange a meeting with the inventor. Such a meeting, if it occurred, was delayed as a consequence of Bailey having addressed his letters to Menlo Park and of both Edison and the inventor traveling to the Philadelphia Exposition. The matter was still pending at the end of September; the editors have not determined its outcome. Bailey to TAE, 1, 10, 16, and 30 Sept. 1884, all DF (*TAED* D8403ZGH, D8403ZGJ, D8403ZGO, D8403ZHB).

3. In his 1 September letter (see note 2), Bailey sent his regards to Stockton L. Griffin and Charles Batchelor. A telegraph operator, Griffin (b. 1841) had served the Union Army in that capacity and had worked with Edison in Louisville. He was in charge of Western Union's eastern wires at the firm's New York offices when Edison hired him as a personal secretary in June 1878. He left Edison's employ in February 1881, after which the editors have little information about his activities. In 1900, Griffin was working as a clerk in a Los Angeles telegraph office. U.S. Census Bureau 1982? (1900) roll T623_89, p. 13B (Los Angeles Ward 3, Los Angeles, Calif.); Plum 1882, 253; Docs. 169 n. 1, 1322 n. 1; *TAEB* 6 chap. 9 introduction.

–2730–

From William Curtis Taylor

[Philadelphia, c. September 9, 1884][1]

My Dear Mr. Edison:

Hearing that darkie at the gate every night repeat his set speech to the people entering,[2] about the direction to take where to leave their watches,[3] etc., makes me think of your phonograph. I was going to suggest to Mr. Hammer last night, if I had seen him, to have the same fellow impress his speech in a phonograph and then grind it out, for the remainder of the exhibition. It would make a great deal of fun, and the instrument would afterwards make a valuable relic for the Franklin Institute.[4] Very truly Ys

W Curtis Taylor[5]

I think Mr. Gilliland's[6] negative will be a success. None are finished yet   W.C.T.

ALS, NjWOE, DF (*TAED* D8464V1).

1. The International Electrical Exhibition opened in Philadelphia on 2 September. Taylor likely sent this letter between 5 September, when he reintroduced himself in a letter to Edison, and 11 September, when Edison arrived in Philadelphia.

2. In an illustrated article on the Edison exhibits at the International Exposition, *Scientific American* described the scene to which Taylor referred in a single long paragraph that read, in its entirety:

> One of the electrical comicalities of the Exhibition was the illuminated colored gentlemen who politely distributed cards to astonished visitors. The Edison Company conceived the idea of so locating one of their lamps that it could be seen by all, and to do this most effectually they placed it upon a helmet surmounting the head of the colored party. Two wires led from the lamp under his jacket, down each leg, and terminated in copper disks fastened to his boot heels. Squares of copper of a suitable size for him to stand naturally upon were placed at intervals in the floor, and were electrically connected with the dynamo. So with each heel in contact with a plate he was enabled to make and break the circuit leading to his lamp, the movement required being so slight as not to attract attention, and his hands being free to handle the cards. Many nervous per-

Scientific American*'s
illustration of "the Edison
electrical darky . . . the
illuminated colored
gentlemen [sic] who
politely distributed cards
to astonished visitors"
at the Edison displays,
which the journal called
"One of the electrical
comicalities" of the
International Electrical
Exposition.*

sons were startled by the sudden flashing of the light, and so great were the crowds that continually surrounded this individual that he was frequently obliged to change his quarters in order to keep the passages open. As a further improvement it was the intention to place copper strips under a carpet and provide the heels with sharp points, so that each step would be illuminated. This simple exhibition led many folks from the rural districts to inquire as to the cost of such an appliance, as it was just the thing they wanted "to carry around the house." ["The Edison Exhibit at the Philadelphia Electrical Exhibition, *Sci. Am.* 51 (18 Oct. 1884): 246]

3. Another published report observed that because of the large number of operating dynamos on display, "At many places placards are hung, reading: 'The dynamos will magnetize your watches. Keep away!' The management has provided a safe at the office where watches can be kept, though a piece of silk (not half cotton) well wrapped around the watch answers the same purpose. There are about two hundred dynamos exhibited, the largest in Edison's department." "News and Miscellany," *Medical and Surgical Reporter* 51 (4 Oct. 1884): 388.

4. The phonograph had continued to generate a trickle of written inquiries and suggestions, including several about the time of exhibition. One such letter that Edison received in October asked about connecting a phonograph with a typewriter to transcribe speech. Edison commented on it: "I'm getting old & the problem is too tough for me." A formal reply, prepared by Alfred Tate, left out the part about being too old. TAE marginalia on H. C. Faulkner to TAE, 23 Oct. 1884, DF (*TAED* D8469D); TAE to Faulkner, 30 Oct. 1884, Lbk. 19:327 (*TAED* LB019327).

5. William Curtis Taylor (1825–1905), a prominent portrait and art photographer in Philadelphia, had corresponded with Edison in 1878 about astronomical photography and the tasimeter. Having secured from the Franklin Institute a commission for "making Photographs by Electric Light" at the International Exposition, Taylor set up an electrically lighted atelier visited by thousands of visitors, and he invited Edison to sit for a portrait. After Edison left the city, Taylor obtained William Hammer's approval to hang the resulting print in one of the Edison exhibits. Obituary, *Philadelphia Inquirer,* 18 Feb. 1905, 7; U.S. Census Bureau 1970 (1880), roll T9_1125, p. 143.1000, image 0387 (Ridley, Delaware, Pa.); ibid. 1882? (1900), roll T623_1406, p. 1A (Ridley Park, Delaware, Pa.); Franklin Institute 1884, 60; *Report of Examiners* 1886, Section 26, p. 5; Taylor to TAE, 2, 5, and 7 Sept. 1878; 5 Sept. 1884; undated [Sept. 1884?]; all DF (*TAED* D7835M, D7835N, D7835O, D8464S, D8403ZKE).

6. Electrical manufacturer and inventor Ezra Torrance Gilliland (1848–1903) was a former telegraph associate and business partner with Edison, most recently in promoting the phonograph in 1878. Gilliland had recently been a telephone office manager in Indianapolis, but at this time he superintended the mechanical department of the American Bell Telephone Co. in Boston. See Docs. 543 n. 8, 622, 1334, and 2751; Obituary, *New York Athletic Club Journal* 12 (June 1903): 27; U.S. Census Bureau 1970 (1880), roll T9_295, p. 296.2000, image 0294 (Indianapolis, Marion, Ind.).

Phil Sep. 10, 1884

Dear Sir

Mr. Benty[1] called on Me the other day and said the small Dynamo running the call bells on Telephone was out of order, also their gess engine[2] was broke down, so they employed a boy to turn the crank and run two small Magnetos.

I ameditely went to their work shop, and took one of their men, and had him make the nessary repairs.

Would you please hurry up the Phonograph as that seems to be the most atracting feature to the public.[3]

Also hurry along the rest of the exhibit that we may place them and get the place straightend ought   Yours truly

J. F. Ott

⟨OK— E H Johnson will be there tonight & phono go Saturday I think— E⟩

ALS, NjWOE, DF (*TAED* D8464W).

1. Henry Bentley of the Philadelphia Local Telegraph Co.
2. That is, a gas engine.
3. The official exhibition catalog listed among Edison's exhibits both a "Speaking Phonograph" and a "Phonograph in combination with Telephone." Some time shortly before the Exposition, Edison reportedly told the *New York World* that he was close to completing a "double-grooved" phonograph that would record and reproduce two voices simultaneously. The instrument was powered by electricity so that, Edison claimed, it would turn at a uniform speed. The editors have no other information about this device. Franklin Institute 1884, 26; "Edison's New Phonograph," *Grand Forks (N. Dak.) Daily Herald,* 13 Dec. 1884, 2.

[Menlo Park?][2] Sept. 17th. 1884.

Dear Madam:—

I must apologize for not replying earlier to yours of Sunday evening,[3] but inasmuch as I spend most of my time down here now, the letter was delayed, owing to its having been sent to 65 Fifth Avenue.[4]

Dottie is going back to Madam Meares' school as soon as Mr. Edison and his family settle in the City.[5]

He has taken a flat at 39 East 18th. Street, and I expect they will be all settled there some time next week.[6] Dottie is in Philadelphia just now with her father, attending the Electrical Exhibition there.[7]

I have written Madam Meares to-day.[8] Very truly yours,

Sam'l Insull

TLS (carbon copy), NjWOE, Lbk. 18:383 (*TAED* LB018383).

1. Harriet Clarke gave half-hour music lessons to Marion Edison from at least late 1883. Edison's voucher system recorded payments to her for music instruction and also specifically for piano lessons. Clarke was a singer herself, and her pupils' vocal recitals were occasionally noticed in the *New York Times*. The editors have not identified her connection with the Madam C. Mears School. Voucher nos. 226 (1883); 52, 272, 640 (1884); 65 (1885); "Amusements: Mrs. Clarke's Concert," *NYT,* 29 Apr. 1884, 4; "Mrs. Clarke's Concert," ibid., 18 May 1887, 8; "Notes of the Stage," ibid., 3 Feb. 1888, 5.

2. See note 4 below.

3. Insull probably meant an undated note from Clarke that referred to her unsuccessful attempt to see him on Saturday, which would have been 13 September. Clarke asked if Marion would be returning to the Mears school, where she stated Marion had been happy and which was Mary Edison's choice for her daughter. Clarke to Insull, n.d. [14 Sept. 1884?], DF (*TAED* D8414ZBJ).

4. Insull may have been working at Menlo Park and returned to New York when Edison went to Philadelphia on the sixteenth; Edison telegraphed him at the Fifth Ave. office on this day. It is also possible that he had merely been conducting business at the Edison Machine Works on Goerck St., whose letterhead he used several times for general correspondence in late August. Insull to John Tomlinson, 20 Aug. 1884; Insull to TAE, both 22 Aug. 1884; TAE to Insull, 17 Sept. 1884; all DF (*TAED* D8465ZAI6, D8431ZAL, D8431ZAM, D8465ZAJ2).

5. In December 1884, Edison paid $130.18 for Marion's tuition at the Mears school from October 1884 to February 1885. The fees covered "instruction in English, French and German," as well as meals, paper, and piano use. Marion remained at the school through at least June 1885. Vouchers nos. 554 (1884) and 49 (1885).

6. See Doc. 2721 n. 2.

7. Edison left early on 16 September. He probably returned on 19 September, later paying the Continental Hotel in downtown Philadelphia $31.00 for "Board & Bath   3 days' Lunch." Samuel Insull to H. Alabaster, 16 Sept. 1884; Lbk. 18:379 (*TAED* LB018379); Voucher no. 533 (1884) for Continental Hotel.

8. Insull wrote less definitively to Mme. Mears, explaining that he had the "impression that it is his [Edison's] intention to send Miss Edison to your school" after they settled in New York, "some time towards the end of the month." Insull to A. C. Mears, 17 Sept. 1884, Lbk. 18:381 (*TAED* LB018381).

–2733–

Francis Upton to Samuel Insull

EAST NEWARK, N.J., Sept. 22, 1884[a]

Dear Sir:

I take pleasure in calling to your attention the present financial standing of this Co.

There has been paid in cash into the treasury of this Co. by the various individual stockholders

	$163,011.41
Interest to date	29,809.25
Loan by F. R. Upton.	22,172.91
paid back[1]	$214,993.57

In the year 1883 we made a profit on sales of about $35,000[2]

This year I hope that ~~you~~ we can make about $30,000.

We shall not do as well as last year as we have brought down our prices for lamps sold abroad.

If we have a good business next year I hope to be able to pay 10%.[3] Yrs. Truly

Francis R. Upton   Treas

ALS, NjWOE, DF (*TAED* D8429ZAZ). Letterhead of Edison Lamp Co. ª"East Newark, N.J.," and "188" preprinted.

1. Upton probably meant the total "paid back" in the sense of the company's capital indebtedness (as opposed to capital "paid in"), rather than obligations already paid off. His loans to the business were still outstanding in October, but with only about $12,600 due him at this time. The interest owed by the firm had increased since April from $24,919.19. Edison Lamp Co. to TAE, 1 June, 1 Aug., 1 Sept., 1 Oct. 1884; Philip Dyer to Upton, 10 Apr. 1884; all DF (*TAED* D8429ZAN1, D8429ZAS, D8429ZAW, D8429ZBB, D8429Y); cf. Doc. 2536.

2. In calendar year 1883, the factory had reduced its running loss since inception by almost $38,000 (see Doc. 2590 n. 2); retrospective statements from 1884 and 1885 appear to give conflicting information about gross sales in 1883. In 1884, the firm recorded a profit of $25,138.93 on gross sales of $161,863. Its assets were also increased by the purchase from the Edison Electric Light Co., probably during 1884, of the small Canadian lamp factory at Hamilton, Ont. Edison Lamp Co. statements, 2 Jan. 1884 and 2 Jan. 1885; Edison Lamp Co. agreement with Edison Electric Light Co., undated 1885; all DF (*TAED* D8429D, D8529A, D8528ZAB).

3. Evidently anticipating a 5 percent dividend for 1884 to be paid in March 1885, Charles Batchelor created an entry in an account book but did not enter an amount into his totals. The first dividend payment (1.5 percent) seems to have come in October 1885, followed by 2.0 percent payments in January and April 1886. Cat. 1318:30–31, Batchelor (MBA001 [image 13]); TAE cash book (1 Jan. 1881–30 Mar. 1886): 362, Accts., NjWOE.

–2734–

To William Lloyd Garrison, Jr.

[New York,] Sept. 23rd. 1884.

Dear Sir:—

I have your favor of the 15th. inst., which I should have acknowledged earlier but for my absence from New York.[1]

I shall be glad to see you when you are in New York, with relation to the Brockton Co's. account. I may mention, however, that inasmuch as I have closed out the Construction business at a considerable loss, every cent that is owed me is so much of my working capital locked up, and inasmuch as I am not a big capitalist I cannot afford to have this go along for any considerable length of time, and I shall be glad if you will think over some means of effecting a settlement with me, if not all, certainly of the greater part of the account.[2]

I have written to Mr. Johnson with relation to the H machines. Very truly yours,

Thos. A. Edison   I[nsull]

TL (carbon copy), NjWOE, Lbk. 18:396 (*TAED* LB018396). Signed for Edison by Samuel Insull.

1. Garrison's letter has not been found, but this is probably the letter that Insull summarized for Edward Johnson on 23 September. Insull ascribed to Garrison this paragraph: "In your letter of August 5th. you state that the new 'H' machines are now being built. When Mr. Johnson was at Brockton he seemed to question the need of changing the 'S' dynamos. I should like to know his recommendation regarding them." Insull to Johnson, 23 Sept. 1884, Lbk. 18:395 (*TAED* LB018395).

The question of an H dynamo for the Brockton station had come up as early as April. Garrison's wish for that machine may have stemmed from problems of insufficient power in the Brockton system, about which he and Edison had corresponded in the spring and summer. Edison initially suggested augmenting the plant's boiler capacity with a blower, then later promised that he was "perfectly prepared to stand by my contract with you, and will hold myself responsible if your engines and dynamos will not carry exactly the load which the contract provides for" (TAE to Garrison, 3 and 17 Apr., 7 June 1884, all DF [*TAED* D8416BCE, D8416BII, D8416BTF]). He conceded that the plant could not operate at its full capacity because of an error in the original determination for the feeder lines. He discussed various remedies and, in early August, confirmed that the Machine Works was building H dynamos for Brockton. The underlying causes of the problem seem to have remained obscure, and, in late August, Edison dispatched John Lieb to investigate (TAE to Garrison, 11 June 1884, DF [*TAED* D8416BUJ]; TAE to Garrison, 24 July, 5 and 23 Aug. 1884; Lbk. 18:157, 224B, 264 [*TAED* LB018157, LB018224B, LB018264]).

2. The balance due on the Brockton station remained outstanding into October, when Edison again reminded Garrison about it. Before the end of that month, Edison acknowledged receipt of two notes from Garrison totaling about $3,032 to settle the debt, except for $319 withheld for unspecified repairs that Edison planned to recover from the Electric Tube Co. In February 1885, however, Garrison threatened to disavow the notes because Johnson, contrary to earlier promises Garrison believed he had made, held the Brockton company liable for the new H dynamos. Garrison protested that Edison should bear the expense, as "The contract for construction was made with you personally & legally I presume we must look to you for redress." Frank Hastings, treasurer

of the Edison Co. for Isolated Lighting, became involved in the matter and secured Edison's promise to pay for the new machines and the labor of installing them. The issue was not fully resolved until May 1885, when Edison sent a check covering these costs, and Garrison acknowledged settlement of the Brockton company's claims. TAE to Garrison, 20 Oct. 1884, Lbk. 19:300 (*TAED* LB019300); Garrison to TAE, 3 Feb. and 18 May (with TAE marginalia) 1885; Hastings to TAE, 10 Feb. 1885; both DF (*TAED* D8523I, D8523ZAX, D8523L).

-2735-

Samuel Insull to Francis Upton

[New York,] Sept. 23rd. 1884

Dear Sir:—

Mr. Edison requested me to write to the London Co. and ask them whether they would be inclined to buy Swan lamps from you with a guaranteed life greater than those they now manufacture themselves and with our style of sockets on, or else the same style as they now put on Swan lamps.

Mr. Edison told me to quote a price of forty cents in lots of 2000 or more.

Inasmuch at you are now in direct communication with the Edison & Swan United Electric Light Co. of London, I think it would be far better if this letter was written by you. I do not think that Mr. Edison anticipated that they would accept this proposition, but I think he wanted to fish for the present cost of the manufacture of the Swan lamp to the Edison & Swan Co.[1] Yours truly,

Saml Insull

TLS (carbon copy), NjWOE, Lbk. 18:411 (*TAED* LB018411).

1. The editors have found no further correspondence on this subject.

-2736-

To Frank Hastings

[New York,] 30th Sept [188]4

Dear Sir,

I must again draw your attention to the accounts of the Construction Department against your Company[1] for Canvassing & Engineering Expenses.[2]

I would remind you that these accounts are for cash out of pocket expended by me and do not include any profit whatever as you well know from my examination of my Books. The nonpayment by your Company of these accounts very seriously embarrasses me & I must urge you to make an immediate settlement of same[3]   Yours truly

Thos. A Edison

L (letterpress copy), NjWOE, Lbk. 18:434 (*TAED* LB018434). Written by Samuel Insull.

1. Samuel Insull addressed this letter to Hastings as treasurer of the Edison Electric Light Co. Hastings held the same office in the Edison Electric Illuminating Co. of New York.

2. See Doc. 2704.

3. In January 1885, Edison pressed the company to pay a balance of $20,871 for engineering and other expenses directly related to the work of the Construction Dept. In April, Edison and his principal partners reached a comprehensive settlement with the Edison Electric Light Co. and the Edison Co. for Isolated Lighting covering nearly all expenses, including those of the Construction Dept. The Edison Electric Light Co. agreed at that time to reimburse Edison $66,755.70. TAE to Eugene Crowell, 29 Jan. 1885, Lbk. 20:59 (*TAED* LB020059); TAE agreement with Edison Electric Light Co., 23 Apr. 1885, Miller (*TAED* HM850252).

–2737–

To Edward Johnson

[New York,] 30th Sept [188]4

Dear Sir,[1]

With relation to the Central station Plant installed at Sunbury and the alterations now contemplated and your request[2] that I should state what I am willing to contribute towards the expense of same I think it well to lay before you the exact facts relating to my transactions with the Sunbury Co.

In May 1883 Mr P B. Shaw brought me a Contract signed by the Edison Electric Illuminating Co of Sunbury which called for the installation by me of a 500 Light Plant (ten candle lamps).[3] No definite price was stated in the contract nor was any estimate made by me. It was simply stipulated that I should not charge the Sunbury Co any more than any other customer. Mr. Shaw assured me that the Sunbury Co had ample funds to pay for the Plant. The total cost of the Plant without any profit whatever was $13 335.69. To this I was of course entitled to add a reasonable profit but instead of doing so I rendered a bill for $12,000.68 $11,968 being $1367.69 less than the actual cost of the labor and material. I did this because the Sunbury Plant was the first installed and because I was very anxious to give the business of your Company a fair start in Pennsylvania.

Instead of getting my account settled immediately the Contract called for payment I was unable to get from the Sunbury Company any payment beyond $5,500 notwithstanding that, at the time I took the Contract, Mr Shaw informed me that the Company was quite able to pay for the Plant. Finding that the Company had not got the funds to satisfy my claim I agreed to take the balance due me in the Capital Stock of the Company.[4] I was entitled to $6,500 of Stock besides a small amount of the Promoters Stock. The Company delivered me $4100 of Stock and promised delivery of the balance as soon as the Capital Stock could be increased from $16,000 to $20,000. The Stock was increased and I applied for the balance but was put off with the excuse that the Directors were trying to sell the Stock so as to pay me (about $2400) in cash. I have never been able to get this balance either in Cash or Stock.

I will not comment upon the treatment of myself by the Sunbury Company after the liberal manner in which I acted towards them when installing their Plant but will leave you to draw your own conclusions

With reference to your inquiry as to what amount I will allow (in consideration of the bad state of the present Central Station) towards building another Station in a more central location I would state that providing the Sunbury Co are prepared to pay me in cash for the balance of my account I will make an allowance of an amount equal to what it would cost to put the present building in a thorough State of repair.[5] Yours truly

Thos. A Edison   I[nsull]

L (letterpress copy), NjWOE, Lbk. 18:444 (*TAED* LBo18444). Written by Samuel Insull.

1. This letter was addressed to Johnson as vice president of Edison Electric Light Co.

2. Not found

3. Agreement not found.

4. See Doc. 2498 n. 3.

5. See Doc. 2709. Edison reportedly agreed with Frank McCormick in October to allow $750 toward construction of a new station. As part of a settlement with the company, he was to receive an additional $400 of "promoters stock," then held by the Edison Electric Light Co. Before these provisions could be carried out, however, most of the company's Williamsport investors (including McCormick), not wishing to advance more money, agreed to sell out to Charles Story (of the Edison illuminating company in Harrisburg, Pa.), and a new group of Sunbury investors at fifty cents on the dollar. Edison reluctantly agreed to participate at this price. He reached a full settlement with the Sunbury company, on unknown terms, and signed a release to that effect in September 1885. Story to TAE, 6 Apr. 1885; McCormick to TAE, 13 Apr. and 20 June 1885; all DF (*TAED* D8523ZAD, D8523ZAK, D8523ZBF); TAE to Story, 9 Apr. 1885, Lbk. 20:232A (*TAED* LB020232A); TAE agreement with Edison Illuminating Co. of Sunbury, 22 Sept. 1885, Miller (*TAED* HM850268).

Edison's primary focus over the last three months of the year was his telephone-related research for American Bell. He directed his attention to two key problems: long-distance transmission and selective signaling. Edison's work on improved transmitters and repeaters was designed to improve the operation of the Bell Company's new long-distance lines. At the same time, the company's growing local networks required better methods of signaling individual subscribers. Edison's long-time assistant John Ott was his primary assistant on his telephone research, but daughter Marion also helped in the laboratory during some of the transmitter experiments in early October. As Edison worked on these problems for American Bell, he was also negotiating the terms of his contract with the company, which included resolving Western Union's rights to his telephone inventions. The contract issues remained unresolved at the end of the year even though Edison traveled to Boston just before Christmas in an effort to finalize his agreement with the company. This was one of several trips Edison made to Boston during the fall in connection with his telephone work. It was also in December that Edison and Ezra Gilliland began to discuss developing the system of railway telegraphy that would become a subject of sustained research for Edison in the next year.[1]

During the fall, Edison's love of theater led him to work with theater impresarios to develop specialized lighting effects for their theaters, which were also lighted by Edison isolated plants. In October, he asked the lamp factory to treat their 100-candlepower lamps so that they would give 200 candlepower for "stage Effect" for Koster & Bial's recently remodeled Concert Hall on Twenty-third Street near Sixth Avenue.[2] Then in November and December, he had the lamp factory produce focusable 500-candlepower calcium lights for Steele MacKaye, who was planning on introducing several novel mechanical and lighting effects at his new Lyceum Theatre in New York. Drawings from November and December also show Edison thinking about other special lighting effects. At the end the year, he asked the lamp factory to produce high-candlepower incandescent lamps to replace calcium lamps at McKaye's Lyceum Theatre and at Edward Holland Hastings's Bijou Theatre in Boston.

Edison's recreational activities during the fall included two notable events. On 31 October he joined the officers and employees of the Edison Electric Light Company as part of the massive parade down Fifth Avenue in support of Republican presidential candidate James G. Blaine. Edison and two assistants rode at the front of the procession, each carrying a 500-candlepower lamp. They were followed by 350 uniformed men with electric lights on the plumes of their hats, each carrying a 16-candlepower electric torchlight. The lights were powered by a dynamo in a horse-drawn wagon. The parade culminated at Madison Square where the grandstand opposite the Fifth Avenue Hotel at the intersection of Broadway was lighted by Edison lights.[3] On 10 November, Edison was one of several prominent men in the crowd of 5–10,000 people at Madison Square Garden to see the fight between John L. Sullivan and John M. Laflin, which Sullivan won in a knockout.[4]

In order to focus his attention on inventive work, Edison, aided by Samuel Insull, sought to elect a new board of directors for Edison Electric. As he told the *New York Sun* prior to the annual stockholder's meeting, "we want a Board with less law and more business. . . . I have given a perfect system, and I want to see it sold. I have worked eighteen and twenty hours a day for five years, and I don't want to see my work killed for want of proper pushing."[5] In particular, Edison and Insull wanted to replace the company's president and general counsel Sherburne Eaton and trustee Grosvenor Lowrey, the lawyer who had played such an important role in the formation of Edison Electric. Toward this end, they sought to acquire enough proxies from other stockholders to enable Edison, the company's largest stockholder, to elect his slate of candidates in place of those proposed by the existing board.

Before the vote took place on 28 October, a compromise was reached in which Eaton resigned as president, although he re-

*Illustration from the
15 November 1884*
Scientific American
*of the electrical torchlight
parade for presidential
candidate James Blaine.*

mained as general counsel, and a combined slate was put forward without either Eaton or Lowrey.[6] In addition to this slate, Eugene Crowell was elected president in place of Eaton. However, it was Edward Johnson, who was reelected vice president, who took charge of the day-to-day affairs of the company. At Edison's behest, Johnson also became president of the Edison Company for Isolated Lighting in November.[7] Soon afterward, Uriah Painter, perhaps reflecting on his own prior efforts to run the Edison Speaking Phonograph Company, cautioned Johnson "Do not get over sanguine as to your ability to achieve the results. . . . Edison gets people away up in the skies, then their disappointment is so terrible that they put him down as a fraud and in the reaction they give him no credit for what he really does do."[8]

Having put Johnson and others that he trusted in charge of the Edison lighting business, Edison spent little of his own time on either the business or technology of electric light and power during the next year. Freed of these concerns, he returned full time to the laboratory to work on other inventions.

1. Gilliland's and Edison's Testimony on Behalf of Edison, pp. 4–7, 34–36, *Edison and Gilliland v. Phelps* (*TAED* W100DKB [images 4–5, 19–20]).

2. Doc. 2742.

3. "A Wonderful Display of Electric Lights," *New York Tribune,* 1 Nov. 1884, 2; "Miles on Miles of Them," *New York Sun,* 1 Nov. 1884, 1; "An Electric Torchlight Procession," *Sci. Am.* 51 (1884): 310.

4. "Mr. Sullivan Has a Picnic," *NYT,* 11 Nov. 1884, 1; "Sports and Pastimes," *Salt Lake Herald,* 16 Nov. 1884, 11 (reprinted from the *New York Herald*).

5. "Edison on the Warpath," *New York Sun,* 25 Oct. 1884, 4, Cat. 1140:51, Scraps. (*TAED* SB017051a).

6. Edison's slate consisted of James H. Banker, Charles Batchelor, Charles E. Chinnock, Charles H. Coster, Eugene Crowell, Robert L. Cutting, Jr., Thomas A. Edison, William Lloyd Garrison, Edward H. Johnson, J. P. Morgan, Francis Upton, Erastus Wiman, and James Hood Wright. The slate proposed by the existing board included Edwin D. Adams, Banker, Coster, Robert L. Cutting, Jr., Crowell, Edison, Sherburne B. Eaton, Garrison, Johnson, Grosvenor P. Lowrey, José de Navarro, Spencer Trask, and Wright. The elected board included Adams, Banker, Batchelor, Chinnock, Coster, Crowell, Edison, Garrison, Johnson, Trask, Upton, Wiman, and Wright. "Mr. Edison's Activities," *New York Tribune,* 26 Oct. 1884, 10.

7. About the same time Johnson became president of the Sprague Electric Railway and Motor Co. to exploit Frank Sprague's patents. Edison's attorney, John Tomlinson, was one of the Sprague Co.'s trustees and William Hammer served as the company's secretary. Dalzell 2010, 68–69.

8. Painter to Johnson, 14 Dec. 1884, UHP (*TAED* X154A4AI).

-2738-

To Thomas Handford[1]

[New York,] 1st Oct [188]4

Dear Sir,

Referring to yours of 2nd Sept[2] as I am anxious to cut short my expenses in connection with English Light Patents please do not incur any more expenses in this connection on my account unless you[a] receive instructions from me subsequent to this date.

Please let me know if there are any further Govt Fees falling due shortly on Patents you have taken out for me[3]   Yours truly

Thos. A Edison   I[nsull]

L (letterpress copy), NjWOE, Lbk. 18:448 (*TAED* LB018448). Written by Samuel Insull. [a]Obscured overwritten text.

1. In 1861, Thomas John Handford (1842?–1890), a native of Surrey, was working in London as a draftsman's clerk. He was a mechanical draftsman in his own right ten years later but had become a clerk in a

London patent agency by 1881. In March 1882, at Edward Johnson's suggestion and with the concurrence of Theodore Waterhouse, Edison engaged him as his patent agent in Great Britain. United Kingdom General Register Office, 1B:35 (Pancras, London, July–Sept. 1890); United Kingdom 1861, GSU roll 542596, class RG9, piece 229, folio 125, p. 30; ibid. 1871, GSU roll 824599, class RG10, piece 218, folio 28, p. 50; ibid. 1881, GSU roll 1341073, class RG11, piece 337, folio 121, p. 51; Handford to Waterhouse, 3 Mar. 1882; Waterhouse to Johnson, 4 Mar. 1882; both DF (*TAED* D8248Y, D8248X).

2. Not found.

3. In acknowledging this letter, Handford gave Edison his preliminary impression that no patent fees would come due soon. Edison prompted him about the matter in November, pointing out that if he could not reach an understanding with the Edison & Swan United Electric Light Co., he would "consider the advisability of letting the patents lapse." After receiving Handford's confirmation that no fees would fall due in the coming year, Edison asked him to notify the company that it should either purchase the patents or indicate their intentions otherwise. See Doc. 2719; Handford to TAE, 21 Oct. and 18 Nov. 1884; both DF (*TAED* D8468ZBP, D8468ZBV); TAE to Handford, 6 Nov. 1884 and 26 Jan. 1885; Lbk. 19:354, 20:52G (*TAED* LB019354, LB020052G).

–2739–

From George Prescott

New York, Oct. 7. 1884.

Friend Edison.

I send you herewith a copy of my new work on dynamo-electricity.[1] Please let me know what you think of it. Yours truly

Geo. B. Prescott[2]

⟨I had already bought a2 copies from Brentano[3]— I think it a first class work ~~and am quite~~ surprised that you have done so well—[4] TAE⟩

ALS, NjWOE, DF (*TAED* D8410O).

1. Prescott 1884a was a comprehensive work, the third chapter of which was devoted to Edison's lighting system.

2. George Bartlett Prescott (1830–1894), the former chief electrician of Western Union, had already written several important books on electricity. He dedicated all his time to writing after retiring from Western Union in 1882. Prescott's professional relationship with Edison had been strained in the 1870s by the protracted controversy over their respective roles in developing multiple telegraphy, particularly the quadruplex. Prescott went on to work with Edward Weston on dynamo design in 1885. Docs. 148 n. 4, 1692 n. 1; Woodbury 1949, 159–60.

3. August Brentano (1829–1886) had set up a newsstand in front of the New York Hotel in 1853. He later relocated the thriving business to the old Revere House, at Broadway and Houston St., expanded his inventory, and renamed it Brentano's Literary Emporium. Brentano moved his business into an actual store in 1860 and, ten years later, to

its contemporary location at 33 Union Square, down the block from Tiffany and Co. Brentano developed a busy cosmopolitan trade in English- and foreign-language books and periodicals; he sold out in 1877 to nephews who continued their uncle's business practices. *Ency. NYC*, s.v. "booksellers"; "Death of August Brentano," *NYT*, 3 Nov. 1886, 4.

4. In the formal reply, which he wrote in Edison's name, Samuel Insull changed this phrase to: "I think the work is a first class one and beg to congratulate you on the successful completion of same." TAE to Prescott, 8 Oct. 1884, Lbk. 19:295A (*TAED* LB019295A).

–2740–

Leslie Dodd Ward[1] to Samuel Insull

Newark Oct 7th 84.

Dear Sir:—

The enclosed is a copy of a bill I sent to Mr Edison several months ago and afterwards at his request forwarded an itemized account[2]

I have attended his family for the past ten years and rec'd in cash only \$195. but took two shares of "Light Stock" for which I allowed \$1200.[3]— Since then I have taken all the stock in the other Co.'s that was allotted me amounting to several hundred dollars.[4]

I do not complain because they would bring but little now for I took my chances with the rest but to have the last bill ignored is more than I expected.—

The amount is very small when compared with the personal inconvenience and loss of time I was subjected to and my regard for the family I had attended so long was the only thing that caused me to respond so promptly at all hours day or night. Mr Edison has always dealt fairly with me but he knows so little of such matters that I have taken the liberty of explaining the case to you and ask as a favor that you will give it your personal attention[5]

With thanks for past[a] courtesies   I remain yours truly

Leslie D. Ward

ALS, NjWOE, DF (*TAED* D8403ZHK). [a]Interlined above.

1. Leslie Dodd Ward (1845–1910) was a Newark physician who had attended the Edison family since at least 1878. A native of Columbia, N.J., Ward was descended from English settlers of colonial Connecticut and New Jersey. He graduated from the College of Physicians and Surgeons in 1868, and his experience included membership on the staff of St. Michael's and St. Barnabas's hospitals, as well as several years serving as County Physician for Essex County; he had also operated a Newark pharmacy. Ward was one of the cofounders and original investors (1875) in what became the Prudential Life Insurance Co. and was named the company's first medical director. He held that position

for several decades, gradually withdrawing from private practice as his responsibilities grew, and became the company's first vice president in 1884. Officially a resident of Newark, Ward spent much of his time at Brooklake Park, his 1,000-acre estate in Columbia, an affluent community that was renamed Florham Park in 1899. Doc. 2213 n. 3; Obituary, *NYT,* 14 July 1910, 7; "Necrology," *Proceedings of the New Jersey Historical Society* 7 (1913): 155–56; Carr 1975, 18–19; Weis 1988, 41–45, 52; Cunningham 1999, 36–37.

2. The editors have found neither the enclosed bill nor the subsequent itemization.

3. In March 1881, Ward accepted one share of the Edison Electric Light Co., valued at $600, from Edison in exchange for $58 cash and the cancellation of his bill for $542 (Insull to Ward, 10 Mar. 1881, Lbk. 8:39A [*TAED* LB008039A]). In August 1882, Ward agreed to purchase another share for $600, the sum to be paid "in professional services rendered or cash" by 1 January 1883 (Ward to TAE, 14 and 24 Aug. 1882; Ward to Insull, 28 Aug. 1882; all DF [*TAED* D8204ZFP, D8204ZFX, D8204ZFZ]; TAE to Ward, 15 Aug. 1882, Lbk. 7:919A [*TAED* LB007919A]).

4. As of 1 July 1882, Ward was credited with owning one and a half shares of the Light Co., two of the Edison Electric Illuminating Co. of New York, and one of the Edison Co. for Isolated Lighting. His holding in the Light Co. increased to two full shares by September 1883. Edison Electric Light Co. lists of stockholders, 1 July 1882, DF (*TAED* D8224ZAP1) and 29 Sept. 1883, Miller (*TAED* HM830194).

5. Insull replied that he knew nothing about the bill and suggested that Ward's letter "must have been mislaid at Mr Edisons House & so failed to reach me. I will send you a check at an early date" (Insull to Ward, 9 Oct. 1884, Lbk. 19:296, NjWOE). On 19 February 1885, Edison paid Ward a total of $300 from his "House" account for professional services from January 1883 until August 1884; Ward signed receipts for payments of $280 and $20 two days later. Edison's delay in paying was not atypical (Voucher nos. 315 and 543 [1884]; Cash Book [1 Jan. 1881–30 Mar. 1886]: 305, Accts., NjWOE).

–2741–

To Rupert Schmid

[New York,] 8th Oct [188]4

Dear Sir,[1]

A number of my friends have seen the Bust of myself which you executed and they all state that it is a very striking likeness of me.[2] Personally I am very pleased indeed with it & have great pleasure in testifying to my satisfaction   Yours truly

Thos A Edison

LS (letterpress copy), NjWOE, Lbk. 19:295B (*TAED* LB019295B). Written by Samuel Insull.

1. Schmid had asked Edison to write a general letter of reference and a note of introduction to George William Childs, co-owner of the *Philadelphia Public Ledger* and a friend and business associate of Anthony

Drexel. Samuel Insull sent Schmid a reply dated 7 October enclosing this document but explaining that Edison would not write to Childs, whom he did not know. Schmid to TAE, 5 Oct. 1884, DF (*TAED* D8403ZHJ); Insull to Schmid, 7 Oct. 1884, Lbk. 19:294B (*TAED* LB019294B); *ANB*, s.v. "Childs, George William."

2. After the International Exposition in Philadelphia, Schmid's bust of Edison (see Doc. 2722) was included in the annual Black and White art exhibition of the Salmagundi Sketch Club at the National Academy of Design in New York (Schmid to TAE, 11 Dec. 1884, DF [*TAED* D8403ZIW]; "Notes at the Salmagundi," *Studio* 1 [20 Dec. 1884]: 115–17; Shelton 1918, 46). Schmid later delivered to Edison a copy of the casting of the inventor's hands that had been displayed in Philadelphia, but he had to sue to collect the $40 owed for it (Charles Lexow to TAE, 19 Jan. 1885; Robert Greenthal to Ecclesine & Tomlinson, 17 Feb. 1885; New York City District Court to TAE, 17 Feb. 1885; all DF [*TAED* D8512A, D8512D, D8512E]).

–2742–

To Francis Upton

New York, Oct 8 1884[a]

⟨Quick⟩[b]
Upton,

Mr Bergmann wants for Koster & Bials[1] for stage Effect a hundred Candle lamp treated so it will give 200 candle[2] at __ volts which is volts used by K&B ~~also~~ These Lamps must be coiled not with[c] one coil but with two so nearly the whole of the Carbon will be in the coil. The way you do it now is insufficient thus

old way

new way

⟨These have been run through—⟩[d]

He want ½ dozen= [~~th?~~][e] There will be a demand for these. K&B only use them 6 minutes ~~a~~daily—

E

ALS, NjWOE, Vouchers (1885), no. 68 (*TAED* VC85068). Letterhead of Bergmann & Co. Electrical Works. ª"New York," and "188" pre-printed. ᵇMultiply underlined. ᶜObscured overwritten text. ᵈMarginalia probably written by John Marshall. ᵉCanceled.

1. Located at this time on Twenty-third St. near Sixth Ave. in New York, Koster & Bial's Concert Hall was recently remodeled and redecorated. It reopened on 16 August 1884 with an Edison isolated lighting plant of 240 lamps (taking the place of gas); it also used an electric motor to run a ventilating fan in the vestibule. Violinist Eduard Reményi, an Edison acquaintance, had performed at Koster & Bial's in June 1884. "Amusement Notes," *NYT,* 15 Aug. 1884, 4; "Koster & Bial's," ibid., 2 June 1884, 4; "Advertisement," *New York Herald,* 16 Aug. 1884, 1; "Electricity in Theatres," *New York Tribune,* 5 Oct. 1884, 4.

2. Edison may have wanted to treat 100-candlepower filaments by applying a "coating reflective of light" of "Silicon or equivalent material" (see Doc. 2587 n. 2). The Lamp Co. billed Edison on 1 November for six "200 c[andlepower] Special lamps . . . For 'Reflectors.'" He noted that these were "no good to me" (Voucher no. 68 [1885]). Edison may also have been referring to a process for coating carbon filaments with another high-resistance conductor, again such as silicon. Doing so would presumably increase the filament's radiating surface and physical durability. Edison had tried to obtain a patent on this process in 1883 (Case 529) but abandoned it (Patent Application Casebook E-2538:32, PS [*TAED* PT022032]).

TELEPHONE RESEARCH   Docs. 2743–2745, 2750–2751, 2759, 2763, and 2766

In September 1884, Edison returned to research on telephone technology as a consultant to the American Bell Telephone Company. Edison had done almost no work on telephones since the summer of 1879, although he filed a joint patent application with Sigmund Bergmann for an improved transmitter in November 1883 (U.S. Patent 337,254). His renewed work in this field was stimulated by a conversation in September 1884 with his old friend Ezra Gilliland, who was responsible for experimental work for American Bell as head of its Mechanical Department. Gilliland later recalled that when they met at the Philadelphia Electrical Exhibition that September,

> Edison mentioned that his electric light was completed and practically off his hands and he was talking of what would be a good thing to take up next. He asked me to suggest something. I suggested taking up the phonograph and reducing it to a practical instrument. I also suggested a patent train signal that I own and the Smith Induction car

telegraph in which I own a half interest and a long distance telephone transmitter on which I was at that time at work for the Bell Telephone Company. The telephone transmitter seemed to strike him as the best thing to turn his attention to at that time and soon after made an arrangement with the American Bell Telephone Company whereby he was to carry on some experiments with the view of improving the transmitter.[1]

By the end of September, Edison had begun working on telephones for American Bell.[2] His first telephone patent application, executed on 24 September but not filed until 15 December (U.S. Pat. 438,304), was concerned with the problem known as selective signaling between subscribers and central stations. He executed two other patents for selective signaling on 19 December (see Doc. 2763).

Long-distance telephony was becoming a subject of considerable interest at this time. In 1883, American Bell constructed the first truly practical long-distance line between Boston and New York using a copper wire in a metallic circuit. This line spurred the company's interest in long-distance transmission and its support for Edison's work. By the time of the company's annual meeting in March 1885, the Boston–New York line had "proved entirely successful" and soon thereafter American Bell general manager Theodore Vail formed the long-distance subsidiary, American Telephone and Telegraph Company. The rival Long Distance Telephone Company had also been formed in New York in 1884.[3]

The success of the copper-metallic circuit telephone line for long distance service spurred other innovations to improve the service. According to A. S. Hibbard, superintendent of the Wisconsin Telephone Company, who examined the state of long distance telephony in 1885, the "most natural and general remedies" for improving long-distance transmission were "a louder transmitter, and a less sensitive receiver; a receiver which will be affected only by the particular current governed by the transmitter worked in conjunction with it."[4] Edison also completed a joint patent application with Gilliland for a method of electrically isolating the receiver in order to improve its performance (see Doc. 2766). However, he focused his experimental work primarily on improving the carbon transmitter (see Docs. 2744 and 2750) and on using a transmitter as a repeater in order to overcome the attenuation of the signal over a long line (see Doc. 2759).

1. Gilliland's Testimony on Behalf of Edison, *Edison and Gilliland v. Phelps*, pp. 3–4, 24 (*TAED* W100DKB [images 3–4, 14]).

2. Edison's telephone drawings, dated 28 and 29 September 1884, are in Unbound Notes and Drawings (1884), Lab. (*TAED* NS84ACN1, NS84ACN2). His telephone experiment account for 1884 is in Ledger #5:570 (*TAED* AB003 [image 282]). An undated memorandum (probably from late 1884) identified five areas in which Edison was experimenting for the American Bell Telephone Co. In addition to selective signaling, they were: a repeater for long-distance lines, an improved transmitter, elimination of induction from electromagnets in telephone circuits, and means of suppressing crosstalk and induction (TAE memorandum to United Telephone Co., Ltd., n.d., DF [*TAED* D8547ZBN1]).

3. Hall 1887, 23; "Annual Meeting of the American Bell Telephone Co.," *Electrician and Electrical Engineer* 4 (June 1885): 234; Fagen 1975, 202–204; Garnet 1985, 74–77; Long Distance Telephone Co. 1884.

4. Hibbard 1885, 65.

–2743–

Notebook Entry: Telephony[1]

[New York,] Oct. 8, 1884

Telephone Experiments.
Telephone.[2]

Blake box[3]

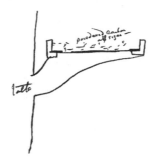

We now try with trans on 187.[4] with 2 B[ergmann] & Cos. cells[5] a regular Edison coil, hand phone; then we change by Switch over to a reg[ular] Edison ~~p~~secondary but have 10 different primaries that we can slip in running from No 16 17 18 20 21 22 24 26 28 30 wire, to obtain best resistance for this trans & battery=[6]

Edison Reg ~~lowerud~~—
64 ohm coil 5 times weaker
34 ohms   3ª   "   "

24	"	3	"	"
12	"	2	"	"
7 ohms		[1/4]^b ½ loudness as reg		
3.25	"	³/₄		"
1.28	"	⁴/₅		"
.44		nearly same		
.53^c		⁵/₆		"

Lower coils not so loud   We are now winding lower re-s[istance] Coils

The Coil twice as long is about same as regular perhaps not quite so loud^d

We try Experiment of adding battery to each coil of dif res until its same loudness as regular Edison

⟨1st 2 alwys B & Co cells⟩

1.28 primary coil. with 2 B & Co. & 1 Carbon[7] as loud as regular   with 2 B & Co & 3 Carbon hurts Ear but only the [jumps?]—[8e] its very loud—

3.25 ohms. with^a ~~34~~ cells ~~not~~ 2 of B & Co. 2 Carbon loud as regular   with 6 louder & clearer   with 8. no louder^d

7 ohms. 5 cells loud as regular.

12 ohms. 8 cells ½ as loud as regular^d

24 ohm poor with 8 cells^d

12^f ohms 8 cells ½ a loud as regular^d

24 ohms poor will 8 cells^d

Tried rubber ring with ₱ two papre Diaphragm ¹/₃₂ inch apart filled with fine coal carbon   the Articulation was good but not loud except in one position   This was lost after talking a little wile   John Ott is now putting some lamp black between them[9]

TAE

J F Ott
Marion[10]

X, NjWOE, Lab., N-82-05-15:187 (*TAED* N203187). Document multiply signed and dated. [a]Obscured overwritten text. [b]Canceled. [c]Decimal point added for clarity. [d]Followed by dividing line. [e]Illegible. [f]From here to end of document written by Marion Edison.

1. See headnote above.

2. Figure labels are "sheet carbon," "mica dia," and "plat."

3. Figure labels are "powdered carbon   all sizes—" and "talk." The "Blake box" refers to the arrangement of the standard telephone transmitter used by the Bell Telephone Co. (Doc. 1740 n. 5). Devised by Francis Blake in 1878, it used a form of inertia transmitter with two spring-mounted electrodes—a platinum bead and a carbon disk—that remained in contact with each other while being vibrated by the diaphragm. This arrangement eliminated the need for frequent adjustment of the pressure on the carbon button, and it also reduced the mechanical vibration from the transmitter mounting. The transmitter was attached to a heavy brass ring and placed inside the front door of a small box ($5^9/_{16} \times 6^{11}/_{16} \times 2^{13}/_{16}$ inches) with an external trumpet-shaped mouthpiece to focus sound waves on the diaphragm. Also inside the box was an induction coil used to amplify the signal (an arrangement originally devised by Edison; see note 5 below). While the Blake transmitter worked adequately for short lines, it was practically useless on long-distance lines. Fagen 1975, 70–71; Prescott 1884b, 368–70; Abbot 1903, 156–59.

4. That is, a line to Western Union Telegraph Co.'s main office at 187 Broadway in New York.

5. This reference was probably to the Bergmann & Haid battery, advertised as "the greatest open circuit battery in the world," supplying the "Great Desideratum" for telephone work. Bergmann & Co. also claimed that its battery was cheaper and more durable, lasting twice as long as the Leclanché and other batteries used for telephones. Berly 1884, American section, 254; Dana 1884.

6. John Ott's drawing of an "Induction Coil with changeble primarys" is in 1884 Unbound Notes and Drawings, Lab. (*TAED* NS84ACN5). Edison had introduced the use of induction coils for telephone transmission in 1877 (see Docs. 1112, 1121, 1139, and 1146; U.S. Pat. 203,019; Prescott 1884b, 116–21; Abbot 1903, 223–24; Miller 1905, 21–22; and Shepardson 1917, 231). He probably referred to the induction coil in the transmitter circuit as the regular Edison coil because, in 1879, he introduced what he called a tertiary coil in the receiver circuit (see Docs. 1759, 1784 [headnote], and 1835; and Brit. Pat. 5,335 [1879], Batchelor [*TAED* MBP021]).

7. Edison meant a modified Grove cell, also known as a Bunsen cell, in which the anode was a carbon rod.

8. Edison may have meant when the voice "jumps" up an octave. See Helmholtz 1954 [1885], 35.

9. Figure labels are "$^1/_{32}$ $^1/_{32}$."

10. Edison's daughter Marion, who wrote the last few items in this notebook entry.

Notebook Entry:
Telephony[1]

I think I have Clearly in my mind the proper theory of constructing a loud telephone & the most powerful utilization of the whole voice to produce the maximum change of resistance. It is this. Use a hard button of Carbon crosshatched to give innumerable points   have this button 2 to 4 inches in diameter resisting on a hard ~~m~~flat metallic surface connecting with one pole— over the button have soft yielding sheet such as a chamois Oiled Cotton etc faced with platina foil or Carbon paper foil— secure the edges only to prevent moving but do not streatch in the least   let it lay dead on the button. now have a very low resistance primary & talk direct to the soft sheet, the strength of the sound waves is the same in every part of the soft armature hence unlike a diaphragm equal pressure will be placed in every part of the Carbon & the whole power of the voice used to make the initial Contact, while with heavy foil or diaphragm the pressure is great at one spot, but one does not get the full benefit of all the pressure as the Carbon makes its greatest Change on the first part of the wave   doubling the pressure scarcely makes ⅕ change further & so on while with my arrangement the ~~f~~ initial sensations of the Carbon only is used.

⟨Didnt turn out so well as expctd   Oct 11 1884⟩

very thin Carbon—

much finer than this 80 @ 100 to inch.

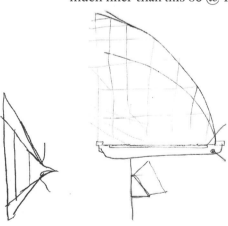

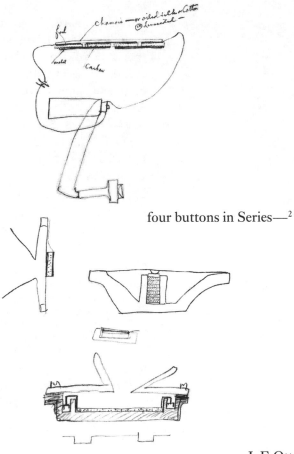

four buttons in Series—[2]

TAE   J. F. Ott

X, NjWOE, Lab., N-82-05-15:199 (*TAED* N203199). Document multiply signed and dated.

1. See Doc. 2743 (headnote).
2. Figure labels are "foil," "chamois— or oiled silk or Cotton i.e. Linseeded—," "metal," and "Carbon."

–2745–

Theodore Vail Memorandum to American Bell Telephone Co. Executive Committee[1]

Boston Oct. 15th 1884[a]

Gentlemen,

In accordance with your wishes, I had an interview with Mr. Edison when last in New York. After a long consultation, Mr. Edison finally agreed to the following as an arrangement that he would be willing to make:—

He would give about half of his time to the Co. to be paid therefor $6000.00 per annum. He has a Laboratory, the expense of which is as follows:—

```
Rent . . . . . . . . . . 2500.00
Mechanic  . . . . . . 1500.00
Mechanic  . . . . . . 1200.00
Boy . . . . . . . . . . .  600.00
                        $5800.00
```

All the experimenting done in this Laboratory, to be so done by him or his assistants, this Company to be charged for only its proportion of time occupied. In addition to this, there will be expense for models made in Bergmanns Manufactory— with which the Laboratory is connected—these models to be paid for at the rate of 60¢ an hour for work— Mr. Edison however agrees that, including half the time of the Laboratory, the total expense outside of his retainer, shall not exceed $4000.00 a year, unless he is called upon by this Co. to make some special expenditures, or do work in such a way as to require extraordinary expenditure.[2]

I explained to Mr Edison, that the above was conditional upon our making some[b] arrangement with the W[estern]. U[nion]. T[elegraph]. Co.[3] I called upon Dr Green[4] the same day, and stated to him our desire, and he said, that he would consent to an assignment by the G[old]. &. S[tock]. T[elegraph]. Co. of the contract between that Co. and Mr. Edison, the G. &. S. T. Co. however assuming the payment of the $6000.00 a year, which it now pays him.[5]

He stated further, that he would bring the matter before the Law Committee, and let me know the result.[6]

In regard to inventions or improvements that are not patentable, the same will belong to us without compensation— All that are patentable, are to be patented, and we are to have the option of taking them at a price to be agreed upon. If a price can not be so agreed upon, the matter to be submitted to arbitration,[b] we reserving the right to refuse to accept any invention at the price fixed by the arbitrators, in which event, the same will belong to Mr. Edison. In case any invention is adopted for other use than Telephony, Mr. Edison is to have a license from us therefor. Respectfully

Theo N Vail[7]   Gen'l Man'r.

ALS, NjWAT, Box 1221 (*TAED* X012F2C). Letterhead of the American Bell Telephone Co. [a]"Boston" and "188" preprinted. [b]Repeated at bottom of one page and top of next page.

1. The American Bell Telephone Co. was created in 1880 as part of the recapitalization of the National Bell Telephone Co. The reorganization allowed National Bell to fulfill its 1879 agreement to purchase

Western Union's telephone rights and instruments. National Bell was itself the product of an 1879 combination of the New England Telephone Co. and the original Bell Telephone Co. Smith 1985, 55–56, 103–105; Reich 2002 [1985], 135.

2. These are the terms of a draft agreement dated October1884 (DF [*TAED* D8472ZAO1]). That agreement was probably the one that Vail sent to Edison on 25 October with the comment that it "is practically your present contract, with a few amendments," meaning Edison's telephone contract with Western Union (see note 3 and Vail to TAE, 25 Oct. 1884, DF [*TAED* D8472ZAM]). Vail also asked Edison to include Canada in the contract, the rights to which were held by Gold & Stock Telegraph Co. (see note 4). Edison agreed to this provision in a 29 October reply (Lbk. 19:314 [*TAED* LB019314]). Although Edison's attorney, John Tomlinson, traveled to Boston in connection with the negotiations in late 1884 and Edison went to Boston at the beginning of January to complete the arrangement, it was apparently never consummated (Tomlinson's Testimony on Behalf of Edison, *Edison and Gilliland v. Phelps*, p. 30 [*TAED* W100DKB (image 17)]; Edison to John Tomlinson, 3 Jan. 1885, Miller [*TAED* HM850231D]; Samuel Insull to Tomlinson, 3 Jan. 1885, and undated draft agreement, both DF [*TAED* D8541A, D8472ZCA]; Vail to American Bell Telephone Co., Executive Committee, 9 May 1885, NjWAT [*TAED* X012F2K]).

3. Western Union owned Edison's existing and future telephone patents under an agreement signed on 31 May 1878. The company was to pay Edison an annual salary of $6,000 and reimburse his experimental and patent costs for a term of seventeen years (Miller [*TAED* HM780045]; see Doc. 1317 esp. n. 5). An undated draft agreement between Edison, American Bell, and Western Union provided for Western Union to give American Bell a general license for patents already taken out by Edison and for his future patents as if American Bell been part of the 1878 contract, agreeing that "ABT Co takes the advantages so far as relating to Edison's inventions hereafter and assumes the obligations arising thereout" (WUTAE [*TAED* X099AZC]). This draft may be the one enclosed by Oscar Madden, American Bell's assistant general manager, in a 25 October 1884 letter to Vail (WUTAE [*TAED* X099AZB]). For the views of Western Union attorney Charles Buckingham on the proposed agreement, see his 1 November 1884 letter to Clarence Cary, another Western Union attorney (WUTAE [X099AZD]). Edison sent another draft agreement with his 3 January note to Tomlinson (see note 2 and Tomlinson to TAE, 5 Jan. 1885, DF [*TAED* D8547C]). It is unclear whether or when an agreement between Western Union and American Bell regarding Edison's telephone patents was concluded. However, by the beginning of 1887, American Bell was paying Edison a salary of $500 per month ($6,000 per year) and continued to make such payments through 1907 (American Bell Telephone Co. to TAE, 31 Jan., 28 Feb., 30 Nov., and 24 Dec. 1887, all DF [*TAED* D8754AAC, D8754AAG, D8754AAO, D8754AAQ]; Ledger #5:100; Ledger #6:167, 190, 253, 267, 270, 272, 274; Ledger #7:238, 325, 330, 338, 416; all Accts. [*TAED* NL011A1: image 75, NL012A1: images 104, 116, 123, 130, 132–34; NL015A1: images 133, 176, 179, 183, 222]); General Ledger #6: 353, 366, 470; General Ledger #7:335, 371, 389, 391, 457, 515; General Ledger #8:443 [*TAEM* 180:61, 673, 675; 181:189, 207, 216, 21, 251, 280,

394]; see also Norvin Green to William Forbes, 1 Nov. 1886, WUTAE [*TAED* X099CX]).

4. Norvin Green (1818–1893) had been president of Western Union since 1878 and also served as first president of the Edison Electric Light Co. *ANB*, s.v. "Green, Norvin"; Docs. 1168 n. 3 and 1576.

5. The Gold and Stock Telegraph Co., a subsidiary of Western Union, had been in charge of Western Union's telephone business, and it paid Edison's telephone royalties. The company also controlled Edison's Canadian telephone patents. See Docs. 1126 and 1417; agreement with Gold & Stock, 2 Nov. 1880; Clarence Cary to TAE, 4 Sept. 1879; G. M. Taylor to TAE, 3 Dec. 1884; all DF (*TAED* D8046ZCA1, D7937ZCE, D8472ZBK); for American and Canadian royalty accounts, see Ledger #8:193–95, 202–203 (*TAED* AB004, images 80–81, 85).

6. Norvin Green reported on his meeting with Vail to Western Union's Law Committee on 15 October. The committee voted to refer the matter of assigning "the option upon Mr. Edison's telephone inventions and improvements secured to Gold and Stock Telegraph Co. by its contract with Mr. Edison dated May 31st 1878" to Green and general manager and vice president John Van Horne "with power." Western Union Law Committee Minutebook No. 5:154, WUTAE.

7. Theodore Newton Vail (1845–1920), cousin of telegraph pioneer Alfred Vail, was superintendent of the U.S. Railway Mail Service before becoming general manager of the Bell Telephone Co. in 1878. He served in that capacity until 1885. He became the first president of American Telephone & Telegraph Co. when it was incorporated in 1885 as the Bell system's long-distance provider. He retired in 1889 but reassumed the presidency of AT&T in 1907. *ANB*, s.v. "Vail, Theodore Newton."

–2746–

To Uriah Painter

NEW YORK, 16th Oct 1884[a]

My Dear Painter,[1]

Please sign & return enclosed form of proxy[2]

I propose to endeavor to get rid of S. B. Eaton & so give E[dward]. H. J[ohnson]. a fair show at running the business

Yours truly

Thomas A. Edison

LS, PHi, UHP (*TAED* X154A4AA). Letterhead of Thomas A. Edison; written by Samuel Insull. [a]"NEW YORK," and "188" preprinted.

1. Uriah H. Painter (1837–1900), a Washington lobbyist and former Philadelphia journalist and railroad promoter, had been instrumental in forming and running the Edison Speaking Phonograph Co. Painter held thirty-two shares of Edison Electric Light Co. stock as of September 1883. See Docs. 672 n. 2, 1190, 1583; Edison Electric Light Co. list of stockholders, 29 Sept. 1883, Miller (*TAED* HM830194).

2. Edison's proxy for the upcoming election of directors of the Edison Electric Light Co. has not been found, but see Docs. 2748, 2752, and 2753. Edison evidently sent the forms and a printed letter (Doc. 2752 n. 4) to a number of shareholders, some of them with quite small hold-

ings (Leslie Ward to TAE, 18 Oct. 1884; Chester Pond to TAE, 21 Oct. 1884; Lee Higginson & Co. to TAE, 21 Oct. 1884; Spencer Borden to TAE, 22 Oct. 1884; Edward Hackett to TAE, 24 Oct. 1884; J. G. Chapman to TAE, 25 Oct. 1884; all DF [*TAED* D8472ZAI, D8403ZHX, D8472ZAJ, D8427ZBV, D8427ZBW, D8427ZBX]; TAE to William Dwelly, Jr., 27 Oct. 1884, Lbk. 19:303 [*TAED* LB019303]). With 1,256 shares to his name at the previous annual meeting, Edison was by far the largest single holder of the roughly 6,800 outstanding shares (Edison Electric Light Co. list of stockholders, 29 Sept. 1883, Miller [*TAED* HM830194]).

According to a newspaper report published just before the election,

> Mr. Edison has decided that the Edison Electric Light Company is not doing the business it should do. He is not manager of the company and never has been. He has done the inventing and left to others the introduction of his system. For some reason the system has not made its way into public use as rapidly as the inventor would wish, and he has come to the conclusion that it is for lack of push and energy on the part of the management. Accordingly a change has been suggested in the Board of Trustees. ["Mr. Edison's Activities," *New York Tribune*, 26 Oct. 1884, 10]

–2747–

Samuel Insull to Caroline Schenck[1]

[New York,] 16th Oct [188]4

Madam

Referring to your favor of 5th Mr Edison desires me to state that all his children are at Private Schools & he has therefore no need of a Governess for them.[2] Yours respectfully

Saml Insull   Private Secy

ALS, NjWOE, Lbk. 18:479 (*TAED* LB018479).

1. Caroline Augusta Schenck (b. 1842?) was born in New York but reportedly moved to San Francisco in 1882 and lived with her father. When he was committed to an asylum at the end of 1883, she remained in the city despite having no other family there. On 5 October 1884, she sent Edison a rambling inquiry about becoming a governess for his children, offering several New York City references; she signed it as "Miss C. Aug. Schenck." U.S. Census Bureau 1965 (1870) roll M593_868, p. 463B, image 342 (Jersey City Ward 13, Hudson, N.J.); "Geer Surrenders," *San Jose (Calif.) Evening News*, 23 Mar. 1896, 4; "Crazy Witnesses," ibid., 24 Apr. 1896, 4; Schenck to TAE, 5 Oct. 1884, DF (*TAED* D8403ZHI).

2. Insull wrote his reply from Edison's notation on Schenck's 5 Oct. letter (see note 1): "children all at Private school." Marion Edison was back at Madame Mears's School (see Doc. 2732). Thomas, Jr., was enrolled from September to June 1885 at M. W. Lyon's Collegiate Institute, on Twenty-second St. at Broadway in New York. That institution advertised itself in an 1878 directory with this description: "Government the minimum of authority, the maximum of kindness and confi-

dence. Best assistants obtainable employed. French and German taught. Rooms—light and well ventilated; teaching—earnest and thorough." The editors do not know the activities at this time of the youngest Edison child, William Leslie, who was nearing his sixth birthday. Voucher nos. 62 and 111 (1885); Steiger's 1878, 49.

—2748—

From Grosvenor Lowrey

[New York,] October 19th/84

My dear Edison:

I have been confined to my room since Saturday last but hope to be out tomorrow.[1]

Mr. Insull was at my office a few days ago to say that he had seen my letter to you about the arc-light contract ~~and suggested lying~~ lying on your desk had suggested that I should see you before giving any time to that—[2] Accordingly I should notwithstanding I had no answer from you have called on you last week had it not been for my illness—

By the time McGowan had returned from his vacation the time for my moving had arrived and I was kept very busy for nearly two weeks in supervising that & the making various alterations in my apartments & just as I was done this sickness caught me

I suppose you will be at the Board meeting tomorrow[a] & if you will then state to me your points I will sit down at once to redraw that very long paper

I hear that you have been collecting proxies with a view to electing new directors in the place of some of the old ones[b] including me   That is all right— If any stock holder thinks I do try properly to serve his interest as a stock-holder it is quite correct for him to vote accordingly [-][c] As to Major Eaton, however other considerations are involved.

You know he[d] has a contract and any change looking to a termation of his connection with the Company should be treated civilly and as a matter of agreement. Besides the new agreements[3] were authorized only because they[b] were part of a re-organization[b] scheme in which Major Eaton, remaining President was to with draw from active business and put his energies entirely into prosecuting suits etc[4]

I do not believe you have reflected that if you or persons under[b] you manage the Light Company as you do the shops, then both sides of that important business will be in one control with the natural risks that it will be carried on as to pay the largest profit to the party in control.[5] The law not only frowns upon but forbids it. Such a state of things is manifestly wrong. You are certainly too clear sighted to think that after all this wrangling the other members of the board would have passed those contracts if they had known or suspected that their complicated provisions were to be interpreted by a new board controlled by the interests on the other side of the contract.

I hope to see you tomorrow at the Board meeting and shall be glad to discuss these matters openly and with a sincere desire to aid in establishing such arrangement as will most conduce to the common interest   Very truly Yours

G. P. Lowrey

LS, NjWOE, DF (*TAED* D8427ZBT). [a]Interlined above. [b]Obscured overwritten text. [c]Canceled. [d]Followed by two heavy horizontal lines.

1. Lowrey may have suffered a flare-up of the gout, which occasionally left him incapacitated. See Doc. 1711.

2. Lowrey wrote Edison on 20 September that, despite the absence of stenographer Frank McGowan, he hoped to dictate "a substitute to the Contract respecting arc lights. . . . (although I understood you to say you did not care for any contract)—because I assume, that eventually you will prefer to use your own dynamo." He invited Edison to give his opinion on a version of "that paper" drafted some time ago, which Lowrey admitted he had not yet read, "not having time this year." That document may have been an understanding with Otto Moses, some form of which had been under discussion since at least May 1882 (Lowrey to TAE, 20 Sept. 1884, DF [*TAED* D8427ZBS]; see Docs. 2286 n. 7 and 2501 n. 15). Alternatively, he may have been referring to a proposed arrangement with the Thomson-Houston Electric Co. to work together when installing systems in New England towns where investors wanted a combined arc and incandescent plant (Thomas Conant to TAE, 9 Aug. 1884; Frank Hastings to TAE, 14 Feb. 1885; both DF [*TAED* D8452J, D8526G]; American Electrical Directory 1886, 314).

3. See Doc. 2725.

4. Eaton reportedly told a newspaper a few days before the contested election that

> everything is now adjusted. There is nothing personal in the issue. Mr. Edison desires a change. Some of the present board, including myself, are objectionable to him and he wants them left out at the coming election. To avoid any unpleasantness and looking to the interest of the company, knowing moreover, that Mr. Edison is worth more to the company than any other man in it, I have offered my services to bring about a satisfactory adjustment of the differences that have arisen. Yesterday I sent in my resignation and offered to give up my contract which binds me to serve the company for three years in legal capacity. . . . In regard to my contract, which is mutually binding on myself and the company, I leave it to the discretion of the Board as to whether my services shall be dispensed with or not. ["Mr. Edison's Activities," *New York Tribune*, 26 Oct. 1884, 10]

Although not reelected to the board, Eaton did retain a significant role with the Edison Electric Light Co. as its legal counsel (Eaton's testimony, 5:3856, *Edison Electric Light Co. v. United States Electric Lighting Co.*, Lit. [*TAED* QD012F3856]).

5. The Edison Electric Light Co. adopted this line of argument in support of its slate of director candidates. Doc. 2752 n. 4.

–2749–

From James Connelly

New York, Oct. 23d 1884[a]

Dr. Sr.

As we rode together over to Philadelphia a few weeks ago, we talked of bleaching cotton cloth by electricity and you said that as soon as you returned to N.Y. you would get up a working model of the invention for me to place in Eastern mills. I make this little recapitulation for fear that in the press of your other business this may have been lost sight of. Has it been remembered? What have you accomplished yet? Have you progressed far enough for me to be of any use? If you send me word when you would like to see me, it will afford me pleasure to call upon you. Very truly Yours

J. H. Connelly[1]

⟨I had forgot all about it being very busy with our Annual Election[2]   I will try & remember it when I go to work next week in Laboratory⟩[3]

ALS, NjWOE, DF (*TAED* D8403ZIA). Letterhead of Connelly & Curtis. [a]"New York," and "188" preprinted.

1. James H. Connelly (1840–1903) was a newspaper reporter who worked for many of the New York newspapers. He also wrote several novels, as well as magazine articles on Theosophical subjects. At this time, he was president of the Cook Publishing Co. and partner with David A. Curtis, secretary and treasurer of Cook Publishing, in the firm of Connelly & Curtis. The partnership's letterhead identified Connelly with the *New York Sun.* In 1885–1886, Connelly and Curtis received a loan from Edison for their proposed magazine, *The Cook.* "James H. Connelly Dead," *NYT,* 16 Mar. 1903, 9; Connelly to TAE, with TAE marginalia, 4 Jan. 1886, DF (*TAED* D8603B).

2. Edison referred to the election of Edison Electric Light Co. directors.

3. Edison's marginal notes served as the basis for a reply to Connelly on 30 October (Lbk. 19:326 [*TAED* LB019326]). Hugh De Coursey Hamilton conducted bleaching experiments in early March 1885 (N–82–12–21:181, Lab. [*TAED* N150181A]); account records are in Ledger #5:574, Accts. (*TAED* AB003 [image 284]).

Notebook Entry:
Telephony and
Electric Lighting[1]

Telephone
Try Square diaphragm in all the telephone Experiments[2a]
[A]

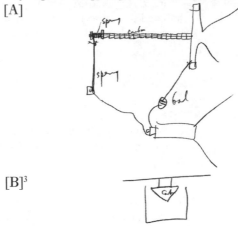

[B][3]

Try following—Galena—Calcopyrite,[4]
phosphide Iron, Conducting[b] Sulphides   peroxide Mangan-
eese art[ificial] & native selected,[5]= peroxide Lead, Peroxide
Silver= Silicon—[6]
[C][7]

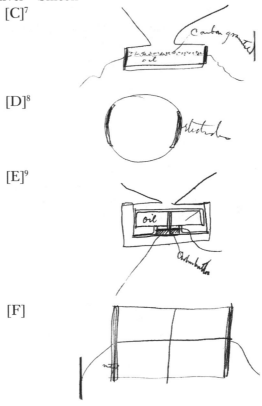

[D][8]

[E][9]

[F]

plate which is insulator covered with granulated Carbon held on by adhesive mixture   talk to this square plate—

[G][10]

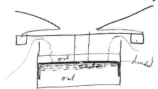

grind up conducting Lamp black with oil etc see if can make a conducting contact Transmitting liquid—

[H]

[I][11]

[J][12]

[K][13]

[L][14]

[M][15]

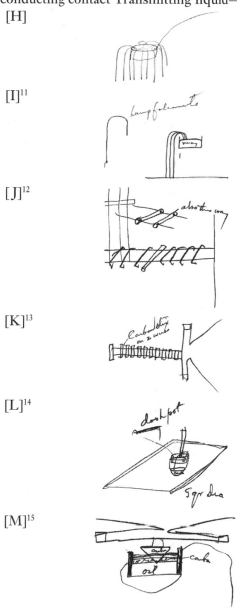

[N][16]

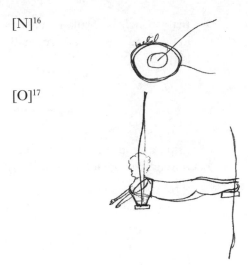

[O][17]

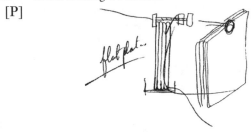

Send to Lamp factory for some Carbonized Cotton[a]

Mix anthracite Carbon[18] with Rubber—also with Copal softened with boileding Linseed

Try that mixture of plaster paris & Carbon—[a]

Try Lampblack & granulated anthracite Carbon with minim oil— ⟨good⟩[c]

Get .40 primary & 100 to 300 secondary Coil wound by Fred Mayer[19] to give ¼ @ ½ inch spark   try this against Regular— think talking be clearer—[a]

[P]

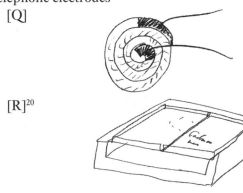

flat plates

Carbonize some cloth at Lamp factory various kinds for telephone electrodes=[a]

[Q]

[R][20]

[S]

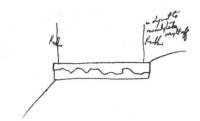

Naturally broken[21]

TAE

X, NjWOE, Lab., N-80-08-09:25 (*TAED* N134025). Document multiply signed and dated. [a]Followed by dividing mark. [b]Obscured overwritten text. [c]Line drawn partially around this paragraph.

1. See Doc. 2743 (headnote).

2. Figure labels are "spring," "carbon," "spring," [—], and "bal." This design is related to Edison's 13 October drawing of the "square plate telephone," on which he noted that it was composed of "peices to prevent p[ac]k[in]g" (N-82-05-15:217, Lab. [*TAED* N203217]). The so-called "long distance" telephone used at this time was adapted from British inventor Henry Hunnings's patented transmitter with granulated carbon. This form of carbon enabled the use of higher battery power and produced clearer talking on long-distance lines. However, "the carbon granules would 'pack' —that is, would cohere, forming a solid mass, which blocked the passage of sound." Users would find it necessary to jar or shake the transmitter in order to loosen the granules and allow normal operation. Various solutions were sought to this problem, including Ezra Gilliland's horizontal diaphragm (U.S. Pat. 384,201) and Edison's improved granulated anthracite carbon (see note 18). It was finally solved in the early 1890s by the adoption of American Bell engineer Anthony White's "solid-back transmitter" (Vaughn 1940, 372; Lockwood 1887, 81–82, Fagen 1975, 73–79).

3. Figure label is "Carbon."

4. Chalcopyrite, a copper iron sulfide mineral.

5. Manganese occurs most commonly as pyrolusite, also known as manganese dioxide or peroxide, which is a good electrical conductor. Because it is difficult to refine, most commercial forms of manganese, including manganese dioxide, are produced artificially (Hawley 1987, s.vv. "manganese" and "manganese dioxide"; Richardson 1863, 1:343–50; Watts 1882, 3:801–16). In a notebook entry of 10 October, Edison mentioned the use of "fine magnesium" (N-80-08-09:19, Lab. [*TAED* N134016]).

6. These are among the semiconducting materials with which Edison experimented in 1877 while developing the carbon button for his telephone transmitter. According to a nearly contemporary account, Edison applied the term "carbon button"

> to the various substances evincing the characteristics so strongly
> displayed by compressed carbon when subjected to electrical action.
> The basis of this, the parent discovery, is the principle that when a
> current of electricity is passed through a quantity of finely divided

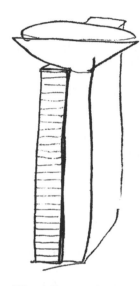

Edison's "square plate Telephone" transmitter drawing of 13 October.

conducting matter, such as finely divided metals, conducting sulphides, oxides, graphite or carbon in its various forms, the slightest
pressure on such finely divided conductor varies the strength of the
electric current by diminishing the resistance offered to its passage.
["Edison's Inventions II. The Carbon Button and Its Offspring,"
Scribner's Monthly 18 (July 1879): 446–47]

See also Prescott 1878, 224; Engler 1880, 14; and Docs. 885, 945, 981,
1005, 1034–35, 1041, 1100.

7. Figure labels are "Carbon granulated" and "oil." For other uses of
oil, see below and Edison's 10 October notebook entry (N-80-08-09:16,
Lab. [*TAED* N134016]). In his joint U.S. Patent 337,254 with Sigmund
Bergmann, Edison used liquids (including mercury), but preferrred oil
to prevent "excessive movement of the diaphragm" and limit the movement of the electrodes "so that they quickly return to their normal position."

8. Figure label is "electrodes."

9. Figure labels are "oil" and "Carbon button."

10. Figure labels are "oil," "Carbon," "divided," and "oil." See also
Edison's notebook entry of 25 October (N-80-08-09:23, Lab. [*TAED*
N134023]).

11. Figure labels are "Lamp filaments," and "mercury." Presumably
the movements of the filaments in the cup of mercury would alter the
resistance of the circuit.

12. Figure label is "also this way." Edison's notebook entry of 10 October may have a design related to Figures J and K that used foil instead
of wire (N-80-08-09:19, Lab. [*TAED* N134016]).

13. Figure label is "Carbon stix on 2 wires."

14. Figure labels are "dash pot," "~~mercury~~," and "sq[ua]r[e] dia-
[phragm]."

15. Figure labels are "carbon," "carbon," and "oil."

16. Figure label is "metal."

17. Edison occasionally drew random male and female figures in
his notebooks. See, for example, Cat. 1174:185; Cat. 297:13, 62; Cat.
1169:7; Vol. 8:225, 322; Vol. 13:49; Vol. 16:43; all Lab. (*TAED* NE1678
[image 41], NM003 [images 63, 210], NM012 [image 10], NV08211
[image 222], NV08320, NV13049, NV16043). He sometimes used such
figures to show the operation of a technology, as in telephone drawings
in Docs. 968 and 971. Edison occasionally also doodled names, phrases,
or quotations. On a looseleaf drawing of a telephone transmitter dated
28 October, he wrote "Mary," presumably his late wife, and several
other unidentified names (Unbound Notes and Drawings [1884], Lab.
[*TAED* NS84ACN4]).

18. This reference marks the first evidence of Edison's use of anthracite coal for a telephone transmitter; he had experimented with anthracite in batteries and fuel cells the previous July (see Docs. 2703 and
2706). He filed a patent on the use of carbonized anthracite in early
1886, noting that it "will not by use become finely powdered and liable to
pack" (cf. note 2). To manufacture this telephone carbon, he subjected
anthracite coal "to a high heat in a manner similar to the carbonization
of materials for the conductors of incandescent lights," and then powdered it "into granules which will pass through a twenty or thirty mesh

screen." This type of carbon proved to be a significant improvement for long-distance transmitters. U.S. Pat. 406,567; Fagen 1975, 74–75.

19. Alfred Marshall Mayer (1836–1897), a veteran teacher and prolific researcher, had been professor of physics at the Stevens Institute in Hoboken since 1871. Mayer worked on a wide variety of topics, including electricity and magnetism, but was best known for his studies on the nature of sound. Mayer had become acquainted with Edison in 1878 in connection with the phonograph; in 1880, he coauthored a study of the electrical characteristics of the Edison carbon lamp. *ANB*, s.v. "Mayer, Alfred Marshall"; see Docs. 1175 and 1927.

20. Figure label is "Carbon on here."

21. Figure labels are "Rubber," "adjust to nearly take weight off," and "Rubber." This drawing (without the figure labels) and text were copied on the same page by Edison's daughter Marion, who made her own doodles and an unrelated list at the back of this notebook (N-80-08-09:140, 218, 280, Lab. [*TAED* N134025: images 33–35]). On the pages preceding this drawing are a freehand map of Europe by Edison, possibly for Marion, and columns of numbers that appear to be unrelated to the notebook entry (N-80-08-09:57–58, Lab. [*TAED* N134025 (images 29–30)].

–2751–

From Ezra Gilliland

Boston, Oct. 25th 1884.[a]

Dear Sir:—

I have received the papers of our joint application for patent for improvement in electro-magnets, and have duly executed the same, and returned them to Mr. Dyer.[1] Am I correct in supposing that Mr. Dyer will continue as our attorney in this matter, or shall we take the case up at this point? Don't forget that arrangement which you are about to enter into with this Company; it is an important question and should be clearly understood. Yours respectfully,

E. T. Gilliland   Supt. Mech. Department[2]   D[3]

⟨Gilliland   What are the wishes of the Co in respect to the prosecution of my cases in the Pat[b] office   I should like Dyer to prepare them but after that the Co could prosecute the appn's or Mr Dyer[4]   pls ask Vail= Dyer is Smart=⟩[5]

ALS, NjWOE, DF (*TAED* D8468ZBR). Letterhead of American Bell Telephone Co. [a]"Boston," and "188" preprinted. [b]Interlined above.

1. Gilliland referred to a joint patent application that he signed on 24 October, three days after Edison had done so. It was filed on 1 December but did not issue until 4 March 1890 as U.S. Patent 422,577. The specification covered a method for preventing the electromagnets used for operating signal bells on individual telephone sets and the annunciators at the central station switchboard, which were always in circuit, from interfering with speech transmission. Edison and Gilliland ac-

complished this goal by using copper or other nonmagnetic conducting material to absorb the inductive force of these "signal-magnets." Pat. App. 422,577.

2. In 1883, Gilliland had established an experimental shop in Boston for American Bell to supplement the company's Electrical and Patent Department. Located at 101 Milk St., it was renamed the Mechanical Department in June 1884. Adams and Butler 1998, 51–52; Fagen 1975, 37–38.

3. Possibly Frank E. Donohoe, one of the witnesses for Gilliland on the patent application.

4. There is no further correspondence with Gilliland regarding Dyer's role, but he continued to act as Edison's patent attorney in relation to this and other telephone patent applications, including filing and guiding the applications through the U.S. Patent Office. See Pat. Apps. 329,030; 340,707; 340,708; 340,709; 422,578; 347,097; 348,114; 378,044; 422,577; 422,579; 438,306; and 478,743.

5. This text formed the basis for a formal reply of 6 November that Samuel Insull wrote and signed on Edison's behalf. Lbk. 19:353 (*TAED* LBo19353).

–2752–

From George Barker

Philadelphia Oct. 26. 1884

My dear Edison:—

I have today received a letter from Lord Rayleigh[1] saying that he should leave Boston tomorrow (Monday)[a] and reach Hotel Brunswick,[2] New York in the evening. He says he will be very glad to see what he can of your electric lighting system. I hope you will be able to show him some attention for in my opinion he is one of the highest type of men of science and as worthy of your effort as any man who has been here this summer.[3] As my classes will detain me here, I shall not be able to get over until Thursday afternoon. So you will arrange with him to suit his convenience. I hope you will take him to see the lamp factory certainly.

I notice in the Tribune of today a statement with reference of the business affairs of your companies. I need not say that as a friend of yours and a stockholder I am sorry at this collision. If you win, your capitalists are alienated; and if they win you are disaffected. But what I notice particularly is that on your ticket and on the ticket proposed as a compromise, the names of both Major Eaton and Mr. Lowrey are omitted.[4] Will you allow me to say how much I regret this. There never was a man who worked harder for the Edison interest or who threw his whole soul into the work of the Edison companies more entirely, than Major Eaton, as it seems to me. While therefore I grant that he is not a business man and that methods of doing business other than his, might have produced better results, I feel very strongly that he is entitled to very great consideration for what he has done and should at least be retained in the Board. His legal ability has been of the greatest service to the Light Company in making its contracts. And now when as I am informed, the Company is about to bring infringement suits against its rivals, and hence needs the very best legal counsel, both Major Eaton and Mr. Lowrey disappear from the Board of Direction. Further, I am quite surprised to see the name of Mr. Chinnock upon your list. Is he such a business man that he can replace such men as Major Eaton or Mr. Lowrey? I said when I sent you my proxy (which I hope you received though I have not heard) that I was glad you purposed putting new life into the concern.[5] But I am afraid you are going to the other extreme now. I do not quite see that those who are to come in are first class business men any more than those who go out. Excuse my freedom of speech and believe me[6]   Cordially yours

George F. Barker

ALS, NjWOE, DF (*TAED* D8403ZID). [a]Interlined above.

1. The experimental and mathematical physicist John William Strutt (1842–1919), the third Baron Rayleigh, had held the Cavendish chair of experimental physics at Cambridge since 1879. He resigned at the end of 1884, the year in which he was also president of the British Association. Rayleigh had researched in acoustics and optics prior to his tenure at Cambridge, but there he undertook a rigorous redetermination of the absolute electrical units. After leaving Cambridge, he maintained a varied research program at his home laboratory. He and William Ramsay isolated atmospheric argon in 1894, for which he won the Nobel Prize for physics in 1904. *Oxford DNB*, s.v. "Strutt, John William"; *DSB*, s.v. "Strutt, John William."

2. The luxurious Hotel Brunswick, at 225 Fifth Ave. between Twenty-sixth and Twenty-seventh Sts., opened in 1871. Its restaurant was designed and operated to rival Delmonico's, located across the street. The Brunswick's superlative reputation was seriously damaged in December 1882, however, when the restaurant unwittingly hosted a bacchanalia by Billy McGlory, the notorious promoter of an antithetical underworld establishment. The house never recovered from the publicity of that event. Its business dropped precipitously, it entered bankruptcy in February 1885, and it closed in July 1885. Williamson 1975, 271–72; Batterberry and Batterberry 1999, 156–58.

3. Rayleigh was among the members of the British Association to visit the United States, particularly Philadelphia, after their August meeting in Montreal. According to Barker, Rayleigh was "very much interested to see the Central Station of which Sir Wm. Thomson had said so much to him." In urging Edison to show him the Pearl St. station and the manufacturing shops, Barker held Rayleigh apart from his contemporaries:

You know of course that he is President of the British Association &, the successor of Maxwell in the University of Cambridge. Such men as he and [James] Dewar represent the highest science of Great Britain, and you can trust everything to their honor. Will you allow me to say that I was not a little surprised in your laboratory on Monday of last week to see the freedom with which you gave your views of the third pole [Edison Effect lamp] business to W[illiam]. H[enry]. P[reece]. after the experience of a few years ago. He is a good fellow as we all know. But dont tell him too much. Your excessive good nature leads you sometimes to do an injustice to yourself in your desire to do a service to your professed friends. [Barker to TAE, 20 Oct. 1884, DF (*TAED* D8403ZHV)]

Edison was still trying to make an appointment with Rayleigh on 29 October. The editors have not determined if they met before Rayleigh's planned departure for England on 1 November (TAE to Barker, 29 Oct. 1884, Lbk. 19:315 [*TAED* LB019315]).

4. According to the *New York Tribune* article that Barker saw ("Mr. Edison's Activities," *New York Tribune,* 26 Oct. 1884, 10), the current board issued a statement expressing the opinion "that it cannot recommend the stockholders to place the management of this company in the hands of parties whose interests may not always be the same as your own." It accordingly presented a slate of directors consisting of: Edison, Grosvenor Lowrey, Sherburne Eaton, Edward Adams, James Banker, Charles Coster, Robert Cutting, Jr., Eugene Crowell, William Lloyd Garrison, Jr., Edward Johnson, José de Navarro, Spencer Trask, and James Hood Wright. The board's concern echoed the argument made by Grosvenor Lowrey in Doc. 2528.

Edison nominated himself, J. Pierpont Morgan, Wright, Coster, Erastus Wiman, Cutting, Crowell, Banker, Garrison, Johnson, Francis Upton, and Charles Batchelor. In soliciting votes, he reportedly issued his own circular letter stating "that only three of these gentlemen (Messrs. Johnson, Upton and Batchelor) and myself have any interest whatever in the manufacturing companies, and this I consider sufficient refutation of the charges made against me" by the company.

As a compromise, Sherburne Eaton arranged to put forward the names of Charles Chinnock, Edison, Adams, Banker, Coster, Crowell, Garrison, Johnson, Upton, Batchelor, Trask, Wright, and Wiman. Eaton's slate was chosen. "Mr. Edison's Directors," *NYT,* 29 Oct. 1884, 8.

5. In the letter enclosing his proxy form (not found), Barker expressed his agreement with Edison that "the Edison Light has suffered for the want of a more vigorous business management; though of course I cannot say what reasons have prevented it so far. Business matters should in my judgment be managed by business men on a business basis. I trust you may succeed in securing a live element in the Board that will enliven things." Barker to TAE, 20 Oct. 1884, DF (*TAED* D8403ZHV).

6. Acknowledging Barker's concerns a few days later, Edison (or more likely Samuel Insull) replied "that the whole question was settled before the meeting took place & when I see you I shall be glad to discuss the matter with you." TAE to Barker, 29 Oct. 1884, Lbk. 19:315 (*TAED* LB019315).

*Samuel Insull to
Alfred Tate*

My Dear Tate:—[1]

Your various letters to me came duly to hand.[2] I have been so terribly busy that it has been absolutely impossible for me to answer you.

Things have been going so badly in general in our business, and the outlook has been so uncertain, that I have even had to deprive myself of the services of the charming and illustrious co-worker "Mrs. Gallagher," whose valuable aid I have called in this devotional Sunday morning.[3]

I will just post you as to the general run of things here. The renowned and illustrious S. B. Eaton retires on this Tuesday next. The enclosed circulars will explain the matter to you.[4] The controversy has been compromised by the Light Co. cutting out all their objectionable men and electing the men nominated on Mr. Edison's ticket, with the exception of Mr. Cutting, in whose place Mr. Spencer Trask[5] is going to be placed.[6] But just stop a moment, I am a little too previous, I should say the election takes place next Tuesday, but, as things look at present, I think we may consider it is all finished up. Thus you see I am at last getting even. I have often told you that I would, and that the great "Mogul" in the back room would have to clear out. Edison, Tomlinson, myself and W. S. Perry have been working like "Trojans" for the last two weeks getting proxies. We have upwards of 3000 proxies now in the safe, but it requires 3500 to control the Co., and Drexel, Morgan & Co. have, in the face of this support which we have obtained, come to the conclusion that the most graceful thing to do is to give Edison what he wants. Eaton disappears from the business entirely, and has given Edison his word that he will not accept the nomination as a Director in any Edison Co. at all in the future, unless such nomination comes from Thomas A. Edison, himself. You can imagine my feelings. There is no one more anxious after wealth than Samuel Insull, but there are times when revenge is sweeter than money, and I have got mine at last.

Johnson will now have full control of the business. The President of the Light Co. will be more a figurehead than anything else, and that President will be Dr. Eugene Crowell,[7] in all probability. I do not know that this change will benefit me in any manner, except that it will certainly put Mr. Edison in a much better humor, and enable him to go ahead on other things, and so I may look forward to making money in other directions. I do not think it is hardly probable that I shall make much money out of the electric light.

Johnson wants me to take charge, in my spare time, of a literary bureau which he is going to start in connection with the Isolated Co. That is he wants me to handle the newspapers, and see that the Edison interests have a fair shake in any electric light paragraphs that the press may give utterance to. Whether I shall take it up I do not know, but I fancy I shall, as $100. or so a month is not to be despised by a "busted Britisher."

I am expecting, with Tomlinson's aid, to form a Co. for the Stock Ticker.[8] We expect to get the Seligmens[9] into it. Of course this means another Secretaryship for Sammy, and another contribution to his fast diminishing "boodle."

When I tell you that I am compelled to desert the society of even such friends as we dined with a Sunday or so before you left here, you will understand how near busted I am, but I am perfectly happy, owing to the turn that electric light affairs seem likely to take.

I have not been to the Machine Works for a week, owing to the active part I have had to take in the Light Co.'s agitation. I understand, however, that the arc light affairs are getting along very well. I may mention, strictly entre nous, that, providing everything is settled in the Light Co. to Mr. Edison perfect satisfaction, that the arc light will, in all probability, be turned over to the Isolated Co. You must not, however, mention this to a soul, as it is a matter that even all our agents know anything about, and I simply send it to you, because I wish you to be first posted. My impression is that if this arrangement takes place, you will be better off for it, inasmuch as I think we shall then be able to make a deal by which you can control the incandescent interests in Canada. Anyway you may rest assured that if I see an opportunity to do you a turn in this direction I shall not forget it, and I think I can flatter myself that I have the ear of those who will have the deciding of such matters. Meantime, you will go ahead just as if the arc light business would remain in the same state as the present, and do not even mention this matter in writing to Batchelor, as it might cause me trouble. One thing you may be sure, that I will take care of your interests just the same as if you were right on the ground here reminding me of it. If arc light affairs take the turn that I expect, which for your interests I hope they will, I fancy you will be called back to New York for consultation.[10]

There are many things that I wish to see you on, and which

I can hardly write about, and if you should be requested to report at headquarters, please send your gripsack to 247 Fifth Avenue, and share my "bugwalk"[11] with me.

When I tell you that Tomlinson took part in a Democratic parade yesterday,[12] you will understand how thoroughly depraved he has become. I think that he has sunk away down beyond redemption.

Hoping to see you when you come to New York, and reminding you to be sure to put up at my hotel, believe, me, Ever Sincerely Yours,

TL (carbon copy), NjWOE, DF (*TAED* D8427ZBY).

1. This letter was addressed to Tate in Peterborough, Ont. Tate had been traveling through Ontario and Michigan, canvassing towns for isolated plants and central station lighting, and appointing sales agents for the Edison Machine Works. He sent several dozen reports on localities and individuals to Charles Batchelor between late September and early November, eventually using forms that he had printed for this purpose in early October. His reports were collected in a scrapbook (Canvassing Reports [1884], Misc. Scraps. [*TAED* SB015]). In July, Tate had received a written testament jointly from Edison and the Edison Machine Works of his competence to install isolated and central station plants. Soon after, he was making plans to go to Argentina (TAE and Machine Works statement, 19 July 1884; Tate to Insull, 3 and 6 Aug. 1884; all DF [*TAED* D8431ZAK, D8465ZAD, D8465ZAE]; see also Doc. 2711 n. 7).

2. Not found.

3. Not identified.

4. The circulars have not been found, but presumably Insull referred to the letters (discussed in Doc. 2752 n. 4) issued by Edison and the Edison Electric Light Co. board on behalf of their rival slates of candidates for the board of directors.

5. An investment banker, philanthropist, and member of the New York Stock Exchange, Spencer Trask (1844–1909) was not listed among the stockholders of the Edison Electric Light Co. in 1883, but he owned shares of the Edison Electric Illuminating Co. of New York and served as president of that company from 1884 to 1889 and again from 1891 to 1899. He (or one of his offices) had an Edison isolated lighting plant in Albany at this time (Edison Electric Light Co. list of stockholders, 1 July 1882; Sherburne Eaton to TAE, 13 June 1882; both DF [*TAED* D8224ZAP1, D8226W]; Jones 1940, 168). After graduating from Princeton in 1866, Trask briefly worked for his uncle, Henry G. Marquand, and formed a short series of partnerships. These arrangements evolved into a partnership with George Foster Peabody, which became Spencer Trask & Co. in 1881 (Worth 2008, 17; Advertisement, *Wall Street Daily News*, 2 May 1881, 2; ibid., 9 June 1881, 2; "Mr. Trask's Career," *New York Tribune*, 1 Jan. 1910, 7; *WWW*, s.v. "Trask, Spencer"; *NCAB* 11:444).

6. Doc. 2752 n. 4.

7. Eugene Crowell (1817–1894) earned a medical degree from the

University of the City of New York (later New York University) and practiced medicine before moving to California in 1849. There he made a fortune in the wholesale drug business and pursued political interests, including terms as president of the state's Know-Nothing Party (1854) and Supervisor for the City of San Francisco (1861). After Crowell returned to New York in 1868, he became noted as a spiritualist, and he published several books on the subject (Obituary, *Appletons' Annual Cyclopaedia and Register of Important Events* 19 (1895): 572, "Dr. Eugene Crowell Dying," *New York Herald,* 29 Oct. 1884, 9; Obituary, *Brooklyn Eagle,* 29 Oct. 1894, 1). He held stock in the Edison Electric Light Co., the Edison Electric Illuminating Co. of New York, and the Edison Co. for Isolated Lighting, and about this time was considering taking a financial interest in the Edison Lamp Co. Crowell served as president of the Edison Electric Light Co. from 1884 to 1886 and became president of the Edison Electric Light Illuminating Co. of Brooklyn in 1887 (list of stockholders, 1 July 1882, DF [*TAED* D8224ZAP1]; Crowell to Insull, 20 Sept. 1884 and 2 Oct. 1884, both DF [*TAED* D8429ZAY, D8429ZBC]; Insull to Crowell, 6 Oct. 1884, Lbk. 18:460A [*TAED* LB018460A]; Jones 1940, 252).

8. Nothing seems to have come of this venture in the near future, as the financing that Insull anticipated (see note 9) withdrew in the face of a subscription-rate cut by the well-established Gold and Stock Telegraph Co. in early 1885. Insull to John Tomlinson, 3 Mar. 1885, DF (*TAED* D8546I); Insull to TAE, 12 Mar. 1885, Lbk. 20:165 (*TAED* LB020165).

9. In February 1885, Edison granted an exclusive sixty-day option on the rights for an improved chemical telegraph to the preeminent international banking house of J. & W. Seligman & Co. He also promised to give the bankers one-half of the stock in any new company they might form to exploit the invention for "the quotation of securities and general financial news" (TAE agreement with J. & W. Seligman & Co., 20 Feb. 1885, DF [*TAED* D8546E]). Founded in 1862, J. & W. Seligman & Co. had New York headquarters in the Mills Building, and its Paris branch (Seligman Frères) had been among the original investors in Edison's electrical companies in France (*Ency. NYC,* s.v. "J. and W. Seligman"; *Finance and Industry* 1886, 78). Henry Seligman, son of one of the founding partners, had been a classmate with John Tomlinson at the University of the City of New York; he joined the New York house in 1880 after a tour of duty with the Anglo-Californian Bank in San Francisco (*General Alumni Catalogue* 1906, 48; Obituary, *NYT,* 23 Dec. 1933, 15).

10. See Doc. 2748 n. 2 regarding arc lighting arrangements. In his reports on canvassing towns for lighting plants (see note 1), Tate made a point of mentioning efforts by competing electrical companies to install arc light systems.

11. Bug-walk, or bug walk, was popular slang for a bed (c. 1850–1930), implying the infestation of a mattress (Partridge and Beale 2002, 148, 337). Insull's personal address was listed as 247 Fifth Ave., near Twenty-eighth St., in a city directory covering the year from May 1884 to May 1885; for the prior year, he was listed at 38 W. Eleventh St. (*Trow's* 1885, 848; ibid. 1884, 823).

12. A parade of businessmen in support of Grover Cleveland's presi-

dential candidacy had occurred along lower Broadway on 25 October. "Cleveland Men in Line," *NYT,* 26 Oct. 1884, 1.

–2754–

From Francis Upton

Dear Mr. Edison:

About three months since by special arrangement with Mr. Insull, The Edison Lamp Co. left $4232.77 ~~left~~ in your hands that you had collected as their agent, on the express agreement that you would meet a note for $2750.00 that would come to your office for payment. on Nov. 3.[1] Mr. Insull [~~shack? schook~~][b] shook[c] hands on the agreement. It was also understood that we would never call on you for the balance of the account or for money due on

Mfg Acct.	351.23
Con's[2] "	787.90
	1,039.13

Making a total of $5,371.90 left in your hands to meet a note of one half the amount.

Today ~~we~~I received a formal note saying that you would not fulfill your obligation to us.[3]

This compels us to pass a large portion of our payroll or all of it and to destroy the credit we are trying to build up at the bank by keeping a small deposit.

There is ~~also~~ in addition to the sums mentioned due us from the Machine Works

| 370.50 |
| 566.60 |
| $937.10 |

or a total of $6309—

We have always been ready to help you and intended to aid you when we left the balance mentioned in your hands, yet we feel that you should not throw us overboard without a word of help now. Yrs. Truly

Edison Lamp Co. By Francis R. Upton   Treas

ALS, NjWOE, Upton (*TAED* MU071). Letterhead of Edison Lamp Co.; monetary expressions standardized for clarity. [a]"EAST NEWARK, N.J.," and "188" preprinted. [b]Canceled. [c]Interlined above.

1. Upton enclosed the note, which was dated 1 November 1883 and payable to himself. At some later time, the maturity date was altered from "3" to "4" November 1884.

2. The Edison Construction Dept.

3. Samuel Insull, who managed Edison's finances, wrote in Edison's name that "I have been unable to make collections I had anticipated getting & hence my failure to keep my promise to you." TAE to Upton, 30 Oct. 1884, DF (*TAED* MU070).

–2755–

To Thomas Mendenhall[1]

[New York,] 6th Nov [188]4

Dear Sir,

Referring to your of 27th ult. I am much obliged to you for the enclosures about Japan.[2]

With relation to Carbon Buttons do you mean that you would like some with the surface of the Lamp Black plated. This is somewhat difficult as the buttons will absorb the solution by capillarity[3]   I will try it however if you say so   Yours truly,

Thos. A. Edison   I[Insull]

L (letterpress copy), NjWOE, Lbk. 19:368A (*TAED* LB019368A). Written by Samuel Insull.

1. Thomas Corwin Mendenhall (1841–1924), a physicist and mathematician, was teaching at the Ohio Agricultural and Mechanical College (later Ohio State University) in 1878 when he took up an offer to teach physics at Tokyo's Imperial University. After three years there, where he also conducted research in meteorology and seismology, Mendenhall returned to Ohio State in 1881 to teach physics and run the Ohio Meteorological Bureau. He left both Ohio positions in 1884 to join the U.S. Signal Corps as professor of electrical sciences and head of its Instrumentation Division. Mendenhall became the president of the Rose Polytechnic Institute in 1886 and then, after a stint as superintendent of the U.S. Coast and Geodetic Survey, of the Worcester Polytechnic Institute. Mendenhall had received a retainer for unspecified purposes on behalf of the Edison Electric Light Co. in 1882. *ANB*, s.v. "Mendenhall, Thomas Corwin"; George Barker to TAE, 30 May 1882, DF (*TAED* D8204ZCF).

2. Edison's marginalia on Mendenhall's 27 October letter formed the basis for this reply. Samuel Insull also acknowledged Mendenhall's correspondence on 6 November, explaining that Edison had been sick but would answer soon (Mendenhall to TAE, 27 Oct. 1884, DF [*TAED* D8472ZAN]; Insull to Mendenhall, 6 Nov. 1884, Lbk. 19:343 [*TAED* LB019343]). Mendenhall had enclosed a letter from Takuma Dan, a prominent Japanese businessman and official of the Industry Ministry who was posted at the Miike mine, a major coal mine opened in the late 1860s and nationalized in 1872 (*PMJHF*, s.v. "Dan, Takuma" [accessed 13 Apr. 2010]; Hoshino and Iijima 1992, 133). Dan provided details about the mine's operation and his wish to light it electrically. In May, Dan had used Mendenhall as an intermediary to request an estimate for an Edison lighting plant, having already obtained figures from the

Brush electric light interests. Edison prepared a proposal (not found) in July, which he asked Mendenhall to forward to Dan with his own personal endorsement of the Edison system (Dan to Mendenhall, 21 Aug. 1884, DF [*TAED* D8420ZAC]; TAE to Mendehall, 21 July 1884, Lbk. 18:142 [*TAED* LB018142]).

3. Mendenhall had corresponded with Edison since at least 1882 about the electrical properties of the carbon buttons used in telephone transmitters, carbon microphones, and Edison's tasimeter (*TAED*, s.v. "Mendenhall, Thomas"). The variability of lampblack carbon's resistance under changing pressure had been widely recognized but poorly understood; the question of whether the cause should be understood as a change in the carbon itself or merely better contact between carbon particles and the electrodes had stoked the 1878 transatlantic controversy over credit for inventing the microphone. Edison had generally argued the former case, with his British antagonists taking the latter position (see Doc. 1367). After experimenting with Edison's tasimeter in 1882, Mendenhall had put forward his own cautious belief that improved physical contact could not entirely account for the phenomenon (Mendenhall 1882). British electrical engineer Sylvanus Thompson, among others, attacked this conclusion, pointing out that "with the exception of Professor Mendenhall, all who have investigated the point are agreed in their verdict" on the importance of physical contact (Thompson 1882, 434). Mendenhall persisted, however. He presented to the American Association for the Advancement of Science, convened at Philadelphia in September 1884, results of tests on hard carbon that seemed "to point conclusively in favor of the theory that the resistance of the carbon itself is altered by pressure. The experiments made by him on soft carbon are open to criticism, though they also point to the change taking place in the carbon" ("Proceedings of the Section of Physics," *Nature* 4 [19 Sept. 1884]: 296).

Mendenhall briefly described in his 27 October letter (see note 2) new experiments, which he hoped would be definitive. He asked if Edison could provide carbon buttons lightly electroplated with copper on the top and bottom surfaces that would guarantee good electrical contact. Although Insull promised that Edison had assigned someone to the task, Mendenhall had not received any by February 1885, when he inquired again. Vacationing in Florida at the time, Edison instructed John Ott to try to "plate Copper over the Ends of the soft Carbon buttons for Mendenhall." These efforts were apparently unsuccessful, and Edison instructed Ott, at Mendenhall's request, to send a dozen plain carbon buttons instead (Insull to Mendenhall, 14 Nov. 1884 and 2 Mar. 1885, Lbk. 19:388, 20:142A [*TAED* LB019388, LB020142A]; Mendenhall to TAE, 14 Feb. and 7 Apr. 1885; TAE to Ott, 9 Apr. 1885; all DF [*TAED* D8518F, D8503ZAD, D8503ZAF]; TAE to Ott, 6 Mar. 1885, MiPhM [*TAED* X059AA]). Mendenhall was able to devise a new experiment for comparing the effects of pressure on hard carbon with those on soft carbon, such as lampblack. In 1886, he concluded that while some of the observed change in resistance could "be attributed to change in surface contact between the carbon and the electrodes through which the current is introduced, . . . by far the larger part . . . is due to a real change in the resistance of the carbon itself" (Mendenhall 1886, 228).

*To Alexander
Outerbridge*[1]

My Dear Sir,

Referring to yours of "no date"[2] I have never found time to go much into the aesthetic part of my work & have therefore practically nothing of interest with relation to the matter you speak of, but the tripolar incandescent lamp has, I am told a very important bearing on some laws now being formulated by the "bulged-headed fraternity" of the "Savanic world"!![3]

If you want to have a little amusement I will send you half a dozen of the tripolar lamps   Yours truly

Thomas A. Edison

ALS (letterpress copy), NjWOE, Lbk. 19:351 (*TAED* LB019351). Written by Samuel Insull.

1. Alexander Ewing Outerbridge (1850–1928), trained in chemistry and mathematics, had been an assistant to Henry Morton at the Franklin Institute in 1867. He then spent ten years from 1868 as an assistant in the assay department of the U.S. Mint in Philadelphia, where he published several important discoveries. After a brief stint at the Mint in New Orleans, Outerbridge returned to Philadelphia in 1880 and soon joined William Sellers & Co., a car-wheel foundry, as its metallurgist. He continued an active research program, publishing several prize-winning papers, and was an active member of the Franklin Institute. Outerbridge was introduced to Edison in December 1879 and, after a visit to Menlo Park, made a report to the Franklin Institute on the electric light. *DAB*, s.v. "Outerbridge, Alexander Ewing"; Isaac Norris to TAE, 31 Dec. 1879, DF (*TAED* D7903ZMO); "The Edison Electric Light," *Journal of the Franklin Institute* 109 (Mar. 1880): 145 ff., reprinted in *Edison Electric Light Co. v. United States Electric Lighting Co.*, Complainant's Rebuttal—Exhibits [Vol. VI]: 4212–26, Lit. (*TAED* QD012G4212).

2. Outerbridge had written an undated letter regarding a report he was preparing for the Board of Examiners of the recent International Exposition at Philadelphia. Noting that it would include William Crookes's "Radiant Matter tubes," Outerbridge thought "it would be well to include an account of the interesting tri-polar incandescent lamp which you exhibited and which I unfortunately omitted to include in my lecture on 'Radiant Matter.'" He requested information on Edison's experiments with the bulb, an illustration of it, and the number of amperes obtained through the third electrode. Edison's marginalia on that letter was the basis for his reply. Outerbridge to TAE, n.d. (Nov. 1884), with TAE marginalia, DF (*TAED* D8403ZIJ1); Outerbridge 1881.

3. Edison may have had in mind an informal paper by Edwin Houston and subsequent discussion at the first meeting of the American Institute of Electrical Engineers, held in Philadelphia in October. Houston spoke on "Phenomena in Incandescent Lamps," particularly the mysterious electric current produced in the Edison tripolar ("Edison Effect") lamp. Houston very tentatively suggested that the effect could be caused by a "Crookes' discharge from one of the poles [that] might produce an electrical bombardment against the plate, each molecule

taking a small charge that might produce the effect of a current." William Preece, who evidently had showed considerable excitement over the effect when Edison showed it to him (see Doc. 2752 n. 3), was in attendance. After returning to London, Preece made his own experiments with the lamp. He described his research in an 1885 paper titled "On a Peculiar Behavior of Glow-Lamps," in which he coined the phrase "Edison Effect." Houston 1884, 2–3; Preece 1885, 229–30; Hong 2001, 122–27.

–2757–

Samuel Insull to
Edward Adams

[New York,] 11th Nov. [188]4

Dear Sir,

I have your letter of this date[1] relating to the use of Mr Edisons name, by Colonel Gouraud, as a Director of the Oriental Telephone Co.[2]

Upwards of a year ago Mr Edison requested Mr Gouraud to withdraw his name from the Boards of all English Companies for the Directorates of which Mr Gouraud had nominated him. Mr Edison's reason for this was that he did not wish to have his name used as the Director of Companies, with the management of which [---][a] he had not the slightest connection.[3]

Mr Gouraud is therefore entirely without any authority for the use of Mr Edisons name in the connection spoken of.

As to Mr Hubbard[4] I cannot speak as I have had no communication with him on the subject. Yours truly

Saml Insull   Private Secy

ALS (letterpress copy), NjWOE, Lbk. 19:372 (*TAED* LB019372).
[a]Canceled.

1. Adams reported having received "a further letter from London" to the effect that "Col. Gouraud uses the names of Messrs. Edison and Hubbard among his Directors" for the Oriental Telephone Co., "notwithstanding the report that Mr. Edison has warned him not to do so." He inquired, "What is the situation in this respect?" Edison was listed as a director in a circular letter prepared by a discontented shareholder of the Oriental Co. Adams to Insull, 11 Nov. 1884; Thomas Lloyd to Oriental Telephone Co. shareholders, undated Nov. 1884; both DF (*TAED* D8472ZAS, D8472ZBQ).

2. George Edward Gouraud (1842–1912), an American business agent in London, had an association, not always harmonious, with Edison going back to the early 1870s (Doc. 159 n. 7), including varied roles in the commercialization of automatic telegraphy, the phonograph, telephone, and electric light. As chairman and president of the Oriental Telephone Co., Ltd., since May 1884, Gouraud was preparing for a proxy fight to install a new board of directors at this time ("Notes," *Teleg. J. and Elec. Rev.* 14 [24 May 1884]: 442; "City Notes, Reports, Meetings, &c.," ibid.

14 [24 May 1884]: 380; "Miscellaneous Companies," *Investor's Monthly Manual* 14 [31 Dec. 1884]: 665; Gouraud to TAE, 4 Aug. 1884, DF [*TAED* D8472S]). The company was incorporated on 4 February 1881 as the result of agreements among Edison, Alexander Graham Bell, the Oriental Bell Telephone Co. of New York, the Anglo-Indian Telephone Co., Ltd., and others. It was authorized to sell telephones in India, Ceylon, Java, Japan, China, South Africa, Australia, New Zealand, Egypt, Turkey, Greece, Malta, and the Hawaiian Islands (Doc. 2056 n. 7). Edison was identified as one of its directors at the end of 1882, but he did not appear in subsequent published lists (Oriental Telephone Co., Ltd., report to directors, 18 May 1883, DF [*TAED* D8374U]; Skinner 1882, 359; Berly 1883, 320). He officially retired as a director at the conclusion of a three-year term at the end of 1884 ("Company Meetings and Reports," *Electrician* 14 [25 Apr. 1885]: 504).

3. Insull referred to a letter Edison had written to Gouraud in March 1882, after several instances in which Edison's name appeared on lists of directors (see Docs. 2258 n. 7, 2187, and 2229).

4. Gardiner Greene Hubbard (1822–1897), father-in-law of Alexander Graham Bell and a key figure in the development of Bell's telephone and control of his patents, had helped to organize the Oriental Co. Like Edison, he retired from the board in late 1884 at the conclusion of his three-year term. Hubbard had moved in 1879 to Washington, D.C., where he devoted his time to various civic projects. See Doc. 1945; *ANB*, s.v. "Hubbard, Gardiner Greene"; "Company Meetings and Reports," *Electrician* 14 (25 Apr. 1885): 504.

–2758–

To William Brewster

[New York, c. November 15, 1884[1]]

Brewster,

Something like this I think will answer

A syndicate to be formed to [control?][a] electric lighting business in [various?][a] parts of the world with a <u>callable</u> capital of say $100,000. Say they trade under the name of Brewster & Co[2]

One member of the Syndicate to be chosen to decide if the Syndicate shall advance on bill lading for each specific order or refuse to do so as the form of payment is satisfactory or unsatisfactory. If satisfactory members are called upon to advance their proportion. In this manner the money is called [--][b] up as needed.

If unsatisfactory and they refuse to advance on bill of lading Edison has right to go to other parties on this specific order—

Edison & shops to furnish all appliances at same price as is charged to Edison Electric Light Co in US.

Of profits derived Edison is to receive 25 pct. the Syndicate 75 pct from which they pay Mr Brewster, ~~profits~~. Arrangement to continue 3 years.[3]

AL (letterpress copy), MiDbEI(H), EP&RI (*TAED* Xoo1A1CD). [a]Illegible. [b]Canceled.

1. This document evidently predates November 18, when Edison personally explained to the writer and traveler William Henry Bishop that his "friend Mr Brewster desires to present a scheme to you for working the Electric Light in various foreign countries. Please hear his plain and simple refrain." TAE to Bishop, 18 Nov. 1884, EP&RI (*TAED* Xoo1A1CB).

2. Sometime in November (perhaps the 21st), Edison told Charles Batchelor that he had "arranged with Brewster to work up the outside countries Corea Japan etc as he can find outside Capital. please give him the correspondence & estimates & other data for Maritana [presumably the Maritima region of Peru, from the Andes to the coast], Japan, Salvador, Guttemala etc." Edison also told Samuel Insull that he had "decided to put the outside country biz into the hands of Brewster to work please have Johnny [Randolph] hunt up all letters regarding it, Shanghai Cotton mill, Honlulu etc." TAE to Batchelor, [21?] Nov. 1884, EP&RI (*TAED* Xoo1A1CC); TAE to Insull, 26 Nov. 1884, DF (*TAED* D8427ZCC); *WGD*, s.v. "Maritima, Cordillera."

3. Years later, Brewster remembered having Edison's handwritten contract for this matter. It is not clear if the present document was the one he recalled; the editors have not found a separate agreement. In January 1885, Brewster made a proposal to the Edison Co. for Isolated Lighting for conducting business in South America, but the company declined his terms. "Synopsis of Important Events," in "Brewster, William," Pioneers Bio.; Edward Johnson to Brewster, 6 Feb. 1885; Brewster to TAE, n.d. [6 Feb. 1885]; both DF (*TAED* D8522B, D8522C).

–2759–

Memorandum to Richard Dyer[1]

[New York,] Nov 20 1884

Dyer—
    Patent this—[2]

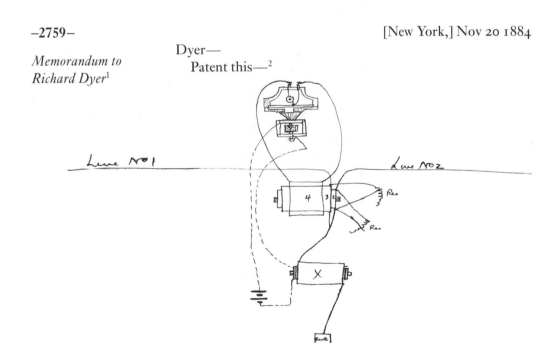

2 & 3 are coils but instead of being wound one over the other the two wires are wound together the current passing through differentially so as to neutralize each other on the No 4 that works the receiver:

X is the Regular coil. When a wave is sent from the action of its primary the same passes through 3 & 2 & neutralizing Each other relatively to 4   the receiver gets none of the current it sends by its action on the trans= The trouble with the old device was the reaction which made[a] a squeak like a piccolo— this stops it.

The Res. are for balancing when one line is very long and the other short= but they will probably be unnecessary in most Cases & switch will be over= This <u>reaction</u> squeak is a fundamental thing & getting rid of it is a <u>big thing   it wks nicely</u>—[3] Edison

ADS, NjWOE, Lab., Unbound Notes and Drawings (1884) (*TAED* NS84ACO5). [a]Obscured overwritten text.

1. See Doc. 2743 (headnote).

2. Figure labels are "4," "3," "2," "Res," "Res," "X," and "Earth." This is one of three patent applications for telephone repeaters that Edison executed on 9 December and filed on 15 December. This one issued as U.S. Patent 340,707 on 27 April 1886; the others issued on 4 March 1890 as U.S. Patents 422,578 and 422,579. For Edison's work on repeaters related to these applications, see drawings dated 5, 17, 19, 21–24, and 29–30 Nov. 1884, Unbound Notes and Drawings (1884), all Lab. (*TAED* NS84ACN6, NS84ACO2, NS84ACO4, NS84ACO7, NS84ACO8, NS84ACO9, NS84ACP1, NS84ACP2, NS84ACP3).

3. In designing telephone repeaters, Edison, like other inventors, relied on the ability of a carbon transmitter to amplify an acoustic signal moving a diaphragm (Doc. 1277 n. 4). The arrangement shown here marked a significant improvement in repeater circuitry. It enabled two-way transmission with a single amplifying device "without the use of switches for changing the line and local circuits in the relay apparatus, and will at the same time produce clear articulation free from the confusion produced by the reciprocal action of the repeating-instruments." In addition, Edison could use "a single combined receiver and transmitter . . . to relay in either direction without the use of switching apparatus, and also to provide means whereby my devices for relaying in either direction without switches can be used effectively with connected lines of widely-different resistances." With this circuit arrangement, Edison was able to prevent feedback in which the amplified output current returned to input, thus causing a "squeak" or "singing" sound. U.S. Pat. 340,707; Fagen 1975, 253–54, 264–65; Lockwood 1896, 596–97, 626–27.

Technical Note:
Electric Locomotion

NEW YORK, Nov 24 1884[a]

Electro Locomobile[1]

TAE

X, NjWOE, Lab., Unbound Notes and Drawings (1884) (*TAED* NS84ACP). Letterhead of Thomas A. Edison. [a]"NEW YORK," pre-printed.

1. Nothing further is known regarding Edison's idea for an electric vehicle that did not run on rails. Presumably the motor, which encompassed the entire span of the carriage, was battery powered.

–2761–

To Samuel Mullen[1]

[New York,] 25th Nov [188]4

Dear Sir,

Referring to your favor of 15th inst in reply to "application No 2953 I beg to state that I have not, so far as I can remember, requested space at your exposition[2]

I have within the last few years spent from $65,000 to $75,000 in exhibiting my inventions at various[a] Expositions and in as much as I gain no pecuniary advantage by so doing I have been compelled to refuse to make [Exhibits?][b] unless [the?][b] expenses are defrayed by Exposition Authorities.

If your management desire an exhibit from me I shall be glad to loan my collection of Instruments provided the expenses of shipment & return & the erection of the exhibit and the expenses of one of my assistants isare guaranteed me.

Of course you understand that my exhibit is a Scientific one—simply showing [each?][c] collection of my inventions & has no financial benefit to me   Yours truly

Thos. A Edison   I[nsull]

L (letterpress copy), NjWOE, Lbk. 19:412 (*TAED* LB019412). Written by Samuel Insull. [a]Obscured overwritten text. [b]Copied off bottom of page. [c]Illegible.

1. Samuel Mullen (b. 1844?) was the secretary and chief of installation for the World's Industrial and Cotton Centennial Exposition, to be held in New Orleans. Apparently a native of Indiana, Mullen was identified in the 1880 federal census as a store clerk; two years later, he held an unspecified position representing a New Orleans association of rice growers. U.S. Census Bureau 1970 (1880), roll T9_458, p. 123.1000, image 0248 (New Orleans, Orleans, La.); Mullen statement to Tariff Commission [n.d.] printed in U.S. Tariff Commission 1882, 2:2537.

2. Mullen had completed and signed a printed "permit for space" at the World's Industrial and Cotton Centennial Exposition. The form, dated 15 November, allocated a $24' \times 24'$ space in the main building (New Orleans Cotton Exhibition to TAE, 15 Nov. 1884, DF [*TAED* D8464ZAK]). Although Edison declined the space, the Edison Co. for Isolated Lighting was one of four electric companies that won bids to provide electric lighting. (The others were the Louisiana Electric Co., the Fort Wayne Jenney Co., and the Brush Electric Light Co.) The Edison Co. provided 6,000 lamps to light various structures, including 1,786 16-candlepower lamps for the Main Building and 885 of the same type lamps for the Government Building. Edison incandescent lamps were also used in the Administration Building, the Art Gallery, and the Music Hall. The Isolated Co.'s plant at the exhibition, which was said to be the largest in the world at the time, included six automatic steam engines and twelve dynamos. According to the company brochure, the exhibition allowed for a favorable comparison of Edison's incandescent light with other companies' arc lighting for large indoor spaces. The exposition ran from 16 December 1884 until 2 June 1885; Edison visited in late February (Barrett 1894, 3; Edison Co. for Isolated Lighting Bulletin, 3:6–7, 6 June 1885, CR [*TAED* CC003]; Perkins 1885, 63; Edison Co. for Isolated Lighting to New Orleans Cotton Exhibition, 2 Mar. 1885, UHP [*TAED* X154A4CG]; Fairall 1885, 405; Kendall 1922, 1:462; Watson 1984; Israel 1998, 237).

–2762–

To Steele MacKaye

NEW YORK, [November, 1884][1a]

Friend Mackaye[2]

Impossible for me to come to the dinner tonight   Saturday I do the conservative citizen act by taking my family to alleged theatres.[3] Is not this bearing the cross. I have six calcium lamps of 500 C Power each.[4] They were not made properly to get a good focus. I went over to the lamp factory Thursday and explained exactly what is wanted. Will try those I have monday   Yours

Edison.

ALS, NhD, MacKaye (*TAED* X009AA). Letterhead of Thomas A. Edison. [a]"NEW YORK," preprinted.

1. At least two meetings between MacKaye and Edison were scheduled in November 1884. One was planned for Tuesday evening, 13 No-

vember at 18 West Twenty-third St.; the other for the afternoon of Friday, 28 November (the day after Thanksgiving), with MacKaye, *New York Times* theater critic William Winter, and additional friends. The party expected to visit Edison's New York laboratory. The latter engagement may have conflicted with a 4 p.m. meeting of the directors of the Edison Electric Light Co. of Europe at the laboratory. Charles Hughes to TAE, 12 Nov. 1884, DF (*TAED* D8403ZIL); MacKaye 1927, 1:428–29; see Doc. 2667 esp. n. 10.

2. James Morrison Steele MacKaye (1842–1894), the son of prominent New York business and civic leader Col. James Morrison McKay, was presently planning the new Lyceum Theatre in New York City. A versatile theatrical influence, MacKaye had studied in Paris and become a champion of François Delsarte's acting system, which he imported to the United States. His career as an actor, lecturer, playwright, and acting teacher got underway during the 1870s. By the end of that decade, he had also patented an innovative double stage (U.S. Pat. 222,143), which was installed at New York's Madison Square Theatre under his management. This stage improved the mechanics of scene changes and permitted overhead, indirect stage lighting. Another MacKaye invention, a folding chair (U.S. Pat. 300,617), enhanced theater fire safety by increasing aisle space, and MacKaye had it installed at the Lyceum. *ANB*, s.v. "MacKaye, Steele"; "Movable Theater Stages," *Sci. Am.* 50 (5 Apr. 1884): 207; Brown 1903, 2:415–17; MacKaye 1882; "Folding Chairs" (advertisement) *American Architect and Building News*, 18 (7 Nov. 1885): 1; MacKaye 1911.

Edison and MacKaye were clearly well acquainted at this time and would later consort together on a number of occasions, but it is not clear how they met. Years later, Edison recalled that MacKaye "first came into 65 5th Ave office to ask of the possibilities of Electric Lighting." TAE diary, 12 July 1885, Cat. 117, Diary (*TAED* MA001); TAE questionaire responses, n.d. [Sept. 1925], MacKaye (*TAED* X009AF); MacKaye 1927, 1:483.

3. The editors have not identified Edison's plans for amusement.

4. See Docs. 2764 and 2768.

Notebook Entry:
Telephony[1]

Selective Signalling Telephone[2]
[A]

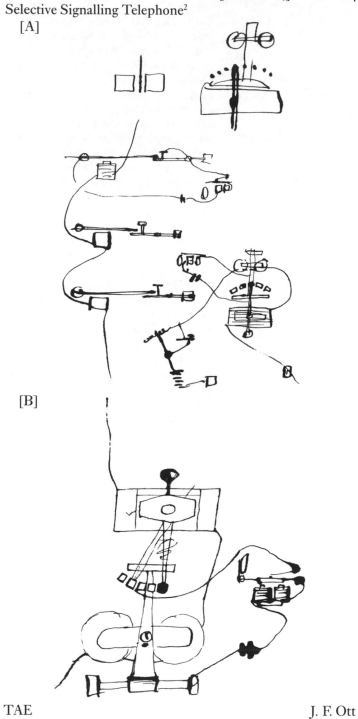

[B]

TAE   J. F. Ott

X, NjWOE, Lab., Unbound Notes and Drawings (1884) (*TAED* NS84ACQ). Letterhead of Thomas A. Edison.

1. See Doc. 2743 (headnote).

2. Telephone systems with multiple subscribers required some means of signaling a user for an incoming call. In some systems, all parties on the line heard an audible signal, usually a bell, in the form of a different code for each subscriber. A preferred method was selective signaling, in which only the party being called heard the bell. There were two primary methods used to achieve selective signaling. The first employed currents of different polarity to actuate the local signals, while the other method used instruments that responded only to specific frequencies. The first evidence of Edison's work on the problem of selective signaling is a patent application that he signed on 24 September. The application described the use of vibrating pendulums or reeds to make and break a circuit at different rates of vibration or current pulsations, a variation of using different frequencies. Edison filed the application on 15 December 1884; after substantial revisions in 1886, it issued in October 1890 as U.S. Patent 438,304. Fagen 1975, 121–22; Lockwood 1892; Miller 1905, 423–24; Pat. App. 438,304.

This document has two of the few drawings related to Edison's research on this subject during the fall of 1884; others date from mid-October and early December (N-82-05-15:210–15, Unbound Notes and Drawings [1884], both Lab. [*TAED* N203211, N203214, NS84ACR]). Figure A appears to be related to Edison's U.S. Pat. 347,097, one of two applications on selective signaling that he executed on 19 December 1884. Edison used a local circuit controller to operate an audible or visual signal indicator for each telephone. Each circuit controller responded to a different amount of current from the signaling instrument. These controllers, two of which are shown on the right side of Figure A, consisted of "a galvanometer whose needle plays in front of a series of contacts and whose coils are in the line." Each circuit controller would have either as many contacts as there were instruments on the line or a single contact placed so as to respond only to a particular current. The signaling station used apparatus such as an adjustable resistance to control the strength of current sent over the line, and the signal device only activated when the local circuit was closed by the galvanometer needle touching the proper contact.

In the other application that Edison executed on 19 December (issued as U.S. Patent 340,708), he replaced the galvanometer with an electromechanical device that used a polarized relay (like that shown in Figure B) to move a contact until it touched the correct stop, closed the circuit, and activated the signal device.

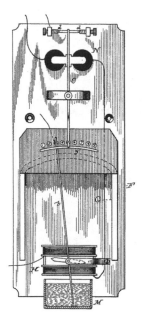

Selective-signaling circuit controller device, with galvanometer, from Edison's U.S. Patent 347,097.

Selective-signaling device using polarized relay, from Edison's U.S. Patent 340,708.

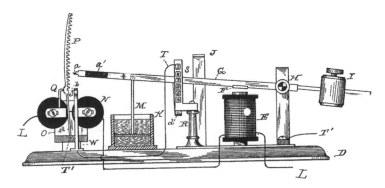

Technical Note:
Electric Lighting

[A][1]

[B][2]

[C]

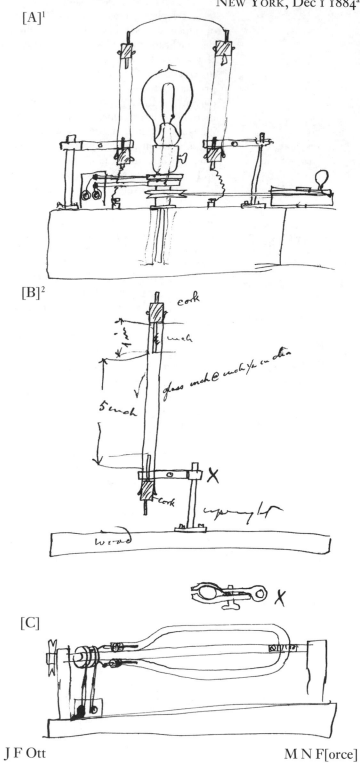

J F Ott   M N F[orce]

X, NjWOE, Lab. (*TAED* NS84ACQ1). Letterhead of Thomas A. Edison; document multiply signed and dated. ᵃ"NEW YORK," and "188" preprinted.

1. The editors have found scant context from which to infer the operation or purpose of this enigmatic lighting device. Edison likely intended it as a practical rather than an experimental piece of equipment, one capable of creating unusual lighting effects suitable for theatrical or other forms of spectacular lighting. The design could have been prompted by his conversations with theatrical inventor Steele MacKaye (see Docs. 2762 and 2768) or perhaps by his visits to the International Electrical Exhibition.

The functional components seem to include an ordinary incandescent lamp on a vertical shaft, a belt drive leading from a horizontal hand crank at the far right, and two tubes forming an electrical circuit independent of the carbon lamp. The tubes are stoppered with corks (shown in Figure B), suggesting that they were not meant to be evacuated, their length being too great for a spark in any case (see also Figure C). The corks could confine a conductive substance, such as a gas, that would become luminous when electrified. Some sort of circular switch appears to lie at the base of the bulb on the shaft. Rotating the shaft would move the contact points to turn the bulb on or off; or, if the switch included variable resistance contacts, to various dim states in between. Such switches were a standard element of Edison's repertoire, but it remains unclear why he would want the bulb itself to rotate, too. One speculative suggestion is that he wished to take advantage of the nonspherical distribution of light characteristic of the carbon lamp. Because each leg of the filament effectively shadowed the other, the lamp radiated less light in the plane of the horseshoe than elsewhere (see Doc. 1927). Considered a disadvantage for ordinary lighting, this phenomenon could have produced an unusual effect in conjunction with rotation and a variable switch.

On 13 November, Edison made a drawing that, although even less clear than those in this document, could have some relation to this device. The purpose of the glass enclosure is not apparent, but perhaps Edison thought to fill it with a conductive gas or other substance. Alternatively, the device could be related to experiments he was doing around this time on the problem of carbon carrying. Unbound Notes and Drawings (1884), Lab. (*TAED* NS84ACO).

Two incandescent bulbs (their electrical connections not shown) are contained in a glass vessel cradled between two sets of rollers and rotated by friction wheels.

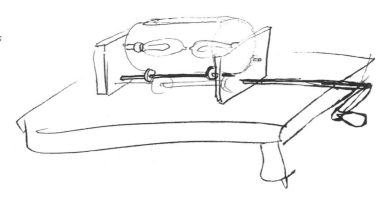

2. Figure labels are "cork," "1 in," "inch," "glass inch @ inch ½ in dia," "5 inch," "X," "cork," "upright," "wind," and "X."

-2765-

Ezra Gilliland to Theodore Vail

Boston, Dec. 9th 1884 84[a]

Dear Sir:

The enclosed bills from Bergman and Co. are for work done on models used by Mr. Edison in his experiments for our company, during the months of October and November, under the arrangement which we have entered into with him.[1]

I have carefully looked over the experiments, and tested the apparatus, and am pleased to report that he is making great progress in the direction of improving our long-distance transmitter, and has also produced a very satisfactory telephone repeater. He has also produced an individual bell, working upon an entirely new principle, and in a laboratory[b] test it works admirably, being simple in its construction, and working quickly.[2]

Any of ten subscribers placed upon one circuit can be selected and rung up, within five seconds. The apparatus of course, requires a practical test upon a working line and I recommend that he be instructed to make enough instruments to equip a line, so that it may be thoroughly tested.[3] If you desire, I will give you a complete description of the apparatus and explain the principle upon which it works. By the terms of the contract between our Company and Mr. Edison, he is limited in the amount which he is to expend in models and experiments. No definite plan has been made as to how this work shall be carried on. I think that it should come under the supervision of this Department, and would like instructions upon this point. Yours respectfully,

E. T. Gilliland   Supt Mech Dept

ALS, NjWAT, Box 1221 (*TAED* X012F2E). Letterhead of American Bell Telephone Co.; "Enclosed bill" written on preprinted subject line. [a]"Boston," and "188" preprinted; "84" interlined above. [b]Repeated at bottom of one page and top of the next.

1. See Doc. 2745 regarding Edison's arrangement with the company. The bills have not been found. In April 1885, Gilliland indicated that payment of Bergmann's bills had been approved, but it is not clear if he referred to the same bills mentioned in this document. Gilliland also indicated Vail's wish that all experimental expenses in New York be stopped until a final arrangement was reached with Edison. Vail to TAE, 28 Apr. 1885, DF (*TAED* D8547ZAA).

2. See Doc. 2743 (headnote) and related documents.

3. In mid-January 1885, Edison requested and received permission to conduct experiments on five miles of Western Electric cable stored in Jersey City. The editors have not determined whether those experiments were related to the tests proposed by Gilliland. Gilliland to Vail, 13 Jan. 1885; William Ross to Gilliland, 19 Jan. 1885; both Box 1221, NjWAT (*TAED* X012F2F, X012F2G); Gilliland to TAE, 21 Jan. 1885, DF (*TAED* D8547F).

–2766–

*Memorandum to
Richard Dyer*[1]

NEW YORK Dec 11 1884[a]

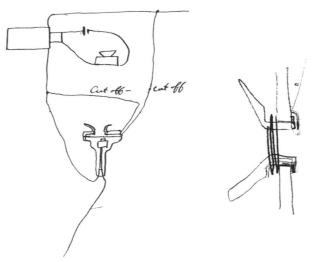

Dick=

New app[licatio]n= You know that in transmitting its necessary to have the sSecondary ie[b] the induction coil in circuit but not essential in Receiving   The induction coil by its Extra Current knocks the talking down nearly $\frac{1}{2}$ now if it could be Cut out when receiving it would be good thing= I arrange the mouth or rather ear piece of the receiver Loose & have an extra movement given it when the party presses it harder on the ear the ear piece will go inward & close a circuit[2]

AD, NjWOE, Lab., Cat.1150 (*TAED* NM019AAP). Letterhead of Thomas A. Edison. [a]"NEW YORK" and "188" preprinted. [b]Circled.

1. See Doc. 2743 (headnote).
2. This memorandum was the basis for a patent application that Edison and Ezra Gilliland executed on 2 January 1885. It was not filed until 14 October 1885 and issued as U.S. Patent 340,709 on 27 April 1886. In the standard telephone set, the receiver was in circuit with the secondary coil of the transmitter's induction coil, and induced currents in the coil sometimes caused low volume in the receiver. Edison and Gilliland sought to overcome this problem by placing a shunt around the induction coil that would cut it out of the receiving circuit when a user's

ear pressed against the earpiece. The user would need to remove this pressure when transmitting so that the induction coil could be brought back into the circuit. Edison and Gilliland had executed another patent application in October (filed in December) that covered a method for preventing electromagnets used with signaling apparatus such as bells from interfering with the receiver. It issued in March 1890 as U.S. Patent 422,577.

–2767–

Memorandum to Richard Dyer

Boston, [c. December 20–24, 1884][1a]

Dyer—=

Gilliland has got out best receiver I ever saw but the Cos ~~man~~ patent man says nothing patentable   now I think there is

So please draw up a brief Spcfn & claims & send in to gilliland—[2] Theres no money in it for you directly but it will help you

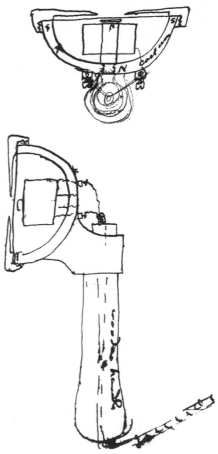

X cast iron hardened & magnetized—or cast steel—or dropped steel=[3]

Claim wooden handle hollow for cord go through— also comb[inatio]n claims

Gill thinks "Ader" had something like this but dont think it was a complete cup[4]

I thought of claim—

In telephone Recvr a [coil?][b] cup of iron or steel permanent the edges of which form one permanently magnetized pole upon which rests the diaphragm rests   The bottom of the cup forming the other pole & to which the Soft iron core surrounded by the helix is secured & presses upwards so its pole is in close prox to the centre of the diaphragm

AD, NjWOE, Lab. (*TAED* NM020AAB). Letterhead of American Bell Telephone Co. [a]"Boston," preprinted. [b]Canceled.

1. Edison probably wrote this document during his visit to Boston the week before Christmas. According to Ezra Gilliland's subsequent testimony, "Edisons work for the American Bell Telephone Company brought him to Boston frequently. While in Boston he stopped with me at my residence. He visited Boston the week before Christmas 1884, remained with us three or four days, leaving there the day before Christmas to return to New York." Edison also attended the Boston Electrical Exhibition on 23 December. While the front of this document was subsequently dated "Jan 85," probably by someone in Richard Dyer's office, the back carries the date "Dec 84" under a list of three drawings, presumably for the patent application, that were intended to show side, vertical, and horizontal elevations. Gilliland's Testimony on Behalf of Edison, pp. 5–7, *Edison and Gilliland v. Phelps* (*TAED* W100DKB [images 5–7]); "Electrical Exhibition," *Boston Journal*, 24 Dec. 1884.

2. This request was probably the basis for Gilliland's U.S. Patent 343,449, executed 25 August 1885 and filed on 14 September 1885. The application was not prepared by Richard Dyer or his partner.

3. Figure labels are "S," "S," "N," and "cast iron."

4. Clement-Agnes Ader (1841–1925) is known more for his steam-driven airplane experiments than for his work on the telephone. He installed the first telephone network in Paris in 1879, and the following year he founded the Société Générale des Telephones, which operated telephone networks throughout France. Many of these networks employed his telephone receiver, which became standard in France by the early 1890s and was still used there and elsewhere in Europe more than a decade later. The Ader receiver employed a single nickel-plated magnet bent into the shape of a ring to serve as a handle. A small ring of soft iron placed in the end of the mouthpiece acted to intensify the magnetic field and produce louder articulation. Huurderman 2003, 169; Bennett and Webb 1974, 162; Miller 1905, 39–40; Lockwood 1883, 303–4; Homans 1901, 90–91; U.S. Pat. 241,580.

To Edison Lamp Co.

[New York,] 26th Dec [188]4

Dear Sir,

Referring to your[a] Bill of 24th inst I do not think the lamps you sent me are what I want.[1]

What I desire to do is to get a lamp of high candlepower to take the place of Calcium Lights in Theatres.[2] Mr Steele Mackaye of the Lyceum Theatre[3] & Mr Hastings of the Boston Bijou Theatre[4] are both pressing me to get up something   I explained to Mr. Holzer what I wanted. Yours truly,

Thos. A Edison   I[nsull]

L (letterpress copy), NjWOE, Lbk. 19:470 (*TAED* LB019470). Written by Samuel Insull. [a]Obscured overwritten text.

1. The bills have not been found, but see Docs. 2460, 2762, and 2764.

2. Calcium light (known also as lime light or Drummond light) exploited the luminous qualities of calcium-oxide when heated to high temperatures, often by a hydrogen flame. It offered an intense and focusable light source for spotlights as well as special effects. Bergman 1977, 273; Hocking and Lambert 1987, 306.

3. New York's new Lyceum Theatre was planned by architect Philip Gengembre Hubert (whose earlier work included the Navarro Flats). It opened on 6 April 1885 as an acting conservatory and theater on Fourth Ave. between Twenty-third and Twenty-fourth Sts. (Brown 1903, 3:419; "The New Lyceum," *New York Tribune*, 1 June 1884, 9; Price 1914, 74–76). The Edison Co. for Isolated Lighting prepared an installation contract for the Lyceum Theatre Co. as early as 3 December 1884, and an undated revised estimate for the job (amounting to $10,040) indicates it would provide lighting and manual regulators for the stage and auditorium, including wiring for 500 lamps (Edison Co. for Isolated Lighting, unsigned installation contract, 3 Dec. 1884; "Revised Estimate for the New Lyceum Theatre," n.d.; both Box 8, Folder 11, MacKaye). The Lyceum featured electric lighting throughout the lobby and house; lights for both the stage and the house reportedly were capable of changes in color and intensity. Other features included fixtures by Tiffany and Co. and electrically powered ventilation and stage machinery (Marienthal 1966, 85–86, 97, 99, 102, 106). The overall design of the Lyceum installation has been attributed to Edison and Luther Stieringer, whose experience included work on the Brockton City Theatre. An 1885 news report attributed to Edison the claim of having "perfected an apparatus" to provide the stage with "more natural shadows" than those from footlights or calcium lights. Years later, Edison recalled that he went to the Lyceum "on & off" to supervise the installation process (*"Edisonia"* 1904, 147; "Gossip of the Theatres," *NYT,* 22 Mar. 1885, 14; TAE questionaire responses, n.d. [Sept. 1925], MacKaye [*TAED* X009AF]).

4. Boston's Bijou Theatre on Washington St. was dedicated in December 1882; its Edison electric lighting system provided illumination for the foyer, entrances, galleries, dressing room, and the stage (see Doc.

2460). The New England Dept. of the Edison Co. for Isolated Lighting managed the installation. Edward Johnson and Verity and Co. (London) worked on decorative chandeliers for the auditorium, and Edison personally supervised the final testing of the system. Boston Bijou Theatre Co. to Johnson, 17 Jan. 1883; Verity and Co. to Edward Hastings, 13 Nov. 1882, both Series 1, Boston Bijou; "The New Theatre: The Bijou, Resplendent in Beauty, to Open Tonight," *Boston Daily Advertiser,* 11 Dec. 1882, 8; Jehl 1937–41, 997–98; King 2005, 71.

The theater company rented the plant from the Edison Isolated Co. at $130 for every week (or fraction thereof) in which the lights were used. According to its original contract, the theater also was responsible for hiring a caretaker for the system and for publicizing its use of the Edison system. Other terms allowed for the removal of the plant if power became available from a central station. Up to 750 incandescent lights were to be available during performances and dress rehearsals, and as many as 175 lights would be operable at any time. Under a revised contract of 1 December 1884 (which terminated in August 1885), the number of lamps required during performances decreased to 500 lights, though the weekly fee remained unchanged. The Edison Electric Light Co. reported 644 lamps installed as of December 1882 (Edison Co. for Isolated Lighting agreements with Boston Bijou Theatre Co., 22 Nov. 1883 and 1 Dec. 1884, both Series 1, Boston Bijou; Edison Electric Light Co. Bulletin, 15:31, 20 Dec. 1882, CR [*TAED* CB015229]). The Bijou also carried accounts for calcium, gas, and arc lighting (Gitelman and Collins 2009, 10; King 2005, 72). The theater participated in a novel display at the 1884 American Electrical Exhibition in Boston's Mechanics Hall, where a miniature illuminated model of the Bijou was connected by Edison's motograph telephone to nightly performances at the theater (Cabot 1885, 7–8; "At the Electrical Exhibition," *Boston Daily Globe,* 13 Dec. 1884, 5).

–2769–

To Mutual Life Insurance Co.[1]

[New York,] 26th Dec [188]4

Dear Sirs,

I enclose the policy on my life no. 163075 & dated Dec 24th 1880 for $10 000. Will you please have all my additions to Policy endorsed on same.[2]

As my wife is dead would it be possible for me to have a new policy issued in favor ~~of such~~ of my children. Yours truly,

Thos. A. Edison   I[Insull].

L (letterpress copy), NjWOE, Lbk. 9:500 (*TAED* LB019500). Written by Samuel Insull.

1. The Mutual Life Insurance Co. of New York was chartered by the state legislature in April 1842 and issued its first policy on 1 February 1843. Its offices were presently located at 140–46 Broadway in New York City. Edison had been using the company for a variety of financial services, including bonds, mortgages, fire insurance, and life insurance,

since at least 1876. He had paid the semi-annual premium ($234.40) on this policy in early July. Clough 1946, 3, 32; Mutual Life Insurance Co. to TAE, 24 June 1884, Miller (*TAED* HM840222A); Mutual Life Insurance Co. to TAE, 2 July 1884, DF (*TAED* D8415B); see *TAED*, s.v. "Mutual Life Insurance Co. of New York."

2. The editors have not identified the specific changes that Edison sought.

Appendix 1

Edison's Autobiographical Notes

From 1907 to 1909 Edison wrote a series of autobiographical notes to assist Thomas C. Martin and Frank L. Dyer in their preparation of his authorized biography.[1] Edison produced Document D, including notes on queries posed by Martin, probably about October 1907.[2] This was followed by the recollections in books A and G, made in September and October of 1908. This material was incorporated into the initial chapters of the biography, which were complete by February 1909; Martin then requested additional personal reminiscences from Edison in order to flesh out the remaining chapters.[3] William Meadowcroft, who was coordinating the project, acknowledged in May 1909 that the continuing lack of Edison's additional material was a "very serious affair," and the next month Edison produced the notes in books E and F.[4] Some of these formed the basis for oral interviews with Martin, the typed transcripts of which became documents B and C; together, these four documents served as the basis for anecdotes related in later chapters of the published biography.

Five of the documents contain sections related to events of the period of Volume Seven; those sections are published here.[5] Edison sometimes referred in the same paragraph to the periods covered by more than a single volume; these paragraphs will be reprinted as appropriate. Each document has been designated by a letter and each paragraph has been sequentially numbered. A few individual items that were inadvertently omitted from previous volumes are presented here. Items that are either solely by the the interlocutor or completely indistinct as to time have not been transcribed.

1. Dyer & Martin 1910. The designations A through F were assigned to these documents in Volume One, which also contains a general editorial discussion of them. See *TAEB* 1 App. 1; document G was discovered later.

2. An Edison notebook entry from this time reads "Martins book take Lab note bk 4, 1 Extra . . . answer Martins immediate notes." PN-07-09-15, Lab. (*TAED* NP077).

3. Martin to TAE, 23 Feb. 1909, Meadowcroft (*TAED* MM001BAP).

4. Meadowcroft to Martin, 24 May 1909, Meadowcroft (*TAED* MM001BAQ).

5. The autobiographical documents designated E and G do not refer to the period of this volume. The sections from A published in Volumes One and Four were drawn from a typed version of Edison's notes prepared by William Meadowcroft. However, a copy of Edison's original manuscript, in a notebook labeled "Book No. 1   September 1, 1908   Mr. Edison's notes re. Biography," was published in Part IV of the microfilm edition. Meadowcroft (*TAED* MM002).

A. BOOK NO. 1

The following is a transcription of a typescript that William Meadowcroft prepared from reminiscences that were originally written by Edison in a notebook labeled "Book No. 1" and dated 11 September 1908. The contents of the notebook pertain to the period covered by Volume One with the exception of the initial five paragraphs, which describe Edison's 1878 Western trip, and a short paragraph (14), which relates Edison's alleged discovery of a Morse diary while renting a house in Gramercy Park in 1882–83.

[14] In 1885 I rented a house in Grammarcy Park New York City which many years ago was the most fashionable quarter of the city. one day I went into the Garrret & found in a drawer the private Diary of SFB Morse the inventor of the telegraph, as now used. This seemed to be a strange Coincidence   The diary is now in the Library of the Soc of Elec Engineers

B. FIRST BATCH

The following is a transcription of a typescript that Edison revised. At the top of the first page is a handwritten note: "First Batch   Notes dictated by Mr Edison to T. C. Martin June, 1909.— Pencil indicates Mr. Edison's revision."

"Honest" John Kruesi

[27] One of the workmen I had at Menlo Park was John Kruesi, who afterwards became from his experience engineer of the lighting stations and subsequently engineer of the Edison Electric Light Works at Schenectady. Kruesi was very exact in his expressions. At the time we were promoting and putting up electric light stations in Pennsylvania, New York and New England there would be delegations of different people who proposed to pay for these stations. They would come to our office in New York at 65 to talk over the specifications, the cost and other things. At first Mr. Kruesi was brought in, but whenever a statement was made which he could not understand or did not believe could be substantiated, he would blurt right out among these stockholders that he didn't believe it. Finally it disturbed these committees so much and raised so many doubts in their minds, one of my chief associates said: "Here Kruesi, we don't want you to come to any of these meetings any longer. You are too painfully[a] honest." I said to him. "We always tell the truth. It may be deferred truth, but it is the truth." He could not understand that.

[29] After the station had been running several months and was technically a success, we began to look after the financial part. We started to collect some bills but we found that our books were kept badly and that the person in charge who was no business man had neglected that part of it. In fact he did not know anything about the station anyway. So I got the directors to permit me to hire a man to run the station. This was Mr. Chinnock, who was then superintendent of the Metropolitan Telephone Company of New York. I knew Chinnock to be [-][b] square and of good business ability and induced him to leave his job. I made him a personal guarantee that if he would take hold of the station and put it on a commercial basis and paid 5 per cent on $600,000, I would give him $10,000 out of my own pocket. He took hold, performed the feat, and I paid him the $10,000. I might remark in this connection that years afterwards I applied to the Edison Electric Light Company asking them if they would not like to pay me this money, as it was spent when I was very hard up and made the company a success, and was the foundation of their present prosperity. They said they "were sorry," that is "Wall Street sorry," and refused to do it. This shows what a nice, genial, generous lot of people they have over in Wall street.

[30] Chinnock had a great deal of trouble getting the customers straightened out. I remember one man who had a sa-

loon on Nassau street. He had his lights burning for two or three months. It was in June, and Chinnock put in a bill for $20; July for $20; August about $28; September about $35. Of course the nights were getting longer. October about $40. November about $45. Then the man call Chinnock up. He said: "I want to see you about my electric light bill." Chinnock went up to see him. He said, "I have the honor." "Well," he said, "my bill has gone from 20 up to 28, 35, 45. I want you to understand, young fellow, that my limit is 60!"

[31] After Chinnock had all this trouble due to the incompetency of the previous superintendent, a man came in and said to him: "Did Mr. Blank have charge of this station?" "Yes." "Did he know anything about the running of a station like this?" Chinnock said: "Does he <u>know</u> anything running a station like this? No, sir. He doesn't even suspect anything."

[32] One day Chinnock came to me and said: "I have a new customer." I said "What is it?" He said "I have a fellow who is going to take 250 lights." He said "~~I have a fellow who is going to take 250 lights~~"   I said "What for?" "He has a place down here in a top loft and has got 250 barrels of "rot gut" whiskey. He puts a light down in the barrel and lights it up and "ages" the whisky. I met Chinnock several weeks after and said, "How is the whiskey man getting along?" "It's all right; he is paying his bill. It fixes the whiskey and takes the shudder right out of it." Somebody went and took out a patent on this idea later.

[33] In the second year we had put the Stock Exchange on the circuits of the station, but were very fearful that there could be a combination of heavy demand and a dark day and that there would be an overloaded station. We had an index like a steam gauge, called an ampere meter to indicate the amount of current going out. I was up at 65 one afternoon. A sudden black cloud came up and I telephoned to Chinnock and^c asked him about the load. He said "We are up to the muzzle, and everything is running all right."^{d} By and by it became so thick we could not see across the street. I telephone down again and felt something would happen; but fortunately, it did not. I said to Chinnock: "How is it now?" He said "Everything red hot, and the ampere meter has made 17 revolutions!"

A NOCTURNAL SURPRISE

[49] One night when I had my laboratory at the top of the Bergmann works, on Avenue B and 17th Street, covering about a quarter of the block, about 2 o'clock in the morning I heard

"tramp, tramp, tramp" on the stairs. Six men walked into the room, six of the engineers. They never looked at me but walked right into my cubby hole, sat down, threw all the apparatus off the table and started a poker[c] game. They never answered me, but stayed there until about six o'clock, when they walked out and never looked at me. The building was six stories high. My father came there when he was 80 years of age. The old man had powerful lungs. In fact when I was examined by the Mutual Life Insurance Company in 1873, my lung expansion was taken by the doctor. The old gentleman was there at the time. He said to the doctor "I wish you would take my lung expansion." The doctor took it and his surprise was very great, as it was one of the largest on record. I think it was five and one-half inches. There were only three or four could beat it. Little Bergmann hadn't much lung power. The old man said to Bergmann, "Let's run upstairs." Bergmann said "Yes" and ran up. When they got there Bergmann was done up, but my father never showed a sign of it. There was an elevator there, and ~~one~~ each[a] day while it was traveling up I held the stem of my watch up against the column in the elevator shaft and it finished the winding by the time I got up the six stories.

UNPROFITABLE LAMP MANUFACTURE.

[54] When we first started the electric light, it was soon seen that we had to have a factory for manufacturing lamps. As the Edison Light Company did not seem disposed to go into manufacturing, with what money I could raise from my other inventions and royalties, and some assistance, we started a small lamp factory at Menlo Park. The lamps at that time were costing about $1.25 each to make so I said to the company "If you will give me a contract during the life of the patents I will make all the lamps required by the company and deliver them for 40 cents." The company jumped at the chance of this offer and a contract was drawn up. We then bought at a receiver's sale at Harrison, N.J. a very large brick factory which had been used for an oil cloth works. We got it at a great bargain and only paid a small sum down, and the balance on mortgage. We moved the lamp works from Menlo Park to Harrison. The first year the lamps cost us about $1.10. We sold them for 40 cents, but there were only about 20,000 or 30,000 of them. The next year they cost us about 70 cents and we sold them for 40. There were a good many and we lost more the second year than the first. The third year I had succeeded in getting up machinery and in changing the processes until

it got down so that they cost us somewhere around 50 cents. I still sold them for 40 cents and lost more money that year than any other because the sales were increasing rapidly. The fourth year I got it down to 37 cents and I made all the money up in one year that I had lost previously. I finally got it down to 22 cents and sold them for 40 cents and they were made by the million. Whereupon the Wall street people thought it was a very lucrative business, so they concluded they would like to have it and bought me out.

[55] One of the incidents which caused a very great cheapening was that when we started one the important processes had to be done by experts, which was the sealing in of the part carrying the filament, into the globe, which was rather a delicate operation in those days and required several months of training before any one could seal in a fair number of parts in a day. When we got up to the point where we employed 80 of the experts, they formed a union, and knowing it was impossible to manufacture lamps without them, they became very insolent. One instance was that the son of one of these experts was employed in the office and when he was told to do anything would not do it or would give an insolent reply. He was discharged, whereupon the union notified us that unless the boy was taken back the whole body would go out. It got so bad that the manager came to me and said he could not stand it any longer; something had got to be done. They were not only more surly, but they were diminishing the output, and it became impossible to manage the works. He got me enthused on the subject, so I started in to see if it was not possible to do that operation by machinery. After feeling around several days I got a clue how to do it. I then put men on it I could trust and made the preliminary machinery. That seemed to work pretty well. I then made a another machine which did the work nicely. I then made a third machine and would bring in the yard men, ordinary laborers, and when I could get these men to put the in parts in as well as the trained experts, in an hour, I considered the machine complete. I then went secretly to work and made 30 of the machines. Up in the top loft of the factory we stored these machines and at night we put up benches and got everything all ready. Then we discharged the office boy. Then the union went out. It has been out ever since.

FIGURING OUT MAINS.

[61] It is true that Sprague figured out mains for us of new stations while he was at Brockton, on a new mathematical ba-

sis, but we already had a good system of determining the size of the mains and of laying them out in miniature in German silver wire. We made a complete survey of the place before figuring them out. This system was so perfect that we could go into a man's store and say: "Your gas bill in December was $62.40." When he looked it up it was usually with in 5 per cent of it. We sometimes found that our estimates were too small, and I soon discovered the cause of this. We went to a place in Sixth Avenue. The man's bill ought to have been $16. It was $32. We took a delicate meter up there and found that there was a leak, which has been going on for fifteen years. Then I found that leakage was very general in New York, and that many complaints of gas bills were due to bad pipes in men's houses. For instance, when we took the factory at Avenue B and 17th street, I told Bergmann he had better test his pipes to see what the leakage was. It was rather extensive factory. Upon testing it from Saturday night to Monday morning we found his leakage bill was about $85 a month. We used a little one foot test meter in this work.

VISITORS TO 65.

[63] I have spoken of Remenyi's visits. Henry E. Dixey, then at the height of his popularity, would come in those days, after theatre hours, and would entertain us with stories—1882-3–4. Another visitor who used to give us a great deal of amusement and pleasure was Capt. Shaw, the head of the London Fire Brigade. He was good company. He would go out among the fire laddies and have a great time. One time Robert Lincoln and Anson Stager, of the Western Union, interested in the electric light, came on to make some arrangement with Major Eaton, president of the Edison Electric Light Company. They came to 65 in the afternoon and Lincoln commenced telling stories—like his father. They told stories all the afternoon, and that night they left for Chicago. When they got to Cleveland, it dawned upon them that they hadn't done any business, so they had to come back on the next train to New York and transact it. They were interested in the Chicago Edison Company, now one of the largest of the systems in the world. I once got telling a man stories at the Harrison lamp factory, in the yard as he was leaving. It was winter and he was all in furs. I had nothing on to protect me against the cold. I told him one story after the other—six of them. Then I got pleurisy and had to be shipped to Florida for cure.

SITTING BULL.

[69] Sitting Bull and 15 Sioux Indians came to Washington to see the Great Father and then to New York, and went to the Goerck Street works. We could make some very good pyrotechnics there so we determined to give the Indians a scare. But it didn't work. We had an arc there that was of a most terrifying character, but they never moved a muscle.

VILLARD'S RALLY.

[73] When Villard was all broken down and in a stuper caused his disasters in connection with Northern Pacific Mrs. Villard sent for me to come and cheer him up. It was very difficult to rouse him from his despair and apathy, but I talked about the electric light to him, and its development, and told him that it would help him win it all back and put him in his former position. Villard did make his great rally, he made money out the electric light, and he got back control of the Northern Pacific. Under no circumstances can a hustler be kept down. If he is only square he is bound to back on his feet. Villard has often been blamed and severely ciriticised, but he was not the only one to blame. His engineers had spent twenty millions too much in building the road and it was not his fault if he had found himself short of money and at that time unable to raise any more.

A HAND DYNAMO.

[79] When we had the factory at Harrison, an importer in the Chinese trade came to us and wanted a dynamo to run by hand power. He explained that in China human labor was cheaper than steam power. I got one of the horsepower forms of machine and put long spokes on it, fitted it up and shipped to China. I never heard of it again.

[81] When I was a young fellow, the first thing I did when I went to a town was to put something into the savings bank and start an account. When I came to New York, I put $30 into a savings bank under the New York Sun office. After it had been in about two weeks, the bank busted. That was in 1870. In 1909 I got back $6.40 with a charge of $1.75 for law expenses. That shows the beauty of New York receiverships.

TD (transcript), NjWOE, Meadowcroft (*TAED* MM003). ªInterlined above in pencil. ᵇCanceled. ᶜ"I telephoned to Chinnock and" interlined above. ᵈQuotation mark added in pencil. ᵉ"r" interlined above in pencil.

C. SECOND BATCH

The following is a transcription of a typescript that includes Edison's revisions. At the top of the first page is a handwritten note: "<u>Second Batch</u>   Mr Edison's notes dictated Mr Martin June 1909   Pencil indicates revision by Mr Edison."

VISIT OF DIAZ.

[14] President Diaz of Mexico, visit this country with Mrs. Diaz, a highly educated and beautiful woman. She spoke very good English. They both took a deep interest in all they saw. I don't know how it ever came about, as it is not in my line, but I seemed to be delegated to show them around. I took them around to railroad buildings, electric light plants, fire departments, and showed them a great variety of things. It lasted two days.

THE EDISON EFFECT.

[18] An effect was shown in connection with the Edison lamps at the Philadelphia Electrical Exhibition of 1884. It became known as the Edison effect—showing a curious current or condition or discharge in the vacuum. It has been since employed by Fleming in England and by DeForest in this country and others as a wireless apparatus. It is really a rectifier of alternating currents, and analogous to those which have since been made on a large scale.

BEATING MORGAN.

[20] The president of the Edison Electric Light Company was a good lawyer but not a business man, and the affairs of the Company suffered. I got interested in this situation around 1884, and took a hand in matters. I am the only man that ever beat Drexel & Morgan Company over an election of directors and officers. I wanted a change so I went around and saw the stockholders and got their proxies. Then the Drexel, Morgan crowd[a] people[b] tried to get them and I had them. My opponents wanted the proxies revoked and new ones made out, but none of the stockholders would break the proxies they had given me. I had their confidence and they believed in my plans. Then Drexel & Morgan agreed to my terms, and we put in a business man—and the Company went ahead. I saw about 75 stockholders and was at it for two whole days. I then went to Mr. Fabbri of the firm & told him if he would put in a business man he could have the proxies. To this he agreed & a business man went in— It was all very friendly.[c]

TD (typescript), NjWOE, Meadowcroft (*TAED* MM003). [a]Interlined above in pencil. [b]Added in pencil. [c]"I then went . . . very friendly." added in pencil.

D. BOOK NO. 2

This undated notebook, labeled "Book No. 2," contains a mix of narrative passages, questions, and notes in Edison's hand. The first two pages are a memo by Meadowcroft, dated 9 January 1920, recounting the preparation and use made of this material between 1907 and 1910. The next sixty-six pages alternately present narrative passages and brief references to various anecdotes, many of which relate to the period covered by this volume. The next nine-page section is labeled "Martin's Questions." The remaining twenty-one pages contain only notes.

[50] House in Grammacy Park found Morses personal Diary—[a]

[103] Remenyi—

[104] Dixie—

[105] Bergmann—boxes nailed to floor[b]

[106] Stewart & Madam Counsend[b]

[107] Bergmann & Galvanometer stopped whistle to save steam[a]

[108] Walking down to Lab ave B & 17 school hours saluted by children thinking I was a priest—[a]

[109] Country Goerck & the Hq—threatened suit—[a]

[110] Try to Chicago Parker—Bergmann thought it was phila—[b]

[111] 65th 5th ave. ofs hours 24— Remenyi H N Dixie Duke of Sunderland—Bull Run Russell—Insull—

[112] W H Vanderbilt[c]—Mrs. Vanderbilt ordering Engines out Tinsel on fire— Insull—ran everything talk shop at [reachers?] funeral[a]

[113] Kiralfy—1st time behind stage—

[114] Chinnock—10 000 if 5% on 600,000 Years after board when about sell out wouldnt reimburse said sorry (wall st sorry)[b]

[115] To assistant Chinnock said dont walk from Yonkers take the train—[b]

[117] Emery & the opposition Steam Heat Explosion Lampblack in Maiden lane E & myself [oughts?] talking over our project[a]

[118] Meters frozen light lighted poured water on—[a]

[119] Casho—dont know anything dont Suspect anything[a]

[120] Saloon mans bill 15 20 25 35 45 Limit was 60[a]

[134] Explosion gas Engines 65 5th ave celler stunned—[a]

[135] Wag got every gas bill ny & laid out mains— also every hoist way—& placed they used power—[a]

[136] 1st starting pearl st Station Engines run away[a]

[137] A&S run 24 hours day 365 days never stopped—[a]

[138] Melted Cobble stone at feeder box—[a]

[139] Aging whisky—shudder out.[a]

[141] Salivation by Mercury—strikers & auto machine that busted them—[a]

[142] Jalop & watchman—

[143] Saluted as Catholic priest walking to Lab Ave B 17

[144] Bergmann stopped whistle wasted steam

[145] Poker 2[c] am [ring?][d] never noticeing me

[146] Father 80 walked 6 stories all puffing old man never puffed—

[147] Country boy mouth full mercury

[148] Jumbo across town to Paris aided by police

[149] Running Lathes in Goerck st—

[150] Tesla dreamer—

[151] Soux indians at Goerck

[152] A&S Engine Crank burst no sleep for seems 3 days— spit blood— Dean

[153] Wkd year $^{1}/_{2}$ raised Economy Lamp from 10 to 15 per hp— how got cigars for many years—

[159] Presdt Diaz & wife NY[b]

[161] Machine for china to be worked by hand—[a]

[162] Steele Mackay & Wm Winters[a]

[166] W H Vanderbilt house lighting—tinsel on fire Etc—

[171] Unanimous decision E E Light Co board nothing nothing in trolly

[267] 1st Jumbo Paris Elec Exh— braking Crank shaft Eng— Volts too low raised by Extra magnets raised by Extra Magnets racing across NY 6 horses all police cleared streets— bbl Beer?

[268] Running Lathes outside Goerck St Tannery Dist Leader— big June & Tannery Leader

[270] Bldg Elec Loco 6 ft drivers

[271] ~~Tors~~ Starting Pearl St Stn torsion shaft—

[274] Christian Herter—Vanderbilts house Vanderbilt coming into 65 5th ave Twombly in WU & telephones

[275] Remenyi— Tinsel on fire WHV house Mrs V ordering out of cellar Eng & B—

[276] Jas G Bennet came into 65 5th ave said go around the world ordered Howland put it in Herald immediately

[277] Elec leaking ann & nassau old Horse—

[278] Emery & Steam heat & myself Laying tubes— Lampblack Opstn—

[279] Smoked my own cigar's—hair horn & gallon—

[294] Laying tubes Hugo O Thompson & inspectors—[e]

[295] Kreuzi Small[c] stations, painfully honest, deferred truth[e]

[296] ~~Chinnock~~

[297] Starting Station Engines run away it Torsion shaft, A & S—[f] Meter red hot. Ampere meter 2 Rev Stock Ex feeder Crossed melted Ton of paving blocks—[a]

[298] Casho, Man having limit to his bill. Emery & Steam Heat Co Lampblack Opposition Co—[a]

[299] Great Cost bulding 1st station surprised—[e]

[303] Johnson & Holbern viaduct[e]

[305] Tesla—Invented Hello for telephn[d]

[306] Puskas & Bailey—failure German Edison Co. now allegman Geshelshaft—English Co never got anything[e]

[319] Lab over Bergmann, poker

[320] Father[c] running up 6 stories wind Waterbury on Elevator[a]

[324] Tube shop in Washn st Pierpont Morgan only investment made repaid 10 times over[e]

[325] Nearly Killed by bricking Machine bolt in Lab—

[344] Duke Sutherland Bull Run Russell— Capt Shaw.

[345] Dixy funny stories—

[346] Glass & Agt of England Ins Co Cherooke Mining Co man & London Ins Co man to get me drunk[a]

[347] Stager & Bob Lincoln tell stories returned & forgot what came for[a]

[351] Loss of English Elec light mistake of Lawyer misleading Fabbri of JPM[a]

[353] 65 5th ave ofs hours 24 daily[a]

[354] Explosion gas Engine in Cellar hurt me—[a]

[356] Big Jim— Dist Tam Leader Lathes run on sidewalk Dean— Tesla here. Sims Engine—broke shaft. All night house lunch 1 clam for season in chowder— [~~6 flys pieces of pie?~~][g] 6 flys for Each pie— Bad neighborhood, theives—accompanied James Russell late night, Spit blood, Deans boy—[a]

[357] Sioux indians— Dog got between belt flattened.[a]

[358] fun on Cor Ann & Nassau leaky pavement, horse.[a]

[362] Invented the multiple tubes & gave plans to the Tannery Boss in 1881,= they adopted it[a]

[364] Villard bust wife sent for me cheer him up— man with 3 lambs, said Wall st men had no forethought best men in [Harwards?][d] Leather   Dynamo order [-----][d]   I had sew bet 100 and nothing several times [------------][d]

[365] Incandescent Lights at Neblos & Iolanthe Bijou Theatre Boston

[367] Ansonia letting us have copper[f] ~~money~~ after seeing our books—terrible time with pay rolls— Selling drafts on London & Cabling[c] money to pay them—[a]

[369] Lamp Contract, of Cutting on telephone wanted know what kind of a Co it was to pay div Every week[a]

[370] When had board meeting I was always the one that was the odd member in voting[a]

[371] ~~We~~ Board unanimous nothing in ~~El~~ Trolly Except Villard[a]

[372] Meadowcroft find Letter signed by all Directors[a]

[376] Moved Lab from Menlo to Ave B & 17 then to Lamp factory

[378] chinese hand Electric Turning

[387] Edison Effect exhibited at Elec Exp phila— Meadowcroft get date, applied by Fleming as wireless receiver & also by De Forrest. Fessenden in Lab.

[391] Wanted change in presdt of E E Light Co, DM & Co wouldn't went around st & got more proxys than they could get   Compromised—

AD (photocopy), NjWOE, Meadowcroft (*TAED* MM005). [a]Paragraph overwritten with a large "X". [b]Paragraph overwritten with a large "X" and followed by dividing mark. [c]Obscured overwritten text. [d]Illegible. [e]Paragraph overwritten with large check mark. [f]Interlined above. [g]Canceled.

F. NOTES (JUNE 1909)

This notebook includes sixteen pages in an unlabeled section in Edison's hand relating to the Dyer and Martin biography. These pages are preceded by a memo to Edison from William Meadowcroft dated "June 28/09" stating that these notes had been copied. Eight of its twenty-four items pertain to the period covered by this volume. There is a typed version of the notes in the William H. Meadowcroft Collection at the Edison National Historic Site. The last fifteen pages are a biographical sketch of Edison's former employee Sigmund Bergmann.

[2] You[a] Left out item of finding Morses Diary in a house I rented in Grammacy Park, Certainly this is not uninteresting—

[8] Soon after this shop was started I sent a man named Stewart down to Santiago Chili to put up[b] a Central Station for Electric Lighting   Stewart after finishing the station returned to N York with glowing accounts of the Country & an order from Madame Cousino the richest woman in Chili for a complete plant with chandiliers for her palace in the suburbs of Santiago. Stewart gave the order to Bergmann, & the price was to be for the chandiliers alone $7,000. Stewart having no place to go generally managed to stay around Bergmanns place recounting the Emmense wealth of Madame Cousino, and Bergmann kept raising the price of the outfit until[b] Stewart realized that these glowing accounts of wealth was running into money when he kept away & the chandiliers went billed for 17 000; cash on bill of Lading, as Bergmann said he wasnt sure Stewarts mind wasnt affected[c] & he wanted to be safe.

[9] At onetime he was making an immense switchboard for the NY Telephone central station the specifications called for mahogany   one day the president called at the shop top to find out what progress was being made, after Explanations Bergmann suggested that it was too bad ~~such~~ that Mahogany should be used with such a beautiful piece of apparatus when for $1000 extra black walnut could be substituted/ The presdt who had been put in his position by influences & ~~not~~ knowing nothing of the business ready assented to this proposition— Bergmann used the ~~inferior~~ walnut at a saving of $1500.

[10] A Jew by the name of Epstein had been in the habit of buying brass chips & turnings from the Lathes & in some way Bergmann found out that he had been cheated ~~so he~~ this hurt his pride & he determined to get even— One day the Jew appeared & said good morning Mr Bergmann have you any chips today, no said B I have none. Thats strange Mr B said the Jew wont you look. no he wouldnt look he knew he had none. finally the jew was so persistent that Bergmann called an assistant & told him to go & see if had any Chips   He returned & reported that they had the finest & largest lot they ever had. The Jew went up to the several large boxes piled full of chips & so heavy that he couldn't lift even one end of a box. Now Mr B said the jew how much for the lot. Epstein said Bergmann you have cheated me and I will no[b] longer sell by the lot but will only sell by the pound. No amount of argument ~~seemed~~ would change Bergmanns determination to sell by the pound

but finally the Jew got up to $250 for the lot & B finally appearing ~~As~~ as if disgusted accepted & made the Jew count out the money & said well Epstein good bye I've got to go down to wall st. the Jew and his assistant then attempted to lift the boxes to carry them out but couldn't & then discovered that calculations as to the quantity had been thrown out because the boxes had all been screwed to the floor & mostly filled with boards with a veneer of brass chips.[d] The Jew ~~was~~ made such a scene that he had to be removed by the police. I met the jew several days afterward he said he had forgiven Mr. B as he was such a smart business man &[e] the scheme was so ingenious.

[11] One day as a joke[f] I filled 3 or 4 sheets of foolscap ~~fill~~ with a jumble of figures, & told Bergmann that they were calculations showing the great loss of power from blowing the factory whistle Bergmann thought it real & never after would he permit the whistle to blow—

[12] Next door to this factory was a Parochial Catholic school & every time I walked pass when the children were out they all saluted with the finger to the head, on inquiry I found they thought I was a priest.

[19] At this Laboratory I ~~have~~ had[a] a series of Vacuum pumps worked by Mercury & used for exhausting Experimental ~~El~~ Incandescent lamps. The main pipe which was full of Mercury was about $7\frac{1}{2}$ feet from the floor along the length of the pipe were outlets to which thick rubber tubing were connected Each tube to a pump. One day while Experimenting with the Mercury pump my assistant an awkward country boy from a farm on Staten Island who had adenoids in his nose & breathed through his mouth which was always wide open was looking up at this pipe at a small leak of mercury when a Rubber tube came off and probably two pounds of Mercury went into his mouth & down his throat[g] & got out through his system somehow. ~~I I had considerable At~~ In a short time he became salivated & his teeth got loose he went home & shortly his mother appeared in the Laboratory with a horsewhip which she proposed to use on the proprietor I was fortunately absent & she was molified somehow by my other assistants— I had given the boy considerable Iodide of potassium to prevent salivation but it did no good in this case

[20] When the first Lamp works was started at Menlo Park one of my experiments seemed to show that hot mercury gave a better vacuum in the lamps than cold mercury I thereupon started to heat ~~the l~~ it—soon all the men got salivated & things looked serious but I found that in mirror factories where Mer-

cury was used Extensively the French govt made the giving of Iodide of potassium compulsory to prevent Salivation. I carried out this idea & made every man take a dose every day but there was great opposition, & hot mercury was finally abandoned

AD (photocopy), NjWOE, Meadowcroft (*TAED* MM005). [a]Interlined above. [b]Obscured overwritten text. [c]"wasnt effected" interlined above. [d]"& mostly filled . . . chips" interlined above. [e]"he was such . . . &" interlined above. [f]"as a joke" interlined above. [g]"down his throat" interlined above.

Appendix 2

Edison Village Plants

A. PLANTS COMPLETED, 1883–1884

This table lists the Edison central station plants built in 1883–1884, in the order in which they went into operation. The editors have not determined with certainty the capacity or cost of each plant because the available documentation is incomplete, especially for the later plants, and sometimes contradictory. Edison frequently provided multiple estimates, some systems were altered in the course of construction, and others were expanded soon after completion. The figures given here reflect the editors' best judgment of what was built and billed to the local illuminating companies.[1] Where it is known, the contract amount is given. The figures have been rounded to the nearest dollar. Most of these plants were constructed, or at least estimated, under the auspices of the Edison Construction Department before the Edison Company for Isolated Lighting officially took over its functions on 1 September 1884. The Isolated Company also had sole responsibility for the Roselle, New Jersey, plant.

City or Town	Operational Date	Canvass Completion Date (approx.)	Contract or Estimate ($)	Underground (U) or Pole (P) System	Initial Capacity (lamps)	Capacity as of 12/31/84
Roselle, N.J.	Jan. 1883	n/a	n/a	P	500[2]	n/a
Sunbury, Pa.	4 July 1883	July 1883	11,968	P	500	900 (as of 4/10/86)
Shamokin, Pa.	c. 20 Sept. 1883	n/a	19,209	P	1,600	1,824
Brockton, Mass.	1 Oct. 1883	Nov. 1883	31,065	U[3]	1,600	3,000
Lawrence, Mass.	20 Oct. 1883	n/a	21,446	U	2,800[4]	2,760
Fall River, Mass.	18 Dec. 1883	May–Sept. 1883	41,810	U	1600	1,650
Tiffin, Ohio	6 Jan. 1884	Aug. 1883	18,000	P	1,000	n/a
Mt. Carmel, Pa.	22 Jan. 1884	n/a	14,560	P	500	543
Bellefonte, Pa.[5]	5 Feb. 1884	July 1883	16,797	P	800	1,773 (as of 4/17/86)
Hazleton, Pa.	11 Feb. 1884	Oct. 1883	20,684	P	1,000	1,736 (as of 3/1/86)
Newburgh, N.Y.	31 Mar. 1884	Aug. 1883	38,000	U	1,600	n/a
Middletown, Ohio	5 Apr. 1884	n/a	13,969	P	500	n/a
Piqua, Ohio	28 Apr. 1884	Aug. 1883	23,000	P	1,000	n/a
Circleville, Ohio	14 June 1884	Dec. 1883	15,291	P	500	n/a
Cumberland, Md.	29 Aug. 1884	Apr. 1884	20,198	P	1,200	2,335 (as of 4/17/86)
Williamsport, Pa.	4 Oct. 1884	June–Dec. 1883	53,226	P	3,200	n/a
Ashland, Pa.	11 Oct. 1884	Feb. 1884	n/a	P	n/a	n/a

1. The principal sources used to construct this table include: Edison Construction Dept. inventory, n.d. (Apr. 1884?), DF (*TAED* D8441N1); Edison Construction Dept. inventories, 19 Dec. 1883 and 7 Mar. 1884; John Campbell to Francis Upton, 23 Jan. 1884, LM 17:179, LM 18:324, 76B (*TAED* LBCD4179, LBCD5324, LBCD5076B); and a notebook used by William Hammer during his tenure as a central station inspector in 1885–86 (Series 1, Box 13, WJH). Most of the listed plants have folders of correspondence specific to them in the Document File at NjWOE and online through the Edison Papers' website (http://edison.rutgers.edu/sernote2.htm). Pertinent correspondence, some of which gives technical or cost specifications, may also be found under the names of particular local companies, or their officers, in Edison, T. A.—Outgoing Correspondence folders for 1883 and 1884, both DF (*TAED* D8316, D8416) and in the eight Construction Department letterbooks, LM 14–21 (*TAED* LBCD1, LBCD2, LBCD3, LBCD4, LBCD5, LBCD6, LBCD7, LBCD8).

2. The plant also supplied 150 street lamps. Doc. 2336 n. 3.

3. Portions of the distribution system may have been on overhead wires.

4. This figure is from the initial estimate; the system may have been built for 1,600 lamps.

5. See also TAE to Phillips Shaw, 25 Sept. 1883, LM 15:171 (*TAED* LBCD2171).

B. OTHER PROJECTED PLANTS CANVASSED OR ESTIMATED, 1883–84

To indicate the scale on which Edison (and Samuel Insull) expected to operate the Edison Construction Department, this table lists projected village plants for which the planning got at least as far as a canvass or estimate. It does not include several dozen others for which a canvass was authorized but not undertaken.[1] The locations are arranged alphabetically within each state. The editors have not been able to find information in all categories for each locale, and in some cases, they have been unable to resolve conflicting figures; monetary amounts have been rounded to the nearest dollar. Many of these plants began operating in 1885 or 1886.[2] Except as noted, the information is compiled from the same principal sources identified in Table A above.

State	City or Town	Canvass Completion Date (approx.)	Estimate ($)	Underground (U) or Pole (P) System	Capacity (lamps)
Alabama					
	Mobile	Dec. 1883			
Connecticut					
	Danbury	July 1883	25,166	U	1,000
	Hartford	July 1883	72,863		3,200
	Meriden	Feb. 1884			
	Norwalk	Feb. 1884			
	Norwich	Aug. 1883	23,576	U	1,000
	Southington	July 1883	17,487	P	600
Georgia					
	Atlanta	Dec. 1883			
Illinois					
	Mendota	Aug. 1883			
Iowa					
	Cedar Rapids	Jan. 1884	37,343	P	1,600
	Davenport	July 1883	25,350	P[3]	1,000
	Des Moines	Feb. 1884	23,745		1,600
Kentucky					
	Cynthiana	Oct. 1883			
	Frankfort	Jan. 1884			
	Lexington	Jan. 1884			
	Louisville	Sept. 1883			
	Maysville	Oct. 1883			
	Paducah	Feb. 1884			
Louisiana					
	New Orleans	Dec. 1883			
Maine					
	Portland	Aug. 1883	45,079	U	1,600
Massachusetts					
	Holyoke	Feb. 1884			
	Lowell	n/a	52,240		3,200
	Natick	Aug. 1883	23,233	U	1,000
	North Adams	n/a	39,988	U	1,600?
	Palmer	Aug. 1883			
	Pittsfield	July 1883	40,127	P	1,600?
	Southbridge	Sept. 1883	20,670	U	500
	Spencer	Aug. 1883			
	Webster	Sept. 1883			
	Westboro	Aug. 1883	16,039	P	500
	Worcester	Aug. 1883	74,006	U	3,200

Michigan					
	Adrian	Dec. 1883			
	Ann Arbor	Dec. 1883			
	Grand Haven	Nov. 1883			
	Hillsdale	Dec. 1883	18,287	P	500
	Ypsilanti	Dec. 1883	14,824	U	500
Minnesota					
	Minneapolis	July 1883			
	St. Paul	June 1883			
Missouri					
	Kansas City	Oct. 1883	43,565	P	1,600 or 2,535
	St. Louis	no canvass[4]	115,625		5,400
New Jersey					
	New Brunswick[5]	May 1884			
	Orange	Feb. 1884			
New York					
	Albany	Feb. 1884			
	Auburn	Oct. 1883	40,844	U	1,600 or 1,985
	Haverstraw	Aug. 1883			
	Hudson	Nov. 1883	32,535	P	1,600 or 1,949
	Irvington	Feb. 1884			
	New York City[6]	Jan. 1884			
	Poughkeepsie	Nov. 1883	40,665	U	1,600 or 2,411
	Syracuse	Oct. 1883	70,056	U	3,200 or 4,327
	Utica	June 1883	63,720	U	3,200
Ohio					
	Akron	Sept. 1883			
	Canton	Nov. 1883			
	Chillicothe	Nov. 1883			
	Columbus	Nov. 1883			
	Georgetown	July 1883			
	Hamilton	July 1883			
	London	Sept. 1883			
	Lima	Aug. 1883			
	Northfield	Aug. 1883			
	Ravenna	Aug. 1883			
	Springfield	Mar. 1884			1,000
	Urbana	Sept. 1883			
	Washington Court House	Aug. 1883			1,600
Pennsylvania					
	Erie	July 1883	37,276		approx. 1,800
	Harrisburg	Feb. 1884			1,132

Johnstown	n/a		P	
Lancaster	Jan. 1884			
Lock Haven	July 1883			
Mauch Chunk	Dec. 1883			700
McKeesport	n/a		P	
Milton	Jan. 1884			
Pottstown	Jan. 1884			
Pottsville	Jan. 1884			
Renova	July 1883			
Union City	Feb. 1884			
Wellsboro	Dec. 1883			
Wilkes Barre	Jan. 1884			

Tennessee

Knoxville				
	Jan. 1884			1,000 or 1,568

Texas

Dallas	Dec. 1883	24,796	P	
Galveston	Dec. 1883			
Sherman	Dec. 1883			1,600

Virginia

Danville	Feb. 1884	23,000(?)		1,600

Wisconsin

Appleton[7]	July 1883	33,579	P	1,600 or 1,968
Madison	Nov. 1883	37,788	P	500
Neenah	July 1883	7,885(?)		1,600
Racine	Aug. 1883	31,030		

1. The uncanvassed locations are included in an undated Edison Construction Dept. inventory, probably from April 1884 (DF [*TAED* D8441N1]).

2. Lists of central stations in operation as of specific dates may be found in Edison Electric Light Co. annual reports, 27 Oct. 1885, 26 Oct. 1886, 25 Oct. 1887, 23 Oct. 1888; all PPC (*TAED* CA001A, CA001B, CA019C, CA019D).

3. Pole lines planned for street lighting; plans for remainder of distribution system are unclear.

4. Although no canvass was ordered for St. Louis, plans for the station were well advanced and Samuel Insull expected a signed contract. See Doc. 2456.

5. See Edison Construction Dept. memorandum, June 1884, LM 19:392 (*TAED* LBCD6392) and Samuel Insull to Charles Hughes, 9 June 1884, DF (*TAED* D8416BTJ).

6. Second ("Madison Square") district.

7. This projected station was in addition to the large isolated plant that began lighting an Appleton paper mill in September 1882 and, later, two private homes. Doc. 2352 n. 3.

Appendix 3

Specifications of Dynamos Produced at the Edison Machine Works, April 1883–December 1884

This table provides basic specifications for the new Edison dynamo models designed and manufactured during the period of this volume. It is intended as a companion to the table in *TAEB* 6 App. 3. It does not include the short-core modifications to older "C" (Jumbo) machines, such as those added to the New York Pearl St. station in 1884.[1]

These models were made for nominally 110-volt central station or isolated plant systems. They were rated by the number of "A" lamps they could operate. ("A" lamps were manufactured to give 16 candlepower at about 100–103 volts.) With increasing frequency, the capacity of these dynamos was also expressed in amperes. Unlike the earlier machines listed in *TAEB* 6 App. 3, they were not rated for the "B" lamps used in 55-volt isolated installations.

These dynamo models embodied the new short-core style that Edison developed in 1883. They proved to be a transitional form that was produced for a relatively short period, less than two years. Even in that brief span, continuing experiments and experience in the field led to innumerable modifications, making it impossible now to state definitively the specifications of any given model. Adding to the complexity is the fact that some information presented here is taken from experimental records, which did not necessarily reflect the design in production at that moment. As pointed out below (see note 3), there is a particularly wide range of values for amperes. This disparity may result from variously measuring a dynamo's output at its terminals or calculating its effective generating power in an actual distribution system having substantial built-in electrical losses. The sources listed below give more-or-less standardized figures, but they represent neither all the exceptions nor all the documents bearing on dynamo specifications.

Model	HP^2	RPM	Field Coils	Height × Base (in.)	Wt. (lbs.)	Amperes (approx.)[3]	"A" Lamps[4]	Shop Price/ Advertised Price ($)[5]
G[6]	3.5	1,670	2 (series)		1,072	[18][7]	25	175/450
R[8]	7	2,000	2 (parallel)	41.5 × 31.5 × 18.1	1,187	50 (40)	50	250/750
T	12.5	1,700	4 (parallel)	48.2 × 44.9 × 23.6	1,645	100 (80)	100	400/1,350
S[9]	[28][10]	1,300	2 (parallel)		3,871	200 (160)	200	725/2,400
Y[11]	[42][12]	1,200	2 (series)		5,450	300 (240)	300	1,075/3,450
H[13]	65	1,100	6 (3 series pairs in parallel)	61 × 77 × 33	7,600	400 (300–345)	400	1,500/4,500

1. The list of dynamo models is adapted from Franklin Institute 1884, 25–26, 61, 75, and TAE to Société d'Appareillage Électrique, 14 Feb. 1884, DF (*TAED* D8416ANU). Except as noted, the principal sources for electrical and physical dimensions are William Hammer Notebook 8, 1882[–1884], pp. 24–42, Ser. 1, Box 13, Folder 2, WJH (*TAED* X098F02), and Frank Sprague to TAE, 16 Apr. 1884, DF (*TAED* D8442ZDG). Also helpful are experimental records that Charles Batchelor made at the Edison Machine Works in May 1884, Cat. 1235:72–78, Batchelor (*TAED* MBN012072–MBN012078).

2. Horsepower figures are taken from William Hammer Notebook 8, 1882[–1884], pp. 24–42, Ser. 1, Box 13, Folder 2, WJH (*TAED* X098F02). As noted in *TAEB* 6 App. 3 n. 2, his figures may reflect the power applied to the armature pulley by the belt, rather than the rating of the steam engine.

3. Contemporary documents by Edison and the Edison Machine Works provide a significant range of amperage ratings for these dynamos, sometimes allowing for the anticipated skill of the operating engineer. It is not clear when those values were based on direct measurement, calculated from design characteristics, or extrapolated from the number of lamps. (In a distribution system, both the current and voltage would be higher at the machine terminals than at the lamps, although Edison's engineers sometimes complained that, in fact, they were not high enough.) The first figures in this column were given to Edison by Frank Sprague on 16 April 1884 (DF [*TAED* D8442ZDG]) and probably represent values at the machine terminals. The second figures (in parentheses) are taken from a 1916 lecture by Edison assistant John Vail (Vail 1916, 4) and may represent effective ratings under actual service conditions. Both Sprague's and Vail's figures present a higher ratio between amperes and the number of lamps than those provided by William Hammer for the earlier generation of machines and presented in *TAEB* 6 App. 3. In general, amperage may be calculated using Ohm's Law ($V=IR$, where V=voltage, I=current in amps, and R=resistance in

Appendix 3

ohms) based on an effective lamp resistance of about 140 to 150 ohms at approximately 99 to 105 volts. In a parallel circuit with a number (n) of identical resistors (R), such as ideal lamps, the aggregate resistance is 1 divided by n/R.

4. Many 110-volt village plants used 10-candlepower lamps; approximately 1.6 times as many of these "C" lamps could operate under the same conditions as the standard 16-candlepower "A" lamps.

5. For a list of shop prices (i.e., for internal sales) and variable profit margins on dynamos, see Edison Electric Light Co.'s memorandum on "Prices of Dynamos and Percentages for Chile," n.d. (1884), DF (*TAED* D8435ZBO); see also prospective pricing of forthcoming models in TAE to Charles Rocap, 5 June 1883, Lbk. 17:58 (*TAED* LB017058). Retail prices are from Edison Co. for Isolated Lighting brochure, [c. Nov. 1883], p. 60, CR (*TAED* CA002B [image 32]).

6. This model was dropped from production at Edison Machine Works some time before the end of 1883, although Edison continued to make improvements to it in the new year. TAE to Deutsche Edison Gesellschaft, 17 Dec. 1883; TAE to Comitato per le Applicazioni dell'Elettricita, 4 Jan. 1884; both DF (*TAED* D8316BTX5, D8416ABJ); see also Cat. 1235:72, Batchelor (*TAED* MBN012072).

7. This value is extrapolated from the number of lamps in the same ratio as John Vail's figures for the other machines (see note 3). Edison rated this model at 22 amperes for use in Henry Rowland's experiments. TAE to Rowland, 5 May 1884, DF (*TAED* LBCD6102).

8. Regarding the R model, see (in addition to the principal sources listed in notes 1–2) Frank Sprague to TAE, 16 Apr. 1884 (*TAED* D8442ZDG).

9. Regarding the S model, see (in addition to the principal sources listed in note 1) Samuel Insull to E. H. Lord, 4 Dec. 1883; Edison Machine Works memorandum, 25 June 1884; both DF (*TAED* D8316BMS, D8431ZAD); and TAE to Société d'Appareillage Électrique, 5 Dec. 1883, LM 2:32C (*TAED* LM002032D).

10. This figure is an approximation based on 7.1 lamps per horsepower, which is similar to the output of the other models.

11. William Hammer did not include this model in his London notebook. In addition to other sources listed in note 1, see Y dynamo test reports from summer and fall 1883 in Electric Light—Edison Electric Light Co.—Testing Department; Insull to TAE, 20 Sept. 1883; TAE to Batchelor, 21 Sept. 1883; all DF (*TAED* D8330, D8316AWB, D8316AWF).

The Y model represented some of the disparities that could exist between test ratings and performance under actual operating conditions. William Andrews, for example, observed that the Y used at the Bellefonte, Pa., central station had produced, under trial in New York, 246 amperes at 112 volts from 1,200 rpms. He pointed out, however, that he was forced to operate the two Y machines in Hazleton, Pa., at 1,370 rpms to carry a smaller load, and he did not think the machines should normally be run at "less than 1380 to 1400 RPMs." Andrews to Insull, 14 Feb. 1884, DF (*TAED* D8442ZBC).

12. This figure is an approximation based on 7.1 lamps per horsepower, which is similar to the output of the other models.

13. See Cat. 1235:46, Batchelor (*TAED* MBN012046). Several variations of the H are identified by Vail 1916 (4). The six-core version reportedly gave 300, 320, or 345 amperes. Four-core versions gave 345 or 460 amps (the latter at 125 volts but 875 rpm). The editors have found no other information specific to these configurations.

Appendix 4

Edison's Patents, April 1883–December 1884

Edison's prolific patent activity diminished as he threw himself into the business of constructing central station lighting plants in 1883 and 1884. Particular lulls in the completion of new applications occurred several times in 1884, due partly to the Edisons' Florida vacation in late winter and early spring and Mary Edison's death that summer. Though confident that his lighting system was reasonably mature and commercially viable, Edison continually tinkered with its components as his experience with their manufacture and operation grew. Frequently, however, his alterations were made in back-and-forth communication with subordinates and associates and were not made public through the Patent Office. Many of the applications that he did file had to do with electrical distribution and the economical operation of the system as a whole, and they arose from hard practical experience in central stations.

Volumes 1–5 of *The Papers of Thomas A. Edison* included a complete list of all patents originated by Edison in the relevant period. However, since this information is now available on the Edison Papers website, as described below, it has not been duplicated here. Instead (as in Volume 6), a chart (A) depicts Edison's applications by month (according to execution or filing date, respectively). It includes unsuccessful or abandoned applications that Edison completed through April 1884. The cumulative totals are shown in a chart (B) by subject area. Finally, a detailed list (C) enumerates the patents taken out by, or with, Edison's associates.

The full text and drawings of Edison's U.S. patents are on the Thomas A. Edison Papers website at http://edison.rutgers .edu/patents.htm. They can be searched by execution date,

A. EDISON'S PATENT APPLICATIONS BY MONTH

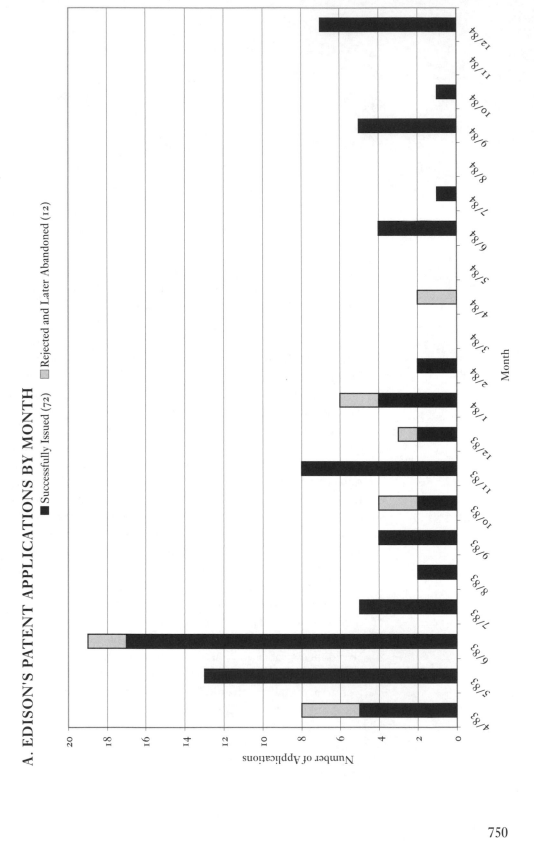

■ Successfully Issued (72)   ▨ Rejected and Later Abandoned (12)

Number of Applications

Month

750

patent number, or subject; patents within particular subject areas are further organized by execution date. The claims and drawings of ultimately unsuccessful applications filed through early 1884 (identified by application case number) are recorded in Patent Application Casebook E-2538, Patent Application Drawings (c. 1879–1886), both PS (*TAED* PT022, PT023, PT004); unsuccessful applications are also described in a separate book of Patent Applications: Abstracts of Edison's Abandoned Applications (1876–85), PS (*TAED* PT004).

B. EDISON'S PATENTS BY SUBJECT

This chart shows the distribution of Edison's successful patent applications by subject area. The broad category of incandescent lamps includes filaments, other components, and manufacturing processes. "Electrical Generation" covers dynamos, their components and operation, and direct conversion or fuel cell processes. The twelve unsuccessful applications completed through April 1884, not represented here, consisted of six for lamps, five for electrical generation, and one (insulated conductors) for the electrical system generally. Three others, each dated 1 July 1884 and listed in the Abstracts of Edison's Abandoned Applications (1876–85), consisted of one for electrical conductors and two for dynamo telegraphy.

B. EDISON'S PATENTS BY SUBJECT

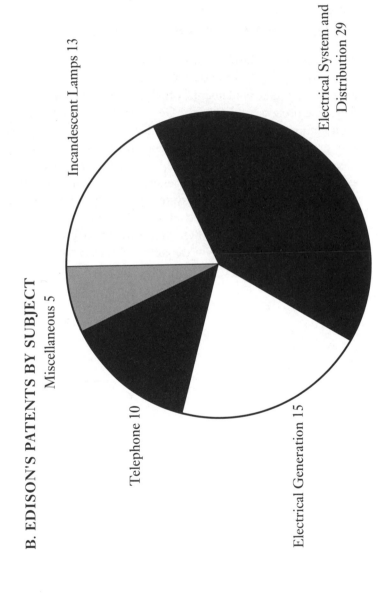

Miscellaneous 5

Incandescent Lamps 13

Electrical System and Distribution 29

Telephone 10

Electrical Generation 15

752

C. U.S. PATENTS RELATING TO ELECTRICITY BY EDISON EMPLOYEES AND ASSOCIATES, EXECUTED APRIL 1883–DECEMBER 1884[1]

This list identifies U.S. patents obtained by individuals whom Edison (or the Edison Electric Light Company) employed, or with whom he had a significant professional relationship in this period. Inventors who left Edison's orbit are identified (with approximate dates) in the notes, but all their known patents from this period are included.[2] These individuals, several of whom were important in their own right, had formative associations with Edison, and the editors have not attempted to disentangle their inventive interests and creativity from his. The names are presented alphabetically; under each name, the patents appear by execution date (where known). This appendix also includes a handful of patents whose applications predate 1 April 1883 that were inadvertently omitted from the corresponding list in *TAEB* 6 App. 5.C. The U.S. Patent and Trademark Office maintains full-text images of all issued patents, searchable by patent number, at: http://patft.uspto.gov/; images may also be obtained through Google Patents at http://www.google.com/patents.

Patentee	Patent No.	Title	Executed	Filed	Issued	Assigned to
Andrews, William	317,610	Electric Lighting System and Switch	12 Aug. 1884	8 Aug. 1884	12 May 1885	Edison Electric Light Co.
Andrews, William	317,700	System of Electrical Distribution	8 Oct. 1884	31 Oct. 1884	12 May 1885	Edison Electric Light Co.
Andrews, William	348,371	Safety-Catch	4 Nov. 1884	20 Nov. 1884	31 Aug. 1886	
Bergmann, Sigmund	341,723	Electrical Switch	18 Dec. 1882	8 Jan. 1883	11 May 1886	Bergmann & Co.
Bergmann, Sigmund	288,916	Galvanic Battery	6 Apr. 1883	11 Apr. 1883	20 Nov. 1883	
Bergmann, Sigmund, with TAE	337,254	Telephone	10 Nov. 1883	13 Nov. 1883	2 Mar. 1886	
Bergmann, Sigmund	337,231	Telephone	1 Dec. 1883	4 Dec. 1883	2 Mar. 1886	Bergmann & Co.
Bergmann, Sigmund	311,100	Socket for Incandescent Electric Lamps	4 Apr. 1884	10 Apr. 1884	20 Jan. 1885	
Bergmann, Sigmund	319,384	Safety-Catch Plug for Electric Circuits	21 Apr. 1884	6 June 1884	2 June 1885	
Bergmann, Sigmund	337,296	Electric Light Fixture	27 Oct. 1884	31 Oct. 1884	2 Mar. 1886	Bergmann & Co.
Bergmann, Sigmund	337,232	Telephone Receiver	11 Dec. 1884	31 Dec. 1884	2 Mar. 1886	
Bradley, Charles[3]	280,563	Electrical Measuring Apparatus	2 Feb. 1883	17 Feb. 1883	3 July 1883	
Bradley, Charles	287,501	Method of Electrical Testing	2 Feb. 1883	17 Feb. 1883	30 Oct. 1883	Edison Electric Light Co.
Bradley, Charles	291,141	System of Electrical Distribution	17 Aug. 1883	18 Aug. 1883	1 Jan. 1884	Edison Electric Light Co.
Bradley, Charles	353,915	Electrical Testing	6 Feb. 1884	5 Apr. 1884	7 Dec. 1886	Edison Electric Light Co.
Bradley, Charles	312,802	Secondary Battery	18 Apr. 1884	19 Apr. 1884	24 Feb. 1885	Bradley Electric Power Co.
Bradley, Charles	312,803	Electric Conducting Material	10 Nov. 1884	11 Nov. 1884	24 Feb. 1885	Bradley Electric Power Co.
Clarke, Charles[4]	284,382	Circuit and Apparatus for Electric Temperature and Pressure Indicators	9 Jan. 1883	19 Jan. 1883	4 Sept. 1883	Robert Hewitt, Jr.
Clarke, Charles	285,572	Telethermometer	15 June 1883	18 June 1883	25 Sept. 1883	Robert Hewitt, Jr.
Clarke, Charles	304,907	Electrical Meter	21 Dec. 1883	26 Dec. 1883	9 Sept. 1884	Telemeter Co.
Clarke, Charles	301,805	Transmitting Device for Primary Electric Clocks and Means for Actuating Secondary Clocks Thereby	11 Jan. 1884	14 Jan. 1884	8 July 1884	Telemeter Co.
Clarke, Charles	304,908	Electric Indicating Device for Elevators	20 Mar. 1884	22 Mar. 1884	9 Sept. 1884	Telemeter Co.

Name	Patent No.	Title				Notes
Clarke, Charles	327,657	Method of Operating Circuits in Telemeter Systems	19 Aug. 1884	20 Aug. 1884	6 Oct. 1885	
Clarke, Charles	327,525	Electric Valve-Operator	28 Nov. 1884	1 Dec. 1884	6 Oct. 1885	
Dyer, Richard	287,757	Thermostatic Fire-Indicator	8 June 1883	11 June 1883	30 Oct. 1883	One-third to Henry W. Seely
Dyer, Richard	305,164	Electrical Cigar-Lighter	5 Dec. 1883	5 Mar. 1884	16 Sept. 1884	
Greenfield, Edwin	316,260	Telephone	Not stated.	6 Oct. 1884	21 Apr. 1885	Half to Holmes Burglar Alarm Telegraph Co.
Holzer, William	293,879	Manufacture of Incandescing Electric Lamps	17 May 1883	7 June 1883	19 Feb. 1884	
Holzer, William	295,398	Manufacture of Incandescing Electric Lamps	17 May 1883	7 June 1883	18 Mar. 1884	Edison Lamp Co.
Holzer, William	323,150	Method of Manufacturing Incandescent Electric Lamps	1 May 1884	5 June 1884	28 July 1885	
Johnson, Edward	339,298	Electrical Switch	18 Dec. 1882	8 Jan. 1883	6 Apr. 1886	
Johnson, Edward	356,689	Telephone	1 Dec. 1883	3 Dec. 1883	25 Jan. 1887	
Kenny, Patrick, and TAE	314,115	Chemical Stock Quotation Telegraph	9 Feb. 1884	19 Mar. 1884	17 Mar. 1885	
Kruesi, John	288,454	Machine for Insulating Electrical Conductors	6 Apr. 1883	11 Apr. 1883	13 Nov. 1883	Electrical Tube Co.
Kruesi, John	296,185	Electrical Conductor and Connecting Device Therefor	28 Sept. 1883	20 Oct. 1883	1 Apr. 1884	Electrical Tube Co.
Kruesi, John	322,385	Electrical Conducting System	1 Nov. 1884	20 Nov. 1884	14 July 1885	
Moses, Otto	343,188	Arc Lamp	26 Jan. 1883	17 Feb. 1883	8 June 1886	
Moses, Otto	310,145	Incandescent Electric Lamp	15 Oct. 1883	18 Oct. 1883	30 Dec. 1884	
Moses, Otto	316,993	Apparatus for Carbonizing Incandescents	21 Dec. 1883	28 Dec. 1883	5 May 1885	
Moses, Otto	321,309	Mold for Making Incandescents	2 Aug. 1884	13 Aug. 1884	30 June 1885	
Moses, Otto	324,038	Incandescent Electric Lamp	21 July 1884	30 July 1884	11 Aug. 1885	
Moses, Otto	324,334	Method of Manufacturing Incandescent Electric Lamps	16 Oct. 1883	29 Oct. 1883	11 Aug. 1885	
Moses, Otto	325,958	Incandescent Electric Lamp	26 July 1884	30 July 1884	8 Sept. 1885	
Sprague, Frank[5]	295,454	Electro-Dynamic Motor	19 Apr. 1883	2 May 1883	18 Mar. 1884	
Sprague, Frank	314,891	Electrical Indicator	22 Jan. 1884	24 Jan. 1884	31 Mar. 1885	

Name	Patent No.	Title	Date 1	Date 2	Date 3	Assignee/Notes
Sprague, Frank	313,247	Regulator for Electro-Dynamic Motors	12 Feb. 1884	21 Feb. 1884	3 Mar. 1885	
Sprague, Frank	309,167	Adjustable Resistance for Electrical Circuits	19 Mar. 1884	2 Apr. 1884	9 Dec. 1884	Samuel Insull
Sprague, Frank	315,179	Electro-Dynamic Motor	23 Apr. 1884	30 Apr. 1884	7 Apr. 1885	
Sprague, Frank	315,180	Electro-Dynamic Motor	23 May 1884	9 June 1884	7 Apr. 1885	
Sprague, Frank	315,181	Electric Motor and Generator	2 July 1884	19 July 1884	7 Apr. 1885	
Sprague, Frank	315,182	Electric Motor and Generator	2 July 1884	19 July 1884	7 Apr. 1885	
Sprague, Frank	313,546	Electro-Dynamic Motor	21 Aug. 1884	4 Nov. 1884	10 Mar. 1885	
Sprague, Frank	318,668	Method of Operating Electric-Railway Trains	12 Dec. 1884	22 Dec. 1884	26 May 1885	Sprague Electric Railway and Motor Co.
Sprague, Frank	323,459	Electric-Railway System	12 Dec. 1884	19 Jan. 1885	4 Aug. 1885	Sprague Electric Railway and Motor Co.
Sprague, Frank	338,313	Electric Railway System	12 Dec. 1884	15 Dec. 1884	23 Mar. 1886	Sprague Electric Railway and Motor Co.
Stern, William[6]	346,388	System of Lighting Railway-Trains by Electricity	9 Nov. 1882	18 Nov. 1882	27 July 1886	Stern and Thomas A. Edison
Stieringer, Luther	281,576	Safety-Catch for Electric-Light Circuits	19 Feb. 1883	16 Mar. 1883	17 July 1883	Edison Electric Light Co.
Stieringer, Luther	294,697	Combined Gas and Electric-Light Fixture	7 June 1882	14 June 1882	4 Mar. 1884	
Stieringer, Luther	307,879	Combined Gas and Electric Light Fixture	27 Mar. 1884	5 Apr. 1884	11 Nov. 1884	
Stieringer, Luther, and Jonathan Vail	343,087	House-Wiring for Electric Lights	21 June 1884	1 July 1884	1 June 1886	
Stieringer, Luther, and Jonathan Vail	317,879	Wiring Interiors for Electric Lights	28 June 1884	14 July 1884	12 May 1885	Half to Charles F. Hanington and Richard N. Dyer
Upton, Francis	315,679	Lightning-Conductor	23 June 1883	25 June 1883	14 Apr. 1885	
Vail, Jonathan[7]	286,350	Electrical Stand-Lamp	21 Oct. 1882	6 Dec. 1882	9 Oct. 1883	
Vail, Jonathan	308,712	Electric-Light Fixture	9 Nov. 1883	24 Nov. 1883	2 Dec. 1884	Bergmann & Co.
Vail, Jonathan	308,713	Electric-Light Conductor for Structures	6 Dec. 1883	12 Dec. 1883	2 Dec. 1884	
Vail, Jonathan; see "Stieringer, Luther"						
Wheeler, Schuyler[8]	335,099	System of Electrical Distribution	22 Sept. 1883	24 Sept. 1883	26 Jan. 1886	

1. Compiled from yearly supplements of the U.S. Patent Office's *Index of Patents Relating to Electricity* (Washington, D.C.: GPO).

2. Two applications that Frank Sprague completed after April 1883, but before beginning his association with Edison, are excluded.

3. Became an independent inventor or engineer sometime in 1883. Doc. 2464 n. 7.

4. Left full-time employment with the Edison Electric Light Co. in early 1884. Doc. 2585 n. 15.

5. Employed by Edison from June 1883 to April 1884. See Docs. 2456, 2656, and 2657.

6. William A. Stern (1860–1914), an acquaintance or associate from Edison's Menlo Park days, was with the Philadelphia printing firm Edward Stern & Co. in 1883. In 1884, he was connected with the Pittsburgh Electric Co. Stern met with George Westinghouse and officials of the Pennsylvania Railroad in early 1883 to try to commercialize the patent for lighting railroad cars by electricity. Apparently unsuccessful, he later offered to sell his interest to Edison. Edison declined and instead offered to sell his portion to Stern. *TAEB* 6 App. 5.C.

7. Jonathan H. Vail (1852–1926), a relative of the telegraph pioneer Alfred Vail and of the telephone executive Theodore Vail, began working in Edison's Menlo Park machine shop in 1880. At the shop and, from 1881, at the Edison Machine Works, Vail gained extensive experience in dynamo assembly and installation. He set up a number of the early Edison isolated lighting plants and, in late 1881, became general superintendent of the Edison Co. for Isolated Lighting. He remained in that position until 1895, supervising construction of central stations in New York, Chicago, Boston, and other cities. "Vail, Jonathan H.," Pioneers Bio.

8. Schuyler Skaats Wheeler (1860–1923) attended Columbia College but left before graduation to begin working for the Jablochkoff and U.S. Electric Lighting Cos. He began working for Edison in 1882 in connection with the Pearl St. central station, and he was later involved with installing central stations in Fall River, Mass., and Newburgh, N.Y. In 1886, Wheeler helped to establish the C. & C. Electric Motor Co., an early and important manufacturer of heavy motors. After years of manufacturing, inventing, and consulting, Wheeler served as president of the American Institute of Electrical Engineers in 1905–1906. "Wheeler, Schuyler Skaats," Pioneers Bio.

Bibliography

Abbott, Arthur V. 1903. *Telephony: A Manual of the Design, Construction, and Operation of Telephone Exchanges.* New York: McGraw Publishing Co.

Abbott, Benjamin Vaughan. 1886. *The Patent Laws of All Nations.* Vol. 1. Washington, D.C.: Charles R. Brodix.

Acheson, Edward G. 1965. *A Pathfinder: Inventor, Scientist, Industrialist.* Port Huron, Mich.: Acheson Industries.

Acts of the Ninety-Fourth Legislature of the State of New Jersey. 1870. Newark: E. N. Fuller.

Acts of the One Hundred and Eighth Legislature of the State of New Jersey. 1884. Camden: Sinnickson Chew.

Adams, Stephen B., and Orville R. Butler. 1998. *Manufacturing the Future: A History of Western Electric.* Cambridge: Cambridge University Press.

Aldrich, Mark. 1997. *Safety First: Technology, Labor, and Business in the Building of American Work Safety, 1870–1939.* Baltimore: Johns Hopkins University Press.

Alglave, Ém[ile], and J. Boulard. 1884. *Electric Light: Its History, Production, and Applications.* Translated with notes and additions by C. M. Lungren and T. O'Conor Sloane. New York: D. Appleton and Co.

American Electrical Directory. (Printed annually.) Fort Wayne, Ind.: Star Iron Tower Co.

Andreas, A. T. 1886. *History of Chicago from the Earliest Period to the Present Time.* Vol. 3, *From the Fire of 1871 until 1885.* Chicago: A. T. Andreas Co.

Andrews, Herbert Cornelius, and Sandford Charles Hinsdale. 1906. *Descendants of Robert Hinsdale.* Edited by Alfred L. Holman. Lombard, Ill.: Privately Printed.

Andrews, W.S. 1924. "The Belt Driven Bipolar Dynamos." *General Electric Review* 27: 163–66.

Annual Report of the President of Johns Hopkins University [11th]. 1886. Baltimore: Johns Hopkins University.

Appleton's Hand-Book of American Travel: Northern and Eastern Tour . . . 1870. New York: D. Appleton and Co.

Appleton's Illustrated Hand-Book of American Winter Resorts for Tourists and Invalids. 1877. New York: D. Appleton and Co.

Arapostathis, Stathis. 2007. "Dynamos, Tests, and Consulting in the Career of John Hopkinson in the 1880s. Researche et Innovation dans L'industrie Électrique." *Annales historiques de l'électricité* 5:9–32.

Aron, Cindy S. 1999. *Working at Play: A History of Vacations in the United States.* New York: Oxford University Press.

Association of the Bar of the City of New York. 1915. *Yearbook.* New York: Knickerbocker Press.

Atkinson, E. 1883. *Elementary Treatise on Physics, Experimental and Applied.* Translated from *Ganot's Eléments de Physique.* New York: William Wood and Co.

Baedeker, Karl. 1890. *Northern Germany as Far as the Bavarian and Austrian Frontiers.* Leipsic: Karl Baedeker.

Baldwin, Neil. 1995. *Edison: Inventing the Century.* New York: Hyperion.

Banes, Charles H. 1885. *General Report of the Chairman of the Committee on Exhibitions.* [International Electrical Exhibition.] Philadelphia: Franklin Institute.

Barrère, Albert, and Charles G. Leland, eds. 1967 [1889]. *A Dictionary of Slang, Jargon & Cant.* Detroit: Gale Research Co. Reprint.

Barrett, J. P. 1894. *Electricity at the Columbian Exposition.* Chicago: R. R. Donelley & Sons.

Bartlett, James H. 1885. "Manufacture of Charcoal Iron in the Province of Quebec." *Journal of the United States Association of Charcoal Iron Workers* 6 (no. 4): 225–36.

Barton, Ruth 1990. "'An Influential Set of Chaps': The X-Club and Royal Society Politics, 1864–85." *British Journal for the History of Science* 23 (no. 1): 53–81.

Bass, Bob. 2008. *When Steamboats Reigned in Florida.* Gainesville: University Press of Florida.

Bassett, Herbert Henry. 1900. *Men of Note in Finance and Commerce, with Which Is Incorporated Men of Office. A Biographical Business Directory.* London: E. Wilson.

Batterberry, Michael, and Ariane Batterberry. 1999. *On the Town in New York: The Landmark History of Eating, Drinking, and Entertainments from the American Revolution to the Food Revolution.* New York: Routledge.

Bazerman, Charles. 2002. *The Languages of Edison's Electric Light.* Cambridge, Mass.: MIT Press.

Beach, Frederick Converse, and George Edwin Rines, eds. 1911. *The Americana: A Universal Reference Library. Biographies.* New York: Americana Co.

Beck, Bill. 1995. *PP&L: 75 Years of Powering the Future.* Eden Prairie, Minn.: Pennsylvania Power & Light Co.

Beecher, Willis J. 1883. *General Catalogue of the Auburn Theological Seminary.* Auburn, N.Y: Daily Advertiser and Weekly Journal Printing House.

Beers, J. H. & Co. 1906. *Commemorative Biographical Record of the County of Lambton, Ontario.* Toronto: Hill Binding Co.

Bell, Herbert C. 1891. *History of Northumberland County Pennsylvania.* Chicago: Brown, Runk & Co.

Benjamin, Park. 1886. *The Age of Electricity: From Amber-Soul to Telephone.* New York: Charles Scribner and Sons.

Bennett, Alfred Rosling, and Herbert Laws Webb. 1974 [1895]. *The Telephone Systems of the Continent of Europe.* New York: Arno Press.

Bergman, Gösta M. 1977. *Lighting in the Theatre.* Stockholm: Almqvist & Wiksell.

Berly, J. A. 1883. *J. A. Berly's British, American and Continental Electrical Directory and Advertiser.* London: Wm. Dawson & Sons.

Bernhardt, Sarah. 1907. *My Double Life: Memoirs of Sarah Bernhardt.* London: William Heinemann.

Biedermann, R., and Otto H. Krause. 1880. "Foreign Patents." *Journal of the American Chemical Society* 2:355–59.

Bishop, William H. 1878. "A Night with Edison." *Scribner's Monthly* 17 (no. 1): 88–99.

Bitmead, Richard. 1910. *French Polishing and Enamelling: A Practical Work of Instruction.* London: Crosby, Lockwood and Son.

Blustein, Bonnie Ellen. 1986. "The Brief Career of 'Cerebral Hyperaemia': William A. Hammond and His Onsomniac Patients, 1854–1890." *Journal of the History of Medicine and Allied Sciences* 41:24–51.

Bolton, Sir Francis. 1886. "Further Historical Notes on the Electric Light." *Journal of the Society of Telegraph-Engineers and Electricians* 15 (no. 63): 391–516.

Bonner, Judith Hopkins. 1982. "George David Coulon: A Nineteenth-Century French Louisiana Painter." *Southern Quarterly* 20 (no. 2): 41–61.

Boston Landmarks Commission. "George Milliken House: Study Report." Edited by Boston Environmental Dept. City of Boston, n.d. [post-2005].

Bowers, Brian. 1969. *R. E. B. Crompton: Pioneer Electrical Engineer.* Science Museum Booklet. London: HMSO.

———. 1982. *A History of Electric Light & Power.* London: Peter Peregrinus, Ltd., in association with the Science Museum.

Boxman, Raymond L. 2000. "Vacuum Arc Deposition: Early History and Recent Developments." Paper presented at the IEEE Nineteenth International Symposium on Discharges and Electrical Insulation in Vacuum, Xi'an, China.

Boyd, Andrew W., comp. (Printed annually.) *Boyd's Directory of the District of Columbia.* Washington, D.C: William H. Boyd.

Bradsby, H. C., ed. 1893. *History of Luzerne County Pennsylvania, with Biographical Selections.* Chicago: S. B. Nelson & Co.

Brearly, Harry Chase. 1916. *The History of the National Board of Fire Underwriters; Fifty Years of Civilizing a Force.* New York: Fredrick A. Stokes Co.

Bright, Arthur A., Jr. 1972 [1949]. *The Electric Lamp Industry: Technological Change and Economic Development from 1800 to 1947.* New York: Arno Press.

Brock, W. H. 1981. "Advancing Science: The British Association and the Professional Practice of Science." In *The Parliament of Science:*

The British Association for the Advancement of Science 1831–1981, edited by Roy MacLeod and Peter Collins, 89–117. Northwood, United Kingdom: Science Reviews Ltd.

Brooks, Herbert B. 1953. "Electrical Indicating Instruments Used in Early Edison Central Stations." *Journal of the Franklin Institute* 256 (no. 5): 401–22.

Brown, Glenn. 1970 [1903]. *History of the United States Capitol.* New York: De Capo Press.

Brown, Rossiter Johnson, and John Howard, eds. 1904. *The Twentieth-Century Biographical Dictionary of Notable Americans.* 10 vols. Boston: Biographical Society.

Brown, T. Allston. 1903. *A History of the New York Stage from the First Performance in 1732 to 1901.* Vol. 2. New York: Dodd, Mead and Co.

Bruce, Robert V. 1973. *Bell: Alexander Graham Bell and the Conquest of Solitude.* Boston: Little, Brown & Co.

Buck, Solon Justus. 1963 [1913]. *The Granger Movement: A Study of Agricultural Organization and Its Political, Economic, and Social Manifestations, 1870–1880.* Lincoln: University of Nebraska Press.

Burk, Kathleen. 1989. *Morgan Grenfell 1838–1988: The Biography of a Merchant Bank.* Oxford, New York: Oxford University Press.

Burns, R. W. 2004. *Communications: An International History of the Formative Years.* London: Institution of Electrical Engineers.

Cabot, F. Elliot. 1885. "The Electrical Exhibition at Boston." *American Architect and Building News* 17 (no. 3): 7–8.

Cahen, Isadore. 1870. "La Quinzaine Politique & Religieuse." *Archives Israélites* 31 (no. 14): 417–25.

Caillard, E. M. 1891. *Electricity: The Science of the Nineteenth Century: A Sketch for General Readers.* New York: D. Appleton and Co.

Carhart, Henry S. 1893. "The Future Ohm, Ampere, and Volt." *Science* 21 (no. 524): 86–87.

Carlson, W. Bernard. *Innovation as a Social Process: Elihu Thomson and the Rise of General Electric, 1870–1900.* Cambridge: Cambridge University Press.

Carosso, Vincent P. 1987. *The Morgans: Private International Bankers, 1854–1913.* Cambridge Mass.: Harvard University Press.

Carosso, Vincent, and Richard Sylla. 1991. "U.S. Banks and International Finance." In *International Banking 1870–1914*, edited by Rondo Cameron and V. I. Bovykin, 48–71. New York: Oxford University Press.

Carr, William H. A. 1975. *From Three Cents a Week . . . : The Story of the Prudential Insurance Company of America.* Englewood Cliffs, N.J.: Prentice-Hall, Inc.

Carroll, Walter F. 1989. *Brockton: From Rural Parish to Urban Center, an Illustrated History.* Northridge, Calif.: Windsor Publications.

Carter, Charles S., ed. 1921. *History of the Class of '70, Department of Literature, Science and the Arts, University of Michigan: Supplement, 1903–1921.* Milwaukee, Wisc.

Cassis, Yossef. 1994. *City Bankers, 1890–1914.* Cambridge: Cambridge University Press.

Ceppi M. de L, Sergio, and Gonzalo Vial Correa. 1983. *Chile, 100 Años de Industria, 1883–1983* Santiago, Chile: Sociedad de Fomento Fabril.

Chandler, Alfred D., Jr. 1990. *Scale and Scope: The Dynamics of Industrial Capitalism*. Cambridge, Mass.: Harvard University Press.

Chappell, Phil Edward, and Eloise Martin Chappell. 1983 [1900]. *A Genealogical History of the Chappell, Dickie, and Other Kindred Families*. F. B. Chappell Family.

Child, Hamilton, comp. 1886. *Gazetteer of Grafton County, N.H., 1709–1886*. Syracuse, N.Y.: Syracuse Journal Co.

Cleveland Directory. (Printed annually.) Cleveland: Cleveland Directory Co.

Clough, Shepard B. 1946. *A Century of American Life Insurance: A History of the Mutual Life Insurance Company of New York, 1843–1943*. New York: Columbia University Press.

Collins, Emerson, and John W. Jordan, eds. 1906. *Genealogical and Personal History of Lycoming County, Pennsylvania*. New York: Lewis Publishing.

The Columbia Gazetter of the World Online. 2005. Columbia University Press.

Commerford, Martin Thomas. 1889. *The Electric Motor and Its Applications*, 2nd ed. New York: W. J. Johnston.

———. 1922. *Forty Years of Edison Service, 1882–1922*. New York: New York Edison Co.

Commissioners of Sewers of the City of London, Select Committee re Electric Lighting. 1883. *Proposals from the Edison Electric Light Company and the Gulcher Electric Light and Power Company*. Charles Skipper & East.

Congressional Directory: Compiled for the Use of Congress by Ben[jamin]. Perley Poore. (Printed annually.) Washington, D.C.: GPO.

Conkling, Alfred Ronald. 1889. *The Life and Letters of Roscoe Conkling: Orator, Statesman, Advocate*. New York: Charles L. Webster & Co.

Conot, Robert. 1979. *A Streak of Luck: The Life and Legend of Thomas Alva Edison*. New York: Seaview Books.

Cooley, Arnold James. 1880. *Cyclopaedia of Practical Receipts and Collateral Information in the Arts, Manufactures, Professions, and Trades*. London: Churchill.

Cordulack, Shelley Wood. 2005. "A Franco-American Battle of Beams: Electricity and the Selling of Modernity." *Journal of Design History* 18 (no. 2): 147–66.

Courtwright, David T. 2001. *Dark Paradise: A History of Opiate Addiction in America*. Cambridge, Mass.: Harvard University Press.

Crisp, Frederick Arthur. 1904. *Visitation of England and Wales*. Vol. 12. [London]: Privately printed.

Crompton, Rookes Evelyn Bell. 1928. *Reminiscences*. London: Constable & Co., Ltd.

Crosby, Oliver Marvin. 1887. *Florida Facts Both Bright and Blue: A Guide Book to Intending Settlers, Tourists, and Investors, from a Northern Standpoint*. New York: privately printed.

Cudahy, Brian J. 1990. *Over and Back: The History of Ferryboats in New York Harbor*. New York: Fordham University Press.

Culbertson, John Norcross. 1926. "Reminiscences of Early Telephone Work, 1878–1913." Unpublished typescript. NjWAT.

Cumbler, John T. 1979. *Working-Class Community in Industrial America: Work, Leisure and Struggle in Two Industrial Cities, 1880–1930*. Westport, Conn.: Greenwood Press.

Cunningham, John T. 1999. *Florham Park*. Charleston, S.C.: Arcadia.

Curry, Wade Chester. 1958. "Steele Mackaye: Producer and Director." Ph.D. diss., University of Illinois at Urbana-Champaign.

Dalzell, Frederick. 2009. *Engineering Invention: Frank J. Sprague and the U.S. Electrical Industry*. Cambridge, Mass.: MIT Press.

Dana, C. L. 1884. "New Portable Faradic Battery." *The Medical Record* 25 (no. 22, Mar. 1884): 334–35.

Davis, William T. 1895. *Bench and Bar of the Commonwealth of Massachusetts*. Boston: Boston History Co.

Dawson, Andrew. 2004. *Lives of the Philadelphia Engineers: Capital, Class, and Revolution, 1830–1890*. Burlington, Vt.: Ashgate.

DeKosky, Robert K. 1976. "William Crookes and the Fourth State of Matter." *Isis* 67:36–60.

Denby, Elaine. 1998. *Grand Hotels: Reality and Illusion*. London: Reaktion Books.

Detwiler, Thomas C. 1881. "Trichinae Spiralis." Thesis, University of Pennsylvania.

Dewey, Fred P. 1885. "The Copper Industry of the United States." *The Chautauquan* 6:95–98.

Dickens, Charles. 1998. *The Letters of Charles Dickens*. Vol. 10 (1862–1864). Edited by Kathleen Tillotson and Graham Storey. Oxford: Clarendon Press.

Dickson, W. K. L., and Antonia Dickson. 1894. *The Life and Inventions of Thomas A. Edison*. New York: Thomas Y. Crowell & Co.

Dilworth, Richardson. 2005. *The Urban Origins of Suburban Autonomy*. Cambridge, Mass.: Harvard University Press.

Directory of Directors of the City of New York. 1915–16. New York: Directory of Directors Co.

Directory of Iron and Steel Works in United States and Canada. 1890. 10th ed. Philadelphia: American Iron and Steel Assoc.

Distinguished Men of Philadelphia and Pennsylvania. 1913. Philadelphia: The Press Co.

Dobyns, Kenneth W. 1994. *The Patent Office Pony: A History of the Early Patent Office*. Fredericksburg, Va: Sergeant Kirkland's Museum and Historical Society, Inc.

Dormanen, Sue. "Major Frank McLaughlin, the Mastermind Behind the Creation of the Golden Gate Villa." Golden Gate Villa of Santa Cruz, Calif. In possession of Thomas Edison Papers.

Dredge, James. 1882–1885. *Electric Illumination*. 2 vols. London: Engineering.

Dublin, Thomas, and Walter Licht. 2005. *The Face of Decline: The Pennsylvania Anthracite Region in the Twentieth Century*. Ithaca, N.Y.: Cornell University Press.

Dudley, Myron Samuel. 1903. *Class of Sixty-Three, Williams College, 1863–1903: Fortieth Year Report*. Boston: Thomas Todd.

Duffy, Eve. 2007. "Oskar Von Miller and the Art of the Electrical Exhibition: Staging Modernity in Weimar Germany." *German History* 25:517–38.

Dyer, Frank, and T.C. Martin. 1910. *Edison: His Life and Inventions*. 2 vols. New York: Harper & Bros.

Edison, Thomas A. 1885. "Electricity: Man's Slave." *Scientific American* 52 (no. 21): 185.

"Edisonia": A Brief History of the Edison Electric Lighting System. 1904. New York: Association of Edison Illuminating Companies.

"Edison System of Electric Railways." In *Second Annual Report*, 27–38. New York: Edison General Electric Co., 1891.

Edlund, E[rik]. 1882 [1881]. "On the Electrical Resistance of Vacuum." *Philosophical Magazine* 13 [5th Series]: 1–20.

———. 1883 [1882]. "Researches on the Passage of Electricity through Rarefied Air." *Philosophical Magazine* 15 [5th Series]: 1–22.

Eduardo, Posada Carbó. 1996. *The Columbian Caribbean: A Regional History, 1870–1950*. Oxford: Clarendon Press.

Edwards, Clement Higgins, and E. W. W. Edwards. 1883. *The Electric Lighting Act, 1882*. London: William Clowes & Son, Ltd.

Eisenman, Harry J. III. 1967. "Charles F. Brush: Pioneer Innovator in Electric Technology." Ph.D. diss. Case Institute of Technology.

Eliezer, Shalom, and Yaffa Eliezer. 2001. *The Fourth State of Matter: An Introduction to Plasma Science*. Philadelphia: Institute of Physics.

Engler, E. A. 1880. "The Carbon Button." *Popular Science Monthly* 17:1–27.

Ernst, Hagen. 1885. *Die Elekrische Beleuchtung*. Berlin: Julius Springer.

Étrangères, Ministère des Affairs. 1882. "Conférence Internationale pour la Détermination des Unités Électriques, 16 Octobre–12 Octobre 1882: Procès-Verbaux." Paris: Imprimerie Nationale.

Evans, Richard, ed. 1884. *New York's Great Industries*. New York: Historical Publishing Co.

Fagen, M. D., ed. 1975. *A History of Engineering and Science in the Bell System: The Early Years (1875–1925)*. [New York]: Bell Telephone Laboratories, Inc.

Fairall, Herbert S. 1885. *The World's Industrial and Cotton Exposition, New Orleans, 1884–1885*. Iowa City, Iowa: Republican Publishing Co.

Feldenkirchen, Wilfried. 1999. *Siemens 1918–1945*. Columbus: Ohio State University Press.

———. 2000. *Siemens: From Workshop to Global Player*. Munich: Piper.

Feuerwerker, Albert. 1958. *China's Early Industrialization: Sheng Hsuan-Huai (1844–1916) and Mandarian Enterprise*. Cambridge, Mass.: Harvard University Press.

Finance and Industry: The New York Stock Exchange. 1886. New York: Historical Publishing Co.

Findling, John E. 1996. "Opening the Door to the World: International Expositions in the South, 1881–1907." *Studies in Popular Culture* 19:29–38.

———, and Kimberly D. Pelle. 1990. *Historical Dictionary of World's Fairs and Exposition, 1851–1988*. New York: Greenwood Press.

Fletcher, W. I. 1909. *The Annual Library Index 1908*. New York: Publishers' Weekly.

Florida State Census. 1971 [1885]. *Schedules of the Florida State Census of 1885*. Washington, D.C.: National Archives, National Archives Microfilm Publication M845.

Fox, Robert. 1996. "Thomas Edison's Parisian Campaign: Incandescent Lighting and the Hidden Face of Technology Transfer." *Annals of Science* 53:157–93.

Francis, John W. 1813. "Futher Observations on Mercury." *The American Medical and Philosophical Register* 3 (no. 4): 476–519.

Franklin Institute. 1884. *Official Catalogue: International Electrical Exhibition*. Philadelphia: Burk & McFetridge.

Frary, Ihna Thayer. 1960 [1940]. *They Built the Capitol*. Freeport, N.Y: Books for Libraries Press.

Frazar, Everett. 1884. *Korea and Her Relations to China, Japan, and the United States*. Orange, N.J.: Chronicle Book and Job Printing Office.

Freeman, A. C. 1901. *The American State Reports*. San Francisco: Bancroft-Whitney.

Friedel, Robert D., Paul Israel, and Bernard S. Finn. 1986. *Edison's Electric Light: Biography of an Invention*. New Brunswick, N.J.: Rutgers University Press.

Frisbie, Louise K. 1980. *Florida's Fabled Inns*. Bartow, Fla.: Imperial Publishing Co.

Füssl, Wilhelm. 2005. *Oskar Von Miller: Eine Biographie*. Munich: C. H. Beck.

Garnet, Robert W. 1985. *The Telephone Enterprise: The Evolution of the Bell System's Horizontal Structure, 1876–1909*. Baltimore: Johns Hopkins University Press.

Gas Commissioners of the Commonwealth of Massachusetts. 1887. *Second Annual Report of the Board*. Boston: Wright & Potter Printing Co.

———. 1917. *Thirty-third Annual Report of the Board*. Boston: Wright & Potter Printing Co.

Gates, William B. 1951. *Michigan Copper and Boston Dollars*. Cambridge, Mass.: Harvard University Press.

Gearhart, Heber G. 1936. "Edison's Experiment in Sunbury." *The Northumberland County Historical Society Proceedings and Addresses* 8: 176–80.

General Alumni Catalogue of New York University. 1916. New York: New York University.

Gibson, Jane Mork. 1984. "The International Electrical Exhibition of 1884 and the National Conference of Electricians: A Study in Early Electrical History." M.A. thesis, University of Pennsylvania.

Gitelman, Lisa, and Theresa M. Collins. 2009. "Medium Light: Revisiting Edisonian Modernity." *Critical Quarterly* 51:1–14.

Gooday, Graeme J. N. 1995. "The Morals of Energy Metering: Constructing and Deconstructing the Precision of the Victorian Electrical Engineer's Ammeter and Voltmeter." In *The Values of Precision*, edited by M. Norton Wise, 239–82. Princeton: Princeton University Press.

———. 2008. *Domesticating Electricity: Technology, Uncertainty, and Gender, 1880–1914*. London: Pickering & Chatto.

Görges, Hans. 1929. *50 Jahre Elektrotechnischer Verein: Festschrift zum Fünfzigjährigen Bestehen des Elektrotechnischen Vereins*. Berlin.

Goss, Charles Fredric, ed. 1912. *Cincinnati: The Queen City, 1788–1912*. Vol. 3. Cincinnati: S. J. Clark Publishing Co.

Great Southern Exposition of Art, Industry, and Agriculture. 1886. Louis-ville.

Grenander, M. E. 1980. "Henrietta Stackpole and Olive Harper: Ema-nations of the Great Democracy." *Bulletin of Research in the Humani-ties* 83 (no. 3): 406–22.

Grieg, James. 1970. *John Hopkinson: Electrical Engineer.* Science Mu-seum Booklet. London: HMSO.

Grove, William R. 1843. "On the Gas Voltaic Batteries Made with a View of Ascertaining the Rationale of Its Action and Its Application to Eudiometry." *Philosophical Transactions of the Royal Society of London* 133: 91–112.

———. 1845. "On the Voltaic Battery. Voltaic Action of Phosphorous, Sulphur and Hyrdocarbons." *Philosophical Transactions of the Royal Society of London* 135: 351–61.

———. 1849. "On the Effect of Surrounding Media on Voltaic Ignition." *Philosophical Transactions of the Royal Society of London* 139: 49–59.

———. 1854. "On the Electricity of Flame." *Philosophical Magazine* 8:403–4.

The Guide to the Charities of New York and Brooklyn. 1885–86. New York: P. F. McBreen

Gutiérrez, Claudio, and Flavio Gutiérrez. 2006. "Physics: Trajectory in Chile." *Historia* [Santiago] 39:477–96.

Hafner, Arthur W. 1993. *Directory of Deceased American Physicians, 1804–1929.* 2 vols. Chicago: American Medical Association.

Hall, E. J. 1887. "Notes on Long Distance Telephone Work." *Meeting of the National Telephone Exchange Association* 9:23–26.

Hall, Henry. 1895–1896. *America's Successful Men of Affairs. An Encyclo-pedia of Comtemporaneous Biography.* New York: New York Tribune.

Harlow, Alvin F. 1936. *Old Wires and New Waves: The History of the Tele-graph, Telephone, and Wireless.* New York: D. Appleton-Century Co.

Harrison, Mitchell Charles. 1902. *Prominent and Progressive Americans: An Encyclopedia of Contemporaneous Biography.* Vol. 1. New York: New York Tribune.

Hartnoll, Phylis, and Peter Found. 1996. *The Concise Oxford Companion to the Theatre.* Oxford: Oxford University Press.

Harvard Law School. 1905. *Quinquennial Catalogue of the Law School of Harvard University 1817–1904.* Cambridge, Mass.: Harvard Law School.

Hasluck, Paul N., ed. 1908. *Wood Finishing, Comprising Staining, Var-nishing, and Polishing.* Philadelphia: David McKay.

Hausman, William, Peter Hertner, and Mira Wilkins. 2008. *Global Electrification: Multinational Enterprise and International Finance in the History of Light and Power, 1878–2007.* Cambridge: Cambridge University Press.

Hawley, Gessner Goodrich. 1987. *Condensed Chemical Dictionary.* 11th ed. Edited by Sr. N. Irving Sax and Richard J. Lewis. New York: Van Nostrand Reinhold Co.

Haywood, William. 1883. *Electric Lighting of the Holborn Viaduct by the Edison System. Report to the Streets Committee of the Honourable Commis-sioners of Sewers of the City of London.* Charles Skipper & East. In CPF.

Heap, David Porter. 1884. *Report on the International Exhibition of Electricity Held at Paris August to November, 1881*. Washington, D.C.: GPO.

Heerding, A. 1986. *The History of N.V. Philips' Gloeilampenfabrieken: A Company of Many Parts*. Vol. 2. Cambridge: Cambridge University Press.

Helmholtz, Hermann von. 1954 [1885]. *On the Sensations of Tone as a Physiological Basis for the Theory of Music*. New York: Dover Publications.

Henshall, James A[lexander]. 1884. *Camping and Cruising in Florida*. Cincinnati: Robert Clarke & Co.

Herfindahl, Orris C. 1959. *Copper Costs and Prices: 1870–1957*. Baltimore: Johns Hopkins University Press.

Hibbard, A. S. 1885. "Some Practical Results in Long Distance Telephoning." *Meeting of the National Telephone Exchange Association* 7:60–66.

Hicks, Maurice J. 1979. *Roselle, New Jersey: Site of Thomas Alva Edison's First Village Plant*. Roselle, N.J.: Roselle Historical Society.

History of St. Clair County, Michigan. 1883. Chicago: A. T. Andreas and Co.

Hocking, M. B., and M. L. Lambert. 1987. "A Reacquaintance with the Limelight." *Journal of Chemical Education* 64 (no. 4): 306–9.

Hodgkins, John B. 1979. "Thomas A. Edison and Major Frank McLaughlin: Their Quest for Gold in Butte County." *Association for Northern California Records and Research*, Research Paper No. 5. Chico, Calif.

Hofman, H. O. 1914. *Metallurgy of Copper*. New York: McGraw-Hill Book Co.

Holbrook, A. Stephen. (Printed annually.) *Holbrook's Newark City Directory*. Newark, N.J: A. Stephen Holbrook.

Hollerith, Herman. 1883. "Statistics of Power Used in Manufactures." In *Tenth Census of the United States*. Vol. 2, *Manufacturing*. Washington, D.C.: GPO.

Holliday, Guy H., ed. 1935. *Quinquennial Catalogue of the Law School of Harvard University, 1817–1934*. Cambridge, Mass.: Harvard University.

Homans, James E. 1901. *A B C of the Telephone*. New York: T. Ardel & Co.

Homberger, Eric. 2000. *Mrs. Astor's New York: Money and Social Power in a Gilded Age*. New Haven: Yale University Press.

Hong, Sungook. 2001. *Wireless: From Marconi's Black-Box to the Audion*. Cambridge, Mass.: MIT Press.

Hopkins, Albert A., and A. Russell Bond, comps. and eds. 1915. *Scientific American Reference Book*. New York: Munn & Co.

Hopkins, Nevil Monroe. 1905. *Experimental Electrochemistry*. New York: D. Van Nostrand.

Horna, Hernan. 1982. "Transportation Modernization and Entrepreneurship in Nineteenth-Century Colombia." *Journal of Latin American Studies* 14 (no. 1): 33–53.

Hoshino, Yoshiro, and Nobuko Iijima. 1992. "The Miike Coal-Mine

Explosion." In *Industrial Pollution in Japan*, edited by Jun Ui, chap. 5. Tokyo: United Nations University.

Houston, Edwin J. 1884. "Notes on Phenomena in Incandescent Lamps." *Transactions of the American Institute of Electrical Engineers* 1:1–8.

———. 1885. *Glimpses of the International Exhibition: The Telephone.* Philadelphia: The Franklin Institute.

Howe, Henry. 1889. *Historical Collections of Ohio, an Encyclopedia of the State.* 2 vols. Columbus, Ohio: Henry Howe & Son.

Howell, John W. 1886. *The Edison System of Central Station Lighting.* New York: Edison Electric Light Co.

Hughes, Thomas P. 1962. "British Electrical Industry Lag: 1882–1888." *Technology and Culture* 3: 27–44.

Hunter, Louis C., and Lynwood Bryant. 1991. *A History of Industrial Power in the United States, 1780–1930.* Vol. 3, *The Transmission of Power.* Cambridge, Mass.: MIT Press.

Huret, Jules. 1899. *Sarah Bernhardt.* London: Chapman & Hall.

Huurdeman, Anton A. 2003. *The Worldwide History of Telecommunications.* New York: Wiley.

Insull, Samuel. 1992. *The Memoirs of Samuel Insull: An Autobiography.* Edited by Larry Plachno. Polo, Ill.: Transportation Trails.

———. "Memoirs." n.d. Typescript on deposit at EP&RI (transcript at NjWOE).

International Publishing Co. 1888. *Illustrated New York: The Metropolis of To-Day.* New York: International Publishing Co.

Israel, Paul. 1998. *Edison: A Life of Invention.* New York: John Wiley & Sons.

Jackson, Dugald C. 1934. "The Evolution of Electrical Engineering." *Electrical Engineering* 53:770–76.

Jacques, William W. 1896. "Electricity Direct from Coal." *Harper's New Monthly Magazine* (December): 144–50.

Jehl, Francis. 1937–41. *Menlo Park Reminiscences.* 3 vols. Dearborn, Mich.: Edison Institute.

Jenkins, Reese V. 1984. "Edison's New Light for Newburgh." *Orange County Historical Society Journal* 13:2–12.

Jenks, William J. 1885. "Description of the Edison Electric Light Plant, of Brockton, Mass." [reprinted from *Electrical Review*]. In PPC (*TAED* CA003A).

———. 1893. *Electricity as a Fire Hazard.* Schenectady, N.Y: General Electric Co.

Johansen, Dorothy O. 1936. "Simeon G. Reed, Pioneer." *Bulletin of the Business Historical Society.* 10:37–43.

Johnson, Christopher. 2006. *This Grand & Magnificent Place: The Wilderness of the White Mountains.* Lebanon, N.H.: University of New Hampshire Press.

Jones, Francis Arthur. 1907. *Thomas Alva Edison: Sixty Years of an Inventor's Life.* New York: Thomas Y. Crowell & Co.

Jones, Hugh A. 1995 [1984]. "Edison's Experiment in Northumberland County." *Northumberland County Historical Society Proceedings and Addresses* 29:69–90.

Jones, H. Macnaughton. 1884. *Practical Manual of Diseases of Women and Uterine Therapeutics: For Students and Practitioners.* New York: D. Appleton.

Jones, Payson. 1940. *A Power History of the Consolidated Edison System, 1878–1900.* New York: Consolidated Edison Co. of New York.

Jordan, D. W. 1990. "The Magnetic Circuit Model, 1850–1890: The Resisted Flow Image in Magnetostatics." *British Journal for the History of Science* 23:131–73.

Josephson, Matthew. 1992 [1959]. *Edison: A Biography.* New York: John Wiley & Sons.

Kandall, Stephen R. 1996. *Substance and Shadow: Women and Addiction in the United States.* Cambridge: Harvard University Press.

Kane, H. H. 1880. *The Hypodermic Injection of Morphia: Its History, Advantages and Dangers.* New York: Chas. L. Bermingham & Co.

Kanigel, Robert. 1997. *The One Best Way: Frederick Winslow Taylor and the Enigma of Efficiency.* New York: Viking.

Kapp, Gisbert. 1882–1883. "Crompton's Coumpound Machine." *Electrician* 10:114–15, 154–55, 200–202, 246–47.

———. 1885–86. "Modern Continuous-Current Dynamo Electric Machines and Their Engines." *Minutes of Proceedings of the Institution of Civil Engineers* 83:123–274.

———. 1891. *Electric Transmission of Energy and Its Transformation, Subdivision and Distribution.* London: Whittaker & Co.

Kendall, John Smith. 1922. *History of New Orleans.* 3 vols. Chicago: Lewis Publishing Co.

Kershaw, Michael. 2007. "The International Electrical Units: A Failure in Standardization?" *Studies in History and Philosophy of Science* 38:108–31.

Ketelaar, J. A. A. 1993. "History." In *Fuel Cell Systems,* edited by Leo J. M. J and Michael N. Mugerwa Blomen. New York: Plenum Press.

Kettenburg, Fritz. 1978. "George Solon Ladd." *Insulators: Crown Jewels of the Wire* 10:4.

Kiddle, Henry, and Alexander J. Schem. 1883. *The Cyclopaedia of Education: A Dictionary of Information for the Use of Teachers, School Officers, Parents, and Others.* 3rd ed. New York: E. Steiger & Co.

[Kimball, W. L.]. 1884. "The Determination of the Ohm." *Science* 3 (no. 48): 10–11; reprinted under Kimball's name as "The Determination of the Ohm," in *Johns Hopkins University Circular* [1885] 3 (no. 29): 56–7.

King, Donald C. 2005. *The Theatres of Boston: A Stage and Screen History.* Jefferson, N.C.: McFarland & Co.

Kingman, Bradford. 1895. *History of Brockton, Plymouth County, Massachusetts, 1656–1894.* Syracuse, N.Y.: D. Mason & Co.

Kinley, David. 1913. *Money: A Study of the Theory of the Medium of Exchange.* London: MacMillan & Co.

Kintner, C[harles]. J. 1884. "Reports of Examiner Kintner." *Transactions of the American Institute of Electrical Engineers* 1:7–9.

Kobrak, Christopher. 2007. *Banking on Global Markets: Deutsche Bank and the United States, 1870 to the Present.* Cambridge: Cambridge University Press.

Kocka, Jürgen. 1999. *Industrial Culture & Bourgeois Society: Business, Labor, and Bureaucracy in Modern Germany*. New York: Berghahn Books.

Kolbe, Hermann, and T. S. Humpidge. 1884. *A Short Text-Book of Inorganic Chemistry*. New York: J. Wiley & Sons.

Krause, Otto H. 1880. "Foreign Patents." *Journal of the American Chemical Society* 2 (no. 7): 355–59.

Landau, Sarah Bradford, and Carl W. Condit. 1966. *Rise of the New York Skyscraper, 1965–1913*. New Haven, Conn.: Yale University Press.

Lanthier, Pierre. 1988. "Les Constructions Électriques en France: Financement et Stratégies de Six Groupes Industriels Internationaux, 1880–1940." Ph.D. diss., 3 vols., University of Paris X (Nanterre).

Lantz, M. S. 1884. *The Acme Cyclopedia and Dictionary*. Philadelphia: Lantz-Lantz Publishing Co.

Lathrop, William G. 1909. *The Brass Industry in Connecticut: A Study of the Origin and Development of the Brass Industry in the Naugatuck Valley*. Shelton, Conn.: W. G. Lathrop.

Latimer, Lewis H. 1890. *Incandescent Electric Lighting: A Practical Description of the Edison System*. New York: D. Van Nostrand Co.

Lawson, Publius Virgilius. 1908. "The Invention of the Roller Flour Mill." *Proceedings of the 55th Annual Meeting of the State Historical Society of Wisconsin* 55:244–58.

Leary, Peter J. 1893. *Newark, N. J. Illustrated. A Souvenir of the City and Its Numerous Industries*. Newark, N. J.: William A. Baker.

Lee, Francis Bazley, ed. 1907. *Genealogical and Personal Memorial of Mercer County, New Jersey*. Vol. 2. New York: Lewis Publishing Co.

Lee, Henry. 1885. *The Tourist's Guide of Florida: Illustrated with Wood-Cuts of Scenes in Florida, Etc., Also Maps of Florida and Jacksonville*. New York: Leve & Alden Printing Co.

Lehlbach, Charles F. J. 1889. "New Phases of Lead and Mercury Posioning." *Medical and Surgical Reporter* 61 (no. 13): 342.

Lemann, Nicholas. 1999. *The Big Test: The Secret History of the American Meritocracy*. New York: Farrar, Straus and Giroux.

Leonard, John W. 1922. *Who's Who in Engineering: A Biographical Dictionary of Contemporaries, 1922–1923*. New York: John W. Leonard Corp.

Levy, Donald M. 1912. *Modern Copper Smelting*. London: C. Griffin & Co., Ltd.

Little, George Thomas. 1909. *Genealogical and Family History of the State of Maine*. Vol. 3. New York: Lewis Historical Publishing Co.

Lloyd, Thomas W. 1929. *History of Lycoming County Pennsylvania*. Topeka and Indianapolis: Historical Publishing Co.

Lock, Alfred G., and Charles G. Lock. 1878. *A Practical Treatise on the Manufacture of Sulphuric Acid*. London: S. Low, Marston, Searle, and Rivingon.

Lockwood, Thomas Dixon. 1883. *Electricity, Magnetism, and Electric Telegraphy; a Practical Guide and Hand-Book of General Information for Electrical Students, Operators, and Inspectors*. New York: D. Van Nostrand.

———. 1887. "Some Recent Advances in Telephony." *Transactions of the American Institute of Electrical Engineers* 3:74–94.

————. 1892. "Selective or Individual Signals." *Transactions of the American Institute of Electrical Engineers* 9:527–45.

————. 1896. "Telephone Repeaters or Relays, and Repeating Systems." *Electrical World* 28:593–97, 626–27, 660–61.

Long Distance Telephone Co. 1884. *The Long Distance Telephone*. New York.

MacDonald, George A. 1896. *How Successful Lawyers Were Educated*. New York: Banks & Brothers.

MacLeod, Roy, and Peter Collins, eds. 1981. *The Parliament of Science: The British Association for the Advancement of Science, 1831–1981*. Northwood, United Kingdom: Science Reviews Ltd.

MacKaye, Percy. 1911. "Steele MacKaye, Dynamic Artist of the American Theatre; an Outline of His Life Work." *The Drama: A Quarterly Review of Dramatic Literature* 1 (no. 4): 138–73.

————. 1927. *Epoch: The Life of Steele MacKaye*. 2 vols. New York: Boni & Liveright.

MacKaye, Steele. 1882. "Safety in Theaters." *North American Review* 135 (no. 311): 461.

Manzotti, Luigi. 1883. *Excelsior: The Great Mimical Dramatic Ballet Spectacle*. [New York?]: Kiralfy Bros.

Marble, Louis M. 1920–21. "Edgar M. Marble." *Journal of the Patent Office Society* 3: 156–58.

Marienthal, Harold Seymour. 1966. *A Historical Study of the New York Lyceum Theatre under the Management of Steele MacKaye, 1884–1885*. Ph.D. diss., University of Southern California.

Marley, David F. 2005. *Historic Cities of the Americas: An Illustrated Encyclopedia*. 2 vols. Santa Barbara, Calif.: ABC-CLIO.

Marquis, James. 1993. *Merchant Adventurer: The Story of W. R. Grace*. Wilmington, Del.: SR Books.

Marshall, David Trumbull. 1930. *Recollections of Boyhood Days in Old Metuchen*. Flushing, N.Y.: Case Publishing.

————. 1931. *Recollections of Edison*. Boston: Christopher Publishing House.

Maver, William, Jr. 1888. "Dynamo Machines in Telegraphy—the New Plant in the Western Union Building, New York." *Electrical World* 11:67, 79–80.

————. 1892. *American Telegraphy: Systems, Apparatus, Operation*. New York: J. H. Bunnell & Co.

Mayo, John. 1987. *British Merchants and Chilean Development, 1851–1886*. Boulder, Colo.: Westview Press.

McClure, James Baird. 1879. *Edison and His Inventions*. Chicago: Rhodes & McClure Publishing Co.

McCullough, David. 1972. *The Great Bridge*. New York: Simon and Schuster.

McCune, George McAfee, and John A. Harrison, eds. 1951. *Korean-American Relations: Documents Pertaining to the Far Eastern Diplomacy of the United States. Volume I: The Initial Period, 1883–1886*. Berkeley: University of California Press.

McMahon, A. Michal. 1984. *The Making of a Profession: A Century of Electrical Engineering in America*. New York: IEEE Press.

Medical Society of the County of New York. 1895. *The Medical Directory of the City of New York.* New York: Stettiner, Lambert & Co.

Mendenhall, Thomas Corwin. 1882. "On the Influence of Time on the Change in the Resistance of the Carbon Disk of Edison's Tasimeter." *American Journal of Science* 24 (3rd series): 43–46.

———. 1886. "On the Electrical Resistance of Soft Carbon under Pressure." *American Journal of Science* 32 (3rd series): 218–28.

Middleton, William D., and William D. Middleton III. 2009. *Frank Julian Sprague: Electrical Inventor & Engineer.* Bloomington: Indiana University Press.

Mighels, Ella Sterling. 1893. *The Story of the Files: A Review of Californian Writers and Literature.* San Francisco: Co-operative Print Co.

Miller, Francis T. 1940. *Thomas A. Edison: An Inspiring Story for Boys.* Philadelphia: The John C. Winston Company.

Miller, Jimmy H. 1991. *The Life of Harold Sellers Colton: A Philadelphia Brahmin in Flagstaff.* Tsaile, Ariz.: Navajo Community College Press.

Miller, Kempster Blanchard. 1905. *American Telephone Practice.* New York: McGraw Publishing Co.

Milliken, John H. 1884. "Dot and Dash Polka." In Library of Congress, Music Division, American 19th-century sheet music, digital id: sm1884 07951.

Minton, Henry. 1884. *Uterine Therapeutics.* New York: A. L. Chatterton Publishing Co.

Moens, Peter Gilles, and Gabriel Moens. 1998. *International Trade and Business: Law Policy and Ethics.* Sydney, Australia: Cavendish Publishers.

Moore, Charles. 1887. "Electric Lighting in the City of Detroit." *Publications of the American Economic Association* (no. 2): 539–50.

Morrell, Jack, and Arnold Thackray. 1981. *Gentlemen of Science: Early Years of the British Association for the Advancement of Science.* Oxford: Clarendon Press.

Morrill, Frank J., William O. Hultgren, and Eric J. Salamonsson. 2005. *Worcester.* Charleston, S.C.: Arcadia Publishing.

Morris, Charles, ed. 1896. *Men of the Century: An Historical Work.* Philadelphia: L. R. Hamersly.

Morris, Ira K. 1898. *Morris's Memorial History of Staten Island, New York.* New York: Memorial Publishing Co.

Morris, Roy Jr. 1995. *Ambrose Bierce: Alone in Bad Company.* New York: Crown Publishers, Inc.

Müller, Falk. 2004. *Gasentladungsforschung Im 19. Jahrhundert.* Diepholz: GNT.

Murray, David. 1972 [1899]. *History of Education in New Jersey.* Port Washington, N.Y: Kennikat Press.

Nam, Moon-hyon. "Early History of Electrical Engineering in Korea: Edison and First Electric Lighting in the Kingdom of Corea." Paper presented at "Promoting the History of EE" conference, Singapore, 23–26 Jan. 2000. In possession of Thomas Edison Papers.

Nash, Michael, John Rumm & Craig Orr. 1985. *Pennsylvania Power & Light Company; a Guide to the Records.* Wilmington, Del.: Hagley Museum and Library.

Neff, Frank A. 1946. "Edison, a Great American." *The Northumberland County Historical Society Proceedings and Addresses* 15:181–99.

Newark Daily Advertiser. 1872. *Hand Book and Guide for the City of Newark, New Jersey*. Newark, N.J: Newark Daily Advertiser.

Newhnam-Davis, Nathaniel. 1908. *Gourmet's Guide to Europe*. New York: Brentano's.

New Jersey Secretary of State. 1914. *Corporations of New Jersey: List of Certificates to December 31, 1911*. Trenton, N.J.: MacCrellish & Quigley.

"New Orleans, Louisiana Birth Records Index, 1790–1899." Provo, Utah: The Generations Network, Inc. (2002). Accessed through Ancestry.com.

"New Orleans, Louisiana Death Records Index, 1804–1949." Provo, Utah: The Generations Network, Inc. (2002). Accessed through Ancestry.com.

New York State Bar Association. 1911. *Proceedings of the Thirty-Fourth Annual Meeting*. Albany, N.Y.: Argus Co.

New York State Board of Railroad Commissioners. 1889. *Annual Report*. Albany, N.Y.

Niaudet, Alfred. 1880. *Elementary Treatise on Electric Batteries*. New York: Wiley.

Nichols, Theodore E. 1954. "The Rise of Barranquilla." *The Hispanic American Historical Review* 34 (no. 2): 158–74.

Norcross, Frank W. 1901. *A History of the New York Swamp*. New York: Chiswick Press.

Novak, Michael. 1978. *The Guns of Lattimer*. East Brunswick, N.J.: Transaction Publishers.

Nutt, John J. 1891. *Newburgh: Her Institutions, Industries, and Leading Citizens*. Newburgh, N.Y.: Ritchie & Hull.

Nye, Bill. 1882. "Entomologists." In *Forty Liars and Others Lies*. Chicago: Bedford Clarke & Co.

Nye, David E. 1983. *The Invented Self: An Anti-Biography, from Documents of Thomas A. Edison*. Odense, Denmark: Odense University Press.

———. 1990. *Electrifying America: Social Meanings of New Technology, 1880–1940*. Cambridge, Mass.: MIT Press.

Orange, A. D. 1981. "Beginnings of the British Association, 1831–1851." In *The Parliament of Science: The British Association for the Advancement of Science, 1831–1981*, edited by Roy MacLeod and Peter Collins, 43–88. Northwood, United Kingdom: Science Reviews Ltd.

Osborn, Col. Norris G. 1907. *Men of Mark in Connecticut*. Hartford: William R. Goodspeed.

Oser, Marion Edison. 1956. "Early Recollections of Mrs. Marion Edison Oser." Typescript of oral history, Box 16, Edison Biographical Collection, NjWOE.

Outerbridge, Alexander E[wing]. 1881. "A Fourth State of Matter." *Journal of the Franklin Institute* 111 (no. 4): 287–97.

Paine, Walter. [n.d.]. *Reminiscences of My Experience in Business-Building*. Typescript in Box 41, EP & RI.

Pancaldi, Giuliano. 1981. "Scientific Internationalism and the British

Association." In *The Parliament of Science: The British Association for the Advancement of Science, 1831–1981,* edited by Roy MacLeod and Peter Collins, 145–69. Northwood, United Kingdom: Science Reviews Ltd.

Paquier, Serge. 1998. *Histoire de L'électricité en Suisse: La Dynamique d'un Petit Pays Européen 1875–1939.* Vol. 2. Genève, Switzerland: Éditions Passe Présent.

Partridge, Eric, and Paul Beale. 2002. *A Dictionary of Slang and Unconventional English.* 8th ed. New York: Routledge.

Passenger Lists of Vessels Arriving at New York, 1820–97. 1962. Washington, D.C.: National Archives Microfilm Publication M237. Accessed through Ancestry.com

Passer, Harold C. 1953. *The Electrical Manufacturers.* Cambridge, Mass.: Harvard University Press.

Pennsylvania General Assembly. 1883. *Laws of the General Assembly of the State of Pennsylvania, Passed at the Session of 1883.* Harrisburg, Pa.: Lane S. Hart.

———. 1885. *Laws of the General Assembly of the State of Pennsylvania, Passed at the Session of 1885.* Harrisburg, Pa.: Lane S. Hart.

Perkins, Daniel W. 1885. *Practical Common Sense Guide Book through the World's Industrial and Cotton Centennial Exposition at New Orleans.* Harrisburg, Pa.: Lane S. Hart.

Piccirilli, Ricardo, ed. 1953. *Diccionario Histórico Argentino.* Buenos Aires: Ediciones Histricas Argentinas.

Pinheiro, Mario J. "Plasma: The Genesis of the Word." (arXiv:physics/0703260v1 [physics.hist-ph]). Accessed January 2009 at http://arxiv.org/PS_cache/physics/pdf/0703/0703260v1.pdf.

Plum, William Rattle. 1882. *The Military Telegraph During the Civil War in the United States: With an Exposition of Ancient and Modern Means of Communication, and of the Federal and Confederate Cipher Systems.* Chicago: Jansen, McClurg & Co.

Polk's Ann Arbor, Ypsilanti and Washtenaw County Directory. (Printed annually.) Detroit: R. L. Polk & Co.

Poor, Henry V. 1889. *Manual of the Railroads of the United States.* New York: H. V. & H. W. Poor.

Pope, Franklin. 1871. *The Telegraphic Instructor: A Hand-Book for Students of Telegraphy.* New York: F. L. Pope.

———. 1892. *Modern Practice of the Electric Telegraph.* 15th ed., New York: Van Nostrand.

Porter, Charles T. 1908. *Engineering Reminiscences Contributed to "Power" and "American Machinist."* New York: John Wiley & Sons.

Preece, William Henry. 1884. "A Visit to Canada and the United States in the Year 1884." *Journal of the Society of Telegraph-Engineers and Electricians* 13 (no. 11): 570–92.

———. 1885. "On a Peculiar Behavior of Glow-Lamps When Raised to High Incandescence." *Proceedings of the Royal Society of London* 38 (Mar.): 219–30.

Prescott, George B. 1884a. *Dynamo-Electricity: Its Generation, Application, Transmission, Storage and Measurement.* New York: D. Appleton and Co.

———. 1884b. *Bell's Electric Speaking Telephone: Its Invention, Construction, Application, Modification and History.* New York: D. Appleton.

Price, C. Matlack. 1914. "A Pioneer in Apartment House Architecture: Memoir on Philip G. Hubert's Work." *Architectural Record* 36:74–76.

The Providence Directory and Rhode Island Business Directory. 1892. Providence: Sampson, Murdock & Co.

Quain, Richard. 1883. *A Dictionary of Medicine Including General Pathology, General Therapeutics, Hygiene, and the Diseases Peculiar to Women and Children.* New York: D. Appleton and Co.

Ramirez, Ron. *The History of Philco.* Chap. 1: "Lamps to Batteries." www.philcoradio.com/history. Accessed May 2010

Ramson, John. 1884. "The St. Johns Region in Florida." *Outing and the Wheelman* 3 (no. 5): 321–27.

Rand's New York City Business Directory. (Printed annually.) New York: Rand Business Directory Co.

Ratzel, Eric. 1998. "Un Aventurier des Temps Industriels: Pierre Eugène Secrétan, 1836–1899." *Cahiers d'Historie de l'Aluminium* 22: 27–48.

Reich, Leonard S. 2002 [1985]. *The Making of American Industrial Research: Science and Business at GE and Bell, 1876–1926.* Cambridge: Cambridge University Press.

Reid, James D. 1879. *The Telegraph in America.* New York: Derby Bros.

———. 1886. *The Telegraph in America.* rev. ed. New York: John Polhemus.

Remsen, Ira. 1903. *A College Text-Book of Chemistry.* New York: Henry Holt and Co.

Report of the Electrical Conference at Philadelphia in September 1884. 1886. Washington, D.C.: GPO.

Report of the Examiners of Section XIX. [International Electrical Exhibition.] "Telegraphic Systems." 1884. Philadelphia: Franklin Institute.

Report of the Examiners of Section XXIX. [International Electrical Exhibition.] "Educational Apparatus." 1885. Philadelphia: Franklin Institute.

Report of the Examiners of Section XXVI. [International Electrical Exhibition.] "Applications of Electricity to Artistic Effects . . . . 1886. Philadelphia: Franklin Institute.

Reynolds, J. Russell, and H. Charlton Bastian, 1879. "Congestion of the Brain." In *A System of Medicine,* edited by J. Russell Reynolds, 844–54. Philadelphia: Henry C. Lea.

Rhode Island General Assembly. 1881. *Acts and Resolves Passed by the General Assembly of the State of Rhode Island and Providence Plantation at the May Session 1881.* Providence: E. L. Freeman & Co.

Richardson, Joe M. 1964. "The Florida Excursion of President Chester A. Arthur." *Tequesta* 24:41–47.

Richardson, Thomas. 1863. *Chemical Technology; or, Chemistry in Its Applications to the Arts & Manufactures.* London: H. Baillière.

Richter, Ernst. 1883. "Experimental Results by Siemens and Halske on Dynamo-Electrical Machines with Constant Terminal E.M.F." *Electrician* 11 (19 May):19–21.

Ricord, Frederick W. 1897. *History of Union County, New Jersey.* Vol. 1. Newark, N.J.: East Jersey History Co.

Riesco, Alejandro Valdés. 1897. *Quiebras: Comentarios al Libro IV Del Código de Comercio.* Vol. 1. Santiago, Chile: Establecimiento Poligráfico Roma.

Ripley, George, and Charles A. Dana, eds. 1883. *The American Cyclopaedia: A Popular Dictionary of General Knowledge.* New York: Appleton & Co.

Rippy, J. Fred. 1945. "The Development of Public Utilities in Colombia." *The Hispanic American Historical Review* 25 (no. 1): 132–37.

Roscoe, Henry E., and Carl Schorlemmer. 1878. *A Treatise on Chemistry.* New York: D. Appleton.

Rosenberg, Robert. 1990. "Academic Physics and the Origins of Electrical Engineering in America." Ph.D. diss., Johns Hopkins University.

Rowland, Henry A. 1902. *The Physical Papers of Henry Augustus Rowland.* Baltimore: Johns Hopkins University Press.

Rule, William, ed. and comp., with George F. Mellen and J. Woolridge. 1900. *Standard History of Knoxville, Tennessee.* Chicago: Lewis Publishing Co.

Samuel, Cartner, III. 1968. *Cyrus Field: Man of Two Worlds.* New York: G. P. Putman's Sons.

Scannell, J. J., ed. 1917. *Scannell's New Jersey's First Citizens: Biographies and Portraits of the Notable Living Men and Women of New Jersey.* Paterson, N.J.: Scannell.

Schaffer, Simon. 1992. "Late Victorian Metrology and Its Instrumentation: A Manufactory of Ohms." In *Invisible Connections: Instruments, Institutions, and Science,* edited by R. Bud and S. E. Cozzens, 23–56. Bellingham, Wash.: SPIE Optical Engineering Press.

———. 1994. "Rayleigh and the Establishment of Electrical Standards." *European Journal of Physics* 15: 277–84.

Schallenberg, Richard H. 1981. "The Anomalous Storage Battery: An American Lag in Early Electrical Engineering." *Technology and Culture* 22 (no. 4): 725–52.

———. 1982. *Bottled Energy: Electrical Engineering and the Evolution of Chemical Energy Storage.* Philadelphia: American Philosophical Society.

Scharf, J. Thomas, and Thompson Westcott. 1884. *History of Philadelphia, 1609–1884.* 3 vols. Philadelphia: L. H. Everts & Co.

Schellen, Heinrich. 1884. *Magneto-Electric and Dynamo-Electric Machines: Their Construction and Practical Application to Electric Lighting and the Transmission of Power.* Translated by Nathaniel S. Keith and Percy Neymann. New York: D. Van Nostrand.

Scott, Charles F. 1934. "The Institute's First Half Century." *Electrical Engineering* 53:645–70.

Secretary's Report Number VII [Harvard College Class of 1880]. 1905. Cambridge, Mass.: Riverside Press.

Shamokin Area Centennial: Then and Now, 1864–1964. 1964. Shamokin, Pa.

Shaw, G. M. 1878. "Sketch of Thomas Alva Edison." *Popular Science Monthly* 13:487.

Shelton, W. H. 1918. *The Salmagundi Club, Being a History of Its Beginning as a Sketch Class, Its Public Service as the Black and White Society, and Its Career as a Club from MDCCCLXXI to MCMXVII.* Boston: Houghton, Mifflin.

Shepardson, George Defrees. 1917. *Telephone Apparatus: An Introduction to the Development and Theory.* New York: D. Appleton and Co.

Sheppard, F. H. W., ed. 1960. *Survey of London.* Vol. 29, pt. 1, *The Parish of St. James Westminster, South of Piccadilly.* London: London County Council [by] the Athlone Press, University of London.

Siemens, Werner von. 2005 [1892]. *Werner Von Siemens Recollections.* Edited by Wilfried Feldenkirchen. Munich: Piper.

Silliman, Benjamin. 1871. *Principles of Physics, or Natural Philosophy.* New York: Ivison, Blakeman, Taylor & Co.

Simmons, Jack, and Gordon Biddle. 1997. *The Oxford Companion to British Railway History.* Oxford and New York: Oxford University Press.

Simonds, William Adams. 1934. *Edison: His Life, His Work, His Genius.* Indianapolis: Bobbs-Merrill Co.

Singer, Isidore, ed. 1910. *International Insurance Encyclopedia.* Vol. 1, *Insurance Men, Past and Present.* New York: American Encyclopedic Library Assoc.

Skinner, Thomas. 1882. *The Stock Exchange Year-Book for 1883.* London: Gassell, Fetter, Galpin & Co.

Smith, Albert W. 1925. *John Edson Sweet: A Story of Achievement in Engineering and of Its Influence Upon Men.* New York: American Society of Mechanical Engineers.

Smith, George David. 1985. *The Anatomy of a Business Strategy: Bell, Western Electric, and the Origins of the American Telephone Industry.* Baltimore: Johns Hopkins University Press.

Sobel, Robert. 1978. *Biographical Directory of the Governors of the United States, 1789–1978.* Westport, Conn.: Meckler Books.

Souvenir Booklet Presented by the United Light & Railways Company on the Occasion of the Public Opening of the Riverside Power Station. 1926. United Light & Railways Co.

Spehr, Paul C. 2008. *The Man Who Made Movies: W. K. L. Dickson.* Eastleigh: John Libbey.

Spencer, Herbert. 1884. *The Principles of Sociology.* Vol. 1, New York: D. Appleton & Co.

Spero, James, and Edmund Vincent Gillon. 2002. *New York at the End of the 20th Century: A Photographic Guide.* Mineola, N.Y.: Dover Publications.

Spies, Stacy. 2000. *Metuchen.* Charleston, S.C.: Arcadia.

Sprague, Frank S. 1883a. "The Edison-Hopkinson Dynamo-Electric Machine." *Engineer* 56:105; also published without notice of transmittal as "Report on the Edison-Hopkinson Dynamo," *Electrician* 11 (11 Aug. 1883): 296–99.

———. 1883b. *Report on the Exhibits at the Crystal Palace Electrical Exhibition, 1882. Office of Naval Intelligence, General Information Series.* Washington, D.C.: GPO.

———. 1904. "Some Personal Experiences." *Street Railway Journal* 24 (no. 15): 566–75.

Squires, William H., ed 1905. "Alumniana with Necrology." *Hamilton Literary Magazine* 40 (June): 237–48.

Standard Corporation Service. 1917. New York: Standard Statistics.

Steiger's Educational Directory for 1878. 1878. New York: E. Steiger.

Stieringer, Luther. 1901a. "The Evolution of Exposition Lighting." *Western Electrician* 29 (no. 12): 187–89.

———. 1901b. "From Christmas Tree to Pan-American." *Electrical World and Engineer* 38:285–90.

Stocking, Charles Henry Wright. 1903. *The Stocking Ancestry: Comprising the Descendants of George Stocking, Founder of the American Family.* Chicago: Lakeside Press.

Stover, Donald L. 1982. *American Sculpture: The Collection of the Fine Arts Museums of San Francisco.* San Francisco: Fine Arts Museums of San Francisco.

Strelinger, Chas. A. & Co. 1979 [1897]. *Wood Workers' Tools: Being a Catalogue of Tools, Supplies, Machinery, and Similar Goods Used by Carpenters, Builders, Cabinet Makers.* In *Trade Catalogues at the Winterthur Museum, Part 2,* edited and compiled by Eleanor McD. Thompson. Bethesda, Md.: University Publications of America.

Stross, Randall. 2007. *The Wizard of Menlo Park.* New York: Crown Publishers.

Strouse, Jean. 1999. *Morgan: American Financier.* New York: Random House.

Sullivan, Joseph Patrick. 1995. "From Municipal Ownership to Regulation: Municipal Utility Reform in New York City, 1880–1907." Ph.D. diss., Rutgers University.

Swan, J[oseph]. 1881. "Electric Lighting by Incandescence." *Electrician* 7 (no. 17): 283–84.

Swan, Kenneth R. 1946. *Sir Joseph Swan and the Invention of the Incandescent Electric Lamp.* London: Longmans, Green and Co.

Swan, Mary Edmonds, and Kenneth R. Swan. 1929. *Sir Joseph Wilson Swan F. R. S.: A Memoir.* London: Ernest Benn, Ltd.

Swanson, Kara W. 2009. "The Emergence of the Professional Patent Practitioner." *Technology and Culture* 50 (no. 3): 519–48.

Swinton, Alan Archibald Campbell. 1884. *The Principles and Practice of Electric Lighting.* London: Longmans, Green.

Taltavall, John B. 1893. *Telegraphers of to-Day; Descriptive, Historical, Biographical.* New York: John B. Taltavall.

Tate, Alfred O. 1938. *Edison's Open Door: The Life Story of Thomas A. Edison, a Great Individualist.* New York: E. P. Dutton.

Taylor, Jocelyn Pierson. 1978. *Grosvenor Porter Lowrey.* New York: Privately printed.

Teal, Harvey J. 2001. *Partners with the Sun: South Carolina Photographers, 1840–1940.* Columbia, S.C: University of South Carolina Press.

Teisch, Jessica B. 1981. "Great Western Power, 'White Coal,' and Industrial Capitalism in the West." *Pacific Historical Review* 70 (no. 2): 221–53.

Telegraphy. 1901. Vol. 2. Scranton, Pa.: International Textbook Co.

Third Report of the Secretary of the Class of 1881 of Harvard College. 1887. Boston: Rand Avery Co.

Thirty Years of New York, 1882–1912, Being a History of Electrical Development in Manhattan and the Bronx. 1913. New York: New York Edison Co.

Thomas de la Peña, Carolyn. 2003. *The Body Electric: How Strange Machines Built the Modern American.* New York: New York University Press.

Thompson, Edgar K. 1954. "The First Light." *U.S. Naval Institute Proceedings* 80 (no.12): 1390–91.

Thompson, Silvanus P. 1882. "Note on the Alleged Change in the Resistance of Carbon Due to Change of Pressure." *American Journal of Science* 24 (3rd series): 433–34.

————. 1884a. *Recent Progress in Dynamo-Electric Machines: Being a Supplement to "Dynamo-Electric Machinery."* New York: Van Nostrand.

————. 1884b. "Recent Progress in Dynamo-Electric Machines." *Electrician* 12 (29 Mar.): 474–78.

————. 1896. *Dynamo-Electric Machinery: A Manual for Students of Electrotechnics.* 5th ed. New York: American Technical Book Co.

————. 1902. *Dynamo-Electric Machinery: A Manual for Students of Electrotechnics.* 8th ed. New York: M. Strong.

Thornall, Jay W. 1982. *Israel Thornell [sic], Planter, Woodbridge, New Jersey, and His Descendants of the Woodbridge–Metuchen, N.J., Area.* Princeton, N.J: Minute Press.

Thorndale, Theresa. 1900. "The Birthplace of Edison Dreams of Her Fallen Greatness." [Firelands Historical Society, Norwalk, Ohio] 13:716–23.

Thorne, Robert. 1984. "Places of Refreshments in the Nineteenth-Century City." In *Buildings and Society: Essays on the Social Development of the Built Environment,* edited by Anthony D. King, 228–53. London: Routledge & Kegan Paul.

Thurston, Robert H. 1939. *A History of the Growth of the Steam-Engine.* Centennial ed. Ithaca, N.Y.: Cornell University Press.

Tidd, Ernest George. 1889. "The Use of Electricity for Theatre Lighting." *Journal of the Society of Telegraph Engineers and Electricians* 17:459–66.

Tolles, Bryant Franklin. 2003. *Resort Hotels of the Adirondacks: The Architecture of a Summer Paradise.* Hanover, N. H.: University Press of New England.

Torrey, David, comp. 1885. *Memoir of Major Jason Torrey, of Bethany Wayne County, Pa.* Scranton, Pa: James S. Horton.

Trachtenberg, Alan. 1965. *Brooklyn Bridge: Fact and Symbol.* Chicago: University of Chicago Press.

The Trow Co-Partnership and Corporation Directory of the Boroughs of Manhattan and the Bronx. 1906. New York: Trow Directory Co.

Trow's New York City Directory. (Printed annually.) New York: Trow City Directory Co. [Cited in *TAEB* 1–5 under the name of H. Wilson, compiler.].

Tyler, Daniel F. 1881. *Where to Go in Florida.* New York: Hopcraft & Co.

United Kingdom General Register Office. *England and Wales Civil Reg-*

istration Indexes England & Wales Death Index: 1837–1915. Accessed through Ancestry.com.

United Kingdom Office for National Statistics. 1861. *Census Returns of England and Wales.* London. Accessed through Ancestry.com.

———. 1871. *Census Returns of England and Wales.* London. Accessed through Ancestry.com.

———. 1881. *Census Returns of England and Wales.* London. Accessed through Ancestry.com.

United Kingdom. Parliament. House of Commons. 1883a (224). *Bill to Confirm Provisional Orders by Board of Trade under Electric Lighting Act (1882) Relating to Bermondsey, Clerkenwell, Hampstead, Holborn, Hornsey, St. George's-in-East, St. Giles (Brush), St. James' and St. Martin's, St. Luke's and Wandsworth.* London: HMSO.

———. 1883b. *Select Committee of Electric Lighting Provisional Orders Bills.* London: HMSO.

———. 1888. *Report of the Royal Commission on the Working of the Metropolitan Board of Works Inquiry, Interim Report.* London: HMSO.

United States. 1901. *Journal of the Executive Proceedings of the Senate of the United States of America.* Washington, D.C.: GPO.

University of Michigan. 1913. *General Announcement of Courses in Engineering and Architecture 1913–1914.* Ann Arbor: University of Michigan.

Urquhart, John W. 1881. *Electro-Typing.* London: Crosby, Lockwood and Co.

U.S. Census Bureau. 1883–88. *Tenth Census of the United States.* Washington, D.C.: GPO.

———. 1963? (1850). *Population Schedules of the Seventh Census of the United States, 1850.* National Archives Microfilm Publication Microcopy M432. Washington, D.C.: National Archives. Accessed through Ancestry.com.

———. 1967? (1860). *Population Schedules of the Ninth Census of the United States, 1860.* National Archives Microfilm Publication Microcopy M653. Washington, D.C.: National Archives. Accessed through Ancestry.com.

———. 1965 (1870). *Population Schedules of the Ninth Census of the United States, 1870.* National Archives Microfilm Publication Microcopy M593. Washington, D.C.: National Archives. Accessed through Ancestry.com.

———. 1970 (1880). *Population Schedules of the Tenth Census of the United States, 1880.* National Archives Microfilm Publication Microcopy T9. Washington, D.C.: National Archives. Accessed through Ancestry.com

———. 1982? (1900). *Population Schedules of the Twelfth Census of the United States, 1900.* National Archives Microfilm Publication Microcopy T623. Washington D.C.: National Archives. Accessed through Ancestry.com.

———. 1982 (1910). *Population Schedules of the Thirteenth Census of the United States, 1910.* National Archives Microfilm Publication Microcopy T624. Washington, D.C.: National Archives. Accessed through Ancestry.com

———. 1992 (1920). *Population Schedules of the Fourteenth Census of the United States, 1920.* National Archives Microfilm Publication Microcopy T625. Washington, D.C.: National Archives. Accessed through Ancestry.com

———. 2002 (1930). *Population Schedules of the Fifteenth Census of the United States, 1930.* Washington, D.C.: National Archives Microfilm Publication Microcopy T626. Accessed through Ancestry.com.

U.S. Department of the Interior. (Printed annually.) *Annual Register of the Department of the Interior.* Washington, D.C: GPO.

U.S. Department of Labor. 1914. *Strike in the Copper Mining District of Michigan.* Washington, D.C.: GPO.

U.S. Department of State. 1951. *Despatches from United States Ministers to Korea, 1883–1905.* Washington, National Archives.

———. *Passport Applications, 1795–1905.* National Archives Microfilm Publication M1372. Washington, D.C.: National Archives. Accessed through Ancestry.com.

U.S. Internal Revenue Service. 1977 (1866). *Internal Revenue Assessment Lists for Pennsylvania, 1862–1866.* (Record Group 58). Washington, D.C.: National Archives. Accessed through Ancestry.com.

U.S. Patent Office. [Annual supplements]. *Index of Patents Relating to Electricity.* Washington, D.C.: GPO.

U.S. Senate. 1884. *Report of the Committee on Post Offices and Post Roads on Postal Telegraph.* Washington, D.C.: GPO.

U.S. Tariff Commission. 1882. *Report.* 2 Vols. Washington, D.C.: GPO.

Usselman, Steven W. 2002. *Regulating Railroad Innovation: Business, Technology, and Politics in America, 1840–1920.* New York: Cambridge University Press.

Vail, J[ohn] H. 1916 "Experiences in Pioneer Electrical Engineering." Typescript of 28 April lecture before Rochester, N.Y., section of AIEE. Box 42, EP & RI.

Van den Noort, Jan. 1993. *Licht Op Het Geb: Geschiedenis Van Het Gemeente-Energiebedrijf Rotterdam.* Rotterdam: NV-GEB.

Van der Feijst, G. 1975. *Geschiedenis Van Schiedam.* Schiedam: Interbook International.

Vargas, Fernando Silva. 2008. "Formas De Sociabilidad En Una Urbe Portuaria: Valparaíso, 1850–1910." *Boletín de la Academia Chilena de Historia* 1 (no. 117): 81–159.

Vaughn, John. 1940. "The Thirtieth Anniversary of a Great Invention." *Scribner's Magazine* 40: 365–77.

Veit, Richard F., and Mark Nonestied. 2008. *New Jersey Cemeteries and Tombstones: History in the Landscape.* New Brunswick, N.J: Rivergate Books.

Verity, John B. 1891. *Electricity Up to Date, for Light, Power, and Traction.* London and New York: Frederick Warne & Co.

Vicuña Mackenna, Benjamín. 1877. *De Valparaiso a Santiago: Datos, Impresiones, Noticias, Episodios De Viaje.* Santiago, Chile: Libreria del Mercurio.

———. 1881. *La Edad del Oro en Chile: O Sea una Demostracion Histórica de la Maravillosa.* Vol. 1. Santiago, Chile: Imprenta Cervantes.

———. 1882. *El Libro de la Plata.* Santiago, Chile: Imprenta Cervantes.

Waits, Robert K. 2001. "Edison's Vacuum Coating Patents." *Journal of Vacuum Science and Technology [A]* 19: 1666–73.

Wasserman, Mark. 2000. *Everyday Life and Politics in Nineteenth-Century Mexico: Men, Women, and War*. Albuquerque: University of New Mexico Press.

Waterhouse, Theodore. 1894. *Theodore Waterhouse, 1838–1891: Notes of His Life and Extracts from His Letters and Papers*. London: Chiswick Press.

Watkins, John Elfreth. 1919. *Famous Mysteries: Curious and Fantastic Riddles of Human Life That Have Never Been Solved*. Philadelphia: John C. Winston.

Watson, Thomas D. 1984. "Staging the 'Crowning Achievement of the Age'—Major Edward A. Burke, New Orleans and the Cotton Centennial Exposition, Pt. II." *Louisiana History* 25 (no. 4): 341–66.

Watt, Alexander. 1887. *Electro-Deposition: A Practical Treatise on the Electrolysis of Gold, Silver, Copper, Nickel, and Other Metals, and Alloys*. 2nd ed. London: Crosby Lockwood and Co.

Watts, Henry. 1882. *A Dictionary of Chemistry and the Allied Branches of Other Sciences*. New ed. London: Longmans Green & Co.

Waugh's Blue Book of Leading Hotels and Resorts of the World. (Printed annually.) Boston: W. Wallace Waugh & Son.

Weis, Eleanor. 1988. *Saga of a Crossroads: Florham Park*. Florham Park, N.J.: Historical Society of Florham Park.

Wells College. 1894. *General Catalogue of the Officers and Students with Historical Sketches of the Founder*. Aurora, N.Y.: Wells College.

Wharton, J. J. S., and J. M. Lely. 1883. *Wharton's Law-Lexicon*. London: Steven and Sons.

Whipple's Electric, Gas and Street Railway Financial Reference Directory. 1890. Detroit: Fred H. Whipple Co.

Wile, Frederic William. 1914. *Men around the Kaiser: The Makers of Modern Germany*. Indianapolis: Bobbs-Merrill Co.

Wilkinson, David J. 2004. "The Contributions of Lavoisier, Scheele, and Priestley to the Early Understanding of Respiratory Physiology in the Eighteenth Century." *Review of Resuscitation* 61: 249–55.

Williams, F. W. 1906. *A History of the Class of Seventy-Nine, Yale College . . . .* Cambridge, Mass.: University Press.

Williams College. 1939. *Obituary Record of the Alumni*. Williamstown, Mass.: Williams College.

Williamson, Jefferson. 1975. *The American Hotel*. New York: Arno Press.

Wilmer, G. W. A. 1915. *Ordinances of the City of Middletown*. Middletown, Ohio: Naegele-Auer Printing Co.

Wilson, J. F. 1988. *Ferranti and the British Electrical Industry, 1864–1930*. Manchester: Manchester University Press.

Wiman, Erastus. 1884. *Competitive Telegraphy, Can It Be Sustained?: An Interview from the Chicago "Inter-Ocean."* New York.

Withers, Hartley. 1933. *National Provincial Bank, 1833 to 1933*. London: National Provincial Bank.

Wolkonowicz, John Paul. 1981. *The Philco Corporation: Historical Review and Strategic Analysis*. M.S. thesis, Massachusetts Institute of Technology.

Wood, H. Truman. 1882. "Seeing by Telegraph." *Frank Leslie's Popular Monthly* 14 (Dec.): 756–59.

Woodbury, David O. 1949. *A Measure for Greatness: A Short Biography of Edward Weston.* New York: McGraw Hill.

Woodings, Calvin, ed. 2000. *Regenerated Cellulose Fibres.* Cambridge: Woodhead Publishing, Ltd.

Woods, John W. (Printed Annually.) *Wood's Baltimore City Directory.* Baltimore: John W. Woods.

Worboys, Michael. 1981. "The British Association and Empire: Science and Social Imperialism, 1880–1940." In *The Parliament of Science: The British Association for the Advancement of Science, 1831–1981,* edited by Roy MacLeod and Peter Collins, 170–87. Northwood, United Kingdom: Science Reviews Ltd.

Worth, David S. 2008. *Spencer Trask: Enigmatic Titan.* New York: Kabique.

Wrege, Charles D., and Ronald G. Greenwood. 1984. "William E. Sawyer and the Rise and Fall of America's First Incandescent Electric Light Company, 1878–1881." *Business and Economic History* 13:31–48.

Wright, Helen Martha. 1939. *The First Presbyterian Congregation, Mendham, Morris County, New Jersey: History and Records, 1738–1938.* Jersey City, N.J.: Helen Martha Wright.

Yale, University. 1909. *Obituary Record of Graduates of Yale University Deceased during the Academical Year Ending in June.* New Haven, Conn.: Yale University.

Young, William C. 1973. *Documents of American Theater History.* Vol. 1, *Famous American Playhouses, 1736–1899.* Chicago: American Library Association.

Zaremba, Charles W. 1883. *The Merchants' and Tourists' Guide to Mexico.* Chicago: Althrop Publishing.

Zhaojin, Ji. 2003. *A History of Modern Shanghai Banking; The Rise and Decline of China's Finance Capitalism.* Armonk, N.Y.: M. E. Sharpe.

Credits

Courtesy of the AT&T Archives & History Center: Docs. 2745, 2765. From the collections of the Rare Book and Manuscript Library (Edison Manuscript Collection) of Columbia University: Doc. 2560. Reproduced with permission of the Rauner Special Collections Library (Percy MacKaye Papers), Dartmouth College: Doc. 2762. Courtesy of the Henry Ford Museum and Greenfield Village: Docs. 2559, 2627, 2630, 2758. From the collections of the Historical Society of Pennsylvania (Uriah Hunt Painter Papers): Doc. 2746. Courtesy of the Library of Congress (American Memory): illustration on p. 479. Reproduced with permission of the Cudahy Memorial Library (Samuel Insull Papers), Loyola University: Doc. 2438. Reproduced with permission of the Metropolitan Opera Archives: Doc. 2445. From the collections of the Manuscript and Archives Division (Frank J. Sprague Papers) of the New York Public Library: Docs. 2575, 2584; also from the New York Public Library: frontispiece. Courtesy of the National Museum of American History (William Hammer Collection), Smithsonian Institution: illustrations on pp. 181–82 (neg. 2003-35552 and Hammer Coll. Series 4).

Reproduced from Thompson 1884a, 48: illustration on p. 31. Reproduced from *Electrical Review* 4 (23 Aug. 1884), 1: illustration on p. 221. Reproduced from Alglave and Boulard 1884, 160: illustration on p. 304. Reproduced from *Scientific American* 51 (18 Oct. 1884), 239 and (15 Nov. 1884), 310: illustrations on pp. 573, 668. Reproduced from United States patents 370,123; 274,290; 280,563; 287,524; 391,595: illustrations on pp. 12, 40, 292, 328, 582.

Index

Boldface page numbers signify primary references or identifications; italic numbers, illustrations. Page numbers refer to headnote or document text unless the reference appears only in a footnote.

Achard, Arthur, 487 nn. 1–2, 488 nn. 5–6
Acheson, Edward Goodrich, 369, 372, **390** n. 1, 466, 468
Acme Cyclopedia and Dictionary, 534
Adams, Edward Dean, **583,** 669 n. 6, 696 n. 4, 705
Adams, Milton F., 313 n. 2
Adams, Theodore, **244**
Ader, Clement-Agnes, 719 n. 1
African-Americans, 145–*46,* 598 n. 5, 656–*57*
Agreements with: American Bell, 666, 680–81; Ansonia Brass & Copper Co., 241–43; Armington & Sims Engine Co., 38 n. 3; Babcock & Wilcox, 38; Brockton Edison, 81–82, 218; central station engineers, 423–24; Compagnie Continentale Edison, 240 n. 7; Eaton, 61–63; Edison Electric, 61–63, 418 n. 2, 460 n. 1, 483, 548 n. 5, 577 n. 2, 583–84, 727, 735; Edison Electric, Ltd., 193 n. 7, 236–38, 635–36; Edison Electric of Europe, 496–97; Lawrence Edison, 218; Lowell Edison, 218; Ohio Edison, 387 n. 12, 433; Piqua Edison, 433; Reed, 61–63; Shamokin Edison, 81–82, 218; Sherman, 577–78; Stephen Field, 61–63; Stewart, 418 n. 2, 532; Sunbury Edison, 218, 615, 664; Tiffin Edison, 311; Utica Edison, 250; Western Union, 681; Williamsport Edison, 382
Alaska (steamship), 128 n. 2, 133

Albany, N.Y.: proposed central station, 4, 26, 28, 108; TAE in, 4, 12 n. 10, 28, 633
Allen, W. H., 145 n. 2
Altenberg, George, **508**
American Association for the Advancement of Science, 313 n. 4, 415 n. 6, 571 n. 1, 703 n. 3
American Bell Telephone Co., **28,** 557; agreement with TAE, 666, 680–81; agreement with Western Union, 681; Electrical and Patent Dept., 694 n. 2; infringement suits, 28; local networks, 666; long-distance lines, 666, 675; Mechanical Dept., 590, 658 n. 6, 674, 693; and patents, 693; TAE's salary, 682 n. 3; TAE's work for, 591, 666, 674–75, 716; transmitters for, 677 n. 3, 691 n. 2; Vail as general manager, 675, 680–81
American Brass Association, 242, 244 n. 6
American District Telegraph Co., 309
American Electrical Exhibition (Boston), 719 n. 1, 720 n. 4
American Institute (New York), 313 n. 3
American Institute of Electrical Engineers, 529, 571 nn. 1–2, 704 n. 3, 724, 757 n. 8
American Novelty Co., 335 n. 2
American Pin Co., 244 n. 6
American Press Association, 278
American Society of Mechanical Engineers, 235 n. 6

American Solid Leather Button Co., 655 n. 1
American Swedes Iron Co., 398 n. 7
American Telegraph Co., 99 n. 1, 478 n. 4, 683 n. 7
American Telegraph Works, 567 n. 7
American Telephone & Telegraph Co., 675, 683 n. 7
Anderson, Hugo, **162**
Anderson, William A., **98,** 233 n. 3
Andrews, William Symes, **23,** 198 n. 4; appointed chief electrician, 4, 41; and boilers, 462; and central station dynamos, 367; and central stations, 151, 186–87, 199 n. 9, 220–22, 264–65, 279, 337, 358, 373 n. 1, 421 n. 2, 462–63, 515–16; and commutator brushes, 30 n. 14, 317; dynamo experiments, 15, 23, 65–66; on economy of conductors, 448–49; and engineer's exam, 209–10, 222; and safety fuses, 120 n. 2; salary, 127, 162; and spare dynamo connections, 186–87; and village plant instructions, 171 n. 1, 187; and voltage indicators, 203 n. 5, 291, 293–94, 378–79; and voltage regulation, 220–22
—letters: to Edison Construction, 448–49; from Insull, 515–16; to Insull, 337; from TAE, 209–10, 394; to TAE, 23, 65–66, 186–87, 367, 378–79
Anglo-American Brush Electric Light Corp., Ltd., 84 n. 2

Bruch, Charles Patterson, **56**

Brunel, Isambard, 357 n. 7

Brush, Charles, 509 n. 2

Brush Electric Light Co., 98, 477 n. 3, 507, 710 n. 2

Brush Illuminating Co., 344 n. 1

Brush Provincial Lighting Co., 95 n. 1

Brush-Swan Electric Light Co. of New-England, 284 n. 3

Buchel, Emile François, **348**

Buchel Machine Works, 348 n. 1

Buckingham, Charles L., **596**, 682 n. 3

Bunn, J. F., 311 n. 1

Burnett, Mr., 162

Burnham, Charles E., **162**

Burrell, David Hamlin, **457** n. 5

Byron, George Gordon (Lord), 140 n. 34

Cable codes, 400

Calcium lamps, 667, 720

Callahan, Edward, 309 n. 1

Cambridge University, 695 n. 1

Campbell, Charles H., **184**, 358, 403, 551 n. 2

Campbell (A.) & Sons, 185 n. 1

Canada: patent law, 465 n. 3; TAE in, 4, 12 n. 10, 57 n. 2, 95; Tate in, 385; telephone in, 683 n. 5. *See also* Edison Lamp Co.: lamp factory: and Canadian factory; Electric lighting: in Canada

Canadian Club (New York), 592 n. 1

Canadian Pacific Railway Co., 63 n. 3

Card, Benjamin, 91 n. 1

Carman, Theodore, 630 n. 3

Carman, William, 630 n. 3

Carranza, Carlos, 195 n. 2

Carvalho, David Nunes, 653 n. 2

Cary, Clarence, 682 n. 3

Cases (patent): No. 187, 228 n. 2; No. 283, 441 n. 12; No. 404, 548 n. 5; No. 579, 118; No. 611, 349; No. 614, 392–93, 559 n. 3; No. 616, 391 n. 2; No. 619, 426 n. 1

Casho, Joseph, 34 n. 42, **108**, 733–34

Cassel, Ernest, 322 n. 3

Cassidy, Edward J., 460

Cassidy, Patrick, 461 n. 2

Catlin, Benjamin Rush, **596**

Catlin, Fred, 343–**44**

Caveats: arc light carbons, 459 n. 4; nonferrous ore separation, 539–43

Ceccarini, Giovanni, 528 n. 5

Centennial Exhibition (Philadelphia), 97 n. 1, 435 n. 8

Central American Cable Co., 334 n. 3

Central Edison Light Co., 148 n. 6

Central High School (Philadelphia), 571 n. 2

Central Park Building Co., 246 n. 1

Central Thomson-Houston Electric Co. (Cincinnati), 201 n. 1

Chadbourne, Edwin Ruthvin, 623 n. 5

Chamberlain, Joseph, 135 n. 3

Chamberlaine, William Wilson, **422**

Chandler, Albert B., **334**

Chase, R. Gardner, 479 n. 6

Chase (R. Gardner) & Co., 478

Chatard, Alfred, 183 n. 12

Chemistry: chemical recording paper, 600 n. 1; electrodeposition, 390, 444; electromotograph, 472 n. 6; French polish, 458, 471; lamp filaments, 230, 353, 454, 456, 466, 468–71, 473–74; primary batteries, 608; storage batteries, 472; telephone, 688–90. *See also* Artificial materials; Fuel cells

Cherbuliez, Antoine, 487 n. 1

Cherokee Gold Mining Co., 734

Chicago: Common Council, 155 n. 4; electric railways in, 67; Haverly Theatre, 327; Palmer House, 262 n. 1; TAE in, 4, 107 n. 10, 124, 128. *See also* Bliss, George; National Exposition of Railway Appliances; Western Edison Light Co.

Chicago Daily Tribune, 75, 219 n. 7, 431 n. 5

Chicago General Electric Co., 538 n. 5

Chicago Inter-Ocean, 592

Chicago Republican, 593 n. 2

Chicago Telephone Co., 155 n. 4

Childs, George William, 672 n. 1

Chile: telephone in, 618 n. 3. *See also* Electric lighting: in Chile

Chinnock, Charles Edward, **108**, 134; and commutator brushes, 269; and fire safety, 283 n. 4; proposed as Edison Electric director, 669 n. 6, 695; as superintendent of first district, 98 n. 3, 108, 123, 358, 439, 531, 725–26, 732

Churchill, Randolph, **100**

Cincinnati: arc lighting in, 477 n. 3; capitalists, 122; exposition in, 114. *See also* Central Edison Light Co.; Ohio Edison Installation Co.

City and Guilds Central Technical Institution, 512 n. 2

Clarendon Hotel (New York), 2, 153, 268 n. 3, 278, 280, 285 n. 1, *309*, 443 n. 1, 521 n. 1, 575

Clark, Edward, **147**

Clark, Henry A., 78 n. 11, 91 n. 1, 146–**47**, 423 n. 3, 650 n. 3

Clark, John Marshall, **154**

Clark, William J., 618 n. 3

Clarke, Charles Lorenzo, **162**; and conductor determinations, 358; and Engineering Dept., 338 39; and meter men's exam, 210 n. 1; and National Conference of Electricians, 636 n. 1; and Pearl St., 289 n. 4; reminiscences, 544 n. 2; resigns from Edison Electric, 372, 385; salary, 162; and voltage indicators, 291

Clarke, Harriet, 566 n. 4, **659**

Claudius, Hermann, 79 n. 14, **128**, 162, 182 n. 8, 361 n. 4

Cleveland, Grover, 699 n. 12

Clifton House, 505 n. 2

Coerper, Carl, 283 n. 3

Colombo, Giuseppe, **120**, 160 nn. 1–2, 373–74, 503–4

Colt's Armory, 235 n. 6

Columbia, 381 n. 3

Columbia College: electrical engineering program, 4, 57, 68–69; Law School, 543 n. 4; School of Mines, 538 n. 4

Columbian University Law School, 598 n. 3

Comitato per Applicazioni dell'Elettricita Edison. *See* Società Generale Italiana di Elettricità Sistema Edison

Commercial Advertiser, 368

Commercial Union Assurance Co., 136 n. 13

Commonwealth Edison (Chicago), 175 n. 12

Compañía de Luz Eléctrica de Edison de Santiago, La, 416–17, 530, 618 n. 3

Compagnie Continentale Edison, 322 n. 3, 390 n. 1, 504 n. 11; and Deutsche Edison, 641 n. 2; Dutch interests, 351 n. 4; and

589, 620–21, 631–32; Newark shops, 54 n. 12; orders fireworks, 152, 589; owns cows, 495; at Philadelphia Electrical Exhibition, 589–90, 653, 656 n. 2, 659, 674, 687, 715 n. 1; and physiognomy, 149; portrait of, 590; reminiscences, 723–38; reputation, 100 n. 3, 565–66, 668; residences, 2, 110, 113 n. 16, 153, 248, 268, 280, 285 n. 1, 309, 443 n. 1, 481, 483–84, 519, 521, 535–36, 562–63, 575, 589, 620, 622, 629–30, 637, 659, 724, 732, 734, 736; in Shamokin, 4, 42, 152, 269, 277 n. 1; sleep habits, 653 n. 2; in Sunbury, 40–41, 151, 153, 196, 358; as telegrapher, 590; testimony on overhead wires, 280, 344 n. 1; as vice-president of AIEE, 529; views on Jews, 736–37; views on scientific community, 704; as Wizard of Menlo Park, 562; in Worcester, 481, 522; work habits, 568 n. 11

—finances: assists Pitt Edison, 218 n. 4; assists Reményi, 262; assists Samuel Edison, 575; and British patent expenses, 669; for cow pasturing, 495 n. 1; Drexel, Morgan loans, 279, 308; and Edison Construction, 76, 123, 251 n. 2, 346–47, 370, 384, 405, 481, 513–14, 609, 614–15, 654–55, 661–65, 725, 735; experimental accounts, 346–47; Florida vacation, 426, 437, 452–53, 485; and Germania Bank, 329–30; and Jennie Stilwell's education, 616 n. 1, 639; lobbying payments, 576; and manufacturing shops, 26, 433, 513–14, 610, 661, 701; medical bills, 671; owed by Edison Ore Milling Co., 396; proposal to Brewster, 706; Seyfert suit, 483, 518–19, 535–36, 603; and South American lighting, 578 n. 1; stock in Edison companies, 610, 651 n. 6

Edison, Thomas Alva, Jr. ("Dash"), 478 n. 3, 564; appearance of, 565; education, 590, 684; mother's death, 589, 626–27, 632, 634; moves to 39 E. 18th St., 637 n. 2; as TAE's beneficiary, 721

Edison, William Leslie, 564, 626; birth, 621, 632; education, 684;

mother's death, 589, 626–27, 632, 634; moves to 39 E. 18th St., 637 n. 2; as TAE's beneficiary, 721

Edison, William Pitt, 217, 443 n. 6, 630 n. 5

Edison and His Inventions, 560

Edison Co. for Isolated Lighting, 33 n. 32, 149 n. 1, 366 n. 4, 538 n. 3; agents, 25, 78 n. 12, 149 n. 10, 644; agreements with Edison Electric, 126 n. 3, 329, 585, 588, 642–50; automatic regulator for, 440; and Brockton station, 662 n. 2; business prospects, 97, 109, 122–23, 158; catalog, 21 n. 28, 654 n. 2; central station licenses, 45 n. 12; Chicago Dept., 31 n. 21, 55 n. 1; defect books, 171 n. 3, 303 n. 4; dynamo orders, 21 n. 19, 25–26, 173, 260, 384, 433; exhibitions, 54, 126 n. 3, 158, 572 n. 3, 710 n. 2; finances, 25–26, 123, 162–63, 278, 384; Goddard offered presidency, 102; Johnson as president, 584, 668, 683, 698; lighting of government buildings, 148 n. 4; lists of plants, 124 n. 6; New England Dept., 25, 31 n. 21, 123, 142, 145, 226, 387 nn. 9–10, 605 n. 1, 720 n. 4; offices, 25; promotes central stations, 43, 75–78; proposed South American business, 707 n. 3; relations with manufacturing shops, 330, 410, 648–49; reorganization of, 409–11, 482, 583–85, 642–50; report on Maxim plants, 91 n. 2; and ship-lighting, 19 n. 4; stock, 308, 646–48, 672 n. 4; takes over central station business, 509–10, 513, 515, 517–18, 583, 588, 605, 643–46, 648; wiring for, 123

Edison Effect, 279, 294–95, 299–300, 325–26, 359, 369, 378–79, 394, 399, 695 n. 3, 704, 731, 735

Edison Electric Illuminating Co. of Ashland, 651 n. 4; central station, 740

Edison Electric Illuminating Co. of Bellefonte: license from Edison Electric, 651 n. 4; payment to TAE, 610, 614

—central station: begins operation, 383, 740; capacity of, 740; conductor determinations, 172; cost of, 336, 740; dynamo, 449 n. 1;

staff, 421 n. 2, 434 n. 6; steam engine, 433, 462; wiring, 406

Edison Electric Illuminating Co. of Boston, 28, 72 n. 1

Edison Electric Illuminating Co. of Brockton: agreement with TAE, 81–82, 218; conductor determinations, 728; investors, 43; license from Edison Electric, 651 n. 4; officers, 48 n. 44; organization of, 33 n. 37, 43, 72; payment to TAE, 277 n. 1, 614, 661–62; price charged customers, 524

—central station, 39; begins operation, 43, 278, 358, 740; broken lamps, 304 n. 2; canvass for, 43; capacity of, 122, 740; construction of, 42, 109, 122, 152, 160 n. 7, 264–65; cost of, 122, 740; customers, 189, 267, 277 n. 2; dynamos, 449 n. 1, 662 nn. 1–2; estimates for, 72, 92 n. 2; first underground three-wire, 43, 278, 358; meters, 304; operation of, 358–59, 662 n. 1; overhead wires, 149; photograph of, 653 n. 2; Siemens visits, 402; spare dynamo, 304–5; staff, 149, 281; switches, 264–65; underground conductors, 43, 307; voltage box, 440 n. 4; voltage indicator, 295, 326 n. 1; voltage regulation, 264–65; wiring, 43, 72, 189–90, 264–65

Edison Electric Illuminating Co. of Brooklyn, 111 n. 6

Edison Electric Illuminating Co. of Circleville: central station, 78 n. 13, 201 n. 4, 433, 477 n. 3, 482, 740; license from Edison Electric, 651 n. 4; street lighting, 475–76

Edison Electric Illuminating Co. of Cumberland, 651 n. 4; central station, 740

Edison Electric Illuminating Co. of Fall River: license from Edison Electric, 651 n. 4; payment to TAE, 614, 654–55

—central station, 360; begins operation, 44, 359, 368 n. 3, 740; capacity of, 44, 740; complaints about, 655 n. 2; conductor determinations, 142; construction of, 44, 306–7, 311; cost of, 336, 740; cost of conductors, 45 n. 9; dynamo, 367; estimates for, 142, 277 n. 1; feeder regulators, 463; Siemens visits, 402; underground

Electric lighting: arc lighting, 43, 88 n. 4, 91 n. 1, 106 n. 8, 143, 148 n. 4, 212, 283 n. 3, 343–44, 441 n. 17, 475–76, 507–8, 559, 571 n. 2, 594, 611, 653, 686 n. 2, 698, 720 n. 4; in Argentina, 36 n. 3, 418 n. 2, 550, 617–18; in Australia, 460 n. 1; in Bolivia, 418 n. 2; in Brazil, 418 n. 2, 439; in British Guiana, 439; Brooklyn Bridge, 91 n. 1, 105; in Canada, 4, 12 n. 10, 57 n. 2, 698; in Central America, 460 n. 1, 482; in Chile, 120 n. 2, 371, 416–17, 530–31, 550, 617, 736; in China, 730, 733, 735; Christmas trees, 363 n. 11; in Cincinnati, 477 n. 3; in Cleveland, 477 n. 3; in Colombia, 87–88; compared to gas lighting, 72, 93, 245, 273–74, 331, 525; dramatic lighting, 116, 152, 215, 572 n. 3, 666–68, 673, 710, 714–15, 720, 735; exhibitions, 4, 12 n. 10, 54, 57 n. 2, 95 n. 1, 105, 420 n. 8, 522 n. 2, 572 n. 3; fire safety, 98, 116 n. 2, 280, 282–83; foreign business proposal, 706; Guatemala, 707 n. 2; infringement suits, 279–80, 319; in Japan, 552, 702, 707 n. 2; in Korea, 552, 707 n. 2; in Mexico, 418 n. 2, 460 n. 1, 550; for mining, 260–61, 702 n. 2; in Peru, 707 n. 2; and photography, 590, 658 n. 5; in Salvador, 707 n. 2; signs, 181–82, 572 n. 3, 573; in South America, 36 n. 3, 195, 416–17, 441 n. 17, 460 n. 1, 482, 550; in Spain, 488 n. 5; street lighting, 87–88, 332–33, 475–76, 507–8
—patent rights: Argentina, 482, 577–78; Austria, 641 n. 2, 642 n. 4; Bolivia, 482; Britain, 30 n. 6, 33 n. 34, 84 n. 2, 138 n. 23, 139 n. 32, 191, 351, 635–36; Canada, 643, 645; Chile, 482; Denmark, 641 n. 2; Europe, 160 n. 8; Germany, 97 n. 1, 158 n. 1, 640–41; Hungary, 642 n. 4; Italy, 486–87; Mexico, 550; Norway, 139 n. 32; Russia, 641 n. 2; South America, 3, 36 n. 1, 139 n. 32, 371, 416, 441 n. 17, 460 n. 1, 482, 530, 550, 618 n. 2; Sweden, 139 n. 32; Switzerland, 486–87
Electric Lighting Act of 1882, 135 n. 3, 357 n. 7

Electric lighting central stations, 699 n. 1; Appleton, 368 n. 3; Argentina, 4; Baranquilla, 87–88; Berlin, 178, 526; Buenos Aires, 195; conductor determinations, 179, 288, 361 n. 4, 448–49, 497 n. 2; copper costs, 45 n. 9, 242–43; copper supply, 43, 241–42; economics of, 206, 267, 272, 274–75, 309–10; estimate for, 438; in Europe, 4; Holborn Viaduct, 30 n. 13, 120 n. 2, 350, 493 n. 9, 544 n. 2, 734; London, 126, 133; maintenance of, 423–24; motors for, 37; Paris, 288, 323 n. 5, 355, 526; Rotterdam, 351 n. 4; Santiago, 371, 416–17, 530, 618, 736; in South America, 4; steam engines, 114, 180, 229; test (resistance) lamps, 170, 267 n. 8; U.S. Electric Lighting Co.'s, 90; Valparaiso, 371, 416, 617–18. See also Electric lighting distribution system; Società Generale Italiana di Elettricità Sistema Edison: Milan central station
—village system: Albany, 4, 28; boilers, 462; compared to Pearl St., 3–4, 39; costs of, 41, 74; customer expectations, 271–76; dangers of electrocution, 343–44; distribution system, 341; economics of, 288, 448–49; first three-wire, 278; operating instructions, 151, 168–71, 271–76; return on investment, 206; Roselle, 39, 73, 94, 119 n. 3, 164 n. 14, 199 n. 9, 263, 276 n. 1, 651 n. 4, 654 n. 2, 740; Sarnia, 95 n. 2; short circuits, 120; spare dynamos for, 151, 186–87; statistics, 534; steam engines for, 462; and street lighting, 475–76. See also central stations under the various Edison Electric Illuminating Cos.; Edison (Thomas A.) Construction Dept.
—patents: ampere meter, 125 n. 11; meter, 5–12; overhead wires, 267 n. 6; regulation, 86 n. 2; regulators, 172 n. 7; three-wire, 29 n. 2, 39–40; voltage conversion, 59 n. 2; voltage indicators, 291, 296, 299–301
Electric lighting distribution system: ampere meters, 123, 359, 600 n. 1, 733–34; feeder regula-

tors, 290 n. 6, 359, 381, 439, 463, 532; high voltage, 579–81; hydroelectric, 58–59; inside wiring, 270 n. 5, 302; lightning protection, 94, 151, 188 n. 5, 197; metallic circuit, 95 n. 3; meters, 5–12, 51, 118, 274, 304–5, 733–34; overhead wires, 3–4, 48 n. 45, 94–95, 267 n. 6, 280, 343–44, 358; power transmission, 511–12; remote control switching, 118–19, 171; safety fuse, 120, 358–59, 471; switches, 187, 202; three-wire, 29 n. 2, 39–40, 43, 58–59, 74, 151, 153 n. 1, 171, 179, 207, 220–21, 222, 288, 290, 293, 299–301, 304–5, 341, 358, 501, 580; two-wire, 43, 119 n. 3, 293, 341, 501; underground conductors, 43, 95, 344 n. 1, 385; use of storage batteries, 284 n. 3; voltage indicators, 42, 202, 221, 290–301, 325–27, 338–39, 359, 369–70, 378–79, 381–82, 394 n. 1, 399, 433, 439, 463, 467, 532; voltage regulation, 42, 151, 172 n. 7, 220–21, 222, 264–65, 279, 290–96, 300, 327, 369–70; voltages, 15, 39–40, 59 n. 2, 86 n. 2, 153 n. 1, 344; volt boxes, 381, 439; water power, 580; wire insulation, 392–93, 655
Electric lighting generators, 502; for arc lighting, 560 n. 1; armatures, 13–18, 24, 65–66, 211–12, 269, 365, 425–26, 441 n. 11; bipolar, 13; C dynamo (Jumbo), 13, 24, 65, 112 n. 12, 160 n. 2, 174 n. 7, 178, 229 n. 2, 270 n. 6, 347, 350, 373–74, 490, 501, 544, 636 n. 4, 733, 745; for China, 730, 735; commutator brushes, 14, 24, 31, 168–70, 269, 279, 316–17, 349; commutators, 365, 463; compound-wound, 181, 207–8, 653; copper disk, 471 n. 5; for determination of ohm, 483, 505; disk dynamo, 20 n. 7, 439; Edison Machine Works experiments, 13, 15, 587; E dynamo, 65, 420 n. 8; for electroplating, 259; exhibitions, 15, 17, 658 n. 3; experimental costs, 15, 346–47; field magnets, 3, 13–18, 23–24, 288, 425–26; Franklin Institute dynamo tests, 571 n. 2; gasoline engines for, 340, 500 n. 2; G dynamo, 746; Gramme's, 435 n. 8;

Hennesey, Jerome, 334 n. 3
Herrick, Albert, 69 n. 4
Herter, Christian, 733
Hewitt, Abram, **57**
Hewitt, Robert, Jr., 387 n. 15
Hibbard, A. S., 675
Highland National Bank, 405 n. 1
Hipple, James, 183 n. 12, 323 n. 2
Hiscox, Gardner Dexter, **597**
Hix, William Preston, 91 n. 1, **143**, 650 n. 3
Hoadley (J. C.) Engine Works, 60 n. 1
Hoboken Land and Improvement Co., 440 n. 7
Hoffbauer, F., 361 n. 2
Holborn Restaurant, 377, 491 n. 1, 492 n. 7, 493 n. 9
Holland, Charles, 335 n. 2
Holloway, James, **491**
Holzer, Alice Stilwell (Mrs. William), 185, **494** n. 1, 567 n. 6; reminiscences, 562 n. 5, 567 n. 8, 623 n. 14; stock transfers, 351 n. 3; and TAE's and Mary's courtship, 566 n. 5; and TAE's and Mary's honeymoon, 635 n. 8; vacations with Mary Edison, 152, 185; works for TAE, 566 n. 5
Holzer, William, 186 n. 3, **248**, 555 n. 2; Edison Lamp Co. stock, 248, 498; and high candlepower lamps, 720; lamp experiments, 353 n. 1, 388–89, 450; at Long Beach Hotel, 152; and proposed contracting company, 510 n. 1; salary, 248
Honness, Ada, **360**
Hopkinson, Edward, 14, 442 n. 19
Hopkinson, John, **23**, 342 n. 5, 442 n. 19, 511; and dynamo design, 3, 12–18, 23–24, 110, 211–12; and Edison Electric Ltd., 133; and Edison & Swan United, 442 n. 19, 492 n. 2; and isolated plants, 131; patents, 14, 24, 29 n. 2, 191, 224; three-wire system, 29 n. 2, 39, 224
Hoskin, John, 91 n. 1, 650 n. 3
Hotel Brunswick (New York), **694**
Houston, Edwin James, **570–71**, 595 n. 2, 704 n. 3
Howard, Henry, **60**, 433, 450
Howell, Ed. P., 63
Howell, John, **104**; Edison Lamp Co. stock, 248; education, 69 n. 3; investigates lamp breakage, 310 n. 3; investment in Edison Lamp

Co., 497–98; and lamp tests, 104–5; and National Conference of Electricians, 636 n. 1; and voltage indicators, 296, 369, 399
Howell, Wilson, 650 n. 3
Howland, Gardiner, 734
Hubbard, Elisha B., 311 n. 1, **525**
Hubbard, Gardiner Greene, **90**, 571 n. 1, 705
Hubert, Philip Gengembre, 720 n. 3
Hughes, Charles T., 387 n. 13, 630 n. 3, 650 n. 3
Hunnings, Henry, 691 n. 2
Hurlbut, H. R., 468 n. 1
Hutchinson, Joseph, **25**, 173, 233 n. 3, 372, 433, 439
Hygroscope, 493

India ink, 438
Indian River, Fla., 428–*30*, 443
Inductorium, 634 n. 5
Ingersoll Rock Drill Co., 599 n. 13
Institution of Electrical Engineers, 491 n. 1
Insull, Samuel, **29**, 164 n. 12, 732; and Acheson, 390 n. 1; and Albany village plant, 26, 28; and arc lighting contract, 685; in Berlin, 375; and canvassers, 142; and carbon buttons for Mendenhall, 703 n. 3; and closing of Edison Construction, 588; complains to Edison Machine Works, 367 n. 2; and Danville village plant, 44 n. 1; dines at Delmonico's, 271 n. 10; and Edison Electric directors' election, 667, 697–98; and Edison Isolated, 698; and Edison Lamp Co., 144 n. 1; and Edison residences, 481, 521; on Edison's dynamo experiments, 15, 24; in Europe, 279, 336, 341, 350–51, 354–55, 375; and foreign lighting business, 707 n. 2; and inside wiring, 302; and lamp manufacturing, 279; and lamps for Edison & Swan United, 663; and lighting of Parliament, 100 n. 2; living arrangements, 29 n. 3; in London, 279, 350–51, 354–55, 375–77; as manager of Edison Construction, 4, 26–28, 41, 72 n. 2, 73–76, 80, 92, 127–28, 155 n. 1, 172–73, 184, 201 n. 2, 269, 311, 336–37, 369, 433, 439–40, 451, 462–63, 515–16, 537; and manufacturing shop finances, 412 n. 7; and

Marion Edison's education, 659; and Mary Edison's death, 624; at Mary Edison's funeral, 629; and Mary Edison's health, 268 n. 1; and merger with Swan United, 107–8, 133, 239 n. 5; in Milan, 375; negotiates with Edison & Swan United, 635; negotiates with Italian Edison Co., 279; negotiates with Paris companies, 371; negotiates with Sunbury Edison, 202 n. 3, 615; negotiations with Armington & Sims, 433–34, 450; in Newburgh, 302 n. 3, 515; in Paris, 350–51, 354–55, 375; personal finances, 27–28; powers of attorney, 74, 80, 330 n. 1; and proposed contracting company, 510 n. 1; in Providence, 433; relations with Eaton, 433, 697; relations with Johnson, 27–29, 109–10, 127–28, 134; relations with Sims, 433; and reorganization of Edison Electric, 667; and reorganization of French Edison companies, 279, 529 n. 10; returns to New York, 351 n. 1; and Santiago station, 618 n. 3; stenographer, 388 n. 17; stock in Electric Tube Co., 651 n. 6; and Swedish lamp patent, 465 n. 6; as TAE's business manager, 25–26, 60 n. 2, 71 n. 6, 109, 173, 243 n. 3, 314, 329–30, 364, 428–29, 433–34, 439–40, 467, 499 n. 3, 544, 701; and TAE's household, 146 n. 2; and TAE's residences, 637; as TAE's secretary, 95, 100 n. 2, 106 n. 8, 176 n. 1, 186 n. 5, 188 nn. 1 & 8, 250 n. 1, 284–85, 309, 335, 485 n. 4, 509 n. 5, 591 n. 8, 602 n. 2, 607, 609, 634 n. 1, 665, 671 n. 4, 672 n. 1, 694 n. 5, 696 n. 6; views on Jews, 354–55; and voltage indicators, 293; and voltage regulation, 266 n. 3; as witness, 577–78
—letters: to Adams, 705; from Andrews, 337; to Andrews, 515–16; to Batchelor, 122–24; to Brown, 104; to Campbell, 184; from C. Clarke, 338–39; to H. Clarke, 659; to Cowles, 614; to Culbertson, 89–90; from Eaton, 451; to Eaton, 467; to Ferrell, 232; from Garrison, 189–90; to Germania

News Reporting Telegraph, 566
n. 5
New York and New England Railroad Co., 102 n. 1
New York Board of Fire Underwriters, 98, 99, 283 n. 4, 344 n. 1,
410–11
New York City: Blaine parade, 215
n. 2; Board of Aldermen, 134, 344
n. 1, 385; Board of Electrical
Control, 598 n. 7; Bowery, 143;
Brooklyn Bridge, 91 n. 1, 105;
Dakota Apartments, 246 n. 3;
Democratic parade, 699; map, 2;
Mills Building, 246 n. 3, 700 n. 9;
Navarro Flats, 244–46; Tammany
Hall, 734; underground wire legislation, 385. *See also* Edison,
Thomas Alva: residences; Edison
Electric Illuminating Co. of
New York; Electric lighting isolated plants; Gramercy Park;
Theaters
New York Daily Graphic, 568 n. 13
New York Daily Tribune, 589
New York Electrical Society, 313
New York Elevated Railroad, 99 n. 1
New York Evening Post, 83, 151, 219
n. 6
New York Foundling Asylum, 623
n. 5
New York Herald, 254, 628 n. 3
New York Hotel, 670 n. 3
New York Loan Improvement Co.,
246 n. 3
New York Mail and Express, 99 n. 1
New York Steam Co., 734
New York Stock Exchange, 368, 699
n. 5, 726
New York Sun, 106 n. 8, 687 n. 1,
730
New York Times, 106 n. 8, 156 n. 1,
283 n. 1, 343, 368, 659 n. 1, 710
n. 1
New York Tribune, 362 n. 6, 694
New York Underground Railway
Co., 102 n. 1
New York World, 145, 560–61, 562–
66, 621, 630–34, 659 n. 3
Niaudet, Alfred, 252
Norfolk Electric Lighting Co., 423
n. 1
Northern Pacific Railroad Co., 73,
380, 411 n. 4, 413 n. 10, 574 n. 3,
730
Norton, Frank Beardsley, 621
Nye, Bill, 128

Oakland Daily News, 568 n. 13
Ocean Magnetic Iron Co., 398 n. 6
Odorscope, 493 n. 2, 572 n. 3
O'Gorman, William, 484 n. 1
Ohio Agricultural and Mechanical
College (Ohio State University),
702 n. 1
Ohio Edison Electric Installation
Co., 50 n. 1, 78 n. 12, 198 n. 6,
387 n. 12; agency contract, 650
n. 3; agreements with TAE, 387
n. 11, 433; central station contracts, 78 n. 13; and Circleville
station, 476 n. 1; investors, 122;
license from Edison Electric, 651
n. 4; and Middletown station,
507–8; offices, 124 n. 4; operations of, 199–201; organization
of, 122; and Piqua station, 516
n. 1
Ohio Meterological Bureau, 702
n. 1
Ohm, determination of, 483, 505
Ohm's law, 221
Olan, Mr., 464 n. 1, 465 n. 3
Omaha Daily Bee, 219 n. 6
Oregon and Transcontinental Co.,
380, 413 n. 10
Oregon Railway and Navigation
Co., 63 n. 3, 380, 574 n. 1
Oregon Steam Navigation Co., 63
n. 3, 381 n. 3
Oriental Bank Co., 136 n. 13
Oriental Telephone Co., 705
Oriole Festival (Baltimore), 215 n. 2
Oser, Marion Edison. *See* Edison,
Marion ("Dot")
Ott, John F., **52**; and ampere meter,
125 n. 11; carbon button plating
experiments, 703 n. 3; discusses
Kellogg, 601 n. 2; and dramatic
lighting experiments, 714; and
duplicating ink, 335 n. 4; and dynamos for telegraphy, 547–48,
587 n. 2; and dynamos for telephony, 659; and Edison Machine
Works experiments, 587; filament
experiments, 353, 371, 388, 440,
468–69; instructions from TAE,
371, 466, 468–69, 547–48; and
isolated plant regulator, 440; laboratory reports, **51**, 51–52; lamp
experiments, 231 n. 3; meter experiments, 10 n. 1; ore separation
experiments, 543 n. 2; and TAE's
exhibit, 659; and telephone experiments, 666, 677, 680, 712;

and voltage boxes, 439; and voltage indicators, 291, 293–95, 301
n. 2, 369, 399, 434 n. 5, 440; voltage regulation experiments, 279;
and wire drawing and annealing,
587 n. 2
—letters: from Batchelor, 587; to
Edison, 51–52; from TAE, 466,
468–69; to TAE, 659
Outerbridge, Alexander Ewing,
704

Page, Homer, 627 n. 4
Page, Marion Edison (sister), 626–
27
Paine, Sidney Borden, 284 n. 3, **604**
Paine, Walter, 48 n. 48
Painter, Uriah H., 668, **683**
Palatka, Fla., 428, *430,* 440 n. 1,
444, 453
Parker, Samuel, 478 n. 2
Parker & Ditson, 478 n. 2
Parliament (Britain): electric light
legislation, 133–34; lighting of,
83, 100, 137 n. 16, 553
Parrish, Dillwyn, 54 n. 9
Patent applications: ampere meter,
125 n. 11; arc lighting, 559; with
Bergmann, 674; chemical stock
quotation, 612 n. 3; coated filaments, 673 n. 2; commutator
brushes, 279, 316–17; dynamo
regulation, 349; dynamos, 425–
26; dynamos for telegraphy, 548
n. 5; filament coatings, 391 n. 2;
fuel cells, 152, 256–58; with Gilliland, 675; high-voltage transmission, 581 n. 1; lamp filaments,
228 n. 2, 230, 231–32; light fixtures, 117 n. 3; meters, 5–12; pulleys, 557; remote control switching, 118–19; repeaters, 707–8;
selective signaling, 675, 713 n. 2;
table of, 750; telephone receiver,
675, 717; telephone transmitter,
674; three-wire system, 224; vacuum deposition, 391 n. 2; voltage
conversion, 59 n. 2; voltage indicators, 291, 294, 299–301; voltage
regulation, 172 n. 7; wire drawing
and annealing, 545–46, 587 n. 2;
wire insulation, 392–93
Patent interferences: with Blake, 35
n. 50; with Drawbaugh, 35 n. 50;
with Field, 63 n. 2; with Freeman, 319 n. 3; with Latimer, 439;
with Sawyer and Man, 227, 324;

Thomson galvanometer, 379
Thomson-Houston Electric Co., 269, 571 n. 2, 686 n. 2
Thornall, Israel, **495**
Tiffany & Co., 670 n. 3, 720 n. 3
Tilley, John, 441 n. 17
Time Telegraph Co., 103 n. 4
Tobey, Edward, 175 n. 11
Tocoi, Fla., 429–31, 453 n. 6
Tokyo Imperial University, 702 n. 1
Tomlinson, John Canfield, **319**; and agreement with station engineers, 425 n. 1, 439; and American Bell agreement, 682 n. 2; in Boston, 682 n. 2; and chemical stock printer, 698; and Democratic parade, 699; as director of Sprague Electric Co., 669 n. 7; and Edison Construction contracts, 383; and Edison Electric directors' election, 697; education, 700 n. 9; and electric light infringements, 279–80, 319; notified of Mary Edison's death, 625; and patent interferences, 439; and Seyfert suit, 518–19, 536 nn. 2–3, 603; and Williamsport payment, 382
—letters: to Insull, 603; from Lozier, 625; to Vroom, 518–19
Torres, Manuel Montt, 416 n. 1
Trask, Spencer, 669 n. 6, 696 n. 4, **697**
Trask (Spencer) & Co., 50 n. 2, 699 n. 5
Trenton, USS, 553 n. 5
Trowbridge, John, 595 n. 2
Truth (London), 83
Truth (New York), 368
Tudor, Frederick, **244**
Tudor (F.) & Co., **244**
Turrettini, Théodore, 487 nn. 1–2
Tyler, George H., 117 n. 2
Tyler, Henry D., 567 n. 8
Tyler, William, 650 n. 3
Typewriter, 483

Ullmo, Simon, 401 n. 2
Ulmann, Carl, 551 n. 2
Unger, Geh, 97 n. 2
Unger, William, 566 n. 5
Union Pacific Railroad, 141 n. 2, 177 n. 2
Union Theological Seminary, 630 n. 2
United States Electric Lighting Co., 33 n. 38; arc lighting, 140; Brooklyn Bridge lighting, 319 n. 1; cen-

tral stations, 90; at Chicago railway exhibition, 105, 141 n. 1; high candlepower lamp, 594; isolated plants, 90, 140–41, 146; rivalry with Edison companies, 596 n. 3; Southern Exposition exhibit, 143; and Weston, 156, 594 n. 1, 596 n. 3
United States Illuminating Co., 106 n. 8, 344 n. 1
United Telephone Co., Ltd., 136 n. 8, 193 n. 7, 376
University College School, 491 n. 1
University of Chile, 420 n. 8
University of Michigan, 598 nn. 3 & 7
University of Pennsylvania, 386 n. 6, 595 n. 1
University of the City of New York (NYU), 313 n. 4, 319 n. 1, 699 n. 1 & 7, 700 n. 9
Upham, James C., **522**
Upton, Elijah, 287 n. 3
Upton, Francis Robbins, **24**, 528 n. 5; and AAAS, 313 n. 4, 416 n. 6; and British lamp manufacturing, 555 n. 2; and dramatic lighting, 116, 673; and dynamos, 16, 24; as Edison Electric director, 669 n. 6, 696 n. 4; and Edison Isolated, 584; and European lamp business, 322 n. 2; and incorporation of Edison Lamp Co., 248–49; investment in Edison Lamp Co., 498; invests in Shamokin Edison, 45 n. 13; and lamp breakage, 310 n. 3; as lamp factory superintendent, 30 n. 8, 70, 451 n. 1; and lamp improvements, 24; and lamp quality, 121 n. 1; and lamps for Edison & Swan United, 663; and lamp tests, 121; loans to Edison Lamp Co., 70, 314–15, 661; and National Conference of Electricians, 637 n. 1; and proposed contracting company, 510 n. 1; and reorganization of Edison Electric, 411 nn. 1 & 3, 584–85; and reorganization of French companies, 529 n. 10; salary, 286; as treasurer of Edison Lamp Co., 30 n. 8, 248–49, 285–86, 314–15, 395, 412 n. 7, 497, 660–61, 701; vacation, 144; visits Newburgh station, 516 n. 3; and Worcester exhibition, 522 n. 2

—letters: from Insull, 663; to Insull, 248–49, 660–61; from TAE, 144, 399, 673; to TAE, 70, 285–86, 304, 314–15, 395, 497–98, 701; to Villard, 516–18
U.S. Bureau of Mines, 262 n. 3
U.S. Circuit Court, 574 n. 1
U.S. Coast and Geodetic Survey, 702 n. 1
U.S. Commissioner of Patents, 194 n. 2, 228 n. 2, 319 n. 3, 324, 596
U.S. Congress, 505 n. 2, 593 n. 3, 595 n. 2
U.S. Mint (New Orleans), 704 n. 1
U.S. Naval Academy, 67 n. 2, 440 n. 7, 506 n. 3
U.S. Patent Office, 749; carbon filament interferences, 227, 319 n. 3, 324; dynamo interference, 441 nn. 11–12; dynamo telegraphy application, 548 n. 5; fuel cell applications, 255 n. 8, 258 n. 4; reform of division of electricity, 589, 596–97; telephone interferences, 194; and telephone patents, 693; vacuum deposition application, 391 n. 2; wire drawing and annealing application, 546 n. 3
U.S. Post Office Dept., 593 n. 3
U.S. Railway Mail Service, 683 n. 7
U.S. Secretary of the Interior, 228 n. 2, 324
U.S. Senate, 486 n. 1
U.S. Signal Corps, 702 n. 1
U.S. Supreme Court, 324
USS *Trenton,* 553 n. 5

Vacuum deposition, 369, 390–91, 444, 466
Vacuum pumps, 388, 737
Vail, Jonathan H., **757** n. 7
Vail, Theodore Newton, 675, 680–81, 693
Valparaiso Sporting Club, 419 n. 5
Van Cleve, Cornelius, **162**, 186 n. 2
Vanderbilt, William H., 93 n. 2, 732–33
Vedder, John, 430
Velocipede, 637 n. 2
Vera, or the Nihilists, 262 n. 2
Verity, John B., **490**–91
Verity, Thomas, 137 n. 16
Verity Bros., 491 n. 1
Verity & Co., 720 n. 4
Verity & Sons, 378 n. 13, 491
Vicuña Mackenna, Benjamin, **415**